"Filled with thoughtful pieces penned by writers from around the world, this handbook covers a variety of topics focusing on theoretical as well as practical facets of tourism experience management and marketing. The topics covered range from traditional frameworks for designing, delivering and managing tourism experiences to more current and emerging topics such as co-creating tourism experiences with customers and using technological advances to enhance and promote tourism experiences. This volume is a valuable resource for tourism researchers and practitioners."

A. Parasuraman, *Professor of Marketing and Holder of the James W. McLamore Chair, University of Miami, USA*

"*The Routledge Handbook of Tourism Experience Management and Marketing* represents an unparalleled collection of thoughtful contributions to the advancement of experience thinking. It will prove an invaluable resource not only within the tourism industry, but for any experience-focused professional, as many of the models, metrics, design considerations, and technology twists put forth provide insights applicable to any experience-minded business. Indeed, in the Experience Economy – where time is the currency of consumption – all customers are tourists; and this volume helps set the agenda as enterprises compete to offer compelling places to spend their time."

B. Joseph Pine II & James H. Gilmore, *co-authors of* The Experience Economy: Competing for Time, Attention, and Money

"*The Routledge Handbook of Tourism Experience Management and Marketing* provides comprehensive coverage of recent experience research and thinking. The handbook provides 52 chapters from authors from around the world. Sections of the handbook that are of topical interest include designing experiences, technology-enabled experiences, and sustainable tourism experiences. The chapters cover experience in diverse contexts such as hospitality, attractions, gastronomy, travel, and accommodation. Theoretical, methodological, and empirical chapters include international case studies. This edited book is a timely collection providing the state of the art of management and marketing of tourism experiences."

Noel Scott, *Professor of Tourism Management, Sustainability Research Centre, University of the Sunshine Coast, Australia*

The Routledge Handbook of Tourism Experience Management and Marketing

The Routledge Handbook of Tourism Experience Management and Marketing offers a comprehensive and thorough inquiry into both customary and emergent issues of tourism experience and co-creation. Drawing together contributions from 83 authors from 28 countries with varied backgrounds and interdisciplinary interests, the handbook highlights multiple representations and interpretations of the theme. It also integrates a selection of illustrative global case studies to effectively present its chapter contents.

Tourism experience drives the contemporary tourist's behavior as they travel in pursuit of experiencing unique and unusual destinations and activities. Creating a memorable and enduring experience is therefore a prerequisite for all tourism business organizations, irrespective of the nature of their products or services. This handbook focuses on conceptualizing, designing, staging, managing, and marketing paradigms of tourism experiences from both supply and demand perspectives. It sheds substantial light on the contemporary theories, practices, and future developments in the arena of experiential tourism management and marketing.

Encompassing the latest thinking and research themes, this will be an essential reference for upper-level students, researchers, academics, and industry practitioners of hospitality as well as those of tourism, gastronomy, management, marketing, consumer behavior, cultural studies, development studies and international business, encouraging dialogue across disciplinary boundaries.

Saurabh Kumar Dixit, PhD is an Associate Professor and Head of the Department of Tourism and Hotel Management, North-Eastern Hill University, Shillong (Meghalaya) India. He holds a Bachelor's degree in Hotel Management and Catering Technology, a Master's degree in Tourism Management, and a Doctorate (PhD) in Hotel Management. His research interests include Consumer Behavior, Gastronomic Tourism, Service Marketing, Experience Management, and Marketing in hospitality and tourism contexts. He has worked more than 20 years in a number of Indian universities and educational institutes and successfully completed different research projects relating to hospitality and tourism management. He has 11 books to his credit including *The Routledge Handbook of Consumer Behavior in Hospitality and Tourism* and *The Routledge Handbook of Gastronomic Tourism*. He can be contacted at saurabh5sk@yahoo.com.

The Routledge Handbook of Tourism Experience Management and Marketing

Edited by Saurabh Kumar Dixit

Routledge
Taylor & Francis Group
LONDON AND NEW YORK

First published 2020
by Routledge
2 Park Square, Milton Park, Abingdon, Oxon OX14 4RN

and by Routledge
52 Vanderbilt Avenue, New York, NY 10017

Routledge is an imprint of the Taylor & Francis Group, an informa business

British Library Cataloguing-in-Publication Data
A catalogue record for this book is available from the British Library

Library of Congress Cataloging-in-Publication Data
Names: Dixit, Saurabh Kumar, 1979- editor.
Title: The Routledge handbook of tourism experience management and marketing / edited by Saurabh Kumar Dixit.
Description: Abingdon, Oxon ; New York, NY : Routledge, 2020. |
Includes bibliographical references and index.
Identifiers: LCCN 2020000381 (print) | LCCN 2020000382 (ebook)
Subjects: LCSH: Tourism–Management. | Tourism–Marketing.
Classification: LCC G155.A1 R687 2020 (print) | LCC G155.A1 (ebook) |
DDC 910.68–dc23
LC record available at https://lccn.loc.gov/2020000381
LC ebook record available at https://lccn.loc.gov/2020000382

ISBN: 978-0-367-19678-3 (hbk)
ISBN: 978-0-367-49275-5 (pbk)
ISBN: 978-0-429-20391-6 (ebk)

Typeset in Bembo
by Swales & Willis, Exeter, Devon, UK

Dedicated to the (loving) memory of my father, Dr. Rajesh Kumar Dixit, who left for the heavenly abode on March 12, 2019.

Contents

Figures

Tables

Contributors

Abijith Abraham is a research scholar at the Department of Tourism and Hotel Management, North-Eastern Hill University, Shillong (Meghalaya), India, and at present he is working on his PhD thesis, titled, "Tourism destination competitiveness through analytic hierarchy process: a case study of Thrissur district of Kerala". His research interests include Destination Competitiveness, Experiential Management, and Marketing. He holds a Bachelor's Degree in Hospitality and Hotel Administration and a Master's Degree in Business Administration (Tourism and Travel Management). He is a recipient of the prestigious Junior Research Fellowship (JRF) awarded by the Union Grants Commission (UGC), New Delhi. He was also a research intern at Equitable Tourism Options (EQUATIONS), Bangalore (India), where he worked on "Report on rapid impact assessment in North Bengal on Gorumara National Park, Jaldapara National Park and Buxa Tiger Reserve".

António Sérgio Araújo de Almeida, PhD (Tourism Sciences) is an Adjunct Professor at the Superior School of Tourism and Technology of the Sea belonging to the Polytechnic of Leiria, Portugal. He is also an integrated member of CiTUR Centre for Tourism Research, Development and Innovation. His interest areas are tourism, tourism experience, and identity contributions to sustainability of territories.

Melissa Baker, PhD is an Assistant Professor at the Department of Hospitality & Tourism Management in the Isenberg School of Management at the University of Massachusetts Amherst, USA. She is a multi-award-winning teacher conducting classes on food and beverage management, service experience management, and human resources management. Her research focuses on service experience management with a concentration on service failure and recovery and appearance and impression formation.

Adela Balderas-Cejudo is a Lecturer and Researcher at Esic Business and Marketing School and Research Fellow at the Oxford Institute of Population Ageing, University of Oxford. Professor at Cámara Bilbao University Business School and of the Bachelor Degree in Gastronomy and Culinary Arts, Basque Culinary Center, University of Mondragon, Dr. Balderas gained her PhD in Business Administration. She also has a Master in Marketing, an Executive MBA, and a Master in Career Coaching, and she has specialized in professional coaching with her studies at the University of New York, Stanford, and Harvard University. Dr. Balderas, who is a consultant in marketing and business skills, has teaching experience at undergraduate and graduate levels. Lecturer at ESIC Business and Marketing School, she is currently Director of Master in Restaurant Innovation and Management and Master in F&B Management at Basque Culinary Center, mentor in Imagine at Silicon Valley and a consultant. Dr. Balderas is Visiting

Professor and Lecturer at the University of Salamanca (Spain), College of Economics of the University of Xiamen (China), L University of Regensburg (Germany), University of Northumbria (United Kingdom), and École hôtelière de Lausanne (Switzerland).

Adiyukh Berbekova received her Bachelor's degree in Legal studies and in Tourism Management in Nalchik, Russia, and a Master's degree in International Tourism Management from the Westcoast University of Applied Sciences, Heide, Germany. Adiyukh is a recipient of a DAAD scholarship. She is a PhD student in the Isenberg School of Management at the University of Massachusetts Amherst, USA. Her main research interest is crisis management in the tourism and hospitality industry.

Jennifer Chan Kim Lian is a Professor of Tourism and Hospitality Management at the Faculty of Business, Economics and Accountancy, and Director of Borneo Tourism Research Centre, Universiti Malaysia Sabah. She received her PhD in Tourism and Hospitality Management from the University of Strathclyde, UK. Her research and publication activities focus on sustainable tourism and hospitality, consumer behavior, tourism marketing, and mixed methods research.

Guo-Wei Chen is a research assistant at the business school in Nanfang College of Sun Yat-Sen University, China. He received his BS in financial engineering from Nanfang College of Sun Yat-Sen University. His research interests focus on bond, financial market, portfolio investment, and data analytics.

Rashmiranjan Choudhury is enthusiastic about nature, a traveler by passion and a researcher by profession, and is presently pursuing his doctoral program at North-Eastern Hill University, Shillong (India). He has completed his Masters in Tourism Management and is a Junior Research Fellow awardee from the University Grant Commission, Govt. of India. His research interests include community-based tourism, technology in tourism, sustainability, and smart tourism. He has presented several research papers at various conferences and published two articles in referred research journals.

Cristiana Compagno is Full Professor of Management at the University of Udine (Italy). Her current research interests are in the field of entrepreneurship and innovation. She has authored or co-authored several international publications on these topics. Since 2000, she has headed the start-up incubator of the University of Udine and has led the Scientific Park "Friuli Innovazione" (North-eastern Italy). From 2008 until 2013 she served as Rector of the University of Udine. She currently serves as a board member in research, managerial, and financial institutions.

Valentina Della Corte is a Full Professor of Business Management. She received her PhD at Ca'Foscari University, Italy. She is the author and reviewer of numerous articles in specialized journals, both national and international, of contributions in books with plural authors and of monographic works. She has coordinated several research activities and cooperates actively in Bachelor's degree, Master's degree, and PhD programs in Italy and Europe.

Saurabh Kumar Dixit, PhD is an Associate Professor and Head of the Department of Tourism and Hotel Management, North-Eastern Hill University, Shillong (Meghalaya), India. He holds a Bachelor's degree in Hotel Management and Catering Technology, a Master's degree in Tourism Management, and a Doctorate (PhD) in Hotel Management. His research interests include Consumer Behavior, Gastronomic Tourism, Service Marketing, Experience Management, and Marketing in hospitality and tourism contexts. He has worked for more than 20 years in a number of Indian universities/educational institutes and has successfully completed different research projects relating to hospitality and tourism management. He has 11 books to his credit including *The Routledge Handbook of Consumer Behaviour in Hospitality and Tourism* (2017) and *The Routledge Handbook of Gastronomic Tourism* (2019). His detailed profile can be seen at https://www.routledge.com/authors/i15903-saurabh-dixit.

Elvisa Drishti is a permanent Lecturer at the Business Administration Department of the University of Shkodra "Luigj Gurakuqi", Albania. She lectures on applied statistics, research methods, and econometrics, and she holds a PhD in Labour Market Economics from Birmingham Business School. Elvisa does research in Labor Statistics, Econometrics, and Labor Economics. She is a Career Integration Fellow at CERGE-EI. Their current projects are "EconBiz" and "WageIndicator Albania" and she is also an InGRID visiting researcher.

Richard Ek is an Associate Professor in Human Geography at Lund University, Sweden. His research and publications cover several areas; place marketing and destination branding, tourism theory and philosophy, spatial theory, biopolitics, the geography of handcraft breweries, and the imagination of spatial touristic futures.

Fabrizio Ferrari is Assistant Professor in Economic Geography in the Department of Economic Studies, University "G. d'Annunzio" of Chieti-Pescara, Italy. His fields of research are about tourism places, local and regional development, urban growth, and population and migration studies.

Jörg Finsterwalder is Associate Professor of Marketing at the University of Canterbury Business School, New Zealand. His interests include co-creation of services, customer-to-customer interaction, group services and consumer tribes, customer experience, wellbeing, refugees and service provision, disasters and services, and food consumption and wastage. His work has been published in *Journal of Business Research*, *Journal of Retailing and Consumer Services*, *Journal of Services Marketing*, and *Journal of Service Theory and Practice.*

Anestis K. Fotiadis, PhD, is a Professor of Tourism at Zayed University, College of Communication and Media Science, Abu Dhabi, UAE. His research focuses on event management, rural tourism, and sustainable development. He has published more than 40 research papers in international academic journals such as *Tourism Management* and *Journal of Business Research*, along with several books and book chapters.

Mariana de Freitas Coelho is a Marketing Professor at Federal University of Viçosa, Minas Gerais, Brazil. A Post-Doctoral researcher in Hospitality at Universidade Anhembi Morumbi, São Paulo Brazil, she holds a Marketing PhD at Federal University of Minas Gerais, Brazil. She is interested in the following research themes: tourism experience, tourism marketing, hospitality, consumer behavior, and entrepreneurship.

Burkhard von Freyberg, PhD is Professor of Hospitality Management for the Faculty of Tourism at Munich University of Applied Sciences, Germany. He is in charge of the Hospitality Management Master program and started his career in hotel management before studying business administration at the Ludwig-Maximilians-University in Munich and the University of Regensburg. His research field is strategic hospitality management and he is a frequent guest lecturer at various institutions. Burkhard is editor and author of various professional hospitality text books and a member of the jury of the "European Hotel of the Year Award".

Warwick Frost is Professor of Tourism, Heritage and the Media at La Trobe University, Melbourne, Australia. His research interests include heritage, events, nature-based attractions, and the interaction between media, popular culture, and tourism.

José-Carlos García-Rosell, PhD, is Senior Lecturer in Responsible Tourism Business at the University of Lapland, Faculty of Social Sciences/Multidimensional Tourism Institute (MTI), Finland. His research interests include corporate social responsibility, ethical consumption, tourism service development, management education, and ethnographic action research. He has published in a range of books and journals including *Management Learning*, *Journal of Business Ethics*, *Journal of Sustainable Tourism*, *Tourism Recreation Research*, and *Society and Leisure*.

Roberta Garibaldi is Professor of Marketing and Tourism Management at the University of Bergamo. She is a Member of the Board of Advisors and Ambassador for Italy of the World Food Travel Association, a Member of the World Gastronomy Institute, and a Board Member of IGCAT – International Institute of Gastronomy, Culture, Art and Tourism. She is also a Member of the Executive Board of SISTUR (the Italian Society of Tourism Sciences) and is associated with ATLAS and AIEST, international scientific associations devoted to tourism.

Giovanna Del Gaudio has a PhD in Business Administration at Federico II University of Naples, Italy. Her research focuses on tourism destinations, value creation, and value capture with the application of the resource-based view. She is the author of a book and of some book chapters and articles published in national and international journals, and she has attended several national and international conferences.

Sandhiya Goolaup is an Assistant Professor in Marketing at the University of Borås, Sweden. Her research interest revolves around tourist/consumer experience, destination development, extraordinary experience, value creation, and phenomenology. Sandhiya has published in journals such as *Journal of Travel Research* and *Scandinavian Journal of Hospitality and Tourism*.

Punam Gupta, PhD, is Principal of Career College, Bhopal (India). She worked as an Associate Professor for over 35 years at Dev Samaj College for Women, Chandigarh. She is a scholar in English literature and linguistics and was awarded a gold medal during her BA Hons (English), and again for obtaining the first position in MA (English) in Panjab University. She has written several papers and has published books on English as well as co-authored a book on Rural Marketing.

Minni Haanpää, PhD, is Senior Lecturer in Tourism Studies at the University of Lapland, Faculty of Social Sciences/ Multidimensional Tourism Institute (MTI), Finland. Her research interests are in the areas of tourism and event experiences; co-creation; ethnographic methodologies, especially videography; and pedagogic practice in the higher education of tourism.

Her work has been published in *Tourism Recreation Research, Society and Leisure*, and in edited, peer-reviewed books.

Farooq Haq is an Associate Professor and the Chair of Marketing & International Business Program at the Canadian University, Dubai (UAE). He received his PhD from Charles Darwin University in Australia in 2011. His research interests are in various types of special interest tourism such as spiritual and heritage tourism. He is also among the early researchers of Islamic Marketing, halal branding, and the marketing of Islamic products and services. He has published in various journal articles, conference papers, and book chapters in his areas of research.

Anne Hardy is the Co-Director of the Tourism Research and Education Network (TRENd) at the University of Tasmania, Australia. She conducts research in three areas: the neo-tribal behavior of tourists; sustainable tourism; and tracking tourists' movement. Anne is the lead investigator on the Tourism Tracer project that has received international acclaim and awards. She is a passionate advocate for knowledge transfer.

Robert J. Harrington, PhD, Associate Director and Professor in the School of Hospitality Business Management at Washington State University, received his PhD and MBA from Washington State University, and a BBA from Boise State University, USA. Prior to joining WSU-Tri-Cities, he was the 21st Century Endowed Chair in Hospitality at the University of Arkansas, faculty member at the University of Guelph, Canada, and Dean of the Chef John Folse Culinary Institute. He has widely published in the areas of hospitality innovation, strategic management, and food and wine, and is author of *Food and Wine Pairing: A Sensory Experience.*

Ann Hindley is a Senior Lecturer in Tourism, Marketing, and Event Management and a researcher in the International Thriving at Work Research Group at the University of Chester, United Kingdom. Her interests lie in the area of sustainability, responsibility, and consumer behavior.

Linda D. Hollebeek, PhD, is a Senior Associate Professor at Montpellier Business School and Full Professor (Adjunct) at Tallinn University of Technology, Estonia. Her research interest centers on customer/consumer engagement and interactive consumer/brand relationships, and her work to date has been published in *Journal of the Academy of Marketing Science*, *Journal of Service Research*, *Journal of Business Research*, *Industrial Marketing Management*, *European Journal of Marketing*, and *Journal of Interactive Marketing*, among others. She was awarded the 2014 ANZMAC Emerging Researcher Award and serves as Associate Editor for the *European Journal of Marketing* and the *Journal of Business Research.*

Jan Huizing is a Research Fellow in City Hospitality and a Senior Lecturer in Strategy Development at Hotelschool The Hague, International Hospitality Business School. He received his Master's degree in Business Administration from Rijksuniversiteit Groningen, The Netherlands. Before joining Hotelschool The Hague Jan Huizing held senior marketing and management positions in leading service and knowledge companies. He contributed to customer value and business growth at ANWB (provider of services in recreation, tourism and mobility) and ADP (global payroll and HR business solutions). Next to academic contributions, Jan Huizing is a keynote speaker at industry conferences and workshops. His research interest is city hospitality design, with a focus on the healthy balance between visitors and residents, and how meaningful interactions contribute to hospitable, sustainable cities and communities.

Mine Inanc is a staff member at Dokuz Eylul University, Turkey. She obtained a BSc degree in Radio, Television, and Cinema from Lefke European University, Northern Cyprus, Turkey, and a Master's degree in Communication Sciences from Yasar University, Turkey. Mine is currently undertaking her PhD in the field of public relations and communication at Ege University, Turkey. Her fields of expertise engage with public relations, political communication, big data, new media, and consumer experiences.

Jens Friis Jensen is teaching Associate Professor at the Department of Arts and Humanities at Roskilde University, Denmark. Before joining the university, he had a professional career in tourism both in the private and the public sector in Denmark and internationally at the UNWTO. Currently he is involved in three national tourism research and development projects focusing on innovation in tourism.

Eva Maria Jernsand is a Researcher in Marketing at the School of Business, Economics and Law and the Centre for Tourism, both at the University of Gothenburg, Sweden. Her research interests revolve around learning, inclusiveness, and co-creation in place branding and destination development. Eva Maria has published in journals such as *Place Branding and Public Diplomacy*, *Tourism Recreation Research*, and *Action Research Journal*.

Lee Jolliffe is Professor of Hospitality and Tourism at the University of New Brunswick, Canada, and a Visiting Professor at Ulster University, Northern Ireland. With an academic background in museum studies, Dr. Jolliffe has written extensively on heritage tourism and in particular on niche areas of culinary tourism, including both tea- and coffee-related tourism.

Yeonjung (Alice) Kang, M.S., is a PhD student in the Hospitality & Tourism Management department of Isenberg School of Management at the University of Massachusetts Amherst. Her research focuses on branding and consumer behavior in the food and beverage industry.

Kawon Kim, PhD, is an Assistant Professor at the School of Hotel, Restaurant, and Tourism Management in the College of Hospitality, Sports and Retail Management at the University of South Carolina, USA. Her teaching focuses on food service and event management, and human resources management. Her main stream of research focuses on social and interpersonal influence in a service setting.

Metin Kozak is Professor in the School of Tourism, Dokuz Eylul University, Turkey. He holds both Master's and PhD degrees in Tourism Management. His research focuses on consumer behavior, destination management, and sustainability. He acts as the co-editor of *Anatolia: An International Journal of Tourism and Hospitality Research* and has been to several universities in the US, Europe, and Asia as a visiting scholar and also many conferences as an invited speaker.

Drita Kruja is Professor of Services Marketing, Marketing Research, and Research Methods at the Management & Marketing Department and also Vice Rector of the European

University of Tirana, Albania. She is involved in a wide range of academic and scientific activities, which are mainly related to marketing and tourism issues. She is author and co-author of three book chapters published by Routledge and some textbooks, and she is a member of editorial boards and a reviewer of many national and international journals.

Dinesh Kumar, PhD, is Associate Professor at Jagran Lakecity University, Bhopal (India). He had corporate experience for over 15 years after which he switched to teaching in 1995. He has taught at leading business schools in India, has traveled widely and has published and presented many papers. He is the author of four major books on Marketing: *Marketing Channels* (2012) and *Consumer Behavior* (2015), both published by Oxford University Press, *The Connected Consumer* published by Business Expert Press, New York, and *Rural Marketing*, published by Sage (2017).

Jennifer Laing is an Associate Professor at La Trobe University, Melbourne, Australia. Her research interests include travel narratives; the role of events in society; rural and regional regeneration through tourism and events; and health and wellness tourism.

Mia Larson is an Associate Professor in business management at Lund University, Sweden. Her research has mainly focused on event management and destination development from different perspectives. Her publications cover topics such as network collaboration, innovation, tourist behavior, gender, social media, and the sharing economy.

Kuan-Huei Lee, PhD, is an Associate Professor in the Design and Specialized Businesses Cluster at Singapore Institute of Technology (SIT). She teaches Organization Behavior, Management in Aviation, F&B Management, Cuisine & Culture and Leadership & Ethics across Accountancy, Hospitality Business, Air Transport Management in SIT and BBA in Food Business Management for the Culinary Institute of America, SG. Dr. Lee's publications include academic journals, book chapters, conference proceedings, and industry case studies, which she developed and used as teaching material. Her research interests include food tourism, organization and management, and cross-cultural studies of tourist traveling behaviors.

Young-Sook Lee is Professor in Tourism Management at UiT the Arctic University of Norway. Before joining UiT in 2014, she held an academic position at Griffith University in Australia for over 15 years. With her academic backgrounds in literature and business, and a PhD in sociology, her research interests cover Asian cultural philosophies in tourism, the meanings of nature and sustainability, tourist behaviors, and Arctic tourism.

George W. Leeson is Co-Director of the Oxford Institute of Population Ageing, University of Oxford. Dr. Leeson's first degree was in Mathematics, followed by a Masters in Applied Statistics, both from Oxford. His Doctoral work was in Demography. Dr. Leeson directs the Institute's research networks in Latin America (LARNA) and in Central and Eastern Europe (EAST) and also the Centre for Migration and Ageing Populations (MAP Centre). Dr. Leeson has directed the Danish Longitudinal Future Study, which elucidates the attitudes and aspirations of future generations of older people in Denmark, and he is Principal Investigator with Professor Sarah Harper on the Global Ageing Study, a survey of 44,000 men and women aged 40 to 80 in 24 countries.

Tiffany S. Legendre, PhD, is an Assistant Professor in the Conrad N. Hilton College of Hotel & Restaurant Management at the University of Houston. Her research focuses on branding and sustainability, and her teaching concentrates on food and beverage management, sustainability, and research methods.

Wilhelm Loibl is a Digital Marketing Lecturer in the Department of Marketing and Sales at the University of Chester, United Kingdom. His research interests include web and social media mining and semantic web technologies, both with a special focus on the tourism industry.

Poh Theng Loo, PhD, is an Assistant Professor at the Department of International Tourism and Hospitality at I-Shou University, Taiwan. Her research interests are consumer behavior and services marketing, particularly in service failure and recovery, coping with complaints and online complaints.

Sandra Maria Correia Loureiro, PhD, has the following research interests: relationship marketing, tourism marketing, and marketing connected to VR, AR, and AI. Her papers have been published in a variety of peer-reviewed journals that include *Journal of Marketing*, *Journal of Retailing*, *Tourism Management*, *Journal of Business Research*, *International Journal of Hospitality Management*, *Journal of Travel and Tourism Marketing*, *Journal of Service Management*, *Journal of Cleaner Production*, *Journal of Brand Management*, and *Online Information Review*. Her work has also been presented at respected international conferences such as EMAC, ANZMAC, and KAMS-GMC. Sandra serves as a reviewer for several international journals and conferences and has participated in several research projects funded by the EU and FCT (Foundation for Science and Technology).

Keith Mandabach is a Professor of Hotel, Restaurant, and Tourism Management at New Mexico State University, USA with an Agricultural Experiment station appointment. He completed his Doctorate and Master's degree at the University of Houston and his Bachelor's Degree at St. Edward's University, USA. His research interests are gastronomy/culinary arts, food and beverage management, hotel management, and the use of technology/social media in tourism and hospitality marketing and customer engagement.

Michela C. Mason is Associate Professor of marketing at the Department of Economics and Statistics at the University of Udine, Italy. Her current research interests are in the area of tourism marketing and consumer behavior. Her papers have been published in *Journal of Business Research*, *Tourism Management*, *Journal of Hospitality Marketing and Management*, and others.

Frans Melissen is Breda University of Applied Sciences' Professor of Sustainable Experience Design. His research focuses on the link between sustainable development and human behavior, with special emphasis on mitigating the social dilemma by means of sustainable experience design. Frans is also the co-author of the books *Sustainable Customer Experience Design* and *Improving Sustainability in the Hospitality Industry*.

Roberta Minazzi is Associate Professor of Tourism Marketing and Management at University of Insubria (Como), Italy, and Lecturer in Marketing at the Master in "Management del Made in Italy" and in "Hospitality and tourism management" at IULM University (Milan).

She received her PhD in Marketing and Communication at IULM University. Her research interests mainly focus on social media marketing, electronic word-of-mouth, quality management, and customer satisfaction in services. She is the author of papers published in national and international journals and of the book *Social Media Marketing in Tourism and Hospitality* (Springer, 2015).

Christy Yen Nee Ng is an Assistant Professor at the Institute for Tourism Studies (IFT), Macau. She holds a PhD in Hospitality Administration from Oklahoma State University, USA and she has research interests in relationship marketing, customer satisfaction, job satisfaction, and service quality.

Can-Seng Ooi is a Sociologist and Professor of Cultural and Heritage Tourism at the University of Tasmania, Australia. His research interests are varied and focus on tourism-in-society issues, including job polarization in the industry, destination branding as social engineering, and community development through tourism. His comparative work covers Singapore, Denmark, Australia, and China.

Michael C. Ottenbacher, PhD, is Department Head and a Mary L. Vanier Endowed Professor of the Hospitality Management Department at Kansas State University, USA. He received his PhD in Marketing from the University of Otago, New Zealand, and his Master and Bachelor of Science in Hospitality Management from Florida International University, USA. Prior to joining Kansas State University he was teaching at Heilbronn University, Germany, San Diego State University, USA, University of Guelph, Canada, and the University of Surrey, UK. In addition to academia, Dr. Michael Ottenbacher has extensive business experience and has worked in senior hospitality positions in the USA, UK, France, and Germany. Professor Ottenbacher has published widely in leading hospitality, tourism, and management journals. He is on the editorial board of several journals and has been the co-editor of the *Journal of Hospitality and Tourism Education* since 2012. Further, Dr. Ottenbacher has been a Visiting Professor at the Institut Paul Bocuse in France, the German-Vietnamese University in Ho Chi Minh City, Vietnam, Taylor's University in Kuala Lumpur, Malaysia, and HTW Chur in Chur, Switzerland.

Alexandros Paraskevas, PhD, is a Professor in Strategic Risk Management, Chair in Hospitality Management, and the Director of ICHARM (International Centre for Hospitality and Aviation Resilience Management) at the London Geller College of Hospitality and Tourism in the University of West London. Alexandros researches the governance and management of risks/crises both in an organizational and tourism destination context. He has co-authored the book *Planning Research in Hospitality and Tourism* (2016, Routledge) and co-edited the *Handbook of Research Methods in Complexity Science* (2018, Edward Elgar Publishing).

Adam Pawlicz is a Research Scientist with 15 years of research experience in the hospitality and e-tourism area. Adam Pawlicz is an Associate Professor at Szczecin University, Poland, and a visiting professor at Klaipeda University, Lithuania, and Vidzeme University, Latvia. He has co-authored over 100 publications and is a member of the scientific committee of various international journals such as *Entrepreneurship* issued by SWU "Neofit Rilski", Bulgaria. His current research focuses on the impact of sharing economy and online travel agencies on the hospitality market.

Pairach Piboonrungroj is Associate Dean of the Faculty of Economics, Acting Associate Dean of the College of Maritime Studies, and Management and Founding Director of the MICE Excellence Centre at Chiang Mai University, Thailand. He is a member of the Chiang Mai Province MICE City Committee and is certified by various MICE institutes including CMP by MPI, CIS by SITE, and CEM by IAEE.

Girish Prayag is Associate Professor of Marketing in the Department of Management, Marketing, and Entrepreneurship at the University of Canterbury, New Zealand. He received his PhD in Tourism Management from the University of Waikato. His research interests are related to the emotional experiences of tourists, tourism market segmentation, place attachment, and organizational resilience. His work has been published in the *Journal of Travel Research*, *Tourism Management*, and *Annals of Tourism Research*.

Francesco Raggiotto is research fellow of marketing at the Department of Economics and Statistics at the University of Udine, Italy. His current research interests are in consumer behavior and tourism management. He has published in the *Journal of Business Research*, *Tourism Management*, and other publications.

Raouf Ahmad Rather, MBA (Marketing), is a Research Associate at the Department of Tourism Studies, Central University of Kashmir, Jammu and Kashmir, India. His research areas are consumer/customer engagement, co-creation, service innovation, and customer loyalty. His work to date is published in journals including *European Journal of Marketing*, *International Journal of Contemporary Hospitality Management*, and *Journal of Hospitality Marketing & Management*, among others.

Hanım Kader Şanlıöz-Özgen is affiliated as an Assistant Professor in the Department of Hotel Management, School of Applied Sciences, Ozyegin University, Turkey. She holds an MBA degree and obtained her PhD degree in Tourism Management at Dokuz Eylul University, Turkey. Her fields of expertise are experience management, accommodation experience in hotels, hospitality marketing, quality management, and sustainable tourism.

Heike Schänzel is an Associate Professor at Auckland University of Technology in Auckland, New Zealand, and program leader for post-graduate tourism. Her research interests include tourist behavior and experiential travel; families, children, and grandparents in tourism; sociality in tourism; femininities and masculinities in tourism research; innovative and qualitative research methodologies; and critical theory development in tourism and hospitality.

Laura Schmidt is currently a PhD student at Universidad de Las Palmas de Gran Canaria. After an apprenticeship she studied Tourism Management (BA) and Hospitality Management (MA) at Munich University of Applied Sciences. Since 2017, she has worked as Communication Manager for Cocoon Hotels in Munich. Laura Schmidt is a Lecturer at Munich University as well as at Baden-Wuerttemberg Cooperative State University, Ravensburg. She is the author of various hospitality publications and her PhD research deals with the topic of succession in the context of familiness and knowledge management in family-owned hotel companies in Germany.

Anke Schneider is a Researcher in the field of emotional color perception. In 2007 she graduated in Computer Visualistics at the University of Koblenz-Landau. She received her doctoral degree in 2016 from the TU Chemnitz and for several years she has worked at WU Vienna.

Sabrina Seeler is a post-doctoral Researcher in experienced-based tourism at Nord University Business School (Norway) and is part of the research group for Marketing, Management and Innovation of Experiences (MMIE). Having experience in qualitative and quantitative research methods, Sabrina's research interests include the creation and consumption of experiences, consumer behavior in tourism, social media marketing, visitor management, and strategic destination management.

Joanna Seraphim is an Assistant Professor and the Head of the Incubation Centre at Amity University, Dubai (UAE). She holds a PhD degree in Anthropology from École des Hautes Études de Sciences Sociales, Paris, France. As part of UAE's 2015 Year of Innovation, Dr. Joanna was selected to be in the first UAE government-funded cohort to study Design Thinking and Entrepreneurship at Stanford University. Joanna's ethnographic fieldwork has taken her from France to Canada, where she has collaborated with the indigenous communities to help preserve their heritage. She also studied how ICTs empowered Emirati women. Dr. Seraphim's research interests are in heritage, cultural preservation and transmission, design thinking, innovation, and gender studies.

Marlusa de Sevilha Gosling is a doctorate Professor at Federal University of Minas Gerais, Brazil. She coordinates the research nucleo NEECIM-Tur and the journal *Marketing and Tourism Review*.

Fabiola Sfodera, PhD, is Assistant Professor of Marketing at the Department of Communication and Social Research, Sapienza, University of Rome. She is an expert in tourism and cultural heritage, and has coordinated research projects and participated in international research groups. Her research interests are focused on consumer behavior in the fields of tourism and cultural heritage, destination marketing, and consumer experience through technology.

Bert Smit, MA, is Principal Lecturer of Consumer Experience and Service Design at Breda University of Applied Sciences. His research focuses on design methods for hospitality and tourism experiences. He has a background in social sciences, but after being involved in the development of tourism attractions he has taken an interest in how engineering design can be applied to intangible concepts such as experiences.

Samuel Spector earned his PhD from the Department of Tourism at the University of Otago in New Zealand and was subsequently a Post-doctoral Research Fellow at the Centre for Sustainability, also at the University of Otago. He is now a Lecturer in the Department of Management, Marketing, and Entrepreneurship at the University of Canterbury, New Zealand. His research focuses on the intersections between tourism, transport, sustainability, and resilience.

Anastasia Ef. Spyridou is currently a PhD candidate at Gdansk University of Technology, Faculty of Management and Economics, Poland. Her research interests are event management, tourism economics, and bankruptcy issues.

Paul Strickland is a Lecturer at La Trobe University, Melbourne, Australia. His research interests include ethnic restaurants, Bhutanese studies, wine tourism, and wine marketing.

Jon Sundbo is Professor in business administration and innovation at Roskilde University, Denmark. He is leader of Roskilde University's Centre for experience and tourism research and has published over 200 articles, book and book chapters, primarily on innovation, service business, experience, and tourism. He has been an advisor to the EU, the Danish government, and several firms and trade organizations within service and tourism.

Karoline Wiegerink holds the Chair in City Hospitality and City Marketing at Hotelschool The Hague and is the Director of a Platform for Customer Centricity. During her professional career, she has always combined academic and practical work. As an economist and marketeer, she gained special experience in the field of live communication and event marketing and has shared that knowledge as a Director at Erasmus Centre of Event Marketing (ECBM), as a Lecturer at Erasmus School of Economics, and while working as a part-time Associate Professor at Nyenrode Business University. As a consultant, co-creator, and keynote speaker she nowadays focuses on creating customer value through hospitality experience. Her focus areas are cities, urban spaces, and neighborhoods wherein different dimensions of hospitality contribute to livability and lovability.

Kim M. Williams is a Lecturer at William Angliss Institute, Melbourne, Australia. Her research interests include human resources, event management, gastronomy, fashion, and wine tourism.

Brian Kee Mun, Wong is an Associate Professor and currently heads the School of Business at the UOW Malaysia KDU University College. He holds a PhD in Tourism Management from the University of Malaya, Kuala Lumpur. He is a marketing, management, and entrepreneurship research enthusiast, particularly in the area of travel, tourism, and hospitality.

Roy C. Wood, PhD, has held Professorships at universities in the UK, China, and The Netherlands. He has been Principal of IMI University Centre, Switzerland; Dean of the Oberoi Hotels Centre for Learning and Development, India; and COO of a private further education provider in Bahrain. Wood is credited on over 200 publications and is currently a Visiting Professor at the University of Cumbria.

Arch G. Woodside is Distinguished Visiting Professor, Yonsei Frontier Lab, Yonsei University, Seoul, Republic of Korea and Honorary Professor of Marketing, Curtin University. He is the Editor-in-Chief of the *Journal of Global Scholars of Marketing Science*; Senior Editor of the *Journal of Global Fashion Marketing*; Series Editor of *Advances in Culture, Tourism and Hospitality Research* book series; and Series Editor of *Advances in Business Marketing and Purchasing* book series. He is a member and Fellow of the Royal Society of Canada; the International Academy for the Study of Tourism; the American Psychological Association; the Association of Psychological Science; Society for Marketing Advances; Global Academy of Marketing and Management Associations (GAMMA); and the Academy of Innovation,

Entrepreneurship, and Knowledge. He is the Founding Editor-in-Chief of the *International Journal of Culture, Tourism and Hospitality Research*. He is the former Editor-in-Chief of the *Journal of Business Research*, 1976–2016. He has also authored and co-authored 350+ research articles appearing in the *Annals of Tourism Research, Tourism Management, Journal of Travel Research, Journal of Marketing, Journal of Applied Psychology, Journal of Business Research, Journal of Marketing Research, Psychology & Marketing*, and 37 additional tourism, hospitality, marketing, management, and operations research journals. He resides in Marietta, Georgia.

Hung-Che Wu is an Associate Professor at the Business School in Nanfang College of Sun Yat-Sen University, China. He holds his PhD in marketing from Lincoln University. He is currently working as an editor-in-chief for *Probe – Business Management* and is a reviewing editor for *Cogent Business & Management*. His research interests include green marketing, hospitality management, service quality, customer satisfaction, and consumer behavior.

Xiang Ying Mei pursued her PhD focusing on tourism innovation, which was completed at the University of Queensland in 2012. She has many years of industry experience from the hospitality industry. Dr. Mei's research interests include government policies, innovation in service industries, experience management, consumer behavior, branding, tourism, and destination management and regional development. She has authored multiple book chapters and scientific papers in peer-reviewed journals. She is an Associate Professor in Marketing at Inland Norway University of Applied Sciences, Norway.

Yan Zhang is from Shenzhen, China. She studied for her BA in Business English at Xiamen University, Tan Kah Kee College in China. Upon completion of her BA she moved to Norway for her Master's in Tourism Studies at UiT The Arctic University of Norway. She is currently a Marketing Manager at AUSDOM HongKong Limited and is passionate about gaining a better understanding of the dynamic Chinese market, specifically related to digital media.

Foreword: Tourists' enacting, interpreting, and evaluating experiences in contexts

Saurabh K. Dixit's *Routledge Handbook of Tourism Experience Management and Marketing* contributes original theory while deeply explaining a vast amount of relevant literature on tourism experience. The *Handbook*'s principal focus is on the following topics: describing/explaining tourists' experience plans, actions, and outcomes; how tourists assess their travel processes, intermediate, and final destinations; and emic and etic descriptions, explanations, and assessments of mega actions, events, and performances relating directly to their trips. An independent researcher's or tourist-related strategist's interpretations of a tourist's plans, actions, and trip outcomes are "emic" explanations. A tourist's own interpretations of antecedents/causes, actions, outcomes, and (re) evaluations are "etic" explanations. Surprisingly, a large hole exists in the relevant extant literature on the understanding, explaining, assessing, predicting, and (less so) planning of tourists' experiences – both hospitality, destination, and event strategists plan tourists' experiences as well as tourists themselves plan tourists' experiences. Dixit and colleagues describe this vacuum and offer a highly useful review of the literature that puts an end to the pervasive meager attention to tourism experience. This handbook is original in identifying an obvious but overlooked truism: describing, understanding, and managing the tourist's experiences are fundamental to highly beneficial outcomes for the tourist; high quality-of-life for local residents; and profitable cash flows for destination, event, and hospitality firms.

Because so many of the topics may be *unfamiliar* to most of us, taking an extra five minutes for reading the table of contents (TOC) closely would be beneficial. Doing so helps to identify three groups of chapters at least – the groupings will be unique to each of us: "must read now for core knowledge" and applications; "key sources relevant for studies that I have in mind later" and "shocking topics for me chapters". My "must read now" chapters include chapters 1, 2, 5, 7, 13, 27, and 47. "Key sources relevant for my research" include chapters 3, 9, 11, 23, 33, 41, and 45. "My shocking topics chapters" include 15, 30, 31, 36, 38, 42, and 50. The chapter titles in the TOC are clues to the reasons of my assignments. Warning: using this three-group selection–assignment protocol is likely to increase your own must-read-now perspectives toward several chapters in this Handbook.

Core tourism experience tenets

"Quantum tourism" (QT) is a useful nascent theoretical concept to adopt in the study of tourism experiences. QT refers to the discretizing and connecting of pinnacle, major-secondary, and minor processes, steps, and events (i.e., parts/milestones) of a completely or in-part discretionary trip. The doing/performing, describing, and evaluating of the three levels impact the functioning efficiency and judged quality of the total tourist experience. Pinnacle processes/events include a potentially and typically limited number of high-point event-experiences occurring during the trip (e.g., your one-time only night at the La Scala Opera House in Milan; catching a 5-lb catfish at a lake with your ten-year-old daughter; going-to-see and viewing flowing molten lava on the "Big Island" of Hawaii) – a pinnacle event occurring that was planned. An unplanned possibly life-changing event can be a pinnacle experience as well (e.g., Hirschman's (1978) near-death mugging during a vacation trip with her husband). Since the 1980s, a large share, possibly half, of American families with children visit either Disneyworld or Disneyland – a ritualized macro-experience expression that confirms middle-class standing. Bicycling road-touring of Alaska, for example, can be a pinnacle experience – macro experiences vary in length of time, monetary cost, and emotional intensity but all are identifiable as being a core reason for the trip or an unexpected enjoyable or unpleasant happening. A pinnacle experience may occur inside another (macro) pinnacle experience. For example, an American tourist spending a one-week lifetime dream vacation in Milan during which she attends an opera at La Scala may count both occurrences as pinnacle events.

Not all pinnacle experiences are planned occurrences or positive. The following story is an example of a negative pinnacle tourist experience.

> When we were about 2 miles from our hotel, I glanced down at the clock on the dashboard to check the time so I could pay the babysitter. It was 10:45 pm. As I looked up, I noticed the lights of another car coming up behind us. We were in no hurry, so Ray slowed down to let it pass. There was an enormous crash. Our car was thrown off the road and into a culvert. "Damn," I thought, "those people are drunk and now we've had a wreck." In the headlights of our car I saw two shadowy figures running toward us. There was a bright flash of light and an explosion. BAM! ... The entire windshield on his side shattered. Ray screamed, "They're shooting at us; they're shooting at us!" I yelled to him "Drive the car! Drive the car!" There was another flash of light, and another explosion on my side of the car and the rest of the windshield fell away. Suddenly the two men were jerking open the doors of the car. One began hitting Ray in the chest with his gun demanding his wallet. The other put his gun to my head and began grabbing my jewelry and my pocketbook. My ears strained for the sound of the gun going off for a third time. I knew it was inevitable. I knew we were dead. But instead of shooting us, the men ordered us to get out of the car. As we were exiting, another pair of headlights appeared on the road; one man shouted to the other "Someone's coming, Mon, let's go." And they jumped back in their car and sped away, screaming, laughing, and waving their guns in the air. I ran into the street and began waving my arms at the oncoming car. It stopped. I yelled to them "We've just been shot at; we've been robbed; our car is wrecked; please help us." It turned out to be 4 tourists from Ohio (never have I been so glad to see tourists from Ohio) and they did help us.
>
> *(Hirschman 1991: 2)*

Major-secondary tourist experiences include the configuration of experiences associated with accommodations (e.g., hotel stays as well as fine-dining events). Tourists likely assess the majority of the events in contexts to be minor experiences, for example, a taxi ride from a railway station to a hotel, lunch at a fast-fast stop via a street vendor, and window shopping on Main Street in a strange city. When something unexpected and/or extraordinary occurs, a seemingly minor experience can become a major-secondary experience or even a pinnacle experience. Two sources are available that include more than two dozen both emic and etic stories about seemingly minor events that are transformed into major-secondary tourist experiences. First, a special issue on service experiences, mostly tourism-related, of the *International Journal of Business and Economics* (IJBE, Volume16, Issues 2 and 3) includes 30 customer-experience research articles – each followed by training for tourists and service providers to avoid the problems described in each story. (This journal is one of two journals entitled *IJBE*; this one is published by Feng Chia University, Taichung.) Second, *Trade Tales* (Woodside 2017) is volume 14 in Emerald Publishing's *Advances in Culture, Tourism, and Hospitality Research* book series; this book includes 25 emic/etic mostly tourism-related stories of minor to major-secondary experiences gone wrong and how they were resolved – with additional discussion in each article on how the problem might have been resolved better.

Detailed research studies of tourists' experiences, problems, and resolutions illustrate the need for deep reporting/interpreting of both the profound and the mundane of real-life tourism experiences/stories. Such studies are necessary to enable researchers to accurately explain, describe, and predict communications, processes, and outcomes in tourism experiences. In Chapter 1 in the *Handbook*, authors Sundbo and Dixit communicate this research tenet by explicating economic, psychological, and sociological perspectives of tourism experiences. A major conclusion is accurate to draw here: Dixit's *Handbook* is a pinnacle event in tourism research because the book's chapters collectively recognize and report on the essence of tourism – tourism experiences as interpreted by tourists and researchers.

The tourism experience focus is necessary to embrace if researchers seek to do more than build symmetric predictive models using 5-point and 7-point scales via multiple regression analysis (MRA) and structural equation models (SEM). Predictive models of tourism experiences and outcomes need to begin with rich descriptions and move beyond modeling by symmetric tools. The theoretical possibility of modeling for predicting outcomes asymmetrically has become reality in tourism experience studies. Woodside (2019) develops this view of accurate predictive modeling of tourist experiences and outcomes and offers empirical comparisons of asymmetric versus symmetric models in a field experiment of tourism experience

Brilliance

While all the chapters in this *Handbook* are worth reading, a few deserve the label brilliant. "Brilliance" in insights and in providing advances toward a theory of tourism experience applies to the following chapters: de Freitas's chapter 7, the essence of memorable experience; Goolaup and Jernsand's chapter 15, learning through extraordinary tourism experiences; chapter 30, Kim and Baker's managing service failures and service recoveries; Kumar's chapter 31, dark tourism: negative experiences as part of modern tourism; and Elk, Larson, and Ooi's chapter 48, the emotional labor of the co-created tourism experience. "Brilliance" applies to additional chapters as well. Read on to find these and your personal choices worthy of the label.

Arch G. Woodside
Department of Marketing
Boston College
Chestnut Hill
Massachusetts, USA

References

Hirschman, E. (1991). Presidential address Secular mortality and the dark side of consumer behavior: Or how semiotics saved my life. In Holman, R. H., and Solomon, M. R. (eds), *North American – Advances in Consumer Research* (pp. 1–4), 18. Provo, UT: Association for Consumer Research.

Woodside, A. G. (Ed.) (2017), *Trade Tales: Decoding Customers' Stories (Advances in Culture, Tourism and Hospitality Research, Volume 14).* Bingley, UK: Emerald Publishing. DOI: 10.1108/S1871-317320170000014030.

Woodside, A. G., (2019). Accurate case outcome models in economics, psychology, and marketing. *Psychology & Marketing*, 36:1046–1061. DOI: 10.1002/mar.21255.

Acknowledgements

The Routledge Handbook of Tourism Experience Management and Marketing could see the light of day due to the enduring assistance, support, and enthusiastic collaboration of many people. Here, I would like to have the opportunity to acknowledge them all.

I would like to thank all colleagues and the researchers in the field of tourism and hospitality, who graciously volunteered their time and efforts to put this amazing handbook together. They integrated the advised comments and alterations with full zeal within the stipulated timelines. I would like to express my sincere gratitude and thanks to all the contributors of the *Handbook*. I would also like to express my sincere appreciation to Arch G. Woodside for writing the foreword to this *Handbook*. I wish to extend my heartfelt thanks to Emma Travis, Commissioning Editor, and Lydia Kessell, Editorial Assistant of Routledge, for their patience and sincere support from the submission of the initial proposal to the full manuscript. I would also like to express a word of appreciation for all the other members of the Routledge editorial, production, and marketing teams for the successful accomplishment of this *Handbook* project. I would also like to acknowledge the constant encouragement and motivation offered by my father, mother, and other family members. Last but not least, I wish to express special thanks to my wonderful wife, Seema, and my lovely kids, Rakshit and Rayaan, for their constant support during the entire *Handbook* project.

Saurabh Kumar Dixit
Shillong
2019

Introduction

Saurabh Kumar Dixit

Our experiences are a bigger part of ourselves than our material goods.

(Gilovich, Kumar, and Jampol 2015)

The above quote of Gilovich, Kumar, and Jampol (2015) followed their 20-year study, which concluded that people derive more happiness from acquiring life experiences than from mere material possessions. Building favorable experiences is therefore crucial for all organizations operating in any industry. As an idea, experience has a broad appeal and has its linkages with different specialized disciplines of social science, such as anthropology, economics, marketing, psychology, sociology, and ethnology (Carù and Cova 2003). The multidisciplinary orientation and usage of experience has given wide-ranging and enigmatic definitions and theoretical meanings to the term. However, the organizations in the contemporary business arena are losing control of experience construction as the customers are actively engaged in the co-creation and shaping of the organizational experiences.

Experience can be traced into the mainstream literature of consumer behavior (where the consumers are deemed to be the sole decision makers) giving rise to the emergence of the experiential approach of consumer behavior way back in the eighties, emphasizing the role of emotions in decision-making (Holbrook and Hirschman 1982). Experience has been defined as "a steady flow of fantasies, feelings, and fun" (Holbrook and Hirschman 1982: 132); "the act of living through an observation of events and also refers to training and the subsequent knowledge and skill acquired" (Hoch 2002: 448); the "result of encountering, undergoing, or living through situations" and "triggered stimulations to the senses, the heart, and the mind" (Schmitt 1999: 25). According to Carlson (1997), an experience can be defined as a constant flow of thoughts and feelings that occur during moments of consciousness.

Tourism experience, first proposed by Boorstin in 1961, is defined as a popular consumer behavior that is a non-spontaneous and pre-fabricated experience gained from travel. Historically, tourism experience has been discussed as a phenomenon since the 1960s and became more popular in the 1970s, with the discussion of the phenomenology of experience by authors like MacCannell (1973) and Cohen (1979), Quinlan-Cutler and Carmichael (2010). A noteworthy publication on tourism experience was the book by Csikszentmihalyi (1975) (Ritchie, Tung, and Ritchie 2011), which offered a further path to the exploration of tourism experience. Initially, in the seventies, tourism experience was treated either as something superficial, or as a search for authenticity, or an effort to escape from an alienated world (Cohen 1979).

In the nineties, the emphasis within experience was to understand the thoughts and feelings of tourists, as reported in diaries or interviews (Andereck, Bricker, Kerstetter, and Nickerson 2006; Quinlan-Cutler and Carmichael 2010). More recently, Pine and Gilmore (1999) and Schmitt (2000, 2004) have emphasized how service management through experiential tools can enhance tourism experience. Andersson (2007) refers to the tourist experience as "the moment when tourism consumption and tourism production meet". There were further advances in the concept of experience, and this has brought about various perspectives on all kinds of experience. Ritchie, Tung, and Ritchie (2011) point out some experience typologies: great experience, quality experience, extraordinary experience, memorable experience and, more recently, the creative experience (Ihamäki 2012; Richards 2011; Tan, Kung, and Luh 2013). Although relevant in the literature, these terms are being used without a commonly agreed definition and interchangeably (Knobloch, Robertson, and Aitken 2016).

The notion of experience has spawned much attention in tourism management as well as marketing. Otto and Ritchie (1996) describe experience in leisure and tourism as "the subjective mental state felt by the participants". Hospitality and tourism businesses predominantly deal with the consumption of products and services, which in turn results largely from the varied nature of experiences (affirmative, neutral, or negative). Due to the intangible and extremely perishable nature of these products/services, the experiences shaped by tourism activities determine the suitability or refutation of these offerings. The "tourism product" is nowadays not only appreciated as the physical entity of the core product, but also the whole experience associated with the product. As a result of which, every hospitality and tourism organization endeavors to market authentic and engaging travel experiences to its consumers. Overall tourism experiences are shaped from the three important components i.e. the service/product itself, the service settings/environment, and the service delivery systems. Each of these three components strives to meet the expectations of the tourists, in order to give them the feeling of a quality tourism experience.

An understanding of the ways in which tourists experience the places and people they visit is, therefore, fundamental to the study of tourism consumption. The tourists fulfill their personal and hedonic desires mostly by immersing themselves in vivid tourism experiences. Tourism is therefore appreciated as a transformative experience that satisfies escapism and hedonic motives amongst the tourists. Tourists' behavioral intentions are also largely dependent upon the variety of authentic experiences a destination offers. These experiences contribute essentially to the generation of pleasure, satisfaction, and loyalty among the consumers through the consumption of tourism products and services.

Tourism experience can be comprehended from both supply and demand perspectives. The supply side embraces tourism by offering a series of experiences attained through an amalgamation of a diverse array of products and services shaped in collaboration with a series of businesses such as hotels, restaurants, travel agents, transport operators, etc. The demand side of tourism experiences involves visualizing a customer centric approach in tourism experience creation. It therefore involves the developing and managing of organizational offerings as per the cravings of the customers. Good customer experience leads to a higher level of customer satisfaction.

There are five models of consumer experience that have been adapted to research into evaluating tourism consumer experience – hierarchical, flow, planned behavior, typological, and insider–outsider models (Dixit 2017: 34; Prentice, Witt, and Hamer 1998). These models are used generally to undertake tourism experiential research. Under the experiential setting, Pine and Gilmore (1999) introduced the term *Experience Economy*, arguing that the economy has moved to a stage where the "competitive battleground lies in staging experiences". This means that, "customer experience itself will be the only truly competitive advantage" ("Gartner surveys confirm customer experience is the new battlefield", 2014) and that customer experience becomes product entity itself (Pine and Gilmore 1999). They classified experience according to the level of customer involvement and customer participation. The results were four realms of tourist experience: *Entertainment, Educational, Esthetic, and Escapist.* For example, travelers are much more likely to stay in a hostel or utilize "Couch Surfing" if they are looking to immerse themselves in local culture, which may be described as "escapist". On the other hand, mass tourists prefer to stay in a globally recognized resort to seek "entertainment". Schmitt (1999: 53) explained that marketers can create for customers sensory experiences (SENSE), affective experiences (FEEL), creative cognitive experiences (THINK), physical experiences, behaviors and lifestyles (ACT), and socio-identity experiences that result from relating to a reference group or culture (RELATE). These experiences provide "sensory, emotional, cognitive, behavioral, and relational" values that are presented in the form of an experience sequence, as illustrated in Figure I.1.

Ritchie and Crouch (2003) suggest that the service experience in tourism contains four dimensions that detail what tourists seek and thus for that reason they should be taken into consideration when creating experiences (see Figure I.2). The first dimension is *hedonics* – for example; doing something memorable, doing something thrilling, or being challenged in some way. Again, here the experience is focused on what the tourist would view as pleasurable. The second dimension is *peace of mind*. This would include physical comfort and safety along with privacy and relaxation. Dimension three is labeled as *involvement*. This contains issues related to having choices and being educated and informed. The fourth and last dimension is *recognition*.

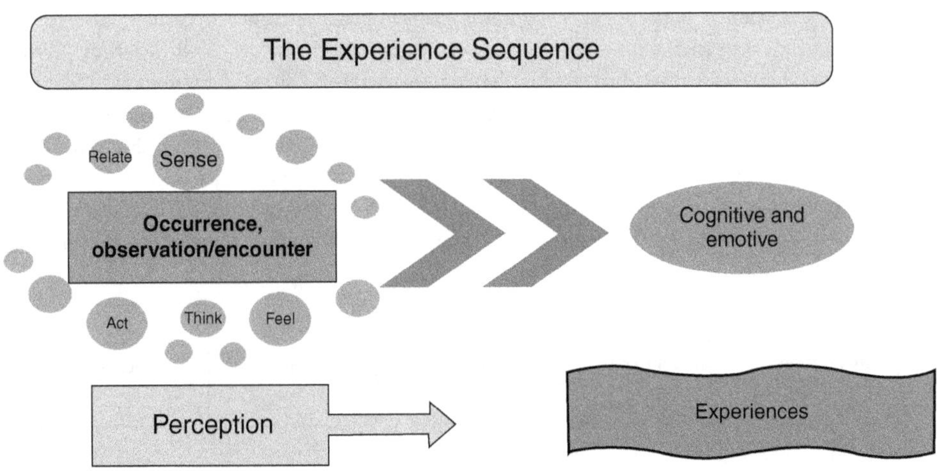

Figure I.1 The experience sequence
(*Source*: Chan, Jennifer Kim Lian and Kasim, Azilah 2017)

Figure I.2 The tourism experience dimensions
(*Source*: Author based on Ritchie and Crouch 2003)

Vacationers usually perceive and evaluate their journey as a total experience. The tourism experience usually encompasses the amalgam of experiences encountered by the tourists from different services/products, including their attitudes and expectations. The components of experience building normally involve the physical, intellectual, emotional, and spiritual traits of the tourists. The travelers therefore observe different phases of experiences when taking their trip. The Clawson and Knetsch (1966) model identified five distinct phases of travel experience: anticipation (planning and information gathering), travel to site, on-site activities, return travel, and recollection (see Figure I.3). The interaction between the tourist and service providers during each phase contributes to shaping the total outcome of a trip experience.

Tourism experience is therefore a complex consumptive attribute shaped by the totality of internal and external impulses, influences, and emotions, as well as the physical environment. The individual dimensions that come together (Shaw and Ivens 2002) within the individual may involve the consumer emotionally, physically, intellectually, and spiritually (Mossberg 2007) and are memorable (Pine and Gilmore 1999). Hultén (2011) proposes that the environmental context in which a tourism experience is obtained has an impact on customers' emotional connection with a brand. Hence, no two individuals will have the same experience because experience dimensions are interpreted individually.

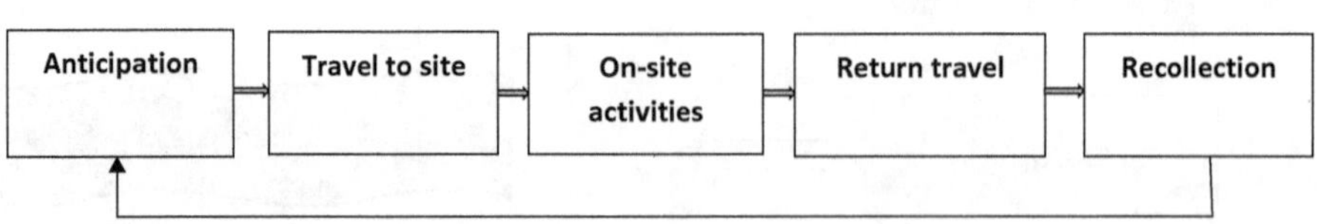

Figure I.3 Stages of tourism experience formation
(*Source*: Author based on Clawson and Knetsch 1966)

Moreover, the organization cannot stage a consumer experience; instead, organizations can only create the environment and the circumstances in which consumers could have an experience (Mossberg 2007).

Researchers have now acknowledged that tourism and hospitality are the business of marketing and selling experiences (Li 2000; Mannell and Iso-Ahola 1987; Ooi 2002; Prentice, Witt, and Hamer 1998; Waitt 2000). Schmitt (1999) who proposes a new approach called *experiential marketing*, provides a strategic framework that stands in contrast to the traditional marketing perspective of that time. *Experiential marketing* and the attempt to form memorable and extraordinary experiences have become significant to contemporary tourism organizations. This approach emphasizes the customer experience and replacing purely functional values by focusing on consumption as a holistic experience. It is the task of the marketing managers/planners to weigh the potential value of each type of experiential tourist and decide which of them should be targeted for their destination offerings. However, understanding tourism experiences is difficult because of their existential nature; experiences are embodied in people, are felt personally, and can only be expressed to and not felt by other people. Experiences are also multifaceted; they arise from activities and the environment as well as the social contexts embedded in the activities. It is therefore recognized that the organizations should include the consumers in the process of the co-production and co-creation of products/services.

In the present *Handbook*, an attempt has been made to construct a comprehensive foundation for an understanding of varied perspectives of tourism experience by means of impartial and rational coverage of up-to-date philosophies, practices, and global success stories. The volume therefore aims at bridging the gap in the contemporary available literature by carefully examining the diverse perspectives of tourism experience. Inside the tourism body of knowledge, tourism experiences have sometimes been synonymously linked to tourist experiences. The present *Handbook* is a prudent effort to bring together chapters on both the demand and supply sides of the tourism experience. The *Handbook* offers the fusion of chapters authored by educators, industry practitioners, and eminent experts involved in experiential tourism research, industrial practices, and operations from different corners of the globe. The *Handbook* will definitely help the researchers and practitioners in studying different paradigms of tourism experience management and marketing with minimal effort.

The handbook organization

The 52 chapters of the *Handbook* are divided into six parts to offer a greater understanding of the facets associated with tourism experience management and marketing. The collection of themes presented in the *Handbook* exemplifies an exceptional scholarly attempt to cover a large number of both theoretical and applied dimensions, in order to nurture synergetic interface between academia and industry. Part I "Tourism experience: theories, structure and frameworks" attempts to develop the critical understanding of the tourism experience by means of state-of-the-art reviews of theory and research; Part II "Crafting tourism experiences" addresses the topics relating to the construction and designing of tourism experience; Part III "Marketing tourism experiences" strives to deliver contemporary approaches to marketing tourism experiences; Part IV "Technology enabled tourism experiences" highlights the modern technological interventions in tourism experience management and marketing; Part V "Sustainable tourism experience" aims to examine the sustainability perspectives relating to tourism experience; Part VI "Emerging avenues of tourism experience and co-creation" looks into the recent and future developments and issues in tourism experience. The multidisciplinary nature of the *Handbook* will enable its readers to appreciate tourism experience from multifarious standpoints.

Part I: tourism experience: theories, structure and frameworks

Part I includes ten chapters offering the comprehensive basic understanding of diverse dimensions of tourism experience. This part therefore builds the theoretical base for the readers to progress their horizon toward the prospering erudite field of tourism experience. Chapter 1 authored by Jon Sundbo and Saurabh Kumar Dixit has endeavored to reveal the theoretical foundation of tourism experience: the core theme of the *Handbook*. Different disciplines' understanding of experience are therefore presented and discussed in this chapter. The chapter also throws light on how experience is considered in relation to tourism – in literature and practice. Drita Kruja and Elvisa Drishti in Chapter 2 critically assess the existing body of research on tourism experience through systematic review techniques. The chapter investigates the main research questions, context, data collection, analysis methods, and main findings of the papers published in major tourism and hospitality journals from 2009–2018. This chapter also provides a reference guide in relation to the main aspects of the published studies in this research field. Fabrizio Ferrari, in Chapter 3, details the dynamic drivers of tourist experiences. He further elaborates on how tourist motivations evolve with the time horizon. Bert Smit and Frans Melissen in Chapter 4 distinguish different types of (customer) experiences, which can be expressed based on two dimensions – duration and outcome. These dimensions can be used to create a typology of experiences consisting of four main types or categories: epiphanies, transformative experiences, peak experiences, and extended experiences. Chapter 5 by Girish Prayag, Samuel Spector, and Jörg Finsterwalder critically evaluates the concept of customer experience as applied to the tourism industry. A case study on customer experience evaluation in Brazilian restaurants in the UK is incorporated in the chapter to highlight the importance of cognition and emotions as drivers of customer experience.

The purpose of Chapter 6 by Girish Prayag is two-fold. First, it offers a review of past topics of research on emotions in the tourism field. Second, it provides an assessment of the different ways of measuring emotions in tourism. Mariana de Freitas Coelho and Marlusa de Sevilha Gosling, in Chapter 7, deliberate the concept of the Memorable Tourism Experience (MTE) and review the results of memorable experience studies. This chapter also helps readers to find gaps in the area and guide future directions on memorable experience researches. Chapter 8, authored by Poh Theng Loo, elucidates the importance of service employees in service delivery, understanding customer experiences in tourism, the impact of service employees on customer experience, and strategies to manage service employees and customer experience. Keith Mandabach in Chapter 9 provides an overview and insight into the role that tourist emotional experiences play in the decision-making process and how they might alter their perceived overall image of the tourism site or experience. The chapter also discusses how the role of emotional experiences reflects on the tourists' reported satisfaction levels as well as developing loyalty for that tourism product and/or the intent to recommend that said product. Roy C. Wood in Chapter 10 posits a critique of the notion of the experience economy in tourism. It is concluded in the chapter that the concept of the experience economy has insufficient theoretical weight or practical utility to render it anything other than a mildly interesting insight into tourism.

Part II: crafting tourism experiences

Part II with 11 chapters highlights the varied constructs of designing tourism experience from different perspectives. In their Chapter 11, Bert Smit and Frans Melissen review a range of approaches, tools, and methods that could be used for designing experiences and their relevance and applicability to designing specific types of experiences within tourism. Jens Friis Jensen and Jon Sundbo in their Chapter 12 deal with how tourists and tourism businesses decide which experiences are good and important. The focus of the chapter is that there are two parties in tourism experiential choices, the tourism businesses and the tourists, and the two parties' choices do not always correspond. The gap between tourism businesses' experience propositions and tourists' realized experiences needs to be bridged. Chapter 13, authored by Melissa A. Baker and Kawon Kim, discusses the experiencescape and its key elements, i.e. the physical servicescape and the social servicescape, as well as the influences each has on the overall customer experience and key outcome factors such as satisfaction, word of mouth, and return intention. Brian Kee Mun Wong and Christy Yen Nee Ng in Chapter 14 discuss destination image, perception of tourism destination, and tourist motivation in relation to tourism experience. The importance of these elements to practitioners, when they strategize and implement positive and memorable tourism experiences, is also highlighted. Chapter 15 by Eva Maria Jernsand and Sandhiya Goolaup aims to understand how tourists learn through extraordinary experience. The chapter also suggests that experiential and transformational learning theories are specifically useful to the understanding of tourists' extraordinary experiences, although the core aspects of these are also found in learning theories on, for instance, education and organization. Karoline Wiegerink and Jan Huizing in Chapter 16 elaborate on the *city hospitality experience model* and describe the value-adding role of programs in which volunteer residents act as city hosts. This value creation is related to the three domains of hospitality, i.e. private, social, and commercial.

Joanna Seraphim and Farooq Haq in Chapter 17 applied design thinking to augment experiential heritage tourism in the UAE. The findings of this study identified authenticity, memorability, use of the five senses, and engagement as key elements for an effective experiential heritage tourism. Chapter 18 authored by Paul Strickland, Jennifer Laing, Warwick Frost, and Kim M. Williams explores trends in experience design implemented by wineries in Victoria, Australia, with the aim of attracting millennials as tourists. The chapter therefore contributes to existing theory by highlighting millennials as a potential market to increase their wine sales. Chapter 19, authored by Lee Jolliffe and Pairach Piboonrungroj, reviews the use of themes and stories in designing tourism experiences. Designers of the tourism experience can derive benefits from understanding the role that themes and stories play in the visitor experience. Minni Haanpää and José-Carlos García-Rosell. the authors of Chapter 20, review how tourism experiences are lived, performed, and co-created through the body by drawing on empirical data from animal-based tourism experiences in Finnish Lapland, namely dog-sledding experiences. Through these embodied performances, memorable tourism experiences are co-created. Jon Sundbo in Chapter 21 presents the theory of experiential dissonance, which is a new concept referring to situations in which tourists are attracted by one experience (called Frame) while another experience (called Occupant) is crucial for the positive value of the total experience and for the Frame experience being consumed again.

Part III: marketing tourism experiences

Part III comprises ten chapters which delve into the various ways of investigating, understanding, and documenting the experiential marketing strategies to be used in tourism businesses and destinations. Chapter 22 by Valentina Della Corte and Giovanna Del Gaudio reviews the issue of the evolution of some traditional marketing tools used by hospitality firms, like Customer Relationship Management and Loyalty Systems, as well as empirically examining the impact that the rise of the sharing economy has on revenue management (RM) techniques as well as the exploitation of CRM. Chapter 23, authored by Kuan-Huei Lee, discusses the different requirements, benefits, and contents of a marketing plan. The chapter further discusses stages in the life cycle of a product from introduction, growth, maturity, and decline, and various marketing strategies to be used in each of the stages. Raouf Ahmad Rather and Linda D. Hollebeek in Chapter 24 create a discourse on the effects of customers' tourism engagement with experiential marketing activities by developing and testing a framework. The authors further study the relationships between customer engagement, experience, identification, and loyalty in the chapter. Chapter 25 by Christy Yen Nee Ng and Brian Kee Mun Wong epitomizes the concept of relationship marketing. This will enable producers and customers to understand each other's needs and develop the willingness to cooperate with each other; to shift their focus from "transactional exchange" to "relational exchange". Robert J. Harrington, Michael C. Ottenbacher, Burkhard von Freyberg, Alexandros Paraskevas, and Laura Schmidt authored Chapter 26, which sheds light on a better understanding of the Oktoberfest tourists' customer journey, perceptions of attribute quality during the beer tent experience, and how attributes might be classified using moderated regression and multiple dependent variables.

Anestis K. Fotiadis and Anastasia Ef. Spyridou in Chapter 27 throw light on different dimensions of service quality. The chapter examines how service quality and customer satisfaction are influencing the industry. Finally, a case study highlighting the dimensions and outcomes of experiential quality relating to the visitors of port wineries is presented. Saurabh Kumar Dixit, who authored Chapter 28, reviews the role of gastronomic tourism in formulating destination-marketing strategies. The chapter further explores the origin of gastronomic experiences, tourism experience co-creation through gastronomy, authenticity in gastronomic tourism experiences, and the understanding of food consumption behavior. Chapter 29, authored by H. Kader Şanlıöz–Özgen and Metin Kozak, explores the concept of brand experience with its distinct relevance to the hotel industry, given its principal role within the tourism industry. Secondly, it presents the relationship between brand experience and reputation management, with reference to the use of social media. Kawon Kim and Melissa A. Baker in Chapter 30 offer a review of the existing literature on service failure, customer responses to service failure, and service recovery in relation to the travel and tourism industry, and suggests strategies that organizations can utilize to manage customers' unsatisfactory service experiences and increase customer satisfaction. Dinesh Kumar and Punam Gupta in Chapter 31 describe "dark tourism", in which negative experiences are marketed as destinations. The chapter also offers examples of the places that have become popular due to their dark history.

Part IV: technology enabled tourism experiences

Part IV consists of seven chapters that offer evidence of how technological interventions can be used to augment an effective tourism experience by fascinating, engaging, and building a lasting bond with the tourists. Chapter 32, authored by Adam Pawlicz, endeavors to identify distinct characteristics of the sharing economy experience, presents differences between the traditional and sharing economy, and highlights the importance of sharing economy platforms in creating value for customers. Roberta Minazzi in Chapter 33 deliberates on the "traveler journey" and explores travelers' experience during the different steps of their travel behavior (pre, during, and post-trip). The analysis of the "traveler journey" shows a gradual convergence of travel behavior activities from the pre-trip and the post-trip stages to the during-trip. Sabrina Seeler and Heike Schänzel in their Chapter 34 offer a holistic understanding of the multifacetedness of co-constructed tourist experiences. The chapter discourses about types of co-construction, the motivations behind digital narratives, and ways in which tourism practitioners are leveraging on first-person voices. Chapter 35, authored by Roberta Garibaldi and Fabiola Sfodera, focuses on the ways in which new technologies have enhanced the tourist experience by adding new values and contents. After a brief overview of scholarly research, in order to outline the general context, a selection of international case studies related to wine tourism is presented in the chapter. Anke Schneider, Wilhelm Loibl, and Ann Hindley Dayna Ortner in their Chapter 36 review the importance of a picture's color characteristics on websites. The chapter uses an automated approach to explore the hue, saturation, and luminance of pictures on UK DMO websites, in order to shape the UK's e-tourism experience. Yeonjung (Alice) Kang and Tiffany S. Legendre aim to understand how tourist experiences are managed at a destination level in their Chapter 37. The chapter further adopts a value co-creation framework to illustrate an exemplary case from the Korea Tourism Organization (KTO). In Chapter 38, Sandra Maria Correia Loureiro offers a review of tourist experience in real and virtual worlds (virtual and augmented reality). The chapter further examines existent definitions and conceptualizations of experience (and related constructs), virtual reality (VR), and augmented reality (AR), and delivers a historical perspective of real and virtual experiences within tourism.

Part V: sustainable tourism experience

Part V contains a cluster of five chapters that examine sustainability as an essential element in tourism experience construction. Frans Melissen and Bert Smit in Chapter 39 explore this multifaceted relationship between tourism and sustainability by focusing on the concept of the visitor experience. The chapter concludes that adapting tourism to becoming more sustainable is not a mission impossible but, by integrating sustainability in the design process for visitor experiences, could even enhance visitor experiences. António Sérgio Araújo de Almeida in Chapter 40 elucidates the meaning, scope, and impacts of identity decomposition in the perspective of a correspondent stratification of the intensifying factors of the tourism experience (IFTE) and its linkages to tourism sustainability. Chapter 41, authored by Jennifer Chan Kim Lian, reviews rural tourism experiences as experiential products; and illuminates the importance of co-creation experiences to achieve sustainable tourism experiences (STEs) within the rural tourism context at Kota Belud, Sabah, Malaysia. Hung-Che Wu and Guo-Wei Chen, who authored Chapter 42, contribute to the theoretical advancement of the field of attraction management by empirically testing the interrelationships of experiential quality, green emotional attachment, green image, green switching experience, green experiential satisfaction, and green experiential loyalty. Chapter 43 by Rashmiranjan Choudhury and Saurabh Kumar Dixit enlightens readers about tourism experience building from the perspectives of tourism stakeholders; specifically the guest and host perspective. The chapter also explicitly investigates the guest–host responsibilities and sustainability issues involved in providing a quality tourism experience.

Part VI: emerging avenues of tourism experience and co-creation

Part VI encompasses nine chapters addressing the evolving themes and issues of tourism experiential management and marketing. Mine Inanc and Metin Kozak in Chapter 44 aim to develop a model in a specific reference to how the co-creation system works. The model also indicates the details of the elements, e.g. inputs (interactions among consumers and between consumers and service providers, etc.), outputs (consumer satisfaction, behavioral intentions, etc.), and instruments (environment, experience, etc.). Chapter 45, authored by Xiang Ying Mei, focuses on exploring the complexity of tourist experiences and the relevance of value co-creation through Service Dominant Logic (SDL) in the tourism industry. The book chapter also explores two case studies in order investigate how co-creation can occur through the SDL in practice. Francesco Raggiotto, Cristiana Compagno, and Michela C. Mason, who authored Chapter 46, deal with the challenges and opportunities for sustainable tourism development related to the employment of place-based business models. The chapter presents the case of Albergo Diffuso, an Italian model of hospitality, as a Place-Based Business Model, examining how the strong bond place-business impacts on the model's ability to organize resources, obtain stakeholder support, and ultimately create value. Yan Zhang and Young-Sook Lee in Chapter 47 investigate face culture and its attendant traits, reflected in the Chinese tourists' consumption behavior. The chapter further brings out diverse aspects in co-creating tourist experiences in the Arctic destinations.

Richard Ek, Anne Hardy, Mia Larson, and Can-Seng Ooi, authors of Chapter 48, offer a review of emotional labor and discuss how it has been elaborated upon in tourism and hospitality management studies. The chapter also sheds light on the sharing economy with the case of Airbnb hosting. Chapter 49 by Abijith Abraham and Saurabh Kumar Dixit deliberates on the question of whether a subjective variable such as tourism experience has an impact on the competitiveness of a destination. To illustrate the chapter content authors presented the case of Kerala (India). Adela Balderas-Cejudo and George W. Leeson in their Chapter 50 highlight the needs and wants of seniors regarding their leisure traveling experiences. The chapter concludes with a discussion of key considerations in the improvement of customer experience in the silver economy. Chapter 51, contributed by Adiyukh Berbekova and Melissa A. Baker, discusses key definitions of crisis management, types of crises, and examples of recent crises. The chapter also provides an overview of crisis management plans, the importance of leadership, and key stages of the crisis management plan. Saurabh Kumar Dixit concludes the *Handbook* with a chapter entitled "Conclusion: Some Reflections on Experiential Tourism Management and Marketing", where the findings and themes of the diverse chapters are reviewed and synthesized, and future research directions for tourism experience research are shown.

The handbook is intended to provide management and marketing implications/ recommendations for tourism businesses and destinations, to enable them to create, manage, and market extraordinary tourism experiences effectively. Therefore, this volume has been created to add to this growing body of research by probing how tourism organizations are marketing experiences in order to lure an increasingly sophisticated traveler. The *Handbook* therefore stresses that marketing is the link between the experiential products/services, service operators, and the needs/wants of consumers. The *Handbook* also attempts to offer a complete reference source as it comprises chapters on both applied and academic interests.

The *Handbook* has been designed for the undergraduate, masters and research students, practitioners, and/or industry consultants. The interdisciplinary organization and diverse content of the volume also makes it of interest for even non-tourism researchers in fields such as Marketing, Consumer Behavior, Management, Psychology, Anthropology, International Business, Sociology, Cultural Studies, etc. As an editor of the *Handbook*, I strongly believe that the diverse and multidisciplinary nature of the *Handbook* will definitely supplement the existing body of knowledge on experiential management and marketing for tourism.

I am indebted to all of the chapter authors for sharing their latest knowledge and research through this handbook. It has been a pleasure and an excellent learning experience working with you all!

References

Andereck, K., Bricker, K., Kerstetter, D., and Nickerson, N. (2006). Connecting experiences to quality: Understanding the meanings behind visitors experiences. In N. Jennings, and G. e Nickerson (Org) (Eds.). *Quality tourism experiences* (pp. 81–98). Oxford: Butterworth-Heinemann.

Andersson, T. D. (2007). The tourist in the experience economy. *Scandinavian Journal of Hospitality and Tourism*, 7(1): 46–58. DOI: 10.1080/15022250701224035

Boorstin, D. J. (1961). *The image: A guide to Pseudo-events in America*. New York, NY: Atheneum.

Carlson, R. (1997). *Experienced cognition*. New York, NY: Lawrence Erlbaum Associations.

Carù, A., and Cova, B. (2003). Revisiting consumption experience a more humble but complete view of the concept. *Marketing Theory*, 3(2): 267–286. DOI: 10.1177/14705931030032004

Chan, J. K. L., and Kasim, A. (2017). Factors affecting tourist buying behavior. In Dixit, S. K. (Ed.). *The Routledge handbook of consumer behaviour in hospitality and tourism* (pp. 32–45). London: Routledge. DOI: 10.4324/9781315659657

Clawson, M., and Knetsch, J. L. (1966). *Economics of outdoor recreation*. Baltimore, MD: John Hopkins Press.

Cohen, E. (1979). A phenomenology of tourist experiences. *Sociology*, 13(2): 179–201.

Csikszentmihalyi, M. (1975). *Beyond boredom and anxiety: The experience of play in work and games*. San Francisco, CA: Jossey-Bass, Inc.

Dixit, S. K. (2017). Introduction. In Dixit, S. K. (Ed.). *The Routledge handbook of consumer behaviour in hospitality and tourism* (pp. 1–3). London: Routledge. DOI: 10.4324/9781315659657

Gilovich, T., Kumar, A., and Jampol, L. (2015), A wonderful life: Experiential consumption and the pursuit of happiness. *Journal of Consumer Psychology*, 25(1): 152–165. DOI: 10.1016/j.jcps.2014.08.004

Hoch, S. J. (2002). Product experience is seductive. *Journal of Consumer Research*, 29(3): 448–454. DOI: 10.1086/344422

Holbrook, M. B., and Hirschman, E. C. (1982). The experiential aspects of consumption: Consumer fantasies, feelings, and fun. *Journal of Consumer Research*, 9(2): 132–140. DOI: 10.1086/208906

Hultén, B. (2011). Sensory marketing: The multi-sensory brand-experience concept. *European Business Review*, 23(3): 256–273. DOI: 10.1108/09555341111130245

Ihamäki, P. (2012). Geocachers: The creative tourism experience. *Journal of Hospitality and Tourism Technology*, 3(3): 152–175. DOI: 10.1108/17579881211264468

Knobloch, U., Robertson, K., and Aitken, R. (2016). Experience, emotion, and eudaimonia: A consideration of tourist experiences and well-being. *Journal of Travel Research*, 56(5): 651–662. DOI: 10.1177/0047287516650937

Li, Y. (2000). Geographical consciousness and tourism experience. *Annals of Tourism Research*, 27(4): 863–883. DOI: 10.1016/S0160-7383(99)00112-7

MacCannell, D. (1973). Staged authenticity: Arrangements of social space in tourist settings. *The American Journal of Sociology*, 79(3): 589–603. DOI: www.jstor.org/stable/2776259

Mannell, R. C., and Iso-Ahola, S. E. (1987). Psychological nature of leisure and tourism experience. *Annals of Tourism Research*, 14(3): 314–331. DOI: 10.1016/0160-7383(87)90105-8

Mossberg, L. (2007). A marketing approach to the tourist experience. *Scandinavian Journal of Hospitality and Tourism*, 7(1): 59–74. DOI: 10.1080/15022250701231915

Ooi, C.-S. (2002). *Cultural tourism and tourism cultures: The business of mediating experiences in Copenhagen and Singapore*. Copenhagen: Copenhagen Business School Press.

Otto, J. E., and Ritchie, J. B. (1996). The service experience in tourism. *Tourism Management*, 17(3): 165–174. DOI: 10.1016/0261-5177(96)00003-9

Pine, B. J., and Gilmore, J. H. (1999). Welcome to the experience economy. *Harvard Business Review*: 97–105 (July–August).

Prentice, R. C., Witt, S. F., and Hamer, C. (1998). Tourism as experience: The case of heritage parks. *Annals of Tourism Research*, 25(1): 1–24. DOI: 10.1016/S0160-7383(98)00084-X

Quinlan-Cutler, S., and Carmichael, B. (2010). The dimensions of customer experience. In M. Morgan, P. Lugosi, and J. R. B. Ritchie (Org) (Eds.). *The tourism in leisure experience: Consumer and managerial perspectives* (pp. 3–26). Bristol: Aspects of Tourism.

Richards, G. (2011). Creativity and tourism: The state of art. *Annals of Tourism Research*, 38(4): 1225–1253. DOI: 10.1016/j.annals.2011.07.008

Ritchie, J. R., and Crouch, G. I. (2003). *The competitive destination: A sustainable tourism perspective*. Cambridge, MA: CABI Publishing.

Ritchie, J. R. B., Tung, V. W. S., and Ritchie, R. J. B. (2011). Tourism experience management research: Emergence, evolution and future directions. *International Journal of Contemporary Hospitality Management*, 23(4): 419–438. DOI: 10.1108/09596111111129968

Schmitt, B. (1999). Experiential marketing. *Journal of Marketing Management*, 15: 53–67. DOI: 10.1362/026725799784870496

Schmitt, B. (2000). *Marketing experimental*. São Paulo: Nobel.

Schmitt, B. H. (2004). *Gestão de Experiência do Cliente*. Porto Alegre: Bookman.

Shaw, C., and Ivens, J. (2002). *Building great customer experiences*. New York, NY: Palgrave Macmillan.

Tan, S. K., Kung, S. F., and Luh, D. B. (2013). A model of "creative experience" in creative tourism. *Annals of Tourism Research*, 41: 153–174. DOI: 10.1016/j.annals.2012.12.002

Waitt, G. (2000). Consuming heritage: Perceived historical authenticity. *Annals of Tourism Research*, 27(4): 835–862. DOI: 10.1016/S0160-7383(99)00115-2

Part I

Tourism experience

Theories, structure and frameworks

1

Conceptualizations of tourism experience

Jon Sundbo and Saurabh Kumar Dixit

Introduction

Experiences are posited at the center of tourism consumption; hence tourists travel in search of novel opportunities to attain memorable and enriching experiences. The scenario of modern society, the needs of consumers and the marketing practices of the organizations are changing dramatically. Sophisticated attitudinal, societal and technological changes have led to the creation of memorable experiences, giving an added benefit to the tourist and a competitive advantage to the business organizations. The tourism experience has been noted by both academics and policymakers as pivotal; however, the disparity in the central meaning of the concept is still evident (Jennings et al. 2009).

Therefore, this introductory chapter of the *Handbook* provides an overview of experience by highlighting the role of experience in tourism and discusses definitions of experience and different approaches to explain the growing interest in experiencing. The chapter deliberates different stages of tourism experience formation and the evolution of different experiential theories. The chapter also elaborates multidisciplinary perspectives of experience.

How experience is considered an important factor in tourism, in literature and practice will be discussed in the section to follow. This includes both what tourists experience as well as why and how tourism firms develop experiences as part of their product portfolio. Experience not only has an academic, conceptual interest, but it is also crucial for tourism firms' and organizations' business and daily work. This association between academic conceptualization and business practice will also be emphasized.

Experience: an overview

Experience is a complex psychological phenomenon, which can be conceptualized in varied manners. The current focus on experience may make it appear as though experience is a new phenomenon. This awareness has particularly been raised by Pine and Gilmore's seminal work *The Experience Economy* from 1999. But experience is a phenomenon that probably is as old as mankind. Cave paintings dating back 14,000 years are proof that mankind, already many years ago, was occupied by experiencing and telling stories about experiences. Even the scientific registration of experience is not new. The American philosopher John Dewey (1929, 1934, 1938) had even in the 1920s and 30s written several books on experience's role in peoples' perception of art, education and nature. What is new, however, is a tremendous increase in experiences measured as human activities, which in economic terms is caused by increased wealth in societies and the satisfaction of material needs, leading people to seek fulfillment of mental and sensory needs. The forms of experience have also changed due to technological development; a great deal of experience is now derived from ICT media (TV, smartphones, the Internet, etc.) and cheap air transport increases long-distance tourism. Scientific attention toward and the conceptual discussion of experience has also developed substantially during the last two decades.

We may start the discussion and explication of the experience concept with a brief general definition:

> An experience is something that happens in peoples' minds, it is determined by external stimuli and elaborated via the mental awareness that people have from earlier motivational experiences, mental needs and personal strategies. Experience can be released by stimuli that affect all the senses.
>
> *(Sundbo and Sørensen 2013: 2)*

Sometimes it is added that to be classified as an experience, an event should be extraordinary and memorable (Pine and Gilmore 1999, Mossberg 2008), but that criterion has been previously discussed (Sundbo and Sørensen 2013). It has also been debated whether experiences should be authentic (e.g. Gilmore and Pine 2007). It might be queried what authenticity is, but also events and stimuli that in no way could be classified as authentic can be called an experience, for example visiting a shopping mall or an "artificial" folk dance performed for tourists. Organizations stage an experience whenever they engage customers, connecting with them in a personal, memorable way. Staging experiences is not about entertaining customers; it is about engaging them (Pine and Gilmore 1999).

Multidisciplinary perspectives of experience

Although the above definition emphasizes the mental and maybe physiological elements, experience has also been illustrated from numerous different disciplinary perspectives. These different perspectives look into different dimensions of experiences among individuals and society, besides illuminating the reasons why comprehending experiences is important for people. The contemporary theoretical understanding of experience is generally conceived in three disciplinary approaches: economics, psychology and sociology. In the following paragraphs, these different perspectives will be presented and discussed.

Economic perspective on experience

Within an economic perspective Pine and Gilmore (1999) looked at experience as a general business phenomenon, which can be used in any business either as core products, add-ons to goods and services or in marketing. They talk about experience as "staging business as a theatre". They argue that experience is a new field with great consumer interest, high added value and profit. Economic growth increasingly comes from experiences, which replace services as the fastest growing economic activity category, just as services in the late 1900s replaced goods. There are also analyses and theories that look at experience as a more delimited and focused economic sector. These analyses have defined a core experience sector which includes, for example: tourism, leisure and sport, entertainment, design, games and other ICT-based experiences (e.g. Nilsen and Dale 2013). Around that sector is often defined an "experience-related" sector containing, for example: the arts, architecture, TV and publishing. These are the traditional cultural or creative industries. Some analyses concern creative industries in which the "experience-related" sector, and sometimes the above defined experience industries, are included (e.g. Department for Culture, Media and Sport 2001, Power 2010). Statistically there is a confusion between "experience industries (or economy)", "culture industries (or economy)" and "creative industries (or economy)" (see Andersson and Andersson 2006, Lorentzen 2013). Tourism may be placed in both "experience industries" and "creative industries", but hardly in "culture industries" although there is an overlap between cultural activities and tourism. Thus, we may conclude that economics is increasingly paying attention to experience, but the exact definition of an economic experience sector is still under discussion.

Psychological perspective on experience

In the previous section we have defined experience as a mental phenomenon. In spite of that, psychology has only demonstrated a limited interest in explaining and defining experience. Csikszentmihalyi (2002) has presented his *flow theory*: flow is when people become very occupied by a certain achievement activity, such as becoming able to climb a mountain, write a novel or become a wine expert. The learning and progress in the achievement makes people forget time and place, and gives them a certain mental satisfaction – an inner peace and a sense of meaningfulness. Csikszentmihalyi calls this sensation flow experience. This is correct, but not all experience is flow (see also Privette 1983). For example, admiring the architecture of an old building that a tourist passes by may be an experience to be remembered but it cannot be classified as flow. The explanation of experience as a stimulus-based mental process has been discussed by Jantzen (2013). He presents a model of experiencing where external stimuli and the individual's perception of these decides whether the individual gets an experience and which type of experience she or he will get. Jantzen suggests two dimensions for analyzing the experience: one is the hedonic tone (whether the stimuli are pleasant or unpleasant) and the other is arousal (how much awareness and energy the stimuli create). This leads to four types of experience: relaxation, boredom, excitement and anxiety. Jantzen's theory is behavioristic and to some degree physiological, which might be discussed by other psychologists, but without doubt an experience is connected with the physiological senses (hearing, seeing, smelling, tasting, feeling with the fingers, tongue, etc.). However, any larger discussion has not yet emerged. It has been suggested that experiences need not only be positive and enjoyable. They can also be sad (such as participating in a funeral) and can include more serious activities, such as learning processes cf. Pine and Gilmore (1999), who operate with four types of experiences: aesthetic, escapism, entertainment and educational. One semantic point of consequence to note is that the term "experience" in English covers two meanings, which in other languages (for example German) are expressed by two different terms: one that characterizes joy (in German: *erlebniss*) and one that characterizes learning (in German: *erfahrung*). This must be taken into consideration when we define and discuss experience utilizing the English language.

Sociological perspective on experience

The sociological side of experience has not been emphasized that much. The German sociologist Gerhard Schulze wrote a large report (in German) in 1992 containing an empirical investigation of peoples' cultural behavior and consumption in the city of Nürnberg. He divided the population into different strata depending on their cultural consumption behavior, based on age, education and general lifestyle. Schulze characterized this consumption as experience ("*erlebniss*") and developed a theory of the experience society ("*erlebniss gesellschaft*") based on the idea that experience consumption has become an important part of peoples' lives. Sundbo and Sundbo (2018) have further developed the sociological explanation of people's search for experiences. They explain the act of seeking out experiences as giving people an identity and being a social status marker. People around the world create certain interest fields, for example sport or gastronomy, and within them specific "paradigms" called Interest Regimes develop, for example interest in bicycling or molecular cuisine. An Interest Regime is a common framework for interaction and discourses.

The most developed sociologically based approach to experience is the marketing one, which concerns people's consumer behavior. Experience marketing has become a large field within the marketing discipline (e.g. Schmitt 2003, Milligan and Smith 2008, Smilanski 2009, Dixit 2017). This field is a continuation of the view that consumption is also emotional or hedonic (Holbrook and Hirschman 1982, Carù and Cova 2003). The idea is that people seek out emotional experiences so much that it is an advantage for firms to connect sales of goods and services with experience. The success of the experience concept within marketing is partly caused by Customer Relationship Marketing having become a dominant approach. The approach stems from service theory, which has emphasized service suppliers' close interaction with customers in the moment of production and consumption of a service (Lovelock and Wirtz 2011). The customers' assessment of the provided service and the provision process is seen as crucial for their repeat purchases and willingness to pay a high price for the service. Customers' assessment is called customer experience. This theoretical view is called Service-Dominant-Logic (Vargo and Lusch 2006) and is considered a general marketing principle for all sectors. Such a use of the experience concept has been discussed and Sundbo (2015) suggests distinguishing between the "learning" ("*erfahrung*") experience of whether the service solved the problem, called instrumental experience, and a "pure" ("*erlebniss*") experience such as watching TV or running a marathon, called expressive experience.

Technology assisted experiences

These experiences are very important today. Radio and TV are old experience technologies. However, currently a large part of experiencing happens on the Internet, mobile phones and other new technology platforms, including social media, artificial intelligence and virtual reality. Surfing the Internet for travel information is a vital part of the tourism experience today. One model distinguishes close experiences, in which the experience production takes place where the experiencing persons are, e.g. a theater, as opposed to distant experiences, where the experience production takes place away from where the experiencing persons are and is delivered via technology, e.g. TV and music streaming on mobile phones (Sundbo and Hagedorn-Rasmussen 2008). The distant, technology mediated experiences are fundamentally based on the same psychological mechanisms as the close ones. However, where close experiences can activate all senses, distant ones cannot yet activate them all – not taste, smell and feeling. Close experiences are growing in number, but distant experiences are growing faster. Conceptual discussions of what distinguishes distant, technology mediated experiences and the design of these from close experiences and their design has already begun (Jensen 2013). What has changed with the emergence of distant, technology mediated experiences production is a larger user focus. Where, for example, traditional arts production – a movie or a novel – is a one-way communication from the director or author to the audience, contemporary technology mediated experiences production takes its point of departure from the users – what they want. This constitutes part of the process of the industrialization of experience production (Sundbo and Sørensen 2013) – a market approach where the customer and her or his willingness to buy is crucial. Experiences are co-created with customers. This development may be criticized as commodization and undermining artistic freedom; or it can be seen as democratization (Von Hippel 2005). However, there is a consensus that technology mediated experiences allow users to select and even produce experiences themselves.

Experience in tourism and tourism research

Experiencing is on the agenda for every tourism firm and organization. Hence, the study of experiences has long been an important part of tourism research. The contemporary tourism body of knowledge places an increasing accent on quality tourism experiences and perceives experience as the core of tourism businesses. Likewise, tourism companies pay added attention to the creation and management of quality experiences for tourists by placing tourism experience at the center of their products and services. Numerous experience innovations are currently being developed and are transforming tourism institutions. Typical examples are museums, which are transformed from scientific storage locations to experience centers, and the many types of festivals introducing new side-activities outside the core theme such as gastronomic meals, games, interactive home pages, lectures, etc. Ooi (2003) summarized tourism experience research into five different categories:

Cognitive psychology of tourism experiences – Under this category tourist perceptions and how these perceptions affect their experience (Mannell and Iso-Ahola 1987) are to be studied. Tourists' preconceived philosophies and expectations have been shown to affect the consumption, evaluation and experience of products.

Effect of tourism experiences – This category focuses on tourists seeking out experiences that benefit them personally in terms of improved mood, feelings of well-being or learning about other cultures (Prentice et al. 1998).

Depth of experiential engagement – This category includes those studies concentrating on "optimal" or "peak" experiences, which are seen as out of the ordinary and especially meaningful and engaging (Quan and Wang 2004).

Phenomenological approach – Under this research attempts to capture the whole range of experiences rather than concentrating on peak experience (Cohen 1979).

The gap between tourism experiences and everyday life – In this approach tourism is seen as distinct from everyday life and tourists are attracted by these differences (Urry 1990).

Technology mediated experiences and ICT as discussed in the previous section have added a crucial dimension to tourism experience. They can be used onsite as a complementary tool to the core attraction. Tourism is by definition close, not distant, experiencing. However, even distant ICT-based communication is part of a tourist experience. People seek information on the Internet about possible tourism experiences before they decide to go to a specific destination. When they return home, they visit web pages or watch videos from the destination. Tourism research talks about experiences *before, during, and after the journey* as a totality of a tourist journey (Mossberg 2007). It is no longer only the core activity or event that the tourist seeks, for example beach and sun or cultural institutions, that forms the impression the tourist returns home with. Also other seemingly peripheral factors and side-activities influence these impressions, for example the meals, the nature, the hotel bed, even the weather. Tourism firms and organizations must emphasize these factors and contribute resources to provide a good total experience. This requires collaboration between tourist firms and organizations, even some outside the traditional tourist sector, for example retailers, nature and environment authorities and farmers and manufacturers, who might contribute to give the tourists a good total experience. That is why the destination is important. A destination may be a place that tourists are going to, but it is also a social and economic community that can be the framework for collaborating and innovating to produce good total experiences for the tourists.

Experience has long been a topic within tourism research (e.g Arnould and Price 1993, Jackson et al. 1996, Uriely 2005). That interest has increased in recent years, which also forms the backdrop for this *Handbook*. Often the concept of experience, as is the case with other popular and widely used concepts such as innovation, democracy and community, is used in scientific literature without any definition or discussion of the concept. As experience increasingly becomes a concept within tourism research and theory, the need for discussing its meaning and definition becomes urgent. This chapter in the *Handbook* is one attempt to fulfill that need.

Experiences in tourism are fundamentally identical to experiences in general. They are a mental journey that may be remembered. The definition and discussions referred to in the previous section are also equally applicable to tourism. Since tourism fundamentally is about experiencing, it raises the question whether the experience concept at all is meaningful to use within tourism research. As a general term it cannot be used to differentiate different types of tourism and may not contribute to specific explanations of different tourist behavior. The earlier suggestion that only extraordinary experiences that are very memorable should be called experiences within tourism (Mossberg 2008) is an attempt to solve that problem. However, the tourism sector does not only provide experiences but also basic services, for example transport and clean hotel rooms, which need not be experiential or memorable. As earlier argued, even ordinary, not very memorable events such as relaxation in a very ordinary sense can be characterized as a tourist experience.

The tourist experience

A tourist's experience may be fundamentally explained from the general psychological approach to experiencing cf. the above discussion. However, tourists' experiences must also be understood in specific tourism contexts. Tourists relocate from their regular domicile in order to experience and we may suppose that the experiences ought to be different from those they could have at home. If not, they would have no interest in traveling and incurring the extra costs. At least some elements of the total experience should be different, such as the environment. People interested in opera may travel to an opera festival in the grounds of a castle, and then the spatial environment is different from the theater at home.

Tourists' motives for experiencing may be hedonic and self-gratifying (cf. the definition at the start; Carù and Cova 2003), but they may also stem from a need for relaxation – to alleviate stress and mentally recover from a busy everyday life. People may further become tourists to learn about other cultures or maybe even live within these cultures. This leads to the observation that research has revealed (Jantzen et al. 2012, Coelho et al. 2018), namely that experiences and tourism increasingly should be meaningful, i.e. that they should procure a meaning of life (e.g. religious, political or mental insight) or create possibilities for charity and helping other people. It also leads to the perspective of immersion (Pine and Gilmore 1999): some tourists have an experience by becoming immersed in tourism activities

(Hansen and Mossberg 2013). They become very absorbed by an activity, for example mountain climbing or building a water supply in a developing country. Immersion has a large element of flow (cf. Csikszentmihalyi 2002).

Some tourists' motives may be just habits or social norms. They are used to going on holiday, perhaps in their own summer house, or they live according to a social norm dictating that you should go on holiday. This raises the question of whether this routine motive for tourism implies experience. For these people, experiencing is not the motive for becoming a tourist, but such a routine tourist may nevertheless encounter some unintended experiences that are memorable.

People's motives for becoming tourists, and which experiences they seek, can be difficult to grasp. The main motives are not always what they seem to be or what the tourists themselves would say. The tourists may have other, maybe unconscious, motives. That can, for example, be revealed by letting them interpret their own photos from holidays (Andersson et al. 2016, Larsen 2006). It also, at times, leads to the primary experience not becoming the most important one to the traveler, but rather that the side-activities are the ones remembered most vividly and fondly cf. chapter 21 on experiential dissonance. It might be the meal in an amusement park or the seat in a cinema.

Tourists' motives for seeking particular experiences, or which side-experiences they seek when they are at the destination, must also be understood from their general Interest Regimes (Sundbo and Sundbo 2018).

Stages of tourists' experience

There has been a significant and diverse effort to gain a better understanding of tourism experiences in a variety of consumption environments, including atmospherics, tourism, leisure, food and beverages and lodging. The tourism experience is the multidimensional activity formed by people's encounters with products, services and businesses. It is considered to be a consumptive process involving several stages. Ahonen (2001) has developed a model to understand tourists' experience. He suggested that tourists' motives for experiences can be divided into four categories: *Getting emotionally affected, getting informed, getting practiced (to learn)* and *getting transformed*. The motivational elements in tourism experiences are physical, mental, moral and social. Based on that, he illustrated a seven stages model of experience processes, which is presented in Figure 1.1. Experience must be seen as a process that includes stages before and after the core experience (cf. Mossberg's, 2007, talk about experiences before, during and after the journey). The stages are stated as:

This model proposes *Orientation* (Awakening of interest and expectations where people test whether this possible experience fits with their general Interest Regime cf. above, Sundbo and Sundbo 2018), *Attachments* (Decision on whether to seek this experience or not cf. chapter 12 on experiential decision choice) as *before the journey* stages, *Visiting* (The actual visit consisting of the travel and the stay at the destination) as *during the journey* stage, *Evaluation* (Comparison with earlier experiences and how the visit fitted into the general Interest Regime), *Storing* (Documentation in the form of photographs, video, memory, presentations, etc.), *Reflection* (Ensuring that the experience will be an element in future mental awareness) and *Enrichment* (Using the experience to further strengthen the individual's position within an Interest Regime and social status). New practices created during the journey are covered in *after the journey* stages in the tourism experience process. The above model involves a complex process from the awakening of interest to the utilization of gained experiences. All touristic experiences do not cover all the seven stages.

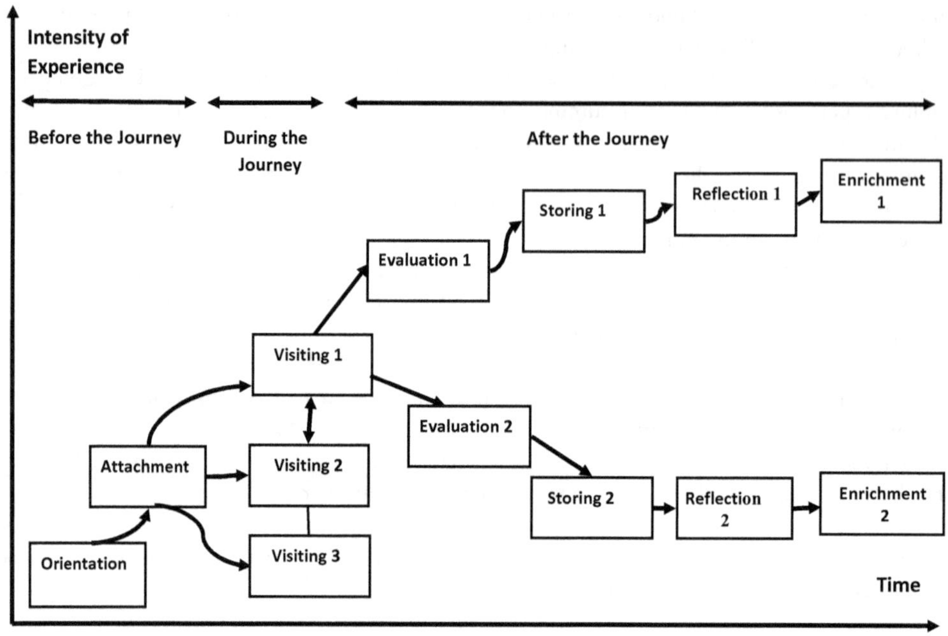

Figure 1.1 The stages of tourists' experience
(Adapted from Ahonen 2001)

Degrees of intensity vary at each stage according to the case, and there is some critical level of intensity establishing the ground for the next stage. Each further stage is an indication of the strength of the touristic experience. The more stages that are covered, the stronger the intensity of the experience is.

Tourism firms' experience proposition

Even though tourism obviously is about experiencing, many tourism firms are still not very focused on that and are quite bad at developing experiences for the tourists (e.g. Sørensen and Jensen 2015). The tourism sector provides two types of products: *services* such as transport, meals and beds and *"pure" experiences* such as nature, gastronomy and storytelling. The first is about delivering a basic experience. This is an absolute necessity for tourism; without a transport system and food there would be no tourism. The latter is about delivering expressive, or hedonic, experiences. This may differentiate the tourism products and destinations into high and low value ones for the tourists.

These two types of products, the services and the "pure" experiences, can be combined and tourism firms can add experiences to their services. Nevertheless, tourist firms often continue providing the same services that they did in the past. This is typically the case with overnight-stay enterprises such as hotels, camping sites and holiday home renting firms. All tourist firms can add experience elements to their services. These can be within the spatial framework, such as the architecture of buildings, the design of furniture or new concepts for designing amusement parks. They can also relate to behavioral aspects, i.e. how the personnel interact with the guests: a hotel receptionist may inform guests about possible experiences in town or about the special architecture of the hotel; the personnel in an amusement park can turn a negative experience, like queueing or a scenario where small children are denied admission to violent roller-coasters, into a positive one by small, kind explanations or by suggesting an alternative attraction (Sørensen and Jensen 2015). ICT technology can also be used. The tourist firm can introduce interactive information websites, YouTube videos or maybe games. Museums could likewise introduce interactive videos or mobile phones that play a small video about how people lived in historical times.

Innovation within experiences is important. Tourist firms have traditionally been low on innovation activities (e.g. Sundbo et al. 2007) but have recently increased their innovation activities. The experience sector has in recent investigations (for example in Denmark, Sundbo et al. 2013) shown itself to be more innovative than manufacturing and service sectors. However, the tourism industry is not the most innovative experience industry.

The tourism firm cannot be sure of which experiences the tourists create for themselves – that depends on the tourists' mental awareness toward experiences, earlier experiences and general Interest Regime cf. the above discussion. The tourist firms can only deliver experience propositions: they can provide experience elements that they *think* will create a specific experience in the tourists' minds, but they cannot be sure that the intended experience is really created. The tourist may get another experience because the mental awareness system is formed by the tourist's earlier experiences and Interest Regimes and his or her personality. Which is why we maintain that a tourist firm, or any other experience firm, can only provide experience propositions. This concept is inspired by the concept of value proposition that service theory has introduced (Vargo 2011). Service theory claims that service firms cannot be sure of which experience the customers will get and which value the provided service creates for the customer, partly because the customer entangles the instrumental and the expressive experience of the service delivery. Therefore, a service firm can only provide value propositions.

Experience design is a concept which is relevant here. Researchers and practitioners have developed tools and models for experience design (e.g. Benz 2014, Sundbo and Hagedorn-Rasmussen 2008, Hoboken 2013) and methods to investigate what the tourists really experience in their own minds. Following the trend of increased user-orientation in experience innovation, the design of experience elements can advantageously be achieved in interaction and co-creation with the tourists (Pullmann and Gross 2004, Eide and Mossberg 2013, Stare and Krizaj 2018). Storytelling can be an important part of experience elements and the marketing of new experiences (Mossberg 2008).

Even for destinations, the concepts of service proposition, service design and storytelling are important. Tourists often perceive the destination as one total experience. Consequently, the destination should also think in terms of developing this total experience and the storytelling about it.

Conclusions

Experience is important for the tourism sector. Tourism offers the marketplace to create and deliver memorable experiences; this marketplace can occur virtually. In this virtual setting tourists do not simply buy products and services and nor are they engrossed by refined environments or physical settings. There is an increasing scientific understanding of the formation of an enduring tourism experience. But regardless of a large number of works devoted to the domain of experience, there is still not one authorized definition. The concept is associated with a variety of facets (such as extraordinarily memorable, flow, entertainment, learning and so forth). Undoubtedly, the word "experience" is among the most used in the marketing literature. However, there are still challenges such as: How to create experience propositions that appeal to tourists' expectations? How to understand tourists' experience preparedness or Interest Regime? Therefore, it is important to study experiencing processes and to create scientific understanding – models, theories and empirical knowledge. That will be done in this *Handbook*. This will also be of utmost interest to tourism firms, which have the sale of experiences as their business model.

Experience can be viewed from multiple perspectives including the customer, the organization or both in the co-creation process. Experience must be understood as total experience. The concept not only implies the traditional tourism product (i.e. attractions, hotel rooms), but also other experience elements in the destination and peoples' expectations as well as their earlier experiences (their Interest Regimes). Moreover, people seek different types of experiences, e.g. hedonic, educational, meaningful, etc. This too must be addressed by tourism firms and organizations in their business plans and innovation activities.

References

Ahonen, S. (2001). 'Towards a general theory of tourism experiences: modelling experience process in tourism', *Tourism Review*, 56 (3/4), 33–37. DOI: 10.1108/eb058368

Andersson, Å. and Andersson, D. (2006). *The Economics of experiences, the Arts and Entertainment*, Cheltenham: Edward Elgar.

Andersson, T., Getz, D., Vujicic, S., Robinson, R., and Cavicchi, A. (2016). 'Preferred travel experiences of foodies: an application of photo elicitation', *Journal of Vacation Marketing*, 22 (1), 55–67. DOI: 10.1177/1356766715589621

Arnould, E. and Price, L. (1993). 'River magic: extraordinary experiences and the extended service encounter', *Journal of Consumer Research*, 20 (1), 24–45. DOI: 10.1086/209331

Benz, P. (ed.) (2014). *Experience Design, Concepts and Case Studies*, London: Bloomsbury.

Carù, A. and Cova, B. (2003). 'revisiting consumption experience: a more humble but complete view of the concept', *Marketing Theory*, 3 (2), 267–286. DOI: 10.1177/14705931030032004

Coelho, M., Gosling, M., and Araujo de Almeida, A. (2018). 'Tourism experiences: core processes of memorable trips', *Journal of Hospitality and Tourism Management*, 37, 11–22. DOI: 10.1016/j.jhtm.2018.08.004

Cohen, E. (1979). 'A phenomenology of tourist experiences', *Sociology*, 12, 179–202. DOI: 10.1177/003803857901300203

Csikszentmihalyi, M. (2002). *Flow*, London: Rider.

Department for Culture, Media and Sport. (2001). *Creative Industries 2001 – Mapping Document*, London: Department for Culture, Media and Sport.

Dewey, J. (1929). *Experience and Nature*, 1958 publication. New York: Dover publications.

Dewey, J. (1934). *Art as Experience*, New York: Capricorne.

Dewey, J. (1938). *Experience and Education*, 1963 publication. New York: Collier.

Dixit, S.K. (ed.) (2017). *The Routledge Handbook of Consumer Behaviour in Hospitality and Tourism*. London: Routledge. DOI: 10.4324/9781315659657

Eide, D. and Mossberg, L. (2013). 'Towards more intertwined innovation types: innovation through experience design focusing on customer interactions'. In Sundbo, J. and Sørensen, F. (eds.), *Handbook on the Experience Economy* (pp. 248–268). Cheltenham: Edward Elgar. DOI: 10.4337/9781781004227.00019

Gilmore, J. and Pine, J. (2007). *Authenticity: What Consumers Really Want*, Boston, MA: Harvard Business School Press.

Hansen, A. and Mossberg, L. (2013). 'Consumer immersion: a key to extraordinary experiences'. In Sundbo, J. and Sørensen, F. (eds.), *Handbook on the Experience Economy* (pp. 209–227). Cheltenham: Edward Elgar. DOI: 10.4337/9781781004227.00017

Hoboken, N.J. (2013). *Experience design, a Framework for Integrating brand, experience, and Value*, London: John Wiley.

Holbrook, M.B. and Hirschman, E.C. (1982). 'The experiential aspects of consumption: consumer fantasies, feelings and fun', *Journal of Consumer Research*, 9 (2), 132–142. DOI: 10.1086/208906

Jackson, M., White, G., and Schmierer, C. (1996). 'Tourism experiences within an attributional framework', *Annals of Tourism Research*, 23 (4), 798–910. DOI: 10.1016/0160-7383(96)00003-5

Jantzen, C. (2013). 'Experiencing and experiences: a psychological framework'. In Sundbo, J. and Sørensen, F. (eds.), *Handbook on the Experience Economy* (pp. 146–170). Cheltenham: Edward Elgar. DOI: 10.4337/9781781004227.00013

Jantzen, C., Fitchett, J., Østergaard, P., and Vetner, M. (2012). 'Just for fun? The emotional regime of experiential consumption', *Marketing Theory*, 12 (2), 137–154. DOI: 10.1177/1470593112441565

Jennings, G., Young-Sook, L., Ayling, A., Lunny, B., Cater, C., and Ollenburg, C. (2009). 'Quality tourism experiences: reviews, reflections, research agendas', *Journal of Hospitality Marketing & Management*, 18, 294–310. DOI: 10.1080/19368620802594169

Jensen, J.F. (2013). 'IT and experiences: user experience, experience design and user-experience design'. In Sundbo, J. and Sørensen, F. (eds.), *Handbook on the Experience Economy* (pp. 179–208). Cheltenham: Edward Elgar. DOI: 10.4337/9781781004227.00016

Larsen, J. (2006). 'Picturing Bornholm: producing and consuming a tourist place through picturing practice', *Scandinavian Journal of Hospitality and Tourism*, 6 (2), 75–94. DOI: 10.1080/15022250600658853

Lorentzen, A. (2013). 'Post-industrial growth: experience, culture or creative economies?'. In Sundbo, J. and Sørensen, F. (eds.), *Handbook on the Experience Economy* (pp. 45–64). Cheltenham: Edward Elgar. DOI: 10.4337/9781781004227.00008

Lovelock, C. and Wirtz, J. (2011). *Services Marketing*, Boston: Prentice Hall.

Mannell, R.C. and Iso-Ahola, S.E. (1987). 'Psychological nature of leisure and tourism experience', *Annals of Tourism Research*, 14, 314–331. DOI: 10.1016/0160-7383(87)90105-8

Milligan, A. and Smith, S. (2008). *See, Feel, Think, Do: The Power of Experience Marketing*, London: Cyan Books.

Mossberg, L. (2007). 'A marketing approach to the tourist', *Scandinavian Journal of Hospitality and Tourism*, 7 (1), 59–74. DOI: 10.1080/15022250701231915

Mossberg, L. (2008). 'Extraordinary experiences through storytelling', *Scandinavian Journal of Hospitality and Tourism*, 8 (3), 195–210. DOI: 10.1080/15022250802532443

Nilsen, B. and Dale, B. (2013). 'Defining and categorizing experience industries'. In Sundbo, J. and Sørensen, F. (eds.), *Handbook on the Experience Economy* (pp. 65–83). Cheltenham: Edward Elgar. DOI: 10.4337/9781781004227.00009

Ooi, C.-S. (2003). 'Crafting tourism experiences: managing the attention product', paper presented at the 12th Nordic Symposium on Tourism and Hospitality Research, Stavanger, 2-5 October.

Pine, J. and Gilmore, J. (1999). *The Experience Economy*, Boston: Harvard Business School Press.

Power, D. (2010). *Priority Sector Report: Creative and Cultural Industries*, European Cluster Observatory. Bruxelles: The EU Commission.

Prentice, R.C., Witt, S.F., and Hamer, C. (1998). 'Tourism as experience: the case of heritage parks', *Annals of Tourism Research*, 25, 1–24. DOI: 10.1016/S0160-7383(98)00084-X

Privette, G. (1983). 'Peak experience, peak performance and flow: a comparative analysis of positive human experiences', *Journal of Personality and Social Psychology*, 45 (6), 1361–1368. DOI: 10.1037/0022-3514.45.6.1361

Pullmann, M.E. and Gross, M.A. (2004). 'Ability of experience design elements to elicit emotions and loyalty behaviors', *Decision Sciences*, 35 (3), 551–578. DOI: 10.1111/j.0011-7315.2004.02611.x

Quan, S. and Wang, N. (2004). 'Towards a structural model of the tourist experience: an illustration from food experiences in tourism', *Tourism Management*, 25, 297–305. DOI: 10.1016/S0261-5177(03)00130-4

Schmitt, B. (2003). *Customer Experience Management: A Revolutionary Approach to Connecting with Your Customers*, New Jersey: John Wiley.

Schulze, G. (1992). *Die Erlebnis-gesellschaft*, Frankfurt: Campus.

Smilanski, S. (2009). *Experiential Marketing. A Practical Guide to Interactive Brand Experiences*, London: Kogan Paul.

Sørensen, F. and Jensen, J.F. (2015). 'Value creation and knowledge development in tourism experience encounters', *Tourism Management*, 46, 336–346. DOI: 10.1016/j.tourman.2014.07.009

Stare, M. and Krizaj, D. (2018). 'Crossing the frontier between touchpoints, innovation and experience design in tourism'. In Scupola, A. and Fuglsang, L. (eds.), *Services, Experiences and Innovation* (pp. 81–106). Cheltenham: Edward Elgar. DOI: 10.4337/9781788114301.00011

Sundbo, D. and Sundbo, J. (2018). 'Interest regimes in fruit wine tourism', *Journal of Gastronomy and Tourism*, 3 (3), 163–176. DOI: 10.3727/216929718X15410151918027

Sundbo, J. (2015). 'From service quality to experience – and back again?', *International Journal of Quality and Service Sciences*, 7 (1), 107–119. DOI: 10.1108/IJQSS-01-2015-0009

Sundbo, J. and Hagedorn-Rasmussen, P. (2008). 'The backstaging of experience production'. In Sundbo, J. and Darmer, P. (eds.), *Creating Experiences in the Experience Economy* (pp. 83–110). Cheltenham: Edward Elgar. DOI: 10.4337/9781848444003.00009

Sundbo, J., Orfila-Sintes, F., and Sørensen, F. (2007). 'The innovative behaviour of tourism firms – comparative studies of Denmark and Spain', *Research Policy*, 36 (1), 88–106. DOI: 10.1016/j.respol.2006.08.004

Sundbo, J. and Sørensen, F. (2013). 'Introduction to the experience economy'. In Sundbo, J. and Sørensen, F. (eds.), *Handbook on the Experience Economy* (pp. 1–17). Cheltenham: Edward Elgar. DOI: 10.4337/9781781004227.00005

Sundbo, J., Sørensen, F., and Fuglsang, L. (2013). 'Innovation in the experience sector'. In Sundbo, J. and Sørensen, F. (eds.), *Handbook on the Experience Economy* (pp. 228–247). Cheltenham: Edward Elgar. DOI: 10.4337/9781781004227.00018

Uriely, N. (2005). 'The tourist experience – conceptual developments', *Annals of Tourism Research*, 32, 199–216. DOI: 10.1016/j.annals.2004.07.008

Urry, J. (1990). *The Tourist Gaze: Leisure and Travel in Contemporary Societies*, London: Sage Publications.

Vargo, S. (2011). 'Market systems, stakeholders and value propositions. Toward a service-dominant logic-based theory of the market', *European Journal of Marketing*, 45 (1-2), 217–222. DOI: 10.1108/03090561111095667

Vargo, S. and Lusch, R. (eds.) (2006). *The Service-Dominant Logic of Marketing*, Armonk: Sharpe.

Von Hippel, E. (2005). *Democratizing Innovation*, Cambridge Mass: MIT Press.

2

A systematic review of the tourism experience research from 2009 to 2018

Drita Kruja and Elvisa Drishti

Introduction

Tourist experience is increasingly recognized as a key research theme in relation to the rapidly changing motivations of tourism destination choices since the 70s. The 'experience economy' has been pioneered as an object of research by Pine and Gilmore (1999) and narrowed down in tourism service experience by Otto and Ritchie (1996). Recent developments in the field of hospitality and tourism have led to a shift of emphasis from a supply-driven towards a demand-driven tourist experience (Kong and Johnston 2011; Minkiewicz et al. 2014). Modern experiential research in marketing echoes this view and it has since established a linkage with the 'resource-based theory' as a theoretical stand in which the customer-based firm orientation and the co-creation of added value portray the tourist as a 'maker' – and not just a 'taker' – of service experience (Campos et al. 2018; Prahalad and Ramaswamy 2004; Vargo et al. 2014).

From a *consumer (demand)* standpoint, a wealth of literature has provided support for the relationship between consumer behaviours such as tourist experience[1] and revisit intentions/decisions and client/traveller loyalty, word-of-mouth recommendations, and the emotional aspects of tourist satisfaction (Barnes et al. 2014; Biran et al. 2011; Brunner-Sperdin et al. 2012; Kruja 2017; Ma et al. 2013; Mathis et al. 2016; Pearce and Kang 2009; Robinson and Clifford 2012; Sharpley and Jepson 2011; Tussyadiah and Zach 2012). Tourist experience is also a major area of interest within the field of destination image (Kruja 2018; Morrison 2013; Pike 2012, 2002; Pike and Page 2014; Stepchenkova and Mills 2010). From the *firm (supply)* perspective, firms, in their quest for increased profit and market share have been looking for innovative methods to differentiate themselves in order to capture the customer's attention and attain emotional attachments (Binkhorst and Dekker 2009; Brunner-Sperdin et al. 2012). After all, tourism industry enterprises are in the business of selling positive memorable tourist experiences (Kim 2014; Park and Santos 2017; Tung and Ritchie 2011).

Several attempts have been made to provide a structured understanding of tourist experience. A number of researchers have used marketing theoretical frameworks such as 'consumer involvement theory' (Beatty et al. 1988; Pearce and Kang 2009), 'survival theory of tourist consumption' (McIntyre 2007), and 'flow theory' (Csikszentmihalyi 2014) as points of departure. On the other hand, following the perceptive and cognitive nature of experience, other researchers have sought theoretical frameworks from the field of psychology to develop and test different theoretical constructs that serve as antecedents of tourist experience, and/or even a tourist experience scale as such. For example 'appraisal theory' (Scherer 2005; Scherer et al. 2001), 'categorization theory' (Rosch and Lloyd 1978), 'script theory' (Tomkins 1987), 'intimacy theory' (Trauer and Ryan 2005), and 'motivation theory' (Snepenger et al. 2006), or even from the spirituality field a construct such as 'mindfulness theory' (Chen et al. 2017; Moscardo 2009). However, the generalisability of much published research into the general theory of touristic experiences (Aho 2001) is quite problematic due to the unstructured methodologies used in these studies. Of course, central to this issue is also a lack of general agreement on the tourist experience construct per se, since the variety of destinations and its multifaceted nature make it difficult to operationalize and/or inform an exhaustive list of features.

The current elusive state of affairs in tourist research is reflected in the fact that much of the research up to now has been qualitative in nature, i.e. phenomenological, grounded theory, and ethnographic approach. These methodologies, however, while offering thought provoking accounts on the tourist experience dimensions have failed to converge towards a universal concept (Ritchie et al. 2011). Hence, what is now needed is a synthetization of the previous research in terms of types of experience, visited sites, methodological approach, (un) structure, and findings. Following Pike's (2002) approach, this chapter will systematically review 51 articles in order to address the following set of questions:

- How many studies had examined tourist experience in an explicit travel context?
- How many studies utilized qualitative methods either in the measurement of destination image or in the development of attribute lists?
- How many studies used a structured approach?
- What range of methods were used in the data analysis?

Apart from measuring tourist experience, revisit intentions/decisions, and client/traveller loyalty, word-of-mouth recommendations, and the emotional aspects of tourist satisfaction, what other research variables (main or moderating) were included?

Methods

Consistent with previous extensive and comprehensive literature reviews on the tourist experience research topic (Ritchie and Hudson 2009; Tung and Ritchie 2011), this study seeks to narratively review the existing body of research on tourism experience in order to critically assess the most important research themes. Since this is a relatively unestablished research topic in tourism research literatures, this investigation of secondary nature will generate a structured scheme of the main research questions, context, data collection and analysis methods, and main findings of the papers published in three major tourism and hospitality journals along the 2009–2018 time horizon. This time horizon is common convention in past studies.

Several methods were applied in order to identify and analyse the relevant studies. As a first step, following a computerized search of the main databases, e.g. EBSCOhost, ProQuest, etc., where 'tourist experience' was used as the keyword in the title, abstract, and keywords of the articles, two major journals with reputable impact factor and numbers of citations were identified: *Annals of Tourism Research* and *Journal of Travel Research.*

The data analysis method took the form of content analysis to identify common themes in terms of definition, research questions, theoretical framework, research design, antecedents, consequences, moderators, and type of measurement of tourist experience.

Table 2.1 provides information on the orientation (demand or supply oriented), country/countries and site/sites visited, approach/methodology, sampling technique, types of experience, data collection, data analysis, and findings. Additionally, three more aspects were analysed. First, whether the methods were structured or unstructured, reported in column A, and addressed the tourist experience measurement and/or the development of scales to measure attributes associated with tourist experience. For those studies that employed a structured approach (S) a second column, B, was used to indicate the independent variables. Column C indicates the size of the sample.

Findings

- *Conceptualization of tourist experience*: The comprehensive review of the 51 articles revealed that in the majority of cases the conceptual definition of tourist experience includes affective and experiential attributes such as meaningful, not ordinary, extraordinary, multi-sensory, emotional, unique, memorable, personal, subjective, transformational, authentic, sacred, social, intimate, related with self-identity …
- The majority of studies (45) were related to specific tourist sites and that influenced the type of tourism experience. For example: creative experience, medieval tournament, food service experience, literary experience, spiritual experience, film-induced experience, embodied flamenco experience, nudist experience, shared media experience, mindful experience, past reference experience, staging experience, responsible/ethical experience, heritage experience, slum experience, spiritual/religious experience, dark experience, business travel experience, co-creation experience, prior and recent experience, female experience (in Islamic destination), hiking experience, geo-based experience, student experience, host experience, memorable experience, emotional experience, museum experience, culinary experience, cruise experience, transformative experience, etc.
- The main tourist destinations analysed in the papers were in Europe, the USA, and Asia.
- The majority of papers used unstructured techniques to operationalize the tourist experience or its destination-related experience construct.
- Almost all of the studies used qualitative methods, with tourists being either included or observed as a sampling unit, either as a mono-method or as a mixed-method.
- The main methodological approaches were a phenomenological and ethnographic approach followed by survey methodology.
- The most used data collection method was ‘in-depth interview’ followed by ‘questionnaire’.
- The main data analysis method was ‘theme identification’ from the content analysis, followed by ‘factorial analysis’, and ‘structural equation modelling’.
- The majority of studies were demand-oriented, with the primary sampling unit being the tourist.
- The main sampling techniques were ‘purposive sampling’ and ‘snowball sampling’, typical of the exploratory methodological approaches that dominate the field.

Table 2.1 Summary of tourism experience studies 2009–2018

Author (date)	*Orientation*	*Countries/ Sites visited*	*Approach/ Methodology*	*Sampling technique*	*Types of experience*	*A*	*B*	*C*	*Data collection*	*Data analysis*	*Findings*
(Tussyadiah and Fesenmaier 2009)	Demand oriented (Online bloggers)	Worldwide Tourist sites	Ethnographic approach	Purposive sampling	Shared media experience	U		120	Online communities and individual blogs	Content analyses	Online shared videos are valid mediators of tourist experience
(Andriotis 2009)	Demand oriented (Tourist)	Greece (Mount Athos)	Phenomenological approach	Purposive sampling	Religious experience	U		27	Participant observation, written experiences in visitor books, researcher journal notes	Theme identification	Four emergent themes: cultural, environmental, secular, and educational
(Pearce and Kang 2009)	Demand oriented (Tourist)	Australia	Confirmatory approach Survey	Purposive sampling	Prior and recent experience	S	Traveller loyalty & experience	254	Questionnaire	Descriptive analysis; mean comparison; correlation analysis	Consumer involvement theory validation; visitors with no prior and recent experience had the lowest continuing interest scores
(Brown 2009)	Demand oriented (International students)	UK (Universities)	Ethnographic approach	Purposive sampling	Student experience	U		150 13	Participant observation & in-depth interviews	Theme identification	International education increases intercultural competence, and shifts self-understanding, with long-term implications for personal and professional life

(Poria et al. 2009)	Demand oriented (Tourist)	Israel (Jerusalem and Beer-Sheva)	Mixed methodology; Qualitative & Quantitative	Convenience sampling; Systematic sampling	Emotional experience	S	Interpretation	20 166	Semi-structured interviews; face-to-face personal interviews;	Thematic analysis (Mind mapping technique); Reliability analysis; Exploratory factor analysis; T-tests; Correlation analysis	Managerial contribution; need to adopt innovative approaches to the management of heritage tourist attractions and provide different interpretations for different visitors.
(Buchmann et al. 2010)	Demand oriented (Tour participants)	New Zealand (Film-induced tourist destinations)	Exploratory Qualitative	Purposive sampling	Film-induced experience	U		33	pre- and post-tour questionnaires; individual and group interviews, observations, and tour journals	Case study	This study enriches the theoretical debate on the concept of authenticity by examining its relevance to the experiences of film tourists.
(Andriotis 2010)	Demand oriented (Gay vacationers)	Greece (Crete)	Ethnographic approach	Purposive sampling	Nudist experience	U			Non-participant observation notes	Analyses in relation to Foucault's principles of heterotopia	Beach, body, and sexuality related to tourism experience; conceptual contribution to beach heterotopia

(*Continued*)

Table 2.1 (Cont.)

Author (date)	*Orientation*	*Countries/ Sites visited*	*Approach/ Methodology*	*Sampling technique*	*Types of experience*	*A*	*B*	*C*	*Data collection*	*Data analysis*	*Findings*
(Huang and Hsu 2010)	Demand oriented (Tourist)	USA	Confirmatory approach Survey	Convenience sampling; Quota sampling	Cruise experience	S	C2C interaction	613	Online questionnaire	Exploratory and Confirmatory factor analyses; Structural equation modelling	The importance of incorporation of C2C interaction as one component of the relationship marketing theory
(Tung and Ritchie 2011)	Demand oriented (University students)	Canada	Grounded theory approach	Snowball sampling	Memorable experience	U		208	In-depth interviews	Theme identification	Four key dimensions of memorable experiences: affect, expectations, consequentiality, and recollection
(Sharpley and Jepson 2011)	Demand oriented (Tourist)	UK (English Lake District)	Exploratory Qualitative	Purposive sampling	Spiritual experience	U	Religion, spirituality, and rural tourism	18	Focus groups	Theme identification	Although tourists do not purposefully visit the Lake District in search of spiritual fulfilment, their visits frequently embrace a subconscious emotional dimension

(Biran et al. 2011)	Demand oriented (Tourist)	Poland (Auschwitz, Krakow)	Mixed methodology; Qualitative & Quantitative	Purposive sampling; Quasi-probability sampling	Dark experience	U		30 25 198	Semi-structured interviews, questionnaires,	Reliability analysis; Cluster analysis; Exploratory factor analysis	Four motivations of visit: see it to believe it, learning and understanding Famous death tourist attraction, emotional heritage experience
(Frantál and Kunc 2011)	Demand & supply oriented (Tourists & business owners)	Czech Republic	Mixed methodology	Semi-quota & probability sampling	Image experience	U		156 73	Questionnaire & semi-structured interviews	Ranking/preferences; Correlation analysis;	Wind turbines do not affect (negatively) the destination choice
(Robinson and Clifford 2012)	Demand oriented (Tournament visitors)	Australia (Medieval Tournament)	Confirmatory approach Survey	Probability sampling	Food service experience	S	Food authenticity	588	Observation and questionnaire	Principal component analysis	Scale development & testing; conceptual & practical contribution; associations of authenticity & revisit intentions
(Mekawy 2012)	Supply oriented (Craftsmen, business owners, agricultural & land workers, etc.)	Egypt (Slums)	Confirmatory approach Survey	Snowball & probability sampling	Responsible slum experience	S	Responsible tourism	464 89	Two staged questionnaires	Exploratory factor analysis	Slum inhabitants show positive attitudes toward the possibility of benefiting from slum tourism

(Continued)

Table 2.1 (Cont.)

Author (date)	*Orientation*	*Countries/ Sites visited*	*Approach/ Methodology*	*Sampling technique*	*Types of experience*	*A*	*B*	*C*	*Data collection*	*Data analysis*	*Findings*
(Tussyadiah and Zach 2012)	Demand oriented (Tourist)	USA (Midwestern states)	Confirmatory approach Survey	Non-probability sampling	Geo-based experience	S	Geo-based technology, tourism experience	415	Questionnaire	Confirmatory factor analysis; Regression analysis	Geo-based technology allows tourist to gain meaningful experiences & emotional attachment with places
(Wang et al. 2012)	Demand oriented (Tourist)	China (Zhejiang)	Confirmatory approach Survey	Probability sampling	Tourist experience	S	Service quality, and revisit intention	267	Questionnaire	Structural equation modelling	Tourist experience is a mediator between service quality and revisit intention; contributions for experiential marketing and sustainable development
(Shani and Uriely 2012)	Supply oriented (Local residents)	Israeli (Eilat)	Ethnographic approach	Purposive sampling	Host experience	U		51	In-depth interviews	Content analysis & theme identification	Local hosts may experience several sorts of difficulties, including loss of privacy, extra expenditures, hard physical work, and mental stress deriving from conflicting obligations of working and hosting

(Kim et al. 2012)	Demand oriented (University students)	USA	Confirmatory approach Survey	Probability sampling	Memorable tourism experience	S	Hedonism, refreshment, local culture, meaningfulness, knowledge, involvement, and novelty	62 511	In-depth interviews; Questionnaire	Content analysis; Exploratory and confirmatory factor analyses	Development and validation of a scale to measure memorable tourism experiences
(Hosany 2012)	Demand oriented (Tourist)	UK	Confirmatory approach Survey	Probability sampling	Emotional experience	S	Appraisals of pleasantness, goal congruence, and internal self-compatibility	520	Mail questionnaire	Ttest; Canonical correlation analysis.	Conceptual and and managerial contributions; the first emotion-appraisal model
(Tan et al. 2013)	Demand & Supply oriented (Tourist)	Taiwan (Four Creative Life Industry sites)	Grounded theory approach	Purposive sampling	Creative experience	U		32	In-depth interviews & observation	Theme identification	Conceptual contribution; Experience is a mediator of outer and inner creative experience

(Continued)

Table 2.1 (Cont.)

Author (date)	*Orientation*	*Countries/ Sites visited*	*Approach/ Methodology*	*Sampling technique*	*Types of experience*	*A*	*B*	*C*	*Data collection*	*Data analysis*	*Findings*
(Kidron 2013)	Demand oriented (Site visitors)	Sites of atrocity heritage	Exploratory; Qualitative	Snowball sampling	Emotive experience	U		55	In-depth ethnographic semi-structured interviews	Theme identification; Respondent validation	Gradual marginalization of historical memory work in favour of pathos-filled familial sociality
(Ma et al. 2013)	Demand oriented (Theme park visitors)	China (Happy Valley theme park)	Confirmatory approach Survey	Probability sampling	Emotional experience	S	Emotions	645	Face-to-face questionnaire	Structural equation modelling	Practical contribution in designing tourism experience; Identification of delight dimensions
(Massara and Severino 2013)	Demand oriented (Tourist)	Italy (Rome)	Mixed methodology	Non-probability sampling	Heritage experience	S	Psychological distance	114	Questionnaire & structured interviews	Mediation analysis; Content analysis;	Managerial and research contribution; Psychological distance related to tourism experience
(Beckman et al. 2013)	Demand oriented (Locals and Tourists)	USA (three city downtowns)	Confirmatory approach Survey	Quota sampling	Brand experience (behavioural, sensory, affective, and intellectual experience)	S	Place dependence	600	Intercept survey	Confirmatory factor analysis; Structural equation modelling	Antecedents of place dependence vary by visitor type

(Barnes et al. 2014)	Demand oriented (Tourist)	Denmark, Sweden (Three tourist destinations)	Confirmatory approach Survey	Quota sampling	Destination brand experience	S	Satisfaction; Recommendation intent; Revisit intentions	360 179 215	Self-administered questionnaire	Confirmatory factor analysis	DBE is an important determinant of revisit intentions and word-of-mouth recommendations
(Matteucci 2014)	Demand oriented (Tourist)	Spain (Seville)	Grounded theory approach	Purposive sampling	Embodied flamenco experience	U		20	Semi-structured, in-depth interviews; photo elicitation; pre- and post- observational notes	Theme identification	Three dimensions of body usage in the tourist experience of flamenco: the body as disciplined, expressive, and consumptive.
(Quinlan Cutler et al. 2014)	Demand oriented (University students visitors)	Peru (Inca Trail)	Phenomenological approach	Experience sampling method	Hiking experience	U		15	In-depth interviews & questionnaire; In-Situ Tourist Experience Log	Theme identification	Mobile tourist experience holds meaning related to the perception of self-identity and the encounter of corporeal self
(Kim and Ritchie 2014)	Demand oriented (University students)	Taiwan (Kaohsiung)	Confirmatory approach Survey	Quota sampling	Memorable tourism experience	S	Hedonism, refreshment, local culture, meaningfulness, knowledge, involvement, and novelty	593	Personal interview	Exploratory and Confirmatory factor analyses; Structural equation modelling	Cross-Cultural Validation of a Memorable Tourism Experience Scale (MTES)

(*Continued*)

Table 2.1 (Cont.)

Author (date)	*Orientation*	*Countries/ Sites visited*	*Approach/ Method-ology*	*Sampling technique*	*Types of experience*	*A*	*B*	*C*	*Data collection*	*Data analysis*	*Findings*
(Bosangit et al. 2015)	Demand oriented (British bloggers/ backpack-ers)	Worldwide Attractions		Purposive sampling	Tourist experience	U		19	Blog narratives	Theme identification	Conceptual con-tribution; three elements of experience – risk, challenge, and novelty/learning
(Lee 2015)	Demand oriented (Restaurant visitors)	Taiwan (South Tainan Rail-way Station)	Confirma-tory approach Survey	Probability sampling	Memorable experience	S	Reuse heri-tage; Tour-ist motiv-ation; Nostalgia	615	Face-to-face questionnaire	Structural equa-tion modelling	Significantly positive relation-ships between culinary attrac-tion and cultural inheritance and nostalgia, as well as a correlation between nostal-gia and memor-able tourism experiences,
(Chronis 2015)	Demand & supply oriented (Tourism organizers, business owners, & tourists)	USA (Pennsylva-nia)	Ethno-graphic approach	Purposive sampling	Staging experience	U		+14 +12+100	Participant and non-participant observation; interviews	Theme identification	contribute to the strategic staging of the tourist experi-ence; three over-lapping and intertwined sta-ging modes: communicative, material, and body-space

(Kim and Fesenmaier 2015)	Demand oriented (International students)	USA (Philadelphia)	Exploratory Qualitative	Purposive sampling	Emotional experience	U	2	Affectiva Q-sensor to measure electrodermal activity (EDA)& follow-up interviews	Thematic analysis	New tool to measure emotional experience which, in turn, provides a better foundation for designing and managing tourism experiences.
(Van Es and Reijnders 2016)	Demand oriented (Tour participants)	London, Los Angeles, Stockholm (Crime-detective fiction sites)	Ethnographic approach	Purposive sampling	Literary experience	U	20	In-depth interviews	Theme identification	Convergence between personal narratives set in the city with the fictional and factual place narratives which underlie it
(Unger et al. 2016)	Demand oriented (Employees who travel)	Israel	Grounded theory approach	Snowball sampling	Business travel experience	U	24	In-depth interviews	Theme identification	'Moments of recreation' and 'time off' at airports and aircraft rather than at the destination

(Continued)

Table 2.1 (Cont.)

Author (date)	*Orientation*	*Countries/ Sites visited*	*Approach/ Methodology*	*Sampling technique*	*Types of experience*	*A*	*B*	*C*	*Data collection*	*Data analysis*	*Findings*
(Mathis et al. 2016)	Demand oriented (Tourist)	Worldwide Tourist sites	Confirmatory approach Survey	Probability sampling	Co-creation experience	S	Satisfaction, impact on life, loyalty, customer involvement	561	Questionnaire	Confirmatory factor analysis; Structural equation modelling	Co-creation of an experience consists of: enhanced participation, confidence, facilitative environment, social interaction.
(Chen et al. 2017)	Demand oriented (Taiwanese backpackers)	Australia	Exploratory Qualitative	Snowball sampling	Mindful experiences	U		43	Semi-structured in-depth interviews	Theme identification	Three main themes: paying attention to the experience, living in the present, and non-elaborative awareness
(Ingram et al. 2017)	Demand oriented (Tourist)	Worldwide Tourist sites	Mixed methodology Participative inquiry Triangulation	Purposive & Snowball sampling	Responsible/ethical experience	U		16	Surveys; semi-structured, interview; diaries; photographs; tourist narratives; ethnography; participant observation	Narrative analysis; discourse analysis; Theme identification	Four prominent themes: 'Walking the talk', 'Reflexive inertia', 'Rhetorical cognisance', 'Pragmatic utility'
(Brown and Osman 2017)	Demand oriented (British tourist)	Egypt	Phenomenological approach	Purposive sampling	Female experience (in Islamic destination)			14	Semi-structured interview	Theme identification	Three emergent themes: a dominant male presence; unwanted male attention; conformity to local norms

(Knobloch et al. 2017)	Demand oriented (Site visitors)	New Zealand	Exploratory; Qualitative	Convenience sampling	Three Consumption Contexts (memorable, special, and extraordinary experience)	U		21	In-depth interviews (seven participants were interviewed for each of three consumption contexts)	Content analysis & theme identification	Tourism providers need to understand tourist consumption experiences beyond hedonic enjoyment of the moment, and consider their broader implications on well-being and quality of life
(Park and Santos 2017)	Demand oriented (Korean backpackers)	London & Paris	Exploratory; Qualitative	Quota sampling	Memorable experiences	U		28	Participant observation & three- staged (Pre-travel, during, and post travel) semi-structured interviews	Thematic analysis	Unique and unexpected personal experiences differentiate experiences of tourists
(Kim and Fesenmaier 2017)	Demand oriented (Tourist)	USA	Experiment	Probability sampling	Post-trip experience	S	Non-sharing and social media (Sharing experience)	167 161	Two scenarios	Descriptive Statistics; ANOVA; MANOVA	The study found that sharing unsatisfactory travel experiences through social media helps to reduce negative perceptions of the trip, which, in turn, enhances post-trip evaluations.

(*Continued*)

Table 2.1 (Cont.)

Author (date)	*Orientation*	*Countries/ Sites visited*	*Approach/ Methodology*	*Sampling technique*	*Types of experience*	*A*	*B*	*C*	*Data collection*	*Data analysis*	*Findings*
(Kirillova et al. 2017)	Demand oriented (Tourist)		Descriptive Phenomenological Method	Purposive sampling; Convenience sampling	Transformative tourist experiences	U		6 4	Unstructured phenomenological interviews; via Skype interviews	Theme identification	This study explored the essence of a transformative tourist experience and subsequent long-term changes.
(Moufakkir and Selmi 2018)	Demand oriented (Tourist)	Tunisia (Sahara desert)	Exploratory Qualitative	Purposive sampling	Spiritual experience	U		10	In-depth interviews	Theme identification	Feeling of security, emptiness, Nature-based authenticity, existential-authenticity, escaping, seeking, and finding, closeness to God, self-renewal/ refreshment
(Masiero and Qiu 2018)	Demand oriented (Tourist)	USA, UK, Australia	Confirmatory approach Survey	Probability sampling	Past reference experience	S	Travel experience	472 465 480	Questionnaire	Mixed multinomial logit	Theoretical and managerial contributions
(Antón et al. 2018a)	Demand oriented (Museum non-local and foreign visitors)	Spain	Confirmatory approach Survey	Probability sampling	Museum Emotional experience	S	Satisfaction Attainment Emotion	175	Face-to-face interviews	Descriptive statistics; Confirmatory factor analysis; Correlation analysis; Hierarchical regression analysis	Research and managerial (museum curators and managers) contributions.

(Stone et al. 2018)	Demand oriented (Tourist)	Australia, Ireland, United Kingdom, and USA.	Exploratory Qualitative	Purposive sampling	Memorable culinary experi-ences	U		1022	Online panel and email interviews	Theme identification	Conceptual and managerial contributions.
(Jorgenson et al. 2018)	Demand oriented (National park visitors)	USA (Yellow-stone National Park)	Confirma-tory approach Survey	Stratified sampling	Memorable experience	S	Personal memory's impact & frequency of rehearsal of tourists	704	On-site and mail-back survey	Exploratory factor analysis; reliability testing	New Tourism Autobiographical Memory Scale (TAMS)
(Chen and Yoon 2019)	Demand oriented (Tourist)	USA	Confirma-tory approach Survey	Probability sampling	Tourist experience	S	Novelty-seeking	556	Online questionnaire	Confirmatory factor analysis; Structural equation modelling	Applying of posi-tive psychology principles enhance the potential contri-bution of tourism experiences to tourists' well-being
(Antón et al 2018b)	Demand oriented (Domestic and foreign visitors)	Spain (Segovia)	Confirma-tory approach Survey	Quota sampling	Perceived Experience	S	Recom-mendation intent; Revisit intentions; satiety	400	Personal interview	Exploratory factor analysis; confirmatory factor analysis; structural equation modelling	The experience value reinforces the intention to return, the inten-tion to recom-mend and the intensification of the experience; the feeling of satiety reduces them

(Continued)

Table 2.1 (Cont.)

Author (date)	*Orientation*	*Countries/ Sites visited*	*Approach/ Methodology*	*Sampling technique*	*Types of experience*	*A*	*B*	*C*	*Data collection*	*Data analysis*	*Findings*
(Kim 2018)	Demand oriented (Tourists from China, HongKong, Japan, South Korea, and USA)	Taiwan	Confirmatory approach Survey	Quota sampling	Memorable tourism experiences		Destination image, tourist satisfaction, revisit intention, and word-of-mouth.	301	Face-to-face interviews	Confirmatory factor analysis; Structural equation modelling	Theoretical and managerial implication Creating of integrated MTE model that links MTEs, DI, overall satisfaction and future behavioural intentions

Summary

This chapter critically assesses the existing body of research on tourism experience through systematic review techniques. Since this is a relatively new and recent research topic in tourism research literatures, this investigation of secondary nature generates a structured scheme of the main research questions, context, data collection and analysis methods, and main findings of the papers published in major tourism and hospitality journals within the 2009–2018 time horizon.

This chapter will provide a reference guide in relation to the main aspects of the published studies in this research field.

To practitioners and tourism experience groups of interest, this systematic review will provide a comprehensive outlook of the main causes and consequences and theoretical frameworks under which tourism experience has been investigated.

Note

1 In this study the terms 'tourist experience' and 'tourism experience' will be used interchangeably.

References

Aho, S.K., (2001). Towards a general theory of touristic experiences: Modelling experience process in tourism. *Tourism Review*. 56(3/4): 33–37. DOI: 10.1108/eb058368

Andriotis, K., (2009). Sacred site experience: A phenomenological study. *Annals of Tourism Research*. 36 (1): 64–84. DOI: 10.1016/j.annals.2008.10.003

Andriotis, K., (2010). Heterotopic erotic oases: The public Nude Beach experience. *Annals of Tourism Research*. 37(4): 1076–1096. DOI: 10.1016/j.annals.2010.04.003

Antón, C., Camarero, C., and Garrido, M.J., (2018a). What to do after visiting a museum? From post-consumption evaluation to intensification and online content generation. *Journal of Travel Research*. 58(6): 1052–1063. DOI: 10.1177/0047287518793040

Antón, C., Camarero, C., and Laguna-García, M., (2018b). Experience value or satiety? The effects of the amount and variety of tourists' activities on perceived experience. *Journal of Travel Research*. 57(7): 920–935. DOI: 10.1177/0047287517727366

Barnes, S.J., Mattsson, J., and Sørensen, F., (2014). Destination brand experience and visitor behavior: Testing a scale in the tourism context. *Annals of Tourism Research*. 48: 121–139. DOI: 10.1016/j.annals.2014.06.002

Beatty, S.E., Homer, P., and Kahle, L.R., (1988). The involvement—commitment model: Theory and implications. *Journal of Business Research*. 16(2): 149–167. DOI: 10.1016/0148-2963(88)90039-2

Beckman, E., Kumar, A., and Kim, Y.K., (2013). The impact of brand experience on downtown success. *Journal of Travel Research*. 52(5): 646–658. DOI: 10.1177/0047287513478502

Binkhorst, E., and Dekker, T.D., (2009). Agenda for co-creation tourism experience research. *Journal of Hospitality Marketing & Management*. 18(2–3): 311–327. DOI: 10.1080/19368620802594193

Biran, A., Poria, Y., and Oren, G., (2011). Sought experiences at (dark) heritage sites. *Annals of Tourism Research*. 38(820–841). DOI: 10.1016/j.annals.2010.12.001

Bosangit, C., Hibbert, S., and McCabe, S., (2015). 'If I was going to die I should at least be having fun': Travel blogs, meaning and tourist experience. *Annals of Tourism Research*. 55: 1–14. DOI: 10.1016/j.annals.2015.08.001

Brown, L., (2009). The transformative power of the international sojourn: An ethnographic study of the international student experience. *Annals of Tourism Research*. 36(3): 502–521. DOI: 10.1016/j.annals.2009.03.002

Brown, L., and Osman, H., (2017). The female tourist experience in Egypt as an Islamic destination. *Annals of Tourism Research*. 63(12–22). DOI: 10.1016/j.annals.2016.12.005

Brunner-Sperdin, A., Peters, M., and Strobl, A., (2012). It is all about the emotional state: Managing tourists' experiences. *International Journal of Hospitality Management*. 31(1): 23–30. DOI: 10.1016/j.ijhm.2011.03.004

Buchmann, A., Moore, K., and Fisher, D., (2010). Experiencing film tourism: Authenticity & fellowship. *Annals of Tourism Research.* 37(1): 229–248. DOI: 10.1016/j.annals.2009.09.005

Campos, A.C., Mendes, J., Valle, P.O.D., and Scott, N., (2018). Co-creation of tourist experiences: A literature review. *Current Issues in Tourism.* 21(4): 369–400. DOI: 10.1080/13683500.2015.1081158

Chen, C.C., and Yoon, S., (2019). Tourism as a pathway to the good life: Comparing the top– down and bottom–up effects. *Journal of Travel Research.* 58(5): 866–876. DOI: 10.1177/0047287518775282

Chen, L.I.-L., Scott, N., and Benckendorff, P., (2017). Mindful tourist experiences: A Buddhist perspective. *Annals of Tourism Research.* 64: 1–12. DOI: 10.1016/j.annals.2017.01.013

Chronis, A., (2015). Moving bodies and the staging of the tourist experience. *Annals of Tourism Research.* 55: 124–140. DOI: 10.1016/j.annals.2015.09.008

Csikszentmihalyi, M., (2014). Toward a psychology of optimal experience, in: Csikszentmihalyi, M. (ed.), *Flow and the Foundations of Positive Psychology: The Collected Works of Mihaly Csikszentmihalyi.* Springer Netherlands, Dordrecht, pp. 209–226. DOI: 10.1007/978-94-017-9088-8_14

Frantál, B., and Kunc, J., (2011). Wind turbines in tourism landscapes: Czech Experience. *Annals of Tourism Research.* 38: 499–519. DOI: 10.1016/j.annals.2010.10.007

Hosany, S., (2012). Appraisal determinants of tourist emotional responses. *Journal of Travel Research.* 51(3): 303–314. DOI: 10.1177/0047287511410320

Huang, J., and Hsu, C.H., (2010). The impact of customer-to-customer interaction on cruise experience and vacation satisfaction. *Journal of Travel Research.* 49(1): 79–92. DOI: 10.1177/0047287509336466

Ingram, C., Caruana, R., and McCabe, S., (2017). Participative inquiry for tourist experience. *Annals of Tourism Research.* 65: 13–24. DOI: 10.1016/j.annals.2017.04.008

Jorgenson, J., Nickerson, N., Dalenberg, D., Angle, J., Metcalf, E., and Freimund, W., (2018). Measuring visitor experiences: Creating and testing the tourism autobiographical memory scale. *Journal of Travel Research.* 58(4): 566–578. DOI: 10.1177/0047287518764344

Kidron, C.A., (2013). Being there together: Dark family tourism and the emotive experience of co-presence in the holocaust past. *Annals of Tourism Research.* 41: 175–194. DOI: 10.1016/j.annals.2012.12.009

Kim, J., and Fesenmaier, D.R., (2015). Measuring emotions in real time: Implications for tourism experience design. *Journal of Travel Research.* 54(4): 419–429. DOI: 10.1177/0047287514550100

Kim, J., and Fesenmaier, D.R., (2017). Sharing tourism experiences: The posttrip experience. *Journal of Travel Research.* 56(1): 28–40. DOI: 10.1177/0047287515620491

Kim, J.H., (2018). The impact of memorable tourism experiences on loyalty behaviors: The mediating effects of destination image and satisfaction. *Journal of Travel Research.* 57(7): 856–870. DOI: 10.1177/0047287517721369

Kim, J.-H., (2014). The antecedents of memorable tourism experiences: The development of a scale to measure the destination attributes associated with memorable experiences. *Tourism Management.* 44: 34–45. DOI: 10.1016/j.tourman.2014.02.007

Kim, J.H., and Ritchie, J.B., (2014). Cross-cultural validation of a memorable tourism experience scale (MTES). *Journal of Travel Research.* 53(3): 323–335. DOI: 10.1177/0047287513496468

Kim, J.H., Ritchie, J.B., and McCormick, B., (2012). Development of a scale to measure memorable tourism experiences. *Journal of Travel Research.* 51(1): 12–25. DOI: 10.1177/0047287510385467

Kirillova, K., Lehto, X., and Cai, L., (2017). Tourism and existential transformation: An empirical investigation. *Journal of Travel Research.* 56(5): 638–650. DOI: 10.1177/0047287516650277

Knobloch, U., Robertson, K., and Aitken, R., (2017). Experience, emotion, and eudaimonia: A consideration of tourist experiences and well-being. *Journal of Travel Research.* 56(5): 651–662. DOI: 10.1177/0047287516650937

Kong, X., and Johnston, R., (2011). The customer experience: A road-map for improvement. *Managing Service Quality.* 21: 5–24. DOI: 10.1108/09604521111100225

Kruja, D., (2017). Product/service management, in: *The Routledge Handbook of Hospitality Marketing.* Routledge, pp. 103–115. www.routledgehandbooks.com/doi/10.4324/9781315445526.ch9

Kruja, D., (2018). Destination marketing research, in: *The Routledge Handbook of Destination Marketing.* Routledge, pp. 35–48. www.routledgehandbooks.com/doi/10.4324/9781315101163-4

Lee, Y.J., (2015). Creating memorable experiences in a reuse heritage site. *Annals of Tourism Research.* 55: 155–170. DOI: 10.1016/j.annals.2015.09.009

Ma, J., Gao, J., Scott, N., and Ding, P., (2013). Customer delight from theme park experiences: The antecedents of delight based on cognitive appraisal theory. *Annals of Tourism Research.* 42: 359–381. DOI: 10.1016/j.annals.2013.02.018

Masiero, L., and Qiu, R.T.R., (2018). Modeling reference experience in destination choice. *Annals of Tourism Research.* 72: 58–74. DOI: 10.1016/j.annals.2018.06.004

Massara, F., and Severino, F., (2013). Psychological distance in the heritage experience. *Annals of Tourism Research.* 42: 108–129. DOI: 10.1016/j.annals.2013.01.005

Mathis, E.F., Kim, H. (Lina), Uysal, M., Sirgy, J.M., and Prebensen, N.K., (2016). The effect of co-creation experience on outcome variable. *Annals of Tourism Research.* 57: 62–75. DOI: 10.1016/j.annals.2015.11.023

Matteucci, X., (2014). Forms of body usage in tourists' experiences of flamenco. *Annals of Tourism Research.* 46: 29–43. DOI: 10.1016/j.annals.2014.02.005

McIntyre, C., (2007). Survival theory: Tourist consumption as a beneficial experiential process in a limited risk setting. *International Journal of Tourism Research.* 9(2): 115–130. DOI: 10.1002/jtr.598

Mekawy, M.A., (2012). Responsible slum tourism: Egyptian experience. *Annals of Tourism Research.* 39: 2092–2113. DOI: 10.1016/j.annals.2012.07.006

Minkiewicz, J., Evans, J., and Bridson, K., (2014). How do consumers co-create their experiences? An exploration in the heritage sector. *Journal of Marketing Management.* 30(1–2): 30–59. DOI: 10.1080/0267257X.2013.800899

Morrison, A.M., (2013). *Marketing and Managing Tourism Destinations.* Routledge. DOI: 10.4324/9780203081976.

Moscardo, G., (2009). Understanding tourist experience through mindfulness theory, in: Kozak, M. and Decrop, A.. (eds.), *Handbook of Tourist Behavior: Theory & Practice.* New York: Routledge. 99–115. DOI: 10.4324/9780203881804.

Moufakkir, O., and Selmi, N., (2018). Examining the spirituality of spiritual tourists: A Sahara desert experience. *Annals of Tourism Research.* 70: 108–119. DOI: 10.1016/j.annals.2017.09.003

Otto, J.E., and Ritchie, J.R.B., (1996). The service experience in tourism. *Tourism Management.* 17(3): 165–174. DOI: 10.1016/0261-5177(96)00003-9

Park, S., and Santos, C.A., (2017). Exploring the tourist experience: A sequential approach. *Journal of Travel Research.* 56(1): 16–27. DOI: 10.1177/0047287515624017

Pearce, P.L., and Kang, M., (2009). The effects of prior and recent experience on continuing interest in tourist settings. *Annals of Tourism Research.* 36: 172–190. DOI: 10.1016/j.annals.2009.01.005

Pike, S., (2002). Destination image analysis—a review of 142 papers from 1973 to 2000. *Tourism Management.* 23(5): 541–549. DOI: 10.1016/S0261-5177(02)00005-5

Pike, S., (2012). *Destination Marketing.* Routledge. DOI: 10.4324/9780080568935.

Pike, S., and Page, S.J., (2014). Destination marketing organizations and destination marketing: A narrative analysis of the literature. *Tourism Management.* 41: 202–227. DOI: 10.1016/j.tourman.2013.09.009

Pine, B.J., and Gilmore, J.H., (1999). *The Experience Economy: Work Is Theatre & Every Business a Stage.* Harvard Business Press, Brighton, MA.

Poria, Y., Biran, A., and Reichel, A., (2009). Visitors' preferences for interpretation at heritage sites. *Journal of Travel Research.* 48(1): 92–105. DOI: 10.1177/0047287508328657

Prahalad, C.K., and Ramaswamy, V., (2004). Co-creation experiences: The next practice in value creation. *Journal of Interactive Marketing.* 18(3): 5–14. DOI: 10.1002/dir.20015

Quinlan Cutler, S., Carmichael, B., and Doherty, S., (2014). The Inca trail experience: Does the journey matter? *Annals of Tourism Research.* 45: 152–166. DOI: 10.1016/j.annals.2013.12.016

Ritchie, B.J.R., Wing Sun Tung, V., and Ritchie, R.J., (2011). Tourism experience management research: Emergence, evolution and future directions. *International Contemporary Hospitality Management.* 23(4): 419–438. DOI: 10.1108/09596111111129968

Ritchie, J.R.B., and Hudson, S., (2009). Understanding and meeting the challenges of consumer/tourist experience research. *International Journal of Tourism Research.* 11(2): 111–126. DOI: 10.1002/jtr.721

Robinson, R.N.S., and Clifford, C., (2012). Authenticity and festival foodservice experiences. *Annals of Tourism Research.* 39: 571–600. DOI: 10.1016/j.annals.2011.06.007

Rosch, E., and Lloyd, B.B.. (Eds.), (1978). *Cognition and Categorization.* Hillsdale, NJ: Lawrence Erlbaum.

Scherer, K.R., (2005). Appraisal theory, in: *Handbook of Cognition and Emotion.* John Wiley, pp. 637–663. DOI: 10.1002/0470013494.ch30

Scherer, K.R., Schorr, A., and Johnstone, T., (2001). *Appraisal Processes in Emotion: Theory, methods, Research.* New York: Oxford University Press.

Shani, A., and Uriely, N., (2012). VFR tourism: The host experience. *Annals of Tourism Research.* 39: 421–440. DOI: 10.1016/j.annals.2011.07.003

Sharpley, R., and Jepson, D., (2011). Rural tourism: A spiritual experience? *Annals of Tourism Research.* 38: 52–71. DOI: 10.1016/j.annals.2010.05.002

Snepenger, D., King, J., Marshall, E., and Uysal, M., (2006). Modeling Iso-Ahola's motivation theory in the tourism context. *Journal of Travel Research.* 45(2): 140–149. DOI: 10.1177/0047287506291592

Stepchenkova, S., and Mills, J.E., (2010). Destination image: A meta-analysis of 2000–2007 research. *Journal of Hospitality Marketing & Management.* 19(6): 575–609. DOI: 10.1080/19368623.2010.493071

Stone, M.J., Soulard, J., Migacz, S., and Wolf, E., (2018). Elements of memorable food, drink, and culinary tourism experiences. *Journal of Travel Research.* 57(8): 1121–1132. DOI: 10.1177/0047287517729758

Tan, S.-K., Kung, S.-F., and Luh, D.-B., (2013). A model of 'creative experience' in creative tourism. *Annals of Tourism Research.* 41: 153–174. DOI: 10.1016/j.annals.2012.12.002

Tomkins, S., (1987). Script theory, in: Arnoff, J., Rabin, A., and Zucker, R.A. (eds.), *Script Theory. The Emergence of Personality.* Springer, New York, NY, pp. 147–216.

Trauer, B., and Ryan, C., (2005). Destination image, romance, and place experience—an application of intimacy theory in tourism. *Tourism Management.* 26(4): 481–491. DOI: 10.1016/j.tourman.2004.02.014

Tung, V.W.S., and Ritchie, J.R.B., (2011). Exploring the essence of memorable tourism experiences. *Annals of Tourism Research.* 38: 1367–1386. DOI: 10.1016/j.annals.2011.03.009

Tussyadiah, I.P., and Fesenmaier, D.R., (2009). Mediating tourist experiences: Access to places via shared videos. *Annals of Tourism Research.* 36: 24–40. DOI: 10.1016/j.annals.2008.10.001

Tussyadiah, I.P., and Zach, F.J., (2012). The role of geo-based technology in place experiences. *Annals of Tourism Research.* 39: 780–800. DOI: 10.1016/j.annals.2011.10.003

Unger, O., Uriely, N., and Fuchs, G., (2016). The business travel experience. *Annals of Tourism Research.* 61: 142–156. DOI: 10.1016/j.annals.2016.10.003

Van Es, N., and Reijnders, S., (2016). Chasing sleuths and unravelling the metropolis: Analyzing the tourist experience of Sherlock Holmes' London, Philip Marlowe's Los Angeles and Lisbeth Salander's Stockholm. *Annals of Tourism Research.* 57: 113–125. DOI: 10.1016/j.annals.2015.11.017

Vargo, S.L., Lusch, R.F., and Lusch, R.F., (2014). Evolving to a new dominant logic for marketing [WWW document]. The Service-Dominant Logic of Marketing. DOI: 10.4324/9781315699035-9

Wang, W., Chen, J.S., Fan, L., and Lu, J., (2012). Tourist experience and Wetland parks: A case of Zhejiang, China. *Annals of Tourism Research.* 39: 1763–1778. DOI: 10.1016/j.annals.2012.05.029

3

Key drivers of tourism experience

Fabrizio Ferrari

Introduction

Exploring tourist motivations is a difficult task that has been undertaken, over a long period and under various aspects, by many scholars. The first studies on the motivations of tourists often highlighted only one main motivation to undertake travel. From the first tourist studies on the motivations of tourists, authenticity certainly emerges as the most important concept, the first fundamental pillar. As the tourism phenomenon grows and evolves into increasingly complex forms, the studies begin to focus on the factors that stimulate demand, often called push factors. In this phase, the stimuli coming from the offer, from the places and from the actors of the tourism sector, although recognized, do not reveal much. Yet, the originality and uniqueness of the tourist experience, the attempt to continually shape and improve the offer, become crucial at the present time.

New paradigms arise, still in many ways to be observed in their different aspects, such as experience, co-production and, finally, the co-creation of the tourist system, in which the local community and visitors continuously interact, stimulating mutual knowledge with an adaptive process that aims at maximum satisfaction both for demand and for supply.

Some early attempts to frame the motivations of tourists: the quest for authenticity

One of the earliest and most influential attempts to examine tourist motivations is the concept of 'authenticity' developed by MacCannell (1973). The basic premise of his position is that tourist behavior is widely shaped by an implicit search for authentic experiences as an antidote to the inauthentic and superficial qualities of modern life.

The notion that tourists are essentially motivated by a quest for authentic experiences has been significantly challenged, starting from the consideration that tourists are more and more informed, more recently thanks to the spread of social media, and are aware of what they expect to see in the place they will visit.

Wang (1999) states that the issue about the authenticity of the tourist destination is irrelevant or less relevant. The author classifies three types of authenticity: '*objective authenticity*' (the true, material and intrinsic authenticity of the place); '*constructed authenticity*' (the so-called staged authenticity according to MacCannell); finally, '*existential authenticity*', subjective and personal, generated by the interaction of the visitor with the vacation place. Therefore, the author postulates that the concept of existential authenticity can explain a wider spectrum of tourist phenomena than the conventional concept of authenticity.

Wang (2000) offers a useful elaboration on the themes of modernity and a continued discussion of authenticity in tourism. Lamenting the weaknesses of economic and psychological explanations of tourism motivation, Wang (2000) frames tourist motivation in terms of wider social changes and structures. Tourism, especially mass tourism, is not a universal phenomenon but a contemporary one that reflects what Wang characterizes as the ambivalence of modernity. Although modernity has a 'love side' in the form of material, intellectual and organizational miracles, it also has a 'dark side' by way of inauthenticity, loss of nature, helplessness, routinization, stress and exclusion. The author, for instance, explains that nature-based tourism is a response to the deterioration of the physical environment or the loss of nature, a dark side of the intellectual dimension (technology and sciences) of modernity.

Olsen (2002) reverses the perspective of authenticity, observing how in many studies the role of the tourist represents modernity that seeks to destroy the past that is the true essence of places. According to the author, the very role of the traditional tourist must be questioned, in order to be able to reach a real experience of places. Authenticity thus becomes a cultural value that is constantly created, reinvented and negotiated through social processes. Therefore, in many situations there is no single, authentic form of experience but a plurality of expressions and perceptions of the authentic.

The dual motivational hypotheses

Looking for a single factor to explain visitors' desire to travel over time has become less feasible, considering the complexity of reality. Thus, different dual motivational models have been proposed, such as those formed by: psychocentric and allocentric factors (Plog 1974); escape and seeking factors (Iso-Ahola 1982); and push and pull factors (Dann 1981). Many other dichotomies have been elaborated over the years, not always providing an adequate motivational understanding of experiential reality.

One of the first works, but still to be considered one of the most interesting, in the context of dual motivational models, is that of Plog (1974). The model conceived examines two categories of visitors, completely different and opposed to each other, but essentially two ends of a continuum: *psychocentric* and *allocentric*; between these two extreme positions, there are two moderate variants of each one (called 'near') and the so-called mid-centric. The author has revised his position more recently (Plog 2001), in an attempt to grasp the personality traits of the different types of tourists. Therefore, Plog (2001) makes *psychocentrics* coincide with *dependable personalities*, cautious and conservative in their daily life, restrictive in spending discretionary incomes, preferring to be surrounded by relatives and friends and liking the routine of their lifestyle. On the contrary, *allocentrics* coincide with *venturers' personalities*; they are intellectually curious, liking to choose new products in everyday life, preferring a daily life with varying activities and challenges and preferring to be alone and somewhat meditative. Many authors argued that the concept fails to account for the fact that tourists travel with different motivations on different occasions (Litvin 2006). As argued by Chen Y. *et al.* (2011), travel personality seems to be a meta-motivation or a premise by which other motivation paradigms and tourist behavior patterns can be discussed. Travel personality can be applied to other forms of consumption and importantly can be viewed as a fundamental force in determining holistic consumer behavior patterns from motivation through to decision.

The dual motivational model proposed by Iso-Ahola (1982) starts from the observation that the satisfaction that individuals expect to derive from involvement in a leisure activity is linked to two motivational forces: approach (seeking) and avoidance (escape). In other words, individuals perceive a leisure activity as a potential satisfaction-producer for two major reasons: it provides certain intrinsic rewards, such as feelings of mastery and competence and it helps them leave the routine environment behind them. According to the author, tourists may escape the personal world (with its troubles and problems) and/or the interpersonal world (formed by relatives and friends) and they may seek personal rewards (simple rest and relaxation, desire to learn new cultures, increase of personal prestige) and/or interpersonal rewards (varied and increased social interaction). In a subsequent reflection on this model, it is observed that recent trends toward more frequent, but shorter, vacations suggest that the escape dimension is a more important motivational force than the seeking dimension for tourism (Mannell and Iso-Ahola 1987). Jamal and Lee (2003) criticize Iso-Ahola's model because it is too tied to psychology at the micro-level to permit the necessary inclusion of a macro-sociological analysis; the authors aim to study both the micro and macro factors that influence tourist motivations. Thus, studies on macro factors focus on the broad social forces that motivate people to take vacations, and the studies on micro factors focus on the internal psychological forces that also motivate them. According to Jamal and Lee (2003), tourist motivation study has to be situated within the wider society and historically situated social changes including modernization, industrialization and urbanization, as well as within spheres of everyday life such as work, family and neighborhood.

Dann (1981) offers an interesting analysis of tourist motivations, classifying them as push and pull factors. In his early studies (Dann 1977) the author focuses on push factors, identifying two types of tourist, the anomic tourist and the ego-enhancement tourist; in both cases, among the factors that stimulate tourism, the author emphasizes the central role of fantasy. On one side, Dann claims that a possible push factor for travel, therefore, lies in the desire to transcend the feeling of isolation and anomie in everyday life, where the tourist simply wishes to get away from it all. On the other side, the tourist, the mirror image of the former, can enhance his/her desire of going to a place where his/her social position is unknown and where he/she can feel superior thanks to this lack of knowledge in those around him/her. Additionally, on return, a further boost can be given to the tourist ego in the recounting of holiday experiences. Pull factors represents the specific attractions of the destination, which induce the traveler to go there once the prior decision to travel has been made (Dann 1981). This theory suggests that people travel because they are 'pushed' by internal forces and 'pulled' by external forces. In other words, these forces describe how individuals are pushed by motivational variables into making a travel decision and how they are pulled or attracted by the destination area.

Uysal (1998) examines the reciprocal interaction between the two phenomena of push and pull items of motivation in the same context to better understand tourist behavior that, in its simplest form, consists of an origin and a destination. The push factors are the behavioral results of an inner emotional state and pose opportunities for interaction and participation; the level of satisfaction, the level of expectation and the eventual need deficiency create a situation of equilibrium or disequilibrium that stimulates the push factors (like motivations, demographic factors, awareness of opportunities, social trends), initially forming a potential demand and, therefore, the current demand, filtering from the first the opportunities and impediments due to economic, sociopsychological and other exogenous determinants (linked to environmental contexts). The responses to the demand side or pull factors, including benefits sought at the destination or desired features in a hotel, would then naturally represent the supply side of travel experience. The quality and availability of tourism supply resources are a critical element in meeting the needs of the everchanging and growing tourism market. So, it is important that destination managements monitor visitors' satisfaction with pull factors such as facilities, programs and services, in order to maintain a sustained and expanding business.

Uysal et al. (2008) focus on a review of the different studies that have tried to identify the push and pull factors that affect the motivation of tourists. In general, according to the authors, all the empirical studies reviewed support and confirm the hypothesis that the factors of push and pull share a common variance that can easily be captured and capitalized on to develop 'product bundles' as implied by the shared variance and create segments by assigning individuals to the common variance that exists between the two. Examples of market segments that have been identified include the studies of Oh H.C. et al. (1995) and Baloglu and Uysal (1996). The first study highlights four overlapping market segments, that is: safety/comfort seekers, culture/history seekers, novelty/adventure seekers and luxury seekers. The second one, similarly, groups tourists into another four overlapping segments: sports/activity seekers; novelty seekers; urban-life seekers; beach-resort seekers. Thus, the examination of push and pull motivation studies also revealed that the number of pull factors included in the studies seem to be larger than the number of push factors, probably due to the fact that pull items could be as large as the scope and nature of attributes that are associated with destinations and amenities.

The multi-motivational hypotheses

As can be noted, the spectrum of motivations has been increased more and more by different elements and features, parallel to the complexification of tourist systems. Several authors in recent times have tried to establish multi-motivational hypotheses to try to frame the drivers of travel in the light of the greater complexification of tourist systems.

In one of the earlier studies of this kind, Cohen (1979) distinguishes five main modes of touristic experiences: recreational mode; diversionary mode; experiential mode; experimental mode; existential mode. Cohen conceptualizes the search for meanings in terms of the 'quest for a center', while stressing the notion of 'center' as the zone of sacred moral values that exists in every society. Accordingly, the five modes of tourist experiences suggested by Cohen are characterized by the meanings assigned by tourists to both the 'center' of their own societies in everyday life and their quest for the 'centers' of other cultures during their excursions. The first two modes involve people committed to the 'center' of their own societies, seeking only mere pleasure; on the contrary, the last three modes relate to people who are alienated from their own society's center, so they search for meaning in other cultures centers' while traveling. The recreational mode serves the need for taking a break from the pressures of daily living in order to restore the strength needed to cope with entertaining, but shallow activities. The recreational mode is associated with those who perceive their daily life as meaningful, while the diversionary mode refers to those alienated from the goals and values of their everyday existence. The experiential mode is compatible with MacCannell's notion regarding the quest for authenticity. This type of tourist enjoys observing the 'authentic life' of 'others' without any attempt to be converted or even engaged in their life. The experimental mode refers to those who do try out the authentic life of 'others', as part of their pursuit for an alternative to the 'center' of their own culture. Finally, the last mode, the existential one, refers to individuals already committed to an 'elective center' that is culturally and geographically external to their own society; while these individuals live their daily routine in a 'spiritual exile' for practical reasons, their travel to a remote 'center' serves their desire to actualize and sustain their spiritual existence.

Uriely et al. (2002) add a sixth mode to the scheme suggested by Cohen, the humanistic mode, which includes those who may seek meaningful experiences in the centers of other cultures without being alienated from their own.

The model elaborated by Crompton (1979) states that the essence of a vacation is a 'break from routine'; it appears to resolve different types of disequilibrium as 'short-term disequilibrium', reflecting a particular set of circumstances or events which are temporal disruptions (called 'pressures'), or as long-term states of disequilibrium, which cannot usually be satisfied by a single pleasure vacation. An initial overt manifestation of disequilibrium leads to a break from routine, so there are three possible alternative behaviors: stay at home, go on a pleasure vacation or travel for other purposes such as visiting friends and relatives or going on a business trip. According to the author, it is possible to classify primarily either sociopsychological or cultural motives which determine the nature and destination of the pleasure vacation if that alternative is chosen. Among the sociopsychological motives, Crompton counts: escape from a perceived mundane environment; exploration and evaluation of self; relaxation; prestige; regression (meaning an opportunity to do things which were inconceivable within the context of their usual lifestyles); enhancement of kinship relationships; and facilitation of social interaction. Among the cultural factors, the author considers novelty and education. The author, therefore, did not develop in depth the cultural factors as much as the sociopsychological ones; in effect, he states that the tourist industry could usefully give more attention to sociopsychological disequilibrium when developing its product and promotion strategies. Consumers motivated by sociopsychological motives were not looking for uniqueness in the product; that is, some specific attribute it possessed which other destinations did not have. Cultural disequilibrium, on the other hand, refers to the desire to see new places or do things in a different environment. Crompton sustains that 'novelty' implies that there was no desire to return to a previously visited destination, no matter how successful the vacation. There is, therefore, a lack of brand loyalty.

Pearce (1988) proposes the idea of a travel career ladder to interpret tourist motivation that changes over time. The author classified tourist motivation into five hierarchical levels, or a 'ladder', with 'relaxation' needs at the lowest level, followed in sequence by 'stimulation', 'relationship' and 'self-esteem and development' needs and 'self-actualization /fulfillment' at the highest level. The logic of this idea lies in one's recognition that different motivations are the result of different travel experiences, which are determined by an individual's life span. This travel career model suggests that people commence their travel with a relatively low goal such as relaxation and pursue higher goals as they become more experienced travelers, until they reach the highest level of self-actualization/fulfillment. Recently, the author reviewed his model by replacing the 'ladder' with a travel career pattern (TCP) approach in which it is the dynamic, multilevel motivational structure that is seen as critical in understanding travel motivation, and it is these patterns that reflect and define careers (Pearce and Lee 2005).

Some recent theoretical approaches

The theoretical schemes shown, although in some cases criticized for some of their aspects, constitute an important starting point, each highlighting some relevant characteristics in the construction of tourism drivers.

In this sense, Chen Y. et al. (2011) tried to construct an integrated motivation framework to explain various tourist behavior patterns, including tourists' destination choices, travel experiences and reactions to distance (distance decaying effects). According to their scheme, the authors state that if the life span of an individual is considered, the distance decaying effect can be reinterpreted by categorizing short-haul tourists as young and first-time visitors with little travel experience, whereas long-haul tourists are more likely to be the elderly and repeat visitors.

Moreover, in recent years, the observation point is increasingly shifting from the intrinsic motivations of the tourist to the creation of experiences for the visitor. Thus Crompton's 'break from routine' must no longer necessarily be satisfied by the 'authenticity' of the location, but by its 'creativity' in renewing itself, by the implementation of the 'fantasy', Dann's key concept.

Relevant from a theoretical point of view in this sense, are the experiential realms of Pine and Gilmore (1999) as a starting point for new studies on tourism, modulated according to the dual passive/active participation and absorption/immersion dualism: entertainment, esthetic, educational and escapist.

According to Quan and Wang (2004), the tourist experience consists of two dimensions, namely, the dimension of the peak touristic experience and the dimension of the supporting consumer experience. If the attractions that constitute the major motivation to pleasure travel are disappointing, then even the high quality of other consumer services such as accommodation cannot fully compensate for this deficiency. However, the experience of an attraction, as a peak touristic experience, cannot replace supporting consumer experience.

Oh H. et al. (2007) offer a first concrete interpretation of Pine and Gilmore's work, pointing out that once the original 'push factors' coalesce, a set of relevant destinations is evoked along with the tourist's attitude associated with each destination in the set. At this point, the expected values of the experience from a destination ('pull factors') often solidify or weaken the tourist's intention to choose the destination.

Volo (2009) states that a 'tourist experience' can be defined as any occurrence that happens to a person outside the usual environment and the contracted time for which a sequence of the following events happens: energy reflecting the state of the environment impinges on sensory organs, the energy pattern is transmitted centrally and is interpreted and categorized according to one's knowledge acquired through time and is integrated and may be stored in the form of memory under some conditions (and thus some learning will occur).

Tung and Ritchie (2011) tried to find the determinants of a 'memorable experience', focusing on four dimensions: affect, expectations, consequentiality and recollection. The authors state that it is necessary to continue reinventing destination by promoting new experiences that tourists can expect when they return, and new memories that they can take home through the development of new memory points throughout the experience.

Chen J.S. et al. (2014) concentrate on dynamic drivers of tourist experiences, focusing on three types: personal, environmental and interactive ones. These drivers change and evolve during and after the trip, reflecting personal responses to emotional inputs received during the visit period. The authors propose five types of experience that can be illuminated by an experience spectrum: disastrous experience, regretful experience, monotonous experience, memorable experience and extraordinary experience. Thus, the drivers may be exploited, in an individual or collective fashion, to transform unfulfilling experiences into memorable ones.

The interactive drivers especially, which stimulate the sociality of the journey, lead to the concepts of co-production and co-construction of tourist experiences (Chathoth et al. 2013), highlighting how the driver formation process is continuous and even able to shape the output of the visit itself. Co-production means that the customers participate in creating the core offering themselves through shared inventiveness and co-design, but co-creation is closely tied to usage, consumption, value-in-use and the premise that value can be determined only by the customer.

Case study 3.1: War and tourism. The Case of 'Trentino Grande Guerra' in Trentino-Alto Adige (Italy)

War tourism is a peculiar form of memory tourism, which developed mainly following the events of the two world wars, due to the need of relatives and survivors to commemorate the places of the battles, often distant and unknown from those of birth and residence (Dunkley et al. 2011).

The tourist experience of the war landscapes requires the visitors to take an active role, as they have to interpret historical and technical information related to the battlefields, take part in commemorative practices and write their own impressions and memories in visitors' books (Winter 2009).

On the one hand, tourism related to war events becomes a form of heritage tourism (Butler and Suntikul 2013), while on the other it goes back to factors connected with the emotions of individuals, as it is linked with the attractiveness of death and its symbolism, namely *thanatourism* (Seaton 1996) and *dark tourism* (Stone and Sharpley 2008).

In summary, there are pull factors based on the individual emotions of each visitor and push factors of a cognitive and situational nature, linked to the monumental and symbolic artifacts of war (Yuill 2003).

The autonomous region of Trentino-Alto Adige is one of the most important territories for tourism in Italy, thanks to its natural environment, its heritage and its sports facilities, especially winter sports: despite being one of the smallest regions in Italy (with about 2 percent of the national population), it is second in terms of total tourist numbers, with 12 percent of the total.

In an entrepreneurial context marked by small and family-run businesses (Tizzoni 2016), it was possible to develop a very participatory territorial community environment, in which the networks of connections between the territorial actors can orient tourist policies and offer new segments of activities, in which tourists can dynamically participate and interact.

Strengthened by a receptive system already relevant and well-structured for every period of the year, in recent times tourism actors in Trentino-Alto Adige have begun to open up to a new tourism channel, linked to the memories of the First World War.

Trentino and Alto Adige were among the most important battle theaters of the First World War, especially due to the ruggedness of the mountains, with fortification systems and trenches that were positioned above 3,000 metres during the so-called 'White War'.

In order to convey different experiences related to the Great War, in 2010 Trentino-Alto Adige, with a regional law, created a 'Trentino Grande Guerra' network formed by public and private operators to celebrate the First World War Centenary (Irimiás 2014; Tizzoni 2016; Franch et al. 2017).

The network has several channels to elicit an authentic experience for visitors and to immerse them as much as possible in the history of the territory during the First World War.

Certainly the material cultural heritage would in itself have been an important attractor on its own, in part due to the natural landscape context in which the network is located, but given that the memories of the Great War tend to disappear after a century, it was decided that messages that stimulate the immaterial and personal needs of visitors could also be conveyed.

The first aspect of the 'Trentino Grande Guerra' network consists of the territories, grouped into eight areas (mostly valleys, plateaus and the city of Trento itself); the greatest testimony is constituted by the fortification works (with 18 main works, especially forts and batteries) still present today, in which restoration works have been started for the use of visitors.

To connect the material network of forts and trenches, different itineraries have been prepared with excursions (sometimes guided) of different levels of difficulty, to satisfy even adventure and sports tourists. Overall, the network of routes has been grouped into the 'Path of Peace', a route that joins the places of memory, in what was the western sector of the Italian-Austrian war, fought between 1915 and 1918.

The route winds for about 604 km, from the Stelvio Pass to the Marmolada, connecting the peaks of the glaciers to the Alpine valley bottoms and creating a wide network of paths that allows the hiker to reach forts, strongholds, military roads and trenches.

The connection between the physical and the memory substrate is then given by military cemeteries and shrines.

The fundamental pivot is given by the network of 18 museums, which host a number of exhibitions and events (often for families, with the intention of transmitting to children and young people the memory of events and promoting a message of peace). The earliest museum to be concerned with First World War remembrance dates back to the 1920s (the War Museum in Rovereto opened in 1921), but most of them were inaugurated recently. Their permanent collections generally illustrate everyday life in the trenches by exhibiting artifacts and relics and showing reconstructions of battlefields and military high mountain buildings via multimedia devices.

Finally, the activities and services supply often includes guided tours, sometimes in collaboration with local associations; other services include edutainment and research activities, where archive materials and thematic libraries are provided; many sites host temporary exhibitions (Tizzoni 2016).

Ultimately, the construction of the First World War tourism network in Trentino-Alto Adige should be read as the overlapping of different layers of tourist products, usable by different types of visitors, with different individual motivations (memory, education for children, family ties, curiosity, adventure, nature and so on), in order to create an experience that can last over time.

Some concluding remarks

Dwyer et al. (2009) frame six global drivers, political, economic, social, technological, demographic and environmental, to identify the main drivers of future tourism supply and demand. The authors warn that in an increasingly turbulent and rapidly changing world, innovation and development driven by both internal and external circumstances will continue, but destinations and firms not adjusting their strategies as their environments change will have difficulty maintaining competitive advantage.

The classic tourist system was formed by visitors with few expectations and goals in front of a standardized, simple and often repetitive tourist offer. In modern times there is a growing knowledge about tourists especially thanks to the new information and communication technologies, with sophisticated database management systems providing the tools to respond to individual preferences and stimulate tourism purchases.

The advent of technological platforms to engage customers at a superior level prior to travel and during/post-travel phases has led to the emergence of co-creative tourism processes and modalities.

Thus, the debate in literature has shifted from push factors to pull factors, to the feelings that tourists have in visited places and the need to always renew the experience so that it is always memorable and not repetitive.

Experience value is created as a result of mutual exchange throughout the whole experience process. Tourists continually shape their tourism product, becoming interactive with the holiday resort and being an active part of the co-creation of experiences. So, it is necessary that a mature tourist system, like the one briefly shown in the case study, should necessarily evolve to anticipate the needs of potential users.

Tourism organizations and destinations would need to exploit all of their resources in order to move toward a higher level of customer engagement while focusing on tourists' experiences, rather than satisfaction alone.

References

Baloglu, S. and Uysal, M. (1996), 'Market segments of push and pull motivations: a canonical correlation approach', *International Journal of Contemporary Hospitality Management*, 8(3):32–38. DOI: 10.1108/09596119610115989

Butler, R. and Suntikul, W. (2013), 'Tourism and war. An ill wind?'. In R. Butler and W. Suntikul (eds.), *Tourism and war* (pp. 1–11). London: Routledge.

Chathoth, P. et al. (2013), 'Co-production versus co-creation: a process based continuum in the hotel service context', *International Journal of Hospitality Management*, 32(1):11–20. DOI: 10.1016/j.ijhm.2012.03.009

Chen, J.S., Prebensen, N.K., and Uysal, M. (2014), 'Dynamic drivers of tourist experience'. In N. K. Prebensen, J.S. Chen, and M. Uysal (eds.), *Creating experience value in tourism* (pp. 11–21). Wallingford: CABI.

Chen, Y., Mak, B., and McKercher, B. (2011), 'What drives people to travel: integrating the tourist motivation paradigms', *Journal of China Tourism Research*, 7(2):120–136. DOI: 10.1080/19388160.2011.576927

Cohen, E. (1979), 'A phenomenology of tourist experiences', *Sociology*, 13(2):179–201. DOI: 10.1177/003803857901300203

Crompton, J. (1979), 'Motivations for pleasure vacation', *Annals of Tourism Research*, 6(4):408–424. DOI: 10.1016/0160-7383(79)90004-5

Dann, G.M.S. (1977), 'Anomie, ego-enhancement and tourism', *Annals of Tourism Research*, 4 (4):184–194. DOI: 10.1016/0160-7383(77)90037-8

Dann, G.M.S. (1981), 'Tourist motivation: an appraisal', *Annals of Tourism Research*, 8(2):219. DOI: 10.1016/0160-7383(81)90082-7

Dunkley, R., Morgan, N., and Westwood, S. (2011), 'Visiting the trenches: exploring meanings and motivations in battlefield tourism', *Tourism Management*, 32(4):860–868. DOI: 10.1016/j.tourman.2010.07.011

Dwyer, L. et al. (2009), 'Destination and enterprise management for a tourism future', *Tourism Management*, 30(1):63–74. DOI: 10.1016/j.tourman.2008.04.002

Franch, M., Irimiás, A., and Buffa, F. (2017), 'Place identity and war heritage: managerial challenges in tourism development in Trentino and Alto Adige/Südtirol', *Place Branding and Public Diplomacy*, 13 (2):119–135. DOI: 10.1057/s41254-016-0019-5

Irimiás, A. (2014), 'The great war heritage site management in Trentino, northern Italy', *Journal of Heritage Tourism*, 9(4):317–331. DOI: 10.1080/1743873X.2014.908886

Iso-Ahola, S.E. (1982), 'Toward a social-psychological theory of tourist motivation: a rejoinder', *Annals of Tourism Research*, 9(2):256–262. DOI: 10.1016/0160-7383(82)90049-4

Jamal, T. and Lee, J.-H. (2003), 'Integrating micro and macro approaches to tourist motivations: toward an interdisciplinary theory', *Tourism Analysis*, 8(1):47–59. DOI: 10.3727/108354203108750166

Litvin, S.W. (2006), 'Revisiting Plog's Model of Allocentricity and Psychocentricity … One More Time', *Cornell Hospitality Quarterly*, 47(3):245–253. DOI: 10.1177/0010880406289138

MacCannell, D. (1973), 'Staged authenticity: arrangements of social space in tourist settings', *American Journal of Sociology*, 79(3):589–603. DOI: 10.1086/225585

Mannell, R.C. and Iso-Ahola, S.E. (1987), 'Psychological nature of leisure and tourism experience', *Annals of Tourism Research*, 14(3):314–331. DOI: 10.1016/0160-7383(87)90105-8

Oh, H., Fiore, A.M., and Jeoung, M. (2007), 'Measuring experience economy concepts: tourism applications', *Journal of Travel Research*, 46:119–132. DOI: 10.1177/0047287507304039

Oh, H.C., Uysal, M., and Weaver, P.A. (1995), 'Product bundles and market segments based on travel motivations: a canonical correlation approach', *International Journal of Hospitality Management*, 14 (2):123–137. DOI: 10.1016/0278-4319(95)00010-A

Olsen, K. (2002), 'Authenticity as a concept in tourism research. The social organization of the experience of authenticity', *Tourist Studies*, 2(2):159–182. DOI: 10.1177/146879702761936644

Pearce, P.L. (1988), *The Ulysses factor: evaluating visitors in tourist settings*. New York: Springer-Verlag.

Pearce, P.L. and Lee, U.-I. (2005), 'Developing the travel career approach to tourist motivation', *Journal of Travel Research*, 43:226–237. DOI: 10.1177/0047287504272020

Pine, B.J., II and Gilmore, J.H. (1999), *The experience economy: work is theatre & every business a stage*. Boston: Harvard Business School Press.

Plog, S.C. (1974), 'Why destination areas rise and fall in popularity', *Cornell Hospitality Quarterly*, 14 (4):55–58. DOI: 10.1177/001088047401400409

Plog, S.C. (2001), 'Why destination areas rise and fall in popularity. An update of a Cornell Quarterly Classic', *Cornell Hospitality Quarterly*, 42(3):13–24. DOI: 10.1177/0010880401423001

Quan, S. and Wang, N. (2004), 'Towards a structural model of the tourist experience: an illustration from food experiences in tourism', *Tourism Management*, 25(3):297–305. DOI: 10.1016/S0261-5177(03) 00130-4

Seaton, A.V. (1996), 'Guided by the dark: from thanatopsis to thanatourism', *International Journal of Heritage Studies*, 2(4):234–244. DOI: 10.1080/13527259608722178

Stone, P.R. and Sharpley, R. (2008), 'Consuming dark tourism: a thanatological perspective', *Annals of Tourism Research*, 36(2):574–595. DOI: 10.1016/j.annals.2008.02.003

Tizzoni, E. (2016), 'The touristification of Great War heritage in the Province of Trento between European history and local identity', *Almatourism*, Special Issue n. 5:84–104. DOI: 10.6092/issn.2036-5195/6170

Tung, V.W.S. and Ritchie, J.R.B. (2011), 'Exploring the essence of memorable tourism experience', *Annals of Tourism Research*, 38(4):1367–1386. DOI: 10.1016/j.annals.2011.03.009

Uriely, N., Yonay, Y., and Simchai, D. (2002), 'Backpacking experiences. A type and form analysis', *Annals of Tourism Research*, 29(2):520–538. DOI: 10.1016/S0160-7383(01)00075-5

Uysal, M. (1998), 'The determinants of tourism demand: a theoretical perspective'. In D. Ioannides and K. G. Debbage (eds.), *The economic geography of the tourist industry* (pp. 79–98). London: Routledge.

Uysal, M., Li, X., and Sirakaya-Turk, E. (2008), 'Push–pull dynamics in travel decisions'. In H. Oh and A. Pizam (eds.), *Handbook of hospitality marketing management* (pp. 412–439). Oxford: Butterworth-Heinemann.

Volo, S. (2009), 'Conceptualizing experience: a tourist based approach', *Journal of Hospitality Marketing & Management*, 18(2–3):111–126. DOI: 10.1080/19368620802590134

Wang, N. (1999), 'Rethinking authenticity in tourism experience', *Annals of Tourism Research*, 26 (2):349–370. DOI: 10.1016/S0160-7383(98)00103-0

Wang, N. (2000), *Tourism and modernity. A sociological analysis*. Amsterdam: Pergamon.

Winter, C. (2009), 'Tourism, social memory and the great war', *Annals of Tourism Research*, 36 (4):607–626. DOI: 10.1016/j.annals.2009.05.002

Yuill, S.M. (2003), 'Dark tourism: understanding visitor motivations on sites of death and disaster', Dissertation, Department of Parks, Recreation and Tourism Sciences, Texas A&M University.

4

Typology of tourism experience

Bert Smit and Frans Melissen

Introduction

If you type the words 'tourism experience' into your favorite academic search engine, you will find roughly 600 academic papers on all kinds of tourism experiences. Some of them relate to food, others to culture. You will find there is a 'special kind of tourism experience' called dark tourism. Some tourism experiences are technology-enabled, others are not. Tourism experiences can be staged in luxury hotels, theme parks and many other experiencescapes. However, when reviewing this range of papers, one thing becomes apparent quite quickly: there seem to be a lot of different interpretations of what a tourism experience actually is, and there seem to be a lot of different types of tourism experience. Brent Ritchie et al. (2011) have therefore identified creating a taxonomy of tourism experiences as one of the top priorities in the field, and this call seems as relevant today as it was then.

Two years before communicating this call, Brent Ritchie and Hudson (2009) identified six 'streams' of research on tourism experiences. Two of these streams focus specifically on consumer experiences and could provide important insights in relation to answering it. First, a significant amount of tourism research is dedicated to the measurement of how tourists 'experience' their multi-sensory environment at any given time during tourism-related activities. Academics in this field aim to understand experiences as a so-called continuous stream of consciousness, for instance by measuring emotions, stress and brain activity. This stimulus–response interpretation allows us to understand the cognitive and affective reactions to the stimuli that surround an individual. However, the way experiences are interpreted in this stream implies that they do not have a natural beginning or ending and, consequently, this makes it hard to distinguish between the tourism experience provided to consumers by tourism providers and elements of these tourism experiences that are not part of a particular value proposition of a particular provider. This brings us to the second stream of research. This stream interprets a tourism experience as the more holistic outcome of having 'gained experience in' or 'having experienced something'. Baddeley (2000) points out that our brain copes with the continuous stream of consciousness that the first stream focuses on by dividing it into experiential episodes. This interpretation refers to learning and the creation of memories. This interpretation also allows us to distinguish the tourism experience as a separate entity. Within business contexts, this interpretation actually allows us to purposely design, develop, market, sell and manage such tourism experiences, for instance in the shape of experiences staged by hotels, restaurants, guided tours and more. Therefore, this chapter focuses on this particular interpretation of tourism experiences, but without ignoring the stimulus–response interpretation from the first stream. The reason for the latter is obvious; measuring experiences could be especially interesting, also within the context of interpreting them as separate entities that can be purposely designed, staged and managed, from the perspective of understanding them better so the provider can know which ones to offer and how they can be sold to which customers and when – and for what price. This chapter therefore explores different types of tourism experiences reported on in literature, as well as their main characteristics, in order to establish a (useful) typology of tourism experiences, which allows for distinguishing between different market offerings.

A bird's eye review of tourism experience research

Research specifically dedicated to experiences, outside of a religious context, first emerged in the 1960s. Economic affluence and scientific and societal development triggered interest from psychologists in personal development and wellbeing. Thorne (1963) and Maslow (1964) more or less simultaneously published their research on so-called peak experiences. Thorne refers to a peak experience as one of the most exciting, rich and fulfilling experiences a person has ever had, which is often described as the highlight of one's life. The opposite of a peak experience would be a nadir experience: an experience that was extremely unpleasant and that represents one of the low points in one's life. Maslow describes a peak experience as an experience leading to a person transcending ordinary reality and truly perceiving 'being' or ultimate reality. He explains that peak experiences are short in duration and accompanied by intense positive emotions and feelings of affection. Maslow, controversially at the time, claims that any person can have a peak experience, whether that person is religious or not. This opened the experience construct up to research beyond theology. Economists and sociologists also became interested in experiences, as they were recognized as an element of the hedonic consumption patterns of more and more households, who increasingly started spending time and money on experiences provided to them by theme parks, holiday resorts and the like. Cohen (1972) was among the first to develop theory on how a typology of tourist experiences could help explain their choices and preferences. It could possibly help us understand the behavior of tourists, who up until that moment were actually viewed as a homogenous group. Based on the work of, amongst others, Boorstin (1964), Turner (1973) and MacCannell (1973), Cohen (1979) distinguished five tourist experience modes: (1) recreational; (2) diversionary; (3) experiential; (4) experimental; and (5) existential. Each of these modes contrasts tourism activities with activities in everyday life and reflects the extent to which and the form in which the tourist wishes to (temporarily) escape from the strain of everyday life.

A similar approach to understanding tourists' experiences was presented by Abrahams (1986), who distinguished between 'ordinary' and 'extraordinary' experiences. In his view, extraordinary experiences stand out from ordinary experiences because they are 'more intense, framed and stylized practices' (Abrahams 1986: 50). These characteristics also make them more memorable without necessarily meeting the requirements of peak experiences, as described by Maslow (1964) and Thorne (1963). Ordinary experiences, also within a tourism context, do not require the extra conscious attention of the consumer because they still fit the schemata a person uses to understand and interact with his environment. In Abrahams' view, extraordinary experiences can be single events and encounters that increase the memorability of an experience. In service marketing literature, Bitner et al. (1990) use the term memorable encounter or out-of-the-ordinary service for a similar concept, which can relate to staff behavior, for instance in a hotel, but also to the so-called 'Gestalt' evaluation of a service. Arnould and Price (1993) use the term 'extended experience' to refer to extraordinary experiences that are not based on a single event, situation or encounter but on a temporary extended series of events that include elements of novelty and discovery. People that have had such an experience would usually refer to it as a single entity because they were immersed in and absorbed by that specific environment for a longer period of time (e.g., a natural park, city or unfamiliar culture). Their memories of this experience are also strongly linked to the specific environment it took place in. More recently, Kim and Fesenmaier (2015) pointed out that memorability of tourism experiences is determined by their score on seven dimensions: hedonism, novelty, local culture, refreshment, meaningfulness, involvement and knowledge. These dimensions fit well with the concepts presented by Abrahams (1986) and Arnould and Price (1993) and present us with an idea of the elements that can be extraordinary (e.g., hedonism and novelty), and those that contribute to the immersive character of extended experiences (e.g., exposure to local culture and involvement).

Goolaup et al. (2018) explored extraordinary food experiences of travelers to develop a theory of surprise. In this study, they distinguish between ordinary and extraordinary food experiences. This is particularly relevant because food experiences are no longer mere ad hoc experiences for travelers; they have become part of the core tourism product for travelers looking for novelty and uniqueness, and who are sensation-seeking and interested in the culture of the host country (Björk and Kauppinen-Räisänen 2016). Goolaup et al. (2018) stress that extraordinary food experiences are not synonymous with memorable food experiences. Memorable food experiences can be negative, neutral and positive, whereas extraordinary food experiences are characterized as being special, hedonic, intense and intrinsically more enjoyable. In the view of Goolaup et al. (2018: 221) an important element of extraordinary food experience is 'the surprise elicited in rare and exceptional food encounters, followed by intense pleasure, derived from relationship with others (socio-pleasure) or related to people's cognitive and emotional reaction to the usage of a product or to an experience (psycho-pleasures)'. For their phenomenological study, tourists were interviewed to understand and explain the source of the feeling of surprise and the type of pleasure induced in relation to these food encounters. In the interviews, respondents were asked to elaborate in as much detail as possible about their most extraordinary food experiences. These stories were contrasted with theory on so-called food cultural capital, which is based on the concept of habitus and cultural capital (Bourdieu 1986). A person's food cultural capital consists of (1) embodied capital, the taste and knowledge of gastronomy a person develops over time; (2) objectified capital, the artifacts a person buys or owns related to food and gastronomy (e.g., cookbooks, kitchen appliances); and (3) institutionalized capital, which relates to formally tested knowledge and skills, e.g. in the shape of diplomas (Bourdieu 1986).In their study, Goolaup et al. (2018) show that a person with high cultural food capital speaks about extraordinary food experiences in terms related to the surprise in the food itself (ingredients, textures, cooking style and simplicity of the dish), whereas pleasure in this case is mostly related to psycho-pleasure. The extraordinary food experience of a person with lower cultural food capital is mostly related to the perceived authenticity of the food, the restaurant setting and the staff. Pleasure for these tourists is mostly socio-pleasure, derived from surprise related to the context a person has immersed in rather than the food itself.

There is more to experience(s)

A somewhat different and less conscious or cognitive type of experience is the so-called optimal experience or flow (Csikszentmihalyi 1991). Flow represents a mental state in which a person is fully focused on a self-rewarding activity. A person reaches this state if and when challenged by engaging in an activity to a level that is only just within the reach of that person. If the challenge posed by a specific activity is not in balance with the skill level of the person undertaking it, this can result in feelings of arousal, worry or anxiety if engaging in it is perceived as a challenge that is beyond his or her skill level. An activity can also result in feelings of control, relaxation or boredom, if and when the challenge linked to that activity is perceived as lower than the person's skill level. Csikszentmihalyi (1991) based his original research on flow on the way people perceive their work, but later on discovered that reaching a state of flow can also happen outside of work environments. Flow has, for instance, also been reported on in research on hiking, sports and cultural activities. Given that an activity at just the right level requires full attention, people engaged with this type of activity generally lose track of time and place but are also less likely to pay attention to basic bodily needs such as hunger, thirst or sleep. In this sense, flow also resembles some of the characteristics of addiction. In one of his follow-up publications, Csikszentmihalyi (2000) describes flow as a state of peak enjoyment, energetic focus and creative concentration experienced by people engaged in adult play. Children can also reach that same state while playing; play generally is a very serious activity to them. The characteristics of flow are similar to those used by other authors to describe experiences, e.g. O'Sullivan and Spangler (1998) indicate that experiences require the involvement and participation of individuals and those individuals reaching a state of being physically, mentally, emotionally, socially or spiritually engaged. Mossberg (2007) describes experiences as many elements coming together and involving a consumer emotionally, physically, intellectually and spiritually. Contrary to the concepts of extraordinary and extended experiences, these descriptions have in common that they refer to a situation, activity or event requiring involvement and the full attention of the consumer, leading to a form of cognitive immersion or introspection, and are associated with losing sense of time and place. There are also a number of differences in these descriptions of flow-like experiences. The skill level of a person in a state of flow could very well increase as a result of engaging in that activity and thus the challenge involved with that activity will need to increase as well for that state of flow to maintain. In other words, flow often incorporates an element of learning or skill development, which is not necessarily incorporated in definitions proposed by other authors.

Pine and Gilmore (1999) have tried to capture these aspects of experiences in their four realms of experience. They define their four realms based on two dimensions: (1) passive vs active participation, and (2) immersion vs absorption. Based on these dimensions, they propose four types of experiences: (1) entertainment, (2) educational, (3) escapist, and (4) aesthetic. In their view, the ideal experience is a mix of these elements, the so-called sweet spot. However, this typology lacks the integration of the temporal aspect of extended and flow-like experiences, which actually consist of a series of events and encounters. Pine and Gilmore (1999) have also picked up on this aspect. They distinguish between staging experiences versus guiding transformations – often also referred to as transformative experiences. In their view, the outcome of staging an experience is a (positive) memory of that particular moment or period, whereas the outcome of guiding transformations is a sustained change in a person with the term guiding referring to both a steering element and the, generally, longer duration of such experiences.

Within the tourism context, this transformative character of tourism experiences has been described by a number of authors. Fu et al. (2015) describe the transformative experience facilitated by a wellness resort. In this resort, guests are stimulated and facilitated to reflect on everyday life and everyday consumption by creating a tranquil and peaceful environment in which personal health plays a key role (e.g., through food and meditation). Many of the guests change their consumption pattern to a healthier lifestyle after returning home. Noy (2004) describes how backpacking for a longer period of time can lead to (a narrative of) self-change, which is often linked to the perceived uniqueness of the visited destination. Brown (2009) reports on the transformative power of the student sojourn, in which the immersion in a different culture for a longer period of time alters a person's beliefs about their own culture. Although it is debatable whether a sojourn can still be seen as a form of tourism, Brown does point out an interesting aspect of getting a deeper insight into the effects of immersing oneself in another culture. From a marketing point of view, it is interesting to understand that some tourism experiences are developed, marketed and consumed as being transformative, whereas other tourism experiences have a transformational effect on tourists without having that specific purpose or promise. Reisinger (2013), amongst others, reports on (commercial) transformative tourism offers, such as volunteer tourism, eco-tourism and WWOOFing. The backpacking study by Noy (2004) points out the transformational (side-) effect of immersing oneself in an unknown environment. Both types of experiences lead to a transformation, as they alter our perception of what is 'normal' and who we are. Skavronskaya et al. (2017) point out that we are more likely to remember tourism experiences if they are different from what we are used to or from what we expect, because we cannot predict or understand them from the schemata we have developed over the years. As a result, we adapt particular schemata to incorporate new knowledge and insights acquired during the tourism experience.

Other authors have observed similar transformations in people as the result of short events, specific situations and encounters, which could be classified as unguided transformations. Denzin (1992) calls them turning point experiences or epiphanies that 'rupture routines and lives, and provoke radical redefinitions of the self' (p.26). Carù and Cova (2003) observe a similar phenomenon and call it philosophical experiences. In psychology, turning point experiences are sometimes referred to as flashbulb memories, because we recollect them in much more detail than other experiences (Brown and Kulik 1977). These types of life-changing experiences bear a close resemblance to trauma (e.g., surviving a car crash), and are generally induced by an external, unexpected circumstance. In contrast, epiphanies and philosophical experiences are induced by a more cognitive moment of becoming aware of a new insight (see e.g., Csikszentmihalyi and Sawyer 1995) rather than an external event. Although the phenomenon of life-changing experiences has also been reported on in tourism contexts, they are not purposely developed and marketed as a tourism-related offer.

Toward a tourism experience typology

From the previous sections, it is clear that even if we narrow the construct down by excluding literature on 'experiencing' and focusing on experiences that are part of particular value propositions in the particular context created by the tourism sector, the words 'tourism experience' can still refer to many different things and are interpreted differently by different authors. However, it is also clear that all of these different interpretations of tourism experiences have some specific elements in common: they all relate to (a series of) events, situations or encounters that demand attention and involvement from people experiencing them and they all lead to some form of memories or learning in combination with specific positive or negative emotions as the result of consuming tourism offerings. This is probably true for experiences in most other contexts as well, with the possible exception of religious contexts. In fact, as Smit and Melissen (2018) point out, by looking at their similarities and differences we can categorize (see Figure 4.1) almost all different types of experiences mentioned earlier based on their position on two dimensions. The first dimension relates to the duration of the experience, ranging from short, one-off events or encounters that could literally only last a few seconds or minutes to a series of events and encounters or specific situations that could last days, weeks or even months. The second dimension relates to the outcome of the experience and ranges from a (positive or negative) memory to a (radical) sustained change in the person(s) involved through learning or skill development.

The four categories of experiences identified in Figure 4.1 can also be used to categorize tourism experiences. Tourism experiences that are short in duration and memorable can be captured by the definitions of memorable encounters and extraordinary experiences. Tourism experiences that are long in duration and that are memorable fit the definition of extended experiences. Epiphanies and turning point experiences are typically short in duration but have a transformational impact on the individual. Finally, transformative experiences and guided transformations also induce a lasting change in the tourist, but these experiences generally last (much) longer.

A third dimension that could be added to Figure 4.1 would be related to the source of the experience. Some types of experiences are mostly induced by external events and the physical context in which they take place, whereas others can be characterized as more cognitive or resulting from subconscious processes. However, these aspects have not been researched in a tourism context.

In today's so-called experience economy, from a tourism business point of view, it makes perfect sense to focus on creating and delivering experiences. However, not all experience types mentioned above are a logical component of value propositions of tourism providers. One could very well argue that epiphanies and turning point experiences cannot – and, maybe, from an ethical perspective actually should not – be the subject of experiences that are purposely designed, staged and managed by tourism providers. It would be difficult to see the 'logic' of developing and marketing tourism epiphanies. Transformative experiences, however, have proven to be interesting tourism value propositions. Given the recent rise in

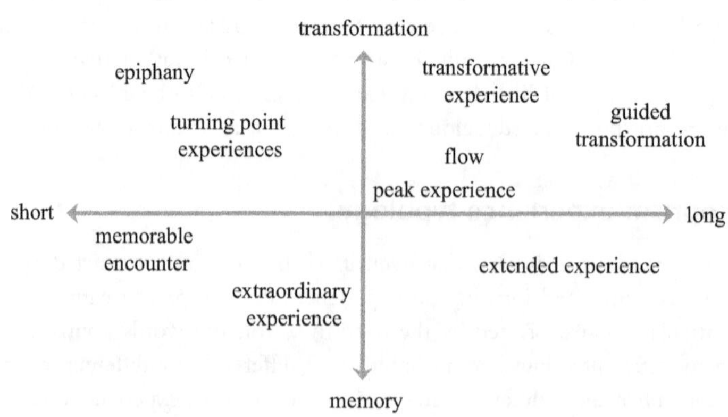

Figure 4.1 Four categories of experiences

health-based, volunteering, food and cultural tourism, it is clear that there is a large group of tourists who are interested in learning, changing their worldviews, personal development and reflection as part of consuming tourism products. The characteristics of transformative experiences, as described above, could thus very well serve as a reference point for purposely developing transformative tourism experiences.

Similarly, flow, and possibly peak, experiences can be an important component of the core of tourism products, for instance in sports tourism, hiking and other forms of physical activities in natural surroundings, that take tourists to their limits with respect to their skills and physical abilities. It is important to distinguish this type of tourism offer from extended experiences, which are not necessarily physically exhaustive but rather can be characterized by their immersive nature in relation to natural environments, other cultures or specific (remote) destinations. Offering this type of tourism experience is based on a value proposition that is not linked to promises of transformation but rather on offering a glimpse of the world outside daily routines, habits and consumption practices. Simultaneously, within the context of formulating new and ambitious value propositions and designing associated tourism experiences, it remains important to realize one thing. Most tourism experiences would never be as impressive, transformative or exhilarating as tourists experience them without the relative simplicity of extraordinary experiences and memorable encounters, in which the genuine hospitality of people you may have never met before, their well-prepared food, their cultural habits, their stories and history and the accommodation they offer, provide those perfect snapshot memories that make traveling so interesting and rewarding.

Conclusion

When we speak about and debate experiences in a tourism and hospitality context, we are actually using the same word for very different concepts and phenomena. In the previous sections we have explored how these are related on two dimensions, duration and outcome, with a possible but unexplored third dimension: source. Each type of experience identified on these dimensions can be valuable and relevant in itself. Especially if it fits the overall purpose of the design. Not every experience should be designed to be transformational, nor should every encounter or interface in tourism and hospitality be designed to be extraordinary. However it can be beneficial to align them in such a way that they create extended experiences. Doing so requires purposive design that starts with the choice for the specific experience and its intended impact on consumers.

References

Abrahams, R.D. (1986). *Ordinary and extraordinary experience. The anthropology of experience, 45–72.* Urbana: University of Illinois Press.

Arnould, E.J. and Price, L.L. (1993). 'River magic: Extraordinary experience and the extended service encounter', *Journal of Consumer Research*, 20(1): 24–45. DOI: 10.1086/209331

Baddeley, A. (2000). 'The episodic buffer: A new component of working memory?', *Trends in Cognitive Sciences*, 4(11): 417–423. DOI: 10.1016/S1364-6613(00)01538-2

Bitner, M.J., Booms, B.H., and Tetreault, M.S. (1990). 'The service encounter: Diagnosing favorable and unfavorable incidents', *Journal of Marketing*, 54(1): 71–84. DOI: 10.2307/1252174

Björk, P. and Kauppinen-Räisänen, H. (2016). 'Local food: A source for destination attraction', *International Journal of Contemporary Hospitality Management*, 28(1): 177–194. DOI: 10.1108/IJCHM-05-2014-0214

Boorstin, D. (1964). *The image: A guide to pseudo-events in America.* New York: Harper & Row.

Bourdieu, P. (1986). 'The forms of capital'. In Richardson, J.G. (ed.), *Handbook of theory and research for the sociology of education* (pp. 241–258). New York: Greenwood.

Brent Ritchie, J.R., Wing Sun Tung, V., and Ritchie, J.B.R. (2011). 'Tourism experience management research: Emergence, evolution and future directions', *International Journal of Contemporary Hospitality Management*, 23(4): 419–438. DOI: 10.1108/09596111111129968

Brown, L. (2009). 'The transformative power of the international sojourn: An ethnographic study of the international student experience', *Annals of Tourism Research*, 36(3): 502–521. DOI: 10.1016/j.annals.2009.03.002

Brown, R. and Kulik, J. (1977). 'Flashbulb memories', *Cognition*, 5: 73–79. DOI: 10.1016/0010-0277(77)90018-X

Carù, A. and Cova, B. (2003). 'Revisiting consumption experience: A more humble but complete view of the concept', *Marketing Theory*, 3(2): 267–286. DOI: 10.1177/14705931030032004

Cohen, E. (1972). 'Toward a sociology of international tourism', *Social Research*, 164–182. WOS: A1972N276500009.

Cohen, E. (1979). 'A phenomenology of tourist experiences', *Sociology*, 13(2): 179–201. DOI: 10.1177/003803857901300203

Csikszentmihalyi, M. (1991). *Flow: The psychology of optimal experience*. New York: Harper Perennial.

Csikszentmihalyi, M. (2000). *Beyond boredom and anxiety*. San Francisco: Jossey-Bass.

Csikszentmihalyi, M. and Sawyer, K. (1995). 'Creative insight: The social dimension of a solitary moment'. In Sternberg, R.J. and Davidson, J.E. (eds.), *The nature of insight* (pp. 329–363). London: MIT Press.

Denzin, N. (1992). 'The many faces of emotionality: Reading persona'. In Ellis, C. and Flaherty, M. (eds.), *Investigating subjectivity: Research on lived experience* (pp. 17–30). London: Sage.

Fu, X., Tanyatanaboon, M., and Lehto, X.Y. (2015). 'Conceptualizing transformative guest experience at retreat centers', *International Journal of Hospitality Management*, 49: 83–92. DOI: 10.1016/j.ijhm.2015.06.004

Goolaup, S., Solér, C., and Nunkoo, R. (2018). 'Developing a theory of surprise from travelers' extraordinary food experiences', *Journal of Travel Research*, 57(2): 218–231. DOI: 10.1177/0047287517691154

Kim, J. and Fesenmaier, D.R. (2015). 'Measuring emotions in real time: Implications for tourism experience design', *Journal of Travel Research*, 54(4): 419–429. DOI: 10.1177/0047287514550100

MacCannell, D. (1973). 'Staged authenticity: Arrangements of social space in tourist settings', *American Journal of Sociology*, 79(3): 589–603. DOI: 10.1086/225585

Maslow, A.H. (1964). *Religions, values, and peak-experiences*. Columbus: Ohio State University Press.

Mossberg, L. (2007). 'A marketing approach to the tourist experience', *Scandinavian Journal of Hospitality and Tourism*, 7(1): 59–74. DOI: 10.1080/15022250701231915

Noy, C. (2004). 'This trip really changed me: Backpackers' narratives of self-change', *Annals of Tourism Research*, 31(1): 78–102. DOI: 10.1016/j.annals.2003.08.004

O'Sullivan, E.L. and Spangler, K.J. (1998). *Experience marketing: Strategies for the new millennium*. State College: Venture Publishing Inc.

Pine, B.J. and Gilmore, J.H. (1999). *The experience economy: Work is theatre & every business a stage*. Boston: Harvard Business Press.

Reisinger, Y. (ed.). (2013). *Transformational tourism: Tourist perspectives*. Wallingford: CABI.

Ritchie, J.B. and Hudson, S. (2009). 'Understanding and meeting the challenges of consumer/tourist experience research', *International Journal of Tourism Research*, 11(2): 111–126. DOI: 10.1002/jtr.721

Skavronskaya, L., Scott, N., Moyle, B., Le, D., Hadinejad, A., Zhang, R., ... Shakeela, A. (2017). 'Cognitive psychology and tourism research: State of the art', *Tourism Review*, 72(2): 221–237. DOI: 10.1108/TR-03-2017-0041

Smit, B. and Melissen, F. (2018). *Sustainable customer experience design: Co-creating experiences in events, tourism and hospitality*. London: Routledge.

Thorne, F.C. (1963). 'The clinical use of peak and nadir experience reports', *Journal of Clinical Psychology*, 19(2): 248–250. DOI: 10.1002/1097-4679(196304)19:2<248::aid-jclp2270190236>3.0.co;2-d

Turner, V. (1973). 'The center out there: Pilgrim's goal', *History of Religions*, 12(3): 191–230.

5

Customer experience in tourism

An overview

Girish Prayag, Samuel Spector and Jörg Finsterwalder

Introduction

This chapter critically evaluates the concept of customer experience as applied to the tourism industry. The chapter starts with a review of concepts, such as customer experience and service experience. The continued conceptual shift toward focusing on the experience economy (Pine and Gilmore 1998, 1999) is particularly relevant for the study of tourism (Adhikari and Bhattacharya 2016; Oh et al. 2007). Specifically, the notion of tourism experience and its synonymous relationship with customer experience can be seen in existing studies as both arguably have cognitive, affective, and behavioral aspects. Customer experience is complex in nature, as it depends on the individual, the situation and the product/service offering. Often researchers have borrowed from the customer experience literature in marketing to inform the conceptualization and measurement of the tourism experience. However, it has been argued that the tourism experience often involves heightened emotional experiences and far more social interactions than the 'customer' experience. In this chapter we focus on customer experience rather than the tourism experience per se.

While we recognize that customer experience management (CEM) has emerged as a higher-order resource of cultural mindsets toward customer experiences, strategic directions for designing customer experiences, and firm capabilities for continually renewing customer experiences, with the goals of achieving and sustaining customer satisfaction and loyalty (Homburg et al. 2017), we do not focus on examining facets of CEM in this chapter. Rather, the chapter outlines some of the ways proposed in the literature to measure customer experience. Customers, but also tourists, interact with firms through a myriad of channels and media and customer experiences are becoming more social in nature with peer customers also influencing the experience (Lemon and Verhoef 2016).

One significant development in recent years is the recognition that other facets of customer experience such as emotional aspects, internal senses, and spiritual facets may be of more importance and relevance to the tourism experience specifically. A case study on customer experience evaluation in Brazilian restaurants in the UK is embedded within the chapter to show the importance of cognition and emotions as drivers of customer experience. The chapter concludes with some suggestions on future research areas that would advance our understanding of customer experience in the tourism field.

The concept of customer experience

The concept of customer experience, as well as experience related research and practice, has increasingly gained importance over the last few years (De Keyser et al. 2015; Jain et al. 2017), and understanding and creating customer experiences have become important research priorities (Marketing Science Institute 2016). Customer experience related research has also crossed disciplines (Finsterwalder and Tombs 2018), for example, into earthquake- and disaster-related work (Finsterwalder and Grey 2016), yet it has also permeated into the tourism realm (Oh et al. 2007). Although, increasingly, efforts have been made to research the construct, it remains elusive, at least in part. This is due to the different notions and conceptualizations as to what customer experience comprises. Two constructs in particular have been heavily discussed in the literature, i.e. the concepts of customer experience and service experience. An initial keyword search utilizing the search parameters 'service experience' and 'customer experience' for the timeframe of 1990–2015 across three different databases (Emerald, Ebsco Host, and Science Direct) undertaken by Jain et al. (2017) returned a total of 1,850 hits. Upon differentiation, the search showed that customer experience returned a total of 963 articles and there were 887 for service experience (duplicates included in results), pointing toward an equally strong utilization of the constructs in literature. While customer experience tends to focus on any encounter a customer might experience during an interaction with a provider, the term service experiences has been specifically utilized to denote actors' experiences in a service context.

Researchers have aimed at classifying both customer and service experience concepts (for example, Helkkula 2011; Jaakkola et al. 2015; Jain et al. 2017; Lipkin 2016). For example, Helkkula (2011) distinguishes outcome-based, process-based, and phenomenologically-based perspectives and comparably, Lipkin (2016) differentiates stimulus-based, interaction-based, and sense-making-based perspectives. Whereas the outcome-based perspective (for example, Holloway et al. 2005) highlights the experience as antecedent or the result of other constructs (Helkkula 2011), stimulus-based approaches (for example, Meyer and Schwager 2007) focus on the S-O-R paradigm in the generation of experience (Lipkin 2016). Process-based or interaction-based approaches (for example, Prahalad and Ramaswamy 2004) focus on social interaction, individual processes and interpretation (Lipkin 2016), understanding it as a sequence of events (Helkkula 2011). Finally, phenomenological or sense-making based approaches (for example, Helkkula et al. 2012; Vargo and Lusch 2004, 2016) hold the view that individuals subjectively experience their lifeworld, 'involving subjective, active, collective and dynamic sense-making' (Lipkin 2016, p. 865).

Drawing on multiple streams of literature, three basic tenets of customer experience can be identified (De Keyser et al. 2015), which are presented below: 1) interaction, 2) uniqueness, and 3) multidimensionality. An actor's customer experience eventuates from an interaction between the actor, for example a customer, and one or multiple other market actors via several interfaces which can be human, for example an employee, or non-human, for example technology (De Keyser et al. 2015; Ramaswamy and Ozcan 2018; Vargo and Lusch 2016). The prerequisite of an experience is interaction (De Keyser et al. 2015). Helkkula et al.'s (2012) research shows that experience is created via real-life interactions but it can also be facilitated by imagined ones. These interactions can be of the past, present, or future as human actors are capable of re-experiencing the past (retrospection), and experiencing the present, as well as pre-experiencing the future by simulating it in the mind (prospection) (Gilbert and Wilson 2007). Furthermore, these interactions elicit 'internal and subjective response[s] (…) to any direct or indirect contact' with another actor (Meyer and Schwager 2007, p. 118). These responses are deemed experiences.

Uniqueness of experiences describes the specific degree of distinctiveness of the customer experience. It can vary from ordinary experiences, for example routine experiences in everyday life, to extraordinary experiences which mark exceptional customer encounters (Arnould and Price 1993). Lastly, multidimensionality relates to the different but interrelated levels of an experience which are cognitive, emotional, sensorial, physical, and social (Gentile et al. 2007; Schmitt 2014). Therefore, based on these three tenets, and drawing on De Keyser et al. (2015) and Lipkin (2016), customer experience can be defined as the cognitive, emotional, sensorial, physical, and social elements that mark a customer interaction. This interaction leads to a process of sense- and meaning-making by the actor. Customer experiences can occur in the search (pre-consumption), purchase/consumption, and after-sale (post-consumption) phases (Arnould et al. 2002).

Measurement of customer experience

As argued by Palmer (2010), the increasing use of the term customer experience is a substitute for the traditional notion of customer relationship in marketing. Palmer (2010) also notes that the greatest problem in operationalizing customer experience as a construct is the complexity of context-specific variables. At best, efforts to conceptualize and operationalize customer experience remain exploratory (Lemon and Verhoef 2016). The experience is conditioned by various emotional states of individuals, different knowledge schemas, and differences between individuals and over time, as well as a variety of situation-specific factors. Another problem relates to the non-linearity of customer experience. For example, from the same roller coaster ride at a theme park, different customers will have varying levels of pleasure, attention focus, anxiety, and feelings of control. Therefore, the experience is non-linear and how it is captured remains challenging unless physiological measures, for example of emotions, are used. In addition, another problem seems to be the identification of the optimal level of experience (Palmer 2010). It is assumed that more intense experiences will be desired by customers. This may be problematic, specifically in the context of dark tourism, for example, where visitors may experience a level of negative emotions that may not be conducive to positive behavioral intentions. Thus, the question remains, at what level should the experience be designed to avoid negative evaluations by customers?

Different methods have been suggested to capture customer experiences, ranging from experiential sampling methods to content analysis. A more thorough review of these methods can be seen in the work of Hwang and Seo (2016). It has been suggested that despite these complexities, multi-dimensional scales are desirable. Schmitt (1999, 2014) introduced five dimensions of customer experience that can inform measurement; namely, sense, feel, think, act, and relate. For example, the 'sense' dimension would imply measuring sensorial experiences while the 'feel' dimension would entail capturing customers' emotional responses. The 'think' dimension would primarily capture cognitive facets of the experience and the 'act' dimension would capture how the product/service is consumed/used. Finally, the 'relate' dimension would involve capturing the social aspects of the experience. These five dimensions have been applied in tourism studies to measure tourist experience (for example, Le et al. 2019; Lee and Chang 2012). Verleye (2015) developed a multi-dimensional experience scale to capture hedonic, cognitive, social, personal, pragmatic, and economic experience dimensions.

Touchpoints seem to be the favored approach to measure and understand the customer journey and thus the experience (Pöppel et al. 2018; Verhoef et al. 2009). For example, Homburg et al. (2017) as part of CEM identify a number of aspects relating to touchpoints.

First, as outlined above customer journeys involve pre-purchase, purchase, and post-purchase situations and should be evaluated holistically. Second, connectivity of touchpoints and the context sensitivity of touchpoints must be understood for strategic reasons. Third, consistency in delivering the experience at each touchpoint is critical for loyalty. Fourth, core touchpoints can be branded and themed to reflect a certain activity or lifestyle of the customer. By extension, these facets imply that not only can touchpoints be designed for a customer journey but they also need to be prioritized, monitored, and adapted. Lemon and Verhoef (2016) identify four broad categories of touchpoints in the customer journey; namely, brand-owned, partner-owned, customer-owned, and social/external and independent. They argue that customer journey analysis is the starting point for understanding customer experiences that feed into CEM. However, Meyer and Schwager (2007) argue that this approach treats the firm and the customer as a monolithic entity, and therefore researchers collect data at only a few touchpoints, typically examining gaps between customers' expectations and their actual experience.

Rather than simply considering interactions at various touchpoints, examining engagement and co-creation may also be desirable (Zolkiewski et al. 2017). Klaus and Maklan (2012) suggest that four facets of experience should be measured, namely product experience, outcome focus, moments-of-truth, and peace-of-mind. Others have suggested the use of Net Promoter Score (NPS), which is a customer experience-centric metric, via a word-of-mouth proxy measure. However, this measure fails to explain where the customer experience issues are (Zolkiewski et al. 2017). Dixon et al. (2010) have proposed the Customer Effort Score (CES) as another metric for assessing customer experiences. However, often these new metrics lack strong theoretical foundations (Lemon and Verhoef 2016). There is a suggestion in the literature to move away from input–output measures to outcome-based measures that better capture, for example, co-creating value. As Hwang and Seo (2016) point out, various scales have been applied from the marketing literature in tourism studies.

Emerging dimensions of customer experience

This section discusses four emerging, under-researched facets of customer experience in the context of tourism. First, we highlight research topics related to tourists' emotions that are likely to provide additional insights, such as measuring the intensity and duration of emotions and evaluating emotions in real-time. Second, we identify the 'internal senses' – including proprioception, interoception, and the vestibular system – as integral to gaining a deeper understanding of customer experiences. Third, we consider spiritual dimensions, a particularly pertinent topic in tourism given the centrality of pilgrimages, spiritual journeys, and transcendental experiences. Finally, we discuss the role of cutting-edge technologies, including the development of artificial intelligence and hyper-customized experiences, in shaping the future contours of the tourism industry.

Emotions have been increasingly identified as exerting a significant influence on customer experiences in a wide range of settings, and several streams of research opportunities can be identified in this area (see also Cohen et al. 2014). Further research is needed to comprehend the interplay between emotional states, affect, and personality traits (Faullant et al. 2011; Gountas and Gountas 2007). This could entail, for instance, endeavoring to gain a deeper understanding of the relationship between transient emotional states and market segmentation tools, such as psychographics, that are based on personality traits and other non-transient characteristics (Lin et al. 2014). Further study is also required regarding the effects of mixed emotions, negative emotions, and anticipated emotions (Sherman et al. 1997; Watson and

Spence 2007). The presence of an extensive information search stage and complex buying behavior in tourism substantiates the importance of examining such factors specifically in this context. Additionally, research methods are needed that elucidate the intensity and duration, rather than only the type, of emotions felt during tourism experiences (Bilewicz and Wojcik 2018). Longitudinal studies, or even measuring emotions in real-time, would be useful in this regard, allowing destinations and attractions to develop 'emotional maps' of tourist experiences (see Kim and Fesenmaier 2015).

The multi-sensory nature of tourism has received significant attention in the literature (for example, Adhikari and Bhattacharya 2016; Ballantyne et al. 2011). However, the effects of the so-called 'internal senses' are in need of further exploration (Agapito et al. 2013; Cohen and Cohen 2017). Proprioception involves awareness of one's muscles and joints. For many forms of adventure activities, such as rock climbing, proprioception plays a central role in the experience. Aspects of health and wellness tourism, such as meditation and yoga classes, encourage participants to gain improved awareness of and connectedness to their proprioceptive systems. Interoception is responsible for sensations related to hunger, thirst, and body temperature. Understanding the role of interoception would offer insights into a range of topics, including food tourism, travel into extreme environments, and tourists' interactions with servicescapes (Bitner 1992). Finally, the vestibular system entails one's sense of balance and spatial orientation and significantly influences a plethora of experiences, such as bicycling, traveling via boat, or engaging in carnival rides. Immersive tourism experiences could offer a particularly fruitful avenue for investigating these internal senses in addition to their interplay with the traditional senses, emotions, and other aspects of customer experiences.

The spiritual dimensions of tourism experiences constitute another area of emerging interest. While tourism related to the 'established' religions has long been investigated, there is space for improved understanding regarding how spirituality affects tourism in secular settings (Sharpley 2009; Timothy and Conover 2006). The experience of tourism is interwoven with one's personal beliefs and worldview and, for some individuals, is seen as a means of attaining meaning and purpose (Willson et al. 2013). Indeed, tourism is perceived to be an integral component of a fulfilling life in some cultures (Sharpley and Jepson 2011). Even in contexts typically considered to be predominantly secular, investigating the spiritual dimensions of tourism may elucidate new aspects of the tourist experience (Sharpley 2009). For instance, long-distance travel to remote locations can be seen as a contemporary incarnation of pilgrimages. Both share similarities including a willingness, or even a desire, on the part of the traveler to endure hardship in the hopes of attaining some form of enrichment, transformation, or transcendence (Sharpley 2009).

There is also space to greatly enhance understanding regarding the role of technologies in tourism experiences. Topics such as social networking have been extensively studied in tourism (for example, Dickinson et al. 2017; Sotiriadis 2017), but technological developments are occurring rapidly. Whether enabling tracking luggage via RFID tags, permitting quick access to information at attractions via QR codes, or allowing guests to pre-customize their hotel room's temperature, emerging technologies are clearly significantly affecting customer experiences (Neuhofer et al. 2015). For example, some Hilton properties now allow guests to upload or select their own artwork, which will then be displayed in their room. As technologies in this realm are progressing quickly, studies regarding the effects of 'hyper-customization' on customer experiences will continue to be pertinent. Additionally, artificial intelligence and other 'smart technologies' are expected to increasingly infiltrate the tourism industry. Indeed, artificial intelligence is predicted to continue replacing some service-sector positions, particularly lower-level jobs (Huang and Rust 2018). Again using Hilton as an example, the hotelier recently introduced a robotic concierge which is capable of carrying out many customer service tasks, including answering questions regarding travel, dining, and attractions. These types of development have important implications for customer experiences as human–human interactions are augmented or replaced by human–machine interactions. Yet, even in a heavily digitized world, many tourism experiences are still co-created. Former research has indicated that tourism experiences are simultaneously individual and social (Cohen et al. 2014, cf. Helkkula 2011; Helkkula et al. 2012), and there is now a growing need to understand the triadic relationship between the individual, social, and technological. Research on encounters with robots and artificial intelligence systems, particularly when such encounters occur in co-created social environments, will offer insights into the tourist experience.

Case study 5.1: Emotions and customer experience in Brazilian restaurants in the UK

The case outlines the role of cognition and emotions in driving customers' evaluations of customer experience in Brazilian restaurants in the UK. The popularity of Brazilian restaurants has surged in the UK following the 2014 FIFA World Cup and the 2016 Rio de Janeiro Olympics. Brazilian restaurants typically are part of the ethnic restaurant segment and usually their uniqueness stems from the Brazilian culture as depicted by the service atmospherics, food, and service quality (Ribeiro and Prayag 2019). As argued earlier, customer experiences are shaped by a variety of drivers and in the restaurant context, social interactions, the core service and the physical environment matter most to customers (Walter et al. 2010). Customers are active creators of the physical environment in the sense that it is they who experience it and give meaning to it (Venkatraman and Nelson 2008). For example in the three Brazilian restaurants surveyed in this study, Bossa Nova, a typical Brazilian music style is played and the food style is Rodizio, which is reminiscent of a typical Brazilian steakhouse (Ribeiro and Prayag 2019). These are cognitive aspects of the experience that customers evaluate.

Data were collected from diners on their perceptions of food quality, the physical environment, and service quality through an email survey. These factors are treated as cognitive aspects of a restaurant experience in this study. At the end of their visit to the restaurant, diners were requested to give their email address so that the survey could be emailed to them. The survey also captured a range of positive and negative emotions that are usually elicited from the dining experience. The emotions represent the affective aspects of the customer experience. We define emotions as the affective or emotional responses generated from a consumption experience, including both functional (for example, food temperature) and hedonic (for example, staff empathy) experiences (Jani and Han 2011). Positive emotions included excitement, joy, positive surprise, and peacefulness, and negative emotions included anger, disgust, shame, and distress (Ribeiro and Prayag 2019). Previous studies have shown that the physical environment, food quality, and service quality can elicit positive emotions (Prayag et al. 2015). However, whether these facets of the restaurant experience can elicit negative emotions remains unclear. Overall customer experience was captured rather than attribute specific (for example, food quality) experiences. At the end of data collection, 665 questionnaires were retained for data analysis purposes (Ribeiro and Prayag 2019). The diners surveyed were mainly females (54.6%), married (50%), and between the ages of 35 to 44 years old (26.8%).

The results showed that both positive (β=0.127) and negative (β=-0.084) emotions have a significant influence on overall customer experience. More interestingly, though, the strongest influence on overall customer experience comes from restaurant atmospherics (β=0.550). Food quality has some influence on overall customer experience (β=0.165), similar to the influence of service quality (β=0.019). The results clearly highlight that negative emotions and service quality are significant but not strong predictors of overall customer experience (Ribeiro and Prayag 2019). In essence, the results suggest that cognition precedes emotion, which in turn shapes the overall evaluation of an experience. For restaurant managers, the findings suggest that service design principles can be applied to magnify the influence of atmospherics, food quality, and service quality on customers' overall perceptions of the experience. In fact, it would be worthwhile for managers to identify the touchpoints that generate heightened emotional responses from diners and manage those to encourage positive post-consumption evaluations.

Conclusion

Customer experience is difficult to define and measure (Palmer 2010). Achieving consensus on its dimensions is still elusive in both the marketing and tourism fields. Customer experience management has emerged as an important topic of research in both fields and it is not surprising that tourism researchers have often applied models and theories developed in non-tourism contexts (Cohen et al. 2014), such as those related to the customer journey and touchpoints to understand the customer experience in the tourism field. However, tourism's distinct attributes – such as the prevalence of group and joint decision-making, the presence of complex purchase behavior, and the interplay between individual and social factors in creating experiences – indicate a need for approaches developed specifically for tourism. We posit that the topics discussed in this section – under-research aspects of emotions, internal senses, spiritual dimensions, and emerging technologies – exert distinct effects in tourism and should be investigated specifically in that setting. Within the experience, co-creation aspects and engagement levels with touchpoints as well as social capital aspects of the relationship need to be researched further (Zolkiewski et al. 2017).

Given the different perspectives that exist on the conceptualization of the concept of customer experience such as outcome-based, process-based, and phenomenologically-based, each with its own theories and models, it is perhaps time for an integrative and holistic approach to be adopted. Likewise, engagement, co-creation, and flow aspects of customer experience are other facets that will need more attention from researchers. A multi-disciplinary approach to researching the concept will advance not only theorization but also operationalization of the concept in both the marketing and tourism fields.

References

Adhikari, A. and Bhattacharya, S. (2016). 'Appraisal of literature on customer experience in tourism sector: Review and framework', *Current Issues in Tourism*, 19 (4): 296–321. DOI: 10.1080/13683500.2015.1082538

Agapito, D., Mendes, J., and Valle, P. (2013). 'Exploring the conceptualization of the sensory dimension of tourist experiences', *Journal of Destination Marketing & Management*, 2 (2): 62–73. DOI: 10.1016/j.jdmm.2013.03.001

Arnould, E., Price, L., and Zinkhan, G. (2002). *Consumers*, New York, NY: McGraw-Hill. DOI: 0071214267

Arnould, E.J. and Price, L.L. (1993). 'River magic: Extraordinary experience and the extended service encounter', *Journal of Consumer Research*, 20 (1): 24–45. DOI: 10.1086/209331

Ballantyne, R., Packer, J., and Sutherland, L.A. (2011). 'Visitors' memories of wildlife tourism: Implications for the design of powerful interpretive experiences', *Tourism Management*, 32 (4): 770–779. DOI: 10.1016/j.tourman.2010.06.012

Bilewicz, M. and Wojcik, A.D. (2018). 'Visiting Auschwitz: Evidence of secondary traumatization among high school students', *American Journal of Orthopsychiatry*, 88 (3): 328–334. DOI: 10.1037/ort0000302

Bitner, M.J. (1992). 'Servicescapes: The impact of physical surroundings on customers and employees', *Journal of Marketing*, 56 (2): 57–71. DOI: 10.2307/1252042

Cohen, S.A. and Cohen, E. (2017). 'New directions in the sociology of tourism', *Current Issues in Tourism*, 22 (2): 153–172. DOI: 10.1080/13683500.2017.1347151

Cohen, S.A., Prayag, G., and Moital, M. (2014). 'Consumer behaviour in tourism: Concepts, influences and opportunities', *Current Issues in Tourism*, 17 (10): 872–909. DOI: 10.1080/13683500.2013.850064

Dadwal, S.S. and Hassan, A. (2016). 'The augmented reality marketing: A merger of marketing and technology in tourism', In *Mobile Computing and Wireless Networks: Concepts, Methodologies, Tools, and Applications* (pp. 63–80), Hershey, PA: IGI Global. DOI: 10.4018/978-1-4666-8751-6.ch004

De Keyser, A., Lemon, K.N., Klaus, P., and Keiningham, T.L. (2015). '*A framework for understanding and managing the customer experience*', Marketing Science Institute Working Paper Series, No. 15–121. Cambridge, MA: Marketing Science Institute.

Dickinson, J.E., Filimonau, V., Hibbert, J.F., Cherrett, T., Davies, N., Norgate, S., Speed, C., and Winstanley, C. (2017). 'Tourism communities and social ties: The role of online and offline tourist social networks in building social capital and sustainable practice', *Journal of Sustainable Tourism*, 25 (2): 163–180. DOI: 10.1080/09669582.2016.1182538

Dixon, M., Freeman, K., and Toman, N. (2010). 'Stop trying to delight your customers', *Harvard Business Review*, 88 (7/8): 116–122.

Faullant, R., Matzler, K., and Mooradian, T.A. (2011). 'Personality, basic emotions, and satisfaction: Primary emotions in the mountaineering experience', *Tourism Management*, 32: 1423–1430. DOI: 10.1016/j.tourman.2011.01.004

Finsterwalder, J. and Grey, H. (2016). 'Customer relationships and experiences during times of disaster: A case study of ballantynes of christchurch', In Hall, C.M., Malinen, S., Vosslamber, R., and Wordsworth, R. (eds.), *Business and Post-Disaster Management: Business, Organisational and Consumer Resilience and the Christchurch Earthquakes* (pp. 132–154), Milton Park, Abingdon, Oxford: Routledge. DOI: 10.4324/9781315640211

Finsterwalder, J. and Tombs, A. (2018). 'Services marketing and customer experience in a post-disciplinary era', *Journal of Service Theory and Practice*, 28 (4): 406–409. DOI: 10.1108/JSTP-07-2018-306

Gentile, C., Spiller, N., and Noci, G. (2007). 'How to sustain the customer experience: An overview of experience components that co-create value with the customer', *European Management Journal*, 25 (5): 395–410. DOI: 10.1016/j.emj.2007.08.005

Gilbert, D.T. and Wilson, T.D. (2007). 'Prospection: Experiencing the future', *Science*, 317 (5843): 1351–1354. DOI: 10.1126/science.1144161

Gountas, J. and Gountas, S. (2007). 'Personality orientations, emotional states, customer satisfaction, and intention to repurchase', *Journal of Business Research*, 60 (1): 72–75. DOI: 10.1016/j.jbusres.2006.08.007

Helkkula, A. (2011). 'Characterising the concept of service experience', *Journal of Service Management*, 22 (3): 367–389. DOI: 10.1108/09564231111136872

Helkkula, A., Kelleher, C., and Pihlström, M. (2012). 'Characterizing value as an experience: Implications for service researchers and managers', *Journal of Service Research*, 15 (1): 59–75. DOI: 10.1177/1094670511426897

Holloway, B.B., Wang, S., and Parish, J.T. (2005). 'The role of cumulative online purchasing experience in service recovery management', *Journal of Interactive Marketing*, 19 (3): 54–67. DOI: 10.1002/dir.20043

Homburg, C., Jozić, D., and Kuehnl, C. (2017). 'Customer experience management: Toward implementing an evolving marketing concept', *Journal of the Academy of Marketing Science*, 45 (3): 377–401. DOI: 10.1007/s11747-015-0460-7

Huang, M.H. and Rust, R.T. (2018). 'Artificial intelligence in service', *Journal of Service Research*, 21 (2): 155–172. DOI: 10.1177/1094670517752459

Hwang, J. and Seo, S. (2016). 'A critical review of research on customer experience management: Theoretical, methodological and cultural perspectives', *International Journal of Contemporary Hospitality Management*, 28 (10): 2218–2246. DOI: 10.1108/IJCHM-04-2015-0192

Jaakkola, E., Helkkula, A., and Aarikka-Stenroos, L. (2015). 'Service experience co-creation: Conceptualization, implications, and future research directions', *Journal of Service Management*, 26 (2): 182–205. DOI: 10.1108/JOSM-12-2014-0323

Jain, R., Aagja, J., and Bagdare, S. (2017). 'Customer experience – A review and research agenda', *Journal of Service Theory and Practice*, 27 (3): 642–662. DOI: 10.1108/JSTP-03-2015-0064

Jani, D. and Han, H. (2011). 'Investigating the key factors affecting behavioral intentions:evidence from a full-service restaurant setting', *International Journal of Contemporary Hospitality Management*, 23 (7): 1000–1018. DOI: 10.1108/09596111111167579

Kim, J. and Fesenmaier, D.R. (2015). 'Measuring emotions in real time: Implications for tourism experience design', *Journal of Travel Research*, 54 (4): 419–429. DOI: 10.1007/978-3-319-03973-2_21

Klaus, P. and Maklan, S. (2012). 'EXQ: A multiple-item scale for assessing service experience', *Journal of Service Management*, 23 (1): 5–33. DOI: 10.1108/09564231211208952

Le, D., Scott, N., and Lohmann, G. (2019). 'Applying experiential marketing in selling tourism dreams', *Journal of Travel & Tourism Marketing*, 36 (2): 220–235. DOI: 10.1080/10548408.2018.1526158

Lee, T.H. and Chang, Y.S. (2012). 'The influence of experiential marketing and activity involvement on the loyalty intentions of wine tourists in Taiwan', *Leisure Studies*, 31 (1): 103–121. DOI: 10.1080/02614367.2011.568067

Lemon, K.N. and Verhoef, P.C. (2016). 'Understanding customer experience throughout the customer journey', *Journal of Marketing*, 80 (6): 69–96. DOI: 10.1509/jm.15.0420

Lin, Y., Kerstetter, D., Nawijn, J., and Mitas, O. (2014). 'Changes in emotions and their interactions with personality in a vacation context', *Tourism Management*, 40: 416–424. DOI: 10.1016/j.tourman.2013.07.013

Lipkin, M. (2016). 'Customer experience formation in today's service landscape', *Journal of Service Management*, 27 (5): 678–703. DOI: 10.1108/JOSM-06-2015-0180

Marketing Science Institute. (2016). *Research Priorities 2016–2018*, Cambridge, MA: Marketing Science Institute. Retrieved from: www.msi.org/uploads/articles/MSI_RP16-18.pdf.

Meyer, C. and Schwager, A. (2007). 'Understanding customer experience', *Harvard Business Review*, 85 (2): 117–126.

Neuhofer, B., Buhalis, D., and Ladkin, A. (2015). 'Smart technologies for personalized experiences: A case study in the hospitality domain', *Electronic Markets*, 25 (3): 243–254. DOI: 10.1007/s12525-015-0182-1

Oh, H., Fiore, A.M., and Jeoung, M. (2007). 'Measuring experience economy concepts: Tourism applications', *Journal of Travel Research*, 46 (2): 119–132. DOI: 10.1177/0047287507304039

Palmer, A. (2010). 'Customer experience management: A critical review of an emerging idea', *Journal of Services Marketing*, 24 (3): 196–208. DOI: 10.1108/08876041011040604

Pine, J. and Gilmore, J. (1998). 'Welcome to the experience economy', *Harvard Business Review*, 78 (1): 97–105.

Pine, J. and Gilmore, J.H. (1999). *The Experience Economy – Work Is Theatre and Every Business A Stage*, Boston, MA: Harvard Business School Press.

Pöppel, J., Finsterwalder, J., and Laycock, R.A. (2018). 'Developing a film-based service experience blueprinting technique', *Journal of Business Research*, 85 (April): 459–466. DOI: 10.1016/j.jbusres.2017.10.024

Prahalad, C.K. and Ramaswamy, V. (2004). 'Co-creation experiences: The next practice in value creation', *Journal of Interactive Marketing*, 18 (3): 5–14. DOI: 10.1002/dir.20015

Prayag, G., Khoo-Lattimore, C., and Sitruk, J. (2015). 'Casual dining on the French Riviera: Examining the relationship between visitors' perceived quality, positive emotions, and behavioral intentions', *Journal of Hospitality Marketing & Management*, 24 (1): 24–46. DOI: 10.1080/19368623.2014.859114

Ramaswamy, V. and Ozcan, K. (2018). 'What is co-creation? An interactional creation framework and its implications for value creation', *Journal of Business Research*, 84 (March): 196–205. DOI: 10.1016/j.jbusres.2017.11.027

Ribeiro, M.A. and Prayag, G. (2019). 'Perceived quality and service experience: Mediating effects of positive and negative emotions', *Journal of Hospitality Marketing & Management*, 28 (3): 285–305. DOI: 10.1080/19368623.2018.1517071

Schmitt, B. (1999). 'Experiential marketing', *Journal of Marketing Management*, 15 (1-3): 53–67. DOI: 10.1362/026725799784870496

Schmitt, B. (2014). 'Experiential marketing: A new framework for design and communications', *Design Management Review*, 25 (4): 19–26. DOI: 10.1111/j.1948-7169.1999.tb00247.x

Sharpley, R. (2009). 'Tourism, religion and spirituality', In Jamal, T. and Robinson, J. (eds.), *The SAGE Handbook of Tourism Studies* (pp. 237–253), London: Sage. DOI: 10.4135/9780857021076

Sharpley, R. and Jepson, D. (2011). 'Rural tourism: A spiritual experience?', *Annals of Tourism Research*, 38 (1): 52–71. DOI: 10.1016/j.annals.2010.05.002

Sherman, E., Mathur, A., and Smith, R.B. (1997). 'Store environment and consumer purchase behavior: Mediating role of consumer emotions', *Psychology & Marketing*, 14 (4): 361–378. DOI: 10.1002/(SICI)1520-6793(199707)14:4%3C361::AID-MAR4%3E3.0.CO;2-7

Sotiriadis, M.D. (2017). 'Sharing tourism experiences in social media: A literature review and a set of suggested business strategies', *International Journal of Contemporary Hospitality Management*, 29 (1): 179–225. DOI: 10.1108/IJCHM-05-2016-0300

Timothy, D.J. and Conover, P.J. (2006). 'Nature religion, self-spirituality and new age tourism', In Timothy, D. and Olsen, D. (eds.), *Tourism, Religion and Spiritual Journeys* (pp. 139–155), London: Routledge. DOI: 10.4324/9780203001073

Vargo, S.L. and Lusch, R.F. (2004). 'Evolving to a new dominant logic for marketing', *Journal of Marketing*, 68 (1): 1–17. DOI: 10.1509/jmkg.68.1.1.24036

Vargo, S.L. and Lusch, R.F. (2016). 'Institutions and axioms: An extension and update of service-dominant logic', *Journal of the Academy of Marketing Science*, 44 (1): 5–23. DOI: 10.1007/s11747-015-0456-3

Venkatraman, M. and Nelson, T. (2008). 'From servicescape to consumptionscape: A photo-elicitation study of Starbucks in the New China', *Journal of International Business Studies*, 39 (6): 1010–1026. DOI: 10.1057/palgrave.jibs.8400353

Verhoef, P.C., Lemon, K.N., Parasuraman, A., Roggeveen, A., Tsiros, M., and Schlesinger, L.A. (2009). 'Customer experience creation: Determinants, dynamics and management strategies', *Journal of Retailing*, 85 (1): 31–41. DOI: 10.1016/j.jretai.2008.11.001

Verleye, K. (2015). 'The co-creation experience from the customer perspective: Its measurement and determinants', *Journal of Service Management*, 26 (2): 321–342. DOI: 10.1108/JOSM-09-2014-0254

Walter, U., Edvardsson, B., and Öström, Å. (2010). 'Drivers of customers' service experiences: A study in the restaurant industry', *Managing Service Quality: An International Journal*, 20 (3): 236–258. DOI: 10.1108/09604521011041961

Watson, L. and Spence, M.T. (2007). 'Causes and consequences of emotions on consumer behaviour: A review and integrative cognitive appraisal theory', *European Journal of Marketing*, 41 (5/6): 487–511. DOI: 10.1108/03090560710737570

Willson, G.B., McIntosh, A.J., and Zahra, A.L. (2013). 'Tourism and spirituality: A phenomenological analysis', *Annals of Tourism Research*, 42: 150–168. DOI: 10.1016/j.annals.2013.01.016

Zolkiewski, J., Story, V., Burton, J., Chan, P., Gomes, A., Hunter-Jones, P., O'Malley, L., Peters, L.D., Raddats, C., and Robinson, W. (2017). 'Strategic B2B customer experience management: The importance of outcomes-based measures', *Journal of Services Marketing*, 31 (2): 172–184. DOI: 10.1108/JSM-10-2016-0350

6

The role and measurement of emotion in tourism experiences

Girish Prayag

Introduction

While the marketing field abounds in studies integrating affective experiences in a holistic understanding of consumer and/or consumption experiences, the emotional turn in tourism is recent (Cohen, Prayag, and Moital 2014). Emotions play a vital role in shaping tourism experiences (Kim and Fesenmaier 2015) and determining tourists' behaviors and actions. The majority of existing studies attempt to understand felt emotions in a post-consumption context (e.g. Bigné, Andreu and Gnoth 2005; Faullant, Matzler, and Mooradian 2011; Hosany and Gilbert 2010; Prayag et al. 2017; Prayag, Hosany, and Odeh 2013). An emerging research strand has attempted to understand the role of anticipated emotions in the tourism experience (Lee, Bruwer, and Song 2017; Lee et al. 2018; Song et al. 2014). As will be discussed in the chapter, there are several gaps in emotion research in the tourism field. Also, from the traditional self-reported measurements (Hosany and Gilbert 2010), other ways of measuring emotions using psychophysiological methods such as electrodermal analysis, facial muscle activity, heart rate response, eye-tracking system, and vascular measures have been suggested (Kim and Fesenmaier 2015; Li, Scott, and Walters 2015). The purpose of this chapter is two-fold. First, it offers a review of past topics of research on emotions in the tourism field. Second, it provides an assessment of the different ways of measuring emotions in tourism. Accordingly, the chapter starts with a review of past and current topics of emotion research in tourism.

Emotion research in tourism

The word "emotion" has been defined differently in the marketing and tourism fields. For example, Lazarus (1991) suggests that emotion is a complex reaction arising from appraisals of self-relevant interactions with the environment. Another definition suggests that emotions are commonly understood to result from an evaluation of an experience in relation to goals, motives, or concerns (Frijda 1988). These two definitions suggest that emotions are drawn from cognitive appraisals, which has received some attention in the tourism literature (Breitsohl and Garrod 2016; Hosany 2012). Tourism researchers often borrow and adapt definitions from either marketing or psychology to describe emotions arising from tourism experiences. Unlike the marketing field, whereby everyday emotions are distinguished from emotions arising specifically from the consumption process by using the term "consumption emotions", tourism researchers have implicitly assumed that emotions arising from tourism and leisure experiences are similar to "consumption emotions". This assumption may not be tenable given that tourism and leisure experiences are often higher in emotional intensity compared to everyday mundane consumption activities. Also, emotional responses from tourism and leisure experiences have been shown to vary significantly over the duration of the consumption process (Nawijn et al. 2013).

A Scopus search of all articles using the keywords "emotion" and "tourism" reveals that 1,072 articles in English were published between 1980 to March 2019, including peer-reviewed articles (969), book chapters (64), and articles in the press (39). As shown in Figure 6.1, there has been a rapid increase in the use of the words "emotion" and "tourism" according to those search criteria from the year 2000 onwards.

An examination of the 1,072 articles reveals a number of research themes related to emotion research in tourism. These themes are: the relationship between emotion and destination image (Hao, Xu and Zhang 2019; Prayag et al. 2017); emotions in the context of festivals, meetings, and exhibitions (Fu et al. 2019; Mason and Paggiaro 2012); emotions and tourism advertising (Hadinejad et al. 2019; Soscia, Prayag, and Hesapci 2019); emotions and food/gastronomy (Organ et al. 2015; Williams, Yuan, and Williams 2019); emotions and the tourism experience itself (Bastiaansen et al. 2019; Rahmani, Gnoth, and Mather 2019); emotional labor of different service providers including tour guides (Heimtun 2016); the relationship between emotion and post-consumption behaviors such as satisfaction, place attachment, behavioral intentions, and loyalty (Ali et al. 2018; Hosany et al. 2017); emotions and how they influence the co-creation of experiences (Malone, McKechnie, and Tynan 2018); emotions in the context of artificial intelligence, virtual reality, and robotics (Huang et al. 2013; Tung and Au 2018); emotions and sustainable tourism (Zhang, Moyle, and Jin 2018); emotions and dark tourism (Boateng, Okoe, and Hinson 2018; Nawijn et al. 2018);

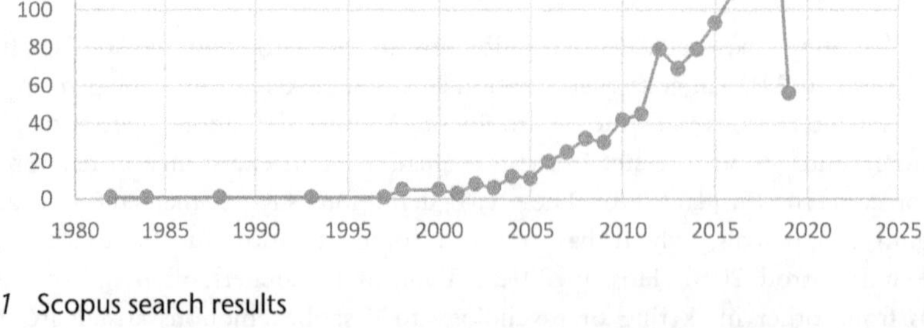

Figure 6.1 Scopus search results

emotions and social media or tourism websites (Hudson et al. 2015; Yan, Zhou, and Wu 2018); emotions and medical tourism (Kemp, Williams and Porter 2015; Loureiro 2017); emotions and well-being (Knobloch, Robertson, and Aitken 2017); emotional solidarity of tourists (Woosnam, Dudensing, and Walker 2015); emotions and rural tourism (Jepson and Sharpley 2015); and emotions and heritage tourism (Smith 2014; Su and Hsu 2013).

These themes highlight not only an interest in research on the evaluation of emotions from special interest tourism segments such as dark tourism, medical tourism, and food/gastronomy but also highlight a number of gaps. For example, the limited attention devoted to, for example, the emotions of residents arising from tourism development (Jordan, Spencer, and Prayag 2019). In fact, very few studies examine emotions in the context of sustainable tourism, despite many studies suggesting a "feel good" effect from supporting sustainable tourism activities (Andereck et al. 2007). While an understanding of how emotions are aroused and primed in the tourism context is burgeoning, there are several aspects such as the core features or components of emotion that are poorly understood. Table 6.1 provides a summary of the various functions that emotion can serve. Tourism researchers have primarily sought to understand emotions arising from evaluations of objects and events but less so, for example, as a way for system regulation. Nevertheless, across disciplines there is agreement that the modalities of emotion include expression, bodily symptoms and arousal, and subjective experience (Scherer 2005). The majority of studies in tourism tend to evaluate the subjective feelings associated with the experience. Scherer (2005) argues that the inclusion of a cognitive, information-processing component is required but most researchers would argue that emotion and cognition are two independent systems that interact. The communication of reaction and behavioral intention evaluated using motor expression has received scant attention in the tourism literature (e.g. Hadinejad et al. 2019).

Table 6.1 Function and emotion component

Emotion function	*Emotion component*	*Examples of tourism studies*
Evaluation of objects and events	Cognitive component (appraisal)	Breitsohl and Garrod 2016, Hosany 2012, Ma et al. 2013
System regulation	Neurophysiological component (bodily symptoms)	Kim and Fesenmaier 2015, Stadler, Jepson, and Wood 2018)
Preparation and direction of action	Motivational component (action tendencies)	Sánchez-García and Currás-Pérez 2011, Nawijn and Fricke 2015
Communication of reaction and behavioral intention	Motor expression component (facial and vocal expression)	Hadinejad et al. 2019
Monitoring of internal state and organism–environment interaction	Subjective feeling component (emotional experience)	Nawijn et al. 2013, Prayag, Hosany, and Odeh 2013, Hosany and Gilbert 2010, Nawijn et al. 2018

Adapted from Scherer (2005)

Emotion measurement

From Table 6.1, it is clear that different components of emotion can be measured, from bodily symptoms to subjective feelings. A vast majority of studies have measured subjective feelings using either the valence (dimensional) or the discrete emotions (emotion-specific) approach. The valence approach measures either positive and negative emotions or pleasure and arousal (Prayag, Hosany, and Odeh 2013). This approach gives a more limited account of emotional experiences (Lazarus 1991), given that there are few fundamental dimensions that organize emotional responses (Mauss and Robinson 2009). The valence approach contrasts states of pleasure (e.g., joy) with states of displeasure (e.g., unhappy), and the arousal approach contrasts states of low (e.g., quiet) and high (e.g., surprise) arousal. However, researchers disagree about which of the dimensional schemes should be used and how the different dimensions are related (Mauss and Robinson 2009). The second approach measures idiosyncratic affective states such as joy, anger, guilt, and shame. This perspective suggests that each dimension corresponds to a unique experience, physiology, and behavior. Over the years, a consensus has emerged that the two approaches do not necessarily conflict (Prayag, Hosany, and Odeh 2013). The emotion component of subjective experience has been measured primarily using self-reported measures in tourism studies. One major criticism of self-reported measures is that there are marked individual differences in awareness of and willingness to report on emotional states, which affects the reliability of the reported emotional experience (Mauss and Robinson 2009). The tourism literature has more recently turned to physiological measures, which are discussed next.

Physiological measures of emotion

Often referred to as physiological measures of emotion, autonomic measures are based on electrodermal (i.e. sweat gland) or cardiovascular (i.e. blood circulatory system) responses (Mauss and Robinson 2009). For example, Stadler, Jepson, and Wood (2018) use electrodermal activity (EDA) tracking technology, through Empatica E4 wristbands, to capture the "unadulterated" emotional response both during the experience and in reliving or remembering it. These authors argued that EDA tracking can enhance qualitative research methodologies in three ways: through its use as an "artifact" to prompt reflection on feelings, through identifying peaks of emotional response, and through highlighting changes in emotional response over time. Kim and Fesenmaier showed how objective emotions of tourists visiting a destination can be captured by tracking their EDA over time and space. EDA is typically quantified in terms of skin conductance level (SCL) or short-duration skin conductance responses (SCRs) (Mauss and Robinson 2009). The major limitation of EDA is that it cannot determine the valence of the emotion. As suggested by Li, Scott, and Walters (2015), this limitation can be addressed by combining EDA with other physiological or self-report methods. Also, the idea that discrete emotions have distinct autonomic signatures is contested in the literature. Studies tend to show that relationships exist between dimensions, particularly those of valence, arousal, and the autonomic nervous system (ANS) response (Mauss and Robinson 2009). Thus, ANS measures typically capture the dimensional aspects of emotional states.

The most commonly used cardiovascular measures include heart rate (HR), blood pressure (BP), total peripheral resistance (TPR), cardiac output (CO), pre-ejection period (PEP), and heart rate variability (HRV) as indicated by Mauss and Robinson (2009). These measures have been sparsely used in tourism studies given the preference for EDA. Facial muscle activity is an important physiological indicator of internal psychological activities (Li, Scott, and Walters 2015). Electromyogram (EMG) measurement is commonly used to assess muscle activity from electrodes placed over the orbicularis oculi muscle, just beneath the lower eyelid (Mauss and Robinson 2009). There seems to be some consensus among researchers that EMG captures the valence dimension of emotional states and works more reliably in the context of high-arousal negative stimuli rather than high-arousal positive stimuli. The measure, however, does not capture discrete emotions (Mauss and Robinson 2009).

Eye movement or eye-tracking measures often supplement other physiological methods as it is important to identify which specific stimuli a subject is looking at when facial muscles move or HR changes (Li, Scott, and Walters 2015). For example, Muñoz-Leiva et al. (2019) analyzed advertising effectiveness in social media in terms of customers' visual attention and self-reported memory. Using eye-tracking technology, participants were instructed to visit three Travel 2.0 websites (T2W), including a hotel's blog, a social network profile (Facebook), and a virtual community profile (TripAdvisor). They found that advertising on Facebook is more effective, followed by the blog, and lastly by Tripadvisor, with a greater amount of information. In another study, Kong et al. (2019) used eye tracking to examine tourism advertisement effectiveness. The results showed that advertisements containing both image and price were the least effective while those containing text and price were the most effective at capturing participants' attention. Babakhani, Ritchie, and Dolnicar (2017) used eye tracking to test the effectiveness of five different messages in voluntary carbon offsetting schemes. The results show low levels of traveler awareness of carbon offsetting schemes, the primary need for any message to attract the air passengers' attention, the superiority of pictorial and short textual information as well as the identification of particularly effective content. Li, Huang, and Christianson (2016) used eye tracking to evaluate advertising effectiveness through an assessment of consumers' visual attention toward tourism photographs with text naturally embedded in landscapes. Their findings show that text embedded in landscapes draws the visual attention of the majority of viewers, irrespective of whether or not they understand the text language.

It is clear from tourism studies that eye tracking as a method is primarily used to assess advertising effectiveness rather than be employed as a supplementary method to understand emotional responses. In marketing some studies are using eye tracking as a complementary method to understand consumer emotional responses. For example, Hernández-Fernández, Mora, and Hernández (2019) used a multi-method approach to evaluate a new technology product. Facial recognition of emotions, including eye tracking, was carried out to understand participants' emotional responses to the presentation of the new product. This was followed by a nethnographic qualitative approach consisting of sentiment analysis of users eWOM generated by product video reviews. Guerreiro, Rita, and Trigueiros (2015) used eye tracking coupled with skin conductance measurements to understand levels of emotional arousal, pleasure, and attention in evaluating the effectiveness of emotionally charged marketing campaigns.

Researchers have argued that physiological correlates of discrete emotions are likely to be found in the brain rather than in peripheral physiological responses (Izard 2007). As a result, electroencephalography (EEG) and neuroimaging methods have been used to capture discrete emotions. EEG measures typically contrast activation in fairly large regions of the brain (i.e. front of brain) versus posterior (i.e. back of brain) in combination with the distinction between left-sided and right-sided hemispheric activation (Mauss and Robinson 2009). For example, Michael et al. (2019) used EEG to assess emotional and cognitive responses to destination images (Dubai, Abu Dhabi, Hong Kong, New York, and London). The EEG calibration was done using functional localizer tests based on the ABM B-ALERT calibration process. General emotional and cognitive differences were identified between the channels through which travel destinations are presented. Words about and names of travel destinations cause higher cognitive loads, which may not be surprising, given the greater associative load that words have than images. Bastiaansen et al. (2018) evaluated the emotional responses of participants in relation to the marketing content of movies using EEG based neuromarketing techniques with implications for destination marketing and destination image management.

Neuroimaging studies using fMRI (functional magnetic resonance imaging) or PET (positron emission tomography) technologies can locate activation in far more specific parts of the brain than EEG (Mauss and Robinson 2009). Therefore, neuroimaging methods may be capable of capturing discrete emotions better than EEG measures. The uptake of oxygen in the blood is captured by fMRI, with the assumption that a greater signal reflects greater blood flow to a particular part of the brain, which in turn is thought to reflect activation of that region (Mauss and Robinson 2009). Because emotional states are complex, neuroimaging studies are purported as the Holy Grail to examine the interrelated activity among multiple brain regions, which can contribute to a better understanding of whether and how emotional specificity is instantiated in the brain (Mauss and Robinson 2009). It remains clear that in tourism studies physiological measures are often used in tandem with other methods such as self-reported, content analysis, and netnography amongst others. For example, a combination of physiological measures of emotion and photo elicitation-based discussions has been used to understand the effects of how reflecting, reliving, and reforming participatory art events enhances long-term effects of travel and tourism (Stadler, Jepson, and Wood 2018). Others have sought to compare the efficacy of self-reported measures of emotion against physiological measures in the context of tourism advertising (Li et al. 2018a). Methods such as facial electromyography and skin conductance have been used to measure emotional responses to tourism advertising (Li et al. 2018b). Shoval, Schvimer, and Tamir (2018) combined spatio-temporal data with physiological measures of emotion and semantic contextual information to obtain an integrative understanding of tourists' experience in time and space.

It is also possible to infer people's emotional states from facial expressions, vocal characteristics, and whole-body behaviors (Mauss and Robinson 2009). For example, prototypical facial behaviors of some basic emotions such as happiness, sadness, anger, fear, and surprise can be identified across different cultures (Fridlund, Ekman, and Oster 1987). Facial behaviors can be identified using facial EMG, which involves measuring electrical potential from facial muscles via the placement of electrodes on the face (Mauss and Robinson 2009). Hadinejad et al. (2019) used image recognition software, FaceReader, to analyze tourists' emotional responses to advertising and associated stimuli. The results showed that participants had low arousal levels and positive emotions when watching a tourism video. Hetland et al. (2016) used the same software to evaluate emotions elicited from tourism commercial films. Their results were similar to Hadinejad et al. (2019) in that the tourism films elicited hardly any facial expressions, which is indicative of low arousal. Vocal characteristics remain sparsely used in tourism studies to assess emotional responses. An exception is the study of Matsuda et al. (2018), which assesses both behavioral cues (eye and head/body movement) and audio-visual data (facial/vocal expressions) during visitors' sightseeing tours to understand emotional status and satisfaction levels. Their study uses a combination of an eye-gaze tracker, physical-activity sensors, and a smartphone (Mauss and Robinson 2009). The most common measures of vocal characteristics are voice amplitude (ie. loudness) and pitch. Higher levels of emotional arousal are generally associated with higher-pitched vocal samples (Scherer 2005). Whole-body behaviors associated with body posture have not received much attention in both the marketing and tourism literature as markers of emotional states. However, this approach might be appropriate for measuring discrete emotions (Mauss and Robinson 2009). Next, a case study based on using the measurement of discrete emotions using the self-reported measure is presented.

Case study 6.1: Motivation and discrete emotions in heritage tourism

Heritage tourism is a well-established area of research. Within heritage tourism studies, visitors' motivation (e.g., Poria, Reichel, and Biran 2006) and post-consumption behaviors (Chen and Chen 2010) are well studied. Yet, limited attention has been devoted to emotional responses from heritage consumption, even though nostalgia has been shown to be a significant emotional driver of heritage tourism experiences (Halewood and Hannam 2001). In the context of heritage tourism, it remains to be ascertained which discrete positive emotions (e.g., joy, happiness, and pleasure) and negative emotions (e.g., guilt, sadness, and regret) are elicited by different heritage sites. In this case study, tourists were surveyed at the Su Nuraxi UNESCO World Heritage Site (WHS), in Sardinia, Italy, as they exited from the site after their visit. The site is the most important monument of the old civilization in Sardinia. Several discrete emotions such as joy, pride, gratitude, love, positive surprise, guilt, regret, and sadness were adapted from previous studies (Holbrook and Batra 1987; Hosany and Gilbert 2010) and measured on a 7-point Likert scale (1=Not at all, 7=Very Much). At the end of the data collection period, 500 usable questionnaires were retained for data analysis. The sample comprised 55% females, with more than half (50.8%) above the age of 45 years old, well-educated (56.8% having an undergraduate degree or more), and with almost an equal split between domestic (49 %) and international visitors (51%).

Figure 6.2 shows that the most strongly felt positive discrete emotions were on average fascinated (5.96), admiration (5.85), amazement (5.84), astonishment (5.66), connected (5.59), and surprise (5.58). Visitors did not generally feel any discrete negative emotions. However, it must be noted that this site is not a dark tourism heritage destination and therefore low arousal of negative emotions is not uncommon. Interestingly, there are no significant differences in the arousal of the positive and negative discrete emotions by gender. However, when domestic versus international visitors were compared, significant differences emerged on the emotions of joy, pleasure, peaceful, fulfilled, serenity, and connected. On all these emotions, domestic visitors had higher arousal compared to international visitors. This is not surprising as this site would elicit strong emotional responses from visitors who had a personal connection to the site.

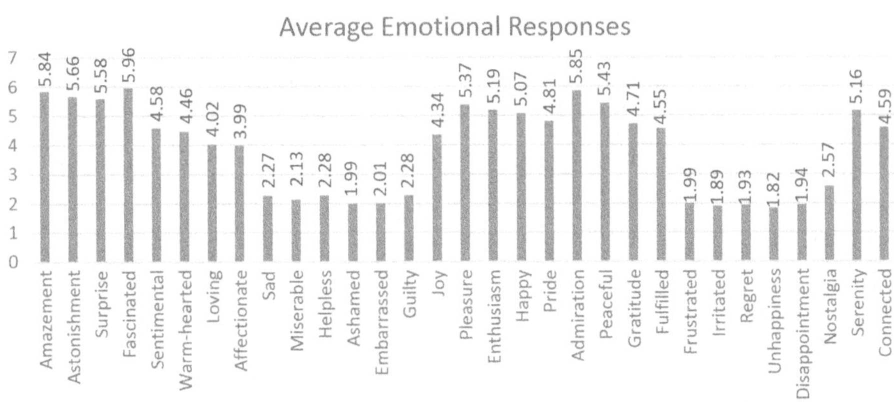

Figure 6.2 Emotional response of visitors to *Su Nuraxi*

Conclusion

Despite the emotional turn in tourism studies, researchers continue to use mainly self-reported measures of emotions, which primarily assesses the component of subjective experience. The other components of appraisal – bodily symptoms, action tendencies, and facial/vocal expressions – remain sparsely researched. A burgeoning literature on physiological measures can be seen, but often the cost associated with the use of such measures remains prohibitive for many researchers. However, physiological measures provide an opportunity to better assess the arousal of discrete emotions. As highlighted in the chapter, there are several measures that can be used to better capture the emotional responses of tourists and there are several gaps that exist in the application of these measures in the tourism literature. It is hoped that future research will endeavor to better understand discrete emotional responses as well as the valence of the experiences of both domestic and international visitors to advance tourism scholarship.

References

Ali, F., Kim, W.G., Li, J., and Jeon, H.M. (2018). 'Make it delightful: Customers' experience, satisfaction and loyalty in Malaysian theme parks', *Journal of Destination Marketing & Management*, 7: 1–11. DOI: 10.1016/j.jdmm.2016.05.003

Andereck, K.L., Valentine, K.M., Vogt, C.A., and Knopf, R.C. (2007). 'A cross-cultural analysis of tourism and quality of life perceptions', *Journal of Sustainable Tourism*, 15(5): 483–502. DOI: 10.2167/jost612.0

Babakhani, N., Ritchie, B.W., and Dolnicar, S. (2017). 'Improving carbon offsetting appeals in online airplane ticket purchasing: Testing new messages, and using new test methods', *Journal of Sustainable Tourism*, 25(7): 955–969. DOI: 10.1080/09669582.2016.1257013

Bastiaansen, M., Lub, X.D., Mitas, O., Jung, T.H., Ascenção, M.P., Han, D.I., Moilanen, T., Smit, B., and Strijbosch, W. (2019). 'Emotions as core building blocks of an experience', *International Journal of Contemporary Hospitality Management*, 31(2): 651–668. DOI: 10.1108/IJCHM-11-2

Bastiaansen, M., Straatman, S., Driessen, E., Mitas, O., Stekelenburg, J., and Wang, L. (2018). 'My destination in your brain: A novel neuromarketing approach for evaluating the effectiveness of destination marketing', *Journal of Destination Marketing & Management*, 7: 76–88. DOI: 10.1016/j.jdmm.2016.09.003

Bigné, J.E., Andreu, L., and Gnoth, J. (2005). 'The theme park experience: An analysis of pleasure, arousal and satisfaction', *Tourism Management*, 26(6): 833–844. DOI: 10.1016/j.tourman.2004.05.006

Boateng, H., Okoe, A.F., and Hinson, R.E. (2018). 'Dark tourism: Exploring tourist's experience at the Cape Coast Castle, Ghana', *Tourism Management Perspectives*, 27: 104–110. DOI: 10.1016/j.tmp.2018.05.004

Breitsohl, J. and Garrod, B. (2016). 'Assessing tourists' cognitive, emotional and behavioural reactions to an unethical destination incident', *Tourism Management*, 54: 209–220. DOI: 10.1016/j.tourman.2015.11.004

Chen, C.-F. and Chen, F.-S. (2010). 'Experience quality, perceived value, satisfaction and behavioural intentions for heritage tourists', *Tourism Management*, 31(1): 29–35. DOI: 10.1016/j.tourman.2009.02.008

Cohen, S.A., Prayag, G., and Moital, M. (2014). 'Consumer behaviour in tourism: Concepts, influences and opportunities', *Current Issues in Tourism*, 17(10): 872–909. DOI: 10.1080/13683500.2013.850064

Faullant, R., Matzler, K., and Mooradian, T.A. (2011). 'Personality, basic emotions, and satisfaction: Primary emotions in the mountaineering experience', *Tourism Management*, 32(6): 1423–1430. DOI: 10.1016/j.tourman.2011.01.004

Fridlund, A.J., Ekman, P., and Oster, H. (1987). 'Facial expressions of emotion', In Siegman, A.W. and Feldstein, S. (Eds.), *Nonverbal Behavior and Communication* (pp. 143–223). Hillsdale, NJ: Lawrence Erlbaum Associates, Inc.

Frijda, N.H. (1988). 'The laws of emotion', *American Psychologist*, 43(5): 349–358. DOI: 10.1037/0003-066X.43.5.349

Fu, X., Yi, X., Okumus, F., and Jin, W. (2019). 'Linking the internal mechanism of exhibition attachment to exhibition satisfaction: A comparison of first-time and repeat attendees', *Tourism Management*, 72: 92–104. DOI: 10.1016/j.tourman.2018.11.002

Guerreiro, J., Rita, P., and Trigueiros, D. (2015). 'Attention, emotions and cause-related marketing effectiveness', *European Journal of Marketing*, 49(11/12): 1728–1750. DOI: 10.1108/EJM-09-2014-054

Hadinejad, A., Moyle, B.D., Scott, N., and Kralj, A. (2019). 'Emotional responses to tourism advertisements: The application of FaceReader™', *Tourism Recreation Research*, 44(1): 131–135. DOI: 10.1080/02508281.2018.1505228

Halewood, C. and Hannam, K. (2001). 'Viking heritage tourism: Authenticity and commodification', *Annals of Tourism Research*, 28(3): 565–580. DOI: 10.1016/S0160-7383(00)00076-1

Hao, X., Xu, S., and Zhang, X. (2019). 'Barrage participation and feedback in travel reality shows: The effects of media on destination image among Generation Y', *Journal of Destination Marketing & Management*, 12: 27–36. DOI: 10.1016/j.jdmm.2019.02.004

Heimtun, B. (2016). 'Emotions and affects at work on Northern Lights tours', *Hospitality and Society*, 6(3): 223–241. DOI: 10.1386/hosp.6.3.223_1

Hernández-Fernández, A., Mora, E., and Hernández, M.I.V. (2019). 'When a new technological product launching fails: A multi-method approach of facial recognition and E-WOM sentiment analysis', *Physiology & Behavior*, 200: 130–138. DOI: 10.1016/j.physbeh.2018.04.023

Hetland, A., Vittersø, J., Fagermo, K., Øvervoll, M., and Dahl, T.I. (2016). 'Visual excitement: Analyzing the effects of three Norwegian tourism films on emotions and behavioral intentions', *Scandinavian Journal of Hospitality and Tourism*, 16(4): 528–547. DOI: 10.1080/15022250.2015.1116405

Holbrook, M.B. and Batra, R. (1987). 'Assessing the role of emotions as mediators of consumer responses to advertising', *Journal of Consumer Research*, 14(3): 404–420. DOI: 10.1086/209123

Hosany, S. (2012). 'Appraisal determinants of tourist emotional responses', *Journal of Travel Research*, 51(3): 303–314. DOI: 10.1177/0047287511410320

Hosany, S. and Gilbert, D. (2010). 'Measuring tourists' emotional experiences toward hedonic holiday destinations', *Journal of Travel Research*, 49(4): 513–526. DOI: 10.1177/0047287509349267

Hosany, S., Prayag, G., Van Der Veen, R., Huang, S., and Deesilatham, S. (2017). 'Mediating effects of place attachment and satisfaction on the relationship between tourists' emotions and intention to recommend', *Journal of Travel Research*, 56(8): 1079–1093. DOI: 10.1177/0047287516678088

Huang, Y.C., Backman, S.J., Backman, K.F., and Moore, D. (2013). 'Exploring user acceptance of 3D virtual worlds in travel and tourism marketing', *Tourism Management*, 36: 490–501. DOI: 10.1016/j.tourman.2012.09.009

Hudson, S., Roth, M.S., Madden, T.J., and Hudson, R. (2015). 'The effects of social media on emotions, brand relationship quality, and word of mouth: An empirical study of music festival attendees', *Tourism Management*, 47: 68–76. DOI: 10.1016/j.tourman.2014.09.001

Izard, C.E. (2007). 'Basic emotions, natural kinds, emotion schemas, and a new paradigm', *Perspectives on Psychological Science*, 2(3): 260–280. DOI: 10.1111/j.1745-6916.2007.00044.x

Jepson, D. and Sharpley, R. (2015). 'More than sense of place? Exploring the emotional dimension of rural tourism experiences', *Journal of Sustainable Tourism*, 23(8-9): 1157–1178. DOI: 10.1080/09669582.2014.953543

Jordan, E.J., Spencer, D.M., and Prayag, G. (2019). 'Tourism impacts, emotions and stress', *Annals of Tourism Research*, 75: 213–226. DOI: 10.1016/j.annals.2019.01.011

Kemp, E., Williams, K.H., and Porter, M., III. (2015). 'Hope across the seas: The role of emotions and risk propensity in medical tourism advertising', *International Journal of Advertising*, 34(4): 621–640. DOI: 10.1080/02650487.2015.1024385

Kim, J. and Fesenmaier, D.R. (2015). 'Measuring emotions in real time: Implications for tourism experience design', *Journal of Travel Research*, 54(4): 419–429. DOI: 10.1177/0047287514550100

Knobloch, U., Robertson, K., and Aitken, R. (2017). 'Experience, emotion, and eudaimonia: A consideration of tourist experiences and well-being', *Journal of Travel Research*, 56(5): 651–662. DOI: 10.1177/0047287516650937

Kong, S., Huang, Z., Scott, N., Zhang, Z.A., and Shen, Z. (2019). 'Web advertisement effectiveness evaluation: Attention and memory', *Journal of Vacation Marketing*, 25(1): 130–146. DOI: 10.1177/1356766718757272

Lazarus, R.S. (1991). 'Progress on a cognitive-motivational-relational theory of emotion', *American Psychologist*, 46(8): 819–834. DOI: 10.1037/0003-066X.46.8.819

Lee, S., Bruwer, J., and Song, H. (2017). 'Experiential and involvement effects on the Korean wine tourist's decision-making process', *Current Issues in Tourism*, 20(12): 1215–1231. DOI: 10.1080/13683500.2015.1050362

Lee, S., Song, H., Lee, C.K., and Petrick, J.F. (2018). 'An integrated model of pop culture fans' travel decision-making processes', *Journal of Travel Research*, 57(5): 687–701. DOI: 10.1177/0047287517708619

Li, Q., Huang, Z.J., and Christianson, K. (2016). 'Visual attention toward tourism photographs with text: An eye-tracking study', *Tourism Management*, 54: 243–258. DOI: 10.1016/j.tourman.2015.11.017

Li, S., Scott, N., and Walters, G. (2015). 'Current and potential methods for measuring emotion in tourism experiences: A review', *Current Issues in Tourism*, 18(9): 805–827. DOI: 10.1080/13683500.2014.975679

Li, S., Walters, G., Packer, J., and Scott, N. (2018a). 'A comparative analysis of self-report and psychophysiological measures of emotion in the context of tourism advertising', *Journal of Travel Research*, 57(8): 1078–1092. DOI: 10.1177/0047287517733555

Li, S., Walters, G., Packer, J., and Scott, N. (2018b). 'Using skin conductance and facial electromyography to measure emotional responses to tourism advertising', *Current Issues in Tourism*, 21 (15): 1761–1783. DOI: 10.1080/13683500.2016.1223023

Loureiro, S.M.C. (2017). 'Medical tourists' emotional and cognitive response to credibility and servicescape', *Current Issues in Tourism*, 20(15): 1633–1652. DOI: 10.1080/13683500.2015.1050363

Ma, J., Gao, J., Scott, N., and Ding, P. (2013). 'Customer delight from theme park experiences: The antecedents of delight based on cognitive appraisal theory', *Annals of Tourism Research*, 42: 359–381. DOI: 10.1016/j.annals.2013.02.018

Malone, S., McKechnie, S., and Tynan, C. (2018). 'Tourists' emotions as a resource for customer value creation, cocreation, and destruction: A customer-grounded understanding', *Journal of Travel Research*, 57(7): 843–855. DOI: 10.1177/0047287517720118

Mason, M.C. and Paggiaro, A. (2012). 'Investigating the role of festivalscape in culinary tourism: The case of food and wine events', *Tourism Management*, 33(6): 1329–1336. DOI: 10.1016/j.tourman.2011.12.016

Matsuda, Y., Fedotov, D., Takahashi, Y., Arakawa, Y., Yasumoto, K., and Minker, W. (2018). 'EmoTour: Estimating emotion and satisfaction of users based on behavioral cues and audiovisual data', *Sensors*, 18(11): 3978. DOI: 10.3390/s18113978

Mauss, I.B. and Robinson, M.D. (2009). 'Measures of emotion: A review', *Cognition and Emotion*, 23(2): 209–237. DOI: 10.1080/02699930802204677

Michael, I., Ramsoy, T., Stephens, M., and Kotsi, F. (2019). 'A study of unconscious emotional and cognitive responses to tourism images using a neuroscience method', *Journal of Islamic Marketing*. DOI: 10.1108/JIMA-09-2017-0098

Muñoz-Leiva, F., Hernández-Méndez, J., and Gómez-Carmona, D. (2019). 'Measuring advertising effectiveness in travel 2.0 websites through eye-tracking technology', *Physiology & Behavior*, 200: 83–95. DOI: 10.1016/j.physbeh.2018.03.002

Nawijn, J. and Fricke, M.C. (2015). 'Visitor emotions and behavioral intentions: The case of concentration camp memorial Neuengamme', *International Journal of Tourism Research*, 17(3): 221–228. DOI: 10.1002/jtr.1977

Nawijn, J., Isaac, R.K., Gridnevskiy, K., and van Liempt, A. (2018). 'Holocaust concentration camp memorial sites: An exploratory study into expected emotional response', *Current Issues in Tourism*, 21 (2): 175–190. DOI: 10.1080/13683500.2015.1058343

Nawijn, J., Mitas, O., Lin, Y., and Kerstetter, D. (2013). 'How do we feel on vacation? A closer look at how emotions change over the course of a trip', *Journal of Travel Research*, 52(2): 265–274. DOI: 10.1177/0047287512465961

Organ, K., Koenig-Lewis, N., Palmer, A., and Probert, J. (2015). 'Festivals as agents for behaviour change: A study of food festival engagement and subsequent food choices', *Tourism Management*, 48: 84–99. DOI: 10.1016/j.tourman.2014.10.021

Poria, Y., Reichel, A., and Biran, A. (2006). 'Heritage site management: Motivations and expectations', *Annals of Tourism Research*, 33: 162–178. DOI: 10.1016/j.annals.2005.08.001

Prayag, G., Hosany, S., Muskat, B., and Del Chiappa, G. (2017). 'Understanding the relationships between tourists' emotional experiences, perceived overall image, satisfaction, and intention to recommend', *Journal of Travel Research*, 56(1): 41–54. DOI: 10.1177/0047287515620567

Prayag, G., Hosany, S., and Odeh, K. (2013). 'The role of tourists' emotional experiences and satisfaction in understanding behavioral intentions', *Journal of Destination Marketing & Management*, 2(2): 118–127. DOI: 10.1016/j.jdmm.2013.05.001

Rahmani, K., Gnoth, J., and Mather, D. (2019). 'A psycholinguistic view of tourists' emotional experiences', *Journal of Travel Research*, 58(2): 192–206. DOI: 10.1177/0047287517753072

Sánchez-García, I. and Currás-Pérez, R. (2011). 'Effects of dissatisfaction in tourist services: The role of anger and regret', *Tourism Management*, 32(6): 1397–1406. DOI: 10.1016/j.tourman.2011.01.016

Scherer, K.R. (2005). 'What are emotions? And how can they be measured?', *Social Science Information*, 44 (4): 695–729. DOI: 10.1177/0539018405058216
Shoval, N., Schvimer, Y., and Tamir, M. (2018). 'Real-time measurement of tourists' objective and subjective emotions in time and space', *Journal of Travel Research*, 57(1): 3–16. DOI: 10.1177/0047287517691155
Smith, L. (2014). 'Visitor emotion, affect and registers of engagement at museums and heritage sites', *Conservation Science in Cultural Heritage*, 14(2): 125–132. DOI: 10.6092/issn.1973-9494/5447
Song, H., You, G.J., Reisinger, Y., Lee, C.K., and Lee, S.K. (2014). 'Behavioral intention of visitors to an oriental medicine festival: An extended model of goal directed behavior', *Tourism Management*, 42: 101–113. DOI: 10.1016/j.tourman.2013.11.003
Soscia, I., Prayag, G., and Hesapci, O. (2019). 'Advertising guilt-laden vacations: The cross-cultural efficacy of a guilt decreasing appeal', *Journal of Hospitality and Tourism Management*, 39: 57–64. DOI: 10.1016/j.jhtm.2019.02.005
Stadler, R., Jepson, A.S., and Wood, E.H. (2018). 'Electrodermal activity measurement within a qualitative methodology: Exploring emotion in leisure experiences', *International Journal of Contemporary Hospitality Management*, 30(11): 3363–3385. DOI: 10.1108/IJCHM-11-2017-0781
Su, L. and Hsu, M.K. (2013). 'Service fairness, consumption emotions, satisfaction, and behavioral intentions: The experience of Chinese heritage tourists', *Journal of Travel & Tourism Marketing*, 30(8): 786–805. DOI: 10.1080/10548408.2013.835228
Tung, V.W.S. and Au, N. (2018). 'Exploring customer experiences with robotics in hospitality', *International Journal of Contemporary Hospitality Management*, 30(7): 2680–2697. DOI: 10.1108/IJCHM-06-2017-0322
Williams, H.A., Yuan, J., and Williams, R.L., Jr. (2019). 'Attributes of memorable gastro-tourists' experiences', *Journal of Hospitality & Tourism Research*, 43(3): 327–348. DOI: 10.1177/1096348018804621
Woosnam, K.M., Dudensing, R.M., and Walker, J.R. (2015). 'How does emotional solidarity factor into visitor spending among birders in the Lower Rio Grande Valley of Texas?', *Journal of Travel Research*, 54(5): 645–658. DOI: 10.1177/0047287514522884
Yan, Q., Zhou, S., and Wu, S. (2018). 'The influences of tourists' emotions on the selection of electronic word of mouth platforms', *Tourism Management*, 66: 348–363. DOI: 10.1016/j.tourman.2017.12.015
Zhang, Y., Moyle, B.D., and Jin, X. (2018). 'Fostering visitors' pro-environmental behaviour in an urban park', *Asia Pacific Journal of Tourism Research*, 23(7): 691–702. DOI: 10.1080/10941665.2018.1487457

7

The essence of memorable experience

Mariana de Freitas Coelho and Marlusa de Sevilha Gosling

Introduction

Experience is discussed in many fields of knowledge, including marketing, consumption (Holbrook and Hirschman 1982; Schmitt 2011) and economy (Pine and Gilmore 1998, 2011). Meanwhile, authors from disciplines such as psychology, anthropology, and sociology have also discussed experience (Hennes 2002).

Experience is subjective and private (Trigo 2010); therefore, two individuals cannot live the same experience (Pine and Gilmore 1998). General experience is unique for the individual (Walls et al. 2011b), and it is essential for socialization, personal satisfaction, learning, and professional articulation (Trigo 2010).

Although one cannot live and experience an event for another individual, one may design the experience to enhance it. Experience design evokes rituals, ceremonies, theater, and architecture (Brunner-Sperdin and Peters 2009).

Many elements are associated with an experience, and it is possible to use management and marketing lenses to understand it. The contribution of the marketing perspective of the experience involves the identification of strategies and possibilities to manage the customer experience through the customization and co-creation of activities (Pine and Gilmore 2011) and the generation of emotions and memorable moments (Schmitt 2004).

In this chapter, the main discussion is devoted to finding the essence of a memorable tourism experience. We provide a synthesis on the results of the leading researches of MTE, which indicates a lack of discussion on the central characteristics of memorable experiences, such as evoking emotions and providing a meaningful experience to the tourist. Also, it reinforces the necessity of providing novelty to the tourist, which means managers should deliver innovative products and understand the tourist's background and travel habits.

The chapter is structured into eight topics. Experience, Tourism experience and memorability, MTE studies classification, The essence of MTE and framework, Case study, Conclusion remarks, and References. It should help readers to find gaps in the area and guide future directions on memorable experience researches. Furthermore, our proposed framework facilitates an understanding of the central characteristics of memorable tourism experience using a holistic approach.

Experience

The term "experience" reflects distinct matters and regularly evokes at least two different definitions. The first expresses experience as knowledge or expertise and the second is connected to living a new situation which generates uncomfortable emotions (Schmitt 1999; Hennes 2002).

This first type of experience relates to regular and routine life experiences. It is relatively unconscious because it is a regular process, flowing continuously between one moment and another in an uninterrupted flow. It also relates to the knowledge acquired over time (Schmitt 1999), leading to expertise, or the know-how of a given task, activity, or automatic experiences that flow continuously between one time and another without interruption.

A person may have experience in simple or complex activities, ranging from making coffee and driving at night to managing sales teams. In the tourism context, one may have experience in working as a host or hotel manager, for example. This experience evokes a set of competencies and know-how, after experiencing a determined event. In other words, experience in this sense comes down to the process of knowing how to do something or knowing something in depth (Coelho 2017).

The second definition determines experience as a process of experiencing a different, strange situation, generating uncomfortable emotions or pleasure (Schmitt 1999). An experience can be characterized as a constant stream of thoughts (cognitive) and feelings (emotional) which occur during moments of consciousness, concerning the elements of experience (Walls et al. 2011a). The second type of experience arises when the normal flow is interrupted by something that attracts attention (an event, a change, something out of place, something unusual, among others) (Hennes 2002).

Turner (1974) and Pezzi and Santos (2012) support this notion of experience as being a transformative and formative agent which begins with shocks of pleasure or pain experienced by a subject, who then seeks to make sense of that situation. However, if one takes Schmitt's (1999) ideas into account, pain and pleasure are not the only outcomes that experience can provide the subject, but also a range of feelings, emotions, and ways of acting and thinking. These emotions in tourism experiences are expressed mainly positively and outwardly through happiness, enthusiasm, joy, and excitement (Coelho, Gosling, and Almeida 2018).

The second type of experience is highlighted in tourism experiences that involve a new and changing context. Meanwhile, daily and everyday experiences are also part of it (Walls et al. 2011a) but are not what is salient in memorable tourism experiences (MTE).

Tourism experience and memorability

Memorable tourism experiences are strongly associated with the tourist's feelings and the emotions experienced during a specific activity (Lee 2015). Tourists evaluate their experience individually (Kim, Ritchie, and Mccormick 2012), but memorable experiences evoke, overall, positive experiences (Morgan 2010; Tung and Ritchie 2011a).

One may ask if every tourism experience is memorable. Besides the set of experiences, which may include transportation, lodging, information, food and beverage, entertainment and other related services, the memorable experience is the one that crosses the memory generation process. A regular tourism experience is defined as the mental state of a person during the time they encounter the service (Kim, Ritchie, and Mccormick 2012). However, not every tourism experience is recorded in memory. Tourists tend to remember the experience when it evokes shared experiences, unexpected events, abundant service options, and notable services and when they feel it was worth it (Morgan 2010).

Another relevant discussion is how strong a memory needs to be in order to evoke MTEs. "A strong memory is not required from every experience, but the greater (positive) memory created and the longer it lasts, the more value is created." (Pine and Gilmore 2011: XXII). The authors stress that what matters most for an experience to be memorable is how it concludes, meaning experiences perceived as enjoyable by tourists are translated into a more memorable event.

Memorable tourism experiences tend to enhance novelty and emotions. Besides, ambience (physical environment) and refreshment (the state of feeling renewed from experience) are salient variables of MTEs (Coelho 2017). Therefore, when researchers ask tourists to select a memorable tourism experience, they tend to share defined experiences, which help on MTEs' categorization:

- Recent experiences (mainly the last three years)
- Not very short trips (from 4 to 30 days)
- Trips they did with someone else (friends, relatives or partner)
- Diverse purposes (mainly vacation, but also study, work, visiting friends and relatives)
- Either domestic or international trips (Coelho 2017).

The experience as a memorable event depends on many elements due to its multiple dimensions. Pine and Gilmore (2011) state the memory from experience depends on its multisensory nature, the level of personal meaningfulness, how and if it is shared, complexity or simplicity, intensity and duration of experiential elements, cultural consideration, and prior life experiences.

Also, studies reinforce that self-transformation is the ultimate stage in experiences (Pine and Gilmore 2011; Coelho 2017). That means, the higher the transformation from experience the more memorable it is for the tourist.

MTE studies classification

MTE studies have ranged from a more rational view of travel experience to the inclusion of emotional and psychological aspects. MTE studies are classified into three major types of studies, 1) the expansive; 2) the managerial/economic; 3) the modeling perspective (Coelho, Gosling, and Almeida 2018).

The expansive perspective is composed of theoretical, exploratory, in-depth, and qualitative studies of MTE. The authors cite Ballantyne, Packer, and Falk (2011) Tung and Ritchie (2011a), Tung and Ritchie (2011b), Knobloch, Robertson, and Aitken (2014, 2016) as part of the expansive perspective.

These studies serve as a basis for advances in concepts; for instance, Knobloch, Robertson, and Aitken (2014) differentiate memorable experiences from extraordinary, special, and peak experiences. They affirm that memorable experiences are related to connecting and shared experiences, and to special places; being less frequent than extraordinary and peak experiences.

Both Tung and Ritchie (2011a) and Knobloch, Robertson, and Aitken (2016) emphasize MTE domains. The older publication highlights affection, expectations, consequentiality, and recollection as MTE dimensions. The latest research lists emotions, personal outcomes, professional staff, scenery, surprise, new experience, adrenaline, and perceived risk.

Thematic researches can be seen in Tung and Ritchie (2011b) research, with memorable experiences for the elderly, and Ballantyne et al. (2011) with destinations featuring wild animals. Coelho, Gosling, and Almeida (2018) are part of the expansive perspective, and at one point they also used an in-depth approach and focused on the cultural aspects and the contextual particularities of Brazilians' memorable experiences.

The studies from the second perspective can help in providing solutions for experiential marketing managers. The managerial or economic perspective follows two directions. The first evokes studies based on the four experiential realms proposed by Pine and Gilmore (1998), which are entertaining, educational, escapist, and aesthetic.

Oh, Fiore, and Jeoung (2007) confirmed that only the aesthetic dimension affects the memory of bed-and-breakfast consumer experiences. Besides the four experience realms, Oh, Fiore, and Jeoung (2007) added memory, satisfaction, and arousal as study variables, pointing out the need to adapt Pine and Gilmore's (1998) theory for tourism studies. Similarly, Pezzi and Vianna (2015) indicated memory, hospitality, safety, and novelty as MTE dimensions.

Another direction of managerial studies is to emphasize the enterprises' and managers' point of view. Coelho, Gosling, and Almeida (2018) only insert Bharwani and Jauhari (2013) as an MTE study that followed this direction. Authors propose a theoretical framework for employees' competencies required to co-create a memorable customer experience. For the authors, hospitality intelligence contributes to a memorable engagement through employee's cultural, experiential, and emotional intelligence. Hence, a lack of the managerial perspective of memorable tourism experiences seems evident in the literature.

The third type of MTE studies, named as the modeling perspective, uncovers MTE dimensions and its causes and effects (Coelho, Gosling, and Almeida 2018). They present two particularities: 1) the focus on final consumers, and 2) structured equation modeling as a method for these researches. The use of SEM as a method indicates an evolution of MTE theory and concepts, but it requires the use of previous theories to establish and test relationships among the constructs.

Coelho, Gosling, and Almeida (2018) assume many MTE studies are part of this perspective, such as Kim (2010), Kim, Ritchie, and Mccormick (2012), Kim (2014), Kim and Ritchie (2014), Lee (2015), Song et al. (2015), Aroeira, Dantas, and Gosling (2016), Kim and Jang (2016), Manthiou et al., (2016); Tsai (2016).

Kim (2010) has carried out one of the first studies to test MTE dimensions. The author proposes hedonism, involvement, novelty, meaningfulness, refreshment, local culture, and knowledge as antecedents of recollection (memory) and vividness. The confirmed relationships were: refreshment, involvement, and local culture affecting recollection and involvement and refreshment and knowledge affecting the vividness of the experience.

Kim, Ritchie, and Mccormick (2012) move forward on Kim's previous studies and propose an MTE scale based on the same seven dimensions: novelty, hedonism, refreshment, meaningfulness, knowledge, involvement, and local culture. Further, Kim and Ritchie (2014) also attest the seven MTE dimensions as antecedents of behavioral intentions. In other words, memorable experiences determine the intention to recommend the destination or to revisit it.

Kim (2014) develops a scale of ten tourism destination's attributes anteceding a memorable tourism experience. The proposed MTE antecedents are: 1) Infrastructure, 2) Accessibility, 3) Local Culture/History, 4) Physiography, 5) Activities and Events, 6) Destination Management, 7) Quality of Service, 8) Hospitality, 9) Place Attachment, 10) Superstructure.

For Lee (2015) nostalgia, culinary attraction, and cultural inheritance precede tourism experiences in heritage attractions. Knowledge and personal emotions were not significant predictors of MTE.

Aroeira, Dantas, and Gosling (2016) confirm that novelty, hedonism, involvement, local culture and knowledge, and refreshment are dimensions of MTE. The authors then propose a fusion of the local culture with knowledge. Besides, meaningfulness was not a significant MTE dimension in the Brazilian study.

Similarly, Tsai (2016) reduces MTEs to five significant dimensions 1) hedonism, 2) involvement, 3) novelty, 4) local culture and 5) knowledge and refreshment. The author confirms some MTE outputs, meaning MTE precedes place dependence, place identity, and behavior intentions. Besides, in the author's tested model the variable place identity impacts behavioral intention. Meanwhile, it does not confirm the relationship between place dependence and behavior intentions.

Kim and Jang (2016) confirm that animation, personality traits, auditory suggestion, olfactory suggestion, and memorabilia impact both memory retrieval and the liveliness of events.

In synthesis, MTE studies are advancing, but still lack a broader understanding of the phenomena which go beyond its psychological and emotional aspects. MTEs have more to do with the experience as a whole and its evaluation, and more can be extracted from tourism experience discussions (Coelho and Gosling 2018). There is also a gap in memory as a variable of MTE. Researches should focus more on memory triggers and the process of sense-making by the consumer.

Also, further studies on the perspective of firms must be encouraged, including employees', hosts', managers', and owners' strategies for providing and designing experience. Hence, new MTE studies should be developed, and further perspectives rather than the three that are discussed here should contribute to understanding the essence of tourism experiences.

The essence of MTE and framework

There is no consensus among authors on what makes some experiences more memorable than others. However, MTE needs to be addressed from a broader perspective, for example by looking for tourism experience studies (Coelho and Gosling 2018). Therefore, Table 7.1 condenses memorable tourism experiences into four main dimensions: 1) Ambience, 2) Culture, 3) Interpersonal Relationships, and 4) Individual and Psychological.

The esthetic dimension of experience involves all aspects of the environment perceived by tourists (Pine and Gilmore 2011). Previous MTE scales did not consider the physical environment, such as Kim, Ritchie, and Mccormick (2012) and Kim and Ritchie (2014). According to Coelho and Gosling (2018), environmental aspects include various physical elements, build-up attractions, and the business environment. They are also described as service systems, hardware, servicescape (Komppula, Ilves, and Airey 2016), layout, design, decoration, physical ambience (Walls 2013), signs, esthetics, and visual communication (Tussyadiah and Fesenmaier 2009).

Experiences enable the development of interactions among individuals and the environment (Davis 2016). Also, external stimuli and environmental observation can arouse a tourist's feelings, sensation, and behaviors (Tung et al. 2016). A relationship between tourists and the environment is built through place attachment and feelings of place belonging (Davis 2016). Hence, a destination attraction or environment can promote emotions and memories for tourists.

Contact with the local culture can enhance memorable tourism experiences, as confirmed by Kim, Ritchie, and Mccormick (2012) and Kim and Ritchie (2014). Cultural heritage and memorable experience have also been previously attested by Lee (2015).

A tourist can experience local culture through many cultural expressions such as culinary, arts, crafts, dance, language, lifestyle, and values (Coelho and Gosling 2018). The adaptation to local manners and the perceived difference between the tourist's culture and the destination's culture are also part of the cultural dimension for MTEs (Coelho, Gosling, and

Table 7.1 – Memorable Experience Dimensions and Related Themes

Memorable Tourism Experience Essence	
Ambience Dimension	
Environment	Tourist contact with physical, constructed, and business attractions
Cultural Dimension	
Culture	Tourist's cultural impressions when opposed to his own culture and experimentation
Interpersonal Relationship Dimension	
Tourist–Companionship	Contact between tourist and travel companions
Tourist–Tourists	Contact between tourist and other tourists
Tourist–Local Agents	Contact between tourist and local agents (residents, service providers)
Individual and Psychological Dimension	
Dream	Personal desire for the experience
Emotion	Emotional state generated from experience
Novelty	Novelty and surprise from the experience
Involvement	Tourist participation in activities
Refreshment	State of feeling rested and renewed
Meaningfulness	To feel great value or meaning
Hedonism	Feelings of pleasure
Knowledge	Known facts or information
Cost-benefit	Evaluation between the costs and tourists' efforts and what was offered for the experience
Outcomes	Behavior intentions Personal transformation Well-being Memory Co-creation Satisfaction
MTE Premise	Sense-making process and personal meaning to the experience

Source: Authors Own

Almeida 2018). A cultural trade can be a means of surprising tourists, who aim to experience novelty by, for instance, living in a local fashion.

MTE studies are not exploring interpersonal influences properly (Coelho and Gosling 2018). Meanwhile, Carù and Cova (2003), Komppula et al. (2016), Quinlan-Cutler and Carmichael (2010) and Walls (2013) stress the importance of interpersonal relationships in tourism experience.

The interaction between tourists and the local community creates opportunities for symbolic exchanges which are enriching for both hosts and guests (Guzmán et al. 2011). Interpretation, sharing, and re-signification are some of the outcomes of trip experiences when one examines interpersonal relations (Tussyadiah and Fesenmaier 2009) through contact or observation in complex services such as tourism (Schmitt 1999). For Coelho and Gosling (2018) for a tourism experience to be memorable the relationships between tourist–travel companions, tourist–tourists, and tourist–local agents are significant. Coelho, Gosling, and Almeida (2018) add travel planning and travel companionship as relevant interpersonal interactions for memorable trips.

Kim, Ritchie, and Mccormick (2012) and Kim and Ritchie (2014) tested and confirmed MTE's psychological factors. In the authors' proposed MTE scale novelty, involvement, refreshment, meaningfulness, hedonism, and knowledge all evoke psychological aspects of memorable experiences. The only omission from the personal and cognitive aspects of the tourists in these scales is culture, which is also evaluated as significant for a memorable experience.

Coelho and Gosling (2018) extend the psychological MTE dimension by adding emotions and dreams as components of MTE scales. "When travel is a result of a desire or dream it has a special meaning for tourists." (Coelho, Gosling, and Almeida 2018: 16). Emotions include pleasure and excitement and impact on visitor's satisfaction and loyalty (Andreu, Gnoth, and Bigne 2005).

Few authors have inserted cost benefit as a critical MTE variable. Kim (2014) tried to assess it and could not confirm it in his studies. However, the cost benefit is part of the experience's evaluation and need to be considered in further studies.

MTE studies should also investigate the outcomes of a memorable experience, rather than its dimensions. Some authors have confirmed MTE dimensions precede tourist's behavioral intentions (intention to revisit and to recommend the experience) (Kim and Ritchie 2014), and satisfaction (Ali, Ryu, and Hussain 2016). Cognitive perception, loyalty, and destination reputation were also confirmed MTE outcomes (Aroeira, Dantas, and Gosling 2016). Coelho (2017) confirms five MTE outcomes; personal transformations, well-being, place attachment, memory, and behavioral intentions. Finally, the connection between tourism experiences and co-creation has been explored (Verleye 2013; Prebensen, Kim, and Uysal 2016), but needs further investigation in MTE studies.

An MTE premise is that there is no memorable experience without personal value. Once the experience depends on the overall personal evaluation of it, how the experience concludes is of critical importance (Pine and Gilmore 2011).

While the MTE occurs following a chronological process (pre-, during and post-trip), these personal meanings can change and enhance new personal meanings. "These moments might become sense-making or express a different role in the tourists' life." (Coelho, Gosling, and Almeida 2018: 22). So memory creation and retention depend on the given meaning and sense-making of particular events and situations of the lived tourism experience.

Case study 7.1

There are many methods and materials that enable the understanding of memorable tourism experiences. If one considers online reviews and social networks, people voluntarily provide feedback about their trips. Therefore, the internet can contribute to exploring MTEs and this case study is only a brief indication of one possible analysis.

This paper briefly recounts two different memorable experiences in order to depict the MTE variables. The first one refers to a Brazilian actress's trip. Her name is Bruna Marquezine and her Instagram has more than 35 million followers. She went on a missionary trip to Mozambique, Africa, in March 2019. On April 1st, the 23-year-old woman posted the following narrative on Instagram, alongside a black and white photo of her smiling next to eight children.

> brunamarquezine I hope I have done for you at least a little of what you have done for me. I'm not coming home like I left. I was transformed. Made new. It healed me through the lives of everyone who crossed my path. I will always be grateful for every affectionate look, for every genuine smile, for every home embracing, for every edifying word, for all generosity, and for honoring my life even when I do not find myself worth it. Thank you for showing me the true character of Christ. I learned so much. And nothing I do or have done will be enough to repay everything I have received. See you later! With love, Bruna

Bruna emphasizes the personal relationships created during her experience (the lives of everyone who crossed my path) and the emotions the encounters have evoked (affection, love). She also demonstrates how she found personal meaning in the experience through learning (I learned so much), new religious meaning (showing me the true character of Christ) and personal transformation (I'm not coming home as I left. I was transformed. Made new. It healed me ...).

The second example is from a nutritionist called Fernanda Scheer. She has 263 thousand followers and traveled with one of her daughters to Paris, France, in April 2019. She posted a photo of her daughter and herself seated and looking at each other and smiling, with the Eiffel Tower in the background.

> Fernandascheernutri We've just arrived, happy to be back, but crying (for real) and already missing a memorable trip.
>
> Paris has something that moves me deeply. It is not glamor, it is not the pomposity ... on the contrary – it is simplicity, the old, the bucolic, the history.
>
> It could be another city alike, but it is special for me because it marked an important phase of my life.
>
> I spent a season (in Quartier Latin) when I was 17 years old, which left beautiful and deep marks.
>
> I have the feeling that I have lived in some other life in this place.
>
> And this love for what the city represents passed as a legacy. My daughter fell in love, of course, and so she cried today.
>
> She knew the city with my eyes – felt, understood the magic that a place with so much history has.
>
> Good travel is so – it leaves its mark on our soul. And this @fotosemparis record says it all
>
> Now we return to normal programming with content and recipes here.
>
> But with much more inspiration and a (still) sweeter look at life.

Fernanda's post is a demonstration of how past experience can be an important variable on MTE. Also, she gave a new meaning to her trip once she took her daughter to experience the city.

Final remarks

Experience can be understood either as acquired knowledge (through regular and routine life experiences) or as a process of experiencing a different situation which generates uncomfortable emotions. Not every tourism experience is to be considered memorable. But there is no memorable experience without personal value, which means that tourists themselves give meaning to their experiences and, therefore, make them remarkable and memorable. This notion also confirms that the ultimate degree of Memorable Tourism Experiences (MTE) is self-transformation.

This chapter presents a research framework which goes beyond the psychological factors (novelty, refreshment, knowledge, and meaningfulness), and also considers cultural, environmental, and inter-relational dimensions to determine the memorability of the tourism experience. Memorable Tourism Experiences (MTE) outcomes are also future research directions, including well-being, co-creation, satisfaction, behavioral intentions, and personal transformation.

References

Ali, F., Ryu, K., and Hussain, K. (2016) 'Influence of experiences on memories, satisfaction and behavioral intentions: A study of creative tourism', *Journal of Travel and Tourism Marketing*, 33(1): 85–100. DOI: 10.1080/10548408.2015.1038418

Andreu, L., Gnoth, J., and Bigne, J.E. (2005) 'The theme park experience: An analysis of pleasure, arousal and satisfaction', *Tourism Management*, 26(6): 833–844. DOI: 10.1016/j.tourman.2004.05.006

Aroeira, T., Dantas, A.C., and Gosling, M. (2016) 'Experiência Turística Memorável, Percepção Cognitiva, Reputação E Lealdade Ao Destino : Um Modelo Empírico', *Revista Turismo Visão e Ação*, 18(3): 584–610.

Ballantyne, R., Packer, J., and Falk, J. (2011) 'Visitors' learning for environmental sustainability: Testing short- and long-term impacts of wildlife tourism experiences using structural equation modelling', *Tourism Management*, 32(6): 1243–1252. DOI: 10.1016/j.tourman.2010.11.003

Bharwani, S. and Jauhari, V. (2013) 'An exploratory study of competencies required to co-create memorable customer experiences in the hospitality industry', *International Journal of Contemporary Hospitality Management*, 25(6): 823–843. DOI: 10.1108/IJCHM-05-2012-0065

Brunner-Sperdin, A. and Peters, M. (2009) 'What influences guests' emotions? The case of high-quality hotels', *International Journal of Tourism Research*, 11: 171–183.

Carù, A. and Cova, B. (2003). 'Revisiting consumption experience: A more humble but complete view of the concept', *Marketing Theory*, 3(2): 267–286. DOI: 10.1177/14705931030032004

Coelho, M.D.F. (2017) *Viagens De Brasileiros: Um Modelo De Relações Entre Experiência Turística Memorável, Mindfulness, Transformações Pessoais E Bem-Estar Subjetivo* (doctoral thesis). Belo Horizonte: Universidade Federal de Minas Gerais. Retrieved from: https://cepead.face.ufmg.br/btd/files/432/aid432n2a1.pdf

Coelho, M.D.F., Gosling, M.D.S., and Almeida, A.S.A.D. (2018) 'Tourism experiences: Core processes of memorable trips', *Journal of Hospitality and Tourism Management*, Elsevier Ltd. 37(May 2017): 11–22. DOI: 10.1016/j.jhtm.2018.08.004

Coelho, M.F. and Gosling, M.S. (2018) 'Memorable Tourism Experience (MTE): A scale proposal and test', *Tourism & Management Studies*, 14(4): 15–24.

Davis, A. (2016) 'Experiential places or places of experience? Place identity and place attachment as mechanisms for creating festival environment', *Tourism Management*, 55: 49–61. DOI: 10.1016/j.tourman.2016.01.006

Guzmán, S.J.M., Astor V., Jr. and Santos, I.J. (2011). 'Turismo de Experiência: Uma Proposta Para o Atual Modelo Turístico Em Itacaré - Bahia', *Revista de Cultura e Turismo*, 5(1): 98–113.

Hennes, T. (2002) 'Rethinking the visitor experience: Transforming obstacle into purpose', *Curator: The Museum Journal*, 45(2): 109–121.

Holbrook, M.B. and Hirschman, E.C. (1982) 'The experiential aspects of consumption: Consumer fantasies, feelings and fun', *The Journal of Consumer Research*, 9(2): 132–140.

Kim, J., Ritchie, J.R.B., and Mccormick, B. (2012) 'Development of a scale to measure memorable tourism experiences', *Journal of Travel Research*, 51(1): 12–25. DOI: 10.1177/0047287510385467

Kim, J.-H. (2010) 'Determining the factors affecting the memorable nature of travel experiences', *Journal of Travel & Tourism Marketing*, 27(8): 780–796. DOI: 10.1080/10548408.2010.526897

Kim, J.-H. (2014) 'The antecedents of memorable tourism experiences: The development of a scale to measure the destination attributes associated with memorable experiences', *Tourism Management*, 44: 34–45. DOI: 10.1016/j.tourman.2014.02.007

Kim, J.-H. and Jang, S. (2016) 'Memory retrieval of cultural event experiences: Examining internal and external influences', *Journal of Travel Research*, 55(3): 322–339. DOI: 10.1177/0047287514553058

Kim, J.-H. and Ritchie, J.R.B. (2014) 'Cross-cultural validation of a Memorable Tourism Experience Scale (MTES)', *Journal of Travel Research*, 53(3): 323–335. DOI: 10.1177/0047287513496468

Knobloch, U., Robertson, K., and Aitken, R. (2014) '(Mis)Understanding the nature of tourist experiences', *Tourism Analysis*, 19(5): 599–608. DOI: 10.3727/108354214X14116690097891

Knobloch, U., Robertson, K., and Aitken, R. (2016) 'Experience, emotion, and eudaimonia: A consideration of tourist experiences and well-being', *Journal of Travel Research*, 1–12. DOI: 10.1177/0047287516650937

Komppula, R., Ilves, R., and Airey, D. (2016) 'Social holidays as a tourist experience in Finland', *Tourism Management*, 52: 521–532. DOI: 10.1016/j.tourman.2015.07.016

Lee, Y.J. (2015) 'Creating memorable experiences in a reuse heritage site', *Annals of Tourism Research.*, 55: 155–170. DOI: 10.1016/j.annals.2015.09.009

Manthiou, A. et al. (2016). 'Investigating the effects of memorable experiences: An extended model of script theory', *Journal of Travel & Tourism Marketing*, 33(3): 362–379. DOI: 10.1080/10548408.2015.1064055

Morgan, M. (2010) 'The experience economy: 10 years on: Where next for experience management?', in Morgan, M., Lugosi, P., and Brent Ritchie, J.R. (eds) *The Tourism and Leisure Experience* (pp. 63–77). Bristol: Channel View Publications.

Oh, H., Fiore, A.M., and Jeoung, M. (2007) 'Measuring experience economy concepts: Tourism applications', *Journal of Travel Research*, 46(2): 119–132. DOI: 10.1177/0047287507304039

Pezzi, E. and Santos, R.J.D. (2012) 'A experiência turística e o turismo de experiência: aproximações entre a antropologia e o marketing' (conference paper), in *Anais do VII Seminário de Pesquisa em Turismo do Mercosul* (pp. 4–13). Caxias do Sul, 2012: Universidade de Caxias do Sul.

Pezzi, E. and Vianna, S.L.G. (2015) 'A Experiência Turística e o Turismo de Experiência : um estudo sobre as dimensões da experiência memorável', *Turismo em Análise*, 26(1): 165–187.

Pine, B.J. and Gilmore, J.H. (1998) 'Welcome to the experience economy', *Harvard Business Review*, (July-August): 97–105.

Pine, B.J.P. and Gilmore, J.H. (2011) *The Experience Economy*. Boston: Harvard Business Review Press.

Prebensen, N.K., Kim, H.L., and Uysal, M. (2016) 'Cocreation as moderator between the experience value and satisfaction relationship', *Journal of Travel Research*, 55(7): 1–12. DOI: 10.1177/0047287515583359

Quinlan-Cutler, S. and Carmichael, B. (2010) 'The dimensions of customer experience', in Morgan, M., Lugosi, P., and Ritchie, B. (eds) *The Tourism in Leisure Experience: Consumer and Managerial Perspectives* (pp. 3–26). Bristol: Aspects of Tourism. Available at: http://books.google.com.br/books/about/The_Tourism_and_Leisure_Experience.html?id=52ReGta3HDIC&pgis=1 [Accessed April 15, 2015].

Schmitt, B. (1999) 'Experiential marketing', *Journal of Marketing Management*, 15(1–3): 53–67.

Schmitt, B. (2011) *Experience Marketing: Concepts, Frameworks and Consumer Insights*. Hanover: Now Publishers Inc. Available at: https://books.google.com/books?id=X7Ya16wdXxkC&pgis=1 [Accessed April 25, 2015].

Schmitt, B.H. (2004) *Gestão De Experiência Do Cliente*. Porto Alegre: Bookman.

Song, H.J. et al. (2015). 'The influence of tourist experience on perceived value and satisfaction with temple stays: The experience economy theory', *Journal of Travel & Tourism Marketing*, 32(4): 401–415. DOI: 10.1080/10548408.2014.898606

Trigo, L.G.G. (2010) 'A viagem como experiência significativa', in Panosso Netto, A. and Gaeta, C. (eds) *Turismo De Experiência* (pp. 21–41). São Paulo: Editora Senac São Paulo.

Tsai, C.S. (2016) 'Memorable tourist experiences and place attachment when consuming local food', *International Journal of Tourism Research*, 548(January): 536–548. DOI: 10.1002/jtr

Tung, V.W.S. et al. (2016). 'A framework of memory management and tourism experiences', *Journal of Travel & Tourism Marketing*, 1–14. DOI: 10.1080/10548408.2016.1260521.

Tung, V.W.S. and Ritchie, J.R.B. (2011a) 'Exploring the essence of memorable tourism experiences', *Annals of Tourism Research*, 38(4): 1367–1386.

Tung, V.W.S. and Ritchie, J.R.B. (2011b) 'Investigating the memorable experiences of the senior travel market: An examination of the reminiscence bump', *Journal of Travel & Tourism Marketing*, 28 (March 2015): 331–343. DOI: 10.1080/10548408.2011.563168

Turner, V.W. (1974) *O Processo Ritual: Estrutura E a Antiestrutura*. Petrópolis: Vozes.

Tussyadiah, I.P. and Fesenmaier, D.R. (2009) 'Mediating tourist experiences. Access to places via shared videos', *Annals of Tourism Research*, 36(1): 24–40. DOI: 10.1016/j.annals.2008.10.001

Verleye, K. (2013) 'The co-creation experience from the customer perspective: Its measurement and determinants', *Journal of Service Management*, 26(2): 321–342. DOI: 10.1108/MRR-09-2015-0216

Walls, A. et al. (2011a). 'Understanding the consumer experience: An exploratory study of luxury hotels', *Journal of Hospitality Marketing and Management*, 20(2): 166–197. DOI: 10.1080/19368623.2011.536074

Walls, A.R. et al. (2011b). 'An epistemological view of consumer experiences', *International Journal of Hospitality Management*, 30(1): 10–21. DOI: 10.1016/j.ijhm.2010.03.008

Walls, A.R. (2013) 'A cross-sectional examination of hotel consumer experience and relative effects on consumer values', *International Journal of Hospitality Management*, 32(1): 179–192. DOI: 10.1016/j.ijhm.2012.04.009

8

Service employees and customer experience

Poh Theng Loo

Introduction

Service employees play an important role in delivering service and also crafting customer experience. At the beginning of this chapter the importance of service employees in customer experience was explained. Service employees are the company brand and one of the key resources that determine the sales and profits of the company. Regardless of the advancement of technology and service delivery facilities, employees are still one of the main ways of ensuring happy customers and generating higher profits. Therefore, understanding customer experience is necessary as it will help service employees know how to craft memorable experiences, which is covered in the second section of this chapter. Further explanation of memorable experience covers experience-involvement and four main aspects of memorable experience.

A memorable experience includes the co-creation of tourists' involvement in the service experience, whereby tourists and employees are both involved and interact throughout the experience. Co-creation experience enables customers to have even more fruitful experiences. For example, the interactions between tourists and local people can be made more meaningful and memorable with the help of a tour guide who explains the values and culture of the local people.

The third section of this chapter explains three critical impacts of service employees on customer experience: (a) customer satisfaction, (b) e-WOM, and (c) loyalty and revisit intentions. This section emphasizes the significant role of service employees and their contribution to customer experience and how this impacts on a company's sales and profits. Last but not least, the fourth section highlights three key strategies in managing service employees and customer experience. These strategies enable leveraging and maximizing the positive impacts of service employees on customer experience.

Importance of service employees in customer experience

Service employees are recognized as the brand of a company and the marketer (Zeithaml, Bitner, and Gremler 2009). In fact, they represent the organization in the eyes of customers (Smith et al. 2010). Service employees have a substantial influence on an organization's sales and also its business productivity. They are the link between an organization and its customers (Wirtz, Chew and Lovelock 2012). In other words, they fulfil a boundary-spanning role as they have frequent contact with customers and they are an important source of information about customers (Bitner et al. 1994). The information about customers obtained by employees can be used to improve service quality and is also helpful for organizational decision-making. For example, in the study of Bitner and her colleagues, it is reported that the types of satisfactory and unsatisfactory incident outcomes which provided useful information to hotels, restaurants and airline companies included employee response to service delivery system failures, employee response to customer needs, unprompted and unsolicited employee and problematic customer behavior. This information enables hospitality companies to know and understand their employees better in order to provide them with necessary training and support. On the other hand, for customers, companies can understand the problems and needs of customers better after their service consumption at their companies and find out ways to provide better service delivery.

Service employees are not only the information source; they also play critical roles in the idea generation stage of the service development process, in particular those employees who are able to read customer needs. Therefore, organizations should invest in and create positive working environments that motivate and reward employees who contribute their flow of ideas for service improvement (Lages and Piercy 2012).

Larivière and his colleagues (2017) have developed a concept of service encounter that highlights how employees and customers play the role of enabler, innovator, coordinator and differentiator in the era of technology. They believe that in the changing interdependent roles of technology, employees and customers exist together in the process of service experience (Larivière et al. 2017). In summary, service employees play a significant role in the process of service delivery as well as the customer experience in the tourism and hospitality industry. Service employees are the ones who shape tourist experience and create circumstances and service environments that are part of the customer experience (Zatori et al. 2018).

Understanding customer experience in tourism

In the tourism industry, there exist lots of experiences of people with similar interests, goals and motivations who meet and interact with each other (Rihova et al. 2015). Customer experience in tourism has been studied and discussed; nowadays, it is no longer merely talked about as an experience but is treasured as a memorable and authentic experience. There are four dimensions of on-site tourist experience introduced by Zatori et al. (2018). The dimensions are emotional, mental, flow-like and social experience-involvement in the context of the guided tour. These dimensions are the indicators of experience-involvement: (a) *emotional experience-involvement* refers to the emotions that are measured as excitement, fascination, inspiration, enjoyment and surprise; (b) *mental experience-involvement* relates to the cognitive senses and it is measured through indicators, including learning, awakening interest, experiencing something thought-provoking and activating the desire to learn; (c) *flow-like experience-involvement* is described as 'a cognitively perceived experience with an emotional value which is conative and/or creative in nature, while its immersive power leads to a higher level of experience-involvement due to active engagement uniqueness or meaningfulness' (Zatori et al. 2018: 119) – the measurement indicators are the perceptions of uniqueness, escapism, meaningfulness and getting lost in the story created during the tour; and (d) *social experience-involvement* results from the social experience which happened during the tour and it is determined by interactions between the customers. The indicators are the degree of interaction within the group, the group atmosphere, the perceived quality of group members and the nature of interaction with service employees.

Another four dimensions have evolved from the memorable aspects, which are affect, expectations, consequentiality, and recollection (Tung and Ritchie 2011): (a) *affect* refers to the positive emotions and feelings related to experiences (e.g. happy, excited) and negative feelings are seldom recalled; (b) *expectations* are the fulfillment of intentions or a description of the surprises encountered during the trip, which were beyond the agendas planned by tourists; (c) *consequentiality* is the personally perceived importance of responses from the trip (e.g. self-discovery, intellectual development, overcoming physical challenges and enhancing social relationships); and (d) *recollection* refers to the deeds done by tourists in reflecting on and/or remembering their tourism experience (e.g. storytelling, souvenir purchases, photographs showed).

According to two different groups of dimensions on memorable tourism experiences, affect and cognition are closely associated with a tourist's experience and memory. The difference is that when a tourist travels with and without a tour guide, they could have different journeys of tourism experiences. Tour-guided experiences emphasize social experience-involvement, whereas traveling without a tour guide focuses on consequentiality and recollection.

In relation to memorable customer experience, co-creation is also critical as part of the experience because tourists become active co-creators of service experiences. They engage in value-creating activities, in which tourists create positive value as outputs for themselves and for other tourists (Rihova et al. 2015). Indeed, the interactions between service employees and tourists and the active participation of tourists are the antecedents of experience co-creation (Buonincontri et al. 2017). Regardless of whether their tourism experiences are with or without a tour guide, tourists can have co-creation experiences between themselves and the locals at the destinations, such as stepping into the daily life of a local, living like a local, tasting, seeing, listening to and experiencing the local culture as well as destination interactive workshops such as the locals teaching tourists about local specialties in terms of food, arts, pottery, painting, dancing, language learning and so on (Binkhorst and Den Dekker 2009). Therefore, Binkhorst and Den Dekker (2009) highlighted that a significant step forward to the creation of memorable tourism experiences is being sensitive to the needs, desires and wishes of individuals and being empowered to act alike. The next section explains how service employees have an impact on customer experience.

Impacts of service employees on customer experience

(a) Customer satisfaction

Experience concepts are associated strongly with the spirit of the destination and its people (Binkhorst and Den Dekker 2009). Interactions between customers and employees have a significant influence on shaping the overall customer experience (Luk and Lloyd 2011). The behavior of service employees in the interaction process with customers during their dining experience would increase customer satisfaction in restaurants (Kong and Jogaratnam 2007); in particular, the human aspects of service such as friendliness, politeness, appearance and reliable behavior of service employees are the most important determinants in customer satisfaction. A recent piece of research investigated the functional aspects (including technical quality of service, food quality and efficiency of service) and personal aspects (include behavior and characteristics of employees such as body language, facial expressions, customer greeting and friendliness) of employee service behavior associated with the customer satisfaction experience, in which personal aspects have a greater impact than functional ones (Alhelalat et al. 2017).

In other words, the behavior of service employees during service delivery is the key determinant of customer satisfaction and revisit intentions; in particular, the employee's helpfulness exerts a greater influence on customer satisfaction than physical attractiveness or displayed positive emotion (Keh et al. 2013). Nevertheless, employees who display positive emotions also have a significant effect on customer satisfaction. This is especially so when this positive emotion is displayed in conjunction with employee helpfulness. Therefore, merely hiring polite or good-looking employees may not be sufficient, as cheerful and helpful employees are absolutely essential in the recruitment requirements.

Customers with positive emotions and who are in the right mood will be better enabled to enjoy a flow of experiences and this leads to customer satisfaction (Brunner-Sperdin et al. 2012). The authenticity of emotional input displayed by service employees has a direct effect on customers' emotional states (Hennig-Thurau et al. 2006). This means employees' genuine positive emotional display is powerful and has a great influence on customer emotions, which will have a later effect on their service consumption satisfaction. However, merely having an employee smiling is not sufficient to influence customer emotions. Therefore, hiring service employees who are good in emotional management and have high emotional intelligence will help to offer a desirable service experience to customers.

(b) Electronic word-of-mouth (e-WOM)

Service employees in tourism, as well as their managers and companies, employ experience-focus management perspective as they recognize that memorable and authentic experiences have an influence on gaining greater competitive advantage in the market (Zatori et al. 2018). In addition, when enjoying a good and enjoyable dining experience crafted and provided by restaurant employees, restaurant customers are motivated to disseminate e-WOM as the customers want to help restaurants and to share their positive experience with other people in an online environment (Jeong and Jang 2011). One of the reasons customers are motivated to help restaurants is based on their satisfaction with employee behavior, in terms of how they are being served and treated (Sundaram et al. 1998). Furthermore, customer-perceived service quality influences customer satisfaction on both a positive and negative emotional basis and subsequently also affects customers' future revisit intentions (Ladhari et al. 2008).

(c) Loyalty and revisit intentions

From the organizational perspective, customer experience can be formed by offering physical environment and social interactions elements during service delivery and service encounters (Carbone and Haeckel 1994; Gupta and Vajic 1999) and this is proven in a later study of Cetin and Dincer (2014) on these elements in relation to customer loyalty and word-of-mouth in hospitality operations. The physical environment elements include safety and security, quality of materials, location of the facility and architectural design; and the social interactional elements are staff showing personal care and the technical knowledge of staff (Cetin and Dincer 2014). A recent study conducted by Brun and her colleagues (2017) on the impact of customer experience on loyalty revealed that the affective dimension (negative) has a substantial influence on customer experience and their loyalty towards companies.

Therefore, hotels, restaurants and tourism product companies should continuously put more effort and investment into enhancing the physical environment (such as the selection of unique and creative designs for facilities and artistic decoration) and social interactions (such as intensive employee training on customized personal interactions and increasing employees' skills and knowledge in respect of service and product standards). Also, it is very important to train service employees in managing customers who experience negative affective during service consumption as well as minimize their negative affective as far as possible to enhance customer loyalty following the service experiences.

Strategies to manage service employees and customer experience

There are different research studies that have recommended strategies to manage and improve service employee and customer experience:

(a) Compatible customer mix management

For tour-guided customers, a compatible customer mix is very important, which means homogeneity of customer group in terms of their preferences, sought benefits, past experiences, physical characteristics and attitude. This would enable managers and service employees to be able to anticipate and prevent some service issues from arising. On the other hand, it is also necessary to avoid putting together incompatible customers such as smoking and non-smoking customers or families with children and couples or single adults. Incompatible customer mix could trigger unavoidable (this should be 'avoidable') conflicts and gaps in customer needs and wants. Therefore, having a compatible customer mix would help to avoid spoiling the customer experience. Introducing a code of behavior, a dress code, smoking rules, punctuality requirements, and so on could help to enhance the overall customer experience throughout their trip if customers agree to follow it (Wu 2007). Service employees are important in helping to run the compatibility management programs by recruiting and training employees who are proactive and willing to make an effort to manage customer behavior as well as promote positive customer-to-customer interactions. Furthermore, service employees can employ games, pre-arranged seating or other approaches to enhance the customer-to-customer interactions.

(b) Manage the customer experience of pre-, during and post-tourism experience phases

It is important that service employees have direct interactions with customers and actively participate with them throughout their experience. Moreover, employees should improve their interactions continuously with existing and potential tourists in the whole service consumption process pre-, during and post-tour or trip by offering direct communication channels and adopting co-creation strategies to involve customers in organizing and planning their trips (Buonincontri et al. 2017). It is also necessary to employ smart technology in the co-creation of customer service experience in tourism, which may also enable the service employees to create interactive, personalized and unique tourism experiences such as mobile apps with augmented reality, interactive and real-time maps, gaming services, tourism destination information and other benefits. During the visit phase, virtual reality tools, the internet such as QR codes, tags, and sensors can be used to support and enhance the customers' tourism experiences in all three phases of their tourism experiences.

Incorporating smart devices and elements into the tourism experiences could help to enhance the relationships between employees and customers as well as between customers themselves. Prior to incorporating and adopting smart technologies and elements into the co-creation experiences, companies and managers have to be careful and should identify different needs and expectations of different types of tourism customer needs and behavior. This would help tourism organizations to engage and interact better with customers in designing and incorporating co-creation with technological elements throughout their customers' tourism experiences, thereby making them especially memorable.

(c) Tourism organizations and employees should be alert and sensitive to customer needs in their experience

Both tourism organizations and service employees need to be sensitive and alert to all types of individual needs, wishes, desires and expectations. They should be empowered to use appropriate strategies and actions to invest resources for creating memorable tourism experiences (Binkhorst and Den Dekker 2009). Detailed data collection and surveys about each customer profile and their preferences, behavior, needs, desires and expectations should be done before planning and carrying out trips. The idea is not only to fulfill customers' expectations but, more importantly, to create a memorable and unique experience. Hence, organizations and managers should plan, allocate and assign sufficient resources in terms of time, money and manpower to support the service employees in delivering and executing strategies to create memorable experiences. More importantly, managers and even service employees should be empowered by their superiors to make decisions and take action in order to offer instant customized services when appropriate for customer needs to be fulfilled.

In addition, organizations should recruit employees who are sensitive and passionate toward customer needs, wishes, desires and expectations and who are self-motivated to be creative and innovative so they can develop plans and strategies for the creation of memorable tourism experiences for different groups of customers at different points of time. Organizations also need to consider the price charged for these memorable and special experiences so that customers feel that their trips are worth the money they have paid.

(d) Investigation into and continuous improvement in service employee and customer experiences

Tourism and hospitality managers should take further action to investigate how different types of employee behavior are likely to affect customers' perceptions and emotions (Luk and Lloyd 2011). For example, the socio-emotional aspects of the behavior of employees could strengthen and enhance interpersonal relationships by showing care and empathy; furthermore, it is important to know to what extent being friendly and enthusiastic to customers would affect the perceptions and emotions of customers.

Hospitality managers and human resource managers should implement procedures for enhancing employees' service behavior, which can be formal or on the job training of employees to equip them with service and human skills to improve the overall customer experience. Training sessions are needed for service employees to improve their attitudes, responses to customers, appearance and service performance because employee–customer interaction would help to improve customer satisfaction (Wu and Liang 2009). In order to foster these skills, managers must carry out reinforcement programs from time to time to evaluate and monitor if employees are committed to the delivery of proper and desirable personal and functional service practices (Alhelalat et al. 2017). Human resource managers should include a 'Desired Personal Qualities' checklist and assessments in recruitment procedures in order to hire service employees who are able to display appropriate and desirable service behavior.

Case study 8.1

In the past three years, ABC hotel, which is a four-star hotel, has had its occupancy rate drop drastically compared to the statistics of previous years. From January 2017, the occupancy rate of 90% dropped to 75% in January 2018 and was recorded as 50% in January 2019. These figures made the hotel owner, Mr. Steven Stone, worried and concerned about the future cost of running this hotel. Thus, he had a meeting with the hotel general manager, Mr. Brown Ryan, to find out the causes of the drastic drop of occupancy rate and he was provided with a detailed report on causes and proposed strategies to overcome and manage this problem.

After compiling some statistics and information, investigations and analysis were done to find out the possible causes of the drastic reduction of occupancy rates in the past three years. It was revealed that several new hotels, ranging from budget hotels to five-star hotels, have opened in the past three years around the ABC hotel area. These new hotels are located within a radius of about five kilometers from ABC hotel. In addition, according to customer feedback and comments in the past twelve months, many negative comments had been made about the behavior of service employees. Further analysis of the negative comments was done by an external research company and the analysis revealed that the negative comments about the behavior of service employees had caused customer dissatisfaction and customer complaints throughout their stay. During the stay, the majority of the complaining customers found that during check-in at the front desk the service employees were not alert and did not try to speed up the check-in process, especially when there was a group check-in or when customers asked for more information about tourist-related information close to the hotel. Customers complained that the information they were given was incomplete and employees were not able to explain further or provide detailed information as needed. There were also comments about rude and inattentive service employees while the customers were eating breakfast and dinner. About 30% of the negative feedback showed that customers did mention their frustration and their decision not to choose ABC hotel again for their future stays and 15% of the complaining customers mentioned that they would definitely tell their friends and family about their bad experiences.

After collecting all of the valid facts and information, the hotel general manager, Mr. Brown Ryan, called for a special team meeting to discuss and brainstorm on strategies to tackle the issues and problems in order to minimize greater negative impacts on the hotel's sales and profits. This special team consists of the middle managers from each department in the hotel. Mr. Ryan informed all special team members from different departments to propose their ideas on ways and strategies to manage and improve the situation, particularly what each department could do and how they would like other departments to work together with them to execute the ideas and strategies of improvement. These coming meetings are very important for this hotel as the outcomes of meetings will affect subsequent actions to be taken to improve and solve the problem of the decreasing occupancy rate.

Questions:

1. Based on the information and facts about the poor occupancy rate issue of ABC hotel, can you as the middle manager of the department explain what your department can do, and what are the ideas and strategies? Choose any two of the departments (human resource, marketing, front desk, restaurant operations) and explain your answers.

2. If you were the general manager and the special team had reported to you with their ideas and strategies, how would you convince the owner of the hotel to increase the budget for improvement? And why do you use those approaches?
3. As mentioned in the case study, as a few new hotels have opened nearby, propose three strategies focusing on how ABC hotel can gain a better competitive advantage in terms of managing the service employee and customer experience.

Conclusions

Service employees and customer experiences are closely related and interchangeable in terms of the connection between their relationships and interactions. Service employees indeed play very significant roles in customer experiences regardless of which company it is in the tourism and hospitality industry, as human interactions cannot totally be replaced by technologies and good relationships and experiences will help companies to gain a competitive advantage and generate more future profits. Nowadays, due to the competitive business environment, tourism and hospitality organizations should offer more than a good experience, but instead a memorable service experience. Therefore, organizations should put greater focus on and apply resources to supporting and enhancing their efforts in creating and developing memorable experiences by offering more service training, and scrutinizing recruitment procedures and requirements that would contribute to customer-positive and memorable experiences.

The impact of service employees on customer experience includes customer satisfaction, e-WOM and revisit intentions. This impact has a significant subsequent effect on tourism and hospitality organizations as satisfied and happy customers tend to spread positive e-WOM and also tend to repeat their visits and service consumption. Subsequently, organizational sales and profits should be increased too. By knowing and understanding how service employees affect customers' experience and their satisfaction, organizations should take proactive and preventive actions to recruit, select, hire, motivate and train service employees. More importantly, it is necessary to support and reward service employees for their efforts in making good and memorable service experiences for customers. Hence, strategies to manage service employee and customer experience are important and necessary.

There are four main strategies mentioned above, (a) compatible customer mix management; (b) managing the customer experience of pre-, during and post-tourism experience phases; (c) tourism organizations and employees to be alert and sensitive to customer needs in experience; and (d) investigation and continuous improvement on service employee and customer experiences. All of these four strategies require many different departments in an organization to work and achieve together. Departments such as quality management, marketing, human resource, operations, and top management should have regular meetings or even form a special team by having representatives from each department to discuss and brainstorm on issues and customize strategies and approaches to manage and improve service employee and customer experience continuously.

References

Alhelalat, J.A., Habiballah, M.A., and Twaissi, N.M. (2017). 'The impact of personal and functional aspects of restaurant employee service behaviour on customer satisfaction'. *International Journal of Hospitality Management*, 66: 46–53. DOI: 10.1016/j.ijhm.2017.07.001

Binkhorst, E. and Den Dekker, T. (2009). 'Agenda for co-creation tourism experience research'. *Journal of Hospitality Marketing and Management*, 18(2-3): 311–327. DOI: 10.1080/19368620802594193

Bitner, M.J., Booms, B.H., and Mohr, L.A. (1994). 'Critical service encounters: The employee's viewpoint'. *The Journal of Marketing*, 58(4): 95–106. DOI: 10.1177/002224299405800408

Brun, I., Rajaobelina, L., Ricard, L., and Berthiaume, B. (2017). 'Impact of customer experience on loyalty: A multichannel examination'. *The Services Industries Journal*, 37(5–6): 317–340. DOI:10.1080/02642069.2017.1322959

Brunner-Sperdin, A., Peters, M., and Strobl, A. (2012). 'It is all about the emotional state: Managing tourists' experiences'. *International Journal of Hospitality Management*, 31(1): 23–30. DOI: 10.1016/j.ijhm.2011.03.004

Buonincontri, P., Morvillo, A., Okumus, F., and van Niekerk, M. (2017). 'Managing the experience co-creation process in tourism destinations: Empirical findings from Naples'. *Tourism Management*, 62: 264–277. DOI: 10.1016/j.tourman.2017.04.014

Carbone, L.P. and Haeckel, S.H. (1994). 'Engineering customer experiences'. *Marketing Management*, 3 (3): 8–19.

Cetin, G. and Dincer, F.I. (2014). 'Influence of customer experience on loyalty and word-of-mouth in hospitality operations'. *Anatolia: An International Journal of Tourism and Hospitality Management*, 25: 181–194. DOI: 10.1080/13032917.2013.841094

Gupta, S. and Vajic, M. (1999). 'The contextual and dialectical nature of experiences'. In J. Fitzsimmons and M. Fitzsimmons (eds.) *New Service Development* (pp. 33–51), Thousand Oaks, CA: Sage.

Hennig-Thurau, T., Groth, M., Paul, M., and Gremler, D.D. (2006). 'Are all smiles created equal? How emotional contagion and emotional labor affect service relationships'. *Journal of Marketing*, 70(3): 58–73. DOI: 10.1509/jmkg.70.3.058

Jeong, E. and Jang, S.S. (2011). 'Restaurant experiences triggering positive electronic word-of-mouth (eWOM) motivations'. *International Journal of Hospitality Management*, 30(2): 356–366. DOI: 10.1016/j.ijhm.2010.08.005

Keh, H.T., Ren, R., Hill, S.R., and Li, X. (2013). 'The beautiful, the cheerful, and the helpful: The effects of service employee attributes on customer satisfaction'. *Psychology and Marketing*, 30(3): 211–226. DOI: 10.1002/mar.20599

Kong, M. and Jogaratnam, G. (2007). 'The influence of culture on perceptions of service employee behavior'. *Managing Service Quality: An International Journal*, 17(3): 275–297. DOI: 10.1108/09604520710744308

Ladhari, R., Brun, I., and Morales, M. (2008). 'Determinants of dining satisfaction and post-dining behavioral intentions'. *International Journal of Hospitality Management*, 27(4): 563–573. DOI: 10.1016/j.ijhm.2007.07.025

Lages, C.R. and Piercy, N.F. (2012). 'Key drivers of frontline employee generation of ideas for customer service improvement'. *Journal of Service Research*, 15(2): 215–230. DOI: 10.1177/1094670511436005

Larivière, B., Bowen, D., Andreassen, T.W., Kunz, W., Sirianni, N.J., Voss, C., Wünderlich, N.V., and De Keyser, A. (2017). '"Service Encounter 2.0": An investigation into the roles of technology, employees and customers'. *Journal of Business Research*, 79: 238–246. DOI: 10.1016/j.jbusres.2017.03.008

Luk, S.T.K. and Lloyd, A.E. (2011). 'Interaction behaviors leading to comfort in the service encounter'. *Journal of Services Marketing*, 25(3): 176–189. DOI: 10.1108/08876041111129164

Rihova, I., Buhalis, D., Moital, M., and Gouthro, M.-B. (2015). 'Conceptualising customer-to-customer value co-creation in tourism'. *International Journal of Tourism Research*, 17(4): 356–363. DOI: 10.1002/jtr.1993

Smith, J.S., Fox, G.L., and Ramirez, E. (2010). 'An integrated perspective of service recovery: A sociotechnical systems approach'. *Journal of Service Research*, 13(4): 439–452. DOI: 10.1177/1094670510375104

Sundaram, D.S., Mitra, K., and Webster, C. (1998). 'Word-of-mouth communications: A motivational analysis'. In J.W. Alba and J.W. Hutchinson (eds.) *NA – Advances in Consumer Research* vol. 25 (pp. 527–531), Provo, UT: Association for Consumer Research. http://acrwebsite.org/volumes/8208/volumes/v25/NA-25

Tung, V.W.S. and Ritchie, J.B. (2011). 'Exploring the essence of memorable tourism experiences'. *Annals of Tourism Research*, 38(4): 1367–1386. DOI: 10.1016/j.annals.2011.03.009

Voon, B.H., de Jager, J., Chitra, K., Kueh, K., and Jussem, P.M. (2013). 'Human service matters: A cross-national study in restaurant industry'. *Asian Journal of Business Research*, 13(2). https://magscholar.com/

Wirtz, J., Chew, P., and Lovelock, C. (2012). *Essentials of Services Marketing* (2nd ed.). Singapore: Pearson.

Wu, C.H.-J. (2007). 'The impact of customer-to-customer interaction and customer homogeneity on customer satisfaction in tourism service—the service encounter prospective'. *Tourism Management*, 28 (6): 1518–1528. DOI: 10.1016/j.tourman.2007.02.002

Wu, C.H.-J. and Liang, R.-D. (2009). 'Effect of experiential value on customer satisfaction with service encounters in luxury-hotel restaurants'. *International Journal of Hospitality Management*, 28(4): 586–593. DOI: 10.1016/j.ijhm.2009.03.008

Zatori, A., Smith, M.K., and Puczko, L. (2018). 'Experience-involvement, memorability and authenticity: The service provider's effect on tourist experience'. *Tourism Management*, 67: 111–126. DOI: 10.1016/j.tourman.2017.12.013

Zeithaml, V.A., Bitner, M.J., and Gremler, D.D. (2009). *Services Marketing: Integrating Customer Focus Across the Firm* (5th ed.). New York: McGraw-Hill Irwin.

9

Relationships between tourists' emotional experiences, perceived overall image, satisfaction, and loyalty

Keith Mandabach

This chapter's purpose is to provide an overview and insight into the role that tourists' emotional experiences play in the decision-making process and how they might alter their perceived overall image of the tourism site or experience. The chapter also discusses how the role of emotional experiences reflects on tourists' reported satisfaction levels as well as developing loyalty for that tourism product and/or intent to recommend that said product. Another topic discussed is how the emotional experiences of Millennials and members of Generation Z are impacted by connectivity as well as the effect of the fear of missing out. Finally, suggestions are offered for developing strategies to better understand and utilize the relationship between customers' emotional experiences and improving the image as well as satisfaction and marketing approaches. Two case studies for the application of these strategies are also presented.

Overview

Tourism businesses provide superior experiential activities and service to their customers by creating a product, maintaining a positive image of the product, and improving customer satisfaction and loyalty through customer engagement. Connecting and engaging customers positively while providing an excellent product and service is sometimes difficult because customer interactions occur at multiple times at multiple levels. Studying the role of tourists' or customers' emotional experiences in this process can help operators of tourism businesses better understand how to serve their customers.

A tourist's image of a destination starts during the research process prior to the visit. Even then experiential components have an emotional impact on our perceptions. Upon arrival, we may be surprised by the ambiance of the hotel, the love we feel from the caring attitude of the people, and the joy we feel because of the incredible sunset we viewed. As we recall and form our perceptions of our travel these emotional recollections may sometimes supersede or alter the perception of the experience.

Satisfaction, perceived image, and loyalty

The premise of this chapter relates to the relationship between tourists' emotions and their perspective on the experience. Thus, it is important when opening or planning a tourism business to consider how the business will deliver satisfaction, develop a positive image, and build loyalty from its customers. A narrative and/or executive summary is usually written that includes answers to questions demonstrating how these important operational goals will be accomplished.

To write the narrative, the operator of the business might want to think of questions that will explain how the business will be successful. One question they might ask is: "What makes my product novel?" Also: "How can this novelty positively surprise my clients and deliver joy, happiness, and a great experience?" Next, we might ask the question: "Does this earn a positive perceived image through customer satisfaction?" Finally, we need to ask: "Will our customers recommend us and why?" The answers to the above questions are sometimes hard to understand but they are interconnected and affect the success of the business.

The best measure of customer satisfaction is whether a customer will recommend your product to others (Reichheld 2006). Most businesses try to do it all and they want to build loyalty from their customers for their product and services. Fantastic food or room, views and/or great ambiance, properly paired beverages, and great human interaction are usually the recipes for success. A great line I was taught by the late R. Scott Morrison, a storied hotel executive, is: "Our business is the business of creating magic and delight and there is nothing better when the guests are satisfied."

The delight creates positive perceived images, and outstanding perceived images usually are the result of high performance standards. It is important to conceptually integrate customer loyalty into the operational framework. Start this integration with customer satisfaction, which is defined as an emotion based on a direct experience, thus referred to as experiential. Improving customers' perceptions by engaging them in the experience may lead to a more positive emotional experience and satisfied customers are more likely to recommend an experience to friends by positive word of mouth (Dick and Basu 1994).

Emotions

Emotions are affective states where intense feelings associated with a specific event cause specific response behavior. Emotions have been studied by psychologists cognitively with dimensional and categorical approaches. Dimensional methods are classified either as positive or negative but do not specify the emotion that caused the reaction. Categorical methods measure specific emotions such as joy, fear, or surprise. Regardless of how they are measured, emotions impact our perceptions of a tourist destination's image and our satisfaction with the experience. Both perceived overall image and satisfaction have a direct impact on our intention to recommend a destination or an activity, which is an excellent method to judge loyalty.

The role of emotions in consumer decision-making, consumption, and product appraisal has been studied in detail, most often using cognitive emotion appraisal theory. Appraisal theory is designed to explain how a wide range of emotions may come to the forefront at the same event or in different people and at different times. They are often highly complex. Appraisal of experiences creates multi-componential response patterns or emotions. The nature of and number of appraisal variables or factors depend on what one can or might want to explain (Moors, Ellsworth, Scherer, and Frijda 2013).

Another study probed the cognitive basis for emotions by asking respondents to recall 15 types of emotional experiences and charted them on three grids. Pleasant and unpleasant emotions were paired with high and low effort, low and high attention with uncertain and certain, and responses with situational and human control. In discriminant analysis, the "15 emotions were correctly identified 40% of the time on the basis of corresponding patterns of cognitive appraisal" (Smith and Ellsworth 1985). It was also interesting how the research addresses both positive and negative emotions. Just as research in other disciplines analyzes the emotional impact of marketing, hotel, restaurant, and tourism management researchers also study the topic.

Tourism applications of emotional cognitive appraisal analysis

Tourists plan their experiences by seeking pleasurable activities. Later, in reflection, our emotions will come out, but during the experience the pleasantness is the antecedent of the emotion. We often seek activities that are not only pleasant but serve to help us accomplish a goal of some sort we have established. An important study by Sameer Hosany, examining the emotional experiences of tourists and their impact on tourists' reflections on their visits, documented that "appraisals of pleasantness, goal congruence, and internal self-compatibility are the main antecedents for tourist determinants of joy, love, and positive surprise" (Hosany 2011). As one discusses emotions it is important to understand where they come from and that these dimensions, referred to as antecedents, are part of our appraisal experience and the antecedent experience is integrated into the emotional determination process.

In this study, pleasantness was statistically the most important antecedent of emotions. This is because tourists determine whether a vacation experience is pleasant or unpleasant and then differentiate the emotional experience. Goal congruence was also found to be a key antecedent of emotional response. This was not surprising because the goal of a vacation is usually to enjoy oneself and create memorable experiences. The third appraisal antecedent of internal self-compatibility relates to whether the destination image matches the tourist's self-concept. The study also focused on the importance of understanding the goals of tourists when planning marketing as they develop marketable destination images. The researcher also encourages marketing strategies that address the tourist's self-concept in an effective way to influence tourist behavior (Hosany 2011).

The term "bucket list" is an example of a travel achievement list but is also influenced by goals and self-compatibility. These lists were once considered by marketers to be primarily for older tourists, but today, in our connected world, tourists of all ages create travel goals as they search for experiences. Goals are important and when we reflect on our tourist experience goal congruence is another antecedent of emotion. The idea of internal self-compatibility as an antecedent to the tourist's emotional response is interesting and there is evidence that travelers seek destinations with images that have congruence or match their self-image (Beerli, Meneses, and Gil 2007).

Other antecedents of emotions include novelty which often leads to surprise (Roseman 1984). Tourists in the same locations often have different goals and expectations. Novelty or finding something new and unique can transcend these differences and impact different markets' emotional responses positively as well as building on goal congruence and self-compatibility. In another study, the emotional experiences that lead to joy, love, and positive surprise were found to have a relationship with tourists' satisfaction and intention to recommend. The authors recommend marketing activities that impact the affective dimensions of tourist experiences by marketing the emotional experience (Hosany and Gilbert 2010).

Destination image and place attachment

It is important that tourists' descriptions of joy, love, and positive surprise receive further attention. Positive surprise includes amazement, astonishment, inspiration, surprise, and fascination. Joy is another emotion that might be cheerful, full of pleasure, joyful, enthusiastic, or delightful. Love can be tender, caring, affectionate, warm-hearted, or simply loving. These are positive emotions and subsets but there are also negative or unpleasant emotions. Unpleasantness includes the components of disappointment, displeasure, regret, sadness, and unhappiness. Further analysis allows the classification of tourists into subgroups to better understand their behavior. The groups include delighted, unemotional, negatives, mixed, and passionates (Hosany and Prayag 2013). This study recommended a "holistic destination experience" and that marketers should try to understand the negatives and the mixed component while building on the strength of the findings of the passionate.

Every hotel, restaurant, or tourism business has a goal to feature something unique in its product that will separate the tourist from previous experiences to drive the passionate. This has always been about creating a destination image, customer satisfaction, and developing loyalty and attachment. Another interesting idea is how on-site experiential reflections or memories bring out strong emotions which drive behavior, specifically satisfaction with the tourist experience, and can also drive place attachment. Destination image, place attachment, and personal involvement all drive tourists' loyalty to a destination (Prayag and Ryan 2012).

One study found that love and pleasure coupled with joy, amazement, caring, and inspiration are important factors central to adding attachment and meaning to the tourist's experiences. The study also found that negative emotional experiences do not lessen the tourist's levels of attachment to a destination and tourists' reflections sometimes have a positive bias (Prayag, Hosany, Muskat, and Chiappa 2015). The way tourists form attachments to places and how they emotionally and personally get involved is changing. In the past, the actual destination was one component and the tourist's experiential involvement began with promotional print or video material. Occasionally news articles might impact a tourist's view and on arrival the experience was in real time. Today, the actual experience is impacted by online connectedness prior to, during, and after the experience. While all tourists are utilizing online resources, the immediate future of tourism is the Millennial market.

Tourist emotional experience for Millennials and beyond

Millennials are generally classified as people born between 1980 and 2000, though Generation Z, which starts in 2001, clearly follows similar behaviors. They are very active users of technology and Asian consumers have the highest activity. Millennials' technological engagement has changed their consumption habits of traditional goods and services but even more so as consumers of hotel, restaurant, and tourism products. The Millennial tourist expects and demands virtual connectivity to facilitate tourism delivery. Prior to the tourism experience, they begin to emotionally connect to the destination and the actual experience is co-created with the online component. This is experiencing a development that is different than traditional travelers. Millennial travelers are dependent on connectivity in the execution of their daily activities and the tourism experience is mediated by technology. Thus, a Millennial's reflection on the experience and corresponding emotional states is impacted by pre, during, and post online activities, and satisfaction, destination image, and decision to recommend are intertwined with the online experience (Mohd, Ismail, Isa, and Jaafar 2019). The need for connectivity, information accessed before and during, as well as photos and comments posted during the experience appear to be a very positive force for the industry. That is because it is driving demand by sharing the experience. The Millennial and beyond segments are more aware of tourism options because of social media.

Fear of missing out (FOMO)

Connectivity has also fueled a phenomenon in today's fast-paced society, which is always connected to social media. It is a new experiential need called the fear of missing out, FOMO, which can produce emotions of irritability, anxiety, and inadequacy, and which drives the tourism market. This is most likely due to the massive amount of knowledge we accumulate about our virtual community friends' activities and the wish that our lives were as full as those of our friends or celebrities. While this might appear to have a negative effect on the tourist's experience the connectivity can be a positive force if understood. Studying how emotional experiences impact on the individual cognitive appraisal process can assist in developing positive tourism marketing plans. Awareness and understanding of FOMO can be used as a tool in understanding the tourist's decision processes, both positive and negative, and lead to effective marketing strategies for those afraid of missing out. (Abel, Buff, and Burr 2016)

FOMO is also important as the tourist market is transitioning from Boomers to Millennials to Generation Z. New loyalty models stress the importance of building loyalty through emotional experiences (Bowen, Chen, and Che 2001). These new target markets seek constant virtual connectivity, building on-site technological engagement that develops tourists' emotional attachment. Traditional guide books and guiding systems are evolving and improving as connectivity becomes the norm. Thus, Millennial destination engagement is improved and positively manipulated by the emotional experience (Mohd, Ismail, Isa, and Jaafar 2019).

Case study 9.1: Tavora valley mystical outing: Cisterns, monasteries and sparkling wine

Tim and Judy had just returned from a Viking River Cruise on the Douro River in Portugal and were reflecting on an optional tour they had been on which was titled: Cisterns, Monasteries, and Sparkling Wine. The tour started in Regguea and followed the above river valley to a historic agricultural area that was developed in the 13th century. Judy laughed that they had been so busy on the cruise that they never researched anything about the tour, and it was amazing how it turned out. Tim remembered feeling nervous when meeting their tour guide, Claudia, because he had no idea what the tour was about when arriving. He was relieved and pleasantly surprised by her Portuguese-accented English because it was easy to understand. Judy reflected that she was really a local person, for all of her family except her (hopefully sometime) had been married in the church at the monastery and her niece would be married there in August. Tim said the joy in her eyes when she talked about her family gave him a chill.

They had been on tours almost daily on the river cruise and were tired of the constant lectures about the history and flora but it was different when Claudia told them who planted and harvested each cork, apple, and almond tree, or recited the names of the plants, flowers, and other trees growing on the side of the road. She also said that she would not make up stories so if she did not know the answer, she would tell them so.

As the Mercedes finished its climb up the mountain, the morning sun filtered through the mountains/trees producing amazing patterns of light and after a while they arrived at the village. Judy and Tim had been amazed when they left the van and then pivoted to survey the vista of soaring mountains, terraced vineyards, and fields, all with ancient granite aqueducts. A communal clothes washing area was being used by villagers just below the parking area. Claudia recounted the story of how the King of Portugal in the 12th century invited the Cistercian monks from Burgundy to found a monastery on the very fertile lands of the Varosa-Tavora Valley. These monks shaped the landscape by farming the land, planting vines, and building imposing monasteries and convents, and more importantly impressed the Pope. Because of the King's dedication to the Church he officially proclaimed him King.

The morning had been a blur as they were invited into the town's café for pastry and coffee because the lady who had the keys to the monastery was a little late. Claudia's stories about the pastries, the local breads, and the hams made the food taste a little better and they moved on to the monastery. The history of the Church and the surrounding ruins came alive as generations of stonecutters' marks and legacies of painted tiles were viewed. This era ended with the 19th-century expulsion of the monks and nuns from Portugal and Claudia told them how her home in the hills had been built with stone bought from the seized monastery. Judy felt that what was really interesting was that the morning just seemed to flow magically as they moved from the monastery to the ballroom of a 19th-century estate, complete with an open fire burning sumac and slash from the vineyard.

The female owner of the estate welcomed our small group with a great white wine made from historic vines and freshly baked bread with local ham, pastry, and chocolate. She regaled them with stories and small talk about her life and family history and how she and her husband had returned to the estate to refurbish it after years abroad. Tim and Judy were excited to hear about and see the renovated bed and breakfast accommodation. They ended their time of reflection thinking about how surprising it was that they had fallen in love with the tiny Tavora Valley and felt joy when they were thinking about their day.

Case study 9.2: How did a Millennial couple survive a vacation without a phone?

Jessica and Noah are a Millennial couple who have recently relocated to Brooklyn, New York, USA. Both are recent university graduates, working in the financial industry in managerial positions and trying to build their careers. It was just before the American Thanksgiving holiday (six days, Wednesday to Tuesday) and they planned on a quiet week at their apartment. Then Noah received a text from his mom offering an all-expense paid vacation to a new resort in the Bahamas. At first they were not sure if they wanted to go, but when they found out that air fare, golf, and excursions were also included they could not resist. A cousin had won the trip on a game show with taxes paid and it was set to expire in a week and the cousin had already reserved the room. They went online and booked the flight for the next day, began to pack and got online to find out about Nassau, Bahamas. But it was already late at night and their flight was early. Jessica worried she was unprepared for the trip but Noah felt they could figure it out as they traveled. They both had unlimited data phone plans. They would be able to look everything up as they traveled.

As they boarded the plane, they were excited but a little nervous. More upsetting was the announcement that the WiFi on the plane was not working. All they could do was watch movies and play games on their phones. The person sitting next to them asked if they were going to the golf tournament and they were excited to learn that the top 20 golfers in the world were playing a charity tournament that weekend. They wondered if tickets were available. When they arrived at the Bahamas airport, they breezed through customs but were disappointed when they discovered that Uber and Lyft did not operate in the Bahamas. They took a cab and googled how much the fee should be and were disappointed when the price went up because they had golf clubs. As they drove to the hotel, the driver explained that the island was having power problems and blackouts were occurring in certain areas and that also affected the cell phone service.

When they arrived at the hotel, the power was out, and they were not able to check in. Their phones did not have service either. After a few minutes, Noah got the attention of the front office manager and explained he was a contest winner and should get special treatment. They were pleasantly surprised when they were assigned a suite with an ocean view, two rooms, balcony, kitchen, and washing machine. The power was still out so they hit the beach. The hotel emergency generator ran the WiFi and the service even worked at the beach, even though there was no cell service. Jessica and Noah celebrated with drinks and then the power came on. They decided to get in the jacuzzi and because they felt their phones were safe they took pictures as they frolicked in the pool. The guy who had talked about the golf tournament gave them two tickets and they were excited, happy, and surprised. But things got really unpleasant when a young boy pulled their phones from the bottom of the jacuzzi. They thought the phones were waterproof but neither was working. They panicked and went into the lobby and asked the concierge for help. She gave them a bowl with raw rice and suggested that they put the phones in the rice and then dry them out in the sun. The concierge called their cell company who said the phones were insured but they could not send a phone to the Bahamas. The concierge said they could get a local phone if they wanted but the price was very high.

They were on a wonderful vacation with no phones. This was not good. Neither had ever been without a phone for more than an hour since they were in grade school. Everything they did and almost every minute of every day they were connected. Their goal was to use their phones to enhance their vacation experience. This was a big change from their self-image. They went back to the room and watched the infomercials about the hotel, the beach, shopping, restaurants, and more. Jessica decided to go back to the concierge to get help planning the next few days. Noah was very upset about the phone and was sitting in the room watching a movie and she knew she had to find a way to turn it around.

A new concierge named Nelda was very energetic and positive when Jessica approached the desk. "Sorry about the phone, you might be able to get a new one downtown cheaply but let's plan a marvelous vacation," said Nelda. She was full of ideas and poured Jessica a complimentary Bahama Mama cocktail as she detailed the experiences she could enjoy. This was very different for she was unable to research, read reviews of the activities, and logically select what was best. But Jessica got into the narrative and in 30 minutes had an incredible itinerary for the rest of the trip which Nelda printed. She also used maps and flyers describing the activities and provided two more Bahama Mamas for Jessica to take to the room.

Noah was still in an unpleasant temper, but the drink helped improve their mood. Jessica explained the grand plan and started to get dressed for the beach-side dinner and they would finish the night trying their luck in the casino. She and Noah strolled across the property to the gourmet restaurant and were surprised by the magnificent view of the surf crashing on the beach from the lucky table they were seated at. The food and service were wonderful. But as they watched the sun set across the ocean they felt real joy. There was romance in the air as well as live jazz as they walked away from the dinner to the dance club. The music energized the night and they stayed until the band stopped playing. They were pleasantly surprised that the entertainers were so good and wished they could look them up online. But because they had no connectivity the experience was different.

The next day they slept in and had a leisurely breakfast on their patio. They took a free shuttle to the golf tournament from their hotel and discovered their ticket included VIP access. Noah had been to golf tournaments before, but this was unique because this was the Tiger Woods Hero charity tournament. Five minutes after arriving, they were walking on the third hole when Tiger walked up to hit a shot five feet from where they were standing. It was so amazing/exciting, and the VIP experience allowed them the opportunity to network with celebrities and powerful business executives. This experience had been a goal of Noah's for a long time and both he and Jessica were comfortable rubbing shoulders with the VIPs.

The concierge had told them they might be able to pick up a cheap phone on the way back to the hotel, but they were hungry and stopped at a local bar and restaurant to listen to Island Music while feasting on local delicacies. The beat was captivating and the clientele was primarily Bahamian. The sunset was even more beautiful than the first night. Jessica and Noah really enjoyed the experience and loved their vacation. They decided that they would be brave and forget about getting a phone and experience their travel the old-fashioned way. They were learning to create their experience conventionally without it being co-created by constant connectivity. They were really enjoying not having a technology-enhanced experience. They felt their goal of a great surprise vacation had worked and they were proud of themselves. The rest of the vacation flew by. They did go to the business center and connect and whenever they made a friend they borrowed their phone just to check something that needed to be looked up. Living unconnected was fun but they were looking forward to their new phones when they arrived home.

Summary

This chapter describes the relations between tourists' emotional experiences, perceived overall image, satisfaction, and loyalty. Satisfaction can best be defined as whether tourists will recommend your product to others. Emotions are experiential and often result in intense feelings and cause specific responses. They are most often studied in tourism utilizing cognitive emotion theory. There are three primary antecedents of the important emotions of joy, love, and positive surprise, which are pleasantness, goal congruence, and internal self-compatibility. Pleasantness is statistically the most important antecedent of a successful tourist experience. Tourists determine pleasantness or unpleasantness in concert with whether the experience is aligned with their goals and compatible with their self-image and then differentiate the emotional experience.

Novelty is another factor that leads to joy, love, and positive surprise and has been found to be related to tourist satisfaction and the intention to recommend. All positive emotional experiences have multiple sub-categories such as delightful, tender, amazement. Negative emotional experiences or unpleasantness might be differentiated by tourists as disappointment, displeasure, regret, sadness, or unhappiness. What is fortunate about the negative experiences is positive emotional experiences will often override the negatives.

The first case study provided examples of the relationship between tourists' emotional experiences and their perceived image, satisfaction, and loyalty and detailed a tour guide changing tourists' preconceptions by building a positive connection to a unique destination. Instead of hearing the tour guide recite details, the guide created a positive emotional experience, building on her personal connection to the cisterns, monasteries, and sparkling wine she was showcasing by making the tourists connect emotionally with the location. The experience built a love for the novelty of the Tavora Valley as well as loyalty to the destination.

The second detailed the Millennials' emotional journey when they were disconnected from the internet due to the resort's loss of electricity. They also dropped their phones in the jacuzzi, making them even more out of touch. The lesson of the case is that they eventually found joy in living in the moment while unconnected. Loving the sunset, the food, and the ocean coupled with fine music and dancing tweaked their emotions and built romance. Being fortunate enough to fall into an incredible celebrity experience at a golf tournament was an awesome emotional experience. Getting lost in the local culture and living for the moment, coupled with sampling unique local cuisine, was the finishing touch that led the Millennials to form an uplifting attachment and love for the location.

Suggested marketing strategies include marketing the affective dimensions of the tourist experience by emphasizing the emotional responses and the holistic experience; marketing to meet prospective tourists' self-goals, self-compatibility, and develop place attachment. Millennials demand connectivity and co-create their emotional experience combining the actual experience, what they are experiencing online, and what they record of the experience. The fear of missing out, or FOMO, can be used as a tool to understand the tourist decision process. New loyalty models stress the importance of building loyalty through emotional experiences. As the boomer generation becomes a smaller part of the market, finding connected methods to market while understanding the emotional experience components will lead to a positive impact on perceived overall image, satisfaction, and loyalty. It will also lead to a stronger tourism industry.

References

Abel, J.P., Buff, C.L., and Burr, S.A. (2016). "Social Media and the Fear of Missing Out: Scale Development and Assessment," *Journal of Business and Economics Research*, 14(1): 33–42. DOI: 10.19030/jber.v14i1.9554

Beerli, A., Meneses, G.D., and Gil, S.M. (2007). "Self-Congruity and Destination Choice," *Annals of Tourism Research*, 34(3): 571–587. DOI: 10.1016/j.annals.2007.01.005

Bowen, J.T., Chen, S.L., and Che, S. (2001). "The Relationship between Customer Loyalty and Customer Satisfaction," *International Journal of Contemporary Hospitality Management*, 13(5): 213–217. DOI: 10.1108/09596110110395893

Dick, A.S. and Basu, K. (1994). "Customer Loyalty: Toward an Integrated Conceptual Framework," *Journal of the Academy of Marketing Science*, 22(2): 99–113. DOI: 10.1177/0092070394222001

Hosany, S. (2011). "Appraisal Determinants of Tourist Emotional Responses," *Journal of Travel Research*, 51(3): 303–314. DOI: 10.1177/0047287511410320

Hosany, S. and Gilbert, D. (2010). "Measuring Tourists' Emotional Experiences toward Hedonic Holiday Destinations," *Journal of Travel Research*, 49(4): 513–526. DOI: 10.1177/0047287509349267

Hosany, S. and Prayag, G. (2013). "Patterns of Tourists' Emotional Responses, Satisfaction, and Intention to Recommend," *Journal of Business Research*, 66(6): 730–737. DOI: 10.1016/j.jbusres.2011.09.011

Mohd, N.S., Ismail, H.N., Isa, N., and Jaafar, S.M.R.S. (2019). "Millennial Tourist Emotional Experience in Technological Engagement at Destination," *International Journal of Built Environment & Sustainability*, 6(1–2): 129–135. DOI: 10.11113/ijbes.v6.n1-2.396

Moors, A., Ellsworth, P.C., Scherer, K.R., and Frijda, N. (2013). "Appraisal Theories of Emotion: State of the Art and Future Development," *Emotion Review*, 5(2): 119–124. DOI: 10.1177/1754073912468165

Prayag, G., Hosany, S., Muskat, B., and Chiappa, G.D. (2015). "Understanding the Relationship between Tourists' Emotional Experiences, Perceived Overall Image, Satisfaction, and Intention to Recommend," *Journal of Travel Research*, 56(1): 41–54. DOI: 10.1177/0047287515620567

Prayag, G. and Ryan, C. (2012). "Antecedents of Tourists' Loyalty to Mauritius: The Role and Influence of Destination Image, Place Attachment, Personal Involvement, and Satisfaction," *Journal of Travel Research*, 51(3): 342–356. DOI: 10.1177/0047287511410321

Reichheld, F. (2006). *Ultimate Question: Driving Good Profits and True Growth*. Cambridge, MA: Harvard Business School Press.

Roseman, I.J. (1984). "Cognitive Determinants of Emotion: A Structural Theory," *Review of Personality & Social Psychology*, 5: 11–36.

Smith, C.A. and Ellsworth, P.C. (1985). "Patterns of Cognitive Appraisal in Education," *Journal of Personality and Social Psychology*, 48(4): 813–838.

10

Tourism and the experience economy

A critique?

Roy C. Wood

Introduction

This chapter assumes familiarity with the central arguments of Pine and Gilmore (1999, 2011) in their elaboration of the "experience economy". Certainly, to researchers in the tourism and hospitality fields the concept needs little introduction, it having been adopted readily by many researchers as a vehicle of analysis (a plausibly representative sample of such analyses might include Andersson 2007; Oh, Fiore and Jeoung 2007; Hosany and Witham 2010; Mehmetoglu and Engen 2011; Loureiro 2014; Sidali, Kastenholz and Bianchi 2015; Kirillova, Lehto and Liping 2017; Hwang and Heesup 2018; Chang 2018a, 2018b). If the literature is to be believed, one core attraction of the concept is its capacity for practical, policy, applications, together with its character as a useful and valid description of reality. In respect of the latter, Pine and Gilmore (1999: 97, cited in Smith, MacLeod and Hart Robertson 2010: 62) themselves state that "customers unquestionably desire experiences." With gung-ho relish, Smith, MacLeod and Hart Robertson (2010: 63) add that: "The tourism industry, traditionally concerned with the provision of standardized travel, accommodation and catering services, is of course embracing the contemporary experience economy and providing more and more tailored and unique experiences for visitors ...[.]" No evidence is presented either for Pine and Gilmore's original assertion or for that made by Smith, MacLeod and Hart Robertson (2010). The remainder of this chapter critically explores the idea of the experience economy with a view to demonstrating its limited utility as a means of exploring tourism and related phenomena.

A brief critique of the "experience economy" concept

Critique is easy, but is barely evident in many studies of the experience economy that simply assume both its conceptual veracity and utility. In an analysis of the influence of Pine and Gilmore's (1998) *Harvard Business Review* article, Ferreira and Teixeira (2013: 21) note that:

> … publication activity on "customer experience" and related concepts has been increasing, mainly in the most recent years (2008–2012) … [.] Such trends seem to convey that the customer experience topic is solidifying its importance in terms of research, receiving increasing attention from marketing academics [.] The multidisciplinary nature of customer experience is quite apparent, ranging from tourism, sport and leisure to hospitality, albeit with prominence in the areas of management and business.

Of course, "customer experience" covers a wide range of perspectives in addition to that of the experience economy. Further, Ferreira and Teixeira (2013) quite reasonably do not differentiate between those studies that are "about" the experience economy and those that merely namecheck it. Freire-Gibb (2011: 1840–1841) reviews the application of experience economy ideas in local economic development in Denmark where the concept has proved especially popular (see also Lorentzen and Hansen 2015). He cites a wide range of studies the general conclusion of which is that the concept is flawed and notes that "it has been possible to find structural limitations of TEE [the experience economy], starting with questionable methodology and statistical 'evidence' from the founders of the term."

More generally, Kociatkiewicz and Kostera (2009: 262) suggest that there is little novel about the concept of the "experience economy" that cannot be traced to earlier managerial and social scientific theories. Referring to diverse previous studies (e.g. Hochschild 1983; Van Maanen and Kunda 1989; Leidner 1993) they correctly suggest that important elements and practices necessary to "successful" creation and management of the experience economy such as emotional labor and performative authenticity are well-established, going as far as to note that: "Perhaps most famously, Ritzer … criticized the transformations of contemporary economy, largely paralleling Pine and Gilmore's description of experience economy, under the name of McDonaldization. His argument highlighted duplicity, irrationality and basic inhumanity of the solutions offered by contemporary business trends."

To those cited we can note other investigations that appear to contain supportive antecedents of the Pine and Gilmore thesis. In the field of consumption studies (Campbell 1987: 89–90) anticipates the concept of the experience economy, writing that:

> The spirit of modern consumerism is anything but materialistic. The idea that contemporary consumers have an insatiable desire to acquire objects represents a serious misunderstanding of the mechanism which impels people to want goods. Their basic motivation is the desire to experience in reality the pleasurable dramas which they have already enjoyed in imagination, and each 'new' product is seen as offering a possibility of realizing this ambition.

Finkelstein (1989) is similarly anticipative when she argues that people visiting restaurants delude themselves, thinking they are acting from largely individual choice and harboring expectations that both their "objective" desires for good food and service and their deeper, emotional need for status and belongingness will be satisfied by such establishments. In reality, far from exercising individual choice, diners comply with imitative and mannered behavior and prevailing fashions, what Finkelstein calls "uncivilized sociality". Lupton (1996: 99) is not so sure, arguing that the dining experience is a dynamic rather than fixed concept, and how we feel about such experiences depends on our state of mind at a given time and in a given context, seemingly implying that it is impossible to *universally* capture for any individual the range of factors that bear on their dining (or any other) experience.

If some of these sources are to be believed, there is, to borrow a biblical assertion, nothing new under the sun. Yet a *fundamental* objection to Pine and Gilmore's concept – the unexplicated nature of "experience" in their work – is rarely articulated. For these authors, the history of western or any other kind of philosophy, a preoccupation of which is the nature of experience, does not exist. It is certainly not as if considerations of existentialism are unknown in consumer studies (e.g. Thompson, Locander and Pollio 1989) nor is it going too far to suggest that all existence is experience. Experience is, quite literally, commonplace. Yet, as psychology tells us, individuals may share specific experiences, but not necessarily in the same way (in connection with this, see Sundbo 2005). An objective to provide a specific and positive experience (and Pine and Gilmore's model is precisely about the creation of positive experiences) does not – nor could not – guarantee that the experience intended will in fact be the one experienced.

Of course, there are businesses – including tourism and hospitality businesses, for example theme parks and themed restaurants – that may consciously set out to create "experiences" (Bryman 2004; Weiss, Feinstein, Hale and Dalbor 2004 – see also the case study in this chapter). There is, however, no universal formula that allows these intentions to be successfully realized – put simply, too much can go wrong. To illustrate this point we can give an example drawn from Pine and Gilmore's (2002: 87–88) application of their concept to the hospitality sector which opens with a question about how timekeeping can be made into a memorable experience (there is no explanation as to why timekeeping should be memorable) and the following "reveal":

> The MGM Grand Hotel & Casino in Las Vegas does just that, by awakening guests with recorded voices of celebrities who have performed there. Guests wake up to a different star on each visit – Rodney Dangerfield one time, say, and Carrot Top the next. Granted, a morning wakeup call represents just one dimension of a guest's overall experience. But what better way to make a great impression?

"What better way to make a great impression?" the authors ask. Given the multiple problems much of the hotel industry has in delivering the basic and essential products and services expected of them (Wood 2018a, 2018b, 2018c) one response to this question is, obviously, to focus on getting those basics right. However, some nearly two decades since the above words were written, who remembers Rodney Dangerfield or Carrot Top? Who, in 2002, from outside of the USA and likely to be a guest at the hotel in question would have heard of these people? What is the logic of the presumption that people are so witless as to be impressed by celebrity wake-up calls? In the revised edition of their book, Pine and Gilmore (2011: xi) utilize a further hospitality example, writing:

> From furnishing sociable lobby spaces to providing beds that promise better sleep experiences (… credit Westin's "Heavenly Bed" for the latter), the hospitality industry now clearly creates new value on the basis of experiences.

One can promise better sleep experiences, but one cannot guarantee them and it seems to have escaped the authors that the "delivery vehicle" – a bed – for such promises is a physical product, not a service. Put very simply, the more one interrogates the notion of the experience economy, the more it is revealed to generate troubling questions rather than reassuring answers.

Case study 10.1: meal experiences

A number of the limitations of the "experience economy" concept referred to in this chapter are evident in discussions of the hospitality meal experience. Hemmington (2007: 749) argues that: " … customers do not buy service delivery, they buy experiences; they do not buy service quality, they buy memories; they do not buy food and drink, they buy meal experiences." The meal experience is an established concept in the field of hospitality management. It originated in the work of Campbell-Smith (1967) whose core assertion was that when dining out, customers' experiences were influenced by a wider range of factors than the quality of food and drink alone. These included, for example, the quality of service, décor, atmosphere and ambience of the restaurant. This recognition of the importance of both tangible and intangible factors in influencing customer experiences seems commonplace today but at the time it was something of a radical departure. A year earlier the UK National Catering Enquiry (1966: 12) had asserted that: "In spite of the trend towards novel décor and a generally more 'swinging' atmosphere, people still go to restaurants primarily for a good meal."

Over time Campbell-Smith's idea of the meal experience has been progressively misunderstood and come to be read as suggesting that intangible elements of the meal experience are more important than tangible ones. Johns, Tyas, Ingold and Hopkinson (1996: 15) capture this nuance, writing that: "It is generally held that the 'product' of a foodservice outlet is a 'meal experience' consisting (like other hospitality 'products') of an amalgam of tangible and intangible components." They assert, however, that there are substantial difficulties in establishing the factors that constitute the meal experience although their own research clearly demonstrated that "food was the most important element of the majority of meal experiences" (Johns, Tyas, Ingold and Hopkinson 1996: 23). Stierand and Wood's (2012) meta-review of the research evidence supports this conclusion suggesting that the tangible elements of the meal experience (food type, quality, range and price) are usually most important to the consumer. Intangible elements of the meal experience are not unimportant, merely less so, which is in keeping with Campbell-Smith's original notion.

Despite the obviousness of the notion that food type, quality, range and price might be the most important factors influencing meal experiences, there are, of course, material advantages to practitioners in the hospitality industry believing in the importance, indeed dominance, of intangible elements in the meal experience. Such a notion implies that the intangible elements of the meal experience can be manipulated to compensate for the limitations of the tangible elements. The quality and cost of these tangible elements can be deliberately limited while performative and intangible elements drive the meal experience. The most common form of this realization of "meal experiences" is the themed or concept restaurant (e.g. Weiss, Feinstein, Hale and Dalbor 2004) and indeed, it is possible to argue that the concept-driven restaurant has become the dominant form of popular commercial catering around the globe. However a thematic concept may present the prospect of a particular type of experience but again, it cannot guarantee that experience, nor can it, beyond perhaps a certain basic level, influence the direction of that experience.

Contextualizing the experience economy

How then are we to understand the success of the concept of the experience economy, both generally and in the field of tourism studies? The academic study of tourism (and indeed hospitality) evidences at least two fundamental weaknesses. The first of these is a widespread and systematic failure to locate the study of tourism phenomena in the wider discourse of the history of ideas within the humanities and social sciences (Wood 2017). For the most part this means that tourism is a pragmatic pursuit, a scholarship of practice. The second weakness is the disciplinary reliance on frequently unexplicated and unevidenced "myths", many with far-reaching consequences which as McKercher and Prideaux (2014: 26) observe:

> … promote ideological beliefs that have little basis in fact, entrench flawed methodologies that produce 'good' but ultimately wrong findings … [.] Many have become elevated to the stature of received doctrines. The promulgation of some of these myths has also had a detrimental effect on industry, with many myths about special interest tourism resulting in the expensive pursuit of phantom markets.

Both weaknesses lend tourism a ready susceptibility to fashionable tropes in mainstream research of diverse kinds, in what often appear as short-cuts to (a) acquiring legitimacy within the scholarly mainstream and/or (b) demonstrating the practical, administrative, "relevance" of tourism scholarship. For example, in terms of legitimacy, those committed to a hospitality "studies" as opposed to hospitality "management" approach (see Lashley 2015; Wood 2015) have engaged extensively with the notion of "McDonaldization" as proposed by Ritzer (1993) while others have sought to harness the slight contributions of Jacques Derrida to "theories" of hospitality (Derrida and Dufourmantelle 2000). In tourism, John Urry's (1992) theory of the tourist gaze has had a wholly disproportionate influence on sociologically informed tourism research.

Of course, no academic discipline is entirely immune from bandwagon effects in knowledge development. Miles (2001) interrogated sociological examples of such effects, several of which have a tangible presence in management and business studies (they were the idea of a mass society; post-industrial society; consumer society; post-modern society; McDonaldized society; risk society; and global society). However, the study of management and business appears particularly prone to generating ideas that are seized upon, developed and within a comparatively short time, forgotten, discarded or ghettoized at the margins of practice. Charles Handy (2007: 185) appears to imply that many of these emerge as part of the academic management field's unsuccessful quest for scientific legitimacy, writing that: "Re-engineering, Core Competences, JIT, Six Sigma, 360-degree Feedback, CRM, Social Network Analysis, Globalization, Format Competition and ROI Marketing are just some of the pseudo-technical terms that make the obvious seem clever." In the case of management and business research, as opposed to simply practice, Griseri, Schipper, Laurie and Diben (2010: 80) take a perhaps more fundamentalist view arguing that: "There are just so many variables, so many different fields and factors, all interacting, that it is for all practical purposes impossible to generate conclusions that carry any significant weight" (see also Griseri 2013). What Griseri, Schipper, Laurie, and Diben (2010) term "railway station" literature, i.e. populist and popular books on business and management, often appear to advocate action in the absence of what would pass for evidence, but appeal to their audience because they are pragmatic. The very idea of evidence-based management is, of course, not only an academic fashion in its own right, but a contested one. Pfeffer and Sutton (2006: 13) suggest that:

> Evidence-based management proceeds from the premise that using better, deeper logic and employing facts to the extent possible permits leaders to do their jobs better. Evidence-based management is based on the belief that facing the hard facts about what works and what doesn't, understanding the dangerous half-truths that constitute so much conventional wisdom about management, and rejecting the total nonsense that too often passes for sound advice will help organizations perform better.

However, many commentators in the "critical management" tradition have pointed to the ideological difficulties attendant on the determination of the nature of evidence (see Tourish 2012 for an interesting review).

While fashionableness in pursuit of legitimacy may explain some or all of the success of concepts such as the "experience economy", there is at least one other perspective to be considered. This is encapsulated in what Roberts (2005) has called the "Ritzerization" of academic management texts, alluding to the work of George Ritzer (e.g. 1993) and his coining of the term "McDonaldization" to describe a contemporary organizational archetype representative of Weberian rationalization predicated on efficiency, calculability, predictability and control. As with the concept of the experience economy, there was little that was new – other than the neologistical coinage – about McDonaldization even when it was launched upon the academic world. One critic savaged Ritzer's Ritzer 1993) book stating that it was " … too simplistic and sloppy to have much educational value. It is not sociology and it would certainly not teach students how to reason" (Gilling 1996: 24). Others noted that there exists a similarity between the essence of McDonaldization and Levitt's (1972) "industrialization of service" thesis and between Ritzer's analysis of the consequences of McDonaldization and the arguments of Harry Braverman (1974), who employed a more astringent Marxist view of work and the economy (Wood 1998).

The arguments of Roberts (2005: 58) are somewhat more general, however, as can be seen from her central claim that

> … there is a formula present in Ritzer's book, which can be found in many academic texts. Such formulae are not primarily designed to create a highly effective tool for learning and developing the knowledge of readers, but, rather, to maximize the commercial success of the book.

Roberts (2005) draws parallels between Ritzer's work and that of Bryman (2004) on "Disneyization" (an extremely popular concept, as one might imagine, in tourism studies). Blackshaw (2010) appears to have coined the term "Ikeaization" and Ahuvia and Izberk-Bilgin (2011) the idea of "ebayization" as specific modifications or mutations of McDonaldization though as yet neither term appears to have caught on. A Ritzerization process in academic publishing "consists of various techniques that are used to present the book's knowledge content and to facilitate ease of marketing" being directed toward making reading easier, a process Roberts (2005: 56) terms "knowledge-lite." There are several such techniques (Roberts 2005: 58–59). The first entails the introduction of a new concept over which the author establishes clear ownership. Secondly, there are associations with a well-known brand and/or company, or some kind of commonly understood activity. Thirdly, the arguments presented evidence a high degree of structure related to the core example (and, one might add, are highly memorable). Fourthly, these arguments tend to be generalized and instanced using less obvious examples. Fifthly, a Ritzerized text will be of a manageable size which renders it attractively accessible in terms of both form and pricing.

To this imperfect summary of Roberts' arguments could no doubt be added other characteristics of Ritzerization. For example, there is a tendency in such literature to offer far-reaching promises of one form or another of "success" if the prescriptions of the book/concept are applied. In reality, as we saw earlier in the case of the experience economy and local economic development, such successful application is, as one would expect, somewhat variable (as it has been with a concept almost as popular as that of the experience economy, Richard Florida's "creative class" – see Florida 2002, and for a critique; Krätke 2010). In other respects, the concept of the experience economy fits well with Roberts' model. There is a clear and clearly-owned concept based on an essential notion – experience – with numerous business examples provided by Pine and Gilmore.

Conclusion

Roberts' (2005) arguments concerning Ritzerization are compelling and expressed with an elegant simplicity. As a framework for understanding the popularity of concepts such as the "experience economy" Ritzerization highlights forms of knowledge production that embrace from the very start an easy, manufactured, populism. Yet, whatever framework is employed to contextualize the idea of the experience economy it remains the case that, as with any other concept, whether intended as revolutionary or not, the nature of the assumptions on which it is based, and its veracity as measured by the quantity and quality of favorable evidence remain the principal criteria by which such a concept should be judged. In this regard, the idea of the experience economy is a weak child indeed. Lived life is experience, and the empirical evidence "for" the experience economy is limited in quantity and quality, not least because of the difficulties inherent in meaningfully operationalizing the concept. Simply put, as a perspective for analyzing tourism and related industries, the utility and value of this particular idea remains undemonstrated.

References

Ahuvia, A. and Izberk-Bilgin, E. (2011). 'Limits of the McDonaldization Thesis: eBayization and Ascendant Trends in Post-industrial Consumer Culture', *Consumption Markets and Culture*, 14 (4): 361–384. DOI: 10.1080/10253866.2011.604496

Andersson, T. D. (2007). 'The Tourist in the Experience Economy', *Scandinavian Journal of Hospitality and Tourism*, 7 (1): 46–58. DOI: 10.1080/15022250701224035

Blackshaw, T. (2010). *Leisure*. London: Routledge. DOI: 10.4324/9780203855959

Braverman, H. (1974). *Labor and Monopoly Capital*. New York: Monthly Review Press.

Bryman, A. (2004). *The Disneyization of Society*. London: Sage. DOI: 10.4135/9781446220122

Campbell, C. (1987). *The Romantic Ethic and the Spirit of Modern Consumerism*. Cambridge: Polity Press. DOI: 10.1007/978-3-319-79066-4

Campbell-Smith, G. (1967). *The Marketing of the Meal Experience*. Guildford: Surrey University Press.

Chang, S. (2018a). 'Experience Economy in Hospitality and Tourism: Gain and Loss Values for Service and Experience', *Tourism Management*, 64: 55–63. DOI: 10.1016/j.tourman.2017.08.004

Chang, S. (2018b). 'Experience Economy in the Hospitality and Tourism Context', *Tourism Management Perspectives*, 27: 83–90. DOI: 10.1016/j.tmp.2018.05.001

Derrida, J. and Dufourmantelle, A. (2000). *Of Hospitality*. Palo Alto: Stanford University Press.

Ferreira, H. and Teixeira, A. A. C. (2013) 'Welcome to the Experience Economy': Assessing the Influence of Customer Experience Literature through Bibliometric Analysis, *FEP Working Papers*, University of Porto.

Finkelstein, J. (1989). *Dining Out: A Sociology of Modern Manners*. Cambridge: Polity Press.

Florida, R. (2002). *The Rise of the Creative Class. And How It's Transforming Work, Leisure and Everyday Life*. New York: Basic Books.

Freire-Gibb, L. C. (2011). 'The Rise and Fall of the Concept of the Experience Economy in the Local Economic Development of Denmark', *European Planning Studies*, 19 (10): 1839–1853. DOI: 10.1080/09654313.2011.614391

Gilling, A. (1996). 'Where There's Mc There's Brass', *The Times Higher Education Supplement*, 19 August: 24, Retrieved from www.timeshighereducation.co.uk/books/where-theres-mc-theres-brass/162237.article, last accessed 04.02.19.

Griseri, P. (2013). *An Introduction to the Philosophy of Management*. London: Sage. DOI: 10.4135/9781446270042

Griseri, P., Schipper, F., Laurie, N., and Diben, M. (2010). 'Real Knowledge Managers', *The Philosophers' Magazine*, Issue 49: 77–80.

Handy, C. (2007). *Myself and Other More Important Matters*. London: Arrow.

Hemmington, N. (2007). 'From Service to Experience: Understanding and Defining the Hospitality Business', *The Service Industries Journal*, 27 (6): 747–755. DOI: 10.1080/02642060701453221

Hochschild, A. R. (1983). *The Managed Heart: Commercialization of Human Feeling*. Berkeley, CA: University of California Press.

Hosany, S. and Witham, M. (2010). 'Dimensions of Cruisers' Experiences, Satisfaction and Intention to Recommend', *Journal of Travel Research*, 49 (3): 351–364. DOI: 10.1177/0047287509346859

Hwang, J. and Heesup, H. (2018). 'A Study on the Application of the Experience Economy to Luxury Cruise Passengers', *Tourism and Hospitality Research*, 18 (4): 478–491. DOI: 10.1177/1467358416682068

Johns, N., Tyas, P., Ingold, T., and Hopkinson, S. (1996). 'Investigation of the Perceived Components of the Meal Experience Using Perceptual Gap Methodology', *Progress in Tourism and Hospitality Research*, 2: 15–26. DOI: 10.1002/(SICI)1099-1603(199603)2:1<15::AID-PTH22>3.0.CO;2-E

Kirillova, K., Lehto, X. Y., and Liping, C. (2017). 'Existential Authenticity and Anxiety as Outcomes: The Tourist in the Experience Economy', *International Journal of Tourism Research*, 19: 13–26. DOI: 10.1002/jtr.2080

Kociatkiewicz, J. and Kostera, M. (2009). 'Experiencing the Shadow: Organizational Exclusion and Denial within Experience Economy', *Organization*, 17 (2): 257–282. DOI: 10.1177/1350508409341114

Krätke, S. (2010). '"Creative Cities" and the Rise of the Dealer Class: A Critique of Richard Florida's Approach to Urban Theory', *International Journal of Urban and Regional Research*, 34 (4): 835–853. DOI: 10.1111/j.1468-2427.2010.00939.x

Lashley, C. (2015). 'Hospitality Studies: Escaping the Tyranny?', *Quality Assurance in Education*, 23 (4): 364–377. DOI: 10.1108/QAE-04-2015-0014

Leidner, R. (1993). *Fast Food, Fast Talk: Service Work and the Routinization of Everyday Life*. Berkeley, CA: University of California Press.

Levitt, T. (1972). 'Production Line Approach to Service', *Harvard Business Review*, 50 (5): 41–52.

Lorentzen, A. and Hansen, C. J. (Eds.) (2015). *The City in the Experience Economy*. London: Routledge. DOI: 10.4324/9780203723449

Loureiro, S. M. C. (2014). 'The Role of the Rural Tourism Experience Economy in Place Attachment and Behavioural Intentions', *International Journal of Hospitality Management*, 40: 1–9. DOI: 10.1016/j.ijhm.2014.02.010

Lupton, D. (1996). *Food, the Body and the Self*. London: Sage.

McKercher, B. and Prideaux, B. (2014). 'Academic Myths of Tourism', *Annals of Tourism Research*, 46: 16–28. DOI: 10.1016/j.annals.2014.02.003

Mehmetoglu, M. and Engen, M. (2011). 'Pine and Gilmore's Concept of Experience Economy and Its Dimensions: An Empirical Examination in Tourism', *Journal of Quality Assurance in Hospitality and Tourism*, 12 (4): 237–255. DOI: 10.1080/1528008X.2011.541847

Miles, S. (2001). *Social Theory and the Real World*. London: Sage. DOI: 10.4135/9781446218778

National Catering Inquiry. (1966). *The British Eating Out*. Glasgow: National Catering Inquiry.

Oh, H., Fiore, A. M., and Jeoung, M. (2007). 'Measuring Experience Economy Concepts: Tourism Applications', *Journal of Travel Research*, 46: 119–132. DOI: 10.1177/0047287507304039

Pfeffer, J. and Sutton, R. L. (2006). *Hard Facts, Dangerous Half-Truths and Total Nonsense: Profiting from Evidence-Based Management*. Boston: Harvard Business Press.

Pine, J. and Gilmore, J. (1998). 'Welcome to the Experience Economy', *Harvard Business Review*, 76 (4): 97–105.

Pine, J. and Gilmore, J. (1999). *The Experience Economy - Work Is Theatre and Every Business Is a Stage*. Boston: Harvard Business School Press.

Pine, J. and Gilmore, J. (2002). 'Differentiating Hospitality Operations via Experiences: Why Selling Services Is Not Enough', *Cornell Hotel and Restaurant Administration Quarterly*, 43 (3): 87–96. DOI: 10.1177/0010880402433009

Pine, J. and Gilmore, J. (2011). *The Experience Economy: Updated Edition*. Boston: Harvard Business School Press.

Ritzer, G. (1993). *The McDonaldization of Society. Beverley Hills*. Thousand Oaks, CA: Sage Publications, Pine Forge Press.

Roberts, J. (2005). 'The Ritzerization of Knowledge', *Critical Perspectives on International Business*, 1 (1): 56–63.

Sidali, K. L., Kastenholz, E., and Bianchi, R. (2015). 'Food Tourism, Niche Markets and Products in Rural Tourism: Combining the Intimacy Model and the Experience Economy as a Rural Development Strategy', *Journal of Sustainable Tourism*, 23 (8–9): 1179–1197. DOI: 10.1080/09669582.2013.836210

Smith, M., MacLeod, N., and Hart Robertson, M. (2010). *Key Concepts in Tourist Studies*. London: Sage. DOI: 10.4135/9781446251027

Stierand, M. and Wood, R. C. (2012). 'Reconceptualising the Commercial Meal Experience in the Hospitality Industry', *Journal of Hospitality and Tourism Management*, 19 (1): 143–148. DOI: 10.1017/jht.2012.1

Sundbo, J. (2005). 'Experience Economy'. In S. M. Dahlgaard-Park (ed.) *The Sage Encyclopedia of Quality and the Service Economy* (pp. 219–222). Thousand Oaks: Sage.

Thompson, C. J., Locander, W. B., and Pollio, H. R. (1989). 'Putting Consumer Experience Back into Consumer Research: The Philosophy and Method of Existential-phenomenology', *Journal of Consumer Research*, 16 (2): 133–146. DOI: 10.1086/209203

Tourish, D. (2012). '"Evidence Based Management", or "Evidence Oriented Organizing"? A Critical Realist Perspective', *Organization*, 20 (2): 173–192. DOI: 10.1177/1350508411435281

Urry, J. (1992). *The Tourist Gaze*. London: Sage.

Van Maanen, J. and Kunda, G. (1989). '"Real Feelings": Emotional Expression and Organizational Culture'. In B. Staw and L. L. Cummings (eds.) *Research in Organizational Behavior* (vol. 11, pp. 43–102). Greenwich, CT: JAI Press.

Weiss, R., Feinstein, R., Hale, A., and Dalbor, M. (2004). 'Customer Satisfaction of Theme Restaurant Attributes and Their Influence on Return Intent', *Journal of Foodservice Business Research*, 7 (1): 23–41. DOI: 10.1300/J369v07n01_03

Wood, R. C. (1998). 'New Wine in Old Bottles: Critical Limitations of the McDonaldization Thesis – The Case of Hospitality Services'. In M. Alfino, J. S. Caputo, and R. Wynyard (eds.) *McDonaldization Revisited: Critical Essays on Consumer Culture* (pp. 85–104). Westport, CT: Praeger.

Wood, R. C. (2015). *Hospitality Management: A Brief Introduction*. London: Sage. DOI: 10.4135/9781473920521

Wood, R. C. (2017). 'The Unspoken Question: A Response to Thomas and Ormerod', *Tourism Management*, 62: 390–393. DOI: 10.1016/j.tourman.2017.02.008

Wood, R. C. (2018a). 'Managing Hotel Accommodation: An Overview'. In R. C. Wood (ed.) *Hotel Accommodation Management* (pp. 1–14). London: Routledge.

Wood, R. C. (2018b). 'Hotel Accommodation Design'. In R. C. Wood (ed.) *Hotel Accommodation Management* (pp. 46–56). London: Routledge.

Wood, R. C. (2018c). 'Housekeeping Management'. In R. C. Wood (ed.) *Hotel Accommodation Management* (pp. 70–82). London: Routledge.

Part II
Crafting tourism experiences

11

Understanding the experience design process

Bert Smit and Frans Melissen

Introduction

Experiences can be successfully designed, staged and managed. Popular visitor attractions such as Disney World and SixSenses Resorts, but also very different types of tourist experiences provided by, for instance, the National Museum of Natural History in Paris, are excellent examples of well-designed tourism experiences. What makes the experiences that these providers offer to and stage for tourists so successful? What makes them well-designed? To assist in answering these questions, it is important to understand how these experiences were designed in the first place and the main reference points for ending up with a "good" design. Therefore, this chapter focusses on the experience design process. In essence, design processes focus on translating customers' needs and wishes into products that can fulfill these needs and wishes. Within these processes, it is not only important to account for what potential customers want but also to establish how to make that happen within the boundaries set by what the specific business or organization that is going to stage the experience actually wants and is willing and able to provide. A successful design process delivers the design for an experience that can be successfully staged and managed.

In the field of tourism, there is increasing attention paid to the role of design in tourism experiences. A search in the EBSCO database on experience design in the tourism and hospitality field provides 206 articles, most of which were written in the last five years. A quick review of these articles clearly highlights a number of differences between the design methodologies, principles, approaches and methods reported on or discussed, but also some similarities. Most of those differences relate to differences in interpretation of the term design and differences in interpretation of the term experience. As Tussyadiah (2014) points out, design in relation to tourism experiences can have three different meanings. First, design can refer to a unique, aesthetically appealing proposition. Within this context, design processes mostly focus on sensory aspects of user experiences (e.g., Ross and Wensveen 2010; Sanders and Stappers 2008). In tourism, these processes are mostly applied to designing websites and apps but also dishes and interior design. Second, design can be a 'pervasive practice' (Tussyadiah 2014: 544) of human-centered innovation and problem solving, often referred to as design thinking. Within this context, design relates to applying descriptive design models (e.g., Eekels and Roozenburg 1995; Martin and Hanington 2012), which are applied to develop new concepts for, for instance, hotels and restaurants. Third, design can refer to the process of creating (a detailed plan for) all the touch points between service providers and customers. Within this context, design seems mostly related to engineering design processes (e.g., Cross 1994; Pahl and Beitz 2013; Patrício, Fisk, Falcão e Cunha, and Constantine 2011), which are generally applied to designing the physical aspects of tourism service systems such as airports, resorts and hotel rooms.

A similar difference can be observed regarding the way experiences are defined and interpreted (see e.g., Helkkula and Holopainen 2011; Smit and Melissen 2018), resulting in different views on how to design them. Some define experiences as the cognitive, behavioral, emotional, sensorial and social responses to the stimuli offered by a firm throughout the purchase journey. These authors conclude that experiences cannot be designed as a set of stimuli that lead to the same responses from every consumer (e.g., Lemon and Verhoef 2016). In their view, firms can only try to model the experience for their customers in the way their customer journey and touch points are designed (Kim and Fesenmaier 2016; Teixeira, Patrício, Nunes, Nóbrega, Fisk and Constantine 2012). Others define experiences in terms of their ability to create a positive memory related to an experience provider (e.g., Tung and Ritchie 2011), intention to recommend (e.g., Hosany and Witham 2010) or self-development (e.g., Richards and Wilson 2006). These authors conclude that experiences can have a similar outcome across (groups of) consumers and could be purposely designed for that purpose. However, as Tussyadiah (2014) points out, even though there are a number of publications on designing experiences, there is no consensus on which steps need to be included in the design process for tourism experiences.

Case study 11.1

Emotions play a key role in tourism experiences (see e.g., Bastiaansen et al. 2019). Their ability to delight, arouse and stimulate other (positive) sensations is important in creating memories, intention to recommend and customer loyalty. For experience design, this means that one of the purposes of any design is to stimulate and facilitate particular customer emotions. To be able to assess and evaluate the design of an experience, we need to measure (amongst others) the emotions solicited and identify a possible gap with the intended emotions. Kim and Fesenmaier (2015) have explored measuring emotions in real time. Other measurements, e.g. through interviews and questionnaires, have a number of downsides, for instance due to so-called slippage in reporting them but also due to consumers not always being consciously aware of their feelings. Moreover, recording this emotional (response) data without recording the (stimulus) context in which it took place creates a major problem for design evaluation. Obviously, measuring emotions in real time using reliable equipment can also have an effect on the respondents' experience itself. Kim and Fesenmaier, for this reason, use Electro Dermal Activity (EDA) sensors in combination with GPS, which have a limited impact on respondents. In their case study, two respondents were sent on the same tour of Philadelphia during which their EDA was measured. After the tour, both respondents reported on their experience using diary notes. The combination of diary and EDA data showed, for instance, that the respondent who had the window seat on a bus tour enjoyed it more than the respondent in the aisle seat because she could barely look outside. Similarly, the respondent who was not an art fan was more bored during a street art tour than the respondent who was into art. Visiting a park that reminded one respondent of a park in her home-town triggered stronger emotions than it did for the other respondent. Kim and Fesenmaier point out that EDA in combination with GPS provides experience designers in a tourism context the opportunity to create so-called emotion maps of destinations, which in turn can assist in pinpointing meaningful touch points, attractions and activities for particular groups of tourists. The benefit of this approach over using diary notes is that a diary only gives insights into separate moments that were noted by the respondents, whereas the EDA supplements this data with data on respondents' emotional states throughout the daytrip and the sequence in which these occur.

Design processes

Design processes come in many forms and can include many different steps, tools and methods. The realm of design science is at least as diverse as the social sciences. Similar to social sciences, design science has provided a variety of theories, models and methods to assist in creating new products, services and interactions with their users. These are linked to various and very different approaches to design. Some approaches could be described as very precise, quantifiable and calculative, for instance linked to the design of IT equipment. Others resemble a more free-flowing process, usually focused on quickly identifying opportunities for (co-creation of) new or improved products or services. This type of design could for instance be linked to the design of websites and usually involves rapid prototyping processes.

Despite these very different approaches to design, most professionals and academics in design agree that successfully creating new or improved products, services and technologies usually requires a step-by-step design process. Together, these steps make up so-called design models. Two main types of design approaches can be distinguished: those based on descriptive models and those based on prescriptive models.

For most descriptive models (for an example, see Figure 11.1), the starting point is an ill-defined problem or opportunity. This type of design process is generally applied to generate (ideas for) new concepts and product innovation opportunities. According to the second interpretation of design put forward by Tussyadiah (2014), this approach is usually referred to as design thinking. In descriptive models, the focus is on creating a solution concept early in the process, and to then use that solution to gather feedback on the concept, analyze this feedback and evaluate the prototype in order to further refine and define the problem or opportunity and

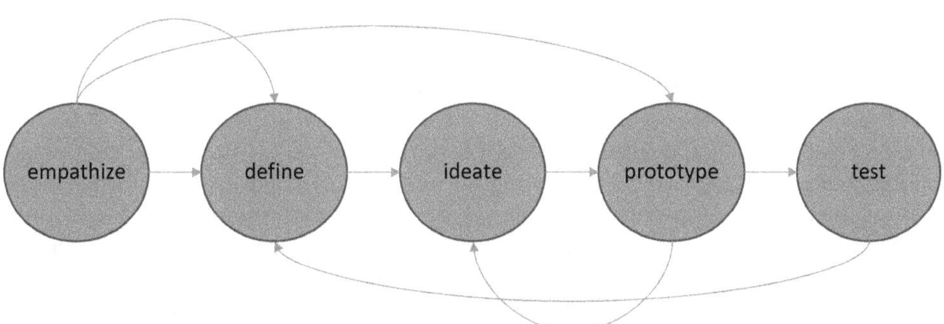

Figure 11.1 Example of a descriptive design model
Adapted from (Interaction Design Foundation 2018)

create a second (version of the) solution concept. One of the benefits of this approach to design is that it is relatively fast. Applying it, for instance in the development of digital services by means of applying the Google Design Sprint method (Knapp, Zeratsky, and Kowitz 2016), allows a design team to generate a first prototype in a matter of weeks or even days. A second benefit of descriptive design models is that they allow for users and other stakeholders to be involved in co-creating and co-designing tangible prototypes. Engaging with the target audience in this way can help build commitment for the final solution. One of the major disadvantages of applying descriptive design models is that zooming in on a potential solution in an early stage of the design process implies that designers are at an increased risk of creating 'a beautiful solution to the wrong problem' (Smit and Melissen 2018: 75). Those involved in the design process might easily get blinkered by the opportunity their initial idea creates. Using this approach to design can undoubtedly be beneficial in organizations that foster a culture of open innovation. However, successfully applying this approach to design requires 'a very experienced designer or design team that is able to empathize with users while relying on previous experiences, rules of thumb and assumptions' (Smit and Melissen 2018: 75).

Marriott, one of the biggest hotel chains in the world, is actively experimenting with this approach to design and innovation. They run a particular hotel in which all kinds of prototypes are tested on guests. This so-called M Beta hotel, a living lab in the shape of a hotel, changes regularly, with new prototypes for hotel rooms, beds, bars, services, but also menus, dishes and drinks and guests are continuously stimulated to rate and review their new ideas. Successful ideas are then standardized and 'extrapolated' to other hotels run by Marriott. Obviously, for Marriott, applying this approach does not mean that they actually start their design process with an entirely ill-defined problem or opportunity; they are already in the business of developing and operating new hotels and hotel brands, and have been for some while. Consequently, they can rely on previous experiences and reference points following from those experiences in coming up with new ideas. Using this approach is probably not advisable for design teams that are 'new' to the 'hotel world'. They would be exposed to a significant risk of ending up with beautiful solutions for the wrong problem or the wrong solutions for the right problem.

One of the ways to prevent investing time and resources in wrong solutions is to consciously invest more time and effort into defining the particular problem or opportunity at hand. This is a typical characteristic of prescriptive design models, which form the basis for an approach to design in line with the third interpretation of design referred to in the introduction section. According to Tussyadiah (2014), design in this case refers to the creation of a detailed plan. Prescriptive design models incorporate a more analytic, systematic and/or mathematical approach to design. To avoid the main disadvantage of descriptive design models, prescriptive design models assist designers in focusing on identifying and understanding the actual problem, and its sub problems, and determining the components and requirements possible solutions must incorporate and meet before generating alternatives for those solutions. This prevents designers creating solutions for the wrong problem or creating unfeasible solutions. After analyzing the problem, prescriptive models then focus on defining (sometimes together with other stakeholders) the scope within which possible (sub) solutions must fit. This allows designers to focus on elements that have been identified as being really important when they start generating potential (sub) solutions, which are subsequently combined into complete designs (detailed plans). To pick the best complete design, each design alternative is then evaluated against the (performance) requirements set in the first stage of the design process. Prescriptive design processes (e.g. Pahl and Beitz 2013) are broadly applied in fields such as manufacturing design, mechanical engineering and building design. The main benefit of applying this type of design process is that it is more precise, whereas one of the main downsides is that it takes more time and effort to get to a solution.

These various benefits and downsides involved with following different approaches to design also imply that it is wise to carefully consider which approach to follow in which specific situation before starting your design process. This most certainly also applies to the specific context of designing tourism experiences. Depending on the particular type of tourism experience an organization or business would like to design, different design processes could, and probably should, be applied to increase the chances of ending up with a successful tourism experience design. For instance, if the aim is to design an AR-application that distributes historic facts of a particular destination to tourists, the user interface might very well be successfully developed using a design process based on a descriptive model. However, if the aim is to design an amazing new visitor center that takes tourists back to life on the streets during the French Revolution, applying a prescriptive model probably is the way to go.

Towards a generic experience design process

Although there are many differences between descriptive and prescriptive design models, Cross (1994) points out there are also many similarities. In his influential book *Engineering Design Methods: Strategies for Product Design*, Cross (1994) proposes seven generic design steps, based on an elaborate analysis of a wide range of design models and processes. These seven steps do not provide the ultimate solution for each design challenge, but rather they provide an overview of the generic stages of a design process and how these (ought to) inform each other. Smit and Melissen (2018) have taken these seven generic design steps to create a comprehensive yet hands-on approach to experience design, highlighting the specific tools and methods that could be applied in each step of the process to (co-)create tourism and hospitality experiences, both in the shape of overall customer journeys and detailed individual touch points. This approach might prove helpful for tourism experience designers and therefore it is presented below.

Figure 11.2 shows an adapted version of the generic design process suggested by Cross (1994). The first three steps in this process are primarily focused on analyzing the problem or central aim of the design task at hand to ensure that the final design will include the appropriate sub solutions to fulfill that aim.

The first step, clarifying objectives, focuses on answering the question: what is the purpose of the (tourism) experience we are going to design? The objectives identified at the start might still change during the process as a result of creating a better understanding of customer needs and wishes, as well as competing experiences, or through feedback on possible (sub) solutions. However, to avoid losing time and resources on irrelevant aspects, it is important to have a clear overview of all relevant objectives at this stage of the process. Smit and Melissen (2018) propose to capture customer objectives by creating personas representing the target audience(s) as well as the 'host organization', destination or brand and potential other stakeholders (e.g., local community). For each persona, objectives related to the experience can be formulated in terms of feelings, memories and enthusiasm. For instance, Persona A will remember ... when asked about us; Persona B will associate us with ... after visiting us. These objectives and sub-objectives should then be defined and ordered hierarchically, for instance in the shape of an objectives tree (Cross 1994). This way a common understanding of words such as 'fun', 'relaxing' and 'safe' can be created between

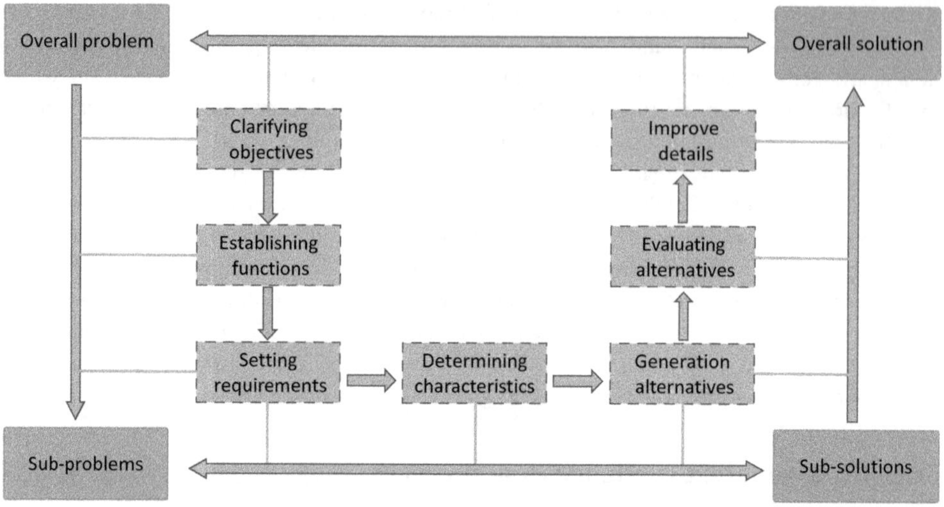

Figure 11.2 A generic model of the design process
Adapted from (Cross 1994)

various stakeholders, and this will tell designers to what extent 'fun' is more important than 'safe' (or vice versa).

In the second step, establishing functions, the focus is on what the design should be able to do. These aspects generally refer to the process elements customers go through as part of a touchpoint or customer journey. In most cases, establishing these functions is quite straightforward. For instance, for a hotel these functions would include aspects such as booking, sleeping and paying. However, in relation to specific types of experiences, such as transformational experiences or peak experiences, functions might not always be related to specific moments in time but rather to a function of the overall experience. For instance, a function could be that the experience should be physically exhausting or facilitate a state of flow. Designers could use tools such as system mapping and reverse engineering to visualize and analyze all relevant functions and their interdependencies.

The third and final step of analyzing the problem or central aim of the design focusses on the performance specifications for the design. In this step, the range of acceptable solutions is determined for a number of criteria. These criteria can be related to customer personas (e.g., price, duration) but also to operational (e.g., capacity), technical (e.g., ISO norms), spatial (e.g., location), financial (e.g., investment budget) or legal (e.g., privacy) specifications. Performance specifications are different from objectives and functions because they set the limits for possible (sub) solutions rather than specifying what the design must do.

These first three steps provide experience designers with information from various disciplines within the organization. They need this information to come up with suitable solutions, but before they can do that they need to translate this information into designable elements. Cross (1994) identifies this as the fourth step in a design process: determining characteristics of the end product. In this step, persona objectives such as 'relaxing' are linked to engineering characteristics such as sound, smell, spatial design, food and drinks. This step identifies the most important elements designers can create, adapt or manipulate to reach the design objective through incorporating them in possible solutions generated in the next step of the process. If step 1 was executed properly, they already know whether 'relaxing' for our personas is linked to, for instance, physical activity in nature or to doing as little as possible while being pampered. Smit and Melissen (2018) propose to use a so-called House of Quality for this step, a method derived from Quality Function Deployment, or to use a Kelly Repertory Grid. Both approaches can help to translate customer needs into design characteristics, while accounting for the reference points set out in the hierarchical model of service quality (Brady and Cronin 2001).

Completing these first four steps of the process sets the stage for the 'more creative' part of design, which is represented by step 5: generating alternatives. This step focusses on creating design alternatives for experiences that fulfill persona needs and wishes, meet the (other) objectives established in step 1 and perform all the functions that need to be performed, while staying within the boundaries set in step 3. In order to do so, specific sub solutions for each function are combined into overall solutions for the experience. In this stage, designers focus on how 'fun' or 'relaxation' can be created, for instance in the shape of activities and facilities. Listing possible sub solutions (touch points) for sub functions in a so-called morphological chart (see e.g., Smit, van Prooijen-Lander and Melissen 2018) can help identify potentially successful combinations of sub solutions, especially when focusing on designing specific types of experiences (e.g., peak experiences) and creating specific memories of the experience (e.g., sipping champagne in a rooftop pool overseeing Central Park). Generating some of these sub solutions might require creativity, inspiration and so-called eureka moments. However, although creativity is undoubtedly an important aspect of design, Cross (1994: 105) also indicates that 'making variations on established themes is [...] an important feature of design' as well. Consequently, combining existing or known (sub) solutions in innovative and smart ways is sometimes more productive and effective than trying to reinvent the wheel through elaborate creative processes. This is especially true for the overall design of the overall experience, which usually combines a(n existing) physical environment, the behavior of staff (already employed by the experience provider) and the behavior of customers while simultaneously performing a range of functions, of which many may already exist. Then again, for specific elements of that overall experience, for instance key touch points, a more creative approach, for instance based on rapid prototyping, might very well add value to the ultimate solution.

At the end of step 5, a number of design alternatives (detailed plans) will have been created. The sixth step then aims to identify the best (combination of) alternatives based on specific criteria that are derived from the objectives (step 1) and requirements (step 3). If all previous steps have been executed correctly, all design alternatives should already meet these criteria. However, it is very likely that not every design alternative meets them in the same way and to the same extent for each of the criteria. Step 6 also serves as a possibility to review the design process, reflecting on, and sometimes improving, some of the decisions taken in earlier steps, possibly together with relevant stakeholders. This could for instance lead to adding or removing particular objectives or functions or to changing the investment budget, now that the consequences of these decisions have become clear. These types of iterations are typical for any design project but are more likely to happen in projects that go through steps 1, 2 and 3 rather quickly, such as those adopting a descriptive design model. Decision-making in step 6 could benefit from applying tools such as multi-criteria analysis and value analysis, also in relation to comparing the ultimate design to competing experience offers in the market.

After finalizing step 6, the design (detailed plan) is ready to be implemented. Step 7 focusses on doing so effectively and efficiently, but does allow further improvement of details. These could relate to materials and technologies used in construction and to the selection and training of staff, but also to further improvements based on customer and staff feedback once the designed experience is actually staged and has been 'consumed' by its first visitors/guests/customers. It is important to note that step 7 might seem unimportant to those mostly interested in the overall concept and the solution principles that have been applied but it is actually a crucial step in ensuring that an experience can successfully be staged and managed and not remain a 'wonderful' idea that never materialized. This is the stage in the design process in which the design team 'hands over' the experience design to those involved with staging and managing the experience. Usually, representatives of management have been involved in the design process, especially in the final design stages. In the hospitality business, for instance, it is common to have a pre-opening team in place in the final months of hotel construction or remodeling. This team usually also stays 'in place' for the first few months or years after formally opening. This team consists of design specialists but also specialists in writing standard operating procedures, recruiting and training staff, negotiating contracts with local suppliers, and so on. Very often, they are the ones that ultimately dot the i's and cross the t's that are needed to truly make the experience come alive. As such, step 7 represents the linking pin between the design cycle and the management cycle of experiences. This is where creativity meets the Deming cycle and Lean management. For these management tools to be effective, it is crucial that management is fully aware and remains aware of the (experience) objectives set in step 1 and that constitute the reasoning behind and underlying logic for the ultimate experience design. As soon as they start prioritizing profit maximization or other types of management objectives over experience objectives, the experience is likely to be downgraded to a mere service (Smit and Melissen 2018). Obviously, this does not mean that profit cannot be combined with designing, staging and managing experiences. The examples with which the chapter kicked off, i.e. Disney, SixSenses and the Natural Museum of Natural History in Paris, prove that this is quite possible. However, doing so needs to account for the following: economic value in tourism generally follows from the experience value that is created for tourists, not the other way around!

Conclusion

The previous sections have illustrated how following a design process and adopting design science methods can be beneficial in designing for tourism and hospitality experiences with a similar outcome across its consumers, without trying to control or manage the individual responses to multi-sensory stimuli. Some of the best practices mentioned in this chapter prove how this outcome focus is beneficial for their business. This chapter has also illustrated some of the opportunities and risks of using particular methods, and how they do assist in creating beautiful solutions for the right problem if applied consistently and also after the implementation of such a design.

References

Bastiaansen, M., Lub, X.D., Mitas, O., Jung, T.H., Ascenção, M.P., Han, D.I., … Strijbosch, W. (2019). 'Emotions as core building blocks of an experience', *International Journal of Contemporary Hospitality Management*, 31(2): 651–668. DOI: 10.1108/IJCHM-11-2017-0761

Brady, M.K. and Cronin, J.J., Jr. (2001). 'Some new thoughts on conceptualizing perceived service quality: a hierarchical approach', *Journal of Marketing*, 65(3): 34–49. DOI: 10.1509/jmkg.65.3.34.18334

Cross, N. (1994). *Engineering design methods: strategies for product design*. New York: John Wiley.

Eekels, J. and Roozenburg, N.F.M. (1995). *Product design: fundamentals and methods*. New York: John Wiley.

Helkkula, A. and Holopainen, M. (2011). 'Service innovation as an experience: differences between employee and user narratives'. In Sundbo, J. and Toivonen, M. (eds.) *User-based innovation in services*, (pp. 281–302). Cheltenham: Edward Elgar Publishing.

Hosany, S. and Witham, M. (2010). 'Dimensions of cruisers' experiences, satisfaction, and intention to recommend', *Journal of Travel Research*, 49(3): 351–364. DOI: 10.1177/0047287509346859

Interaction Design Foundation. (2018). 'The basics of user experience design', [online] Available from www.interaction-design.org/ [Accessed April 26, 2019].

Kim, J. and Fesenmaier, D.R. (2015). 'Measuring emotions in real time: implications for tourism experience design', *Journal of Travel Research*, 54(4): 419–429. DOI: 10.1177/0047287514550100

Kim, J. and Fesenmaier, D.R. (2016). 'Tourism experience and tourism design', In Fesenmaier, D.R. and Xiang, Z. (eds.) *Design science in tourism*, (pp. 17–29). Berlin: Springer.

Knapp, J., Zeratsky, J., and Kowitz, B. (2016). *Sprint: how to solve big problems and test new ideas in just five days*. London: Simon and Schuster.

Lemon, K.N. and Verhoef, P.C. (2016). 'Understanding customer experience throughout the customer journey', *Journal of Marketing*, 80(6): 69–96. DOI: 10.1509/jm.15.0420

Martin, B. and Hanington, B. (2012). *Universal methods of design: 100 ways to research complex problems, develop innovative ideas, and design effective solutions*. Beverly, MA: Rockport.

Pahl, G. and Beitz, W. (2013). *Engineering design: a systematic approach*. Berlin: Springer Science and Business Media.

Patrício, L., Fisk, R.P., Falcão e Cunha, J., and Constantine, L. (2011). 'Multilevel service design: from customer value constellation to service experience blueprinting', *Journal of Service Research*, 14(2): 180–200. DOI: 10.1177/1094670511401901

Richards, G. and Wilson, J. (2006). 'Developing creativity in tourist experiences: a solution to the serial reproduction of culture?', *Tourism Management*, 27(6): 1209–1223. DOI: 10.1016/j.tourman.2005.06.002

Ross, P.R. and Wensveen, S.A. (2010). 'Designing aesthetics of behavior in interaction: using aesthetic experience as a mechanism for design', *International Journal of Design*, 4(2): 3–13. [online] Available from: http://ijdesign.org/index.php/IJDesign/article/view/765/297 [Accessed December 4, 2019].

Sanders, E.B.N. and Stappers, P.J. (2008). 'Co-creation and the new landscapes of design', *Co-Design*, 4 (1): 5–18. DOI: 10.1080/15710880701875068

Smit, B. and Melissen, F. (2018). *Sustainable customer experience design: co-creating experiences in events, tourism and hospitality*. London: Routledge.

Smit, B., van Prooijen-Lander, M., and Melissen, F. (2018). 'Designing hospitality experiences'. In Melissen, F., van der Rest, J.P., Josephi, S., and Blomme, R. (eds.) *Hospitality experience: an introduction to hospitality management* (pp. 155–185). Groningen: Noordhoff Uitgevers.

Teixeira, J., Patrício, L., Nunes, N.J., Nóbrega, L., Fisk, R.P., and Constantine, L. (2012). 'Customer experience modeling: from customer experience to service design', *Journal of Service Management*, 23 (3): 362–376. DOI: 10.1108/09564231211248453

Tung, V.W.S. and Ritchie, J.B. (2011). 'Exploring the essence of memorable tourism experiences', *Annals of Tourism Research*, 38(4): 1367–1386. DOI: 10.1016/j.annals.2011.03.009

Tussyadiah, I.P. (2014). 'Toward a theoretical foundation for experience design in tourism', *Journal of Travel Research*, 53(5): 543–564. DOI: 10.1177/0047287513513172

12

Experiential choice in tourism

Jens Friis Jensen and Jon Sundbo

Introduction: tourism businesses and tourists choose experiences from different logics

This chapter deals with how it is decided which experiences are valuable within tourism. The focus is on the decision process. The main approach of the chapter is that there are two parties in tourism experiential choices; the supplier – tourism businesses, and the buyer – the tourists, and the two parties' choices do not always correspond. Tourism businesses do not provide experiences, but only stimuli that may give tourists an experience (Sundbo and Sørensen 2013) – experience propositions (c.f. Chapter 1). Experiences are mental phenomena created in the minds of the tourists, who have different fields of interest and interpret the stimuli in different ways. The chapter will focus on experiential choice by the two parties. The approach is that each of these two parties makes their own choice. Tourists have varied choices, depending on their fundamental interests (interest regime), contextual needs and preferences, etc. Tourism businesses have certain values, business models and competencies that determine the experiences they offer and they try to satisfy the demand and provide the experiences that tourists want. However, tourism businesses cannot be sure of which experiences tourists create – that depends on tourists' mental awareness toward experiences, earlier experiences and general interest regime (Sundbo and Sundbo 2018, see also Chapter 21 on experiential dissonance). The tourism business can only deliver experience propositions (c.f. the theory of service value proposition, Skålen et al. 2015; see also Chapter 1).

The two parties, tourism businesses and tourists, thus choose experience elements from different perspectives. It must be assumed that the tourism businesses decide their choice of offered experiences based on planning, while the choice for tourists will often be more spontaneous and driven by more or less conscious mental filters and values. There is a dilemma between tourism businesses' and tourists' choice of experience, which must be discussed theoretically and understood empirically. We will therefore first discuss experience design theoretically to create a basis for understanding tourism businesses' choice of experience propositions. Next, we will theoretically discuss tourists' values and interest regimes as a basis for their choice of experience propositions and interpretation of the experience propositions.

Then we will illustrate the choice differences between the two parties with an example of food experiences in an attraction – a family amusement park that has introduced a restaurant serving local food. Qualitative and survey data from a case study in the attraction will be the basis for the illustration. First we will describe how this attraction chooses experience propositions – the single experiences and the total experience of visiting the attraction. Further, how tourism businesses decide which experience proposition to offer and their observations of the guests' reactions to the experience propositions. Next, we will discuss tourists' decisions based on their value systems and interests. Often tourists behave in groups (e.g. families) whose members have different experience preferences. The groups' desires and behavior become a negotiation process within the group, where group structure, power relations, etc., play a role.

Experience design and decision on experience proposition

The experience creation process is an interplay between external stimuli on the one side and inclinations, needs, motivations or mental awareness coming from the individual (organism) itself on the other (Jantzen 2013). Tourists create a holistic experience, for example by visiting an attraction park, which includes many single experiences at different touchpoints.

The experiential choice is an ongoing process starting before the visit to the attraction park, but continues for each touchpoint of the customer journey that constitutes a personalization of the holistic experience. For example, a meal in an attraction park is a unique experience proposition in its own right, but is also a touchpoint in the total customer journey – a visit to a tourist attraction. An experience design model has suggested operating with core and peripheral experiences (Sundbo and Hagedorn-Rasmussen 2008). The former is the fundamental reason for visiting a tourist attraction (for example, that children should have amusement), the latter is the extra service or experience that must be there to fulfill service needs (for example, restaurants to satisfy hunger). Even the peripheral experiences can be important for guests' total experience – sometimes more important than the core experience (c.f. Chapter 21 on experiential dissonance).

Experience design is therefore not about designing an experience, but about designing for experience.

> It is not possible to design experiences since they can never be fully predicted or controlled as they depend on co-producing the performance of visitors and the interrelations of entities. Yet, it is possible to design for experiences: actual experience emerges as visitors (consumers/users) are engaged with spatial designs and with the social situations that ritualize and help shape experiences.
>
> *(Svabo et al. 2013: 321–322)*

So the task of experience design is to design and organize frameworks, develop objects and situations and plan sequences of events which can contribute to interesting and relevant experiences on a physiological, emotional and cognitive level for the tourist (Jantzen et al. 2012). It is important to stress that this definition is based on a psychological understanding of the experience and implies the active involvement of the tourist in experience creation (Jantzen 2013).

From a company perspective, the experiential decision choice implies a multi-layered design approach with two levels that are mutually interdependent. First, the company needs to design an overall experience concept, a meta-concept of tourism experience (Tussyadiah 2014) that translates the strategic intent and positioning into experience value propositions, frameworks and design elements allowing for meaningful interaction and experience creation by the tourist. Second, the concept implies a strategy for the operationalization of the intended experience design into specific design elements, 'experience-facilitating systems and environments' (Tussyadiah 2014) like touchpoints, the physical surroundings, the social environment and the entire customer journey – elements that facilitate, moderate and enable the tourist's mental readiness, understanding and behavior, allowing for interaction and experience creation to take place.

As such, a holistic experience concept contains design principles and a strategy on how to design the operational level elements that translate the overall experience concept into activities, objects, etc., like touchpoints and the whole customer journey from prior imagination and expectations to the recollection of memories. The breakdown of the experience meta-concept into specific design elements also allows a more detailed evaluation of the elements that constitute the concept and continuous work on the improvement of single elements that can hinder the experiential decision choice and cause negative evaluation of the experience proposition.

The experiential choice is also affected by the physical (spatial designs, interactions with tangibles) and social environment (social situations, interactions with employees and group members, other visitors) in the experience scape (O'Dell and Billing 2005, Zomerdijk and Voss 2010, Svabo et al. 2013). Operationalization through the design of touchpoints, customer journey, service delivery, physical spatial design, social encounters are all essential elements that are part of the 'experience facilitating system' and subject to evaluation by both the company and the tourist (Tussyadiah 2014, Ponsignon, Durrieu and Bouzdine-Chameeva 2017).

A gap analysis between the intended (by the company) and the realized (by the tourist) experience could inform the company of why the design fails to enable alignment between the intended and the realized experience.

Interest regimes and values as a basis for tourists' choice of experiences

Tourists' choice of experiences and their subsequent evaluation are decided by their values because these values influence their mental awareness system (c.f. the discussion in Chapter 1). Value creation for customers has always been a core issue in service theory (e.g. Grönroos 2000) to explain customer satisfaction, extra purchase, loyalty, willingness to pay a higher price and other marketing effects. Recently it has even attracted more attention as a general attempt to explain customer behavior and business modeling for firms, among other processes promoted by the theory of service-dominant logic (Vargo and Lusch 2006). Several concepts have been launched to theoretically understand customer value, for example value-in-use (Vargo, Maglio and Akaka 2008), co-creation or co-production (Prahalad and Ramaswamy 2004) and so forth. When looking at the potential for value creation through the introduction of local food in an amusement park, our point of departure is that local food compared to ordinary food has a different experiential value potential due to positive connotations associated with local food (quality, health, local, self-identity and so forth). The experiential value potential meets an increased consumer demand for experiential and symbolic value (Schulze 2013) and a search for meaningful experiences. Following this line of thinking, the challenge for the tourism company is how to organize and design an experience encounter that enables the intended experience and thereby unlocks the experiential values inherent in tourist attractions.

Tourists' values are, to a large extent, decided by their interest regime (c.f. Chapter 1) if they have such a one within the area in question (Sundbo and Sundbo 2018). An interest regime is a specific version of an interest field. The interest field could, for example, be food or sport; the interest regimes could be a particular interest in, for example, healthy food or Japanese food or in mountain biking or football. Many people share the same interest regime around the world. Interest regimes give people an identity and are a basis for communication with other people because they can always talk about their particular interest regimes, for example Japanese food or football. People's choice of tourist experiences are, at least partly, decided by their interest regime. If we talk about restaurants, it might be that people that have an interest regime focusing on Japanese food will choose an attraction park or other tourist attraction where they know there is a Japanese restaurant serving sushi or other Japanese food. The existence of such a phenomenon will probably influence their evaluation of the total experience in a positive way.

For tourism businesses it is an advantage if they can identify their customers' interest regimes because then they can find a systematic value system that decides the customers' choice and evaluation of the experience propositions. Guests might not have an interest regime within the field in question (for example, food or sports activity). It is then much more difficult for tourism businesses to assess guests' values and their choice and evaluation of the experience propositions. People's values may be diffuse and flighty. The values can be decided in the situation where people encounter the experience proposition. Situational and peripheral factors such as a child falling and hurting himself may decide the valuation of the experience proposition and even whether the family will accept the experience offer.

Further, tourists' choice of whether or not they will engage in an experience proposition is not an individual and isolated decision. People choose in social situations and from a social evaluation framework. Tourists are mostly in groups – families, a group of friends or a conducted party. Each individual's choice and evaluation of experience propositions is influenced by the other members of the group, whether these other members accept the experience proposition and their opinion on the proposition. Often the choice is a social negotiation process if there are different wishes, values and interest regimes in the group, for example, a family. This might even develop into a conflict and the conflict in itself may be the reason for a negative evaluation of the experience proposition or a rejection of it, even if the proposition fits with values and interest regimes. People also think of their social reputation and position when they decide upon an experience proposition. Is this type of experience acceptable in my social group (friends, neighbors, at the workplace, etc.)? Will it enhance my position and prestige in the group?

Case study 12.1: Summerland Zealand – an attraction park with changed food supply

The decision choice issues, which were theoretically described above, will be illustrated by a case example, which is the choice of food in an amusement park. The park has been studied in a research project on coastal tourism carried out between 2016 and 2018.

The case is Summerland Zealand in Denmark, which is an amusement park in Odsherred in Zealand, Denmark, 100 km from the capital city Copenhagen. Odsherred is the municipality with the most summer houses in Denmark – about 27,000; most of these are owned by people from Copenhagen, who use them regularly as second homes. The park is owned by one person and there are two other managers, one of whom has responsibility for the restaurants and other food supply; this manager also has other roles, such as responsibility for the running of all the attractions. The park is open from May to September and addresses children's experiences, particularly those of smaller children. The park has many attractions such as switchbacks, merry-go-rounds and so forth and an aquapark. Guests pay an entrance fee and then everything is free. There are many self-managed attractions such as row boats, bicycles, etc. The park is located in the countryside, so there is plenty of space – large greens, small forests and benches and tables where the guests can sit. There are several food offers, almost all of them traditional fast food: burgers, ice cream, pizza, etc. There is also a picnic area with a grill that the guests can use for free. Because of the green spaces, many guests bring their own food and consider the visit a family picnic.

The owner and his staff observed a change in the type of guests some years before this investigation. The traditional guests were typically families with small children from the less wealthy segment of society. However, a new segment emerged – wealthier and more highly educated people, often the summer house owners. This segment has other values and demands for food. Summerland Zealand therefore decided to establish a new restaurant in a tent, café-style, called The Taste of Odsherred, with a menu based on local raw food and local products. The idea of these new food initiatives is to attract and create more value for the new guest segment.

The owner of Summerland Zealand explains that the local food restaurant has created public awareness and branded the park as attractive to the new segment, but has not yet resulted in an increase in terms of profit. Further, practical barriers were encountered, such as the unstable supply of local products from many small producers and the coordination of direct relations with a fairly large number of small producers, rather than having a contract with one wholesale company. Another challenge is difficulties in recruiting staff with the necessary qualifications such as knowledge of local food, professional cooking and guest relations skills. The difficulties in recruiting staff led to the local food restaurant being open for only one month, July, when they can recruit student cooks and other students in their holidays. However, many guests still prefer the fast-food restaurants or bring their own food with them.

The purpose of the study in Summerland Zealand was 1) to investigate the background for and the smartness of this decision to establish a more gastronomic local food restaurant, given that it should attract a new segment and increase guest experience and turnover, 2) to investigate guests' decisions concerning choice of food in the park.

The following methods and data collections were used in the study: The owner and the managers were interviewed several times. The researchers (authors of this chapter) participated in interactive meetings with the owner and managers. Documentary material about the park and the strategy was collected. A survey of guests at The Taste of Odsherred and a burger bar was carried out in 2016. The intention was to describe guests' social characteristics and their motives for buying food in the local food restaurant, their knowledge of the food as local and their opinion about its quality. Further, we asked about their general food and food-price preferences. In the burger restaurant we asked the same questions, including the customers' knowledge of and opinion about local food being served in the amusement park. Informal qualitative interviews were also conducted with selected guests. The interviewers were instructed to select guests randomly, thus different ages, genders and group sizes were represented. The researchers and the survey interviewers made observations in the park. Extensive qualitative interviews were carried out in 2017 with a larger number of guests in the park and an extended focus on the experiential value.

Experience proposition choice in an attraction

The results from the study of Summerland Zealand will be the basis for the following sections, which will describe and discuss the park's and the guests' choices concerning meal propositions. First, the guests' decisions and interest regimes will be analyzed, and then the management's decisions about meal propositions will be described and discussed.

The guests' meal choice

The case study investigated two different meal propositions: the healthy, café-style local meal in the tent and the fast-food propositions (the burger bar). During the study period a third proposition was revealed, namely the grill in the picnic area. The case study investigated which customer segment used the three meal propositions, whether they made their choice from specific values and interest regimes and how they made their decisions. This was done through the surveys, observations and interviews. This section presents selected results from the case study that can tell us about the guests' choice.

The survey shows that the guests at The Taste of Odsherred have a higher social status than the guests at the burger restaurant. They represent more independent business people and managers and they come more from the summer houses in Odsherred, while the guests at the burger restaurant were mainly one-day visitors from Copenhagen and local people from Odsherred. Prices were less important and gastronomic quality more important to guests at The Taste of Odsherred than to guests in the burger bar. The Taste of Odsherred guests said that this restaurant presented a more homely and attractive café atmosphere, but first and foremost they emphasized that the food is healthier than in the fast-food restaurants. Guests at the burger bar said that they chose that because of time, because it is fast and because it has a functional value rather than an experiential value. They were on a family tour where the children's interests are prioritized, and the children generally prioritize the attractions so time spent on meals steals time from the attractions. Some families in the burger bar said that the children, and sometimes the adults, prefer fast-food meals as a food style.

Time was also an issue in The Taste of Odsherred, since even that segment's children generally also had more interest in the attraction's amusements than in food. That raised a decision issue for the guests. Some guests from this segment chose a burger bar because of this time and attraction preference issue. Other families or groups chose to have their meal in The Taste of Odsherred, but the children were allowed to go by themselves to attractions while the parents remain seated in the restaurant. Obviously many parents felt the need to relax in well-known environments away from the attraction razzle-dazzle. The time issue was opposite in the picnic and grill area. There, people emphasized the social get-together, making food and having meals together. This was as important as the attractions, even for the children.

Generally the health aspects of food were emphasized by The Taste of Odsherred segment (73% of interviewees said this was important), but also the burger bar segment (44% said this). The local food aspect was emphasized as important by some of the guests at The Taste of Odsherred, but not by as many as could have been expected (about 16%).

This investigation revealed several values in the guests' choice of meals and food. These values were:

Family values: The visit to Summerland Zealand is a family affair where it is most important that the children are happy and enjoying the visit. Most children's values are:

Enjoying attractions, and perhaps:

Having some fast food (particularly ice cream)

These children's values influence the choice of meal proposition. The adults' own food values also influence the choice of meal proposition. The case study revealed several sets of food values, namely:

Food should be healthy
Café environment is secure and interesting
Food should be fast in an attraction park (speed of service)
Fast food is good and that is what we eat in our social group
The meal should be a social get-together activity

These values are, in some cases, founded in general social values and norms, which are implemented in the situation where guests must decide on the choice of meal and restaurant. To predict guests' meal choice requires knowledge about their general social values, and even then it can be difficult to predict which set of general values guests will follow in the specific situation. In other cases, values are founded in specific food interest regimes. In this case study, we have seen the following food interest regimes represented:

A healthy food interest regime (food is part of a healthy life)
Local food and gastronomic experiences interest regime (meals should be gastronomic experiences)
A social meal interest regime (food is a means for social interaction, which is the real interest regime)

Sometimes interest regimes overlap (such as the healthy and local food/gastronomy ones as a basis for choosing The Taste of Odsherred) and sometimes interest regimes are overruled by other sets of values (such as family interaction values and attention to children's enjoyment).

In some situations, the choice of restaurant is taken spontaneously and coincidentally. Among the guests in the burger bar, 36% said that their choice of restaurant was coincidental, they happened to be there and there was a restaurant. In these situations the food is not an experience, it is just a service – and it must be quick. To others, the choice of place to eat is a reflected choice, mostly at the grill-picnic area where the decision is taken at home before the visit to the park.

For those who more reflexively decide where to eat, food is an experience and values and interest regimes are the basis for choice, or at least they influence the choice. If guests are in groups, which they mostly are in Summerland Zealand, there may be different priorities and different interest regimes and sets of values, for example between children and parents. The observations show that the decision then is often a process of negotiation. This may, for example, end up with the parents sitting in The Taste of Odsherred while the children amuse themselves in the attractions, or the family choose a fast-food restaurant despite the parents preferring a more gastronomic meal; the children and their interest regime win.

The management's decision about meal propositions

The management makes its decisions about food supply from its own perspective, which is an economic one – whether a given activity makes a profit or a loss, and whether the managers find it interesting (it is part of their personal interest regime). Further, the marketing value of the activity also influences the choice of experience propositions.

However, the management also looks at guests' preferences. In Summerland Zealand they attempt to co-create experiences with guests (Kristensson, Matthing and Johansson 2008); they talk to and observe guests and conduct guest surveys every year. Input from external sources such as this case study research is also used.

The assessment from this case research project was that the park did not pay The Taste of Odsherred restaurant sufficient attention. The restaurant was a success, but could have been an even bigger success. The local aspect of the food was not evident enough and staff did not appear sufficiently professional. The management agreed and it led to the engagement of an external chef as a consultant. Further, local food and gastronomy was not part of the managers' interest regimes. The owner was very positive about The Taste of Odsherred because of the PR it gave Summerland Zealand, but he was mostly interested in introducing IT experiences, while the second manager mostly had a production view – this was a loss-making business. He was more focused on the fast-food restaurants and shops; they are profitable and many guests still prefer them.

The result for now is that Summerland Zealand keeps its fast-food restaurants open the whole season and The Taste of Odsherred is open for only one month, the most visited month of the high season. The menu in the latter has also been simplified so that it can be better managed by the staff, while at the same time meeting the requirement from visitors for a kind of 'healthy, local fast-food concept'. A third meal proposition was developed during the case study period. The picnic area with the grill became more important, since many guests used that. The challenge for the park is that the grill makes no profit because the visitors bring their own food. The case study helped the management to develop a new concept where guests can buy a local grill food pack in the park (which combines a need to find new streams of revenue while at the same time meeting a visitor need for combining convenience with the social activity).

Summerland Zealand thus successively made decisions that led to having three kinds of food propositions, each addressing different customer segments and customer food interest regimes: one for those who favor quick and fat-rich fast food, one for those who go for a gastronomic café with a healthy partly fast-food style and one for those on a social get-together picnic.

Intended and realized experiences

A core issue in the discussion of choice of experiences within tourism is to what extent experience providers' propositions fit with tourists' choice decisions, or said in another way: will tourists' realized experience compare with the intended ones? That depends on several factors such as: has the provider interpreted potential guests' interest regimes and values correctly? Is the delivery system appropriate to facilitate the intended experience? What triggers which decision parameters in the concrete choice situations? And, most unpredictably, which coincidental situational factors influence tourists' decisions?

In the case of The Taste of Odsherred, the survey and the interviews showed that the intended and realized experiences fitted to some degree, but not completely. This can be seen in the summary of case results in Table 12.1.

This case thus demonstrates that the choice of the right tourist experience propositions and their implementation is difficult and is a continuous process taking place in a dynamic interaction with the guests.

Table 12.1 Intended experience propositions and guests' realized experiences in the attraction park restaurant The Taste of Odsherred

Management's intended experience	*Guests' realized experience*
• High quality local food experience • The restaurant as an oasis where time, slowness is an integrated part of the meal experience signaling quality, care, freshness of local products • A wish to accommodate both guests' need for a quick meal and the organization's need for a more simple and fast-food-like menu (increase speed of production, reduce independence of skilled staff, reduce food waste) • A feel-good meal experience (contributing to the local community and offering families a healthy good quality meal) • Guests having encountered knowledgeable and skilled staff ensuring engagement and interaction with the guests	• Positive evaluation of the local food, healthy food, good quality, value for money • However, a very simplistic setting/physical environment • Conflicting interests regarding time within the family and between the café experience and the dynamics of being in an amusement park • Conflicting values within the family regarding healthy and hedonic meal consumption (counteractive to the overall amusement park experience) • Misunderstanding of the intended experience, the basic concept of time (almost slow food production) • Problems of internal communication in the restaurant • A number of failures due to lack of competence of the staff. There were expectations of a higher level of staff competence compared to the rest of the seasonal service staff in the other meal options and staff had problems with basic service operations delaying the serving

Conclusion

For people to have experiences is, to a large degree, a matter of choice and decisions. For tourism businesses and institutions it is important to continuously analyze the realized experience and engage in an ongoing development process in interaction with the tourists based on knowledge about their evolving interest regimes. They can never know exactly what the choice will be in concrete situations and they can therefore only offer experience propositions. Tourists' decisions can be coincidental and situational. Their decisions in such situations cannot be exactly foreseen; tourism businesses can only provide propositions that they *think* most tourists will choose and ensure that these propositions have good service elements and no failures. In other situations, tourists' interest regimes and sets of values are decisive for their choice of experience proposition and how they interpret the experience. Tourism businesses can achieve a closer fit between intended and realized experience by knowing about their customers' interest regimes and sets of values.

The case study of an attraction park and its food and meal experience propositions has illustrated some of the challenges in this.

References

Grönroos, C. (2000). *Service Management and Marketing*. Chichester: John Wiley.

Jantzen, C. (2013). 'Experiencing and experiences. A psychological framework'. In Sundbo, J. and Sørensen, F. (eds.), *Handbook on the Experience Economy* (pp. 146–170)), Cheltenham: Edward Elgar.

Jantzen, C., Fitchett, J., Østergaard, P., and Vetner, M. (2012). 'Just for fun? The emotional regime of experiential consumption'. *Marketing Theory*, 12(2): 137–154. DOI: 10.1177/1470593112441565

Kristensson, P., Matthing, J., and Johansson, N. (2008). 'Key strategies in co-creation of newservices'. *International Journal of Service Industry Management*, 19(4): 474–491.

O'Dell, T. and Billing, P. (eds.) (2005). *Experience-scapes*. Copenhagen: Copenhagen Business School Press.

Ponsignon, F., Durrieu, F., and Bouzdine-Chameeva, T. (2017). 'Customer experience design: A case study in the cultural sector'. *Journal of Service Management*, 28(4): 763–787. DOI: 10.1108/JOSM-01-2017-0016

Prahalad, C.K. and Ramaswamy, V. (2004). *The Future of Competition: Co-creating Unique Value with Customers*. Boston: Harvard Business School Press.

Schulze, G. (2013). 'The experience market'. In Sundbo, J. and Sørensen, F. (eds.), *Handbook on the Experience Economy* (pp. 1–17). Cheltenham: Edward Elgar. DOI: 10.4337/9781781004227.00005

Skålen, P., Gummerus, J., Koskull, C., and Magnusson, P. (2015). 'Exploring value propositions and service innovation: A service-dominant logic study'. *Journal of the Academy of Marketing Science*, 43(2): 137–158. DOI: 10.1007/s11747-013-0365-2

Sundbo, D. and Sundbo, J. (2018). 'Interest regimes in fruit wine tourism'. *Journal of Gastronomy and Tourism*, 3(3): 163–176. DOI: 10.3727/216929718X15410151918027

Sundbo, J. and Hagedorn-Rasmussen, P. (2008). 'The backstaging of experience production'. In Sundbo, J. and Darmer, P. (eds.), *Creating Experiences in the Experience Economy* (pp. 83–110), Cheltenham: Edward Elgar.

Sundbo, J. and Sørensen, F. (2013). 'Introduction to the experience economy'. In Sundbo, J. and Sørensen, F. (eds.), *Handbook on the Experience Economy* (pp. 1–17), Cheltenham: Edward Elgar.

Svabo, C., Larsen, J., Haldrup., M., and Bærenholdt, J.O. (2013). 'Experiencing spatial design'. In Sundbo, J. and Sørensen, F. (eds.), *Handbook on the Experience Economy* (pp. 310–324), Cheltenham: Edward Elgar.

Tussyadiah, I. (2014). 'Toward a theoretical foundation for experience design in tourism'. *Journal of Travel Research*, 53(5): 543–564. DOI: 10.1177/0047287513513172

Vargo, S. and Lusch, R. (eds.) (2006). *The Service-Dominant Logic of Marketing*. Armonk: Sharpe.

Vargo, S.L., Maglio, P.P., and Akaka, M.A. (2008). 'On value and value co-creation: A service systems and service logic perspective'. *European Management Journal*, 26(3): 145–152. DOI: 10.1016/j.emj.2008.04.003

Zomerdijk, L. and Voss, C. (2010). 'Service design for experience-centric services'. *Journal ofService Research*, 13(1): 67–82. DOI: 10.1177/1094670509351960

13

The service experiencescape

Melissa A. Baker and Kawon Kim

Introduction

Service experience management is the core of hospitality and tourism experiences. Creating a strong customer experience is a leading management objective and is considered a distinct construct to marketing and management (Lemon and Verhoef 2016). As competition among destinations around the world continues to grow, recognition of their importance to visitors is leading businesses and destinations to market themselves as providing memorable experiences (Andrades and Dimanche 2014). Memorable tourism experiences provide opportunities for destinations and firms to build their identities, increase customer satisfaction, and drive tourist demand.

The customer experience is conceptualized as holistic, comprising multiple touchpoints, in an end to end journey, involving the customers' cognitive, affective, emotional, social, and sensory elements (McColl-Kennedy et al. 2018). There are multiple definitions of customer experience in the literature. Schmitt (1999) identifies five types of experiences: sensory (sense), affective (feel), cognitive (think), physical (act), and social-identity (relate). Verhoef et al. 2009 define customer experience as a multidimensional holistic construct that involves cognitive, affective, emotional, social, and physical responses. In general, scholars and practitioners generally agree that customer experience management is a multidimensional construct that involves cognitive, emotional, behavioral, sensorial, and social components (Schmitt 1999; Verhoef et al. 2009)

The key elements of the experiencescape involve the physical servicescape and the social servicescape. Therefore, this chapter first discusses the experiencescape and its importance in the customer experience. Second, the chapter discusses the physical servicescape including the influence of sight, sound, smell, taste, and touch on customer's perceptions of the experience. Third, the chapter details the elements of the social servicescape, namely the influence of the employees and the other customers. Finally, the key outcomes from the experiencescape are discussed such as how the physical and social servicescape affect customers' perceptions of the overall experience, satisfaction, and future behavioral intentions. This chapter seeks to illuminate the importance of designing the physical and social servicescape and comprehending the strategic importance of these elements in affecting the overall customer experience.

Experiencescape

The experiencescape serves as a canvas for the consumption experience. The experiencescape is a blend of both the physical and social elements that affect the customer's overall experience. This is therefore critical as the heart of hospitality is in the service experiences created by the firm (Baker and Magnini 2016). This will involve both the physical and social elements that the firm creates and manages within their business. The experiencescape is particularly important for its strategic role in effecting favorable customer outcomes such as emotions, memorability, loyalty, and revisit intentions (Mody et al. 2019). The concept of the experiencescape is based on the marketing perspective that recognizes that experiences are highly personal, subjectively perceived, intangible, ever fleeting, and on-going (O'Dell 2005). Therefore, different customers will have different reactions to different experiences. This is why there can be so many different types of experiences targeted at different customers.

From a managerial standpoint, it is important to recognize that the customer experience is influenced by the pre-purchase, purchase, and post-purchase phases (Klink et al. 2018). Value creation elements consist of resources, activities, context, and interactions, and account for the single largest category of service experience comments (McColl-Kennedy et al. 2018). In other words, the design, delivery, and management of the customer experience can be viewed from the firm's point of view, from the customer's point of view, or from a co-creation perspective whereby the customer experience is a culmination of a customer's interaction with other actors in the servicescape (Lemon and Verhoef 2016). While this chapter primarily focuses on the effect of the experiencescape on tourist experiences, it is also important to note that all of these elements affect all constituents within a business from the customers, employees, management, suppliers, and other customers. The explosion of potential customer touchpoints and the reduced control of the experience requires firms to integrate multiple business functions (Lemon and Verhoef 2016) including service management, marketing, human resources, information technology, strategy, and finance in creating, delivering, and assessing positive customer experiences. In other words, the design of the experiencescape can influence how effectively employees are able to perform the necessary functions of their job. If the design is not conducive to efficiency, for example, then it will affect how quickly the employees can serve the guests and subsequent guest satisfaction and the firm's profit. It is therefore important to think about how the experiencescape affects customers, but also management, employees, and the environment.

Physical servicescape

The physical environment in which the experience is created is called the servicescape (Bitner 1992) and can also be described as the physical setting in which a service process takes place. Key components in assessing the servicescape are its role in the service experience, the physical environment, and the influence of atmospherics.

The role of the servicescape

The experiencescape has parallels to the concept of the servicescape and represents an arena in which experiences are staged and consumed (Mody et al. 2019). In other words, the experiencescape and servicescape go hand in hand in ensuring success in tourism destination marketing strategies and patronage (Chui et al. 2010). The servicescape plays many strategic roles, especially in terms of how customers interact with the physical environment as part of the service (Baker and Magnini 2016). Some argue that the physical design is one of the most important elements in determining the experience of the customer. Facility layout, for example, will affect the flow of operations, timeliness, efficiency, production, and overall store performance. Similarly, the design of the physical environment in a hotel can affect the customer's perception of the overall quality of a hotel or the attractiveness of a tourist destination. This is particularly important to note because once the physical servicescape is designed and built, it is not easy to change (Lovelock and Wirtz 2011).

Bitner's (1992) framework describes how the built environment under the service providers' control affects both customers' and employees' cognitive, emotional, physiological, and behavioral responses. The key elements to the servicescape involve the environmental dimensions and how these affect the holistic perceptions of the environment. It is then also important to consider how different customer internal responses such as cognitive, emotional, and physiological responses will affect the experience. Lastly, the social servicescape details how all of these elements will work together to affect customers' behaviors and social interactions.

Physical environment and design

The physical realm includes the arrangement of furniture, fixtures, and equipment that influence a servicescape's functionality, and spatial arrangements, and enhance the ambient elements (Bitner 1992). Cues from the physical environment directly influence how customers act and respond during service encounters (Ballantyne and Nilsson 2017). A sense of place can be facilitated by using cultural artifacts, signs, and symbols that create a sense of belonging to a culture or subculture, thus enriching the customer experience (Bolton et al. 2018). The use of technology, especially smartphones, has shifted from a novelty to becoming an essential part of the consumer experience, with the capabilities of offering personalized information to guests, two-way sharing, context awareness based on smartphone sensors, or communication with other technologies (Dickinson et al. 2014).

Firms also need to consider how digital technologies can serve as a critical element of customer experience management, including such elements as mobile, location-based, artificial intelligence, wearable technologies, neuroscience, and self-serve technologies. Digital technologies can provide a highly personalized and immersive environment that allows for interactivity and rich information exchange between the organization and the customer (Parise et al. 2016) Digital technologies are changing customer expectations and behaviors, how organizations and networks are organized, and the role of frontline human employees (Bolton et al. 2018). Smartphones have changed the communication landscape in the hospitality and tourism sector including room reservations, check-in, keyless technologies, and dedicated apps (Stankov et al. 2019). In other words, everything the guest might need before, during, and after their stay is centralized on a single digital platform.

One key variable that relates to the physical design is authenticity. Establishments seek to construct environments that will be seen as authentic (Wang and Mattila 2013) especially as it relates to their core business offering or tourist destination. Tourists and consumers who seek authentic cultural experiences focus on the authentic elements in the servicescape design (Baker and Kim 2019). Hospitality and tourism managers can enhance customers' perceived authenticity optimizing tangible cues (Kim and Baker 2017a), where authenticity is an important value to the customers. Companies need to embed, execute, and manage these cues to create the desired thoughts and feelings in consumers (Chakravorti 2011). Tangible cues such as furniture, fixtures, signage, and decorations are particularly important as tangible cues in the physical environment (Baker and Kim 2019). Imagine some establishments you have visited and how the design elements such as furniture, decorations, and menus affect your perceptions of authenticity or whether the establishment is seen as inauthentic. Especially as it relates to tourism experience, the level of authenticity is critical in forming tourist perceptions.

Atmospherics

The atmospherics are the sense items within the servicescape that firms can influence. In particular, customers will draw upon the five senses in the servicescape. It is therefore important to consider sight, sound, smell, taste, and touch. For example, elements of sight would include the overall look of a hotel lobby or the colors used in the servicescape design. Colors, for example, are vital in communicating images and themes, but also have immediate perceptual and cognitive significance in human experience of visual consciousness and responsiveness. For example, psychologically, blue tones signify calmness or trust. These colors are used in bank logos or in creating a calm, serene environment in spas. Conversely, red symbolizes boldness and also stimulates hunger and appetite. As such it is often found throughout the physical design of foodservice establishments, especially quick service such as McDonald's, Wendy's, and In-n-Out.

In terms of smell, companies such as Disney pipe in specific smells in order to trigger psychological memories and enhance the experiences. Similarly, spas scent their servicescapes with smells such as lavender, which make customers feel more relaxed. Conversely, realtors use the smell of fresh-baked bread when selling homes as it creates a stronger sense of home and comfort and is found to increase both the price customers pay and the quickness in which a home sells.

From a sound perspective, music or sounds will also influence the experience. Casinos use sounds such as bells when a customer wins on a game. The sound of bells and alarms is heard by the other customers, leading to psychological perceptions that if another customer has won, they can also be a future winner. Therefore, the sounds of bells and alarms increases customers' playing and gambling. In addition, from a foodservice perspective, sounds can also align with the experience. For example, in luxury restaurants, slower tempo music increases the amount customers spend and also increases the length of the dining experience. Conversely, in a quick-service restaurant, fast music is played to encourage customers to eat faster, so they can have more customers moved through the physical environment. Taste is also a significant factor in the servicescape that will affect perceptions, especially in those that have a foodservice component. Lastly, touch is a critical element of the physical servicescape. Everything from the feel of the bedding in a hotel room to the quality of the silverware will be a factor in how customers assess the physical environment.

Case study 13.1

Imagine you are designing a new restaurant. Think about how you would design the physical elements and atmospheric elements of the servicescape to align with your overall brand concept.

1. Create a design board including pictures of what your restaurant would look like. Include a discussion of why you chose the physical element you did and how this might affect the customer experience.
2. Create a playlist of music that would play in your restaurant. Include a discussion of how this music would align with your restaurant concept.

Social servicescape

Service, hospitality, and tourism practitioners are increasingly concerned with the social facets of the service delivery process (Baker and Kim 2018a). The social servicescape refers to the social aspects of the consumption environment (Tombs and McColl-Kennedy 2003) and refers to the customer and employee elements. As interactions are at the heart of the social realm of the customer experience (Bolton et al. 2018), organizations must take account of customers' social environments and their expectations (Bolton et al. 2018). As customers actively co-create experiences, organizations must facilitate customers' interactions with other people, which includes the impact of employees and other customers. A sense of social presence can be experienced in both physical and virtual contexts. The social servicescape model includes how elements of the social servicescape affect customers' affective and cognitive responses. Particularly, the social servicescape involves the two key elements of employees and other customers.

Impact of employees on the social servicescape

One key element that affects the overall experience is the impact of employees. Frontline employees, in particular, are the individuals who have the most amount of direct contact with the guest or tourist (Baker and Magnini 2016). As such, they have the greatest potential to impact the guests' experience through the sheer volume of the interactions. In other words, the frontline employees are the focal agents within the service relationships between the customer and the firm (Chandler and Lusch 2015). Furthermore, these interactions play a key role in the dynamic process of co-creating value and co-creating the experience.

During the service encounter, both tangible and intangible aspects are essential to how customers judge the service experience (Kim and Baker 2019). The language spoken and body language elements such as ethnicity, attractiveness, and courtesy can affect the social interaction and perceptions of the overall experience (Baker and Kim 2018a). Research also finds that the appearance of employees can significantly impact the perceptions of authenticity of the hospitality or tourism experience (Baker and Kim 2018a). The appearance of service providers in service settings is an important element of the social servicescape that effectively conveys an organization's image and core values (Baker and Kim 2019). For example, the level of attractiveness, smile, or other physical characteristics can alter customers' perceptions of service quality and satisfaction.

From an interaction perspective, it is critical that hospitality and tourism employees provide hospitableness to make experiences memorable (Baker and Magnini 2016). An employee is not simply a means to provide service and information, but more importantly, employees' rapport-building behaviors are crucial elements in inducing positive behavioral intentions in customers and tourists (DeWitt and Brady 2003). Developing rapport involves an enjoyable interaction and a personal connection (Kim and Baker 2019).

Impact of other customers

Tombs and McColl-Kennedy (2003) suggest that other customers influence the focal customers' affective and cognitive responses by being a part of the environment or the service experience of individual customers. As such, other customers can either directly or indirectly affect one's service experience (Kim and Baker 2017b). Customer-to-customer interaction is the active or passive interaction between two or more customers inside the service setting (Baker and Kim 2018a). Service experiences are co-created and co-produced through customer-to-customer interactions (McColl-Kennedy et al. 2014). Customers are co-creators, implying that they do not merely have an influence on their own experience, but also have an impact on the experience of other customers (Huang 2008).

Similar to the importance of the influence of employee appearance on the customers' experience, the appearance of other customers also affects how the customer feels emotionally and behaviorally. There is an abundance of social and psychological literature that suggests perceptions are influenced by appearance in both a covert and non-conscious form (Baker and Kim 2018b). Consumers of different demographic and psychographic characteristics seek to infer congruity between themselves and the social servicescape using environmental cues (Baker and Kim 2019). The extent to which other customers may influence the focal customer's service experience and behavior depends on the context of the service encounter and the characteristics of other customers (Kim and Baker 2017b). So customers often seek experiences where the other customers are similar to them in terms of appearance, interests, and lifestyles. For example, different cruises target different types of customers. Carnival cruises target younger individuals seeking a more party, active experience. Princess cruise lines seeks to target middle-aged couples who want a more educational or relaxing experience. Someone seeking to go on a cruise for college spring break would be more disappointed with the experience if most of the other customers are in their 30s to 40s. Conversely, a couple in their 40s would feel more uncomfortable if their cruise was full of 21-year-olds.

Experiencescape outcomes

Finally, it is important to discuss some of the key outcomes related to the experiencescape, servicescape, and social servicescape. An important question is how customer experience management is related to customer outcomes such as satisfaction, rapport, and repeat purchase intention.

Satisfaction

Customer experience management is a crucial element for developing and sustaining customer satisfaction (Chakravorti 2011). The physical environment and social servicescape can affect customer satisfaction for different cultures and subcultures. For example, the preferences of those in the United States may differ from those in Europe. In addition, subculture preferences exist in that those in the Northeast United States vs. the South will also have different subculture preferences (Baker et al. 2013). Furthermore, in service, retail, hospitality, and tourism contexts, the physical servicescape offers customers an immersive experience that can deliver enduring emotional moments and memories (Giraldi et al. 2016). These memories can in turn affect the level of overall satisfaction felt by different customers.

Rapport

Experiences are evaluated not just in relation to the physical environment, but also in terms of the emotional dimensions (Ariffin 2013). From this perspective, firms should focus on the guest–host relationships, such as creating emotional bonds that are developed through the interactions with employees (Baker and Magnini 2016). Similarly, firms should also focus on building rapport with customers as this can have many favorable outcomes to the experience (Kim and Baker 2019). Rapport building behaviors may include an enjoyable interaction and a personal connection. By creating these elements, firms can have greater success in managing the service experience.

Revisit intention

Research finds that customers who are more satisfied with the service experience will have higher levels of loyalty to the destination or brand (Chandler and Lusch 2015) and are more likely to purchase again or recommend the experience to friends and family through traditional word-of-mouth, online through social media such as Instagram, Snapchat or Facebook, or through recommendation websites such as Yelp and TripAdvisor. The economic benefits of competing on customer experience are compelling. For example, companies with superior customer experiences grow revenue five times faster on average than companies with inferior customer experience management (Forrester 2017). In addition, four-fifths of customers are willing to pay more for a better experience (Capgemini 2017). Lastly, providing an outstanding experience for a customer is a hallmark that has ensured the success and strong profit margins of top service companies (Grewal et al. 2017).

Conclusion

In summary, the customer experience is the cornerstone of the entire hospitality and tourism industry. First, firms must understand its importance and design the physical and social elements according to their brand mission. Second, it is important to understand the role of the physical servicescape and atmospherics on the overall customer experience. Third, management must be aware of the vital role both employees and other customers play in affecting customer perceptions and subsequent behaviors. Lastly, effective experience management can have significant effects on tourist satisfaction, rapport, and revisit intention.

References

Andrades, L., and Dimanche, F. (2014). 'Co-creation of experience value: a tourist behavior approach'. In Chen, M. and Uysal, J. (Eds.). *Creating Experience Value in Tourism* (pp. 95–112). London: CABI.

Ariffin, A. A. M. (2013). 'Generic dimensionality of hospitality in the hotel industry: a host – guest relationship perspective', *International Journal of Hospitality Management*, 35: 171–179. DOI: 10.1016/j.ijhm.2013.06.002

Baker, M., Murrmann, S. K., and Green, C. (2013). 'Dining in the city: server behaviors, time preferences, and the effect of urbanization in restaurants', *Journal of Foodservice Business Research*, 16(2): 113–138. DOI: 10.1080/15378020.2013.761042

Baker, M. A., and Kim, K. (2018a). 'The role of language, appearance, and smile on perceptions of authenticity versus rapport', *International Journal of Hospitality Management*, 74: 171–179. DOI: 10.1016/j.ijhm.2018.04.011

Baker, M. A., and Kim, K. (2018b). 'Other customer service failures: emotions, impacts, and attributions', *Journal of Hospitality and Tourism Research*, 42(7): 1067–1085. DOI: 10.1177/1096348016671394

Baker, M. A., and Kim, K. (2019). 'Heritage and authenticity in gastronomic tourism'. In Dixit, S. K. (Ed.). *The Routledge Handbook of Gastronomic Tourism* (pp. 251–258). Oxon: Routledge. DOI: 10.4324/9781315147628

Baker, M. A., and Magnini, V. P. (2016). 'The evolution of services marketing, hospitality marketing and building the constituency model for hospitality marketing', *International Journal of Contemporary Hospitality Management*, 28(8): 1510–1534. DOI: 10.1108/IJCHM-01-2015-0015

Ballantyne, D., and Nilsson, E. (2017). 'All that is solid melts into air: the servicescape in digital servicespace', *Journal of Services Marketing*, 31(3): 26–235. DOI: 10.1108/JSM-03-2016-0115

Bitner, M. J. (1992). 'Servicescapes: the impact of physical surroundings on customers and employees', *Journal of Marketing*, 56(2): 57–71. DOI: 10.1177/002224299205600205

Bolton, R. N., McColl-Kennedy, J. R., Cheung, L., Gallan, A., Orsingher, C., Witell, L., and Zaki, M. (2018). 'Customer experience challenges: bringing together digital, physical and social realms', *Journal of Service Management*, 29(5): 776–808. DOI: 10.1108/JOSM-04-2018-0113

Capgemini. (2017). *The disconnected customer: what digital customer experience leaders teach us about reconnecting with customers*. Retrieved from www.capgemini.com/wp-content/uploads/2017/07/the_disconnected_customer-what_digital_customer_experience_leaders_teach_us_about_reconnecting_with_customers.pdf [Accessed. March 1, 2019].

Chakravorti, S. (2011). 'Managing organizational culture change and knowledge to enhance customer experiences', *Analysis and Framework; Journal of Strategic Marketing*, 19: 123–151. DOI: 10.1080/0965254X.2010.529160

Chandler, J. D., and Lusch, R. F. (2015). 'Service systems: a broadened framework and research agenda on value propositions, engagement, and service experience', *Journal of Service Research*, 18(1): 6–22. DOI: 10.1080/0965254X.2010.529160

Chui, C. T. B., Rahim, F. A., Hassan, F. H., Musa, R., Yusof, J. M., and Hashim, R. H. (2010). 'Exploring tourist experiencescape and servicescape at Taman Negara (national park Malaysia)', *International Journal of Trade, Economics and Finance*, 1(1): 28.

DeWitt, T., and Brady, M. K. (2003). 'Rethinking service recovery strategies: the effect of rapport on consumer responses to service failure', *Journal of Service Research*, 6(2), 193–207. DOI: 10.1177/1094670503257048

Dickinson, J. E., Ghali, K., Cherrett, T., Speed, C., Davies, N., and Norgate, S. (2014). 'Tourism and the smartphone app: capabilities, emerging practice and scope in the travel domain', *Current Issues in Tourism*, 17(1): 84–101. DOI: 10.1080/13683500.2012.718323

Forrester. (2017, June). *Improving CX through business discipline drives growth*. Retrieved from www.forrester.com/report/Improving+CX+Through+Business+Discipline+Drives+Growth/-/E-RES137906 [Accessed: February 27, 2020].

Giraldi, L., Mengoni, M., and Bevilacqua, M. (2016). 'How to enhance customer experience in retail: investigations through a case study', Proceedings of the 23rd ISPE Inc. International Conference on Transdisciplinary Engineering, Vols 3-7, October, pp. 381–390.

Grewal, D., Roggeveen, A. L., Sisodia, R., and Nordfält, J. (2017). 'Enhancing customer engagement through consciousness', *Journal of Retailing*, 93(1): 55–64. DOI: 10.1016/j.jretai.2016.12.001

Huang, W. H. (2008). 'The impact of other-customer failure on service satisfaction', *International Journal of Service Industry Management*, 19: 521–536. DOI: 10.1108/09564230810891941

Kim, K., and Baker, M. A. (2017a). 'The impacts of service provider name, ethnicity, and menu information on perceived authenticity and behaviors', *Cornell Hospitality Quarterly*, 58(3): 312–318. DOI: 10.1177/1938965516686107

Kim, K., and Baker, M. A. (2017b). Koc, E. (Ed.) 'The influence of other customers in service failure and recovery'. In *Service Failures and Recovery in Tourism and Hospitality: A Practical Manual*, Wallingford: CABI (pp. 122).

Kim, K., and Baker, M. A. (2019). 'How the employee looks and looks at you: building customer–employee rapport', *Journal of Hospitality and Tourism Research*, 43(1): 20–40. DOI: 10.1177/1096348017731130

Klink, R. R., Zhang, J. Q., and Athaide, G. A. (2018). 'Designing a customer experience management course', *Journal of Marketing Education*. DOI: 10.1177/0273475318818873

Lemon, K. N., and Verhoef, P. C. (2016). 'Understanding customer experience throughout the customer journey', *Journal of Marketing*, 80(6): 69–96. DOI: 10.1509/jm.15.0420

Lovelock, C., and Wirtz, J. (2011). *Services Marketing: People, Technology, Strategy*, 7th ed., Upper Saddle River, NJ: Prentice Hall.

McColl-Kennedy, J., Cheung, L., and Ferrier, E. (2014). 'Co-creating service experience practices', *Journal of Service Management*, 26: 249–275. DOI: 10.1108/JOSM-08-2014-0204

McColl-Kennedy, J. R., Zaki, M., Lemon, K. N., Urmetzer, F., and Neely, A. (2018). 'Gaining customer experience insights that matter', *Journal of Service Research*, 22(1): 8–26. DOI: 10.1177/1094670518812182

Mody, M., Suess, C., and Lehto, X. (2019). 'Going back to its roots: can hospitableness provide hotels competitive advantage over the sharing economy?', *International Journal of Hospitality Management*, 76: 286–298. DOI: 10.1016/j.ijhm.2018.05.017

O'Dell, T. (2005). 'Experiencescapes: blurring borders and testing connections'. In O'Dell, T. and Billing, P. (Eds.). *Experiencescapes: Tourism, Culture and Economy*. Copenhagen: Copenhagen Business School Press DK (pp. 11–33).

Parise, S., Guinan, P. J., and Kafka, R. (2016). 'Solving the crisis of immediacy: how digital technology can transform the customer experience', *Business Horizons*, 59(4): 411–420. DOI: 10.1016/j.bushor.2016.03.004

Schmitt, B. H. (1999). *Experiential Marketing*. New York: The Free Press.

Stankov, U., Filimonau, V., and Slivar, I. (2019). 'Calm ICT design in hotels: a critical review of applications and implications', *International Journal of Hospitality Management*, 82: 298–307. DOI: 10.1016/j.ijhm.2018.10.012

Tombs, A., and McColl-Kennedy, J. R. (2003). 'Social-servicescape conceptual model', *Marketing Theory*, 3: 447–475. DOI: 10.1177/1470593103040785

Verhoef, P. C., Lemon, K. N., Parasuraman, A., Roggeveen, A., Tsiros, M., and Schlesinger, L. A. (2009). 'Customer experience creation: determinants, dynamics and management strategies', *Journal of Retailing*, 85(1): 31–41. DOI: 10.1016/j.jretai.2008.11.001

Wang, C. Y., and Mattila, A. S. (2013). 'The impact of servicescape cues on consumer prepurchase authenticity assessment and patronage intentions to ethnic restaurants', *Journal of Hospitality and Tourism Research*, 23(1): 1–27. DOI: 10.1177/1096348013491600

14

Exploring image, perception and motivation in tourism experience

Brian Kee Mun Wong and Christy Yen Nee Ng

Introduction

Tourism, an essential industry that contributes about 10% of the world's GDP, is providing one in every ten jobs and accounting for about 30% of the global services trade (Rifai 2017). Despite being a relatively young sector among the global economic industries, the tourism sector plays a role that connects people, culture, society and economy in many nations. In fact, to some nations, tourism is the lifeline of the people. Among the key sustainable factors of the tourism business, either locally or internationally, is the provision of a memorable and positive tourism experience (Kim and Ritchie 2014). In fact, the experience economy is portrayed as the next economic wave after agriculture, industry and services (Pine and Gilmore 1998).

Tourism experience forms through a set of activities in which tourists engage at a tourism destination, allowing them to build their own travel experiences that may or may not fulfill their travel needs and purposes. It is an accumulation of tangible and intangible elements through a process of co-creation, summarized in the mind of a tourist who has a connection on an emotional, intellectual, physical, or spiritual level (Pine and Gilmore 1998; Prahalad and Ramaswamy 2004). It is indeed a rather complex psychological process that includes live experiences (Simon 2015), self-discovery and/or self-transformation (Saunders, Laing and Weiler 2013; Tung and Ritchie 2011), a review of an individual's beliefs and habits (Henning 2012), identity-building (Jaurand 2015), and perhaps a search for better well-being (Panchal 2013). In fact, the multi-faceted tourism experience requires a holistic approach to understand it as it covers a wide range of 'interconnected processes and dynamics involving anticipation, moving to the site, on-site experience, moving back, and post-travel recollection' (Park and Santos 2017: 16).

Tourism provides 'a free area, a mental and physical escape from the immediacy of the multiplicity of impinging pressures in technological society and, as such, holiday experiences provide a scope for the nurturance and cultivation of human identity' (Wearing and Neil 2004: 280). Thus, the importance of tourism experience goes beyond just the benefit of the individual tourist but also extends to tourism businesses and society at large.

In understanding tourists' experiences, academicians and practitioners may discover more of their motivations, behavior and decision-making processes. While there are several methods to understanding tourism experience, this chapter explores destination image, tourist perception and tourist motivation in relation to tourism experience, covering both the demand and supply perspectives.

Destination image

Product or service image is often a major highlight of marketing studies, specifically in the consumer behavior field, as it provides a cue to consumers' choice selection (Baloglu and McCleary 1999). In the context of tourism, destination image plays a similarly essential role in decision making (Kim, Hallab and Kim 2012). The importance of destination image escalates when the competition among destinations is becoming stiffer in this age of globalization. Thus, the role of destination marketers to enhance their destination's competitiveness by creating a unique image is more critical than ever before (Phau, Quintal and Shanka 2014).

Destination image forms through the accumulated impressions, expectations, emotional thoughts, beliefs, knowledge and prejudices of the destination's attributes (Fakeye and Crompton 1991; Kim 2014). In other words, destination image is a multidimensional concept comprising both symbolic and tangible features (Echtner and Ritchie 1993). Despite much focus having been given to the construct in the past decades, a universal definition has yet to emerge.

Destination image was first investigated in a scholarly fashion by Hunt (1975) to study the impression of four Rocky Mountain States' (i.e. Colorado, Montana, Utah, Wyoming) image among the non-resident visitors to the four states, from five preselected geographical subdivisions of the United States. The study attempted to gauge the respondents' impressions of the land, climate, resident population and recreational character of the afore-mentioned states. The respondents also provided feedback on destination attributes such as national parks, national forests, cities, sightseeing, camping, hunting, winter skiing and fishing. The findings concluded that destination image plays an essential role in attracting potential visitors to a touristic region. 'All places have images – good, bad and indifferent – that must be identified and either changed or exploited' (Hunt 1975: 6).

Further research has been done by Fakeye and Crompton (1991), who have differentiated between the destination images of prospective, first time and repeat visitors to the Lower Rio Grande Valley in the United States. The study revealed five factors of destination image: social opportunities and attractions, natural and cultural amenities, accommodations and transportation, infrastructure, foods and friendly people, physical amenities and recreation activities, and bars and evening entertainment. Fakeye and Crompton (1991) conceptualize destination image as three phases of evolution, as follows:

Phase 1: organic image

Prospective tourist has yet to be exposed to any destination promotional messages

Phase 2: induced image

Prospective tourist shows desire to travel and actively search for information as a reaction to the destination promotional messages

Phase 3: complex image

Prospective tourist eventually becomes an experienced tourist at the destination

Destination image is a multidimensional concept. Based on the review of destination image studies from 1975 to 1990, Echtner and Ritchie (1991), Echtner and Ritchie (1993) proposed a three dimensions framework. The first dimension concerns the nature of human information processing while the destination images in this dimension fall within a continuum of attributes–holistic imagery. The second dimension focuses on perceptions and impressions where the images range from functional characteristics (directly observable) to psychological characteristics (less tangible or more difficult to observe). The third dimension provides images that lie in the continuum of common–unique.

The earlier studies clearly indicated that both tourists (demand side) and tourism providers (supply side) play their part in co-creating destination image. Baloglu and McCleary (1999) strengthened the understanding of destination image building among tourists by proposing the forces of stimulus factors (e.g. information sources, previous experience, distribution) and personal factors (e.g. psychological, social). Destination image is measured by perceptual/ cognitive, affective and global elements in the proposed model.

Further studies on destination attributes (i.e. supply side) reveal more variables that represent the mental knowledge, feelings and overall perception of a tourist destination (Chen and Tsai 2007). These include natural and cultural resources, general, tourist and leisure infrastructures, atmosphere, social setting and environment, sun and sand, knowledge, relaxation, entertainment, prestige, destination brand, nature and culture, etc. (Beerli and Martín 2004; Chen and Tsai 2007). Recently, Kim (2014) proposed a ten-dimensional construct that forms tourism experience. These dimensions include infrastructure, accessibility, local culture/history, physiography, activities and events, destination management, quality of service, hospitality, place attachment and superstructure.

A positive destination image leads to higher visit intention (Chen and Tsai 2007; Phau, Quintal and Shanka 2014), particularly the cognitive and affective images (Baloglu and McCleary 1999; Beerli and Martín 2004; Jeong et al. 2009) as compared to the destination physical attributes (Guzman et al. 2012; Martin and Rodriguez del Bosque 2008). The cognitive destination image is based on rational thinking and covers beliefs, impressions and destination attributes. On the other hand, the affective image is derived from the tourist's feelings and emotions towards the tourist destination (Beerli and Martín 2004; Martin and Rodriguez del Bosque 2008).

Some researchers described destination image by the length of the visit that forms the original image initially and the re-evaluation image later (Selby and Nigel 1996). Tourism destination image can also be distinguished via tourists' perceived image and the destination's projected image (Grosspietsch 2006) that sets the tourism experience. Thus, the tourist's perceived image is cognition and impression-based while the latter is defined by the strategies and messages the tourism providers intend to convey to the target tourist segment(s). Tourist perception in relation to tourism experience will be discussed next.

Perceptions of tourism destination

Perception is the method by which people choose, organize and interpret stimuli or inducement into an expressive image of the world (Schiffman and Wisenblit 2015). Tourist perception towards destination image is a crucial element when accessing a destination. Thus, perception in relation to tourism destination is often being used to examine tourism impacts, tourist decision-making and tourism experience. The understanding of tourist perception and tourist experiences allows researchers and practitioners to comprehend the setting on which the tourism industry operates and the involvement of tourists in co-developing the industry together with the tourism providers and policy makers. In general, perception is a process of obtaining and processing information cognitively. Numerous theories have been developed by scholars to explain the perception phenomena. Démuth (2013) categorized perception theories into two groups: bottom-up and top-down. The two groups show an inverse direction of information flow.

The bottom-up theories indicate that an individual acquires and processes data from the lowest sensory level, furthest away from the cognitive levels, but gradually escalates to a more complex and intricate level that happens in the higher-order thinking process (Démuth 2013). When tourists perceive a destination, their sensors start to gather fundamental information (e.g. people, customs, buildings, incidents, etc.) as the individual components of the destination. The pieces of information are then linked together to establish more complex thoughts, which eventually forms the higher-order perception (e.g. safe, dangerous, fun, adventurous, etc.) of the destination. This data-driven processing of perception relies on cues obtained from one or more sources such as news, word of mouth, reports, advertising, promotions, etc. The stimuli may directly influence human perceptions and the individual will correspond to them accordingly. Gibson (1966)'s theory is an example of bottom-up perception formation.

Vice versa, the top-down theories work in the opposite direction. Instead of building perception from the basic individual information of an object, the perception process starts from higher cognitive functions. 'The participation of higher cognitive functions in the process of perception in the form of support of discrimination and interpretation of perceived contents' (Démuth 2013: 30). Thus, the stimulus an individual is exposed to will be processed and organized by the higher cognitive contents that build up upon past experience(s) and/or knowledge. Démuth (2013) further categorized the top-down theories into constructivist (e.g. Gregory's theory (1970)), computational (e.g. Marr's model of perception (1982)) and synthesizing (e.g. Neisser's analysis-by-synthesis model (1967)).

Generally, perception is like image formation. Thus, tourist perceptions are essential in understanding tourism destination image. While the earlier perception studies proposed the bottom-up and top-down approaches, the understanding of perception has been further expanded in the tourism field by means of organic image formation and induced image formation (Pike 2018). Organic or natural image formation resembles the bottom-up theories' approach as tourists are exposed to the surrounding stimuli, gather them and create meaning when they assemble the information together. When new information is received, the perception formed through the previous travel experience may influence the interpretation of the new insights, thus forming another level of perception.

On the other hand, induced or forced image formation is the result of external stimulation by destination marketers, where potential tourists are introduced to persuasive and informative stimuli. In this digital era, tourists are bombarded with substantial marketing communications, and at times can be overloaded. As they are becoming better informed and are making more complex and sophisticated decisions, overselling marketing contents may cause a negative perception among the tourists instead.

In line with that, tourist perception plays an important role for destination marketers as it determines if the perception fits the tourism products or services, and if the tourists even notice the tourism products or services (Solomon 2017). As tourists make decisions and take action based on what they understand to be reality, it is vital for marketers to recognize the idea of perception and its connected impressions, in order to be able to define certain factors that will influence the consumer to respond positively (Schiffman and Wisenblit 2015). As Kaplan and Kaplan (1989) proposed in their model of landscape preference, tourists prefer landscapes that offer a promise of tourist involvement and tourism destinations that will satisfy their needs in the future (Kaplan and Kaplan 1989). The preferences are natural reactions that enhance the perceptual process. In fact, when a destination is able to provide sufficient sense-making and exploratory attributes, tourists are reported to experience positive and holistic perception with a higher level of psychological well-being (Cracolici and Nijkamp 2008; Reitsamer and Brunner-Sperdin 2015).

As there is no universal model for tourist behavior or the destination-selection process (Yoo, Yoon and Park 2018), tourism perception may also be viewed from the internal and external inducement perspective. The perception forms through the internal inducement factors that come from personal demographics and experiences. This includes personal characteristics, country of origin, attitudes, emotions and personality. Emotions or moods can be characterized as a psychological state of arrangement that emerges from the cognitive evaluations of feelings (Bagozzi, Gopinath and Nyer 1999). Emotions are resilient, moderately uncontrolled feelings that influence personal behavior (Hawkins and Mothersbaugh 2010). In fact, studies have suggested that positive tourist emotions lead to positive tourism perceptions (Yang, Gu and Cen 2011; Yüksel and Yüksel 2007).

On the other hand, perception that develops from external inducement factors creates a tourist's expectations of the destination. In most cases, these factors comprise destination attributes such as attraction sites, environment, accommodation, facilities, sun, sand and sea, food, culture, people, commodities, price, etc. (Choi, Law and Heo 2016; Dortyol, Varinli and Kitapci 2014; Hui, Wan and Ho 2007; Kansal, Walia and Goel 2015). Besides these usual destination attributes, the tourism destination marketing communication may also be steered by the destination's themes, events, culture and seasons. The influence of both traditional and digital media plays an essential role in creating these external stimuli that drive destination perception formation among the targeted tourists (Chheang 2011; Šantić, Solic and Vrdoljak 2016).

Cultural diversity also plays its part in influencing tourism destination perception (Hui, Wan and Ho 2007; Kongprasert and Virutamasen 2015). Tourist perceptions can be described in terms of the connection between semantic words and cultural identity. The connection can be grouped into four components: traditional, antique, relax and unique (Kongprasert and Virutamasen 2015). While the traditional component describes tourists' perception of the destination's traditions (i.e. traditional self-defense, market, people), the antique element is associated with tourists' perception of the destination's educative value (i.e. cultural beliefs, cultural performances, architecture). The relax element focuses on the tourist's perception of the destination's ambiance and healthcare availability, while the unique element identifies the integration of traditional and antique, thus known as the unique cultural identity of the destination.

Tourists from different nationalities and with different experiences may perceive the same destination differently. The differences are much influenced by a set of tourist elements (e.g. their profile, anticipation level, expectation, motivation, emotions, etc.) and destination elements (e.g. attitude, offerings, deliveries, etc.). Thus, all stakeholders within the destination's tourism ecosystem need to take the diversity factors (e.g. social, cultural, belief, etc.) of the tourists into consideration when understanding their perception. Competition among tourism destinations is not only on the tourism attributes (e.g. environment, attractions, transportation, hospitality, etc.) but also the experience the tourists encounter at the destination (Buhalis 2000). The holistic experience creates a unified perception of the destination.

Tourism providers need to provide good value and service quality as both assist in creating sustainable customer relationships, thus developing a positive tourist perception (Ariza-Montes et al. 2017; Demirci Orel and Kara 2014). As the tourists interact with the tourism providers, the tourists' value systems, the providers' delivery systems and the actual destination's conditions co-create tourism perception. In other words, space and value systems are the fundamental features shaping the tourist perception. Another key internal factor that affects tourist perception is motivation. Thus, the next section discusses tourist motivation.

Tourist motivation

Need is the key driver in understanding human motivation. It is the inner state of an individual that drives and directs human motivation and behavior (Maslow 1943; Murray 1964) towards self-improvement, pleasure-seeking and spiritual growth (Gilmore 2002). In fact, it does not just satisfy the psychological aspect but also physiological needs (Berkman, Lindquist and Sirgy 1997). Motivation is a crucial element of understanding tourist behavior as it provides insights into tourists' expectations, objectives and requirements when they are involved in tourism activities. Aside from tourists' needs, motivation is also closely linked to the usual way of life of tourists and their personalities when they are forming their expectations (Asbollah et al. 2017) of tourism experience. It is worth noting that while motivation is essential to the understanding of tourist perceptions, it is equally important in assisting researchers and practitioners to understand tourist behavior. In fact, with a better understanding of tourists' motives, destination marketers can tailor their offerings to best facilitate tourist experience and satisfy tourists' needs and wants (Augustyn and Ho 1998).

Inherently, it is rather difficult to understand motivation, particularly for tourists' travel experiences (Pike 2018), as it cannot be observed (Iso-Ahola 1999). In recent tourism studies, it is suggested that besides 'need' as the main driver of motivation, tourists are motivated to take an action due to curiosity and the urge for new experiences and knowledge in an unknown situation (Podoshen 2012; Sharpley and Stone 2009). Thus, Wong and Musa (2014: 143) define motivation as 'a set of psychological and physiological needs that are triggered by curiosity and the urge to gain new experiences and knowledge in a less familiar destination' in their study on retirement tourism. The continuous interest on the motivation subject among tourism scholars has seen an evolution of motivation theories over the years.

Maslow's Hierarchy of Needs (1943) is a widely adopted motivation theory in the social science field. The theory categorizes human needs into five levels, ascending from the most basic physiological needs, to safety, social, self-esteem and lastly self-actualization. While Maslow indicated that the needs are satisfied in stages, the notion has been challenged by many scholars over the years. In fact, the theory has been argued for its non-testability and insufficient empirical evidence (Dye, Mills and Weatherbee 2005), especially in tourism studies (Crompton and McKay 1997). Despite the criticism, many tourism researchers adopted the fundamentals of Maslow's Hierarchy of Needs to further propose new theories within the tourism field. Among the commonly cited ones is Pearce (1988)'s Travel Career Ladder (TCL). As in the Hierarchy of Needs, TCL provides an overview of travelers' motivations through five levels of need: physiological, safety or security, relationship, self-esteem or development and fulfillment. As in Maslow's theory, TCL suggests that lower-level needs must be satisfied before moving on to the next level of need. Recently, Wong and Musa (2014) have provided a different perspective from TCL – the motivations among retirement tourists found to impinge on the various levels of the TCL model and the fact that retirement tourists do not necessarily need to follow the sequence of the hierarchy to be motivated into retiring overseas (refer Figure 14.1). Ryan (1998) and Kim, Pearce, Morrison, and O'Leary (1996) have previously warned about the usage of 'ladder' in the theory, as it can be misleading.

As opposed to Maslow's Hierarchy of Needs, Vroom (1964) proposed the Expectancy theory, which focuses on the outcomes instead. Humans are motivated to do things to reach a specific goal if the individual believes in the worth of the goal and the practicality of doing the things can be visualized. Deci (1975) and Deci and Ryan (1987) have further developed and refined the theory for tourism studies. Motives can also be viewed along a disequilibrium continuum of cultural-socio-psychological aspects (Crompton 1979). Crompton suggested that travelers lie within any point of the continuum prior to traveling, creating certain unmet needs along the way. The unmet

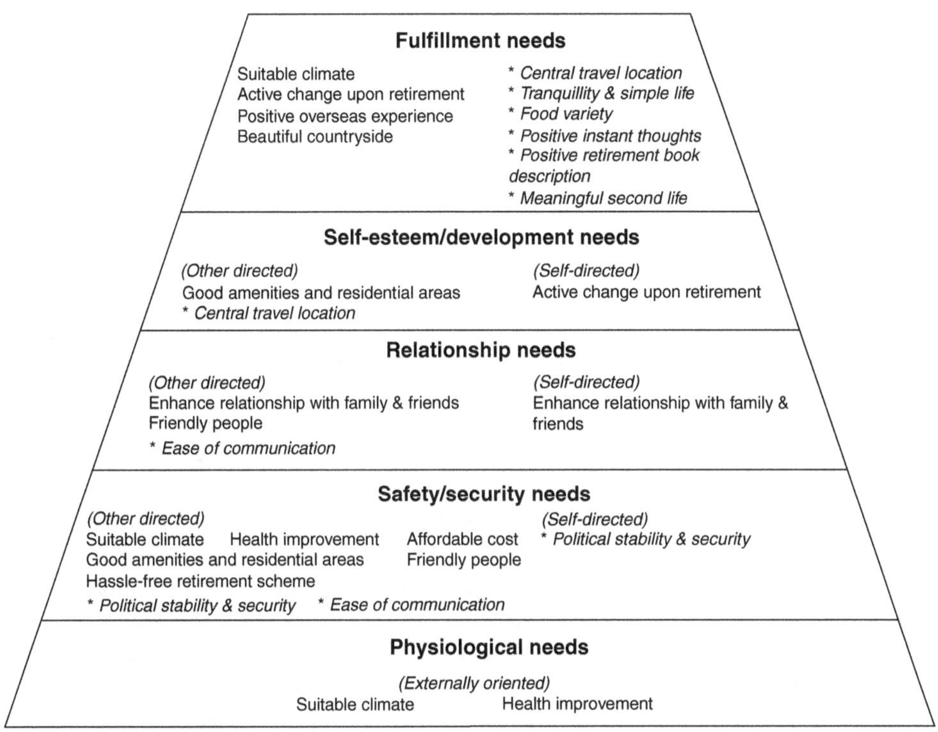

Figure 14.1 Classifications of MM2H Retirees' Travel Motivation Themes in TCL

Source: Wong and Musa 2014: 151

needs motivate them to travel and equilibrium of the needs may be achieved upon traveling. The seven socio-psychological motives for travel in Crompton's theory include 'Escape from a perceived mundane environment', 'Exploration and evaluation', 'Relaxation', 'Prestige', 'Regression', 'Enhancement of kinship relationships', and 'Facilitation of social interaction'.

Based on Tolman's work (1959), Dann (1977) proposed the concept of pull-push motivation in tourism studies. The theory is widely adopted by tourism scholars as it has an important role in comprehending the time and location of an individual's traveling. Push factors are internal desires that drive traveling behavior, which thus explains 'why' travelers travel. On the other hand, pull factors are external forces, mostly destination attributes that attract tourists to the area. While many attempts to explain the relationship between these two factors can be observed, it seems that the convolution of the theory made it difficult to make it clear and detailed. Another motivation theory that is closely linked to the push-pull factor is Iso-Ahola (1982)'s escaping/seeking theory. Escaping refers to the drive of a traveler to leave his or her original location while seeking describes the drive to seek intrinsic rewards by traveling to a different location.

Tourists' motivation sets their expectation on a tourism destination. Thus, travel motivation affects the tourist's experience traveling to and within the tourism destination. In fact, the recent gastronomic tourism study by Agyeiwaah et al. (2019) reported a strong direct impact of motivation on experience. As tourism experience may refer to past or present experience, the former may establish the next level of expectation which eventually may affect the latter. However, some researchers found that tourists' experiences may be just a function of the actual performance, regardless of their expectations (Tse and Wilton 1988). In fact, most satisfactory experiences are not built on expectations (Arnould and Price 1993). Unique tourism experience is context-driven (Björk and Kauppinen-Räisänen 2016) and enhances the level of engagement among tourists (Prebensen et al. 2013). Therefore, it is worth recognizing tourist motivation as an essential element of establishing positive and memorable tourism experience.

Case study 14.1

The aging population is a global phenomenon and it is giving rise to a new breed of tourism product, 'retirement tourism'. For those who intend to retire overseas on a social visa, terminologies such as international retirement home schemes, second home and long stay tourism are also commonly used. In Malaysia, the 'Malaysia My Second Home' (MM2H) programme is the brainchild of the Malaysian government to invite high net worth overseas retirees to spend their retirement years in Malaysia. The programme grants retirees a ten-year multiple-entry social visit pass when making Malaysia their second home retirement destination. However, as the programme has no minimum age requirement, the programme is literally open to any foreigners who fulfill certain criteria laid down by the Ministry of Tourism and Culture Malaysia.

Since the inception of the programme in 1996 (it was known as the 'Silver Hair' programme back then), enrolment in the programme has been a dream come true for many western retirees, as Malaysia provides the right weather, affordable cost and a widely-spoken English environment, enabling them to spend most of their retirement life overseas. Malaysia has a relatively good and positive destination image among foreign tourists. In the past few years, Malaysia has been consistently ranked among the Top 10 in the Annual Global Retirement Index undertaken by *International*

Living magazine and has received multiple international accolades such as Health and Medical Tourism: Destination of the Year Award, World's 5th Best Shopping Destination, Most Popular Tourist Destination, Best Halal Travel Destination and others.

With such international recognition and the right fit of destination attributes, overseas retirees are expected to have a positive organic image formation of Malaysia. Coupled with the marketing efforts done by the MM2H Centre, locally and internationally, the target audiences' induced image formation is further strengthened, creating a stronger perception of the nation as an attractive second home destination. To understand the perception formation further, Wong, Musa and Taha (2017) researched into the MM2H participants' motivations to retire in Malaysia. The study discovered five push motivators (i.e. Prior Overseas Experiences, Overseas Retirement Dream, Unfavourable Political Stability and Security, Escapism, and Health Improvement) and seven pull motivators (i.e. Amenities and Facilities, Leisure Lifestyle, Being Active, Cost and Economics, Conducive Environment, People and Communication and Socialization). While both push and pull motivations have a significant interaction with the overall tourism experience among the MM2H participants in Malaysia, it is the push motivators that portrayed a stronger influence.

Based on the published statistics, a total of 42,271 approved participants were recorded from 2002–2018 (Ministry of Tourism and Culture Malaysia 2014). As shown in Figure 14.2, most of the participants are from China, Japan, Bangladesh, the United Kingdom of Great Britain and Northern Ireland and South Korea. Even though Asian participants showed a growth until 2013, the growth among western participants is a concern. In 2018, the total number of approved applications has sharply reduced by 46%. Thus, it is worth noting that the tourism experience garnered by some of the participants may not necessarily be encouraging, and attention from the MM2H Centre is recommended.

Among the challenges encountered by the MM2H participants while spending time in Malaysia are MM2H operational issues, inconvenient renewal office location, personal safety and security, religious problems and daily life inconveniences (Wong and Musa 2015). The sustainable growth of a tourism product relies on tourism experience. The negative experiences should raise concerns among the relevant organizations and authorities, who should take immediate remedial action and work towards continuous improvement.

Conclusion

Tourism experience sets the sustainable scene of a tourism business. The ultimate experience is not just an outcome of a tourism provider's performance but a co-creation effort together with the environment and the tourists themselves. Although the perception of an experience is a complex process, destination image, perception of tourism destination and tourist motivation play an essential role in influencing tourism experience. Despite the complicated mechanism of perception building and the driving force behind the tourist's actions, it is crucial for tourism providers to use these insights effectively in order to complement the tourist's behavior. Several perception and motivation theories from previous studies have been discussed in order to provide an overview of the constructs in relation to tourism experience. Destination marketers need to identify and develop attractive tourism products and services to cater for different types of tourists who possess different perceptions and motivations. While the destination marketers put much effort in creating a desired destination image, it is the tourism experience that defines the image ultimately.

NO.	COUNTRY OF NATIONALITY	YEAR																	TOTAL (2002–2018)	SHARE (%)
		2002	2003	2004	2005	2006	2007	2008	2009	2010	2011	2012	2013	2014	2015	2016	2017	2018		
1	*People's Republic of China*	241	521	468	502	242	90	120	114	154	405	731	1.337	1.307	719	1.512	2.923	1.495	12,881	30.5
2	*Japan*	49	99	42	87	157	198	210	169	195	423	816	739	428	300	281	352	233	4.778	11.3
3	*People's Republic of Bangladesh*	0	32	204	852	341	149	68	86	74	276	388	285	250	205	283	451	191	4.135	9.8
4	*United Kingdom of Great Britain and Northern Ireland*	108	159	210	199	209	240	208	162	141	153	139	148	117	83	110	200	105	2.691	6.4
5	*Republic of Korea*	5	12	66	60	65	151	86	54	49	64	85	98	137	120	184	693	449	2.378	5.6
6	*Republic of Singapore*	96	143	91	62	94	58	48	61	73	78	83	145	94	67	93	117	56	1.459	3.5
7	*Islamic Republic of Iran*	0	2	8	7	9	59	227	212	227	286	201	51	17	19	8	43	23	1.399	3.3
8	*Taiwan*	38	95	140	186	63	31	16	36	49	70	85	151	83	71	77	134	71	1.396	3.3
9	*Hong Kong Special Administrative Region, China*	17	41	86	56	47	10	6	14	2	15	35	62	44	42	78	339	193	1.087	2.6
10	*Republic of India*	45	123	118	80	51	46	32	35	51	50	56	41	42	46	68	100	63	1.047	2.5
11	*Others*	219	418	484	524	451	471	491	635	484	567	608	618	555	539	653	843	460	9,020	21.3
	TOTAL	818	1.645	1.917	2.615	1.729	1.503	1.512	1.578	1.499	2.387	3.227	3.675	3.074	2.211	3.347	6.195	3.339	42.271	100.0

Figure 14.2 Top 10 Participating Countries in the MM2H Programme (2002–2018)

References

Agyeiwaah, E., Otoo, F. E., Suntikul, W., and Huang, W.-J. (2019). 'Understanding culinary tourist motivation, experience, satisfaction, and loyalty using a structural approach', *Journal of Travel and Tourism Marketing*, 36(3): 295–313. DOI: 10.1080/10548408.2018.1541775

Ariza-Montes, J. A., Arjona-Fuentes, J. M., Han, H., and Law, R. (2017). 'Employee responsibility and basic human values in the hospitality sector', *International Journal of Hospitality Management*, 62: 78–87. DOI: 10.1016/j.ijhm.2016.12.001

Arnould, E. J. and Price, L. L. (1993). 'River magic: Extraordinary experience and the extended service encounter', *Journal of Consumer Research*, 20(1): 24–45. DOI: 10.1086/209331

Asbollah, A. Z., Hassan, N., Muhammad Yusoff, Y., and Idris, H. (2017). 'The tourist behaviour in different environments: A literature review', *Planning Malaysia: Journal of the Malaysian Institute of Planners*, 15(1): 279–288. DOI: 10.21837/pmjournal.v15.i1.244

Augustyn, M. and Ho, S. K. (1998). 'Service quality and tourism', *Journal of Travel Research*, 37(1): 71–75. DOI: 10.1177/004728759803700110

Bagozzi, R. P., Gopinath, M., and Nyer, P. U. (1999). 'The role of emotions in marketing', *Journal of the Academy of Marketing Science*, 27(2): 184–206. DOI: 10.1177/0092070399272005

Baloglu, S. and McCleary, K. W. (1999). 'A model of destination image formation', *Annals of Tourism Research*, 26(4): 868–897. DOI: 10.1016/S0160-7383(99)00030-4

Beerli, A. and Martín, J. D. (2004). 'Tourists' characteristics and the perceived image of tourist destination: A quantitative analysis – A case study of Lanzarote, Spain', *Tourism Management*, 25(5): 623–636. DOI: 10.1016/j.tourman.2003.06.004

Berkman, H., Lindquist, J., and Sirgy, M. J. (1997). *Consumer Behavior: Concepts and Marketing Strategy*. Lincolnwood: NTC Business Books.

Björk, P. and Kauppinen-Räisänen, H. (2016). 'Exploring the multi-dimensionality of travellers'culinary-gastronomic experiences', *Current Issues in Tourism*, 19(12): 1260–1280. DOI: 10.1080/13683500.2013.868412

Buhalis, D. (2000). 'Marketing the competitive destination of the future', *Tourism Management*, 21(1): 97–116. DOI: 10.1016/S0261-5177(99)00095-3

Chen, C. F. and Tsai, D. G. (2007). 'How destination image and evaluative factors affect behavioral intentions', *Tourism Management*, 28(4): 1115–1122. DOI: 10.1016/j.tourman.2006.07.007

Chheang, V. (2011). 'Angkor heritage tourism and tourist perceptions', *Tourismos*, 6(2): 213–240. Available from: www.chios.aegean.gr/tourism/VOLUME_6_No2_art13.pdf

Choi, M., Law, R., and Heo, C. Y. (2016). 'Shopping destinations and trust – Tourist attitudes: Scale development and validation', *Tourism Management*, 54(C): 490–501. DOI: 10.1016/j.tourman.2016.01.005

Cracolici, M. F. and Nijkamp, P. (2008). 'The attractiveness and competitiveness of tourist destinations: A study of southern Italian regions', *Tourism Management*, 30(3): 336–344. DOI: 10.1016/j.tourman.2008.07.006

Crompton, J. L. (1979). 'Motivations for pleasure vacations', *Annals of Tourism Research*, 6(4): 408–424. DOI: 10.1016/0160-7383(79)90004-5

Crompton, J. L. and McKay, S. L. (1997). 'Motives of visitors attending festival events', *Annals of Tourism Research*, 24(2): 425–439. DOI: 10.1016/S0160-7383(97)80010-2

Dann, G. (1977). 'Anomie, ego-enhancement and tourism', *Annals of Tourism Research*, 4(4): 184–194. DOI: 10.1016/0160-7383(77)90037-8

Deci, E. L. (1975). *Intrinsic Motivation*. New York: Plenum. DOI: 10.1007/978-1-4613-4446-9

Deci, E. L. and Ryan, R. M. (1987). 'The support of autonomy and the control of behavior', *Journal of Personality and Social Psychology*, 53(6): 1024–1037. DOI: 10.1037//0022-3514.53.6.1024

Demirci Orel, F. and Kara, A. (2014). 'Supermarket self-checkout service quality, customer satisfaction, and loyalty: Empirical evidence from an emerging market', *Journal of Retailing and Consumer Services*, 21 (2): 118–129. DOI: 10.1016/j.jretconser.2013.07.002

Démuth, A. (2013). *Perception Theories*. Krakow: fftu. Available from: www.academia.edu/30094149/Perception_Theories

Dortyol, I. T., Varinli, I., and Kitapci, O. (2014). 'How do international tourists perceive hotel quality and exploratory study of service quality in Antalya tourism region', *International Journal of Contemporary Hospitality Management*, 26(3): 470–495. DOI: 10.1108/IJCHM-11-2012-0211

Dye, K., Mills, A. J., and Weatherbee, T. (2005). 'Maslow: Man interrupted: Reading management theory in context', *Management Decision*, 43(10): 1375–1395. DOI: 10.1108/00251740510634921

Echtner, C. M. and Ritchie, J. R. B. (1991). 'The meaning and measurement of destination image', *The Journal of Tourism Studies*, 2(2): 2–12. Available from: https://citeseerx.ist.psu.edu/viewdoc/download?doi=10.1.1.89.3276&rep=rep1&type=pdf

Echtner, C. M. and Ritchie, J. R. B. (1993). 'The measurement of destination image: An empirical assessment', *Journal of Travel Research*, 31(3): 1–13. DOI: 10.1177/004728759303100402

Fakeye, P. C. and Crompton, J. L. (1991). 'Image differences between prospective, first-time and repeat visitors to the lower Rio Grande valley', *Journal of Travel Research*, 30(2): 10–16. DOI: 10.1177/004728759103000202

Gibson, J. J. (1966). *The Senses Considered as Perceptual Systems*. Boston: Houghton Mifflin. DOI: 10.1080/10407413.2017.1297680

Gilmore, F. (2002). 'Branding for success'. In N. Morgan, A. Pritchard, and R. Pride (eds.), *Destination Branding: Creating the Unique Destination Proposition* (pp. 57–65). Oxford: Elsevier.

Grosspietsch, M. (2006). 'Perceived and projected images of Rwanda: Visitor and international tour operator perspectives', *Tourism Management*, 27(2): 225–234. DOI: 10.1016/j.tourman.2004.08.005

Guzman, A. B., Castro, B. V., Calanog, J. F. V., Taguinin, A. J. R., Affalla, J. R. S., Aldover, A. L. R., and Gotangco, M. T. C. (2012). 'The Australian tourists' travel motivation and pre-and post-images of the Philippines as their destination', *Asia-Pacific Journal of Innovation in Hospitality and Tourism*, 1(2): 143–164. Available from: www.cabdirect.org/cabdirect/abstract/20143039445

Hawkins, D. I. and Mothersbaugh, D. L. (2010). *Consumer Behaviour: Building Marketing Strategy* (11th Ed.). New York: McGrawHill.

Henning, G. K. (2012). 'The habit of tourism: Experiences and their ontological meaning'. In R. Sharpley and P. R. Stone (eds.), *Contemporary Tourist Experience: Concepts and Consequence* (pp. 25–37). Oxon: Routledge.

Hui, T. K., Wan, D., and Ho, A. (2007). 'Tourists' satisfaction, recommendation and revisiting Singapore', *Tourism Management*, 28(4): 965–975. DOI: 10.1016/j.tourman.2006.08.008

Hunt, J. D. (1975). 'Image as a factor in tourism development', *Journal of Travel Research*, 13(3): 1–7. DOI: 10.1177/004728757501300301

Iso-Ahola, S. E. (1982). 'Toward a social psychological theory of tourism motivation: A rejoinder', *Annals of Tourism Research*, 9(2): 256–262. DOI: 10.1016/0160-7383(82)90049-4

Iso-Ahola, S. E. (1999). 'Motivational foundations of leisure'. In E. L. Jackson and T. L. Burton (eds.), *Leisure Studies: Prospects for the Twenty-first Century* (pp. 35–51). State Gollege: PA: Venture Publishing.

Jaurand, E. (2015). 'Le nudismegay: Uneexpériencetouristiqueidentitaire'. In J.-M. Decroly (ed.), *Le Tourismecommeexpérience. Regards Interdisciplinaires Sur Le Vécutouristique* (pp. 163–174). Québec: Presses de l'Université du Québec.

Jeong, C., Kim, Y. K., Ko, Y. J., Lee, H., and Jeong, R. S. (2009). 'Horse racing image: Re-examination of relations between image and intention to visit', *Journal of Quality Assurance in Hospitality & Tourism*, 10(3): 194–217. DOI: 10.1080/15280080902988139

Kansal, P., Walia, S., and Goel, S. (2015). 'Factors affecting perception of value for money and customer satisfaction for foreign tourists in Goa', *International Journal of Hospitality and Tourism Systems*, 8(2): 93–102. Available from: www.publishingindia.com/ijhts/24/factors-affecting-perception-of-value-for-money-and-customer-satisfaction-for-foreign-tourists-in-goa/433/3073/

Kaplan, S. and Kaplan, R. (1989). *The Experience of Nature: A Psychological Perspective*. New York: Cambridge University Press. Available at: http://willsull.net/resources/270-Readings/ExpNature1to5.pdf

Kim, J. (2014). 'The antecedents of memorable tourism experience: The development of a scale to measure destination attributes associated with memorable experiences', *Tourism Management*, 44(C): 34–45. DOI: 10.1016/j.tourman.2014.02.007

Kim, J. and Ritchie, J. R. B. (2014). 'Cross-cultural validation of a memorable tourism experience scale (MTES)', *Journal of Travel Research*, 53(3): 323–335. DOI: 10.1177/0047287513496468

Kim, K., Hallab, Z., and Kim, J. N. (2012). 'The moderating effect of travel experience in a destination on the relationship between the destination image and the intention to revisit', *Journal of Hospitality Marketing & Management*, 21(5): 486–505. DOI: 10.1080/19368623.2012.626745

Kim, Y. J., Pearce, P. L., Morrison, A. M., and O'Leary, J. T. (1996). 'Mature versus youth travellers: The Korean market', *Asia Pacific Journal of Tourism Research*, 1(1): 102–112. DOI: 10.1080/10941669708721967

Kongprasert, N. and Virutamasen, P. (2015). 'Tourist perceptions to cultural identity: The case of Thai experience', *Procedia – Social and Behavioural Sciences*, 195: 167–174. DOI: 10.1016/j.sbspro.2015.06.428

Marr, D. (1982). *Vision. A Computational Investigation into the Human Representation and Processing of Visual Information.* New York: W.H. Freeman and Company. Available from: http://s-f-walker.org.uk/pubsebooks/epubs/Marr]_Vision_A_Computational_Investigation.pdf

Martin, H. S. and Rodriguez del Bosque, I. A. (2008). 'Exploring the cognitive-affective nature of destination image and the role of psychological factors in its formation', *Tourism Management*, 29(2): 263–277. DOI: 10.1016/j.tourman.2007.03.012

Maslow, A. H. (1943). 'A theory of human motivation', *Psychological Review*, 50: 370–396. Available from: https://psycnet.apa.org/record/1943-03751-001

Ministry of Tourism and Culture Malaysia (2014). 'Malaysia my second home programme'. *MM2H* [online], Available from: www.mm2h.gov.my/index.php/en/home/programme/statistics. [Accessed: May 6, 2019].

Murray, E. J. (1964). *Motivation and Emotion.* Englewood Cliffs: Prentice Hall.

Neisser, U. (1967). *Cognitive Psychology.* New York: Meredith.

Panchal, J. (2013). 'Tourism, wellness and feeling good: Reviewing and studying Asian spa experiences'. In S. and P. Pearce (eds.), *Tourist Experience and Fulfilment: Insights from Positive Psychology* (pp. 72–88). Oxon: Routledge.

Park, S. and Santos, C. A. (2017). 'Exploring the tourist experience: A sequential approach', *Journal of Travel Research*, 56(1): 16–27. DOI: 10.1177/0047287515624017

Pearce, P. L. (1988). *The Ulysses Factor: Evaluating Visitors in Tourist Settings.* New York: Springer-Verlag.

Phau, I., Quintal, V., and Shanka, T. (2014). 'Examining a consumption values theory approach of young tourists toward destination choice intentions', *International Journal of Culture, Tourism and Hospitality Research*, 8(2): 125–139. DOI: 10.1108/IJCTHR-12-2012-0090

Pike, S. (2018). *Tourism Marketing for Small Businesses.* Oxford: Goodfellow Publishers Limited.

Pine, B. J. and Gilmore, J. (1998).'Welcome to the experience economy'. HBR [online], Available from: https://hbr.org/1998/07/welcome-to-the-experience-economy. [Accessed May 6, 2019].

Podoshen, J. S. (2012). 'Dark tourism motivations: Simulation, emotional contagion and topographic comparison', *Tourism Management*, 35: 263–271. DOI: 10.1016/j.tourman.2012.08.002

Prahalad, C. K. and Ramaswamy, V. (2004). *The Future of Competition: Co-Creating Unique Value with Customers.* Boston: Harvard Business School Press.

Prebensen, N. K., Woo, E., Chen, J. S., and Uysal, M. (2013). 'Motivation and involvement as antecedents of the perceived value of the destination experience', *Journal of Travel Research*, 52(2): 253–264. DOI: 10.1177/0047287512461181

Reitsamer, B. F. and Brunner-Sperdin, A. (2015). 'Tourist destination perception and well-being: What makes a destination attractive?', *Journal of Vacation Marketing*, 23(1): 55–72. DOI: 10.1177/1356766715615914

Rifai, T. (2017). 'Tourism: Growth is not the enemy; it's how we manage it that counts'. UNWTO [online]. Available from: https://www.unwto.org/archive/global/press-release/2017-08-15/tourism-growth-not-enemy-it-s-how-we-manage-it-counts. [Accessed: April 2, 2019].

Ryan, C. (1998). 'The travel career ladder: An appraisal', *Annals of Tourism Research*, 25(1): 936–957. DOI: 10.1016/S0160-7383(98)00044-9

Šantić, D., Solic, M., and Vrdoljak, A. (2016). 'Factors affecting the distribution of two Synechococcus ecotypes in the coastal Adriatic Sea', *Acta Adriatica*, 59(1): 51–60. Available from: http://jadran.izor.hr/acta/pdf/59_1_pdf/59_1_4.pdf

Saunders, R., Laing, J., and Weiler, B. (2013). 'Personal transformation through long-distance walking'. In S. Filep& and P. Pearce (eds.), *Tourist Experience and Fulfilment: Insights from Positive Psychology* (pp. 127–146). Oxon: Routledge.

Schiffman, L. G. and Wisenblit, J. (2015). *Consumer Behavior (11th Ed.).* Boston: Pearson Education.

Selby, M. and Nigel, J. M. (1996). 'Reconstruing place image: A case study of its role in destination market research', *Tourism Management*, 17(4): 287–294. DOI: 10.1016/0261-5177(96)00020-9

Sharpley, R. and Stone, P. R. (2009). '(Re) presenting the macabre: Interpretation, kitschification and authenticity'. In R. Sharpley and P. R. Stone (eds.), *The Darker Side of Travel: The Theory and Practice of Dark Tourism* (pp. 109–128). Tonawanda: Channel View.

Simon, G. (2015). 'L'expérienceurbaine des touristes: Uneapprocherelationnelleavec les objets du monde urbain'. In J.-M. Decroly (ed.), *Le Tourismecommeexpérience. Regards Interdisciplinaires Sur Le Vécutouristique* (pp. 93–102). Québec: Presses de l'Université du Québec.

Solomon, M. R. (2017). *Consumer Behavior: Buying, Having and Being*. London: Pearson Education.

Tolman, E. C. (Ed.). (1959). *Psychology: A Study of A Science*. New York: McGraw Hill.

Tse, D. K. and Wilton, P. C. (1988). 'Models of consumer satisfaction formation: An extension', *Journal of Marketing Research*, 25(2): 204–212. DOI: 10.2307/3172652

Tung, V. W. S. and Ritchie, J. R. B. (2011). 'Investigating the memorable experiences of the senior travel market: An examination of the reminiscence bump', *Journal of Travel & Tourism Marketing*, 28(3): 331–343. DOI: 10.1080/10548408.2011.563168

Vroom, V. H. (1964). *Work and Motivation*. New York: John Wiley.

Wearing, S. and Neil, J. (2004). 'Expanding sustainable tourism's conceptualization: Ecotourism, volunteerism and serious leisure'. In S. Williams (ed.), *Tourism: Critical Concepts in Social Sciences* (pp. 280–300). London: Routledge.

Wong, B. K. M., Musa, G., and Taha, A. Z. (2017). 'Malaysia my second home: The influence of push and pull motivations on satisfaction', *Tourism Management*, 61(C): 394–410. DOI: 10.1016/j.tourman.2017.03.003

Wong, K. M. and Musa, G. (2014). 'Retirement motivation among 'Malaysia my second home' participants', *Tourism Management*, 40: 141–154. DOI: 10.1016/j.tourman.2013.06.002

Wong, K. M. and Musa, G. (2015). 'Challenges of international retirees in second home destination: Phenomenological analysis', *Tourism Management Perspectives*, 15: 81–90. DOI: 10.1016/j.tmp.2015.03.010

Yang, J., Gu, Y. K., and Cen, J. (2011). 'Festival tourists' emotion, perceived value, and behavioral intentions: A test of the moderating effect of festivalscape', *Journal of Convention and Event Tourism*, 12 (1): 25–44. DOI: 10.1080/15470148.2010.551292

Yoo, C.-K., Yoon, D., and Park, E. (2018). 'Tourist motivation: An integral approach to destination choices', *Tourism Review*, 73(2): 169–185. DOI: 10.1108/TR-04-2017-0085

Yüksel, A. and Yüksel, F. (2007). 'Shopping risk perceptions: Effects on tourists' emotions, satisfaction and expressed loyalty intentions', *Tourism Management*, 28(3): 703–713. DOI: 10.1016/j.tourman.2006.04.025

15

Learning through extraordinary tourism experiences

Eva Maria Jernsand and Sandhiya Goolaup

Introduction

Tourist experience is a core concept surrounding the hospitality and tourism literature. In today's society, tourists are increasingly looking for experiences that not only provide them with mere fun and satisfaction but can also engage them intellectually through immersion in activities, time and space (Goolaup 2018). In short, they are in search of experiences that are extraordinary in nature. Extraordinary experiences are typically perceived as being emotionally intense, transformative and yielding feelings of personal triumph and a sense of achievement (Arnould and Price 1993). In the study of extraordinary experiences learning is a core aspect, since it makes the experience more meaningful (e.g. Goolaup and Mossberg 2017; Lindberg and Østergaard 2015). Indeed, various scholars have also identified learning new things and personal enrichment as being a key motivating factor to travel (e.g. Shoemaker 1994; Su, Johnson and O'Mahony 2018).

Despite being a critical element in the study of tourist experience, learning has received scant attention. Research has partly focused on learning within the formal education arena, such as study abroad (e.g. Scarinci and Pearce 2012; Stone and Petrick 2013), evaluating the benefits and impact of educational tours (e.g. Cohen 2014), or holidays as a way to contextualize classroom learning (e.g. Bos, McCabe and Johnson 2015). Other streams of research have emphasized the potential of transformational tourism and how learning shapes tourists' behaviors and actions (Pritchard and Morgan 2013; Reisinger 2013; Weaver and Jin 2016).

However, the tourism literature still provides few answers regarding how and in what situations tourists learn. Historically, learning and educational research has been separated from tourism and leisure research (Falk, Ballantyne, Packer and Benckendorff 2012). Even more importantly, the specificity of extraordinariness in tourism experience has not received sufficient consideration in relation to learning. Therefore, the purpose of this study is to understand how tourists learn through extraordinary experiences. Theoretically, it is encased within learning theory, where knowledge and skills are considered as being obtained in an ongoing process (Gipps 1999) and the learner's beliefs are challenged and experiences are made sense of (Shephard 1992). In this study, it is argued that experiential and transformational learning theories are specifically useful to understand tourists' extraordinary experiences, although the core aspects of these are also found in learning theories on, for instance, education and organization. Two illustrative different cases, an oyster experience on the West Coast of Sweden and a guided tour in Kisumu, Kenya are used to demonstrate how learning occurs for tourists engaging in what are perceived as being extraordinary experiences. The understanding of how learning takes place in extraordinary experiences contributes to practical knowledge on how to compose tours and events that enhance learning and transformation.

Learning within the tourism and hospitality literature

Given the complexity involved in defining the term learning, there is no one definition of what it is. Kalinowski and Weiler (1992), for instance, defined it as a lifelong process, which is mostly incidental and natural, whilst Gursoy and McCleary (2004) portrayed it as a process by which experience leads to changes in knowledge, attitudes and/or behavior. According to Romme and van Witteloostuijn (1999), learning is a prerequisite to being able to handle situations and be prosperous in a changing world. Developing such thought, Reisinger (2015b) perceived learning as being instrumental in developing self-confidence, directing actions and helping people understand their own identity, life values and roles in society. Other scholars define learning through the identification of important components. Falk, Ballantyne, Packer and Benckendorff (2012), for instance, see learning as constituting several key features: (1) a lifelong and life-wide process; (2) both a process and an outcome; (3) being highly individual; (4) a process of constructing meaning; (5) context-dependent; (6) a cumulative process and finally (7) fun.

Theories on everyday life learning point to the ambiguity of learning; being both individual and institutional, both policy and practice, and both changing and permanent (Jarvis 2009). When it comes to intentional processes for learning, the literature on student learning (e.g. Gipps 1999; Mathieson 2014; Terenzini 1998) give further understanding of how learning actually takes place. Prior knowledge and beliefs must be challenged, and active participation, social interaction and individual support enhance learning. Furthermore, learning activities and their outcomes must be meaningful for the learner. According to constructivist learning theory, learning is an "ongoing knowledge construction process" (Gipps 1999:372), and this active sense-making of interventions takes place only if the interventions fit into the learner's existing mental framework (Shephard 1992). The process of learning can also be illustrated by learning loops, as in organizational theory (e.g. Argyris and Schön 1996; Flood and Romm 1996). The framework rests on the principle that learning occurs in loops, where higher loops are considered as leading to higher consciousness and awareness. This is dependent on a conversational process, which in many cases is hard since it can reveal difficult, embarrassing and threatening issues (Argyris, Putnam and McLain-Smith 1985). Jernsand for instance, points to the importance of open feedback in development research and practice, which may reveal "uncomfortable matters", which eventually leads to consciousness, awareness and transformation (2017: 97). Regardless of the way it has been conceptualized, it can be argued that learning can occur in planned, unplanned, formal or informal situations.

A review of the tourist experience literature reveals that there are various theoretical frameworks that have been used to explain the relationship between learning and traveling. Falk, Ballantyne, Packer and Benckendorff (2012) use Aristotle's notion of competencies; phronesis, episteme and techne, and discuss how it leads to both passive and active development. As they further argued, studies within tourism have been shown to include all of Aristotle's framework since travel indeed contributes to various practical skills (techne). For instance, Pearce and Foster's (2007) study on backpackers shows how travel leads to generic skills such as problem solving and communication. Likewise, the development of knowledge (episteme) was seen in the work of Moscardo (1998), who emphasizes the incidental learning of sites and species. Lastly, practical wisdom (phronesis) is shown by Lewin (2009) through development of reflexive traits such as self-awareness. Another theoretical framework of learning through tourism experiences is the innovation path. The tourist is seen as co-creator, co-innovator and co-designer in a mutual learning process (e.g. Hall and Williams 2008; Jernsand, Kraff and Mossberg 2015). Tourists engage in activities and, more or less actively, they co-produce knowledge with employees and other actors, which results in different types of innovation (Jernsand 2017; Toivonen and Tuominen 2009).

One of the most widely used theoretical models in tourism and hospitality research is experiential learning theory (ELT), which was first introduced by David A. Kolb in the 1980s. He sees experiential learning as a cycle that occurs through a four-stage model: concrete experience, observations and reflection, formation of abstract concept and active experimentation in new situations (Kolb 1984). The ELT model has mainly been used to understand the learning experience of tourists within various settings (e.g. Ballantyne, Packer and Sutherland 2011; Bos, McCabe and Johnson 2015; Stone and Petrick 2013). Bos, McCabe and Johnson (2015), for example, proposed that experiential learning was typically found within structured holidays abroad with a guided group and a planned route, which made it possible for the participants to socialize in environments they were not used to and where they could acquire knowledge and skills. Ballantyne, Packer and Sutherland (2011) used ELT to explain how tourists experience a wildlife site, what they feel and think, and finally, how they act as an effect of their visit.

Whilst the ELT model has proved to be widely useful, another closely linked model is Mezirow's notion of transformational learning. Transformational learning theory explains the meaning and process of learning experience; to interpret life experiences, examine and make meaning of them, and change thoughts, opinions, and even worldviews (Mezirow 1990, 1997, 2009). According to Mezirow (2009), the learning process includes a disorienting dilemma and critical reflection, which often contain undesirable emotions that trigger the individual towards self-examination, renegotiation of beliefs and assumptions, and engagement in new behavior. However, although human transformational learning is significantly applicable to tourism and important for the future, it is rarely discussed in tourism literature (Reisinger 2013), In most cases where transformation is considered, it refers to merely individual perspectives of experiences, such as wellness (e.g. Bushell and Sheldon 2009) and spirituality (e.g. Wilson, McIntosh and Zahra 2013). Thus, transformational tourism should also deal with how travel and tourism can change human behavior and impact the world in positive ways (Reisinger 2013). As Ballantyne and Packer (2011) appropriately point out, there are few tourism experiences that meet the potential to impact behavior and choices of lifestyles.

Learning in a tourist's extraordinary experience

An extraordinary experience is characterized by immersion, enjoyment, interpersonal interaction, co-creative acts, novelty, surprise, storytelling and transformation (e.g. Arnould and Price 1993; Mossberg 2008; Tumbat and Belk 2011). A review of the literature shows that an extraordinary experience has primarily been conceptualized in two different ways. One stream of research has concentrated on the collaborative interaction and common and shared goals that tourists experience (e.g. Arnould and Price 1993; Celsi, Rose and Leigh

1993) while another stream has highlighted the limited interaction and competitive and individualized goals of the tourists (e.g. Tumbat and Belk 2011). Regardless of the way it has been conceptualized, it can be argued that learning remains a key characteristic at the core of the experience. Falk, Ballantyne, Packer and Benckendorff propose that "[e] xperiences that enable visitors to engage deeply and make personal meaning are likely to have the greatest impact [on learning]" (2012: 920). Arnould and Price (1993:41), for instance, noted that "intense, positive experiences crystallize selfhood, provide life meaning and perspective, confer awareness of one's own mortality, reduce anxiety, and improve fear coping". In conceptualizing memorable experience, Tung and Ritchie likewise identified consequentiality as being a response, which has "some sort of personally perceived importance" (2011:1379). They further classified it as having four sub-dimensions; *enhancing social relationships* through interaction with others, *intellectual development* through eye-opening learning experiences, *self-discovery* through which you "expand your perspective in life" (Tung and Ritchie 2011:1380) and *the overcoming of physical challenges*, where you could develop skills and expertise.

Journeys encountering contrasting cultural or nature and wilderness contexts have significant transformative potential (Morgan 2010). For example, Brymer (2013) highlighted how participants of nature-based extreme sports cared more about the natural environment after being immersed in it. Likewise, Arnould and Price (1993) emphasized how isolation in a natural environment made it easier to connect with people and situations. A further aspect that connects learning theory to extraordinary experience studies is critical moments. Undesirable, threatening issues, anxiety, embarrassment and similar negative aspects are considered crucial in order to reach transformation (Argyris, Putnam and McLain-Smith 1985; Mezirow 2009), and they are also important for a person's meaning-making and identity construction process (Thompson and Hirschman 1995). Studies have shown that interactions that are likely to include accidents and/or fear, which people call "close calls", are more likely to crystallize one's sense of self as well as drive an emergent sense of identity with other people (Arnould and Price 1993; Tumbat and Belk 2011). The experiences, but also the stories, of those close calls made people think of themselves as being capable of mastering difficult situations. It can be argued that these examples contradict a hedonic, pleasurable view of extraordinary experiences, and imply that more attention needs to be paid to challenges and conflicts that occur during experiences (Lindberg and Østergaard 2015; Tumbat and Belk 2011). Thus, negative emotions such as anxiety and identity conflicts are part of extraordinary experiences that may lead to transformation. Last but not least, studies have shown that tourists with different cultural capital (skills, knowledge and experience) are more likely to learn differently (Goolaup, Solér and Nunkoo 2018). For instance, Goolaup, Solér and Nunkoo (2018) showed how tourists with high cultural capital are more likely to learn by interacting with participants that have similar cultural capital (high knowledge and skills), whereas those with low cultural capital learn by interacting with local people.

Contextualizing learning and tourists' extraordinary experiences

To understand how tourists learn through extraordinary experiences, two cases were selected: the Karingo Oyster bar in Sweden and a guided tour in Kisumu, Kenya. These cases were chosen since learning was a key component and the participants conceived the experience as being extraordinary.

Case study 15.1: Karingo Oyster bar

The Karingo Oyster bar, situated on the West Coast of Sweden on the island Käringön, offers small groups of tourists the opportunity to experience Swedish Flat European oysters in a unique environment. To provide insights on how the tourists engage in learning within the extraordinary food experience, observations and 15 interviews were conducted.

The oyster event, which usually lasts for three hours, starts with the guide giving a presentation about the place, the different types of oysters and a description of what types of champagne are suitable to combine with the oysters. Then, the tourists are taken to the oyster farm, where they get to learn about techniques regarding farming and the opening of oysters. Some oysters are collected for tasting, which is the next step of the experience. Unlike a typical restaurant, the visitors get to enjoy the oysters with good champagne in an outdoor hot tub overlooking the sea.

For tourists participating in this extraordinary experiential activity, the element of learning is a key part. For the majority of participants, the discussions and interactions with the environment, other oyster enthusiasts and the guide were perceived as illuminating. The experience enabled them to enhance their knowledge due to the fact that they got to open the oysters themselves, consume the oysters where they were farmed, listen to the stories of the host and discuss their experience with each other. One of the participants eloquently described this:

> The fact that the oysters are grown on the spot and you get to hear the history behind and learn what type of oyster it is and how the farming goes on has been very exciting ...

Interestingly, some of the participants also acknowledged how being part of this experience had enabled them to have a better appreciation of oysters, since they became more aware of their health and environmental benefits. This, according to the participants, was due to the discussions they had with other participants and from the guide's presentation as well as getting to see how oysters are farmed. The quotation below provides an indication of how being part of this experience provided a participant of the event with the opportunity to reflect and adjust his behavior:

> I did not really like oyster before, after learning how to eat it properly and how to pair it with the right champagne, and knowing about its health benefits ... I kind of like it now to tell you the truth

Case study 15.2: Guided tour in Kisumu, Kenya

As part of an ecotourism development project in a small fishing village by Lake Victoria in Kenya, a test tour was held with two Kenyan families. The purpose was to develop the guided tours, and to understand how the tourists experienced the tour. Therefore, seven participants were asked beforehand to write down their reflections of the tour in a notebook during and after the tour. Participatory observations were also conducted during the tour together with a colleague.

The guided tour started by the beach, where the families watched the fish being handled by fishmongers. The children were told that the Nile perch is the queen of the lake, while the hippo is the king, which engaged the kids in the story and made them want to hear its continuation. Continuing the storytelling, the families could easily understand the hazard of the hyacinths when the guide pulled out a water hyacinth from the lake and explained the problem by showing how they float around in the lake and how their long roots entangle with the fishing nets.

The next step was to go on a boat ride in the lake. The mother of one of the families told the story of how they were once chased by a hippo. One of the boys wrote in his notes after spotting a hippo in the lake: "At first I was very scared because I thought the hippo would eat the boat. Then we went near the hippo, we were told that it was friendly." Similarly, the father in the group wrote about the emotional intensity and the thrill of being in the boat: "It was psychologically healing to get into the cold breeze propagated by the water waves. It was also scary to think of what would happen if the wind became boisterous."

After having lunch, the tour continued with the participants making their own craft work, such as necklaces, bangles and earrings out of beads, mussel shells and old fishing lines (Figure 15.1). The guide interacted with the participants, saying "Hold that! Try that!" and "Wow!" Both children and adults were totally immersed in the activities for almost two hours. The dad said, "this is the most enjoyable, because it is an activity". In his notes from the day he wrote:

Figure 15.1 Participants of the test tour in Kisumu learn through extraordinary experience

> The boys were excited to be involved in such activity. It reminded me of the fact that when I do, I do not forget. This together with the souvenirs we made gives us the memory and freshness of the test tour. It is my humble submission that a tour will always be printed in the minds of those involved if activities are included where participants are involved in the actual process

> of doing what that particular tourist site is known for. It is also worthwhile to note that it can make one discover a hidden talent.

One of the girls wrote about the crafts activity:

> This was my best part of the treat – the curio work. I was able to view various products made of things that may be considered as waste. This showed the art of invention in man. Also, we were able to make something for ourselves. Here we were able to learn the art of patience and also had something tangible to remember this great day.

Discussion

This chapter contributes to the current literature by highlighting how extraordinary tourism experiences can be intellectually engaging and transformative. It is argued that tourists involved in extraordinary experiences are more likely to learn by being in an environment where learning is enhanced through (1) engagement. (2) critical moments and (3) reflection. Even though these elements are mutually inclusive, for the sake of clarity, they will be discussed separately.

Engagement

Extraordinary tourism experiences expose people to other environments and different ways of living, which could never be obtained in a classroom. Active engagement, such as interaction with other people, activities and environments is crucial for learning to occur. Engagement allows the tourists to understand and reconsider their existence and find power to act and change themselves and the world (Falk, Ballantyne, Packer and Benckendorff 2012; Gursoy and McCleary 2004; Reisinger 2015a). This is reflected in the Kenyan case, where the anecdote about the "Nile perch and the hippos" engaged the kids in the family tour. The story prompted them to ask questions and listen further, for example about how the roots of the water hyacinths affect the fishermen's ability to fish. Further, the younger tourists learned about patience and being engaged in making crafts, while the older participants reflected on how the children's engagement in different activities improved remembering and learning. The activities also enabled them to develop new skills and discover hidden talents, which they may use in other situations. Likewise, in the case of the Karingo Oyster bar, the activities that the participants were engaged in, and their discussions with other guests and the guide, provided the opportunity to learn by sharing experience, knowledge and skills. In short, these activities and the interaction enabled the tourists to co-create their experiences and enhance their learning. It means that being deeply engaged in activities make the experience more memorable and meaningful and makes it possible to learn from it.

Critical moments

Critical moments reflect encounters that make the tourist question certain circumstances or situations. For instance, the episode where one of the children in the Kenyan case was scared that the hippo would eat the boat indicates a critical or even threatening moment, which is crucial for learning to take place (Argyris, Putnam and McLain-Smith 1985; Mezirow 2009). Likewise, a critical moment was implied by one of the participants, who was nervous about what would happen if there was a storm. Such encounters can have transformative potential, since they trigger consciousness and reactions, helping the tourists to redefine the nature of the situation or the problems they face. This is also in line with Morgan (2010), who highlighted how contrasting cultural or wilderness contexts have transformative potential, especially if they include immersion in nature (Brymer 2013).

Reflection

The learning process includes moments of reflection that occurs after certain actions, conversations or observations (Argyris and Schön 1996; Kolb 1984). Reflection enables individuals to evaluate their own intuitive understanding and provide alternative explanations. As illustrated in the Kenyan case, thoughts on new types of resources and production techniques were triggered as the participants created craft items from waste material. Similarly, some of the participants in the experience at the Karingo bar recognized how the knowledge about the benefits of eating oysters has enabled them to rethink their eating habits. These anecdotes contribute to our understanding of how specific meaningful activities start a sense-making process and result in people acquiring new insights and altering their behavior. Reflection can thus be seen as being closely linked to both engagement and critical moments, since it triggers self-examination and renegotiation.

Conclusion

The purpose of this chapter is to understand how tourists learn through extraordinary experience. The findings indicate interrelated features such as engagement, critical moments and reflection that could contribute to the understanding and conceptualization of extraordinary experiences. It contributes to the current literature by highlighting how and in what situations learning occurs, which has been neglected in previous studies. This understanding can help practitioners in their product development, knowing what specific activities and encounters should be highlighted.

Acknowledgments

We are grateful to The Swedish International Development Agency (SIDA) and Centre for Sustainable Urban Futures (former Mistra Urban Futures) for financial support regarding the Kenyan case.

References

Argyris, C., Putnam, R., and McLain-Smith, D. (1985). *Action science: concepts, methods, and skills for research and intervention*. San Fransisco, CA: Jossey-Bass.

Argyris, C., and Schön, D. A. (1996). *Organizational learning II: theory, method and practice*. Reading, MA: Addison-Wesley.

Arnould, E. J., and Price, L. L. (1993). River magic: extraordinary experience and the extended service encounter. *Journal of Consumer Research*, 20(1): 24–45. DOI: 10.1086/209331.

Ballantyne, R., and Packer, J. (2011). Using tourism free-choice learning experiences to promote environmentally sustainable behavior: the role of post-visit 'action resources'. *Environmental Education Research*, 17(2): 201–215. DOI: 10.1080/13504622.2010.530645.

Ballantyne, R., Packer, J., and Sutherland, L. (2011). Visitors' memories of wildlife tourism: implications for the design of powerful interpretive experiences. *Tourism Management*, 32(4): 770–779. DOI: 10.1016/j.tourman.2010.06.012.

Bos, L., McCabe, S., and Johnson, S. (2015). Learning never goes on holiday: an exploration of social tourism as a context for experiential learning. *Current Issues in Tourism*, 18(9): 859–875. DOI: 10.1080/13683500.2013.790878.

Brymer, E. (2013). Extreme sports as transformational. Tourism. In Y. Reisinger (Ed.) *Transformational tourism: tourist perspectives* (pp. 111–124). Wallingford, UK: CABI.

Bushell, R., and Sheldon, P. (Eds.) (2009). *Wellness and tourism: mind, body, spirit, place.* New York: Cognizant.

Celsi, R. L., Rose, L. R., and Leigh, T. W. (1993). An exploration of high-risk leisure consumption through skydiving. *Journal of Consumer Research*, 20: 1–23.

Cohen, E. H. (2014). Self-assessing the benefits of educational tours. *Journal of Travel Research*, 55(3): 353–361. DOI: 10.1177/0047287514550098.

Falk, J. H., Ballantyne, R., Packer, J., and Benckendorff, P. (2012). Travel and learning: a neglected tourism research area. *Annals of Tourism Research*, 39(2): 908–927. DOI: 10.1016/j.annals.2011.11.016.

Flood, R. L., and Romm, N. R. A. (1996). Contours of diversity management and triple loop learning. *Kybernetes: The International Journal of Systems & Cybernetics*, 25(7–8): 154–163. DOI: 10.1108/03684929610149747.

Gipps, C. (1999). Socio-cultural aspects of assessment. *Review of Research in Education*, 24(1): 355–392.

Goolaup, S. (2018). *On consumer experiences and the extraordinary* (Doctoral thesis), University of Gothenburg, School of Business, Economics, and Law, Gothenburg (2018) (ISBN: 978-91-88623-07-2).

Goolaup, S., and Mossberg, L. (2017). Exploring the concept of extraordinary related to food tourists' nature-based experience. *Scandinavian Journal of Hospitality and Tourism*, 17(1): 27–43. DOI: 10.1080/15022250.2016.1218150.

Goolaup, S., Solér, C., and Nunkoo, R. (2018). Developing a theory of surprise from travelers' extraordinary food experiences. *Journal of Travel Research*, 57(2): 218–231. DOI: 10.1177/0047287517691154.

Gursoy, D., and McCleary, K. W. (2004). An integrative model of tourists' information search behavior. *Annals of Tourism Research*, 31(2): 353–373. DOI: 10.1016/j.annals.2003.12.004.

Hall, C. M., and Williams, A. M. (2008). *Tourism and innovation.* London, UK: Routledge.

Jarvis, P. (2009). Learning from everyday life. In P. Jarvis (Ed.) *The Routledge international handbook of lifelong learning* (pp. 19–30). New York: Routledge.

Jernsand, E. M. (2017). Engagement as transformation: learnings from a tourism development project in Dunga by Lake Victoria, Kenya. *Action Research*, 15(1): 81–99. DOI: 10.1177/1476750316678913.

Jernsand, E. M., Kraff, H., and Mossberg, L. (2015). Tourism experience innovation through design. *Scandinavian Journal of Hospitality and Tourism*, 15(sup1): 98–119. DOI: 10.1080/15022250.2015.1062269.

Kalinowski, K. M., and Weiler, B. (1992). Educational travel. In B. Weiler, and C. M. Hall (Eds.) *Special interest tourism* (pp. 15–26). London, UK: Belhaven.

Kolb, D. (1984). *Experiential learning: experience as the source of learning and development.* Englewood CliÚs, NJ: Prentice Hall.

Lewin, R. (2009). *The handbook of practice and research in study abroad. Higher education and the quest for global citizenship.* New York: Routledge.

Lindberg, F., and Østergaard, P. (2015). Extraordinary consumer experiences: why immersion and transformation cause trouble. *Journal of Consumer Behaviour*, 14: 248–260. DOI: 10.1002/cb.1516.

Mathieson, S. (2014). Student learning. In H. Fry, S. Ketteridge, and S. Marshall (Eds.) *A handbook for teaching and learning in higher education. enhancing academic practice* (4th ed., pp. 63–79). New York: Routledge.

Mezirow, J. (1990). How critical reflection triggers transformative learning. *Fostering Critical Reflection in Adulthood*, 1: 20.

Mezirow, J. (1997). Transformative learning: theory to practice. *New Directions for Adult and Continuing Education*, 74: 5–12.

Mezirow, J. (2009). An overview of transformative learning. In K. Illeris (Ed.) *Contemporary theories of learning: learning theorists … in their own words* (pp. 90–105). New York: Routledge.

Morgan, A. (2010). Journeys into transformation: travel to an 'other' place as a vehicle for transformative learning. *Journal of Transformative Education*, 8(4): 246–268. DOI: 10.1177/1541344611421491.

Moscardo, G. (1998). Interpretation and sustainable tourism: Functions, examples and principles. *Journal of Tourism Studies*, 9(1): 2–13.

Mossberg, L. (2008). Extraordinary experiences through storytelling. *Scandinavian Journal of Hospitality and Tourism*, 8(3): 195–210. DOI: 10.1080/15022250802532443.

Pearce, P. L., and Foster, F. (2007). A "university of travel": Backpacker learning. *Tourism Management*, 28(5): 1285–1298. DOI: 10.1016/j.tourman.2006.11.009.

Pritchard, A., and Morgan, N. (2013). Hopeful tourism: a transformational perspective. In Y. Reisinger (Ed.) *Transformational tourism: tourist perspectives* (pp. 3–14). Wallingford, UK: CABI.

Reisinger, Y. (2013). Preface. In Y. Reisinger (Ed.) *Transformational tourism: tourist perspectives* (pp. xii–xiv). Wallingford, UK: CABI.

Reisinger, Y. (2015a). Personal transformation and travel and tourism. In Y. Reisinger (Ed.) *Transformational tourism: host Perspectives* (pp. 6–13). Wallingford, UK: CABI.

Reisinger, Y. (2015b). Reflections on life purpose. In Y. Reisinger (Ed.) *Transformational tourism: host perspectives* (pp. 3–5). Wallingford, UK: CABI.

Romme, A. G. L., and van Witteloostuijn, A. (1999). Circular organizing and triple loop learning. *Journal of Organizational Change Management*, 12(5): 439–454. DOI: 10.1108/09534819910289110.

Scarinci, J., and Pearce, P. (2012). The perceived influence of travel experiences on learning generic skills. *Tourism Management*, 33(2): 380–386. DOI: 10.1016/j.tourman,2011.04.007.

Shephard, L. (1992). What policy makers who mandate tests should know about the new psychology of intellectual ability and learning. In B. Gifford, and M. O'Connor (Eds.) *Changing assessments: alternative views of aptitude, achievement and instruction* (pp. 301–328). London, UK: Kluwer Academic.

Shoemaker, S. (1994). Segmenting the US travel market according to benefits realized. *Journal of Travel Research*, 32(3): 8–15.

Stone, M. J., and Petrick, J. F. (2013). The educational benefits of travel experiences: a literature review. *Journal of Travel Research*, 52(6): 731–744. DOI: 10.1177/0047287513500588.

Su, D. N., Johnson, L. W., and O'Mahony, B. (2018). Analysis of push and pull factors in food travel motivation. *Current Issues in Tourism*: 1–15. DOI: 10.1080/13683500.2018.1553152.

Terenzini, P. T. (1998) *Research and practice in undergraduate education*. Paper presented at Kansas State University, Manhattan, KS (mimeo). http://link.springer.com/article/10.1023/A:1003709230179

Thompson, C. J., and Hirschman, E. C. (1995). Understanding the socialized body: a poststructuralist analysis of consumers' self-conceptions, body images, and self-care practices. *Journal of Consumer Research*, 22(2): 139–153. DOI: 10.1086/209441.

Toivonen, M., and Tuominen, T. (2009). Emergence of innovation in services. *The Service Industries Journal*, 29(7): 887–902. DOI: 10.1080/02642060902749492.

Tumbat, G., and Belk, R. W. (2011). Marketplace tensions in extraordinary experiences. *Journal of Consumer Research*, 38(1): 42–61. DOI: 10.1086/658220.

Tung, V. W. S., and Ritchie, J. B. (2011). Exploring the essence of memorable tourism experiences. *Annals of Tourism Research*, 38(4): 1367–1386. DOI: 10.1016/j.annals.2011.03.009.

Weaver, D. B., and Jin, X. (2016). Compassion as a neglected motivator for sustainable tourism. *Journal of Sustainable Tourism*, 24(5): 657–672. DOI: 10.1080/09669582.2015.1101130.

Wilson, G., McIntosh, A. J., and Zahra, A. L. (2013). Tourism and spirituality: a phenomenological analysis. *Annals of Tourism Research*, 42(July): 150–168. DOI: 10.1016/j.annals.2013.01.016.

16

Resident-driven city hospitality design and delivery

Karoline Wiegerink and Jan Huizing

Cities are judged by their welcome

In a city where it is attractive to stay, people feel welcome. This applies to residents, visitors and the business community. A hospitable climate leads to engagement, well-being and recommendation. One could say: hospitality pays off. This chapter shows how local voluntary hosts can design specific welcoming experiences and how this contributes to the city as a welcoming community that connects and leads to social added value, even in a situation of overtourism.

"More visitors who stay longer, spend more money, return more often and spread the positive message by word of mouth …" used to be the primary objectives of Destination Marketing Organizations for many years, while aiming for a prosperous tourism industry. However, in recent decades, tourism has experienced a world-wide increase, becoming part of a new global lifestyle (Neuts and Nijkamp 2012). And this has a dark side: today, the discussion about overtourism is emerging. Expansion of urban tourism in European cities is putting increasing pressure on those areas as places to live, work and visit, leading to effects such as tourism gentrification (Pinkster and Boterman 2017).

More and more local authorities and city halls claim that the social costs of mass tourism are now outweighing its economic benefits and are willing to take affirmative action. For example in Rome, the government is considering introducing an automated counting technology to limit the maximum number of visits to the world-famous Trevi Fountain, deeming it "absolutely necessary" in order to protect their city (Horowitz 2017). Or in the case of Amsterdam, an increase in both population and visitors has brought Amsterdam multiple, especially economic, benefits, but the other side of the coin is that residents are experiencing a decline in their quality of life due to an atmosphere that is characterized by terms such as "Urban Jungle", "Disneyfication" and "Venice of the North". Amsterdam's municipality is taking appropriate action and introducing a strategy of "demarketing". The local administration has explicitly expressed its commitment to residents and local business: "First of all, Amsterdam is a city to live, work and undertake in … and Amsterdam as a tourist destination comes second". The DMO Amsterdam Marketing has been transformed into a knowledge partner for cultural promotion, high-end tourism and a program called City in Balance.

The impact of excessive tourism in cities can be related to many aspects such as congestion, cultural heritage, gentrification … but many overtourism issues are related to the perception (often negative) of encounters between different types of tourist groups, residents, entrepreneurs and service providers, due to high tourist numbers at certain times and places. This influences the perceived quality of life of citizens and/or of visitors' experiences in a negative way.

The literature (Neuts and Nijkamp 2012) has concluded that there is no common set of indicators and policies in place, due to the complexity and diversity of the causes and effects of overtourism. Avoiding the negative consequences of mass tourism requires custom-made policies in cooperation with the destinations' stakeholders and policymakers. A recent publication of UNWTO (2017) proposes recommendations for long-term, sustainable tourism. One of them relates to enhancing the integration of local communities in the tourism value chain and promoting their engagement. Understanding, integrating and engaging local communities helps to balance the interests of all stakeholders. And this is where city hospitality comes in. We define city hospitality as *a long-term process aimed at maintaining and enhancing the different target groups' experience of feeling welcome in the city, based on a vision of hospitality that creates added value for the city, with an orchestrated deployment of appropriate strategic instruments that facilitate and support hospitality.* (Wiegerink 2014)

How can the concept of hospitality help to balance interests in overheated tourist areas?

When the Stranger says: "What is the meaning of this City? Do you huddle close together because you love each other?" What will you answer? "We all dwell together to make money from each other?" Or: "This is a community?" (Eliot 1969, quoted in Bell, the "Hospitable City" 2007.)

The value creation of city hospitality (as an answer to the negative effects of overtourism) can be approached through the three domains of hospitality (Lashley 2000): the private, social and commercial domains.

The private domain relates to one person who exhibits hospitality and another who experiences it. Hospitable behavior consists of two components (Dekker et al. 2014): hospitable (professional) service skills and genuine hospitableness. Genuinely hospitable behavior is described by the six basic rules of Gunnarsons and Blohm's (2002) *Hostmanship, the art of making people feel welcome*: serving, The Big Picture, responsibility, caring, knowledge and dialogue.

This type of behavior is connected to personality. People who exhibit this behavior are naturally hospitable, and come across as warm to the guest. Why do people exhibit hospitable behavior? What is the motivation for welcoming behavior? Is it "deriving pleasure oneself from making other people happy", or is it "a sense of duty towards our fellow humans"? In the literature we find no clear answer to this question. The connection between host and guest is a relationship of mutual obligations and reciprocity … depending on situations and circumstances (Telfer 1996).

If we consider the city as the host, then this is a metaphorical term for anyone within the city who fulfills the role of host: any person with guest encounter during the whole visitor journey, ranging from service providers in the hospitality industry, infrastructure, public transport to retailers, cultural institutions and, last but not least, the residents themselves.

A city will be experienced as more welcoming as more hospitable encounters take place between hosts and guests.

From the perspective of the *social domain*, hospitality is a cultural phenomenon, the binding force within and between societies, the social context in which particular hospitality activities take place (Telfer 2000). In early societies, hospitality and the duty to entertain both neighbors and strangers represented a fundamental moral imperative. Nowadays the duty to provide hospitality, to act with generosity as a host, and to protect visitors is more a matter left to the preferences of individuals and communities (Lashley 2000). The level of hospitality of a city can be traced back to cultural and historical roots as well as current political choices and preferences.

A city will be experienced as more welcoming if a culture can be created that is focused on human orientation and cohesion.

The third, the *commercial domain*, views hospitality as a concept reflected in the hospitality industry, but also on a broader scale in all forms of commercial contacts and services. If we consider the discussion on reciprocity and motives that took place earlier, then the hospitality relationship essentially involves exchange: the experience of a friendly welcome, well-being and pleasure in exchange for money, affection or other personal benefits from the host.

The hospitality industry has been building on this welcoming performance within a specific and targeted setting and concept over the centuries. Putting the guest in the central position and designing guest experiences that create long-term benefits are important elements in many service-oriented industries and areas, including a destination.

Hospitality concepts (from a broader perspective) are of important significance in cities and destinations.

The City Hospitality Experience Model – residents can make a difference

We could conclude that city hospitality can contribute to the liveability and lovability of a destination. Which elements are important for creating experiences?

According to the City Hospitality Experience Model (Figure 16.1), the synergy between the city's offering and the needs, motives and expectations of different stakeholders is what creates experiences. Unexpected surprises lead to guest delight and wow experiences, also known as memorable moments, that highlight the entire experience journey (Dekker et al. 2014). These moments are what enhance the degree of lovability perceived towards the city, and are directly related to the value creation of the industry (realized through returning visitors and extended lengths of stay). A city offers its welcoming behavior, hardware and atmosphere to fulfill the needs of multiple target groups, namely visitors, residents and businesses (Wiegerink 2014).

City Hospitality Experiences are created through the city offerings (hardware) and atmosphere, and very often through hospitable behavior: human encounters that make a difference by creating memorable experiences which lead to guest delight, engagement and loyalty.

Human touch (Dekker et al. 2014) is a crucial factor that determines the hospitality experience. Experiencing hospitality is primarily a perception of personal attention (Pijls-Hoekstra et al. 2015). In that sense, residents, the service providers of the city, are important because their welcoming behavior adds a personal touch to the city environment and eventually creates the unexpected wow experiences for the visitors. Thus, they play a crucial role: they experience the city as a resident, while they also contribute to the total city offering through their behavior towards guests. They take on the multi-faceted roles of the city because they live as residents, enjoy as visitors and work as businesses. Therefore, we could consider

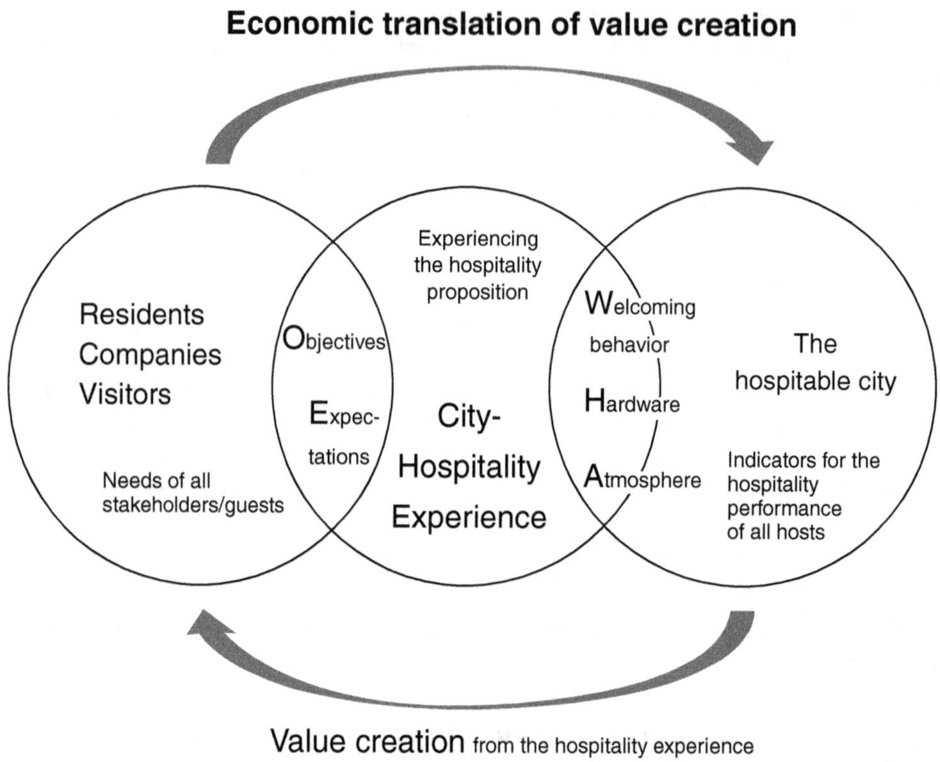

Figure 16.1 City Hospitality Experience Model
(Prooijen and Wiegerink 2012)

residents as the ideal leaders of City Hospitality because they understand the needs of all stakeholder groups, though in a subconscious manner (Park and Wiegerink 2019).

Important touchpoints in the guest journey

The total tourist experience is the sum of the experience of a number of touchpoints before, during and after their stay. The customer journey envisions, designs and visualizes the customer's point of view as a holistic experience and breaks it down into smaller pieces of experiences and service encounters (Lemon and Verhoef 2016). Mapping important touchpoints provides an organization with insights into the journey a customer goes through, identifying the important moments, the so-called touchpoints, that provide a memorable interaction between the "city" and representatives of the target group. Being aware of those moments makes it easier to manage, improve, innovate and work in a city that will become increasingly attractive to visit.

Cities attract a wide variety of consumers who pay a visit for multiple reasons, and by mapping their journey, improvements can be applied to certain touchpoints to increase their overall experience. Touristic product, the atmosphere and hospitable behavior all impact the customer experience and thus the loyalty of guests.

In line with the above-mentioned importance of personal encounters for the guest experience, we consider the contact with a city host as a valuable touchpoint in the visitor journey.

Human encounters can make a major difference by creating memorable experiences, which lead to guest delight, engagement and loyalty. These encounters are provided not only by obvious hosts such as hotel employees, shop personnel and taxi drivers, but also by residents on a volunteer basis.

Interaction between visitors and city hosts – city hospitality value creation

Residents are the main stakeholders in the city experience, either in a providing or in a receiving role. Their leading needs must be fulfilled in an acceptable balance with the needs of other stakeholders. Residents' needs are not only applied in the context of city hospitality, they also concern quality of life, which is what makes them long-term and sustainable.

Gerritsma and Vork (2017) and Ouwehand and Wiegerink (2019) showed that residents who are involved more actively with tourists, e.g. with an economic business interest in tourism or working for a tourism business or any other economic involvement, are more positive towards tourists than residents who are not involved in the tourism industry. It is assumed that greater involvement with tourists leads to a more positive attitude towards them, and hence a better-perceived balance between these stakeholders. Further research is required to see if this is also true for other types of involvement (for example, more contact). The balance and positive perception between residents and visitors are likely to be more positive when both have more interaction and higher economic (or other) interest.

Residents can play an active role in interacting with visitors. When the commercial domain is addressed, i.e. when hospitality is deliberately designed, organized and executed as a concept, a group of residents can have an assigned role of City Host.

What is a city host?

City Hosts represent the city and provide a warm welcome to visitors. City hosts are insightful locals who volunteer as welcoming hosts of the city. They help all visitors with directions and tips on what to do and alleviate language barriers (Park 2018). They are both contact and host at various locations where visitors gather in a city and may need information and assistance (Huizing and Wiegerink 2013). They are approachable, friendly and equipped with extensive local knowledge (City of Melbourne 2013).

A key function of a city host may be to provide information and directions, which would reflect a rather transactional and reactive approach. Hospitable encounters between guest and host are in the private domain of hospitality.

Why do people volunteer as city hosts?

To understand the motives of a city host, it is necessary to distinguish three types (Telfer 1996), each with a specific, dominant orientation. Depending on the main motive, a city host would typically seek and favor a related type of behavior.

Guest-Oriented Motive: the desire to be hospitable, which comes from the need to make the guests feel good and fulfill their needs. For example, by sharing favorite spots. Mutual

Motive: provide hospitality for mutual benefit. The aim is, for example, to be in the company of people, to socialize or to make friends. Self-Directed Motive: being hospitable from the desire to be happier as a host, for example, showing off your knowledge of the city, expressing pride.

A more existential function for a city host would be to be present, visible and easily approachable in the public domain, making a visitor feel comfortable and welcome, possibly safer and more secure; the welcoming feeling is provided either directly or indirectly by "being there". This approach acknowledges the social domain (Lashley 2000), in which the city host is inspired by and co-creates a welcoming culture in the city.

When emphasizing the social domain, city hosts are also referred to as City Ambassadors. City Ambassadors are best described as enthusiastic volunteers who provide a warm welcome to a city visitor, especially during major events when many visitors are in the city. The term Ambassador expresses the professional, representative function of a city host, and relates to feelings of pride and prestige, which can positively add to the job perception of city hosts, and therefore positively influence the job satisfaction and visitor experience. A city host role may thus vary from primarily transactional (answering tourist questions etc.) to more existential and conditional.

The interaction between visitors and city hosts leads visitors to feel greater satisfaction about their city visit as it enhances their journey experience. Research on "Ambassade" and "Amsterdam Welcome Teams" (see cases) indicated that visitors were more satisfied if they had experienced interaction with a city host. It may be assumed – but would require further research – that interaction with a city host leads to a more positive overall journey experience.

How to create sustainable hospitality interaction between city hosts and visitors?

The way city hosts are organized takes different forms; usually city hosts are volunteer residents who partake in an organized structure, either temporarily as a project, possibly in light of a specific sports event, or in a more sustainable scheme if the city adopts a more integrated stance towards city hospitality.

City hosts can be considered professional yet voluntarily based, meaning that they are assigned a specific and organized task which is to interact with visitors. Ultimately, this is a means to an end, like any project or program. It is the aim of a city to have positive interactions between its inhabitants and tourists, regardless of whether it is organized in the form of city hosts or random, informal, spontaneous and not organized at all. The aim of a city host scheme would be not just to enhance the visitors' experience during specific events or at certain periods of time but on a continuous basis as a default enhanced level of service in the city, as in any professional organization. Best practices and a service-oriented, guest-oriented culture are expected to have a positive, leveraged effect.

When the interaction between residents and visitors acknowledges each of the three domains, the role of a city host becomes more effective and sustainable. The interaction is then not limited to providing directions, but seeks to create memorable experiences, by being available and approachable, with an open, interested and attentive attitude. The more city hosts have intrinsic motivation, genuinely want to make the visitor feel welcome, and cherish and are part of a hospitable culture, the more the social and private domains are addressed, which forms the basis for a sustainable approach towards hospitable visitor–resident interactions. With the acknowledgment of the commercial domain, the role of city host is professionally organized in a long-term scheme, with continuous support from the municipality and other stakeholders in the city.

How to engage city hosts?

The participation of residents is voluntary, probably driven more by intrinsic than extrinsic motivational factors. Volunteers form a group with great motivation and job satisfaction that will change their output for the better (Meijs 1997 in van der Vaart 2010). Many organizations increasingly depend on their volunteers and consequently it is important to maintain the level of volunteers connected to your organization or project (Harrison 1995).

Developments over the past few years have threatened to disturb the balance between supply and demand in volunteer work, however. The aging of the population, limited replacement by the younger generation, and a changing lifestyle of the entire population demand a new, up-to-date approach in organizing and offering volunteer work (Bussemaker 2007). It has become harder for organizations to motivate and bind volunteers to their projects, a challenge that various city host projects have faced and will continue to face in the future. According to the principals of the Service Profit Chain, a thorough understanding of the volunteers will lead to the desired low "volunteer turnover", positively impacting the hospitality of a city (Heskett et al. 1994).

Enhancing sustainable co-creation

The key to sustainable experiences is a continuous dynamic of co-creating with the relevant stakeholders. This requires a shared vision, a long-term approach, structural resources (organization, funding) and a strong collaborative culture between all relevant stakeholders.

Creating and managing experiences in a city will be most effective and sustainable if they are the result of solid and ongoing co-creation efforts by the various relevant stakeholders. Who are these relevant stakeholders? They include the municipality, entrepreneurs, shop owners, taxi drivers, public transport, hotel service staff, retailers, terrace waiters, people on the street, inhabitants. The more these stakeholders co-create and collaborate, the stronger the effect will be. Structural effects can best be established by fostering this co-creation, which requires time and effort from all relevant stakeholders, who need to work together not only to execute and implement but also to design and manage – and thus balance – the interests and experiences of all stakeholders. Just creating a blueprint will not do; it is rather a continuous interaction, research and change process. Creating experiences with all these various, diverse stakeholders – not organized as one entity and not directly bound together by one shared purpose – is much more complex than creating changes within a single company. It requires continuous effort, and each stakeholder must think and act beyond individual interests and goals. For example, the general manager of a hotel should be motivated to think and act beyond the interest of his own property and include inhabitants' and other businesses' goals; occasionally, long-term perspectives must prevail over short-term gains if this adds to the balance between all stakeholders.

Projects like "Ambassade" help to create visible successes and best practices, and to move towards a structural co-creation approach within the entire city.

Obviously, a more sustainable approach is achieved if city hospitality is established on a more professional level; the city then has a clear, integrated ambition and vision, formulated SMART long- and short-term goals, and an integrated set of actions in place with allocated resources.

To conclude

Destinations are judged by their welcome. With growing tourist numbers putting increasing pressure on the lovability and liveability of cities, policy makers are seeking effective interventions for long-term sustainable tourism, striving to regain and balance the interests and well-being of all stakeholders.

This chapter elaborated on the value-adding role of programs in which volunteer residents act as city hosts. This value creation is related to the three domains of hospitality: private, social and commercial.

The hospitable encounters between passionate and skilled volunteer city hosts and tourists create memorable experiences for the guests and contribute to the city as a welcoming place.

In their human-to-human approach and continuous and dynamic co-creation, city hosts contribute to a welcoming culture in the city.

When organized and managed professionally, and continuously co-created by relevant stakeholders, city hosting can be seen as a highly valuable hospitality concept in the broader field of service and hospitality provision in destinations.

Case study 16.1: "Ambassade van Den Haag" (Embassy of The Hague)

Inspired by the London Olympics' approach towards visitor hospitality, the municipality of The Hague, The Netherlands, initiated "Ambassade van Den Haag", a volunteer-based city hospitality program aimed at providing visitors with a warm welcome and a memorable stay. More than 300 proud and motivated residents act as city ambassadors and provide high-quality service to the city's visitors at hotspots throughout the city. The program started off as a project dedicated to the Hockey World Cup 2014 and has since developed further into a sustainable city hospitality scheme, in which residents and visitors co-create meaningful experiences in the public space.

Officially launched in November 2013, "Ambassade" (Embassy of The Hague) is part of a longer-term plan by the municipality of The Hague to increase the number of The Hague residents participating in volunteer work from 18% to 25% between 2009 and 2014. The initial assignment was to recruit and train 350 volunteers to be city ambassadors for the visitors to the Hockey World Cup 2014. The dream was to create a showcase for all volunteer projects. It aimed to recruit and train 350 motivated people to deliver high-quality work "without realizing that it was volunteer work". The initial project proposal assumed "passion for hospitality and city pride" as key motives for potential city ambassadors (Huizing and Wiegerink 2015).

So what are a resident's motivations to volunteer as a city ambassador? By analyzing the motivations as formulated in the application process, research (Huizing and van der Wielen 2015) indicated seven different motives amongst the city ambassadors:

1. City Pride (self-directed)
2. Making a contribution (guest-oriented)
3. Reciprocity, doing something in return (self-directed, guest-oriented)
4. Personal development (self-directed)
5. Wanting to be part of an event (mutual)
6. Meeting new people (mutual)
7. Using current competencies (self-directed).

To continuously meet the motives of the city ambassadors and enhance their proud feeling, various programs and activities were developed.

After the application and selection process, a total of 300 volunteer residents were accepted. They received two training sessions at Hotelschool The Hague, a deliberately chosen environment which not only enhanced their hospitality skills but also naturally enhanced their hospitality spirit. In a grand kick-off meeting, all of these residents were officially welcomed and accredited by the mayor of The Hague; each received a certificate and photo of the event, together with a brand new uniform, which were important tangibles to express the achievement and the proud occasion. A website and Facebook page was set up to share hospitality stories between the city ambassadors, in order to continuously feed and cherish a hospitable culture with more and more wonderful compliments from visitors coming in. City hosts with human-oriented, social motivations would find invitations and reward tickets to attend a nice event or sports match – which would at the same time enhance their knowledge of the city; others motivated to use their previous professional experience would find opportunities to apply for a team leader role. A "benefits and rewards program" was set up to offer reward activities. Some of these benefits were co-created with partners in the city, which enhanced the visibility of the city ambassadors and created more interaction with the various stakeholders. At a later stage, once the municipality decided to continue the project, the "Ambassade" became a body of knowledge for other institutions and organized volunteers to interact with visitors and enhance their hospitality skills further. In that role, "Ambassade" has started to provide hospitality training to volunteer hosts in museums. This could lead to a collaborative community in the city in which various hosts share and learn from each other.

Case study 16.2: Amsterdam Welcome Teams: Hospitality pays off!

Welcome teams make the difference! Visitors feel sincerely welcome! There is real contact and attention. Tourists are positively surprised by receiving tips and ideas for experiencing Amsterdam in a different way; they visit places "off the beaten track"... and as a result they feel engaged with the city and share it.

These were the objectives of a unique city hosting pilot project of House of Hospitality Metropol Region Amsterdam (MRA), contributing to valuable tourist experiences as well as to the "City in Balance" project for the dispersion of tourists of the Municipality of Amsterdam as part of the policy to combat the negative aspects of overtourism.

House of Hospitality MRA is a public–private collaboration of the hospitality and tourism-related industry, the municipality of Amsterdam and vocational and higher hospitality education aiming to motivate more people for a career in hospitality and to explore challenges and share knowledge.

The pilot

The setup of the pilot, which took place in 2016 and 2017, was a living lab which tested how city hosts can make a difference for visitors to Amsterdam and how this can become a structural part of vocational and higher hospitality education. The pilot was led by Hotelschool The Hague, Chair City Hospitality, one of the partners of House of Hospitality MRA.

Figure 16.2 Student participant of the Amsterdam Welcome Team in a hospitable encounter with tourists
Source: House of Hospitality MRA

A pool of motivated hosts with an open, welcoming attitude and behavior was created, composed of residents/volunteers experienced in city hosting on the one hand, and motivated vocational and higher education students in tourism and hospitality on the other (Figure 16.2). The students' activities were embedded in the curriculum as part of a practical component in their learning journey.

Before taking up their role as city hosts, the whole group was trained in accurate welcoming appearance and behavior and instructed and informed about current city insights. They were provided with high-quality, recognizable materials and clothing, resources such as iPads and city maps, facilities and support.

The composition of the hospitable teams consisted of a mix of students and skilled city hosts, a nice synergy between experience and young enthusiasm.

Everything was set up to facilitate a top performance of the Welcome teams and lead to positive perceptions, surprise and changed visiting behavior of tourists. The teams worked mainly in busy times during several weeks at the Museum Square, a major landmark in Amsterdam. Their task was: act as a genuine host, show an open posture and appearance ("welcome to Amsterdam"), make contact, start a conversation, provide tourists with local tips and try to offer alternatives to busy mass-tourist attractions.

The outcomes

Research was conducted into how tourists experienced their encounters with the city hosts. How was this exceptional service perceived? What did it lead to, did it pay off?

The majority of the tourists who encountered the welcoming teams were city trippers, who stayed in Amsterdam for a number of days (three days on average). More than 20 nationalities were represented in the sample.

City hosting led to wow experiences, with average scores of over 9 out of 10. Expectations were exceeded, and positive impressions made! The table shows the average scores given to the following questions:

	10-point scale
How was your first impression of the host?	9.4
How friendly was the host during the encounter?	9.7
How enthusiastic was the host during the encounter?	9.2
How hospitable was the host when saying goodbye?	9.5
What is your overall impression of the conversation?	9.7
How likely is it that you would recommend the host to your friends/family?	9.3

In addition: the majority of respondents (65%) reported being happily surprised by the encounter, 14% experienced a "wow moment", and 21% remained neutral. No negative feedback was received. Tourists actually perceived Amsterdam differently after having spoken to the city hosts. More than 70% of tourists considered doing unplanned activities that had been recommended and following tips provided by the hosts. This affects the city of Amsterdam because hosts can actually influence the tourists' behavior, who are more likely to undertake other activities than previously thought. These hosts could actually make a difference and help make Amsterdam a more hospitable place.

Reflection and follow-up

The pilot could be considered successful. Hospitality in the open space provided by Amsterdam hosts (motivated, flexible, capable, enthusiastic residents, students and workers who are happy to volunteer to fill the special host role) can make a difference for a broad range of visitors – the wow factor. They feel sincerely welcome. There is contact and attention. (Welcome to Amsterdam!)

The pilot project showed that there is added value through cooperation between vocational and higher hospitality education students with experienced and voluntary city hosts.

Much attention is required for the recruitment, selection and motivation of participants. This applies to volunteers as well as to students. In the near future, integration in vocational education will become a fact. Furthermore, an incentivization of all local host initiatives has started, striving to create an umbrella approach for the provision of personal hospitality in the public space that pays off.

References

Amsterdam Marketing. (2017). *Metropoolcampagne Amsterdam Bezoeken, Holland Zien.*

Bell, D. (2007). The hospitable city, social relations in commercial spaces. *Progress in Human Geography*, 31 (1): 7–22. DOI: 10.1177/0309132507073526.

Bussemaker, J. (2007). *Beleidsbrief mantelzorg en vrijwilligerswerk: voor elkaar.* Den Haag: Ministerie van VWS.

City of Melbourne. (2013). *City Ambassadors.* Available at: www.thatsmelbourne.com.au/visitors/services/ambassador/Pages/Ambassadors.aspx. [Accessed March 15, 2014].

Dekker, D. M., Lub, X., Tanilon, J., and Groen, B. (2014). Deconstructing the guest experience: what employee behaviours drive customer delight? *32nd EuroCHRIE conference "Hospitality and Tourism Futures"*, Dubai 6–9 October 2014. EuroCHRIE.

Gerritsma, R., and Vork, J. (2017). Amsterdam residents and their attitude towards tourists and tourism. *Philosophy, Communication 2017*, 25: 85–98.

Gunnarson, J., and Blohm, O. (2002). *Hostmanship, De Kunst mensen het gevoel te geven dat ze welkom zijn*, Dialogos Forlag AB.

Harrison, D. A. (1995). Volunteer motivation and attendance decisions: competitive theory testing in multiple samples from a homeless shelter. *Journal of Applied Psychology*, 80 (3): 371–385. DOI:10.1037/0021-9010.80.3.371.

Heskett, J., Jones, T., Loveman, G., Sasser, W., and Schlesinger, L. (1994). Putting the service profit chain to work. *Harvard Business Review*, March–April 1994: 164–174.

Horowitz, J. (2017). No swimming! Rome braces for summer of tourists at its fountains, *The New York Times*, 22 June.

Huizing, J., and Wiegerink, K. (2013). Cherishing the journey wows: fresh approaches for cities to capitalize on their best assets for visitors. *Proceedings of the Service Design and Tourism Conference*, 7th–9th November, Sarasota, FL.

Huizing, J., and Wiegerink, K. (2015). Case: city hospitality – ambassade van Den Haag. In R. Boer, *Brand design* (5th edition, 2015) (pp. 196–208). Amsterdam: Pearson Benelux.

Huizing, J., and van der Wielen, J. (2015). Engaging stakeholders to co-create memorable city hospitality: resident's motives and personas. *EuroCHRIE 2015.*

Lashley, C. (2000). Towards a theoretical understanding. In C. Lashley and A. Morrison *In search of hospitality: theoretical perspectives and debates* (pp. 1–17). Oxford: Butterworth-Heinemann.

Lemon, K. N., and Verhoef, P. C. (2016). Understanding customer experience throughout the customer journey. *Journal of Marketing*, 80 (6): 69–96. DOI: 10.1509/jm.15.0420.

Meijs, L. (1997) in Vaart, I van der (2010). *Motivaties voor vrijwilligerswerk; wat maakt het verschil?* Utrecht: University of Utrecht, p. 8.

Neuts, B., and Nijkamp, P., 2012, *Tourist crowding perception and acceptability in cities: an applied modelling study on Bruges* - 1-s2.0-S0160738312001168-main.pdf.

Ouwehand, M., and Wiegerink, K. (2019). *The attitude and its influential factors of residents towards tourism in the city center of Amsterdam.* HTH Working Paper Series, August, ISSN Number: 2543-2494.

Park, Y., and Wiegerink, K. (2019), Smart resident driven city hospitality. *Conference proceedings international hospitality management symposium, 2019.*

Pijls-Hoekstra, R., Groen, B. H., Galetzka, M., and Pruyn, A. T. H. (2015). Experiencing hospitality: an exploratory study on the experiential dimensions of hospitality. Paper presented at *EuroCHRIE 2015*, Manchester.

Pinkster, F. M., and Boterman, W. R. (2017). When the spell is broken: gentrification, urban tourism and privileged discontent in the Amsterdam canal district. *Cultural Geographies*, 24 (3): 457–472. DOI: 10.1177/1474474017706176.

Prooijen, M., and Wiegerink, K. (2012). *The city hospitality experience model – shaping a hospitable city.* Short paper *EuroCHRIE* 2012.

Telfer, E. (1996). *Food for thought: philosophy and Food.* London: Routledge. DOI: 10.4324/9780203002308.

Telfer, E. (2000). An anthropology of hospitality. In C. Lashley and A. Morrison (Eds.), *In search of hospitality* (pp. 38–55). Oxford: Butterworth Heinemann.

UNWTO. (2017). Overtourism: growth is not the enemy, it is how we manage it. *UNWTO World Travel Market Ministers Summit.*

van der Vaart, I. (2010). *Motivaties voor vrijwilligerswerk: wat maakt het verschil?* (Motivation for voluntary work, what makes the difference?). Utrecht: University of Utrecht.

Wiegerink, K. (2014) *City hospitality marketing – how hospitable is your city? issuu.* Available at: https://issuu.com/hotelschoolthehague/docs/booklet_cityhospitality_engfinal/1 [Accessed March 20, 2019].

17

Experiential heritage tourism designing in the United Arab Emirates

Joanna Seraphim and Farooq Haq

Introduction

This chapter builds on the importance of heritage tourism for the United Arab Emirates. Recently tourism planners realized the commercial-, social- and image-related benefits of heritage tourism (Boussaa 2014; Hoath 2014; Seraphim and Haq 2018). Due to the exponential costs of tourism based on tangible assets and luxury, more meaningful forms of tourism such as cultural and heritage tourism are being sought by the demand side and supply side of tourism. Hence, the search for a form of tourism that could be more profound for tourists was initiated.

The search for meaningful tourism and travel experience leads travelers towards heritage tourism. A detailed review of the history of tourism in the UAE indicated the major goal of establishing the image of a modern and manmade marvel as a tourist destination (Melotti 2014). Following the global trend of a switch of the demand side towards heritage, history and cultural tourism, local heritage has been valorized and tourism strategies have been crafted around it by the supply side. In order to enhance marketability and profitability, the experiential side of heritage tourism is being developed.

As a unique contribution to knowledge and practice, this study applies the design thinking process of innovation to the development of new experiences for heritage tourism in the UAE. Design thinking is "a process for creative solving" (IDEOU 2018). It is a human-centered process that includes five steps: *empathy, defining, ideation, prototyping and testing* (Liedtka, Salzman, and Azer 2017). Through workshops organized with heritage tourism authorities, an effective design for heritage tourism experiences is suggested that will be more customer-centered. Each of the five steps of design thinking has been mapped with the various stages of the research process adopted in this chapter. The "empathy" indicates the intent, motives and benefits of the study. The "defining" illustrates the parameters set to facilitate the research process and reach the findings of the study. The "ideation" and "prototyping" elaborates on the data analysis and triangulation to conclude with the emerging themes. Finally, the "testing" presents interviews and focus group discussions and workshops with respondents as part of data collection and data analysis.

The purpose of this chapter is to: (i) underscore the current state of the experiential nature of heritage tourism in the United Arab Emirates – based on primary and secondary data from public and private resources, (ii) assess different types of tourism experiences developed in the UAE – based on direct interviews and focus groups with tourism experts in the UAE, and (iii) demonstrate how design thinking can be used to develop more customer-centered experiences in heritage tourism for the UAE.

Literature review

The relevant literature studied for this research differentiates heritage tourism from experiential heritage tourism, and it further analyses the different elements that build the heritage tourism experience. The literature review guided this study towards the gap in tourism literature that could be filled by directly questioning related tourists. The direct inquiry was based on face-to-face interviews with tourists who followed a research questionnaire designed for this study. The questionnaire was constructed upon the elements highlighted by the literature that describes the heritage tourism experience. Each of these elements will be discussed in the succeeding sections.

Heritage tourism

Heritage and the branch of tourism emerging from it as heritage tourism have been correlated with culture, history and social norms and practices, under the umbrella of cultural tourism (Chhabra, Healy, and Sills 2003; Yu Park 2010). In recent literature, tourism has been accepted as the successful marketing process for heritage and meanwhile the context and marketed products of heritage and heritage tourism have been intensively argued (Dallen 2011; Dixit 2017; Pera 2017). Graham, Ashworth, and Tunbridge (2000: 3) concluded that heritage is "capable of being interpreted differently within any one culture at any one time, as well as between cultures and through time".

The International Council on Monuments and Sites (ICOMOS) defined "heritage" as a broad concept that includes tangible assets such as natural and cultural environment, encompassing landscape, historic places and build environment as well as intangible assets such as collection, past and continuing cultural practices, knowledge and living experiences (Hosagrahar, Soule, Fusco Girard and Potts 2016). Seraphim and Haq (2018) indicated that the UAE has various recognized heritage sites, specifically those in Al Ain classified as "World Heritage Sites" by UNESCO, including the six oases and the archaeological sites of Hili, Hafeet and Bida bin Saud (UNESCO 2017).

Kumar and Singh (2015) established that heritage tourism includes two main kinds of heritage: natural heritage and cultural heritage. Dallen (2011) further specified that natural heritage includes naturally occurring phenomena such as forests, rivers, mountains, deserts and beaches. Meanwhile, cultural heritage incorporates the history made by people and its physical artifacts, while natural heritage is a naturally created type of tourism where people travel to be close to nature (Dallen 2011). This study is clear of any political motives and focuses on applying design thinking to the improvement of experiential tourism in general and experiential heritage tourism specifically. Fundamental elements derived from the literature to be discussed and utilized as discussion criteria and qualitative data for this paper are detailed below:

Involvement

The outcome of experiential tourism is embedded in the involvement of the traveler with the destination and connected tourism services (Andrades and Dimanche 2014; Meyer and Schwager 2007; Otto and Ritchie 1996). The customer experience has been defined by Meyer and Schwager (2007: 2) as "the internal and subjective reactions and feelings experienced by consumers when they have any direct or indirect contact with the company" and resulting in "customer satisfaction". In experiential heritage tourism, the direct or indirect contact with the people, places and events will be considered as customer involvement and engagement resulting in the tourist's experience.

A different form of tourist experience is the art or tactic of storytelling, which is used as a powerful co-creative behavior in tourism (Pera 2017; Smith and Weed 2007). In order to make a tourist co-creative in a tourism process or enhance their involvement, storytelling has been used successfully (Pera 2017). Narrative inquiry or storytelling has been suggested for theorizing qualitative research, especially in sports tourism (Smith and Weed 2007). However, this study established storytelling as a tool adopted by heritage tourism experts to engage their customers and enrich their experience.

Memorability

Memorability is another significant outcome of a tourist experience resulting in customer satisfaction and destination loyalty (Andrades and Dimanche 2014; Cary 2004; Otto and Ritchie 1996). As discussed previously, tourist experience is an outcome of a set of physical, spiritual, emotional and intellectual pleasures, personally perceived by a tourist (Haq and Medhekar 2019; Otto and Ritchie 1996). The memorability of a tourism experience is based on positive outcomes for a tourist that are more achievable when the tourist is more involved. Cary (2004: 63) further explains that the memorability of a tourist experience depends upon unplanned, incidental, and casual moments for tourists that "simultaneously produces and erases the 'tourist' as a subject". Obviously, the greater the memorability of a tourism experience the greater will be the tourist's loyalty and visit frequency, along with positive word of mouth recommendations. In this research on tourism experience regarding heritage tourism, memorability will be a key exploratory agent. Various questions asked of the tourism operators will be aligned with plans and strategies to enhance the memorability of the travels.

Authenticity and satisfaction

Satisfaction has been commonly understood as the outcome of tourism experiences (Haq and Medhekar 2019; Tung and Ritchie 2011). Based on the research on tourism experiences, Ryan (1997) claims that a tourist is satisfied from the travel experience if there is an obvious equivalence between the need and performance. Researchers agree that the tourism experience is based on two dimensions: existential reality and structural reality (Hamilton-Smith 1987; Otto and Ritchie 1996). Existential reality indicates "high levels of satisfaction, freedom, involvement and (intrinsic) reward", while structural reality involves "the extent to which the activity is task oriented and wherein completion is externally reinforced" (Hamilton-Smith 1987: 334). The satisfaction from heritage tourism experience has been recognized to be heavily dependent upon the authenticity of the tourism experience (Chhabra, Healy, and Sills 2003; Meyer and Schwager 2007; Otto and Ritchie 1996; Redfoot 1984; Ryan 1997).

Authenticity in tourism experiences was classically studied by Redfoot (1984). Based on his qualitative study, Redfoot (1984: 291) concluded that: "the tourist has become the symbol of a peculiarly modern type of inauthenticity". Yi, Fu, Yu, and Jiang (2018) illustrated from their study that authenticity and loyalty are strongly correlated in heritage tourism. In this study, the satisfaction from the heritage tourism experience and its authenticity plays a vital role in preparing the theory and implications. Hence, all respondents were asked about the satisfaction level of their heritage tourists, and how the agencies had measured their tourists' satisfaction.

Research goals and objectives

This chapter aims to fill the gap in the literature by providing triangulated data to discuss experiential heritage tourism in the UAE. Moreover, the design thinking process, which has been successfully accepted and applied in many business scenarios, seems less valued in the tourism industry. This chapter attempts to apply the design thinking process to the study of experiential heritage tourism in the UAE. The overarching research question of this research study is: "How can design thinking be applied to enhance experiential heritage tourism in the UAE"?

Methodology

In this qualitative research, direct interviews and two workshops as focus groups with heritage tourism experts were the research instruments used (Barbour 2008; Bello, Lovelock, and Carr 2018; Micari, Streitwieser, and Light 2005). In order to reach real experts and practitioners in heritage tourism, purposeful sampling was undertaken to effectively utilize the Delphi technique and focus groups (Browne 2005; Pham, Fucci, and Maalej 2018). The analysis of the qualitative data collected was based on the ontology of critical realism since the authors had to take a critical position and realistically report the findings (Lincoln and Guba 2003; Seraphim and Haq 2018).

The respondents interviewed as experts and attendants of the two workshops had some overlap but their roles in the workshops were more focused towards the design thinking process rather than experiential heritage tourism. The list of respondents is attached in Appendix 17.2. The UAE based organizers of heritage tourism who contributed to this study were qualified if they were directly or indirectly designing or organizing heritage tourism, if they were experts in experiential heritage tourism and if they agreed to participate. The interview questions can be found in Appendix 17.1.

The aim of this study was the depth, and not the quantity, of the data. The sample size was not pre-specified and hence the rule of data saturation was adopted: the collected information seemed to be repetitive after achieving an understanding of diverse experiences corresponding to the related topics (Browne 2005). The sample size was finalized when the knowledge provided by tourism experts was similar to previous ones and no new themes could be recognized by both researchers (Lincoln and Guba 2003). The final sample of this study had 13 heritage tourism practitioners.

Findings and discussions

As mentioned earlier, this research is following the five steps of design thinking. Through interviews, the study has accomplished the first step of design thinking: empathy. By having discussions with professionals from the tourism sector, the authors were able to move on to the second step: defining. Several needs were identified.A perception of a lack of authenticity might turn away many potential visitors from a destination. Organizers seem to sell foreign activities as "authentic", because they are more marketable than real authenticity. "*I realized there was a lack of authenticity in the desert. Why is there belly dancing? It is not from here. Why do Emiratis tolerate it?*" (E6).E4 further explained that to attract tourists the UAE has for a long time tried to display a modern image with its glittery skyscraper landscape, to the detriment of its rich history and heritage:

> *A decade or so ago, the city wanted to put Dubai on the map, to become the city of the future. What was overlooked, however, you need to project the city of the future, with an amazing history. A lot of people do not know there is a history; they believe there is no soul.*

Similarly: "*However, people feel it has been staged, orchestrated*" (E5). Many tour guides are frustrated that Emirati history is often ignored by visitors and residents alike.Nevertheless, E6 explained that authenticity alone does not sell; added value is also needed. His authentic experiences gained value with the addition of gourmet meals: "*For example, here we add gastronomy, like our lamb ouzi is a meal the guest will remember for years.*"Indeed, the successful experiential tours in the region added a twist to authenticity. Some tours propose food tours in Dubai, a falconry experience in the desert or a visit to an Emirati house.Experiential tours rely heavily on storytelling. E11 commented that tourists are committing, so they need to "put on a show".E2 insisted that what is important is not the information itself, but the way the information is delivered. For example, he believes it is more interesting to talk about how coffee was discovered rather than listing the steps to do an Emirati coffee. Storytelling techniques and focus were different for each person interviewed. Some concentrated on the stories, other on the tone and hand gestures, others favored jokes: "*You do not want to be an audio guide.*" (E8).In addition, the focus of their storytelling varied depending on the themes of the experiences. The stories could be about a cultural element: "*An Arab trader was traveling in Cambodia. He smelled something, it smells good. People told him the smell came from a forest fire. When the fire was over, the trader takes the tree bark and starts experimenting with it*" (E2).Some focus on the origins of the food and the people behind the food (E3). Falconers talk about their birds: "*One of our birds is called 'Phoenix', because he was injured and came back to health* [...]. *He was reborn from his ashes.*"Experiential tours need the inclusion of the five senses. Some focus mostly on sightseeing and hearing the stories, while others try to combine all of the senses. "*When we walk, we see everything (vision), we are touching and feeling (walking), we are smelling Hindi lane/coffee museum, we hear the birds flying, we are tasting chocolate/dates/pastries/food*" (E7).In the interviews, one issue kept coming back: the need to increase the number of Emiratis involved. Some tourists complained that they barely met any Emiratis during their trip. E1 stated that there are approximately thirty certified Emirati guides in Dubai. E3 asserted that careers in the tourism sector should attract more local youth: "*As ambassadors of their country they* [Emiratis] *play a central role in presenting local culture to visitors.* [...] *the involvement of well-qualified, multilingual UAE nationals needs to be fostered*".A respondent specified that Emirati guides were too expensive: "*For the price of four expatriate guides we can hire one Emirati guide*" (E6).Another dominant issue was that tourists were too shy to talk to Emiratis engaged in tourism. According to E2, elderly ladies hired to showcase the making of handicrafts should have a formal training in welcoming visitors. E8 had a similar experience when he volunteered at the heritage festival of Qasr Al Hosn.

> *I do not know why visitors do not dare to ask questions to locals. I used to gather people every 10 minutes and tell a story to them, for example about how to use the palm fronds, and how they were cut. It started with groups of 3–4 people and ended up with groups of 30.*
>
> *(E8)*

In addition, E2 believes that there should be a volunteering program involving the Emirati youth in the Heritage and Tourism sector, to nourish their interest in these fields. "*We need to teach it from tenth grade in school. We need to set up culture scouts and teach; they will practice with older people and a generation will become more enthusiastic about culture.*"Moreover, unlike expatriate tour guides, the Emirati tour guides live and experience the daily culture. Emirati tour guides insist that they are the only ones who can explain the intricacies of many cultural aspects of Emirati life. "*With tour guides that are not local: they learn things by heart, but they do not know societal realities, such as Emirati marriages*" (E8).E9 reported: "*Many visitors are from outside the country and also expats. Many expats have never visited a local house, even if they have visited a local house, they never have experienced any stories*".In summary, professionals from the tourism sector need to valorize the authenticity of the Emirati heritage. The expatriate tourism professionals are offering exceptional touristic experiences to the guests, such as falconry experiences, walking city tours, food tours, nature and desert tours, photo tours. However, if tourists want to learn about Emirati culture, the Emiratis guides are the ones that possess this specific knowledge.The research accomplished empathy and defining the steps of the design thinking process through heritage and tourism sector professionals' interviews. Based on their feedback, solutions were ideated with the heritage experiential tourism sector's needs in mind, such as the need for an increased Emirati involvement, the inclusion of the five senses, storytelling skills and the valorization of authenticity. The solution selected was to develop a design thinking workshop devoted to heritage experiential tourism, where tourism sector professionals were invited to create their own unique experience for heritage tourists. The authors prototyped and tested this workshop on two occasions. The workshop included the empathy steps, the defining steps and the ideating steps. The participants started by describing who their customers were and they then had to list their customers' feedback on the tour and their experiences of the UAE (empathy steps). In taking into account the empathy step, they had to define their customers' needs (defining step). Finally, they had to design a unique experience for their customers (ideating). After filling up a feedback survey, they were told to prototype and test this new tour on their guests. The results of this testing will be the focus of a future article.The first workshop was conducted with nine people coming from eight different touristic ventures. The participants were tour guides, the founders of an experiential tourism company, the founder of a cultural consulting firm, falconers, an operation manager and a tourism professor. What they really valued in this experience was meeting other professionals from the industry.The second workshop was undertaken for an experiential tourism company. Eight different employees were selected, such as reservationists, sales managers, events sales specialists, a marketing executive, an operations team leader and two conservation guides. Unlike the previous workshop, this event did not offer any opportunity for networking as every participant was from the same company. After each group presented their idea, the supervisors were asked about their thoughts on the workshop. Some ideas seemed creative, but the manager found them difficult to implement. As P7 from the first workshop stated: "*I will think there must be a bridge between the exercise of business aspect and activity aspect.*"For example, a team wanted to extend the duration of their safaris to add more activities but the manager argued that guests would need a second meal, thus increasing the cost of the experience. Another team wanted to add an experience where guests could be welcomed into Emirati families. The manager objected that the company had to confirm the reliability of the family in order to be sure that they would not cancel the tour suddenly, and they would also need to know if they could offer this experience regularly. To implement this experience, the company needs to know and trust the right people.The most important advantage of this workshop, according to the participants was the ability to brainstorm ideas in teams: "*Working in a team means you can bounce ideas off each other.*" Most participants of both workshops appreciated that they had to "think outside the box", and the "analytical breakdown of the process". They liked the fact that design thinking methods guided them through the creation process of the new activity. "*It is something you can apply.*" Participants thought the method could improve the guests' engagement, increase the authenticity of the experience, and add the five senses and storytelling in tours. Most of them affirmed that they would use design thinking methods. The most surprising result was that participants particularly appreciated the social aspect of the workshop. They commended the networking opportunity that might lead to other collaboration possibilities.

Theoretical implications

The qualitative findings and outcomes based on the triangulation of investigators with primary and secondary data identified various themes: authenticity, memorability, use of the five senses and the engagement of tourists. This research implies that experiential heritage tourism in the UAE has great potential and these themes evidently provide a guideline for its growth. The application of design thinking to innovation is also an outstanding contribution of this chapter to the literature of tourism as well as innovation in practice. The themes presented in this research could be used as a research direction for any study regarding the application of design thinking to innovation within experiential heritage tourism in any other country.

Since the authenticity of heritage tourism is a long-standing question for the UAE, discussion with tourism experts delivered a clear finding that any place, group of people or event based on original UAE resources is called authentic. Memorability comes naturally with well-planned and executed experiential heritage tourism. The paper explains that travelers will have a memorable heritage experience due to the unique qualities of the people, places and events, rather than the glamour. Tourism managers need to realize that the five senses build the experience. This paper has demonstrated that seeing, hearing and smelling are overwhelming factors in the creation of a memorable heritage experience.

Engagement for heritage tourism depends on the participation of local people and the role of storytelling. Tourism companies need to train and hire local Emiratis to have first-hand experience with tourists so they get to know about the real Emirati culture, which connects with the heritage. Heritage experience can be built and enhanced by storytelling about the place, people and events.

Conclusion and future research

The conclusion emphasizes the role of local Emiratis as the key to tourist engagement, which would enrich experiential heritage tourism in the UAE. Nonetheless, the Emiratis could add an authentic dimension by transmitting specific cultural knowledge. The design thinking workshops highlighted the importance of having a process to design new interactive touristic experiences. The participants believed that the addition of a business dimension in the design thinking process was required, to make these new ideas more practical to implement.

Although the UAE depends on structural or tangible resources of heritage tourism, this study discovered that the success of heritage tourism, especially in the experiential segment, will depend a lot on intangible resources. The questionnaire that was designed to interact with tourism experts had to be constructed upon key concepts of experiential heritage tourism, which also required an in-depth study of the literature. The literature study confirmed that there was no study that applied design thinking for innovation to experiential heritage tourism.

Looking in the direction of future research, a paradigm shift in research is suggested. The four themes presented in this chapter could be further investigated through heritage tourists coming from overseas, local residents and expats. A quantitative and confirmatory study of experiential heritage tourism in the UAE is a possible subject for future research. A survey questionnaire could be prepared to test and confirm the qualitative findings of this research. Similarly, a survey questionnaire could be designed to study the objectives of public and private tourism operators in the UAE who are involved with experiential heritage tourism. A new method, combining design thinking, socialization and business conditions could be developed and tested again on the tourism sector, as well as on tourists, for future research. Finally, the impact of design thinking on the tourism sector could be studied further. If these institutions used the design thinking process, would they offer more interactive experiences to tourists?

References

Andrades, L., and Dimanche, F. (2014). 'Co-creation of experience value: A tourist behavior approach'. In N. Prebensen, J. Chen, and M. Uysal (eds.), *Creating Experience Value in Tourism* (pp. 95–112). London: Cabi. DOI: 10.1079/9781780643489.0095

Barbour, R. (2008). *Doing Focus Groups*. Thousand Oaks, CA: Sage. DOI: 10.4135/9780857029034

Bello, F. G., Lovelock, B., and Carr, N. (2018). 'Enhancing community participation in tourism planning associated with protected areas in developing countries: Lessons from Malawi'. *Tourism and Hospitality Research*, 18 (3): 309–320. DOI: 10.1177/1467358416647763

Boussaa, D. (2014). 'Cultural Heritage in the Gulf: Blight or blessing? A discussion of evidence from Dubai, Jeddah and Doha'. *Middle East – Topics & Arguments (META) Journal*, 2 (3): 55–70. DOI: 10.17192/meta.2014.3.2176

Browne, K. (2005). 'Snowball sampling: Using social networks to research non-heterosexual women'. *International Journal of Social Research Methodology*, 8 (1): 47–60. DOI: 10.1080/1364557032000081663

Cary, S. H. (2004). 'The tourist moment'. *Annals of Tourism Research*, 31 (1): 61–77. DOI: 10.1177/1468797610382701

Chhabra, D., Healy, R., and Sills, E. (2003). 'Staged authenticity and heritage tourism'. *Annals of Tourism Research*, 30 (3): 702–719. DOI: 10.1016/S0160-7383(03)00044-6

Dallen, J. T. (2011). *Cultural Heritage and Tourism: An Introduction*. Bristol: Channel View Publications. DOI: 10.1080/1743873X.2012.671565

Dixit, S. K. (ed.). (2017). *The Routledge Handbook of Consumer Behaviour in Hospitality and Tourism*. London: Routledge. DOI: 10.4324/9781315659657

Graham, B., Ashworth, G. J., and Tunbridge, J. E. (2000). *A Geography of Heritage*. London: Routledge. DOI: 10.4324/9781315824895

Hamilton-Smith, E. (1987). 'Four kinds of tourism'. *Annals of Tourism Research*, 14 (3): 332–344. DOI: 10.1016/0160-7383(87)90106-X

Haq, F., and Medhekar, A. (2019). 'Challenges for innovative transformation in heritage tourism development in India and Pakistan'. In S. Srivastava (ed.), *Conservation and Promotion of Heritage Tourism* (pp. 127–154). PA: IGI Global. DOI: 10.4018/978-1-5225-6283-2.ch006

Hoath, N. (2014, March 13). 'Call to attract young Emiratis into tourism industry'. *Khaleej Times*. Available from www.khaleejtimes.com/nation/government/call-to-attract-young-8232-emiratis-into-tourism-industry [Accessed: February 26, 2019].

Hosagrahar, J., Soule, J., Fusco Girard, L., and Potts, A. (2016). *Cultural Heritage, the UN Sustainable Development Goals, and the New Urban Agenda*. International Council on Monuments and Sites (ICOMOS). Available from http://www.usicomos.org/wp-content/uploads/2016/05/Final-Concept-Note.pdf [Accessed: February 28, 2020].

IDEOU. (2018). *What is Design Thinking?* [Online]. Available from www.ideou.com/blogs/inspiration/what-is-design-thinking [Accessed October 4, 2019].

Kumar, S., and Singh, R. P. (2015). 'Cultural-heritage tourism in Ayodhya-Faizabad: Scenario and prospects'. *The Geographer*, 1: 66–74.

Liedtka, J., Salzman, R., and Azer, D. (2017). *Design Thinking for the Greater Good: Innovation in the Social Sector*. Columbia University, NY: Columbia Business School Publishing.

Lincoln, Y. S., and Guba, E. G. (2003). 'Paradigmatic controversies, contradictions, and emerging confluences'. In N. K. Denzin and Y. S. Lincoln (eds.), *The Landscape of Qualitative Research: Theories and Issues* (2 ed, pp. 253–291). London: Sage.

Melotti, M. (2014). 'Heritage and tourism. 'Globalization and shifting values in the United Arab Emirates''. *Middle East-Topics & Arguments (META) Journal*, 2 (3): 71–91. DOI: 10.17192/meta.2014.3.2181

Meyer, C., and Schwager, A. (2007). 'Understanding customer experience'. *Harvard Business Review*, 85 (2): 116–126.

Micari, M., Streitwieser, B., and Light, G. (2005). 'Undergraduates leading undergraduates: Peer facilitation in a science workshop program'. *Innovative Higher Education*, 30 (4): 269. DOI: 10.1007/s10755-005-8348-y

Otto, J., and Ritchie, J. R. B. (1996). 'The service experience in tourism'. *Tourism Management*, 17 (3): 165–174. DOI: 10.1016/0261-5177(96)00003-9

Pera, R. (2017). 'Empowering the new traveller: Storytelling as a co-creative behaviour in tourism'. *Current Issues in Tourism*, 20 (4): 331–338. DOI: 10.1080/13683500.2014.982520

Pham, Y. D., Fucci, D., and Maalej, W. (2018). 'A first implementation of a design thinking workshop during a mobile app development project course'. *arXiv preprint* arXiv:1803.10587.

Redfoot, D. L. (1984). 'Touristic authenticity, touristic angst, and modern reality'. *Qualitative Sociology*, 7: 291–309.

Ryan, C. (1997). *The Tourist Experience: A New Introduction*. London: Cassel. DOI: 10.1002/jtr.462

Seraphim, J., and Haq, F. (2018). 'Challenges for promotion of heritage tourism: Case study of the UAE'. In S. Srivastava (ed.), *Conservation and Promotion of Heritage Tourism* (Chapter 2, pp. 35–56). PA: IGI Global. www.igi-global.com/gateway/chapter/213400.

Smith, B., and Weed, M. (2007). 'The potential of narrative research in sports tourism'. *Journal of Sport & Tourism*, 12 (3–4): 249–269. DOI: 10.1080/14775080701737005

Tung, V., and Ritchie, J. R. B. (2011). 'Exploring the essence of memorable tourism experiences'. *Annals of Tourism Research*, 38 (4): 1367–1386. DOI: 10.1016/j.annals.2011.03.009

UNESCO. (2017). 'Cultural sites of Al Ain'. *World Heritage List*. Available from http://whc.unesco.org/en/list/1343 [Accessed: October 4, 2019].

Yi, X., Fu, X., Yu, L., and Jiang, L. (2018). 'Authenticity and loyalty at heritage sites: The moderation effect of postmodern authenticity'. *Tourism Management*, 67: 411–424. DOI: 10.1016/j.tourman.2018.01.013

Yu Park, H. (2010). 'Heritage tourism: Emotional journeys into nationhood'. *Annals of Tourism Research*, 37 (1): 116–135. DOI: 10.1016/j.annals.2009.08.001

Appendix – 17.1

The research questions asked in the interviews:

1. Can you explain the heritage offered by your company? What is unique about your tours (desert safaris, falconry experiences and Bedouin experiences, astronomy, hot air balloon)?
2. How do the tourists experience heritage with their five senses in your tours?
3. What kind of visitors are attending your experiences?
4. How do you reach your visitors? How do they hear about you? What is your marketing strategy?
5. What is the feedback for your tours?
6. Do some of the visitors come back to your tours? Why?
7. What does authenticity mean to you in terms of tourism? What is authentic in the UAE?
8. What are the different experiences offered to tourists in the UAE in terms of heritage?

   a. What activities do you offer to tourists to engage them as active participants?
   b. How can tourists interact more with local Emiratis?

9. What are the weaknesses in experiential tourism? What could be improved?
10. Please suggest your ideas for solutions to the above-mentioned issues related to experiential heritage tourism in the UAE.
11. What do you think is the future of experiential tourism in the UAE?
12. Which element/s of the tourism experience are controlled by your company?
13. How do you measure the satisfaction of your tourism experience customers?

Appendix – 17.2

Experts

E	*Official Description*
E1	The Assistant Head Manager at a Dubai museum
E2	E2 is a businessman who opened more than 40 business ventures in North America and the UAE. In 2016 he founded The Arab Culturalist, a cultural consulting firm which focuses on Arab and Emirati culture and cultural intelligence and spent eight years as the Managing Director and Lecturer of the Sheikh Mohammed Centre for Cultural Understanding.
E3	Project manager
E4	A founder of an experiential tourism company
E5	A founder of an experiential tourism company
E6	A founder of an experiential tourism company
E7	Tour guide and tour manager
E8	Emirati tour guide
E9	Emirati tour guide
E10a	Photographic tour guide
E10b	Art tour company
E11	Falconers/operation manager
E12	A founder of an experiential tourism company

Participants' workshop 1

P1	E2 is a businessman who opened more than 40 business ventures in North America and the UAE. In 2016 he founded The Arab Culturalist, a cultural consulting firm, which focuses on Arab and Emirati culture and cultural intelligence and spent eight years as the Managing Director and Lecturer of the Sheikh Mohammed Centre for Cultural Understanding.
P2	A founder of an experiential tourism company
P3	Falconers/operation manager
P4	Discover Old Dubai with Mr. Street Smart
P5	Tourism professor
P6	Dubai Photo Tours
P7	Tour guide
P8	Tour guide and online tour reservations

Participants' workshop 2

P9	Reservationist – Team Leader
P10	Reservationist
P11	Sales Manager
P12	Events Sales Specialist
P13	Marketing Executive
P14	Operations – Team Leader
P15	Conservation Guide/Astronomer
P16	Conservation Guide

18

Trends in experience design

Strategies for attracting millennials to wineries in Victoria, Australia

Paul Strickland, Jennifer Laing, Warwick Frost and Kim M. Williams

Introduction

Millennials are the newest generation to consume wine legally and it is important to understand their wants and needs in order to encourage future wine sales. They can be defined as being born after 1982 and before 2005, sometimes referred to as Echo Boomers, the Internet Generation, or even a part of Generation Y, but are a group shaped by well-defined events moving through time (Howe and Strauss 2009). Millennials arguably have distinct characteristics, such as being generally well educated, innovative, compassionate and socially conscious (Howe and Strauss 2009). Although some of these characteristics may occur with other generations, millennials appear to combine them all.

Nowak, Thach and Olsen asked the question "how does a winery successfully build its customer base? One method may be to start cultivating relationships with the millennial generation – potentially 76 million new customers" (2006: 316). This seems simple, however Howe and Strauss (2009) argue that marketing strategies often fail to engage with this generation, as most businesses erroneously make the assumption that millennials "will follow blindly along all the life-cycle trends initiated (30 and 40 years ago) by Boomers and confirmed (10 and 20 years ago) by Gen Xers" (p. 148). For the wine industry to continue to grow wine sales, it is essential that millennials are considered as a distinct market segment, as Barber, Dodd and Ghiselli (2008) argue, as this demographic appears to prefers social, hedonic wine marketing campaigns and imagery, whereas Generation X, for example, tend to pay more attention to wine critic opinions.

Alonso et al. (2013) suggest there is a lack of research being conducted into new trends, experiences and technological developments in the wine industry. Further research is therefore required to explore this new market in depth, including millennials' potential as wine tourists in an expanding Australian market. Using a case study approach, this chapter explores trends in experience design and marketing strategies used for attracting millennials to wineries in Victoria, Australia. More exploration of this demographic may assist with encouraging millennials to visit wineries and thus increasing wine sales.

Literature review

Millennials and wine

Winery visitation by all generations is a key factor to increase a winery's customer base (Byrd et al. 2016). The wine industry, however, has a particular interest in understanding the consumption behavior of millennials and encouraging their visits to wineries as they have reached an age where they can make their own purchasing decisions and are now an identifiable demographic (Pomarici and Vecchio 2014). Millennials are becoming a significant market segment for increasing wine sales (MacDonald, Saliba and Bruwer 2013). Nowak and Newton (2008) have highlighted that many millennials like to engage with new, hedonic and social experiences in general, including at wineries and wine events. This generation also appear to have different wine preferences compared to other generations. These include lighter, sweeter wines, and therefore the varietal style of wine offered should be a major consideration when seeking to develop new markets (Saayman, Saayman and Joubert 2012).

Additionally, millennials, in general, are believed to be very tech savvy, have an interest in social issues and be environmentally conscious (Higgins, McGarry Wolf and Wolf 2016), which are all characteristics that the wine industry should consider when seeking to attract this demographic to visit wineries. Millennials have matured using technology in all aspects of their lives, especially smart phones and social media (Deal, Altman and Rogelberg 2010). According to Pelet et al. (2018), this can be exploited by the wine sector, for example with SMS technology, to encourage sales. Research based on millennials has considered website designs for wineries (Nowak and Newton 2008), creating wine brand loyalty (Nowak, Thach and Olsen 2006) and service quality expectations at the cellar door (Charters and O'Neill 2001). Additionally, wine experiences are well documented in other generational studies (Barber, Dodd and Ghiselli 2008; Qenani-Petrela, Wolf and Zuckerman 2007). However, there is a paucity of studies on wine experiences for millennials, and thus a large gap in the literature.

Trends in wine tourism experience design

Dodd (1995) was one of the first researchers to understand the benefits of attracting wine tourists to wineries and identified the importance of offering different tourism products and services to increase wine sales. The focus has now moved on to offering *experiences* to visitors. Wine experiences can include many diverse offerings. The most common in Australia involve wine tasting at the cellar door, wine education, events and wine festivals. These festivals are major tourist attractions and provide strong motivations for travel, increase awareness of destination regions and generate wine sales (Barber, Donovan and Dodd 2010). Indeed, "wine festivals are becoming intertwined with other tourism attractions" and should be considered as a part of every wine region's calendar of events (Stoddard and Clopton 2015: 225). Many wineries are also now offering a variety of novel experiences such as helicopter rides, ballooning and live entertainment such as music or face painting for children, to help bridge the gap of activities for different generations. Some wineries also offer additional products and services such as gift shops, restaurants and cafés and have the ability to host events on site such as weddings, which is a growing market in New World wine countries (Taplin and Nguyen 2016). Although research indicates the importance of wine events for wineries, there are few studies that specifically showcase wine events staged for millennials. This is another important gap in the literature that needs to be addressed.

Furthermore, studies regarding wine experience behavior generally focus on wine purchase behaviors and the reasons why certain wines are purchased according to variety, taste, price and attractiveness (Johnson and Bruwer 2003). Alonso and Northcote (2009) focus on the creation of wine products for different generations, suggesting that each generation has different wine variety preferences. Moreover, evidence regarding the types of experiences at wineries is available but it mainly focuses on the cellar door experience or on other generations than millennials (Fountain and Charters 2010). Research about the types of winery experiences that are specifically targeting millennials and the concerns of management with respect to this gen-eration of wine consumers is scarce. This chapter aims to address this gap in the literature, along with an examination of the implications for the wine industry.

Technology and social media

Tanford, Montgomery and Hertzman (2012) suggest that wine experiences, enjoyment and social recognition are the main driving factors of attendance at a wine event, with technology such as social media becoming more important to promote these events. The emphasis on social media accords with the view that millennials are mostly "technology savvy and use technology in almost every aspect of their lives" (Atkin and Thach 2012: 56). As such, technology is an important platform to communicate with millennials, as it becomes more widely accepted and utilised in daily activities (Szolnoki et al. 2018). Leung et al. (2013: 3) describe technology, in particular the use of social media in the tourism and hospitality industry, as being a "mega trend" and a tool that should be widely embraced by the industry.

However, there is evidence of low adoption rates by the Australian wine industry of some types of technologies, due to time limitations, financial restrictions and lack of industry expertise, including familiarity with and use of certain social media platforms (Strickland et al. 2013). This inability to engage with technology may affect the likelihood of being able to target millennials and thus may have a negative impact on visitation and sales, which is typical of some sectors that do not fully understand how technology can have a positive impact on their business (Cabiddu, De Carlo and Piccoli 2014). This study highlighted that some sectors have an unwillingness for change and suggests a reluctance to accept technology in particular.

Social media is an example of a new form of word-of-mouth communication, which is widely utilised and preferred by millennials (Sigala 2011). Positive word-of-mouth through social media offers a high level of credibility amongst tourists (Bronner and de Hoog 2016). This study found that social media is the main platform that millennials use for finding information on what to do and where to go, especially during their free time. In another study, Hubbard et al. (2012) advocated that most millennials attend festivals because of word-of-mouth or a friend's recommendation and suggested that technology can be very influential in enhancing this effect. In fact, social media is a modern way for many generations to share experiences, including millennials (Sigala 2011). There is a dearth of research, however, as to how wineries are developing and implementing social media marketing strategies to attract millennials as wine tourists. This case study will explore this issue in the context of Victoria, Australia. This case study context was chosen as Victoria has a long-standing and profitable wine industry, as discussed in the next section of this chapter.

Case study 18.1

The state of Victoria is located in the south-eastern corner of the mainland of Australia. Wine making started in Victoria in the 1840s and has continued to grow, making it one of the oldest wine regions in Australia. In 2013, Victoria had 526 wineries offering wine tasting through their cellar doors (tasting rooms), which represents 30% of all Australian wine cellar doors, according to Wine Victoria – Australia (2017a). Victoria offers different microclimates and terroirs, allowing for a variety of regional wine styles. The state produces approximately 30% of all grapes in Australia and is a significant contributor to overall wine production, international wine exports, local wine sales, grape varieties and direct and indirect employment (Wine Victoria – Australia 2017b). In a highly competitive market, the challenge for these wineries is "the need for domestic market growth through a direct route-to-market [which] means regional wineries, particularly small-to-medium producers, are particularly reliant on increasing sales through their cellar door" (Wine Victoria – Australia 2017b: 13). For this reason, 67.8% of Victorian wineries operate a cellar door to sell directly to visitors (Wine Victoria – Australia 2017a).

As millennials are a growing wine consuming market, it is important to explore how wineries are seeking to attract these individuals in order to capitalise on increasing wine sales. In this case study, a variety of wine stakeholders were interviewed to explore how they approached designing experiences for millennials (refer to Table 18.1). Participants were identified by their location and job title, and then approached for a face-to-face interview that lasted generally an hour at their point of business. The snowball technique was used to recruit other wine stakeholders, who were recommended as suitable to be interviewed.

Table 18.1 Participant number and job title/organisation

Participant Number	*Job Title/Organisation*	*Participant Number*	*Job Title/Organisation*
P1	Winery Owner	P9	Winery General Manager
P2	Winery General Manager	P10	Winemaker
P3	Winery Marketing Manager	P11	General Manager, Wine Industry Association
P4	Winery Owner	P12	Representative, Destination Marketing Organisation
P5	Wine Festival Organiser	P13	Director of Festivals and Events, Local Government
P6	Winemaker	P14	Marketing Coordinator, Wine Event Company
P7	Social Media Consultant to Winery	P15	Representative, Destination Marketing Organisation
P8	Social Media Consultant to Winery	P16	Cellar Door Manager

The participants freely discussed the marketing strategies that they used to attract millennials to their winery or the wineries that they represented or serviced. Open questions asked included: *What marketing strategies do you implement? Do you specifically target millennials to visit your winery?* and *What technologies do you use in marketing*? From these discussions, two main themes emerged, being *hedonic and new experiences* and the *use of technology*. This case study discusses each theme in turn, highlighting the viewpoints of the participants.

Hedonic and new experiences

The majority of participants identified that creating hedonic and enjoyable experiences for millennials is an important factor in attracting new and repeat visitors. The participants also indicated that the nature of what constituted hedonic and new experiences could be interpreted in a number of ways. Firstly, visiting a winery for the first time was considered a new experience for some participants. However, P4 commented: "I would say that millennials want to be trendy and try something new. A winery can offer that."

Secondly, the participants indicated that millennials are interested in tasting wines but acknowledged that it needed to be done in a hedonic and enjoyable way, as an important element of the overall experience. For example, P10 stated

> I think millennials want a combination of a lot of things. For example, they want to taste new wines, look at the view, take photos, be educated, make a purchase, have fun and then brag to their friends what a great day they had.

Often this is instant via social media. P12 observed "we have an online presence and social media presence so millennials would easily be able to find us or access information about us and post about us".

Connecting millennials with the winery was seen as very important so as not to alienate this segment and to maintain their interest. P14 commented

> millennials are the next wave of drinkers so I think they really want to get to know the wine, how it's made, how it's grown and have a fun experience with their friends while they are doing it. We make sure that everyone is welcome and feels comfortable tasting wine.

The type of wine available to tourists was also tailored to appeal to millennials, as a new experience. P16 stated

> young wine drinkers have different tastes to older, more experienced wine drinkers so we need to create wines that are appealing. I see millennial wine drinkers as consuming lighter, sweeter wines such as Rosé or fruity Riesling rather than a heavy, full rounded red that is suited to the Baby Boomers' generation.

Additionally, P2 observed that

> I make sure that I have at least two wines that appeal to this generation. For example, I have one red wine that is sweet and another that appeals to older generations. This way, I can offer something for all taste buds.

The fact that these winemakers are producing wines that suit each generation's tastes supports the previous findings of Saayman, Saayman and Joubert (2012).

It is important to recognise that these responses have indicated that some millennials do have a desire to visit a winery, taste wines that suit their palate and be educated about the wine in the process. Wine education is mainly verbal, with the cellar door employees interacting with tourists in a conversational and unscripted way. However, often tasting notes are available and some wineries offer information on their website. P4 commented "I have seen young customers [millennials] access their phones for more information", suggesting that wineries should also provide information pertaining to their wines online, as this is generally millennials' preferred method of obtaining information. This is similar to the findings of Deal, Altman and Rogelberg (2010) with respect to the importance of wineries being online. Information about a winery and their wines may also be obtained through websites, apps, blogs, vlogs or social media, as indicated by the participants. According to P7 "there are so many different ways millennials access information, the wine industry needs to keep up with the changing technology".

Only one winemaker, however, suggested minor changes to their cellar door to accommodate younger customers and the millennial generation. P4 described the changes he had made as creating "more of a lounge room than a traditional cellar door ... we play current music in our tasting lounge so it doesn't seem as stuffy or old fashioned ... it's relaxed". Visiting the winery, a noted point of difference was that the furniture reflected a chilled and relaxed setting that millennials can relate to, offering soft couches and chairs with pillows for the tourist to sit on while tasting wines.

Apart from the interviews, an analysis of the types of activities available at wineries was also conducted through a systematic search of internet sites, brochures and social media, and this unveiled a plethora of additional activities offered by Victorian wineries to create hedonic, enjoyable experiences. For example, wineries may offer customised winery tours of the vineyard, ballooning over the winery region or host a monthly farmers' market. Wineries may also display art installations, operate a restaurant, have an art gallery, offer a cricket pitch, encourage a game of pétanque, conduct winery tours, arrange tractor rides or organise music concerts. These activities appeared to be used to establish a point of differentiation from other wineries. Photographic images of younger people in winery restaurants, at wine events, at music concerts, on winery tours and on Segways were used, suggesting the types of experiences that millennials may enjoy.

Attending a wine event as a new experience at a winery

Each person interviewed was involved in staging an annual wine event, only some of which were specifically aimed at attracting millennial tourists. The majority of participants in this current case study held a similar view that their events were not necessarily just for millennials. For example, P8 stated

> Our calendar of events makes every year fresh and new. There is something for everyone. Even if a visitor has been to a dinner last year, then [the] menu will be different this year. I would consider that a new product offering .... millennials love it.

One winery hosted music events for large numbers of people and sold many local beverage products such as beer, coffee and soft drinks. This particular winery only sells their own wines to visitors to boost their wine sales and they have two particular wines that suit the general taste of millennials (light and sweet). However, there are examples of wine events that sell wines from other wineries in the region. P1 explained that three music events per annum sell approximately half their winery's annual wine production, and therefore hosting music events is a necessary component of their marketing strategy and they must consider the millennial demographic in the future.

Additionally, "millennials can choose which winery they want to go to that is all part of the festival ... they are the ones that like to dress up in a costume and have fun with their friends" (P13). This response reflects the perception that many millennials visit a winery for the social aspect that a winery can provide and consuming wine is a secondary experience consideration. It also highlights that some millennials are outgoing and willing to embrace the atmosphere of the wine event by "dressing up".

Events and functions aimed at millennials

There are specific events aimed at millennials that are not hosted at a winery. They include dinners, wine awards events and functions. For example, there is an annual awards event based on emerging winemakers Australia-wide. This particular evening was hosted in various locations around Australia but was originally hosted in different restaurants with the capacity to cater for 150 attendees. P6 stated "this award is to celebrate young winemakers in the industry". Additionally, P6 commented "we put together a matching dinner and wine package that suit the millennial tastes". The menu reflected modern Spanish tapas food and light, sweet wines, which might be seen as appealing to a younger demographic.

P10 named wine functions and events throughout Victoria that are aimed at the millennial demographic, where their winery was actively involved in the organisation and supply of wine. These included wine tours with lunch and dinner highlighting wine styles that suit millennials and specific sections at wine shows that cater to the millennial market. P12 stated "there is a millennial wine group that organise functions after linking up with the wine industry ... they have a Facebook page, Instagram and even create YouTube videos". Further questions unveiled that this proactive group, also considered a club, specialises in creating functions for millennials and P12 believes the group is "growing in membership".

The use of technology

Social media was specifically mentioned as a technological platform that engages millennials. Wineries in Australia mainly promote themselves via Facebook, Instagram, Twitter and YouTube because "we use the most popular social media in Australia" (P1). Interestingly, while Snapchat is a common social media platform for communicating with groups, no participant mentioned using this platform. Some winery websites have an option of registering a membership, thereby signing an individual up to an email database that may be utilised for future marketing campaigns. However, P3 acknowledged that "we can always improve our marketing" and was the only participant that currently had paid advertisements on Facebook or YouTube. Some wineries did not have an email database. P10 stated that "we have a dedicated person promoting via social media ... but we need to be more engaged", acknowledging that their lack of expertise in social media is an issue. Others did not engage with social media as a major communication tool. P11 noted in relation to a wine festival that "we have been in existence for such a long time that everyone knows about us". Social media, however, was used by some wineries in part because their competitors were using it, rather than being seen as having intrinsic value. P5 explained that, "we really do not need Facebook, but it is all the rage" and further commented that while she thought that social media might be used by all generations, she found that it is more likely to be used by millennials as a source of communication.

For some wineries however, "part of the problem is getting mobile phone reception and the internet. If that problem was solved, we could do more" (P9). P9 is referring to investing in interactive technologies at wineries, such as the ability for wine tourists to instantly upload photos whilst they are at the winery, rather than at a later time. Some wineries do not have adequate internet or mobile phone reception, which is an issue in many regional areas of Victoria. More than half those interviewed indicated that there was a major challenge in not being able to interact with millennials (and other generations) in real time via technology when they were visiting the winery.

It was mentioned that social media could be a substitute for verbal word-of-mouth, which accords with the work of Sigala (2011). Examples given include "millennials use social media everyday so it makes sense to communicate to them in this way" (P6) and "Facebook is a way for millennials to chat" (P13). Most participants commented that much more could be undertaken regarding social media, however only two participants suggested that paid advertising might be a future option, despite the fact that sites such as Facebook are increasingly removing all unpaid commercial advertising in favor of paid advertising. For example, P12 commented that "Facebook can create an event and can invite friends" but has never used that functionality. Similarly, P15 asked "how do I make someone forward an email on or 'like' a Facebook page?" In fact, none of the participants indicated that they had implemented an online electronic word-of-mouth marketing campaign, which could be attributed to their lack of knowledge, time and financial constraints, as many participants acknowledged. P3 also mentioned that more interaction through technology when customers visit could be encouraged, including possible virtual reality tours, posting photos on large screens and online competitions.

Conclusion

This chapter highlighted that millennials are a new but significant contributor to the growing wine tourism sector (Higgins, McGarry Wolf and Wolf 2016). A discussion regarding the marketing strategies currently used to attract millennials to wineries from an industry (supply) perspective occurred, using Victorian wine regions as a case study.

Keeping up with trends in experience design appears to be essential for attracting millennials to wineries in Victoria. Wineries are considering the types of experiences offered and a number are offering new, hedonic, socially engaging and technologically interactive activities that are attractive and on trend for millennials (Ritchie and Hudson 2009). Traditional wine tastings and wine education through the cellar door should continue, however the diffusion of information is now often done in a relaxed, authentic and interactive way. Wine events aimed at millennials are a good example of a current trend involving a hedonic experience. Technology plays an important role in the way millennials communicate and therefore wineries should consider how to engage with social media and all online information at the product design stage of creating experiences. Providing real-time feedback for millennials is also a major consideration, as outlined by Kim and Fesenmaier (2015) and a substitute for word-of-mouth (Sigala 2011).

This chapter contributes to existing theory, highlighting millennials as a potential market to increase wine sales. The practical implications suggest that the wine industry should follow trends and be strategic in experience design to attract millennials to Victorian wineries. Possible ways for a winery to attract millennials as wine tourists involve offering hedonic, enjoyable and new experiences, embracing technology and social media, developing wines that suit the millennial palate and offering wine events specifically for millennials. Further research could explore these strategies from the perspective of millennials as potential wine tourists.

References

Alonso, A. D., et al. (2013). Website and social media usage: Implications for the further development of wine tourism, hospitality, and the wine sector. *Tourism Planning and Development*, 10(3): 229–248. DOI: 10.1080/21568316.2012.747989

Alonso, A. D. and Northcote, J. (2009). Wine, history, landscape: Origin branding in Western Australia. *British Food Journal*, 111(11): 1248–1259. DOI: 10.1108/00070700911001068

Atkin, T. and Thach, L. (2012). Millennial wine consumers: Risk perception and information search. *Wine Economics and Policy*, 1(1): 54–62. DOI: 10.1016/j.wep.2012.08.002

Barber, N., Dodd, T., and Ghiselli, R. (2008). Capturing the younger wine consumer. *Journal of Wine Research*, 19(2): 123–141. DOI: 10.1080/09571260802622225

Barber, N. A., Donovan, J. R., and Dodd, T. H. (2010). Differences in tourism and marketing strategies between wineries based on size and location. *Journal of Travel and Tourism Marketing*, 25(1): 43–57. DOI: 10.1080/10548400802164889

Bronner, F. and de Hoog, R. (2016). Travel websites: Changing visits, evaluations and posts. *Annals of Tourism Research*, 57: 94–112. DOI: 10.1016/j.annals.2017.02.007

Byrd, E. T., et al. (2016). Wine tourism: Motivating visitors through core and supplementary services. *Tourism Management*, 52: 19–29. DOI: 10.1016/j.tourman.2015.06.009

Cabiddu, F., De Carlo, M., and Piccoli, G. (2014). Social media affordances: Enabling customer engagement. *Annals of Tourism Research*, 48: 175–192. DOI: 10.1016/j.annals.2014.06.003

Charters, S. and O'Neill, M. (2001). Service quality at the cellar door: A comparison between regions. *International Journal of Wine Marketing*, 13(3): 7–17. DOI: 10.1108/eb008723

Deal, J. J., Altman, D. G., and Rogelberg, S. G. (2010). Millennials at work: What we know and what we need to do (if anything). *Journal of Business and Psychology*, 25(2): 191–199. DOI: 10.1007/s10869-010-9177-2

Dodd, T. H. (1995). Opportunities and pitfalls of tourism in a developing wine industry. *International Journal of Wine Marketing*, 7(1): 5–16. DOI: 10.1108/eb008636

Fountain, J. and Charters, S. (2010). Generation Y as wine tourists: Their expectations and experiences at the winery cellar door. *Tourism and Generation Y*, 15(5): 47–57. DOI: 10.1079/9781845936013.0047

Higgins, L. M., McGarry Wolf, M. M., and Wolf, M. J. (2016). Wine on Facebook: A look at millennials' wine information search. G. Szolnoki, L. Thach, and D. Kolb. (eds.), *Successful Social Media and Ecommerce Strategies in the Wine Industry* (pp. 13–29). New York: Palgrave Macmillan. DOI: 10.1057/9781137602985_2

Howe, N. and Strauss, W. (2009). *Millennials Rising: The Next Great Generation*. New York: Vintage.

Hubbard, K. W., et al. (2012). Perceptions of quality, satisfaction, loyalty, and approximate spending at an American wine festival. *Journal of Culinary Science and Technology*, 10(4): 337–351. DOI: 10.1080/15428052.2012.735962

Johnson, T. and Bruwer, J. (2003). An empirical confirmation of wine-related lifestyle segments in the Australian wine market. *International Journal of Wine Marketing*, 15(1): 5–33. DOI: 10.1108/eb008749

Leung, D., et al. (2013). Social media in tourism and hospitality: A literature review. *Journal of Travel and Tourism Marketing*, 30(1–2): 3–22. DOI: 10.1080/10548408.2013.750919

Kim, J. and Fesenmaier, D. R. (2015). Measuring emotions in real time: Implications for tourism experience design. *Journal of Travel Research*, 54(4), 419–429. DOI: 10.1177/0047287514550100

MacDonald, J. B., Saliba, A. J., and Bruwer, J. (2013). Wine choice and drivers of consumption explored in relation to generational cohorts and methodology. *Journal of Retailing and Consumer Services*, 20(3): 349–357. DOI: 10.1016/j.jretconser.2013.01.013

Nowak, L., Thach, L., and Olsen, J. E. (2006). Wowing the millennials: Creating brand equity in the wine industry. *Journal of Product and Brand Management*, 15(5): 316–323. DOI: 10.1108/10610420610685712

Nowak, L. I. and Newton, S. (2008). Using winery web sites to launch relationships with millennials. *International Journal of Wine Business Research*, 20(1): 53–67. DOI: 10.1108/17511060810864615

Pelet, J. E., et al. (2018). Winery website loyalty: The role of sales promotion and service attributes. *International Journal of Wine Business Research*, 30(2): 138–152. DOI: 10.1108/IJWBR-01-2017-0003

Pomarici, E. and Vecchio, R. (2014). Millennial generation attitudes to sustainable wine: An exploratory study on Italian consumers. *Journal of Cleaner Production*, 66: 537–545. DOI: 10.1016/j.jclepro.2013.10.058

Qenani-Petrela, E., Wolf, M., and Zuckerman, B. (2007). Generational differences in wine consumption. *Journal of Food Distribution Research*, 38(856-2016-57902): 119.

Ritchie, B. and Hudson, S. (2009). Understanding and meeting the challenges of consumer/tourist experience research. *International Journal of Tourism Research*, 11(2): 111–126. DOI: 10.1002/jtr.721 |

Saayman, M., Saayman, A., and Joubert, E.-M. (2012). Expenditure-based segmentation of visitors to the wacky wine festival. *Tourism Recreation Research*, 37(3): 215–225. DOI: 10.1080/02508281.2012.11081710

Sigala, M. (2011). Special issue on web 2.0 in travel and tourism: Empowering and changing the role of travellers. *Computers in Human Behaviour*, 27: 607–608. DOI: 10.1016/j.chb.2010.04.005

Stoddard, J. E. and Clopton, S. W. (2015). Exploring the differences between new and repeat visitors to North Carolina wineries: Implications for winery marketing strategy development. *Journal of Wine Research*, 26(3): 225–240. DOI: 10.1080/09571264.2015.1057272

Strickland, P., et al. (2013). The acceptance or rejection of social media: A case study of Rochford Winery Estate in Victoria, Australia. *Tourism, Culture and Communication*, 13: 19–27. DOI: 10.3727/109830413X13769180530602

Szolnoki, G., et al. (2018). Using social media for consumer interaction: An international comparison of winery adoption and activity. *Wine Economics and Policy*, 7: 109–119. DOI: 10.1016/j.wep.2018.07.001

Tanford, S., Montgomery, R., and Hertzman, J. (2012). Towards a model of wine event loyalty. *Journal of Convention and Event Tourism*, 13(2): 77–99. DOI: 10.1080/15470148.2017.1296393

Taplin, I. M. and Nguyen, M. T. (2016). Wine versus weddings: Wine tourism in the emerging North Carolina wine industry. K. Lee. (ed.), *Strategic Winery Tourism and Management* (pp. 31–50). New York: CRC Press. DOI: 10.1201/b1959

Wine Victoria – Australia. (2017a). Victorian wine industry snapshot, Victoria [online] Available at www.winevictoria.org.au/images/WVIndustrySnapshot_HR.pdf [Accessed: April 16, 2019].

Wine Victoria – Australia. (2017b). Policy Priorities, Victoria [online] Available at www.winevictoria.org.au/images/Wine_Victoria_Policy_Priorities_July_2015_lowres.pdf [Accessed: April 16, 2019].

19

The role of themes and stories in tourism experiences

Lee Jolliffe and Pairach Piboonrungroj

Introduction

A compelling story can be the basis of a product that the visitor engages with in the form of a tourism experience. Mossberg (2008) observed in the experience industry that a global trend is to build an entire business or parts of it around a story. This chapter examines the identification of themes as the basis of story development and discusses how narratives that form the story and its related storylines are created by different methods. Themes are applicable to destination development where a compelling story has to be articulated, in order to differentiate a location from others and attract visitors, as places compete with each other to create memorable visitor experiences (Campos et al. 2018). As a basis for developing storylines, themes offer a framework for developing opportunities for visitors to co-create their own tourism experience. Themes as the foundation of stories and narratives are in addition relevant for heritage tourism products, at a variety of natural and cultural sites where a thematic heritage identity is maintained and conveyed, or at a museum where a theme and the stories within it is portrayed through exhibitions and programming.

Themes and stories are not so much an emerging facet in the co-creation of tourism experiences as concepts that are often not always fully understood or appreciated. Storytelling in the context of tourism is not without issues and challenges. Designers of tourism experiences are faced with a number of questions to consider. Whose story will be told? How will the story be told? Is it possible to avoid the commodification that may result from the commercial use of the story for tourism? How can authenticity be maintained when the story is being told for the commercial purpose of tourism? How can the story be used to market the tourist experience?

Terminology regarding the development of stories includes the concepts of the theme, storyline, storytelling, storyboard and the story or narrative itself (Table 19.1).

The construction of stories normally begins by assessing resources for the story or narrative to be delivered, grouped into themes which in experience design can be conveyed into storylines to be developed and communicated to the visitor. The art of storytelling, defined both as "the activity of telling or relating stories" and "a design research process"

Table 19.1 Terminology in Story Development

Term	*Definition*
Theme	An idea or a point central to a story, often summed up in a few words.
Storyline	The plot of a novel, play or other narrative form such as an exhibition or an experience.
Storytelling	The activity of telling or relating stories. An interpretive technique. A design research process.
Storyboard	A series of drawings or sketches laying out in sequence how a story will be delivered.
Story	An account or narrative of past people, places or events.
Narrative	A synonym for story, meaning telling of a story. A portrayal of a series of events.

comes into play for the communication of stories through the experience economy. Storytelling is noted as the key to developing extraordinary experiences for visitors (Mossberg 2008). In visitor attractions such as museums stories are often told in a linear fashion within exhibitions. A chronological narrative can help the visitor to make sense of a product or experience and in the process become a co-creator of their own experience. It is noted that involvement and co-creation can facilitate visitor involvement in the story (Mossberg 2008). Techniques typically used in the development of stories include storylines that contain the messages to be delivered, as well as storyboarding in which the method of conveyance is sketched out.

This chapter introduces the role of themes and related stories in the tourist experience and reviews related literature. This draws in part from literature in the museum and interpretation fields where storytelling is recognized as central to the visitor experience (Pera 2017). This process of recognizing themes and developing stories is illustrated through a real-life case study. Beeton (2005) identifies the case study as a useful research method in tourism. The case study choice was one of convenience as the authors were involved in a team-based project with participants at the case study location. It was a form of action research (Veal 2017) which, having the advantage of being goal-driven, may provide innovative approaches to problem-solving and is rewarding for participants (Mason 2002).

Literature review

The tourism literature generally recognizes that tourism experiences can be enhanced by stories, and in addition that a compelling story built by tourists can make for a memorable (Chronis 2012) or extraordinary tourism experience (Mossberg 2008). Moscardo (2010) identified three threads in tourism research on themes and stories as; studies in destination image in marketing and media; research that directly analyses tourist stories and research into the role of narratives; and stories and themes in interpretive settings. This literature review highlights research in those areas.

Destination image in marketing and media

According to Bassano et al. (2019) storytelling about places is a technique to enhance the reputation of regions as they compete for tourism and economic development. Pera (2017) recommended that the value of storytelling in the marketing and communications process should be given increased recognition. Storytelling has therefore been used as a tool within destination development incorporating what Campos et al. (2018) note as the organization and or destination perspective of the co-creation of the tourist experience. The authors furthermore indicate that both the organization and destination perspectives of co-creation are similar in that they share the vision of it being a strategy applied towards competitiveness to be dealt with through management and marketing (Campos et al. 2018). Storytelling is a means of unifying the offerings of the destination.

Tourist stories and research into the role of narratives

Storytelling is a powerful instrument for co-creation of the tourist experience. Moscardo (2010) noted the importance of good storytelling in shaping the tourism experience. Pera (2017) introduced storytelling as an activity that facilitates co-creation. Campos et al. (2018) noted an increasing recognition of the value of the tourist as a co-creator of their experience. This incorporates the tourist perspective of co-creation. Involving tourists in experiences that capture their attention and match their interests has been noted to be important for co-creation (Andrades and Dimanche 2014).

Campos et al. (2018) identify three differing approaches to tourist co-creation reflected in the tourism literature as 1) the tourist contributes to some aspect, phase or the overall tourism experience, 2) the tourist actively participates in on-site experience activities, 3) the tourist interacts with others during the on-site experience, and 4) the tourist engages in an on-site subjective experience. There is thus a range in depth of the experience during co-creation and that can be enriched and encouraged by the use of narratives in interpretive settings that can engage and involve the visitor.

Stories and themes in interpretive settings

One of the foundations of storytelling in tourism, especially relevant for heritage and natural area tourism, can be found within Tilden's (1977) work on interpretation. Moscardo (2010) notes the significance of the use of narratives, stories and themes in interpretive settings that include museums, art galleries, science centres, historic and archaeological sites, zoos and aquaria and both guided and self-guided tours of cultural and natural heritage sites. In *Interpreting Our Heritage* (1977) Tilden portrayed "interpretation" as both an art and a science, being imaginative, meaningful, inspiring and involving while targeting the whole person and accepting that different approaches are needed for different visitors and also cross-cultural audiences.

In the museum context Witcomb (2003) notes that exhibitions set up a situation in which the visitor can create their own narrative. Writing further about storylines and narratives in museum exhibit design Forrest (2014) equates the thematic sections of an exhibit or experience of it from the visitors' perspective as akin to scenes or acts of a play. The author notes exhibit development and design can be guided by a storyline or an approximation of the eventual visitor experience in terms of what the visitor will take away with them (Forrest 2014).

Designers of the destination or site experience can thus use storylines to create opportunities for visitor interactions that will contribute to the visitor experience, both lived on-site and that which the tourist takes away with them. In exhibition design, the storyline may unfold in either a linear or chronological fashion (Forrest 2014) or a more organic spoke and wheel layout that offers the visitor more opportunity for choice. Likewise, for the tourist a standard tour of a destination can occur within the context of mass tourism and provide a linear and standardized, commodified visitor experience of the destination while a tourism product offering more choice, such as a self-guided walk of a town with options, gives more of an opportunity for a unique, customized and original co-created tourism experience.

Case study 19.1: Chiang Mai, Thailand – International MICE City and Doi Pumuen village

The city of Chiang Mai in Northern Thailand is one of the five cities identified by the government of Thailand policy as a MICE (Meetings, Incentives, Conventions and Events) city (together with Bangkok, Pattaya, Khon Kean and Phuket). The designated cities will employ incentive travel programmes, both pre and post tours for business events, convention and exhibition activities in an effort to distribute income to local communities and drive the national economic development plan known as Thailand 4.0. Chiang Mai is a cultural city and province known for its creative economy (Piboonrungroj and Sangkakorn 2013). The city was designated as a UNESCO Creative City for Crafts and Folk Art in 2017. In cooperation with the Thailand Convention and Exhibition Bureau (TCEB in Newswire 2018) and in response to government policy at the national level and a provincial strategic plan on MICE development in Chiang Mai, the MICE Excellence Centre, Faculty of Economics, Chiang Mai University (also known as Chiang MICE) has collaborated with local stakeholders and communities, introducing events and activities aimed at corporate travellers that showcase local lifestyle, history and culture. Involving local communities is seen as a means to promote their communities and neighbourhoods as MICE venues as well as a framework for creating opportunities for sustainable income distribution (Newswire 2018). This initiative directly addresses the TCEB and Thai government goal of developing the Northern Region of the country into a high value creative economy. In 2003 TCEB in cooperation with other government agencies launched "Chiang Mai's Year of MICE" campaign promoting the creative features of the destination for hosting meetings, incentive events, conventions and other types of events. The goal is to promote Chiang Mai as a preferred destination for all four types of international business events or MICE. However, current statistics show that Chiang Mai is now a preferred destination for mainly corporate meetings and incentive travels due to the carrying capacity of Chiang Mai city, which fits well with the small to moderate size group of meetings and incentive travels rather than the large conventions and exhibitions normally held in Bangkok.

Chiang Mai was subsequently notably the first to have an official Provincial MICE development plan, developed by TCEB, Chiang Mai Province and Chiang MICE with a vision under the theme of "Chiang Mai The Splendid City of Culture and Destination for MICE in Asia". This policy outlines key development strategies including upgrading services, improving tourism products and creating opportunities for MICE activities by leveraging cultural capital and developing human resources to support MICE activities in local areas (TCEB in Newswire 2018). The plan states that the vision of the city is to promote Chiang Mai in the MICE market through her culture and comfortable way of life, thanks to several creative and cultural initiatives in the city by various groups of government and non-governmental organizations. In the plan, key strategies and supporting projects to develop MICE infrastructure and marketing campaigns were outlined with designated organizations which will carry out each strategy and its related activities.

Since there is no Convention and Visitor Bureau (CVB) at a provincial level in Thailand, Chiang Mai then needed to identify the role of government organizations which would develop the MICE industry in the city. Whilst in other destinations the trade promotion agency will be responsible for the MICE industry (such as HKTDC in Hong Kong), in Thailand MICE tends to be linked with the tourism industry according to the strengths of Thailand in touristic activities, especially at locations outside Bangkok. In Chiang Mai, the provincial office of the Ministry of Tourism and Sports as well as the office of the province were the main organizations that proposed the budget to develop and promote the MICE industry. However, as MICE is not the top priority for both organizations, therefore Chiang MICE, managed by the university, is now the key public sector player that drives the MICE industry. This is evidenced by it having been the host since 2017 for the pavilion management and set-up to promote Chiang Mai and the northern region in IT&CM Asia, the largest MICE trade show in Asia.

The role of Chiang MICE includes capabilities development and destination marketing, or 4P4X (four P for X) including product, people, place and promotion as well as experience design for visitors. To develop new products for Chiang Mai as a MICE destination, Chiang MICE has a focus on visitor experience design based on the authentic and unique Lanna culture of Chiang Mai. In 2018, Chiang MICE launched the 7 Wonders of Lanna MICE Heritage at IT&CMA2018. The campaign aims to promote new experiential themes for Chiang Mai and the northern region including (1) Altra (Ultra) Adventure, (2) Fantastic Foods, (3) Halal History, (4) Secret Silk, (5) Tea Tradition, (6) Rustic Rice, and (7) Wellness Wisdom. All seven themes were developed with a working group combining key players in the incentive supply chains (Thumachai et al. 2017), Destination Management Companies (DMC), hotels and local community. One of the successful seven wonders themes is "tea tradition", developed by a team of Udomporn Tours as DMC, Phumanee Lahu Home Hotel and Lahu community at Doi Pumuen mountain. Moreover, the tea tradition theme is supported by an academic team from Chiang Mai University and partner academics.

This case therefore examines the work of Chiang MICE with the Lahu tea village of Doi Pumuen, located some distance north of Chiang Mai and Chiang Rai province, to develop tourism to the rural village by constructing a compelling story that could be attractive for the MICE market. The authors were part of the team working with the local community and stakeholders and the local tour operator to identify a viable theme, to develop the overall storyline and to document stories around which hands-on activities for visitors could be constructed. The project goal was to specifically increase the capacity of the village to act as a host for the corporate travellers targeted by the Chiang MICE policy. The six-member team, in addition to the Lahu owner of a local hotel and the manager of the Community Based Tourism (CBT) project and the local tour operator, included academic and practitioner experts in meetings and event management, tour guiding, local culture and tea tourism.

The village

The Lahu are a hill tribe native to China and Mainland Asia. In Northern Thailand, the Lahu village Doi Pumuen has a rich heritage that is a resource for niche tourism. The village diversified from opium production after the Thai King Bhumibol Adulyadej (Rama IX) presented them with a tea tree (*Camellia sinesis var. assamica* variety) in 1969. The village's transition period from an opium and drug-based economy to an agricultural (tea) based economy has seen challenging conditions due to the low added value of the agricultural business in the upstream part of the supply chain. Over the last decade the village has established a vibrant community-based tourism (CBT) programme in which five houses participate in providing activities that highlight the Lahu way of life, including observing local cooking on an open fire hearth, picking and processing tea, making a basket, viewing a cultural performance and staying overnight in a local bamboo house. This programme reflects the principles of CBT specifically developed in Thailand (Suansri 2003), which include the sustainable use of natural resources, the promotion of cultural traditions unique to destination communities, the inclusion of all community groups in planning tourism, profits allocated to a social and economic development community fund, fair and equitable distribution of benefits, and shared educational experiences between tourist and host. Russell (2000) notes CBT is one method of achieving both economic and social regeneration, while at the same time protecting local cultures against the rising tide of globalization.

Through participant observation and discussion with local stakeholders the project team identified a number of resources as a foundation for the further development of the visitor experience here (Table 19.2). In particular (Nishimoto n.d.) argues the Lahu food culture is a case of failed recognition of a "cultural heritage". The team also identified the Lahu tea culture as a resource for further developing tourism activities and experiences here, particularly targeted at the MICE market.

Table 19.2 Resources for Story Development at Doi Pumuen

Resources	*Storylines*
Lahu way of life	Heritage, ancestors
	Self-sufficiency
	Bamboo and thatched-roof houses on log piles
	Nature and walking
	Cultural performance
King's legacy	Diversification from opium and drug trafficking
	The original tea tree
	Thai status for the Lahu
Lahu food culture	Open hearth cooking
	Rice accompanied by side dishes
	Tea, coffee, honey produced in the village
Lahu tea culture	Picking tea
	Processing tea, rolling, drying
	Drinking tea
	Tea nursery
	Bamboo cup making
	Basket-making

The project team also noted that the Lahu language can be used for storytelling as well as interpretation related to the local tea culture, whereby visitors can learn a few words of Lahu during their visit, thus enhancing their village experience (Table 19.3). Lahu is a Tibeto-Burman language spoken by the Lahu people who settled in Thailand through migration from China.

Table 19.3 Storytelling Activities Using the Lahu Language

Activity	*English/Lahu*
Tea Tasting	Have some tea!
Preparing Tea	Brew some tea!
Picking Tea	Pluck tea leaves!
Roast Tea	Roast the tea!
Rolling Tea	Rolling the tea!

Visitors are first introduced to the story of the village through an exhibit at the Phumanee Lahu Home Hotel (PLHH) in nearby Fang district, which acts as a staging area for visits to the rural mountain village. Through illustrative storyboards and artefacts, the exhibit gives visitors a basic introduction to the Lahu culture and way of life before they visit the village. Also, the hotel exhibits the history of tea development with support from the Thai royal family and the village leaders. A relationship with an established tour operator (Udomporn Tours) has also given the village an advantage when arranging for visitors to access the Lahu heritage.

The project

After February 2018 meetings in Chiang Mai, a field visit of several days to the PLHH in Fang and Doi Pumuen and subsequent meetings in Chiang Mai, the working group for this project agreed on the project theme. The objective was to introduce tea-related tourism for the MICE market, with the addition of five new houses each designated as a "Lahu tea heritage home", to the existing CBT program at the village that had previously highlighted the general Lahu way of life. The project aim was to develop the tea tradition at Doi Pumuen as an experiential CSR (Corporate Social Responsibility) activity for a corporate group. The activities in these additional host homes would revolve around the Lahu tea culture of the village in a programme identified as innovative and suitable for the corporate/incentive tourism market in line with the aims of Chiang MICE (Table 19.4).

Table 19.4 Lahu Tea Heritage Theme Storylines

In the Village	*At Lahu Tea Heritage Homes*
Visit King's commemorative tea park, follow King's path to the original tea tree	View the *Camellia sinensis var. assamica* tea drying outside a Lahu home
Pick tea using a traditional basket, experience village life	Tea processing. Learn Lahu for "pick tea" and "process tea"
Observe/participate in tea village lifestyle	Observation of Lahu cuisine. Visitors help brew tea. With Lahu for "Have a cup of tea"
See tea seedlings as the King's legacy	Visit tea nursery to see tea seedlings
Participate in tea-related crafts appreciating the self-sufficient Lahu way of life	Demonstrations of tea basket weaving and bamboo tea cup making

Tourism in tea producing areas has been recognized as having potential positive economic impacts on participating communities. Tea tourism is "motivated by an interest in the history, traditions and consumption of tea" and a tea tourist is described as a tourist experiencing the history, culture and traditions related to the consumption of tea (Jolliffe 2007). This tea-related tourism is a niche tourism activity characterized by small scale activities and tours. In the rural areas where tea is grown there is the opportunity for developing such tourism with the goal of improving local livelihoods. This focus on the tea heritage and culture of Doi Punmen would contribute to the Chiang MICE aim of improving local livelihoods in terms of economic status and social development, including education and healthcare via the CSR programme the MICE visitors can bring to the village.

This focus on Lahu tea heritage also complemented the business aims of the newly established and Lahu owned Ah Pah Tea Company. In addition, the working group and local hosts identified that the village setting and tea-related activities would make it an ideal location for small group meetings, team-building exercises and workshops that could be held before or after events in Chiang Mai, thus meeting the objectives of the Chiang MICE organization. In addition, the tea produced in the village and marketed under the Ah Pah Tea brand would be introduced to meetings, conferences and events participants in Chiang Mai at coffee and tea breaks during these events, extending the benefits of participating in MICE to the local operator and staff from Doi Pumeun. Furthermore, the story of the tea from Doi Pumuen could bring a long-term business to the village resulting from orders from hotels and DMCs to supply organic tea in their hotel or events. Chiang MICE also promoted the Doi Pumuen tea story in their "Winter Wonderland 2018" festival in December 2018, which brought the tea tradition story to the public attention as more than 5,000 visitors were received during the three days of the festival.

This case reflects the environmental, social and economic impacts advocated by both the philosophy of CBT and the policy of Chiang MICE, mirroring the government intention to extend the benefits of such tourism to local communities. The Lahu tea heritage product is respectful of the environment that provides villagers a livelihood and locals are involved in all decisions. In terms of social impacts, sharing with visitors contributes directly to keeping the Lahu tea and culinary heritage alive. From the economic perspective benefits go directly back to the community through the CBT model, thus enhancing local livelihoods.

Analysis

Relating the themes in the literature review to the above case of Chiang MICE and Doi Pumuen Village a number of observations can be made. First, the objectives of Chiang MICE within the overarching Thai government and TCEB policies on the creation of local tourism experiences and opportunities reflect both the organization of Chiang MICE and the destination of Chiang Mai's viewpoint of the co-creation of the tourism experience as a competitive strategy. However, what is unique about the Thai approach is tying this objective to strategies that implement improved livelihoods, especially in the neighbourhoods and rural areas that would not typically benefit from expenditures related to locally hosted city-focused meetings, conventions and events.

Moreover, one of Thailand's strategies for Chiang Mai is to develop the "Food Innopolis", where the local agricultural products will be improved with innovation and technology such as food science and packaging. Therefore, MICE visitor groups that could visit Doi Pumuen could also aim for a business opportunity. The story of Doi Pumuen could not only have a direct economic impact on the village via visitors' expenditure but could also influence new product development for the local tea to gain higher value, such as in the case of teas from other regions such as Puer or Darjeeling tea, or a new packaging design based on the story.

Second, the previous experience of the Doi Pumuen CBT team and their local tour operator with the existing tourism programme allowed the project team to quickly recognize and build on the significant role of tourists and their perspective in the co-creation process. Third, the value of using stories and themes in an interpretive setting is demonstrated by the case of the development of the Lahu tea heritage product in Doi Pumuen village. Through a detailed assessment of story resources, the project team was able to encourage the launch of a new vibrant storyline that will lead to enriched experiences for visitors to the village, with the potential for an associated improvement in livelihoods meeting the goals of sustainable development advocated by the Thai government and both national and provincial MICE agencies.

Third, the story used in the MICE market can be distinguished from the leisure market with a standard and structured protocol since the MICE groups are stricter with the schedule and specific procedures of a particular culture and organization. Chiang MICE provided such preparation for the village by practising with a FAM trip of tourism association leaders who took a day trip to Doi Pumuen. Prior to the trip, Chaing MICE had an expert on protocol and VIP treatment train the service team of the PLHH to learn how to deliver services to the MICE visitors.

Lessons

Lessons related to the development of themes and stories for destination development can be identified from the case of Chiang MICE and Doi Pumuen Village. By establishing a compelling overall theme, the organization is offering a basis for the construction of stories at different participating locations in the village, including the development of products and service delivery. In addition, the case of the Lahu tea heritage programme demonstrates that finding a subtheme and developing a story around it may help a community-based tourism entity to develop a new programme or product to take to market. Within the Chiang MICE initiative in the case of Doi Pumuen village a successful CBT programme needed revitalization so that the operators had something new and engaging to present to the identified incentive tourism market of corporate travellers. Moreover, based on the visitors' feedback, the story development of tea tradition at Doi Pumuen can enhance the experience as well as the commitment to future repeat visits and support for other community products or projects.

Other locations with CBT projects and initiatives in Thailand, and within other countries where this form of tourism is advocated, may therefore be able to use this case to examine the use of stories and themes in diversifying their tourism offer. In effect, diversification of CBT offerings will enhance opportunities for co-creation of experiences by visitors with locals with the potential to fulfil the inherent goal of improving livelihoods at the local level.

The case has therefore shown that establishing themes and associated storylines can serve as a framework for communities to develop, launch and market new experiential tourism products, creating an environment where the tourist can to a certain extent co-create their own experience with local hosts. As demonstrated in the case of Chiang MICE and Doi Pumuen, this new enhanced product development also has a social goal of enhancing local livelihoods by distributing income within the local community.

Conclusion

It is evident from the perspectives of both the organization, the destination community and the tourist that storytelling is a powerful tool for influencing the visitor experience of a destination, site or particular tourism product. It provides a framework for hosting the visitor experience, and as the visitor interacts with the offering they are actively co-creating their own experience with their local hosts and then taking away their own unique visitor experience as the product of tourism.

Developers of the tourism experience can benefit from understanding the role that themes and stories play in the visitor experience. They can also be aware of techniques that can be adopted in effectively developing and implementing themes, narratives and stories, in terms of storyboarding and developing both linear and non-linear paths for the visitor, especially the latter. As a result, each visitor may interact with the components of the story presented by local hosts, therefore co-creating their own unique and meaningful experience.

It is also clear that the tourism experience field can learn from the more developed experience sectors, reflected in particular by museums and interpretive sites as indicated by the literature review. Museums, in terms of exhibit experience design, have established practices that guide the development of an experience, as do sites that benefit from the principles of interpretation, originally put forward by Tilden (1971).

This chapter has a number of limitations. In focusing on the practical issues involved in story identification and delivery it has not delved into the more contentious issues of whose story will be told and the possible dissonance that may occur with the delivery of some messages. It also does not tackle the issues of commodification and the inevitable loss of authenticity when stories and narratives are standardized for tourist consumption. Another limitation is that the chapter does not explore storytelling in the digital age when the presence of digital media also contributes to the provision of narratives and storylines that will influence the visitor experience, either before, during or after the visit (Bassano et al. 2019). With the increased use of technology for storytelling the medium may either enhance or take away from the real life and on-site experience. Developers of stories need to be sensitive to the possible commodification and loss of authenticity and local voices as their product is mediated by digital technologies.

References

Andrades, L. and Dimanche, F. (2014). Co-creation of experience value: A tourist behavior approach. In Prebensen, N. K., Chen, J. S., and Uysal, M. (eds.). *Creating Experience Value in Tourism* (pp. 95–112). Wallingford, Oxfordshire, UK: Cabi. DOI: 10.1079/9781780643489.0095

Bassano, C., Barile, S., Piciocchi, P., Spohrer, J. C., Iandolo, F., and Fisk, R. (2019). Storytelling about places: Tourism marketing in the digital age. *Cities*. 87: 10–20. DOI: 10.1080/0965254X.2018.1464498

Beeton, S. (2005). The case study in tourism research: A multi-method case study approach. In Ritchie, B. W., Burns, P. M., and Palmer, C. A. (eds.). *Tourism Research Methods: Integrating Theory with Practice* (pp. 37–48). Wallingford, Oxfordshire, UK: Cabi. DOI: 10.1079/9780851999968.0037

Campos, A. C., Mendes, J., Valle, P. O. D., and Scott, N. (2018). Co-creation of tourist experiences: A literature review. *Current Issues in Tourism*. 21(4): 369–400. DOI: 10.1080/13683500.2015.1081158

Chronis, A. (2012). Tourists as story-builders: Narrative construction at a heritage museum. *Journal of Travel and Tourism Marketing*. 29(5): 444–459. DOI: 10.1080/10548408.2012.691395

Forrest, R. (2014). Exhibition narrative: The spatial parameters. *Exhibitionist*. 33(1) (Spring 2014): 28–32.

Jolliffe, L. ed., (2007). *Tea and Tourism: Tourists, Traditions and Transformations*. Clevedon: Channel View Publications.

Mason, P. (2002). Action research. In Jafar, J. (ed.). *Encyclopedia of Tourism* (pp. 8–9). London: Routledge.

Moscardo, G. (2010). The shaping of tourist experience. The importance of stories and themes. In Morgan, N., Lugosi, P., and Ritchie, J. R. B. (eds.). *The Tourism and Leisure Experience: Consumer and Managerial Perspectives* (pp. 43–58). Bristol: Channel View Publications.

Mossberg, L. (2008). Extraordinary experiences through storytelling. *Scandinavian Journal of Hospitality and Tourism*. 8: 195–210. DOI: 10.1080/15022250802532443

Newswire. (2018). *TCEB Pushes Chiang Mai into an international MICE city*. News provided by Thailand Convention and Exhibition Bureau, December 3, 2018. Available from www.prnewswire.com/news-releases/tceb-pushes-chiang-mai-into-an-international-mice- [Accessed: March 15, 2019].

Nishimoto, Y. (n.d.). Traditional culture and local society: Lahu food culture as a cultural resource. Available from crs.w3.kanazawa-u.ac.jp/other/doc/3_1_nishimoto_yoichi.pdf [Accessed: February 18, 2019].

Pera, R. (2017). Empowering the new traveller: Storytelling as a co-creative behaviour in tourism. *Current Issues in Tourism*. 20(4): 331–338. DOI: https://doi.org/10.1080/13683500.2014.982520

Piboonrungroj, P. and Sangkakorn, S. (2013). Tourism logistics management: A case of cultural creative tourism in Thailand. *World Conference on Hospitality, Tourism, and Event Research 2013*, Bangkok, Thailand.

Russell, P. (2000). Community-based tourism. *Travel and Tourism Analyst*. No. 5: 89–116.

Suansri, P. (2003). *Community Based Tourism Handbook*. Bangkok: Responsible Ecological Social Tours (REST).

Thumachai, C., Chawalit, P., Chinnakum, W., Chaiboonsri, C., Saosaovaphak, A., Khamhom, T., and Piboonrungroj, P. (2017). A supply chain analysis of incentive travels: A case of Thailand. *International Journal of Supply Chain Management*. 6(1): 155–164.

Tilden, F. (1977). *Interpreting Our Heritage*. Chapel Hill, NC: University of North Carolina Press.

Veal, A. J. (2017). *Research Methods for Leisure and Tourism*. New York: Pearson.

Witcomb, A. (2003). *Re-imagining the Museum: Beyond the Mausoleum*. London: Routledge.

20

Understanding performativity and embodied tourism experiences in animal-based tourism in the Arctic

Minni Haanpää and José-Carlos García-Rosell

Introduction

There is an emerging body of work that explores the role of tourists as active co-creators of their own experiences and thus the places that they visit (EK et al. 2008). These studies challenge the conventional understanding of tourists as passive spectators. As Richard Ek and colleagues (2008) stress, tourists perform and produce tourism experiences actively, corporeally, technically and socially. As a result, when thinking of tourism experiences and their co-creation, we have to understand how tourists engage with multiple stakeholders (employees, locals, animals and nature) in the creation of experiences within a particular temporal and spatial context (see Rantala 2010; Lugosi 2014). Co-creation involves tourists' emotional, affective and embodied engagement with the experiential setting. From this perspective, tourism experiences are conceived as lived and performed through the body in relation to the places where they happen (Veijola and Jokinen 1994). This aspect is particularly emphasized in a nature-based tourism context, in which the tourist performances are influenced by the weather and the natural environment (Rantala et al. 2011).

In this chapter, we reflect on how tourism experiences are lived, performed and co-created through the body by drawing on empirical data from animal-based tourism experiences in an Arctic destination. Namely, we demonstrate the performativity and embodiedness of tourism experiences through observational and autoethnographic data on dog sledding experiences in Finnish Lapland. Dogsledding is not only a popular Arctic tourism activity but a tourist practice in which the human body engages with the natural world and non-human animals. Dog sledding implies corporeal movements in interaction with other bodies (fellow tourists, guides and dogs) and human-made (sleds, clothes and kennel facilities) as well as natural materialities (snow, ice and weather). To understand tourism experiences as embodied performances, we first need to take a closer look at the concepts of embodiedness and performativity in tourism.

Performativity and embodiment in tourism

As sociologists Soile Veijola and Eeva Jokinen (1994) argue in their seminal article 'The body in tourism', tourism is experienced and performed through the body. This argument taps into the wider performance turn in social sciences (e.g. Goffman 1990; Butler 1997). Veijola and Jokinen (1994) propose that the concentration on the tourist gaze, which had been the prevailing way of theorizing tourism, neglects a major part of tourism experiences: the corporeal and embodied performances experienced through the body. This argument has since been developed further in the wider performance turn in tourism studies that began circa two decades ago. The key idea of the performance turn is that, rather than merely 'seeing', tourism is experienced in multisensual ways by being and doing as a body (Edensor 2006; Veijola and Valtonen 2007; Urry and Larsen 2011). Tourist practices are performed in a sensory environment that includes scents, flavours, temperatures and textures (see Valtonen et al. 2010). As tourism geographer David Crouch (2002) argues, tourism places are encountered by bodies and the places surround the bodies. Such theorizing recognizes bodily sensations and affect as essential ways of experiencing tourism (Urry and Larsen 2011: 155).

The performance turn in tourism studies also draws heavily on the ideas of sociologist Erwin Goffman regarding staged performances in tourism places. The places are themed and staged and then experienced through embodied performances (Goffman 1990; Urry and Larsen 2011, 155). The co-creation of tourism experiences thus happens through the active doings of the tourists and different actors taking part in these performances (see Franklin and Crang 2001). According to cultural geographer Philip Vannini (2015: 320–321), performativity as a standpoint puts the focus on action: what is done and what is not done. The moving bodies perceive spaces in particular ways that are connected to the materialities and temporalities of those particular spaces but also the spaces experienced previously. Following this idea, tourism places are multidimensional: the sociality of places involves mobility and connections with other places (Baerenholdt et al. 2004: 33).

Thus, the performance turn recognizes the doings of multiple actors in the tourism place (Urry and Larsen 2011: 155). These are co-created, produced and consumed through the performances of these different human and non-human actors. In the co-creation processes, the performances of guides, for example, choreograph tourists' movement in and through different spaces (see Edensor 2001). In Arctic experiences, nature in itself, as a more-than-human actor, is also a powerful force in guiding the performances (Jóhannesson and Lund 2017). The non-human actors in tourist spaces, such as dogs or weather conditions, can strongly influence tourist experiences (see Rantala et al. 2011; Salmela et al. 2017). Besides the visible guiding, there are certain 'pre-choreographies' guiding the tourism performances (see Parviainen 2011). These different cultural codes and norms influence the co-created performances of tourism. They also relate to questions of power: who is allowed to guide the performances and what kinds of roles different actors can hold in these performances (Urry and Larsen 2011: 155–156).

Even though performances are guided and even regulated, they are never totally predetermined. Tourists are not passive actors who just follow the script; they also actively co-create the experience through their own interpretations, resulting in actions and doings. This makes room for the tourist agency, which can also contest and resist the scripts of tourism places (Urry and Larsen 2011: 156–157). When thinking of nature-based tourism places in the Arctic, the role of guides is substantial in ensuring the successful performances of tourists. Their actions are informed by embodied knowledge, which allows them to assess the weather, the equipment and the safety issues related to being in nature (Rantala 2011). For example, by perceiving the track, snow conditions and places for breaks and sightseeing, a guide becomes more aware of the possibilities for co-creating memorable experiences in nature (Rantala 2010; see also Jóhannesson and Lund 2017). Nevertheless, as tourism scholar Outi Rantala (2010) argues, the actions of the guides and tourists are influenced by global cultural norms, local codes of conduct and the infrastructure of the tourism industry. It is through the embodied practices of hostessing and sightseeing that the meanings related to nature are co-created (see Veijola and Jokinen 2008).

Embodiment also taps into the affectivities of places. Affect as a concept turns the focus to the relational capacities of bodies towards other bodies but also to the surrounding more-than-human actors (e.g. Vannini 2015). Affects are processes in which bodies are 'defined by their capacities to affect and be affected' (Blackman and Venn 2010: 9). They are sensed through embodiment and are psychological and social at the same time (Barbour and Hitchmough 2014: 65). Affects make us do things: they move us, but they can also attach us to places and people (Ahmed 2004: 10–11). In Arctic tourism experiences, the connections of bodies and more-than-human actors create strong affectivities, such as relations to different animals in husky safaris as well as relations to natural phenomena, such as the aurora borealis or the midnight sun (Jóhannesson and Lund 2017; see also Edensor 2015).

Case study 20.1: Dog sledding tours in Lapland

Finnish Lapland is the northernmost province of Finland and the European Union. With 3.5 per cent of Finland's population and about 30 per cent of Finland's total area, Lapland is by far the least densely populated region in Finland and is a rapidly growing wilderness and nature-based tourism destination. The number of annual registered overnight stays in Lapland is nearing 3 million (Regional Council of Lapland 2015). As a winter destination, Lapland attracts tourists from around the world. Although most tourists come from European countries, the number of Asian tourists has increased sharply in recent years. In Lapland, tourists engage in such activities as hunting for the northern lights, driving snowmobiles, visiting Father Christmas and riding reindeer sleighs and dog sleds. The dressing up involved in these winter tourism activities is an essential part of their performative nature. Indeed, the wearing of clothing such as winter overalls, mittens, woollen hats and helmets, which are supplied by the service providers, is central to the articulation of the agency of the body in a Nordic nature-based tourism context (see Cater and Cloke 2007). It is through the safari winter clothing that the tourists embody the liminal identity of an Arctic adventurer.

In particular, animal-based tourism services, such as dog sledding and reindeer sleigh tours, are some of the most popular activities among tourists (García-Rosell and Äijälä 2018). Indeed, these two activities are institutionalized patterns and narratives that tourists follow in the production and performance of Lapland as a tourist destination (see Bærenholdt et al. 2004). As narratives, these activities not only offer an idea of what one can do but also nurture tourists' imagination and expectation towards Lapland. More than 4,000 dogs and 1,000 reindeer work in the tourism industry in Lapland and thus generate more than 15 million euros in annual revenue (García-Rosell and Äijälä 2018). Although Lapland is famous for its reindeer, huskies have been growing in popularity. In 2017, for the first time, husky safaris surpassed snowmobile safaris as the number one tourism activity in Lapland.

A sled dog ride is an adventurous embodied practice, in which tourists have the possibility to be a musher leading their own team of huskies (Figure 20.1). Husky safaris can range from short (0.5–2 km) to medium-length (10–40 km) and multi-day rides (2–8 days). The sleds are pulled by four to six dogs depending on the size and weight of the musher and passenger(s). Almost all of the safaris are organized according to a similar kind of script. Independent of the length of the safari, tourists receive driving and safety instructions before entering the kennel. The driving instructions are provided by a guide demonstrating how to perform the appropriate handling of the sled and the team of dogs. Afterwards, the tourists take possession of the sleds and prepare for the start. Depending on the type of safari, during the ride there are different breaks so that the driver can swap places with the passengers. In that way, all the adult tourists have the possibility to try driving the sled. After the end of the safari, tourists have an opportunity to take a tour of the kennel and to learn about and cuddle the dogs. The physical and socio-cultural context of the husky safari is a good example of a tourist space where the human body engages with the natural world and non-human animals.

Figure 20.1 Dogs ready for the start
Source: Dr José-Carlos García-Rosell

Most of the sled dogs in Lapland are Alaskan huskies, which are a blend of different northern breeds, chosen particularly for their pulling skills. There are also Siberian huskies, which are the breed used in commercial marketing and internet advertisements, contributing to the imagination of potential Lapland visitors.

Performing dog sledding through embodiment

In this chapter, we reflect on our embodied experiences and performances on a dog sledding safari. The vignettes used the results from both authors' (auto)ethnographic fieldwork conducted in a project concerning animal welfare in tourism. During the course of the two-year-long project, the researchers took part in several dog sledding safaris in several kennels located in Finnish Lapland. The experiences reflected in this chapter result from those safaris. To analyse the experiences, we have closely read our fieldwork diaries and used videos and photos taken in the field to elicit the memories of the experience of dog sledding. We have also discussed and reflected on our experiences collaboratively.

Different bodily and performative skills are needed to interact with the dogs and other tourists, cope with the winter conditions and drive the sled. We draw on both authors' accounts of dog sledding experiences.

> We are given instructions by a guide on how to drive the sledge. After that we walk to the kennel, surrounded by guides organizing the dogs in front of the sledges and loud barking. I feel a little intimidated and confused. The guides in my eyes act with a tough guy attitude, organizing dogs, sledges and people in a rapid manner. I feel a little tense and try to follow the instructions. The barking and howling is so loud that it is hard. We walk through the kennel to the departure point and take our places. In the midst of everything I am scared to think about driving the sled and decide not to drive but to sit in the sledge while my colleague takes the driver's place.
>
> I don't dare to pet the dogs although I would like to do so. I feel that I could cause a fight if I start petting one of them. I used to have between 4 and 7 dogs when I lived with my parents, and I know what can happen when a dog becomes jealous. Therefore, I just focus on the sled and on doing everything as we were initially instructed.

The vignettes above are by two different authors and describe the beginning of a safari. The sledding performance is strongly guided by the human and non-human actors participating in it. The guides play a remarkable role in choreographing and orchestrating the performance (see Edensor 2001; Rantala 2010). First, by instructing us, they show what kind of embodied performances and skills are needed in driving the sledge. We are taught how to stand on the skids, how to use the brake and how to help the dogs run uphill by kicking speed. These skills are demonstrated by the guide with a sledge but without a dog pack, as we are observing and trying to memorize her words at the same time. We tourists stand around the sledge. After the instructions, the guides direct us to follow them through the kennel and, as mentioned in the vignette, the loud barking of the dogs makes speaking with other people almost impossible. This makes the interaction between the tourists and the guides more embodied. We are shown where to go through signing accompanied by clear shouting, and accordingly we take our places in the sledges. Besides the loud barking, the dogs are pulling and jumping around, keen to start. Their eager and even slightly aggressive behaviour discourages us from approaching them, as shown in the second vignette. The surrounding human and non-human actors strongly influence our performances as customers (see Crouch 2002) and, as mentioned in the first vignette, the loud and busy atmosphere even makes us a little tense. The purposeful choreographing of the situation by the guides and the unintended guiding of the dogs' action are both strong influencers on the situation.

The physical elements of the kennel are not so much meant to guide tourists but are more or less designed to cater to the needs of the dogs and guides in executing the safaris in an effective manner. Without guidance, we as customers would not receive cues from the physical elements on what to do or where to go. Some guides organize the dogs in front of us as we enter the kennel, and soon we are organized onto the sleds in a similarly effective manner.

> Once the guides open the gate of the kennel, the sleds begin leaving the kennel one after another. The dogs stop barking and focus on running. My body moves from left to right according to the curves of the track. I take my phone to take some pictures and make some videos. Nevertheless, I have to put it back into my jacket pocket from time to time, as my hands get cold. It is because of the wind caused by the speed and the fact that I took thin gloves. My face is getting cold too, kind of freezing. And because of that, I am not able to speak clearly because my skin is frozen.

Driving the sled is a somewhat physically challenging task that requires observation of the surroundings, the actions of the dogs and the condition of their harnesses, the actions of the other sleds in front of and behind the sled and the coordination of one's own embodied actions (see Vannini 2015). As the dogs stop making a noise once they start running, it helps the driver to concentrate. Perhaps the most powerful material guidance for the performance is the track (Figure 20.2). The dogs follow it and it directs where the sleds move. The condition of the snow on the track and its course as nature-based elements affect the ways in which the driver's body senses the movement and experiences the safari. Performing the safari requires adjusting one's own actions to the movement: kicking to help the dogs run uphill and again using the brake downhill or if the sled comes too close to the one in the front, as well as adjusting the body to the lateral movement of the sled and sudden bumps in

Figure 20.2 On the safari track
Source: Dr José-Carlos García-Rosell

the track. The first vignette shows how one of the authors felt that such a 'multitasking' operation was too much to handle at such short notice and decided to sit in the sled instead of driving (see also Rantala et al. 2018).

Besides the materialities affecting the performances, the actions of non-human dogs have a strong influence on the experience. Many times these actions are unpredictable, as one of us describes in the field diary: 'Suddenly something surprising happens. One dog starts defecating. This is a wild scene. It smells and the excrement goes under the sled. I kind of worry that some of it will fly over me.' This unexpected event is revealed as an affective one, and the relation between different actors in the safari is mediated by a detested element. This affectivity also relates to the image of the dog sledding experience as gliding through the pure white Arctic nature, not struggling to sidestep excrement. Besides this, once the dogs begin running, they have a very strong urge to pull. This means that, if a musher falls off the sled for some reason, the dogs will probably continue without her/him. They have their own choreography, mostly ignorant of human wishes or needs. Even though the dogs seem to be independent of human actions, the human musher can be very concerned about the dogs. As the other one of us writes:

> I'm a little bit worried for the dogs. I feel that, by stepping on the brakes, I make the dogs work harder. It is my continuous concern that the dogs will be hit by the sled or will have to do extra work.

The affectivity of the situation builds in the relations between the musher, the dogs and the material environment: the shapes of the terrain and the sled reacting to them.

In addition to the dogs, the weather acts as another non-human actor that strongly influences the bodily performances (see Rantala et al. 2011). Experiencing weather happens through embodiment and senses. In the Arctic winter weather, without proper clothing, one begins to feel chilly quite quickly, as is the case in the vignette. The body begins to react to the coldness, and the fingers and face freeze. This also guides the performance that one is able to make, for example by not being able to take pictures. The coldness makes the experiencing of the safari less pleasant. Besides the unpleasant bodily experience, avoiding becoming too cold is an issue of safety. This is why instructions regarding proper clothing and winter overalls are usually given to the tourists by the guides at the beginning of the safari. Again, the cold weather, as a non-human actor, can also often affect another non-human actor that is significant to the tourist performance: the camera. The coldness makes the camera battery die quickly and in this way prevents the tourists from engaging in the typical performance of taking pictures (Urry and Larsen 2011). The taking of pictures is also affected by the multitasking nature of driving the sled; one simply cannot concentrate on too many multiple actions at the same time.

> Arriving at the kennel, the dogs are very calm. I follow the instructions of the guides, who tell me where to put the sled and when to brake so that they can secure the sled. This is the time when I can finally release the brakes and approach the dogs. It is a moment of gratitude. After the amazing ride, I feel that I have to meet my crew and pet them. One of the guides approaches me and takes a picture of me with one of the dogs. This is a moment of pride. I did it, we did it and this is one of the huskies. I would like to spend more time with the dogs, but they seem pretty relaxed, just lying on the ground and being there. The memories of my own dogs come back. Just being with the animals for a while makes me feel like petting my dogs back home. Feeling their wet noses and tongues licking my fingers and hands, remembering my old friends.

The vignette above describes how the movement stops at the end of the safari and how the atmosphere changes accordingly from hectic to calm. This also produces different affectivities between the different actors. The dogs become calm after the run, and they lie still. Finally, the humans are able to approach the dogs and meet them as a pack but also as individual animals. During the safari, the dogs barely pay any attention to the humans in the sled. This makes them feel quite distant. Once one is able to approach them from the front, they begin to feel closer, since, at the beginning of the safari and during it, the people are usually only behind the dogs. In the vignette, it is also clear how the embodiedness of petting the dogs makes one relate to memories of other places and times, and the company of one's own former pets again highlights the affectivity of the experience.

Discussion and conclusion

In this chapter, we have reflected on the performativity and embodiedness of an Arctic animal-based tourism experience, the husky safari. As a performed tourism experience, a husky safari comes into existence by drawing together dog sledding-related objects, mobilities and proximities (see Bærenholdt et al. 2004; Cater and Cloke 2007). Objects such as snow, ice, sleds and kennel facilities intertwine with different kinds of corporeal mobilities (the trip to the kennel, the driving instructions, the walk towards the dog sleds and the dense choreography of mushing). All this implies close proximity to fellow tourists, guides and dogs. These mobilities and relationalities also result in affective atmospheres between non-human animals and human tourists.

Animal-based tourism experiences have rarely been analysed through embodiment. From the analysis undertaken, it is clear that their co-creation requires a multitude of bodily skills and engages with sensing through the body. The choreography of the performances is guided by human guides as well as by the non-human actors of the tourism place. Dogs, as non-human actors, have powerful agency in creating the affectivities of the experience. In their agency, the familiarity of pet dogs and the unfamiliarity of working animals are a discrepancy that carries strong affectivity. Future research could focus on shedding light on the roles and performances of different actors in animal-based tourism experiences (see Haanpää et al. 2019). Future studies could also explore tourists' embodied performances within a different animal-based tourism experiential setting, such as horse riding, reindeer sledding or whale watching. Finally, there is a need for research that examines tourism experiences co-created through emotional, affective and embodied engagement between humans and non-human agents such as robots, autonomous vehicles and other forms of artificial intelligence.

References

Ahmed, S., (2004). *The cultural politics of emotion*. New York: Routledge.

Bærenholdt, J.O., Haldrup, M., Larsen, J., and Urry, J., (2004). *Performing tourist places*. London: Routledge.

Barbour, K. and Hitchmough, A., (2014). Experiencing affect through site-specific dance. *Emotion, Space and Society*, 12: 63–72. DOI: 10.1016/j.emospa.2013.11.004

Blackman, L. and Venn, C., (2010). Affect. *Body & Society*, 16(1): 7–28. DOI: 10.1177/1357034X09354769

Butler, J., (1997). *Excitable speech: a politics of the performative*. New York: Routledge.

Cater, C. and Cloke, P., (2007). Bodies in action: the performativity of adventure tourism. *Anthropology Today*, 23(6): 13–16. DOI: 10.1111/j.1467-8322.2007.00548.x

Crouch, D., (2002). Surrounded by place: embodied encounters. In S. Coleman and M. Crang (eds.), *Tourism: between place and performance* (pp. 207–218). New York: Berghahn Books.

Edensor, T., (2001). Performing tourism, staging tourism: (re) producing tourist space and practice. *Tourist Studies*, 1(1): 59–81. DOI: 10.1177/146879760100100104

Edensor, T., (2006). Sensing tourist spaces. In C. Minca and T. Oakes (eds.), *Travels in paradox: remapping tourism* (pp. 23–45). Oxford: Rowman and Littlefield Publishers.

Edensor, T., (2015). Introduction to geographies of darkness. *Cultural Geographies*, 22(4): 559–565. DOI: 10.1177/1474474015604807

EK, R., Larsen, J., Hornskov, S.B., and Masfeldt, O.K., (2008). A dynamic framework of tourist experiences: space–time and performances in the experience economy. *Scandinavian Journal of Hospitality and Tourism*, 8(2): 122–140. DOI: 10.1080/15022250802110091

Franklin, A. and Crang, M., (2001). The trouble with tourism and travel theory? *Tourist Studies*, 1(1): 5–22. DOI: 10.1177/146879760100100101

García-Rosell, J.-C. and Äijälä, M., (2018). Animal-based tourism in Lapland. In J. Ojuva (ed.), *Animal welfare in tourism services: examples and practical tips for the well-being of animals used for tourism in Lapland* (pp. 10–24). Rovaniemi: Lapland University of Applied Sciences.

Goffman, E., (1990). *The presentation of self in everyday life*. Harmondsworth: Penguin Books.

Haanpää, M., Salmela, T., García-Rosell, J.-C., and Äijälä, M., (2019). The disruptive 'other'? Exploring human–animal relations in tourism through videography. *Tourism Geographies*, online first. DOI: 10.1080/14616688.2019.1666158

Jóhannesson, G.T. and Lund, K.A., (2017). Aurora borealis: choreographies of darkness and light. *Annals of Tourism Research*, 63: 183–190. DOI: 10.1016/j.annals.2017.02.001

Lugosi, P., (2014). Mobilising identity and culture in experience co-creation and venue. *Tourism Management*, 40: 165–179. DOI: 10.1016/j.tourman.2013.06.005

Parviainen, J., (2011). The standardization process of movement in the fitness industry: the experience design of Les Mills choreographies. *European Journal of Cultural Studies*, 14(5): 526–541. DOI: 10.1177/1367549411412202

Rantala, O., (2010). Tourist practices in the forest. *Annals of Tourism Research*, 37(1): 249–264. DOI: 10.1016/j.annals.2009.09.003

Rantala, O., (2011). An ethnographic approach to nature-based tourism. *Scandinavian Journal of Hospitality and Tourism*, 11(2): 150–165. DOI: 10.1080/15022250.2011.576829

Rantala, O., Hallikainen, V., Ilola, H., and Tuulentie, S., (2018). The softening of adventure tourism. *Scandinavian Journal of Hospitality and Tourism*, 18(4): 343–361. DOI: 10.1080/15022250.2018.1522725

Rantala, O., Valtonen, A., and Markuksela, V., (2011). Materializing tourist weather: ethnography on weather-wise wilderness guiding practices. *Journal of Material Culture*, 16(3): 285–300. DOI: 10.1177/1359183511413646

Regional Council of Lapland. (2015). Lapin matkailustrategia 2015–2018 [Lapland tourism strategy 2015–2018]. *Lapin liiton julkaisuja* Serie A, No. 43.

Salmela, T., Valtonen, A., and Miettinen, S., (2017). An uncanny night in a nature bubble: designing embodied sleeping experiences. In D. Fesenmaier and Z. Xiang (eds.), *Design science in tourism* (pp. 69–93). Cham: Springer.

Urry, J. and Larsen, J., (2011). *The tourist gaze 3.0*. London: Sage.

Valtonen, A., Markuksela, V., and Moisander, J., (2010). Doing sensory ethnography in consumer research. *International Journal of Consumer Studies*, 34(4): 375–380. DOI: 10.1111/j.1470-6431.2010.00876.x

Vannini, P., (2015). Non-representational ethnography: new ways of animating lifeworlds. *Cultural Geographies*, 22(2): 317–327. DOI: 10.1177/1474474014555657

Veijola, S. and Jokinen, E., (1994). The body in tourism. *Theory, Culture & Society*, 11(3): 125–151.

Veijola, S. and Jokinen, E., (2008). Towards a hostessing society? Mobile arrangements of gender and labour. *NORA – Nordic Journal of Feminist and Gender Research*, 16(3): 166–181. DOI: 10.1080/08038740802279901

Veijola, S. and Valtonen, A., (2007). The body in tourism industry. In A. Pritchard, N. Morgan, I. Ateljevic, and C. Harris (eds.), *Tourism and gender: embodiment, sensuality and experience* (pp. 13–31). Oxfordshire: CABI.

21
Experiential dissonance

Jon Sundbo

Introduction

This chapter develops a theory and model of experiential dissonance, which is a new concept. Experiential dissonance occurs when a tourist is attracted by one experience, but another experience is crucial for the total experience. The theory is inspired by an empirical case study of an attraction – a safari park – that presented the researchers with an unforeseen phenomenon that needed to be explained (Sørensen and Sundbo 2014). The study showed that people – primarily families with small children – naturally came to see the animals at the safari park, but the playground turned out to be the decisive factor in making a good day.

This theory concerns situations where people in their role as customers, tourists or carrying out leisure activities sometimes experience something other than the provider intends to create. The "something" in such cases is the more important experience. The question that this theory attempts to answer is why this happens and which tensions it creates in the minds of those who have such experiences. The theoretical challenge is to establish what is the most crucial for the customer's general satisfaction and assessment of the total experience, for example the visit to a safari park, and thus also for whether the customer will become loyal and return and consume the experience again. It might be the core experience, watching animals, but it might also be a "side" activity outside the core one, cf. Chapter 1. An experience can include a multiplicity of factors. For example, a cinema visit may be composed of the experience of the film, the room, the candy store and the candies or popcorn, the café in the cinema, the building and so forth. One may therefore talk about a total experience that includes all such aspects.

The theory developed here thus aims to explain a paradox in experience creation, namely that the experiencing person both experiences the experience the provider intends to create and something else, which may be the main experience that the person can remember. Experience happens in the mind of a person (the customer, consumer) cf. the definition in Chapter 1. The provider provides some stimuli that are intended to create a specific experience in the mind of the receiver (customer, tourist), but the experience that de facto is created may be different. The receiver relates to the stimuli based on her or his former experiences and social habitus (social and cultural capital, norms and values, Bourdieu 1994). However, her or his experience (what she or he remembers as significant, cf. Pine and

Gilmore 1999) is also a social phenomenon, i.e. is influenced by the persons she or he is together with in the situation. If for example the situation concerns a family on an excursion, the children's reactions may have an important influence on the adults' experience. In this situation, the different stimuli, value pressures and norms may create an imbalance or tensions in an individual (for example an adult in a family on an excursion) that the provider is unaware of. The theory of experiential dissonance presents a suggestion for explaining why these tensions emerge and how they influence the experiencing person's future consumption.

That experiencing can create tensions in the customer has not received a great deal of attention in the literature. Thus the theory of experiential dissonance offers a new explanation for an overlooked phenomenon. Dissonance has been observed in tourism, but has not been conceptualized and explained comprehensively. The theory of experiential dissonance contributes to a conceptualization and explanation of the phenomenon.

In order to provide an understanding of the theory and which phenomenon it attempts to explain, the chapter starts with a short presentation of the empirical case. Next, the notion of total experience is discussed. Thereafter the Experiential Dissonance theory and a model will be presented. The model has two core concepts that are called the Frame, which is the main attractor for people and the reason for seeking the total experience, and the Occupant, which gives people the special experience.

Case study 21.1: The Safari Park

The inspiration for the theory of experiential dissonance came out of one specific case study, which was a safari park in Denmark. In contrast to a traditional zoo, the wild animals in the safari park can be experienced in a more "natural" situation from a car. The park concerned is an old manor house with a huge park where the animals are. The guests drive through the park in cars, and they can stop and in some cases get out and touch the animals if these are not too dangerous. Visitors can drive around the park as they wish for a whole day. The primary experience thus is to look at wild animals in a freer environment than a narrow cage and maybe touch them. The park also has other facilities; a café, an ice-cream stall, a playground, a shop, etc.

As part of a research project, the author of this chapter together with other researchers investigated the visitors' behavior in the park and what satisfied them most. The visitors were mainly families with small children. The research was carried out by collecting information about the visitors' behavior by lending GPS sports clocks to 36 families to put in their cars and afterwards tracking their movements in the park by using a software program (Quantum GIS), which can draw maps recording the families' movements (Sørensen and Sundbo 2014). Twenty of the families were interviewed a few weeks after their visit.

The investigation showed that the guests spent most of their time in the playground (Sørensen and Sundbo 2014). It was not an important part of the image of the park, for example in advertising, PR and so forth; the owners thought it was a necessary peripheral service just as is the ice-cream stall. However, the playground proved to be the decisive factor in determining how long people stayed in the park, how satisfied they were with the total visit and the factor that mostly influenced their consideration of whether they would come back or not. Thus the animals in the park were the primary experience that attracted the guests in the first place. However, once the guests were in the park, the playground was the experience that was most decisive for their satisfaction with the visit. This outcome of the investigation could be explained by the fact that the playground was good; it included a nature playground where children (and adults) could climb in trees and so forth, which children greatly enjoy. They also liked to watch and touch the animals, but only for a short time, then it became boring. They spent much more time playing with each other in the playground. It also emerged that although the most important driver behind the guests' visit was to watch animals, when the researchers dug deeper into visitors' motives the most important one was to be together as a family (cf. results of Larsen 2006, who studied families on holidays). This was most successfully carried out in the playground and therefore the experience in the playground was the best.

The visit to the safari park created two important experiences in the minds of the visitors, of which one experience, watching animals, was the primary one, while the other, the playground, was unexpectedly the most important and valuable. Thus two memories or experiences were created in the minds of the visitors and their value may be assessed differently. This makes it difficult for the visitors to make a total assessment of the visit to the park. They are often not aware of the secondary experience being the most successful and valuable because they are led to focus on the primary one that is the rationale for going to the safari park and which is the one promoted by the park.

Experiential dissonance

The principle of the theory is that experiences are not necessarily what they seem to be, which is a consequence of the assumption of separation of the offer of elements from users' experiencing and there is a hidden or missed factor that really decides peoples' reactions and satisfaction. This hidden factor may not characterize all experiences, but it may characterize many. The theory extends and provides a variation on the theoretical understanding of customer satisfaction (Holbrook and Hirschman 1982; Gummesson 1994; Berry and Carbone 2007). Furthermore, it provides a tool for practitioners to better understand their business and emphasizes the factors that really create value for customers. The theory of experiential dissonance provides a better and more sophisticated explanation of customer experiences and can explain more exactly what gives people a memory of a good total experience. The theory can therefore also become a tool in selling experiences and creating good and efficient marketing of experience products or elements (if the experience is an addition to goods or services, cf. Pine and Gilmore 1999, 2011; Schmitt 1999; Schmitt, Rogers, and Vrotsos 2004; Horn and Jensen 2011)

As experiencing happens in peoples' minds, also in their role as customers and tourists, a firm (or public institution) does not provide experiences, only elements that function as external stimuli which might get people to have an experience (Sundbo and Sørensen 2013) – cf. experience propositions Chapter 1. This separation of the stimuli or experience proposition and the receiver's mental experience is the reason why experiential dissonance can emerge in the mind of the receiver; the provider cannot necessarily steer which experiences the receiver builds and is often not aware of any dissonance that is created. The question for the provider, and for a theoretical understanding of why dissonance is created, is how the experience product can be created so that the consumer consumes the experience as it is intended, thus avoiding dissonance.

The following sections discuss general aspects of a theory on experiential dissonance. For purposes of generalization, the experiencing person or group will be termed "user" and the firm or institution that delivers the elements that can give the user experiences is termed the "provider". However, the "user" is a tourist within the framework of this handbook.

Primary and secondary experiences

For the purpose of understanding experiential dissonance it might be useful to define two types of provided experience elements, or experience propositions, which should be called primary and secondary experiences.

The primary experience is the main business of the provider: it is the experience provider's *raison d'être*. The secondary experiences are add-ons. The provider might be aware that a secondary experience may be crucial to the users or not; the users may also themselves seek a secondary experience element and create their own experiences. The provider can emphasize a secondary experience if he is aware of it, when for example marketing a safari park, but if the emphasis is exaggerated then a secondary experience element could become the most important, the main business, or *raison d'être* of the provider. That may not be an advantage to the provider because he will then enter a new business field or industry (the safari park will suddenly be in the amusement park business), which establishes a completely new situation with regards to competition and another focus from the customers.

The total experience

The experiencing person, or tourist, gets a total experience, which consists of all the partial experiences – provoked by the offered primary as well as secondary experience elements or propositions – elaborated in his mind as already stated. This elaboration is influenced by his past experiences and cultural and social capital (cf. Bourdieu 1994) and by the purpose of his experience-consumption (for example being together with the family on holidays). The idea of a total experience with parts or elements is the basis for the theory of experiential dissonance. The concept of experiential dissonance denotes a situation where people come to a provider to experience one phenomenon which is the primary attractor (cf. Mattsson, Sundbo, and Jensen 2005) and that most often is the core experience that the provider offers, but another, secondary, experience turns out to be the most important to the consumers. This special secondary experience element is what occupies people most and is the one on which they spend most time and energy, even if it is not an obvious option; maybe people do not even know beforehand that this experience is possible. This secondary element is the experience phenomenon that makes people most satisfied (what they remember best and would be evaluated by them as the "best of the day") with the visit to an experience destination or event or with the reception of a total experience-provoking package (e.g. a TV channel) and what makes them interested in revisiting the place, event or continue receiving the experience package. In the case that has been the basis for this chapter, the primary attractor is watching animals, the secondary experience is visiting the playground.

People are often not consciously aware of this situation. They may still think that the primary experience is the reason for their visit, or why they want to receive the experience package. However, the secondary experience is nevertheless the crucial determinant of their overall assessment of the total experience.

A model of the total experience: the frame and the occupant

The theory of experiential dissonance which sees experiencing as something that happens in the mind of the user (guest, visitor, customer), must be seen within the framework in which the provider offers elements that may give the user several experiences, which taken together create a total impression, or a total experience, in the mind of the user. This approach helps us analyse 1. Which "pieces" of experience the provider has, and 2. How the user creates the picture and the memory of a total experience which is crucial for how valuable the total experience is to the user; the latter is a decisive factor in whether the user comes back and how high a price he is willing to pay (cf. Pine and Gilmore's 1999, theory of high experience value for users).

We may create a model of the total experience which includes and characterizes both the primary and secondary experiences, which is done in Figure 21.1.

The primary experience is the main attractor, or at least it is seemingly the reason for seeking the total experience (the attraction and other experiences); it becomes the frame for other, added and more secondary experience elements. In the safari park the frame is watching animals. The primary experience could be called the Frame. The Frame (e.g. the excursion to watch animals) contains other, secondary experience elements that may be offered by the provider.

One (or in rare cases more) of these secondary elements gives the visitor a particular experience (such as the playground in the safari park). That part of the experience may be called the Occupant because it is what occupies people, both for a longer time and as an activity that especially absorbs people.

The secondary elements may be "hidden" (sometimes the provider has not seen them as a particular business element) or they may be consciously developed by the provider who may make a large investment in them; however, only as supplementary elements subordinated to the primary element – in this case watching animals – so as not to destroy the anticipated authenticity of the total experience. Thus the Occupant may be intended or unintended from the provider's side.

As said, the Occupant is the most crucial part of a successful total experience. The user will still afterwards think of the Frame as the main experience, but he will, consciously or unconsciously, assess the total experience mostly on the basis of the Occupant. The

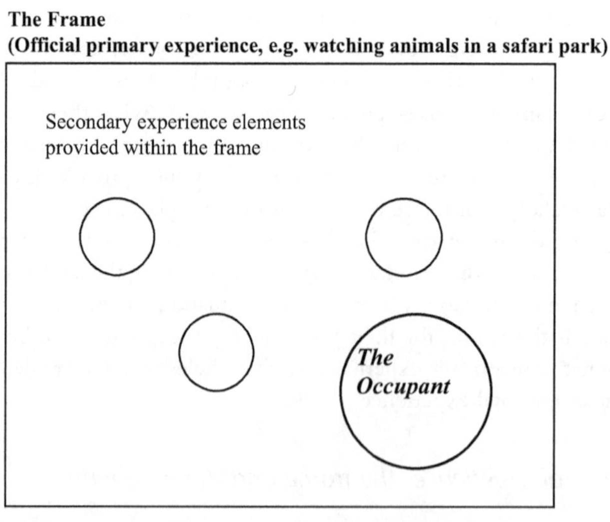

Figure 21.1 The total experience

Occupant is not like a peripheral service, as this is defined in the service management and marketing literature (such as free champagne in the airplane, e.g. Normann 1991; Grönroos 2000). The Occupant is not an inferior add-on, but an integrated part of the total experience. However, the Occupant does not necessarily imply as much absorption and feeling of achievement as expressed by Csikszentmihalyi's (2002) concept of flow. The Occupant might have the character of a flow, but it might as well be just a relaxed and enjoyable activity (such as playing with the children in a playground).

Dissonance

The situation where the user is attracted by the Frame but the Occupant is most successful and crucial for the total experience may create some tension for the user. The total experience is not as expected. There is an inconsistency between which part of the experience is the opportunity to get a total experience and which part is the most successful. This may create a tension in the user's mind about which experience he and his possible co-users (for example a family group) had. This tension can be termed dissonance. In the safari park example the interviews showed that the families, particularly the parents, assessed the playground and related dining places as important and had positive attitudes towards the playground and opposite attitudes towards the dining places. That created some tensions in their general assessment of the safari park, and influenced whether they wanted to come back and mainly why they wanted or did not want to come back. An experiential dissonance – tensions – had been created not only between the assessment of the Frame and the Occupant but also between different members of the family.

The inspiration for the term "experiential dissonance" comes from the psychologist Leon Festinger's (1957) theory on cognitive dissonance (the concept of dissonance has also been applied to organizations, Koolhaas 1982). The theory of experiential dissonance is not an application of Festinger's theory to the experience field. Festinger's theory concerns knowledge and roughly states that if a person has two inconsistent pieces of knowledge or attitudes, dissonance is created in his mind. The individual cannot live with such dissonance and will try to solve it by changing one or both of the attitudes or pieces of knowledge, thus establishing psychic or knowledge harmony in his mind. From Festinger's theory I take the idea of two factors being important for individuals' psychic reaction to a phenomenon and the notion of dissonance as an expression of the fact that the two factors are not in accordance or harmony with each other.

Experiences are generally positive – a pleasure, meaning-creating (gives your life a meaning, as for example when you learn to understand something) or similar. Experiential dissonance does not create psychic displeasure as cognitive dissonance does according to Festinger's theory. However, it creates confusion about which experience is the most positive and why the individual is seeking another part of the experience, the Occupant, rather than the Frame. The experiential dissonance may lead to different reactions from the user. 1. The user may bring order into his mind by suppressing the effect of the Occupant and only remember the experience produced by the Frame. This reaction is not very likely because the user has no unpleasant psychic tension to resolve, but only several positive experiences to remember. 2. He or she may decide that the experience provider provides the Occupant and forget about the Frame. This approach to the total experience may be difficult to sustain since the user is met with the Frame in marketing and at the place of the experience. 3. The user may also turn the dissonance into something conscious and positive: the Occupant is welcome as an addition to the Frame. However, it is an addition that is somewhat irrational and the user must realize that he is not that rational in his choice of experience provisions, but let himself be surprised by the Occupant. The Occupant adds something extra to the Frame, but because it is unexpected and creates some tension it gives a more memorable total experience than a normal visit.

Another example of experiential dissonance is cinemas. Case studies concerning cinemas (Sørensen, Sundbo and Mattsson 2013) have for example shown that watching movies is the Frame, but secondary experiences are the architecture and comfort of the cinema room and the hall, the café, the candy shop and the quality of the popcorn. Popcorn, candy and cola have thus in many cinemas become the Occupant. This Occupant differentiates one cinema from another and the cinema from staying home and watching movies on TV. Experiential dissonance can also emerge when one is watching TV. A learning program (the Frame) can turn out for the watcher to also be amusing (which then becomes the Occupant) and vice versa. Music festivals have for a long time been considered as having the music as the Frame, but the social activities are the Occupant that mostly decides how the visitors assess the total experience of participating in the festival (Sundbo 2013; Lorentzen 2013). In a designer hotel, the overnight stay facilities such as a bed and a restaurant are the Frame, but the furniture design experience is the Occupant that may make guests choose this hotel (Sørensen and Jensen 2015); this Occupant is particularly strong if it is an unexpected experience and thus creates some tensions in the memories of the guests.

Conclusion

This chapter has suggested a theoretical approach to people's experiencing in a customer relationship. Experiential dissonance is a general psychological phenomenon, based on emotions and sensing that people can meet in many situations, for example as tourists. The theory of experiential dissonance is a contribution to understanding peoples' experiencing in general.

The phenomenon also occurs in situations where individuals are tourists and in this situation experiential dissonance is of importance to tourism providers. Experiential dissonance is of importance to tourism firms' business success because it determines the value of the total experience for the tourists and their loyalty, repurchase and general attitude towards the firm.

In this chapter two basic concepts characterizing provided experience elements, the Frame and the Occupant, have been suggested. A theoretical model of experiential dissonance and some hypotheses about experience providers' exploitation of Occupants and experiential dissonance have been established.

The theory of experiential dissonance should be empirically investigated to test its validity. We do not know how general the phenomenon is and how much it means to people. It might also be that a time dynamism in experiencing exists thus the Frame may be more important in the beginning and the Occupant at the end of the experience journey. Such issues should also be investigated further.

References

Berry, L. and Carbone, L. (2007). Build Loyalty through Experience Management, *QualityProgress*, 10 (9): 26–32.

Bourdieu, P. (1994). *Raisons Pratiques*. Paris: Seuil.

Csikszentmihalyi, M. (2002). *Flow*. London: Rider.

Festinger, L. (1957). *A Theory of Cognitive Dissonance*. Stanford: Stanford University Press.

Grönroos, C. (2000). *Service Management and Marketing*. Chichester: John Wiley.

Gummesson, E. (1994). Making Relationship Marketing Operational, *International Journal of Service Industry Management*, 5 (5): 5–20. DOI: 10.1108/09564239410074349

Holbrook, M.B. and Hirschman, E.C. (1982). The Experiential Aspects of Consumption Consumer Fantasies, Feelings and Fun, *Journal of Consumer Research*, 9 (2): 132–142.

Horn, P. and Jensen, J.F. (eds.) (2011). *Experience Leadership in Practice*. Roskilde: MOL Publishers.

Koolhaas, J. (1982). *Organizational Dissonance and Change*. Chichester: John Wiley.

Larsen, J. (2006). Picturing Bornholm: The Production and Consumption of a Tourist Island through Picturing Practices, *Scandinavian Journal of Hospitality and Tourism*, 6 (2): 75–94. DOI: 10.1080/15022250600658853

Lorentzen, A. (2013). The Role of Culture Festivals in Regional Development: The Festival of Tordenskiold in Frederikshavn. In Lindkvist, L. and Lindeborg, L. (eds.), *The Value of Arts for Regional Development* (pp. 43–59). London: Routledge.

Mattsson, J., Sundbo, J., and Jensen, C.F. (2005). Innovation Systems in Tourism: The Role of the Attractors and Scene-takers, *Industry and Innovation*, 12 (3): 357–381.

Normann, R. (1991). *Service Management*, 2 ed. Chichester: John Wiley.

Pine, B.J. and Gilmore, J. (1999). *The Experience Economy*. Boston: Harvard Business School Press.

Pine, B.J. and Gilmore, J. (2011). *The Experience Economy*. Updated Edition. Boston: Harvard Business School Press.

Schmitt, B. (1999). Experiential Marketing, *Journal of Marketing Management*, 15 (1–3): 53–67. DOI: 10.1362/026725799784870496

Schmitt, B., Rogers, D., and Vrotsos, K. (2004). *There's No Business That's Not Show Business Marketing in an Experience Culture*. Upper Saddle River: Prentice Hall.

Sørensen, F. and Jensen, J.F. (2015). Value Creation and Knowledge Development in Tourism Experience Encounters, *Tourism Management*, 46: 336–346. DOI: 10.1016/j.tourman.2014.07.009

Sørensen, F. and Sundbo, J. (2014). Potentials for User-based Innovation in Tourism: The Example of GPS Tracking of Attraction Visitors. In Alsos, G.A., Eide, D., and Madsen, E.L. (eds.), *Handbook of Research on Innovation in Tourism Industries* (pp. 132–153). Cheltenham: Edward Elgar.

Sørensen, F., Sundbo, J., and Mattsson, J. (2013). Organisational Conditions for Service Encounter-based Innovation, *Research Policy*, 42 (8): 1446–1456. DOI: 10.1016/j.respol.2013.04.003

Sundbo, J. (2013). The Roskilde Festival and Its Importance for Regional Industrial and Cultural Development. In Lindkvist, L. and Lindeborg, L. (eds.), *The Value of Arts for Regional Development* (pp. 74–93). London: Routledge.

Sundbo, J. and Sørensen, F. (2013). Introduction. In Sundbo, J. and Sørensen, F. (eds.), *Handbook on the Experience Economy* (pp. 1–17). Cheltenham: Edward Elgar.

Part III

Marketing tourism experiences

22

Traditional versus experiential marketing in tourism

Valentina Della Corte and Giovanna Del Gaudio

Introduction

In recent years, the "sharing economy" has attracted a great deal of attention. There is a trend towards peer-to-peer platforms (Gutiérrez et al. 2017; Karlsson et al. 2017), which allow the customers to access other consumer-owned properties or competencies. These marketplaces enable people to share items like accommodation services with other customers, contributing to the creation of a new asset-light supply paradigm.

The revolution generated by the sharing economy phenomenon is clear evidence that customers are taking over the relationship with companies. Indeed, through the mechanisms introduced by this "peer economy", customers can be a counter-power to companies. The sharing economy can be defined as "a socio-economic system that coordinates" (Wang and Nicolau 2017) "the peer-to-peer-based activity of obtaining, giving, or sharing the access to goods and services" through "community-based online services" (Hamari et al. 2015). Consequently, as this revolutionary and innovative "on-demand economy" evolves, analytical pricing models are going to have a major impact in the future.

Starting from these considerations, this chapter aims to study the evolution of some traditional tools used by hospitality firms, like Customer Relationship Management and Loyalty Systems, as well as to empirically examine the impact that the rise of the sharing economy has on the hospitality offer as well as on such systems in a tourist destination. In more detail, the specific purpose is to study how, in the sharing economy era, knowledge and information sharing between customers has radically changed revenue management (RM) techniques as well as the exploitation of CRM. Additionally, in the analysis we focus on the interactions between the former and the latter, in an overall dynamic process of development. In the study, both hotel firms and other organizations in the hospitality industry (like bed and breakfast establishments, houses, apartments or rooms for rent) are taken into account, since the process involves all types of hospitality firms. This chapter shows how the hospitality industry (including B&B establishments and other types of accommodation), has changed its methods by using marketing policies. Indeed, the transition from the traditional form of marketing i.e. the four Ps of marketing (product, promotion, price and place) to experiential marketing is due to the fact that, in the tourism sector, customer satisfaction happens at the "organic level", the so-called moment of truth. The organic level refers to the phase of service provision and the possible marketing activities during the experience (in particular the servicescape, that is, the general sensorial atmosphere; the relationship between price and quality, connected with the perceived qualitative level of the service; and the productivity level, in term of human resources efficiency while serving) (Della Corte 2013).

One of the limits of traditional marketing is connected with the fact that it does not consider the stage when tourists live the experience. This kind of marketing just acted on the AIDA[1] (Rawal 2013) phases of the potential tourist decision process, while experiential marketing influences the S (satisfaction level) during service provision (Della Corte et al. 2015). Now, in the sharing economy context, this well organized and scheduled process is submitted to a change of direction, because customers, while living their experience, also post comments and pictures of the places they are visiting during their stay, an instantaneous process that exerts a great influence on other potential visitors. Therefore, the organic level tends more and more to assume a relevant role in firms' and destinations' promotions (Della Corte et al. 2018). The idea is then to show which choices firms are making and how they are performing in the sharing economy scenario.

In such a scenario, the non-hotel offer has at an early stage been characterized by a more flexible and open marketing approach, using the internet and social media. These platforms have allowed firms to become known more easily and have enabled them to apply at a lower cost systems that were very complex and costly in the past. CRM, for example, is important for business as it connects and builds a relationship between the firm and its customers. Nowadays, CRM is reinventing itself through three typical "tools" that are an expression of the sharing economy paradigm, which are: mobile, cloud and social media. The use of CRM in the experiential marketing context is of paramount importance since it collects data from and about the customer in order to enhance customer satisfaction and retention.

With reference to the hospitality industry, a very interesting case is Airbnb, which provides a platform for travelers and hosts to interact.

To pursue the research goal, this chapter adopts case study methodology in order to understand whether and how the hotel chain AccorHotels operates through revenue management techniques that use CRM for the creation of personalized offers. The focus is on "Jo&Joe", the new brand launched by AccorHotels, with the purpose of capturing the lucrative millennial demographic and cashing in on the sharing economy.

The choice to proceed with case studies is due to the fact that this is a new specific topic, within which the sharing economy is not just seen as a new, updated innovative idea but as a specific approach driving firms in their practical marketing choices and with reference to the adoption of tools for customer satisfaction and retention. From this point of view, case study analysis allows a deep study of the phenomenon, forming some preliminary implications that can drive to empirical work in the close future.

Interesting results are drawn from both cases, delineating also the pattern that hospitality firms, in their own specificities, are following in the sharing economy.

The first section of the chapter concentrates on a review of the key elements of CRM in the sharing economy era; then the attention shifts to the relationship between CRM and RM in the hospitality industry.

Literature review

In the last two decades, the paradigm of marketing has shifted from transactions to relationships (Bolton 2016; Sheth and Parvatiyar 1993; Vogt 2011). This new vision has caused hotel and non-hotel firms to engage with their customers, considering their relationships with their guests, both actual and potential, as the main reason for the development of commercial strategies.

On the other side, the focus of revenue managers is to maximize revenues, enhancing sales and leveraging in the three areas of pricing, inventory control and customer mix (Wang 2012). However, in revenue management "the customer seems to have been relatively forgotten in this stream of research" (Wirtz et al. 2003: 217). Despite this consideration, during the last few years both scholars and managers have overlapped concepts of customer relationship management and revenue management (Wang 2012). Studies have investigated hotel revenue management by focusing on strategic levers, such as pricing (Baker and Collier 2003; Bayoumi et al. 2013; Choi and Mattila 2004; Kimes and Chase 1998; Noone and Mattila 2009), overbooking (Liberman and Yechiali 1978; Noone and Lee 2011; Phumchusri and Maneesophon 2014; Rothstein 1974), and customer relationship management (Lo et al. 2010; Noone et al. 2003; Wang 2012).

CRM can effectively support the implementation of different allocation methods (Kimes and Thompson 2004) contemplated in the RM philosophy.

The importance of integrating CRM and RM has been studied and expanded upon by several scholars (Belobaba 2002; Dickinson 2001; Jonas 2001; Lieberman 2002; Noone et al. 2003), since their effective implementation can generate profitability for both customers (Buttle 2004) and for the hotel/non-hotel firms (Lo et al. 2010).

The competitive advantage created by the adoption of CRM is connected with the maximization of the revenue from customers over the lifetime of the guest relationship (Wilson et al. 2002). Indeed, without the support of customer data generated by the CRM system, hotel and non-hotel companies are not able to create the right price for the different targets (i.e., leisure and business, or individuals, groups, families, etc.) and/or to achieve different purposes ("encouraging repeat customers, … exceeding customers' expectations, communicating and connecting with customers" – Lo et al. 2010: 146).

The implementation of RM per se is not sufficient because it needs the support of strategic assets, such as Customer Relationship Management (CRM). This mix can determine effective revenue management strategies and influence marketing mix strategies.

The main aim of hotel revenue management is to maximize revenues through the selling of beds (Wang 2012). This requires a clear and effective revenue-optimal strategy. From here, a series of questions emerge: How can revenue managers implement revenue-optimal strategies? What are the tools supporting the development of these strategies?

These questions lead to the development of a precise path for the literature review that takes into account the necessity to study RM in relation to customer relationship management (CRM – Belobaba 2002; Dickinson 2001; Jonas 2001; Lieberman 2002; Noone et al. 2003).

The overlapping perspective between customer relationship management (CRM) and RM is very interesting. CRM has, indeed, the clear goal of creating an alignment between business strategies and the programs of customer loyalty and profitability (Rigby et al. 2002). The classical approach of the implementation of CRM is based on the fact that one of the main strengths is the creation of tailor-made hotel services according to the preferences of the guests, in order to invest in the following communication and promotion marketing policies. The exploitation of CRM allows a differentiation of on-site treatment, (Becerra et al. 2013) such as the creation of ad hoc loyalty programs (Shoemaker and Lewis 1999) in order to retain the most profitable customers, and customer complaint handling (Heung and Lam 2003), etc.

However, little attention is paid to the implications of this integration in a hospitality environment that is increasingly being influenced by the sharing economy.

For example, hotel companies use social media to communicate special offers or other price promotions. These pricing strategies are the result of an attentive exploitation of CRM data. However, the traditional approach of the implementation of CRM is based on the fact that one of the main strengths is the creation of a tailor-made product according to the preferences of the customers in order to invest in communication and promotional marketing policies. Although literature on RM underlines the existence as well as the importance of differences in demand according to customers' preferences and needs (Talluri and van Ryzin 2004; van Ryzin 2005), the existing literature on the theme underemphasizes the use of strategic tools, such as CRM, in the process that leads to the definition of a revenue-optimal strategy.

Hence, from this background the first proposition emerges.

P1: In the hospitality industry, the combined use of CRM and RM can determine effective revenue management strategies

Nevertheless, in the current sharing economy era, the complexity of CRM and RM increases, especially in their interconnections. Considering that one of the main dimensions of the hospitality industry is pricing (Wang and Nicolau 2017), it is interesting to notice that the price policy applied by the hosts of shared accommodation can depend on the so-called "rating visibility" (received stars, Gutt and Herrmann 2015), the location of the houses/rooms (Tang and Sangani, n.d.), and also on the personal background of the hosts (i.e., professional or non-professional hosts – Li et al. 2015).

On the other side, guests looking for this kind of accommodation are more price sensitive, looking for a cost reduction and are more prone to want to live like a local, learn and exchange cultural and social habits (Balck and Cracau 2015).

Differently from the branch of literature (Besanko and Winston 1990) on dynamic models that conceives consumers' choices as rational (Della Corte and Del Gaudio 2012), this chapter assumes that RM has to consider the behavioral approach of consumers based on irrational purchase decisions. In this scenario, the use of CRM plays an important role in the process of revenue-optimal strategy building and monitoring.

CRM can effectively support the implementation of different allocation methods (Kimes 2016). The customers' traits (social variables, propensity to shared accommodation, etc.) and behaviors (first time using sharing accommodation, repeater, etc.) play an important role in the determination of pricing strategies.

Hence, CRM should provide information regarding the choice of accommodation and other preferences (origin and destination, duration of stay – Lin and Huang 2015) that describe the perspective of the demand.

P.2: In the sharing economy era, the interplay between CRM and RM is fundamental to influencing the purchase decisions of customers

Methodology

This research is mainly based on the analysis of the cases of Airbnb and Joe&Joe. The chapter is based on a qualitative case studies analysis, whereas the research approach concerns the understanding of the nature of a phenomenon (Miles and Huberman 1994). According to Aarikka-Stenroos and Sandberg (2012: 201) case studies "allow rich description and comparison" as well as proceeding with cross-case comparisons (Eisenhardt and Graebner 2007).

Consistent with the nature of this work, the case study method allows the authors to take part in the data collection process, as well as to provide very useful context information (Patton 2002), as long as it is being supported by an efficient research design. Case study methodology is in line with the exploratory nature of this chapter as it is a powerful strategy to answer how and why research questions capture contextual richness and the complexity of research issues while understanding the social structures (Yin 2003).

This research is based on an exploratory design where two different cases of shared accommodation are compared. The intent is to examine whether and to what extent these companies implement CRM and RM practices together. Data have been obtained from the websites, reports and other secondary sources.

Case study 22.1

Airbnb

The history of Airbnb began in 2008 when two designers, owning a space to share, decided to host three travelers in their private house.

Since 2008, Airbnb defines itself as "a trusted community marketplace for people to list, discover, and book unique accommodations around the world – online or from a mobile phone or tablet" (Airbnb 2017).

Airbnb has a presence in more than 65,000 cities and 191 countries, with more than 200 million guests. The brand has implemented its CRM strategy with the launch of its new website and mobile app. Airbnb's mobile app is the solution to the customer-facing process as it helps travelers and hosts to book and upload hassle-free. By using multiple CRM software, Airbnb is learning customer behavior and building its strategies starting from the customers' experiences. In addition to CRM, it becomes important to investigate the management of revenue streams from customers.

Indeed, thanks to the implementation of both CRM and RM, Airbnb has:

- Created Instant Booking: Ads with Instant Booking do not require host approval before they can be booked. Guests can simply choose travel dates, book and discuss check-in with the host
- Improved the review system in order to obtain more objective reviews from the guests (Rubin 2014)
- Created "Business Travel Ready": Business Travel Ready lodgings must be complete houses or apartments that are suitable for business guests (Dillet 2015). For example, these lodgings have to supply wireless internet, a place to work with a laptop, self check-in, etc.

Jo&Joe

Jo&Joe is a new brand of the Accor Group and it is said that it "blends the best of private-rental, hostel and hotel formats" (Sell 2016). The aim of the group is to open, by 2020, more than 50 Jo&Joe outlets all around the world. Their location is based on the concept of "less than 15 minutes away from the major points of interest" (Sell 2016).

Jo&Joe is a part of the AccorHotels Group portfolio, a solution designed to compete with the vast international trend-setter and millennial offerings that are revolutionizing the market.

This project is an example of "scale-up" (Beichner et al. 2007).

The Jo&Joe brand is the result of a process of close collaboration with future guests, experts, students and AccorHotels teams, with the ultimate aim of challenging embedded conventions in the hotel industry and imagining the future of hospitality in the sharing economy era (Chierchia 2016), As regards the perspectives of CRM and RM, it is important to outline that this brand exploits the systems and the policy of the Accor Group, which bases its CRM on being more relevant to and personal with the customer.

Thanks to an effective database, offered by the platforms of CRM, the company can keep a record of the customer containing personal details, the degree of satisfaction, loyalty card information, message from smart booking history (first-timer or frequent client) and customer preferences (accommodation preferences, type of bed, air conditioning, personal preferences like sport and culture).

This information helps the company to understand the booking habits of the customer and to launch the right email campaigns, loyalty programs, etc.

This personalization allows an open dialog between the hotel company and the customer and enables the company to anticipate the customer's needs.

Finally, this customer knowledge is used by revenue managers in order to address the right price to the right customer.

Discussion

The above case studies show different histories and ways of doing business (Table 22.1). Indeed, while Airbnb has been based, since its creation, on peer-to-peer platforms and on a precise business model, Jo&Joe on the other hand represents a development of the classical formula of hotels, combining the best of the most popular formats (rents, hostels and hotels) while embracing the philosophy of moments to share.

Table 22.1 – Data on Airbnb and Jo&Joe

	Airbnb	*Jo&Joe*
Year of birth	2008	2017
Social	2.3 mln	3,352
Follower IG	9,163,578	5,535
Follower FB	11,900	313
Followers Twitter	89,062	–
Followers YouTube	3,189,813	–
Human resources – number of employees	1,001–5,000 employees	–
Typology	Private Company	Part of the group AccorHotels group
City	65,000+	1 / Hossegor
Countries	191+	1 / France
Total guests (year 2016)	200,000,000+	–

(Continued)

Table 22.1 (Cont.)

	Airbnb	*Jo&Joe*
Total ads (year 2016)	6,000,000	–
Mission	"Create a world where people can belong when they travel by being connected to local cultures and having unique travel experiences."	"A house open to neighbors and travelers alike. A cool, affordable and caring house, filled with ever so surprising design, events and talents. A house where you feel free to meet and mix, clink glasses, have a bite to eat together, cook, chat, laugh, unwind, dream, work, love, do yoga, play the guitar, explore the city ... and sleep! A house where you will simply live life to the fullest ..."
Development	Airbnb experiences, Los Angeles, 2016	The aim is to open 50 Jo&Joe properties by 2020, with target locations including Paris, Bordeaux, Warsaw, Budapest, Rio and São Paulo.
Development	Airbnb experiences, Los Angeles, 2016	The aim is to open 50 Jo&Joe properties by 2020, with target locations including Paris, Bordeaux, Warsaw, Budapest, Rio and São Paulo.
Guests		
Median Expenditure	45 mln 35$ Italy 50$ United States 34$ France	- -

Both Airbnb and Jo&Joe are the result of the rules imposed and/or created by the players operating in the sharing economy era.

Even if the two propositions are confirmed (P1: In the hospitality industry, the combined use of CRM and RM can determine effective revenue management strategies; P2: In the sharing economy era, the interplay between CRM and RM is fundamental to catch the purchase decisions of the customer), it is important to underline that there is a substantial difference between the two cases.

This difference lies in the managerial competences that are required for both customer relationship management and revenue management.

As regards Airbnb the systems of CRM and RM are implemented through a top-down approach by the central division of Airbnb but, though not always, Airbnb's hosts own the managerial capabilities and insights to determine RM policy based on the CRM data of Airbnb's extranet.

With reference to Jo&Joe, the organizational benefits of belonging to the AccorHotels group allow the managers of the accommodation to exploit the platform in the best way and to take part in training courses on RM and CRM.

Conclusions

The interest in the sharing economy and new formulas of hospitality has led both scholars and managers to focus their attention on this topic, but also to expand their vision on the use and strengths of both CRM and RM.

Indeed, if CRM and RM are implemented separately their effectiveness cannot be fully exploited. Combining them can determine effective revenue management strategies and orientate the marketing mix strategies.

The real revolution, which has already happened in some hotel chains (i.e., Marriot-Kimes 2016), is to incorporate these two aspects. For example, Marriott calls this specific department "Consumer Insights and Revenue Management".

The two cases showed different behaviors in the way in which they use CRM and RM. While Airbnb offers its hosts platforms where the data of the guests are available, revenue management is still unexplored.

Operating in a different way, Jo&Joe exploits the CRM and RM practices of the AccorHotels group. On the other hand, it is important to underline that this is a start-up case and the evolution of the strategy will need to be monitored in future works.

Jo&Joe represents a case of revenue opportunity for hotel companies operating in the sharing economy context. In the last decade, the sharing economy era appeared more as a threat than an opportunity for companies. Nowadays, an overturning of this perspective has come about, since the hotel companies can also exploit the benefits coming from this context. The added value for these companies is the bundle of resources, competences and sedimented knowledge in the fields of CRM and RM applied in this sector.

The chapter shows how CRM can be considered an important junction tool between traditional marketing and experiential marketing. Indeed, CRM can act on the traditional marketing policies as in the case of pricing strategies and the connected revenue management as well as on the development of tailor-made experiences for the guests, thanks to profiling customer information. One of the most important considerations to underline is the fact that the emphasis on experiential marketing for the firms operating in the tourism industry does not imply the avoidance or decline of traditional marketing. Rather, experiential marketing can shape the traditional 4 Ps of marketing and not vice versa.

The work provides important managerial implications showing the potential benefits of CRM in the field of revenue management and highlights the required managerial capabilities to sustain this specific process. Results show that through the use of tools like CRM systems and peer-to-peer platforms, companies can aspire to the implementation of pricing strategies, in terms of RM techniques, and will be able to maximize their own profitability.

Note

1 Known the AIDAS model, this is the decision process characterized by: Awareness, Interest, Desire, Action and Satisfaction.

References

Aarikka-Stenroos, L., and Sandberg, B. (2012). From new-product development to commercialization through networks. *Journal of Business Research*, 65(2): 198–206. DOI: 10.1016/j.jbusres.2011.05.023.

Airbnb. (2017). About us. Airbnb. Retrieved from www.airbnb.ca/about/about-us [Accessed: August 14, 2017].

Baker, T. K., and Collier, D. A. (2003). The benefits of optimizing prices to manage demand in hotel revenue management systems. *Production and Operations Management*, 12(4): 502–518. DOI: 10.1111/j.1937-5956.2003.tb00217.x.

Balck, B., and Cracau, D. (2015). Empirical analysis of customer motives in the shareconomy: a cross-sectoral comparison (Working Paper No. 2/2015). Retrieved from Otto-von-Guericke University Magdeburg, Faculty of Economics and Management.

Bayoumi, A. E. M., Saleh, M., Atiya, A. F., and Aziz, H. A. (2013). Dynamic pricing for hotel revenue management using price multipliers. *Journal of Revenue and Pricing Management*, 12(3): 271–285. DOI: http://doi-org-443.webvpn.fjmu.edu.cn/10.1057/rpm.2012.44.

Becerra, M., Santaló, J., and Silva, R. (2013). Being better vs. being different: differentiation, competition, and pricing strategies in the Spanish hotel industry. *Tourism Management*, 34: 71–79. DOI: 10.1016/j.tourman.2012.03.014.

Beichner, R. J., Saul, J. M., Abbott, D. S., Morse, J. J., Deardorff, D., Allain, R. J., and Risley, J. S. (2007). The student-centered activities for large enrollment undergraduate programs (SCALE-UP) project. *Research-based Reform of University Physics*, 1(1): 2–39.

Belobaba, P. B. (2002). Back to the future? directions for revenue management. *Journal of Revenue and Pricing Management*, 1: 87–89. DOI: 10.1057/palgrave.rpm.5170009.

Besanko, D., and Winston, W. L. (1990). Optimal price skimming by a monopolist facing rational consumers. *Management Science*, 36(5): 555–567. DOI: 10.1287/mnsc.36.5.555.

Bolton, R. N. (2016). *Service Excellence: Creating Customer Experiences that Build Relationships*. New York: Business Experts Press.

Buttle, F. (2004). *Customer Relationship Management: Concepts and Tools*. Oxford: Elsevier Butterworth-Heinemann.

Chierchia, V. (2016) Non solo lusso: Accor punta sul mercato Millennials con il nuovo brand Joe&Joe, prima risposta alla sfida di Marriott. Il sole 24 ore. Retrieved from http://vincenzochierchia.blog.ilsole24ore.com/2016/09/27/non-solo-lusso-accor-punta-sul-mercato-millennials-con-il-nuovo-brand-joejoe-prima-risposta-alla-sfida-di-marriott/ [Accessed: June 12, 2017].

Choi, S., and Mattila, A. S. (2004). Hotel revenue management and its impact on customers' perceptions of fairness. *Journal of Revenue and Pricing Management*, 2(4): 303–314. DOI: 10.1057/palgrave.rpm.5170079.

Della Corte, V. (2013). *Imprese e sistemi turistici: il management*. Milano: Egea.

Della Corte, V., Aria, M., and Del Gaudio, G. (2018). Strategic governance in tourist destinations. *International Journal of Tourism Research*, 20(4): 411–423. DOI: 10.1002/jtr.2192.

Della Corte, V., and Del Gaudio, G. (2012). Dynamic capabilities: a still unexplored issue with growing complexity. *Corporate Ownership and Control*, 9: 327–338.

Della Corte, V., Sciarelli, M., Cascella, C., and Del Gaudio, G. (2015). Customer satisfaction in tourist destination: the case of tourism offer in the city of Naples. *Journal of Investment and Management*, 4 (1-1): 39–50. DOI: 10.11648/j.jim.s.2015040101.16.

Dickinson, C. B. (2001). CRM-enhanced revenue management in the hospitality industry. *Hospitality Upgrade*. Summer, 136–138.

Dillet, R. (2015). Airbnb announces new business travel ready badge, host tools. TechCrunch. Retrieved from http://techcrunch.com/2015/11/12/airbnb-announces-new-business-travel-ready-badge-host-tools/ [Accessed: June 12, 2017].

Eisenhardt, K. M., and Graebner, M. E. (2007). Theory building from cases: opportunities and challenges. *Academy of Management Journal*, 50(1): 25–32.

Gutiérrez, J., García-Palomares, J. C., Romanillos, G., and Salas-Olmedo, M. H. (2017). The eruption of Airbnb in tourist cities: comparing spatial patterns of hotels and peer-to-peer accommodation in Barcelona. *Tourism Management*, 62: 278–291. DOI: 10.1016/j.tourman.2017.05.003.

Gutt, D., and Herrmann, P. (2015). Sharing means caring? Hosts' price reaction to rating visibility. In ECIS 2015 Research-in-Progress Papers (Paper 54). Retrieved from http://aisel.aisnet.org/ecis2015rip/54/ [Accessed: June 13, 2017].

Hamari, J., Sjöklint, M., and Ukkonen, A. (2015). The sharing economy: why people participate in collaborative consumption. *Journal of the Association for Information Science and Technology*. DOI: 10.1002/asi.23552. Advance online publication.

Heung, V. C., and Lam, T. (2003). Customer complaint behaviour towards hotel restaurant services. *International Journal of Contemporary Hospitality Management*, 15(5): 283–289. DOI: 10.1108/09596110310482209.

Jonas, D. (2001) Carriers melding revenue mgmt. and CRM systems. *Business Travel News*. 26th March, 18–19.

Karlsson, L., Kemperman, A., and Dolnicar, S. (2017). May I sleep in your bed? Getting permission to book. *Annals of Tourism Research*, 62: 1–12. DOI: 10.1016/j.annals.2016.10.002.

Kimes, S. E. (2016). The evolution of hotel revenue management. *Journal of Revenue and Pricing Management*, 15(3–4): 247–251. DOI: 10.1057/rpm.2016.27.

Kimes, S. E., Chase, R. B., Choi, S., Lee, P. Y., and Ngonzi, E. N. (1998). Restaurant revenue management: Applying yield management to the restaurant industry. *Cornell Hotel and Restaurant Administration Quarterly*, 39(3): 32–39. DOI: 10.1016/S0010-8804(98)80294-2.

Kimes, S. E., and Thompson, G. M. (2004). Restaurant revenue management at Chevys: determining the best table mix. *Decision Sciences*, 35(3): 371–392. DOI: 10.1111/j.0011-7315.2004.02531.x.

Li, J., Moreno, A., and Zhang, D. J. (2015). Agent behavior in the sharing economy: evidence from Airbnb. Available at SSRN 270827.

Liberman, V., and Yechiali, U. (1978). On the hotel overbooking problem—an inventory system with stochastic cancellations. *Management Science*, 24(11): 1117–1126.

Lieberman, W. H. (2002). Revenue management: what lies ahead? *Journal of Revenue and Pricing Management*, 1(2): 189–195. DOI: 10.1057/palgrave.rpm.5170021.

Lin, Y. H., and Huang, K. (2015). Customer loyalty under the influence of revenue management: the case of Taiwanese hotel customers. *Asia Pacific Journal of Tourism Research*, 20(12): 1374–1388. DOI: 10.1080/10941665.2014.981556.

Lo, A. S., Stalcup, L. D., and Lee, A. (2010). Customer relationship management for hotels in Hong Kong. *International Journal of Contemporary Hospitality Management*, 22(2): 139–159. DOI: 10.1108/09596111011018151.

Miles, M. B., and Huberman, A. M. (1994). *Qualitative Data Analysis: An Expanded Sourcebook*. Thousand Oaks, CA: Sage.

Kimes, S. E. (2016). The evolution of hotel revenue management. *Journal of Revenue and Pricing Management*, 15(3–4): 247–251. DOI: 10.1057/rpm.2016.27.

Noone, B. M., and Mattila, A. S. (2009). Hotel revenue management and the Internet: The effect of price presentation strategies on customers' willingness to book. *International Journal of Hospitality Management*, 28(2): 272–279. DOI: 10.1016/j.ijhm.2008.09.004.

Noone, B. M., and Lee, C. H. (2011). Hotel overbooking: the effect of overcompensation on customers' reactions to denied service. *Journal of Hospitality & Tourism Research*, 35(3): 334–357. DOI: 10.1177/1096348010382238.

Noone, M. B., Kimes, S. E., and Renaghan, L. M. (2003). Integrating customer relationship management: a hotel perspective. *Journal of Revenue and Pricing Management*, 2(1): 7–21. DOI: 10.1057/palgrave.rpm.5170045.

Patton, M. Q. (2002). Two decades of developments in qualitative inquiry: a personal, experiential perspective. *Qualitative Social Work*, 1(3): 261–283. DOI: 10.1177/1473325002001003636.

Phumchusri, N., and Maneesophon, P. (2014). Optimal overbooking decision for hotel rooms revenue management. *Journal of Hospitality and Tourism Technology*, 5(3): 261–277. DOI: 10.1108/JHTT-03-2014-0006.

Rawal, P. (2013). AIDA Marketing Communication Model: stimulating a purchase decision in the minds of the consumers through a linear progression of steps. *International Journal of Multidisciplinary Research in Social & Management Sciences*, 1(1): 37–44.

Rigby, D. K., Reichheld, F. F., and Schefter, P. (2002). Avoid the four perils of CRM. *Harvard Business Review*, 80(2): 101–109.

Rothstein, M. (1974). Hotel overbooking as a Markovian sequential decision process. *Decision Sciences*, 5 (3): 389–404. DOI: 10.1111/j.1540-5915.1974.tb00624.x.

Rubin, B. F. (2014). Airbnb to reviewers: tell us what you really think. CNET. Retrieved from www.cnet.com/news/airbnb-wants-you-to-be-honest/ [Accessed: July 10, 2017].

Sell, G. (2016) Accor launches Jo & Joe brand. Boutique Hotel News. Retrieved from www.boutiquehotelnews.com/home/news/2016/9/27/accor-launches-jo-joe-brand/ [Accessed: June 12, 2017].

Sheth, J. N., and Parvatiyar, A. (1993). *Relationship Marketing: Theory, Methods and Application*. Atlanta, GA: Atlanta Centre for Relationship Marketing.

Shoemaker, S., and Lewis, R. C. (1999). Customer loyalty: the future of hospitality marketing. *International Journal of Hospitality Management*, 18(4): 345–370. DOI: 10.1016/S0278-4319(99)00042-0.

Talluri, K., and van Ryzin, G. (2004). Revenue management under a general discrete choice model of consumer behavior. *Management Science*, 50(1): 15–33. DOI: 10.1287/mnsc.1030.0147.

Tang, E., and Sangani, K. (n.d.). Neighborhood and price prediction for San Francisco Airbnb listings. Retrieved from .stanford.edu",1,0,0>http://cs229</tp:e-path>.stanford.edu/proj2015/236 report. pdf [Accessed: June 13, 2017].

van Ryzin, G. J. (2005). Future of revenue management: models of demand. *Journal of Revenue and Pricing Management*, 4(2): 204–210. DOI: 10.1093/oxfordhb/9780199543175.013.0018.

Vogt, C. A. (2011). Customer relationship management in tourism: management needs and research applications. *Journal of Travel Research*, 50(4): 356–364. DOI: 10.1177/0047287510368140.

Wang, D., and Nicolau, J. L. (2017). Price determinants of sharing economy based accommodation rental: a study of listings from 33 cities on Airbnb. com. *International Journal of Hospitality Management*, 62: 120–131. DOI: 10.1016/j.ijhm.2016.12.007.

Wang, W. L., and Tung, L. (2012). Most feasible strategies for green marketing mix under business sustainable development. *The Business Review, Cambridge*, 20(1): 297–303.

Wilson, H., Daniel, E., and McDonald, M. (2002). Factors for success in customer relationship management (CRM) systems. *Journal of Marketing Management*, 18(1–2): 193–219. DOI: 10.1362/0267257022775918.

Wirtz, J., Kimes, S., Ho, J., and Patterson, P. (2003). Revenue management: resolving potential customer conflicts. *Journal of Revenue and Pricing Management*, 2(3): 216–226. DOI: 10.1057/palgrave.rpm.5170068.

Yin, R. K. (2003). *Case Study Research: Design And Methods*, Third Edition. Applied Social Research Methods series, *5*. Thousand Oaks, CA: Sage.

23

Marketing approach to the tourist experience

Kuan-Huei Lee

Introduction

The marketing of tourism products includes tangible and intangible aspects. Tourists in a destination are using all their senses to experience the destination. They are receiving products and services in the destination, which will form the travel experience that they will take back to their home country. These experiences can become memorable moments of tourists' lives that they will remember and share with family and friends. Destination organizers need to have a plan to market their products and services, starting with the creation of a marketing plan. This chapter discusses the different requirements, benefits and contents of a marketing plan. In the marketing plan should be explained the 8Ps of hospitality and tourism marketing, which are Product, Price, Place, Promotion, Partnership, People, Packaging and Programming.

In the delivery of tourism products and services, many stakeholders are involved in creating the tourist experience (Page and Connell 2014), which can be categorized into Supplier, Carrier, Travel Agency, and DMO (destination marketing organizer). The product offered by these stakeholders follows different stages in the product life cycle. This chapter discusses these stages in the life cycle of a product from introduction, growth and maturity to decline and various strategies to be used in each of the stages. The offering of a product is usually accompanied by a service and it should be monitored using a Service Quality Management system.

A framework of tourist behavior is introduced at the end of the chapter to address the effect of experiential marketing on tourist behavior. The framework explains the mental process that a tourist is undergoing when receiving stimulus in the destination in different forms such as visual, text and audio.

The marketing plan and marketing mix

A marketing plan is a written plan that can be used to guide an organization's marketing activities for a specific period. The period can range from a month to three or five years. The shortest length of marketing plan is called a tactical plan. Longer-term plans are created to govern the direction of the overall marketing goal and are called strategic market plans. Strategic market plans focus on long-term goals and are more oriented to the external competitive environment of the organization.

Requirements of a marketing plan

A marketing plan is similar to a blueprint of a building that determines the actions and steps the management could follow to reach the ultimate goal. There are a number of criteria that the plan must have according to Morrison (2010):

1. Fact-based
2. Organized and coordinated
3. Programmed
4. Budgeted
5. Flexible
6. Controllable
7. Internally consistent and interrelated
8. Clear and simple

Benefits of a marketing plan

An elaborated marketing plan is a useful tool that will allow the organization to assess whether their activities match with the target market. It will also help the organization to be focused on their objectives and the target market; it is a powerful tool that can communicate across different levels internally and externally. The marketing plan can be used as a measurement of success in marketing; it provides the foundation to evaluate all marketing activities launched in the organization. Finally, the marketing plan projects both short-term and long-term targets which allows the organization to stay on track with regard to the longer goals while pursuing their short-term goals.

Contents of a marketing plan

The contents of a marketing plan consist of three main parts: the executive summary, the marketing plan rationale and the implementation plan (see box). The executive summary gives an overview of the plan. with main highlights to enable readers to gain a quick understanding. The marketing plan rationale gives a more detailed analysis of the actual situation of the organization and describes the strategies chosen, the targeted market segment, marketing mixes, position approaches and marketing objectives. The implementation plan is the action plan with step by step activities to achieve the goal and the timeline for each of these planned activities. The implementation plan also specifies the budget associated with these actions, the control procedure and the evaluation processes to make sure that the steps do not stray away from the original path.

Contents of a marketing plan

A. Executive Summary
B. Marketing Plan Rationale
   1. Situation Analysis Highlights (Where are we now?)
      a. Environmental Analysis
      b. Location and Community Analysis
      c. Primary Competitor Analysis
      d. Market Potential Analysis

e. Services Analysis
f. Marketing Position and Plan Analysis
g. Major Strengths, Weaknesses, Opportunities, and Constraints

2. Selected Marketing Strategy (Where would we like to be?)
   a. Market Segmentation and Target Markets
   b. Marketing Strategy
   c. Marketing Mixes
   d. Positioning Approaches
   e. Marketing Objectives

C. Implementation Plan

1. Activities Plan (How do we get there?)
   a. Activities by Target Market for Each Mix Element
   b. Responsibilities for Activities
   c. Timetable and Activity Schedule
2. Marketing Budget (How do we get there?)
   a. Budget by Target Market
   b. Budget by Marketing-Mix Element
   c. Contingency Funds
3. Control Procedures (How do we make sure we get there?)
   a. Results Expected from Each Activity
   b. Progress Reporting and Measures
4. Evaluation Procedures (How do we know if we got there?)
   a. Measurements
   b. Performance Standards
   c. Evaluation Timetable

(*Source*: adapted from Morrison 2010: 295)

The 8 Ps of hospitality and travel marketing

The marketing plan explained how the organization uses the 8Ps of the hospitality and tourism marketing mix. The 8Ps are the well-known 4Ps marketing mix plus four more Ps. These are:

Traditional 4Ps:

- Product
- Price
- Place
- Promotion

Additional 4Ps

- Partnership
- People
- Packaging
- Programming

1. Product: a good, idea or service provided to customers which in the hospitality and tourism context includes tourists.
2. Price: there are different strategies that the organization can choose to set the price. Some pricing strategies are (1) sale-oriented pricing; (2) profit-oriented pricing and (3) status-quo pricing.
3. Place: the combination of different distribution channels that organizations use to communicate their product or service to the customers. A DMO (destination marketing organization) can be involved in this distribution channel.
4. Promotion: a different promotional mix that includes internet marketing, advertising, sales promotion, merchandising, personal selling, publicity and public relations.
5. Partnership: it refers to cooperative marketing from hospitality and travel organizations. Two or more organizations are getting together to launch cooperative promotions that can be short or long term. One example is strategic alliances formed by airlines.
6. People: there are two groups of people, guests and hosts. Guests are the customers or tourists and hosts are those working in different hospitality and tourism organizations. The management of the guest–host relationship is important and will be associated with service quality and customer satisfaction.
7. Packaging: it is all related and additional services in a single transaction. Packaging and programming are related and in many cases they are offered together, such as in the Club Med Package.
8. Programming: it involves the development of special activities or programs associated with the packaging. One example would be parades organized by Disney World for visitors.

Stakeholders in the hospitality and tourism sector

There are four main players in the hospitality and tourism sector: (1) supplier; (2) carrier; (3) travel agency and (4) DMO (destination marketing organizer). Suppliers include lodging, F&B establishments, car rental & sharing, cruises, attractions and casinos. Carriers include airlines, railways, land transport and marine transportation. Travel agencies include all online travel companies, tour operators, event planners and retail travel agencies. DMO refers to national, regional or city destination marketing organizations such as Singapore Tourism Board, Canadian Tourism Commission and Tourism Australia.

Suppliers

1. Lodging sector: this sector includes hotels, hostels, B&B, resorts and Airbnb. The major player in the hotel industry is Marriott International with USD 22.3 billion in sales revenue in 2017. Other major players are Hilton Worldwide, Wyndham Hotel Group,

Host Hotel & Resorts, Hyatt Hotels, Oriental Land, Accor and InterContinental Hotels Group. Marriott International has over 6,500 properties worldwide (Statista 2017). These major hotel players segment their properties according to hotel brand segmentation, targeting different levels of customers.

2. F&B sector: this sector can be divided into two different categories; the commercial foodservice segment and the non-commercial foodservice segment. The commercial foodservice segment includes full service, limited service and quick service restaurants that can be located in different places such as airlines, cruise ships, Ferris wheels and underwater capsules. Non-commercial foodservice establishments are non-profit driven and include school cafeterias, hospital dining and corporate restaurants. The ratio between the commercial and non-commercial segment is 80% and 20% respectively.
3. Car rental & sharing: car rental is growing as well as the car-sharing economy. The major players in car rental are Enterprise, Hertz, National, Alamos, Avis, Budget and Dollar Thrifty. Car sharing is increasing in European and Asian countries. Different players such as Zipcar, Communauto and Maven offer station-based and round-trip car sharing.
4. Cruise line: the global cruise industry revenue was approximately USD 40 billion in 2015, doubled from one decade ago. The passenger capacity grew from 466 thousand in 2015 to 521 thousand by 2018 (Statista 2019). Major players in this sector are Carnival, Royal Caribbean International, Regent Seven Seas, Windstar, Silversea and Princess.
5. Attractions and events: government, non-profit foundation and private operators handle different iconic landmarks and theme parks. World-renowned attractions such as the Great Wall of China, the Grand Canyon and the Taj Mahal are some examples. Private operators usually operate theme parks such as Disney World and Universal Studio. Governmental sectors usually operate national museums and heritage sites. World mega-events such as the Summer and Winter Olympic Games require the effort of all the sectors to organize and execute different programs.
6. Casinos: only adults are usually allowed to join different games in a casino. The largest casino market in the world is in Macau, China. Las Vegas in the United States is the second-largest casino market. There is an increasing growth in the development of casino destinations, especially in the Asia-Pacific region. The major players are Las Vegas Sands, MGM Resorts, Genting, Galaxy Entertainment Group, Wynn Resorts, SJM Holdings, Melco Crown Entertainment and Caesar's Entertainment.

Carrier

1. Airlines: the airlines sector is one of the largest in hospitality and tourism; there are over 5,000 airlines listed with International Civil Aviation Organization (ICAO) codes in the world. The International Air Transport Association (IATA) estimated revenue from passenger business of USD 581billion and from cargo business of USD 59.2billion in 2018.
2. Railways: in large and extended areas, railways are the major carrier linking countries and major cities. Railway systems are important in the United States, Canada, Europe, Japan, China and India. The development of high-speed trains is the increasing trend in passenger railways; the Shinkansen in Japan, Eurostar in Europe and the TGV in France are some of the best-known high-speed trains. The fastest train is the Shanghai Maglev in China, which uses magnetic levitation technology and can travel at 431 km/hr.
3. Ferries and canals: ferries are important for cities near the sea and islands. For example, Washington State Ferries in the United States serves eight counties in Washington and

the Province of British Columbia in Canada has 22 vessels offering ten routes stopping at 20 terminals. The canal sector can be important for tourism and can become a main attraction, such as the Panama Canal that links the Atlantic and Pacific Oceans and is an important cruising destination for tourists.

4. Buses: buses are important land transport that connect cities and major players in guided and sightseeing tours. For long-distance buses, there are significant upgrades in the design to offer bigger, faster and more comfortable seats for passengers.

Travel agency

The travel agency is an intermediary that originally relied on commissions from carriers and other tourist suppliers. Today, with the competition from internet travel and online self-booking by consumers, many travel agencies are diversifying into different service offerings, such as planning and guided tours. The travel agency sector can be distinguished according to the customer base into retail travel agent, tour wholesaler, corporate travel, incentive travel, convention and meeting planner and online travel company. There are four major GDSs (Global Distribution Systems): Amadeus, Galileo, Sabre and Worldspan, which link booking and reservations among travel agencies, airlines, hotel, car rentals and others.

Destination marketing organizations (DMOs)

Most travel destinations have entities assisting in the marketing of the destination. Some are run by government at national, province or city level. Organizations such as Japan National Tourism Organization, Tourism Australia, Singapore Tourism Board and Canadian Tourism Commission are managed at national level with millions of dollars of investment in the planning of tourism marketing and development. In the United States, in residential areas with more than 50,000 inhabitants there is a local DMO called a convention and visitors bureau (CVB). Each CVB is supported by local government funds and their principal objectives are to encourage groups to organize meetings, conventions and trade shows in the city, to assist in the organization of these meetings and to attract tourists from different parts of the world to visit their destination.

Case study 23.1: Singapore Grand Prix

The Singapore Tourism Board (STB) is the organization in charge of the Singapore Grand Prix. The Singapore Grand Prix is part of the FIA Formula One World Championship motor race. The event started in 2008 with the first race in Maria Bay Street Circuit, the first night-time street circuit in Asia for Formula One. The annual event takes place every September for three days and it attracts approximately 250,000 attendees from different parts of the world every year. Parallel events are organized to entertain visitors, ranging from music, food and shopping. All streets involved in the circuit are closed during the race. The contract has been extended to 2021 for STB to continue hosting the event in Singapore.

The last Singapore Grand Prix was held from 20–22 September 2019. Artistic and gastronomic programs were organized, with a new viewing deck called The Cube and the debut of a three-starred Michelin chef for the first time in Singapore (Grand Prix 2019).

Branding and product life cycle

The American Marketing Association defined "brand" as a "name, term, symbol, or a design or a combination of them which is intended to identify the goods or services of one seller or group of sellers and differentiate them from those of competitors". Branding is a process that the organization uses to position itself in the mind of the consumer. There are different strategies that a company can use in branding such as Line Extension, Brand Extension, Multi Branding and New Brands. In line extension, the organization introduces new features into the current product line. In brand extension, the company expands into a new or existing area using the current brand name. Multi branding is when the company introduces more brands into the existing category. New brands are the strategy whereby the company introduces a new brand into the market to compete with existing competitors.

When a product is introduced into the market, there are four stages that we can identify based on the length of time a product is sold for and the sales volume, which is called the Product Life Cycle (Figure 23.1).

Stage I is the introduction stage, when the idea or concept of the product is conceived, developed and introduced into the market. The associated cost at this stage is usually high and sales are low. The company might need to invest in consumer testing and marketing to spread out the launch of this product in the market. The focus is to promote awareness in the market. If there are no, or few competitors, a skimming price strategy can be applied.

Stage II is the growth stage, when the sales start growing as well as the profit with the increase in sales volume. Competitors are attracted and start entering the market. Promotional events and activities can be introduced to maximize the potential of this stage.

Stage III is the maturity stage, when the product reaches the highest market share as well as profit margin. This is the point when the market reaches saturation and some companies might start leaving the market due to poor margins. If companies would like to remain at this stage for as long as possible, some strategies to be introduced are product modifications or improvements in processes to sustain the competitive advantage over similar competitors.

Stage IV is the stage when the product starts to shrink and decline in sales volume. There is pressure for price-cutting. The introduction of new features into the product might delay or prolong the decline, but it might not be able to stop it completely.

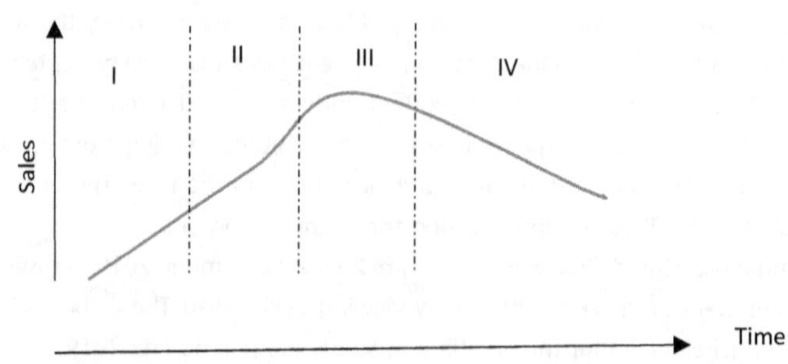

Figure 23.1 Product Life Cycle

Service quality management

It is important for organizations to manage customers' expectations in every hospitality service encounter. The hospitality service encounter is the interaction between the service provider and the customer. Customers usually already have certain expectations and if the fulfillment of their expectations equals or exceeds the service provider's offerings the encounter is therefore successful. If the customer's expectations are not met, there is a gap in the service quality.

There are different tangible and intangible factors that affect the service experience of a customer and the existence of a service gap. It is important for organizations to close all possible gaps in the offering of services. To ensure consistency and eliminate all possible gaps in service offerings, the Total Quality Management (TQM) system is usually implemented in organizations to determine customers' requirements and eliminate service defects. According to Deming's work, there are 14 points associated with TQM (English 1999: 338):

1. Create constancy of purpose for improvement of product and service
2. Adopt the new philosophy. We are in a new economic age
3. Cease dependence on inspection to achieve quality. Eliminate the need for inspection on a mass basis by building quality into the product or service in the first place
4. End the practice of awarding business based on price tag alone. Instead, minimize total cost
5. Improve constantly and forever the system of production and service to improve quality and decrease cost
6. Institute training on the job
7. Adopt and institute leadership with the aim to help people and machines do a better job
8. Drive out fear, so everyone may work effectively
9. Break down barriers between staff areas
10. Eliminate slogans, exhortations and targets for the workforce
11. Eliminate numerical quotas for the workforce and numerical goals for people in management
12. Remove barriers that rob people of pride of workmanship
13. Institute a vigorous program of education and self-improvement for everyone
14. Take action to accomplish the transformation. Put everyone to work to accomplish the transformation.

Another system widely used in the industry is the ISO9000 standard (International Organization for Standards). ISO9000 provides requirements, guidelines, definitions and standard processes for an independent assessment and certification of a service organization's quality management system. The most current version is ISO9000:2015 and there are seven principles of quality management associated with it:

1. Customer focus
    - Understand the needs of existing and future customers
    - Align organizational objectives with customer needs and expectations
    - Meet customer requirements
    - Measure customer satisfaction
    - Manage customer relationships
    - Aim to exceed customer expectations
    - Learn more about customer experience and customer satisfaction.

2. Leadership

    - Establish a vision and direction for the organization
    - Set challenging goals
    - Model organizational values
    - Establish trust
    - Equip and empower employees
    - Recognize employee contributions
    - Learn more about leadership.

3. Engagement of people

    - Ensure that people's abilities are used and valued
    - Make people accountable
    - Enable participation in continual improvement
    - Evaluate individual performance
    - Enable learning and knowledge sharing
    - Enable open discussion of problems and constraints
    - Learn more about employee involvement.

4. Process approach

    - Manage activities as processes
    - Measure the capability of activities
    - Identify linkages between activities
    - Prioritize improvement opportunities
    - Deploy resources effectively
    - Learn more about a process view of work and see process analysis tools.

5. Improvement

    - Improve organizational performance and capabilities
    - Align improvement activities
    - Empower people to make improvements
    - Measure improvement consistently
    - Celebrate improvements
    - Learn more about approaches to continual improvement.

6. Evidence-based decision-making

    - Ensure the accessibility of accurate and reliable data
    - Use appropriate methods to analyse data
    - Make decisions based on analysis
    - Balance data analysis with practical experience
    - See tools for decision-making.

7. Relationship management

    - Identify and select suppliers to manage costs, optimize resources, and create value
    - Establish relationships considering both the short and long term
    - Share expertise, resources, information, and plans with partners

- Collaborate on improvement and development activities
- Recognize supplier successes
- Learn more about supplier quality and see resources related to managing the supply chain.

A framework of tourist behaviors

Experiential marketing is related to how consumers process images (imagery processing) and there is abundant literature in the area of experiential marketing and tourist behavior. Imagery processing is the mental process by which sensory information is represented in working memory (MacInnis and Price 1987: 473). After reviewing extensively the literature in tourist imagery processing, Le et al. (2019) formulated a general framework of tourist behavior to explain the mental process that a tourist is undergoing in a tourist experience (Figure 23.2).

In this framework, three elements are identified as the key to design experiential marketing stimuli, which are (1) representation forms (e.g. picture, text, video), (2) content (levels of details, familiarity, authenticity, narrative themes), and (3) presence or not of an instruction to imagine. Experiential marketing stimuli are designed and introduced to tourists to active their imagery processing with the objective of influencing their behavior. The usage of pictures, video and

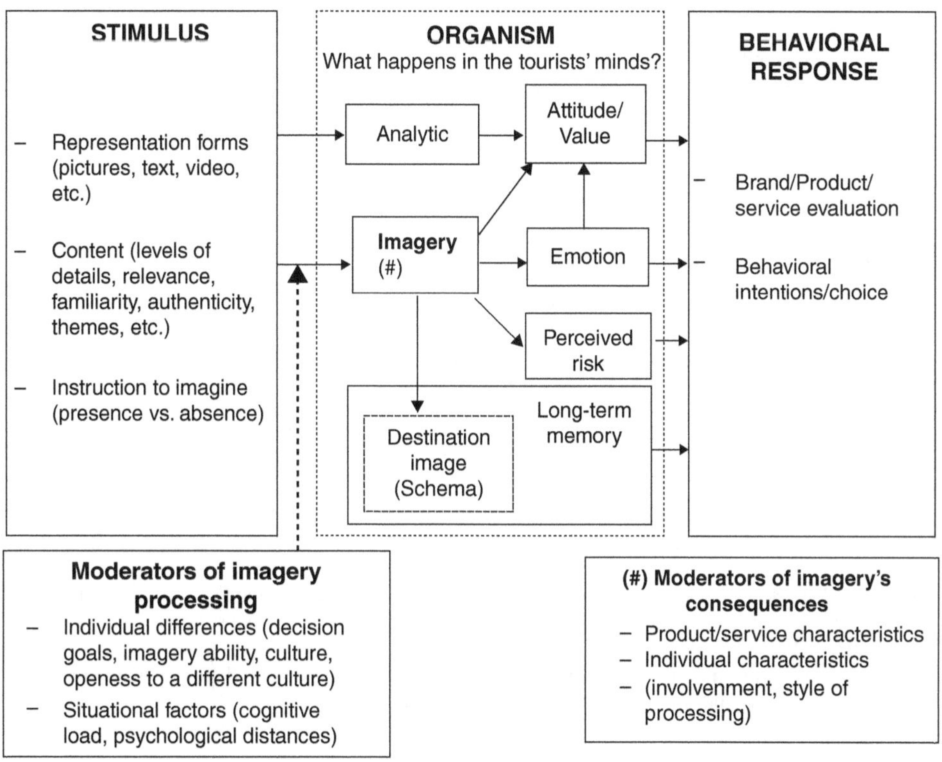

Figure 23.2 Tourist Behavior Framework

Source: adapted from Le et al 2019: 226

audio is important to the generation of emotional imagery. Narrative stories and olfactory triggers are helpful in the creation of dreams and fantasies. The content of the stimuli is also important and it is related to the degree of imagery vividness and elaboration. When tourists perceive authenticity in the story and narrative themes, there is a higher level of imagery processing. The exclusion of analytical and rational information is recommended to be excluded from experiential marketing stimuli as well as direct instructions asking tourists to create images themselves.

There are individual and situational factors as moderators of tourist imagery processing. Individual differences such as imagery ability, cultural background, openness to other cultures and decision goals might affect the same experiential stimulus. Situational factors such as cognitive load and psychological distances also affect imagery processing; it was found that familiar elements help to reduce psychological distance, which will improve tourist imagery processing. Finally, product or service characteristics, the level of involvement of tourists and their decision-making style are also influential in imagery processing.

Conclusion

It is important for the hospitality and tourism sector to know the concept of co-creation of experience with visitors in a destination. The organization-marketing plan, marketing mix, and different stakeholders are working to create peak experiences for tourists that they cannot forget easily. The hospitality and tourism sector industry is not only managing products and services but also the experience of the tourist. Every interaction with the tourist is important in this co-creation, which makes frontline staff some of the key people in this facilitation and management of the process (Campos et al. 2018). The design of this experience will need to actively involve the demand side (tourists) to create with the organization different products and services that tourists are willing to participate in and enjoy. The active participation of the demand side will allow organizations to understand internal aspects such as emotions, perceptions, sensations and other psychological factors that create the feeling and experience of visitors in the destination.

References

Campos, A. C., Mendes, J., Valle, P. O. D., and Scott, N. (2018). Co-creation of tourist experiences: A literature review. *Current Issues in Tourism*, 21(4): 369–400. DOI: 10.1080/13683500.2015.1081158

English, L. P. (1999). *Improving data warehouse and business information quality: Methods for reducing costs and increasing profits* (Vol. 1). New York: John Wiley.

Grand Prix, S. (2019). Retrieved from www.singaporegp.sg/en/news. [Accessed: April 1, 2019].

Le, D., Scott, N., and Lohmann, G. (2019). Applying experiential marketing in selling tourism dreams. *Journal of Travel & Tourism Marketing*, 36(2): 220–235. DOI: 10.1080/10548408.2018.1526158

MacInnis, D. J., and Price, L. L. (1987). The role of imagery in information processing: Review and extensions. *Journal of Consumer Research*, 13(4): 473–491. DOI: 10.1086/209082

Morrison, A. M. (2010). *Hospitality and travel marketing* (4th ed.). New York: Delmar Cengage Learning.

Page, S. J., and Connell, J. (2014). *Tourism: A modern synthesis* (4th ed.). London: Thomson Learning.

Statista. (2017). Sales revenue of selected leading hotel companies worldwide in 2017. Retrieved from www.statista.com/statistics/273064/revenue-of-the-largest-hotel-groups-worldwide/. [Accessed: April 23, 2019].

Statista. (2019). Cruise industry and cruise ships - statistics & facts. Retrieved from www.statista.com/topics/1004/cruise-industry/. [Accessed: April 23, 2019].

24

Experiential marketing for tourism destinations

Raouf Ahmad Rather and Linda D. Hollebeek

Introduction

Over the years, the tourism sector has been marked by rapid growth worldwide, thus revealing its growing economic importance (Indian Brand Equity Foundation 2018; United Nations World Tourism Organization 2014). In line with these developments, the number of tourism destinations has grown to cater for a rapidly developing global tourism demand. The tourism sector has been revolutionized by the increasing use of (new) technology, including social media and sharing economy-based applications such as Airbnb. With the help of these technologies, customers are progressively shaping their own tourism experiences, as substantiated by Pine and Gilmore's (1998) experiential marketing perspective (Le, Scott and Lohmann 2018; Schmitt 1999a, 1999b). Kumar et al. (2019) argue that the focus of business development is moving away from products toward customer-based processes. This leads Prahalad and Ramaswamy (2003: 12) to suggest that "a new point of view is required, one that allows individual customers to actively construct their own consumption experiences through personalized interaction, thereby co-creating unique value for themselves." Likewise, research suggests that people do not want to buy the products, but rather the stories behind them and the experience enabled by the offerings (Hollebeek and Macky 2019).

The reason behind the current tourism boom is a question asked by both academics and the industry. Experiential marketing was derived from the concept of the experience economy (Pine and Gilmore 1998), which is undoubtedly strongly present in the highly intangible experience economy (Le et al. 2018; Quan and Wang 2004; Tsaur, Chiu and Wang 2007), as illustrated by Disneyland's 1955 opening (Hannam 2004). Therefore, what does a customer get from tourism or travel offerings? First and foremost, the customer's perceived benefits lie in experience. While economic offerings such as commodities, goods or services are external to the buyer, experiences are inherently personal and exist only in the mind of a customer who has been engaged on an emotional, physical, intellectual, or even spiritual level (Hollebeek, Glynn and Brodie 2014; Pine and Gilmore 1998). A desirable experience will last long in the minds of its customers and affect their consequent behavior. As a result, managing the customers' experiential environment is a critical issue for the survival and competitive advantage of tourism firms.

Tourism research is shifting from a traditional view of "tourists-as-rational-decision-makers" to a more comprehensive understanding of both experiential and rational determinants of tourist decision-making (Le et al. 2018). Research shows that the demand for experiences is a major trend in the tourism industry and that destinations now increasingly compete by emphasizing the customer experience (Banff Lake Louise Tourism 2005). A key justification for this observation lies in the need for destinations to fight the problem of what Pine and Gilmore (1998) term the *commoditization of services*.

Destinations combine tourism products and intangible experiences to offer an integrated customer experience (Chen et al. 2018). In this vein, Schneider (2004) defines experiential travel as "travel that enriches the soul while broadening the mind." Therefore, tourists are no longer satisfied with traditional services, but instead they desire rich, diversified experiences (Banff Lake Louise Tourism 2005). This increasing demand for experience shows that it is no longer enough for destinations to compete based on their facilities and amenities, but instead they need to create differentiating experiences to continue attracting today's travelers (Banff Lake Louise Tourism 2005; Lemon and Verhoef 2016). Given tourism's nature, relationship marketing and experiential marketing are deemed of paramount importance (Ahn and Back 2018; Rajaobelina 2018). Building long-term relationships with customers and providing them with positive pre-, during and post-service encounter experiences across all touch-points helps tourism firms differentiate themselves from the competition, generate revenue and drive loyalty (Hwang and Seo 2016; Rajaobelina 2018).

Despite these invaluable conceptual works on the experience economy, empirical study remains scarce. One main reason probably lies in the fact that few of these theoretical writings provide an easily operationalizable starting-point for empirical investigation (Brodie and Hollebeek 2011). One exception to this observation is indeed offered by Pine and Gilmore (1998), which explicitly operationalizes the experience economy in terms of four dimensions: education, escapism, esthetics and entertainment. Further, there are few existing studies and none (to our knowledge) in the area of tourism destinations that examine the association between customer engagement dimensions and customer experience and other related variables (e.g. customer identification, behavioral intention) that are deemed to be of interest to tourism firms. Building on the above, this chapter investigates the effects of tourism customers' brand engagement with experiential marketing activities and offers a framework in this area. To do so, we make two specific contributions. First, we develop a model of tourism customer engagement and its effect on customer experience and brand identification and loyalty, which we also test empirically. Second, reflecting on our findings, we offer important implications for tourism marketing academics and managers.

Theoretical framework and hypotheses

Experiential marketing is an emerging marketing management philosophy (Le et al. 2018; Lemon and Verhoef 2016; Tsaur et al. 2007), which has been shown to be effective in driving tourist behavior (Rather 2018c). Experiences are private events that happen in response to the customer's sense(s) being stimulated (e.g. thrill-seeking), which typically result from direct observation or participation in events, whether real, dreamlike, or virtual (Schmitt 1999b). In the tourism context, experience has been described as a subjective mental state felt by participants (Tsaur et al. 2007). Further, experiences are usually not self-generated but induced. They are subordinated or related to the events which they respond to. Moreover, experiences may be viewed as complex, emerging structures. That is, no two experiences are exactly alike (Schmitt 1999b; Tsaur et al. 2007).

The notion of experience entered marketing in the seminal study of Hirschman and Holbrook (1982), which has become a key component in understanding consumer behavior through the overall consumption experience (Tynan and McKechnie 2009). The experiential aspects of consumption progressed at the end of the 1990s when Pine and Gilmore (1998) addressed how economies change. Economic activity aims not for output, but for experience via consumption (Lebergott 1993). Pine and Gilmore (1998) also suggest that experience represents a specific type of economic offering that creates competitive advantage that is difficult to imitate or substitute. Based on these developments, Holbrook and Hirschman (1982) and Lemon and Verhoef (2016)propose customer experience orientation as a prominent trend both for marketing practitioners and researchers. Schmitt (1999a, 1999b) argues that because traditional marketing is engineering-driven, rationaland outdated, a need exists for the development of experiential marketing. The author also suggests that what customers want is products, communications and marketing campaigns that dazzle their senses, touch their hearts and stimulate their minds. They want products, communications and marketing campaigns that they can relate to and incorporate into their lifestyles. They thus want offerings, communications and marketing campaigns to deliver an experience.

Consequently, experiential marketing is increasingly used by businesses to forge experiential connections with customers (Homburg, Jozić and Kuehnl 2017; Schmitt 1999b). Here, Schmitt (1999a, 1999b) proposed the concept of *strategic experiential modules* (SEMs) that managers can use to create different types of customer experiences for their customers. The experiential modules to be managed in experiential marketing include affective experiences (FEEL), sensory experiences (SENSE), behaviors and lifestyles (ACT), creative cognitive experiences (THINK), physical experiencesand socialidentity experiences that result from relating to a reference group or culture (RELATE). The purpose of experiential marketing is to create holistically integrated experiences that possess, at the same time, SENSE, FEEL, THINK, ACT and RELATE qualities.

Customer engagement and customer experience

Although the engagement concept is examined across fields including organizational behavior (i.e. employee engagement), psychology (e.g. task engagement), sociology (i.e. civic engagement) and marketing (e.g. customer engagement or CE; Hollebeek, Srivastava and Chen 2019b; Rather 2018a; Rather, Hollebeek and Islam 2019a), its dynamics in particular settings such as tourism remain nebulous (Ahn and Back 2018; Harrigan et al. 2018). Given its interactive theoretical roots (Brodie et al. 2011), CE has been viewed from relationship marketing (Rather et al. 2019a; Rather and Sharma 2019) and SD logic perspectives (Brodie et al. 2011; Hollebeek et al. 2019b). These authors view CE as customers' resource investments in their interactions (Hollebeek et al. 2019a; Kumar et al. 2019), which is highly relevant in tourism.

CE has been widely viewed to comprise cognitive, affective andbehavioral dimensions (Harrigan et al. 2018; Hollebeek et al. 2014; Rather et al. 2019a; Sharma and Rather 2016). Consequently, the omission of CE's psychological aspects or behavioral activities would likely provide insufficient insight to properly investigate the concept (Ahn and Back 2018; Rather 2018a). However, neither psychological aspects nor behavioral activities alone reveal CE in full. True engagement must reflect an enduring psychological connection, plus ongoing interactive behavioral participation with the object (Ahn and Back 2018; Rather et al. 2019a). Therefore, we employ three-dimensional (cognitive, emotional and behavioral) CE in this study (Ahn and Back 2018; Brodie et al. 2011; Hollebeek et al. 2014).

Engaged customers are expected to play a key role in co-creating customer experience and value (Brakus, Schmitt and Zarantonello 2009). As the direct impact of CE on experience to our knowledge has been validated in only one study (Risitano et al. 2017), we aim to investigate this association in a tourism context. We propose CE's dimensions (i.e. cognitive, affective and behavioral) as important predictors of customer experience with tourism destinations/sites. Thus:

H1: Cognitive CE will positively influence customer experience.
H2: Affective CE will positively influence customer experience.
H3: Behavioral CE will positively influence customer experience.

Customer engagement and customer identification

The concept of customer identification and/or customer brand identification is based on *social identity theory* that defines brand identification as a perceptual construct (Kumar and Kaushik 2018; Rather, Tehseen and Parrey 2018), implying identity matching and identity fit. One reason that customers value psychological and symbolic brand benefits is because they can help construct their self-identity, which denotes the way they present themselves to others (Aguirre-Rodriguez, Bosnjak and Sirgy 2012; Escalas 2004; Rather 2017). Consumers engage in a matching process to identify brands that are congruent with their self-images (Aguirre-Rodriguez et al. 2012). In this vein, Sprott, Czellar and Spangenberg (2009) propose the customer brand engagement in the self-concept construct, in which CE and customers' identification are integrated. Hollebeek et al. (2014) consider customer self-brand connection as a CE consequence, which may develop from customers' specific interactive brand experiences. The links that consumers create between a brand and their own identity are referred to as customerbrand identification. Consequently, brands are perceived to be more important the more closely they link to self (Kumar and Kaushik 2018). Brands with a story that consumers can easily relate to are not only evaluated more favorably but also have a higher likelihood of purchase than brands with fewer self-brand connections (Escalas 2004). Consistent with the literature, we expect that emotional and cognitive brand engagement will facilitate self-brand connections and/or consumer identification. For example, Hollebeek et al. (2014) show that consumers' affective and cognitive brand engagement in social media serves to predict consumers' self-brand connection/identification. In addition, Harrigan et al. (2018) propose that as consumers actively engage with tourism brands, their self-brand connection is reinforced. This links their brand explicitly to the customer's identity or self-image. *Social exchange theory* suggests that consumers will only invest resources when the exchange creates value for them (e.g. status, affiliation; Hollebeek 2011; Rather 2018a; Rather and Camilleri 2019). These arguments suggest that when tourists are engaged, they may identify themselves with a destination brand. In sum, the higher the customer engagement, the higher is the customer's identification. Thus:

H4: Cognitive CE will positively influence customer identification.
H5: Affective CE will positively influence customer identification.
H6: Behavioral CE will positively influence customer identification.

Customer identification and behavioral intention

Behavioral intention is defined as a "stated likelihood to engage in a behaviour" (Oliver 1997: 28). Zeithaml, Berry and Parasuraman (1996) conceptualize four categories of behavioral loyalty: word-of-mouth, purchase intention, price sensitivityand complaining behavior. Understanding the process of how brands influence customer behavioral intention toward the brand is important, because it is an indicator of actual ensuing consumption behavior (Ahn and Back 2018; Hollebeek and Rather 2019; Rather and Sharma 2017; Zeithaml et al. 1996). Previous research on identification revealed sustained, long-term preference for the company's products as one of the major consequences of brand or organization identification (Bhattacharya and Sen 2003). Rather and Hollebeek (2019) advocate that customers who identify with the brand are likely to purchase more and recommend the brands products/services more frequently. Strong customer brand identification can prove vital for developing long-termrelationships (Bhattacharya and Sen 2003; Rather 2018b; Rather et al. 2019b), such as by influencing brand rebuying intentions. Further, Lam et al. (2010) recommend that brand identification increases customers' resistance to switching brands. In a nation brand context, Stokburger-Sauer (2011) establishes that individuals who expressed stronger identification with a nation as a brand were more likely to visit or revisit the destination in the near future. Therefore, we postulate that tourists develop strong loyalty toward a destination by revisiting it in the near future if they manage to develop a formal and enduring relationship with that particular destination. Thus, we hypothesize:

H7: A positive relationship exists between destination customer brand identification and the customer's intent to revisit the destination.

Customer experience and behavioral intention

Existing research has examined customer experience in zoos (e.g. Tsaur et al. 2007), travel agencies (Rajaobelina 2018), wineries (Lee and Chang 2012) and hotel brands (Rather 2018c), to name a few. Experience has also been used in tourism. For example, previous research indicates that experiential marketing positively affects tourist emotions, satisfaction, andbehavioral intentions in zoos and hotel brands (Lee and Chang 2012; Rather 2018c) and customers with a better experience tend to recommend the firm to others. Therefore, we postulate that customers develop strong loyalty (revisit) intentions toward a destination if they have a positive experience with that destination. We suggest:

H8: A positive relationship exists between tourism customer experience and the customer's intent to revisit the destination.

Methodology

A questionnaire was used to measure the constructs contained in our conceptual framework (see Figure 24.1). Minor modifications were made to relevant items to ensure their fit within the tourism destination context. Our five-dimension *customer experience* scale was adapted from Tsaur et al. (2007). A sample item includes: "*I would like to share what I experienced in this destination.*" Second-order *customer engagement* was adapted from Ahn and Back's (2018) scale. A sample item includes: "*Using this tourism destination stimulates my interest to learn more about this destination.*" *Customer identification* was gauged by using the Kumar and Kaushik (2018) four-item scale. A sample item includes: "*If a story in the media criticized this destination, I would feel embarrassed.*" Finally, *behavioralintention* items were sourced from Tsaur et al.'s (2007) five-item tourism-based scale, with a sample item reading: "*Visit this destination more in the next few years.*"

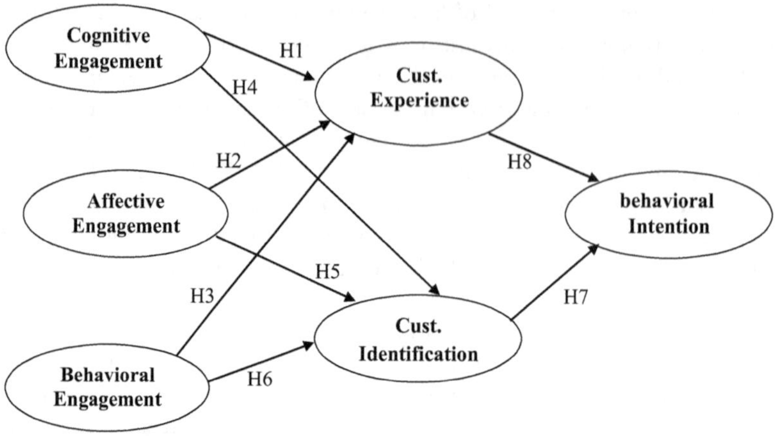

Figure 24.1 Proposed framework

The questionnaire was pilot-tested by using a convenience sample of 40 respondents over a period of ten days. Based on the results, relevant statements were modified as needed to ensure item clarity. The results indicated no issues regarding questionnaire readability. In the main study, respondents were requested to rate each item on seven-point Likert scales (1: strongly disagree through to 7: strongly agree). Due to the lack of a suitable sampling frame, convenience sampling was used to select the respondents for this research (Parrey, Hakim and Rather 2019; Rather et al. 2019a).

We collected data from six famous tourist destinations in India, including Srinagar, Gulmarg, Phalgam, Kokernag, Daksum and Jammu. We selected these particular destinations for several reasons. First, as stated, each of these offer popular tourist destinations and sites in India. Second, these destinations are famous for a range of purposes, including recreation, leisure, adventure and religious purposes, thereby offering a broad insight into customer motivations and behaviors (Indian Brand Equity Foundation 2018). Data was collected mostly during winter, summer and holiday periods, when there were larger numbers of both domestic and international tourists. Our sampling population was limited to those tourists who had visited one or more of these tourist sites at least once. Out of 400 surveys, a sample of 310 completed responses was retained for the final examination, thereby representing a 77.5% usable response rate. Descriptive analyses indicated that 54% of the sample was male. Regarding travel purpose, 65% were leisure and adventure customers, followed by 20% religious and 15% business customers. In addition, 56% of our respondents were from India and 44% were foreigners.

Analysis and results

Measurement model

Confirmatory factor analyses (CFA) were performed on all of the items. The overall goodness-of-fit of our CFA measurement model indicated a reasonable model fit to the data: CFI, TLI, NFI and GFI > 0.90; RMSEA < 0.08 (Bentler and Bonnett 1980). The overall CFA measurement model also achieved satisfactory fit: $\chi2$ = 243.773, $\chi2/df$ = 2.678, CFI = .95,

Table 24.1 Reliability analysis and discriminant validity

Construct	*α*	*CR*	*AVE*	*COGC*	*AFFCE*	*BEHCE*	*SEN*	*FEL*	*ACT*	*THN*	*REL*	*CIDN*	*BI*
COGCE	.84	.81	.54	**.82**									
AFFCE	.88	.80	.53	.78	**.82**								
BEHCE	.91	.83	.55	.71	.69	**.81**							
SEN	.90	.84	.61	.65	.80	.80	**.82**						
FEL	.87	.90	.69	.72	.79	.72	.79	**.80**					
ACT	.92	.84	.65	.71	.74	.63	.80	.79	**.83**				
THN	.94	.87	.52	.78	.73	.74	.81	.75	.79	**.82**			
REL	.84	.86	.55	.79	.63	.72	.64	.71	.81	.75	**.80**		
CIDN	.87	.88	.65	.63	.71	.64	.71	.76	.74	.75	.78	**.81**	
BI	.83	.89	.61	.65	.72	.65	.75	.63	.75	.79	.77	.73	**.82**

α = cronbach's alpha, CR = composite reliability, AVE = average variance extracted, COGCE = cognitive customer engagement, AFFCE = affective customer engagement, BEHCE = behavioral customer engagement, SEN= sense, FEL= feel, ACT = act, THN = think, REL = Relate, CIDN = customer identification, BI = behavioral intention.
The off diagonal factors are the correlations among the study constructs. The bold diagonal factors are the square root of the variance shared among the research factors and its measures.

NFI = .94, GFI = 0.87, SRMR = .053 and RMSEA = .067. Reliability and validity analyses were run to further test our model constructs. To do so, we calculated the average variance extracted (AVE), composite reliability (CR) and Cronbach's alpha for all constructs. We adopted Hair et al.'s (2010) suggested threshold values for convergent validity, including: (i) each item's factor loading should exceed the value of 0.70, (ii) each construct's reliability score should exceed 0.70, and (iii) the AVE should exceed 0.50. Therefore, convergent validity was achieved. For each construct, the square root of the AVE exceeded the correlation between constructs, thereby showing discriminant validity (Table 24.1).

Structural equation model results

We next investigated the data by employing a two-stage structural equation modeling (SEM) technique (Anderson and Gerbing 1988). The overall structural model was tested with maximum likelihood estimation using AMOS (Table 24.2) to assess the model's overall fit on the data. The results are: $\chi2$ = 245.543, $\chi2/df$ = 2.698, CFI = .93, NFI = .91, GFI = 0.85, SRMR = 0.57 and RMSEA = .071, thus reflecting reasonable model fit.

Hypothesis testing results

The results presented in Table 24.2 reveal that all main effects as proposed in H1–H8 were significant. As envisaged in H1, cognitive engagement leads to customer experience, exerting a strong effect (β = 0.60, $p < 0.001$). As hypothesized in H2, affective engagement leads to higher customer experience exerting a very strong effect (β = 0.69, $p < 0.001$). As predicted in H3, behavioral engagement positively affects customer experience, which was found to be moderate in strength (β = 0.21, $p < 0.001$). All the three predictors explain 78% of the observed variance in customer experience. As predicted in H4, the effect of cognitive engagement on customer identification is strong (β = 0.45, $p < 0.001$). As predicted in H5,

Table 24.2 Structural model results

	Relationship	R^2	β	*T value*	*Result*
H1	COGCE → CUSCX	0.60	0.78***	5.44	Supported
H2	AFFCE → CUSCX	0.69	0.78***	9.34	Supported
H3	BEHCE → CUSCX	0.21	0.78***	3.12	Supported
H4	COGCE → CIDN	0.45	0.76***	5.67	Supported
H5	AFFCE → CIDN	0.70	0.76***	10.15	Supported
H6	BEHCE → CIDN	0.23	0.76***	3.89	Supported
H7	CIDN → BI	0.64	0.80***	6.35	Supported
H8	CUSCX → BI	0.71	0.80***	11.23	Supported

Note: *** Significant for $p < 0.001$; COGCE = cognitive customer engagement, AFFCE = affective customer engagement, BEHCE = behavioral customer engagement, CUSCX = customer experience, CIDN = customer identification, BI = behavioral intention.

affective engagement leads to higher customer identification, exerting a very strong effect ($\beta = 0.70$, $p < 0.001$). As predicted in H6, behavioral engagement positively affects customer identification and was found to be intermediate in strength ($\beta = 0.23$, $p < 0.001$). All the three predictors explain 76% of the observed variance in customer identification. Similarly, these findings are also in line with H7, signifying that customer identification positively and strongly influences behavioral intention ($\beta = 0.64$, $p <0.001$). Finally, as predicted in H8, customer experience also strongly affects behavioral intention toward destination ($\beta = 0.71$, $p < 0.001$).

Discussion and implications

Theoretical implications

This research is significant because it extends the tourism literature in various respects. First, we linked CE to customer experience (Schmitt 1999a, 1999b), thereby offering an early empirical study connecting these concepts. Our findings emphasize the significance of a more sophisticated approach to explaining customers' experience and behaviors in the tourism context. These findings are also anticipated to be generalizable to other service settings, for example education, hotel, restaurant and/or professional services.

Second, this research also enhances the understanding of multidimensional CE (Ahn and Back 2018; Brodie et al. 2011). Even though advances have been made in constructing multidimensional engagement scales, there is limited knowledge about the roles of each dimension in tourism customers' experience or behavior. Therefore, the present research extended the literature by empirically testing this theoretical argument in the tourism destination context (Hollebeek and Macky 2019).

Third, this study fits into a growing literature stream that investigates the association between CE, customer experience, customer identification andbehavioral intention (e.g. Le et al. 2019; Rajaobelina 2018). Our empirical findings show that affective, cognitive andbehavioral CE has a positive effect on consumer experience and identification. In particular, affective and cognitive engagement offers the most significant impact on the customer's experience and identification with a destination. Tourists seem to be predominantly driven by affective engagement in their development of customer experience and customer identification. Further, the results reveal that the customer's experience and identification positively affect their behavioral intention, thereby offering a novel insight into the tourism context.

Fourth, our empirical findings serve to further substantiate an insight into the conceptual relationship between customer engagement, customer experience, customer identification andbehavioral intention that has remained largely conceptual so far (e.g. Le et al. 2018). These findings are also likely to be generalizable to other service settings, including hospitality, education or professional services.

Managerial implications

This study also has important practical implications. Le et al. (2018) and Quan and Wang (2004) suggest tourism as a pioneering example of the experience economy. That is, the customer's experience plays a critical role for a tourism firm to attain competitive advantage and survive. Therefore, developing a suitable experiential environment using the concept of experiential marketing is a vital issue for destinations in the context of the tourism industry.

This research proposed a conceptual framework to investigate the influence of customer engagement dimensions on experiences of experiential marketing and customers' identification on the consequent visitors'behaviors. Consequently, destination marketers and managers can develop programs of experiential marketing like visits to tourist sites/ attractions, different holiday packages, tailor-made packages, publicity programs, branding or advertising campaigns and destination attributes in the shape of recreational/leisure, entertainment, educational, cultural, and historical tourism-related activities to increase the onsite experience of tourism destination and to strengthen tourists' loyalty intentions. Programs should also be developed that increase tourists' engagement, for example by providing opportunities for tourists to explore destinations and nearby rural communities. Such programs will increase the tourists' enjoyment and self-identification. Consequently, increasing engagement can strengthen customer experience, identification and ultimately the behavioral intentions of tourists.

Further, the findings suggest that in building and maintaining customers' loyalty, destination managers should create a unique and positive destination experience and identification. From this viewpoint, besides generating unique destination experiences, destination managers should strive hard to build a distinct customers' brand identification. Such a destination brand identification not only resonates with tourists, but also emphasizes the uniqueness of the destination compared with its competitors. Therefore, a sustainable differentiation of the offerings can be achieved through identity distinctiveness. It also helps in enhancing customers' identification with the brand destination (Rather 2017), which further stimulates loyalty among customers (Kumar and Kaushik 2018). Similarly, creating and sharing brand destination stories through YouTube, Facebook, Twitter and Instagram are important in customers' decision-making, along with the consideration of various factors (e.g. narration, color, exterior and interior elements) to increase customers' visual experience. Visual sensory-enabling technologies can facilitate the delivery of product/service information similar to destination brand imaging. Thus, destination brands are shifting to sensory experience-enabling technologies, which offer sensory inputs in the online environment (Ahn and Back 2018). Further, developing online brand communities can encourage CE and promote customers' intellectual experiences with the brand destination (Hollebeek, Jurić and Tang 2017). Tourism marketers may think about offering customized services to increase customer engagement and experience by using big data. Analyzing customers' behavioral data can be applied to develop site services, settings, promotion and travel packages for target customers. For example, online travel agencies such as TripAdvisor.com make it possible for customers to post reviews and ratings about their destination or site experience. Using this (big) data can build new insights by revealing the relationship between marketing activities, customers' experience, identification and their intentions.

Limitations and future research

This study also has several limitations. First, we are aware that this empirical research has relied on cross-sectional data. Therefore, in future, longitudinal study could be used to examine the association of these constructs over a longer time span or through the engagement life cycle (Hollebeek et al. 2019b). Second, the data for this study was gathered from the tourism destination setting. Future research can replicate the findings from this study across different sectors, or in other countries. Third, future research should investigate the role of customerexperience dimensions, namely SENSE, FEEL, THINK, ACT and RELATE in affecting customers'behavioral intention. Finally, scholars may also investigate the role of experiential marketing (customer experience) in influencing other critical visitor outcomes including commitment, satisfaction and so on.

References

Aguirre-Rodriguez, A., Bosnjak, M. and Sirgy, M. J. (2012). 'Moderators of the self-congruity effect on consumer decision-making: A meta-analysis', *Journal of Business Research*, 65 (8): 1179–1188. DOI: 10.1016/j.jbusres.2011.07.031.

Ahn, J. and Back, K. J. (2018). 'Integrated resort: A review of research and directions for future study', *International Journal of Hospitality Management*, 69: 94–101. DOI:10.1016/j.ijhm.2017.10.017.

Anderson, J. C. and Gerbing, D. W. (1988). 'Structural equation modelling in practice: A review and recommended two-step approach', *Psychological Bulletin*, 103 (3): 411–423. DOI: 10.1037/0033-2909.103.3.411.

Banff Lake Louise Tourism. (2005). Retrieved May 03, 2005, from http://www.brand.banfflakelouise.com

Bentler, P. M. and Bonnett, D. G. (1980). 'Significance tests and goodness of fit in the analysis of covariance structures', *Psychological Bulletin*, 88 (3): 588–606.

Bhattacharya, C. B. and Sen, S. (2003). 'Consumer-company identification: A framework for understanding consumers' relationships with companies', *Journal of Marketing*, 67 (2): 76–88.

Brakus, J. J., Schmitt, B. H. and Zarantonello, L. (2009). 'Brand experience: What is it? How is it measured? Does it affect loyalty', *Journal of Marketing*, 73 (3): 52–68.

Brodie, R. J. and Hollebeek, L. D. (2011). 'Advancing and consolidating knowledge about customer engagement', *Journal of Service Research*, 14 (3): 283–284. DOI: 10.1177/1094670511415523.

Brodie, R. J., Hollebeek, L. D., Jurić, B. and Ilić, A. (2011). 'Customer engagement: Conceptual domain, fundamental propositions, and implications for research', *Journal of Service Research*, 14 (3): 252–271. DOI: 10.1177/1094670511411703.

Chen, T., Drennan, J., Andrews, L. and Hollebeek, L. D. (2018). 'User experience sharing: Understanding customer initiation of value co-creation in online communities', *European Journal of Marketing*, 52 (5/6): 1154–1184.

Escalas, J. E. (2004). 'Narrative processing: Building consumer connections to brands', *Journal of Consumer Psychology*, 14 (1): 168–180.

Hair, J. F., Anderson, R. E., Babin, B. J. and Black, W. C. (2010). *Multivariate Data Analysis: A Global Perspective* (Vol. 7). Upper Saddle River, NJ: Pearson.

Hannam, K. (2004). 'Tourism and development II: Marketing destinations, experiences and crises', *Progress in Development Studies*, 4 (3): 256–263.

Harrigan, P., Evers, U., Miles, M. and Daly, T. (2018). 'Customer engagement with tourism social media brands', *Tourism Management*, 59: 597–609.

Hirschman, E. C. and Holbrook, M. B. (1982). 'Hedonic consumption: Emerging concepts, methods and propositions', *Journal of Marketing*, 46 (3): 92–101.

Hollebeek, L. D. (2011). 'Demystifying customer brand engagement: Exploring the Loyalty Nexus', *Journal of Marketing Management*, 27 (7/8): 1–23.

Hollebeek, L. D., Glynn, M. S. and Brodie, R. J. (2014). 'Consumer brand engagement in social media: Conceptualization, scale development and validation', *Journal of Interactive Marketing*, 28 (2): 149–165.

Hollebeek, L. D., Jurić, B. and Tang, W. (2017). 'Virtual brand community engagement practices: A refined typology and model', *Journal of Services Marketing*, 31 (3): 204–217.

Hollebeek, L. D. and Macky, K. (2019). 'Digital content marketing's role in fostering consumer engagement, trust, and value: Framework, fundamental propositions, and implications', *Journal of Interactive Marketing*, 45: 27–41.

Hollebeek, L. D. and Rather, R. A. (2019). 'Service innovativeness and tourism customer outcomes', *International Journal of Contemporary Hospitality Management*, 31 (11): 4227–4246. DOI: 10.1108/IJCHM-03-2018-0256.

Hollebeek, L. D., Sprott, D. E., Andreassen, T. W., Costley, C., Klaus, P., Kuppelwieser, V., ... Rather, R. A. (2019a). 'Customer engagement in evolving technological environments: Synopsis and guiding propositions', *European Journal of Marketing*. DOI: 10.1108/EJM-09-2019-970.

Hollebeek, L. D., Srivastava, R. K. and Chen, T. (2019b). 'SD logic–informed customer engagement: Integrative framework, revised fundamental propositions, and application to CRM', *Journal of the Academy of Marketing Science*, 47 (1): 161–185.

Homburg, C., Jozić, D. and Kuehnl, C. (2017). 'Customer experience management: Toward implementing an evolving marketing concept', *Journal of the Academy of Marketing Science*, 45 (3): 377–401.

Hwang, J. and Seo, S. (2016). 'A critical review of research on customer experience management: Theoretical, methodological and cultural perspectives', *International Journal of Contemporary Hospitality Management*, 28 (10): 2218–2246.

Indian Brand Equity Foundation. (2018). 'Tourism and hospitality industry in India', Retrieved from www.ibef.org/industry/tourism-hospitality-india.aspx [Accessed November 2, 2018].

Kumar, V. and Kaushik, A. K. (2018). 'Destination brand experience and visitor behavior: The mediating role of destination brand identification', *Journal of Travel & Tourism Marketing*, 35 (5): 649–663.

Kumar, V., Rajan, B., Gupta, S. and Dalla Pozza, I. (2019). 'Customer engagement in service', *Journal of the Academy of Marketing Science*, 47 (1): 138–160.

Lam, S. K., Ahearne, M., Hu, Y. and Schillewaert, N. (2010). 'Resistance to brand switching when a radically new brand is introduced: A social identity theory perspective', *Journal of Marketing*, 74 (6): 128–146.

Le, D., Scott, N. and Lohmann, G. (2018). 'Applying experiential marketing in selling tourism dreams', *Journal of Travel & Tourism Marketing*, 1 (16): 220–235. DOI: 10.1080/10548408.2018.1526158.

Lebergott, S. (1993). *Pursuing Happiness: American Consumers in the Twentieth Century*. Princeton, NJ: Princeton University Press.

Lee, T. H. and Chang, Y. S. (2012). 'The influence of experiential marketing and activity involvement on the loyalty intentions of wine tourists in Taiwan', *Leisure Studies*, 31 (1): 103–121.

Lemon, K. N. and Verhoef, P. C. (2016). 'Understanding customer experience throughout the customer journey', *Journal of Marketing*, 80 (6): 69–96.

Oliver, R. L. (1997). *Satisfaction: A Behavioral Perspective on the Consumer*. New York: McGraw-Hill.

Parrey, S. H., Hakim, I. A. A. and Rather, R. A. (2019). 'Mediating role of government initiatives and media influence between perceived risks and destination image: A study of conflict zone', *International Journal of Tourism Cities*, 5 (1): 90–106.

Pine, B. J. and Gilmore, J. H. (1998). 'Welcome to the experience economy', *Harvard Business Review*, 76 (4): 97–105.

Prahalad, C. K. and Ramaswamy, V. (2003). 'The new frontier of experience innovation', *MIT Sloan Management Review*, 44 (4): 12–18.

Quan, S. and Wang, N. (2004). 'Towards a structural model of the tourist experience: An illustration from food experiences in tourism', *Tourism Management*, 25 (3): 297–305.

Rajaobelina, L. (2018). 'The impact of customer experience on relationship quality with travel agencies in a multichannel environment', *Journal of Travel Research*, 57 (2): 206–217.
Rather, R. A. (2017). 'Investigating the impact of customer brand identification on hospitality brand loyalty: A social identity perspective', *Journal of Hospitality Marketing & Management*, 27 (5): 487–513.
Rather, R. A. (2018a). 'Consequences of consumer engagement in service marketing: An empirical exploration', *Journal of Global Marketing*, 31 (1): 1–20.
Rather, R. A. (2018b). 'Exploring customers' attitudes towards the hospitality brands in India: A social identity perspective'. In Camilleri, M. A. (Eds), *The Branding of Tourist Destinations: Theoretical and Empirical Insights* (pp. 207–231). Bingley, UK: Emerald Publishing Limited.
Rather, R. A. (2018c). 'Customer experience, memories and loyalty in Indian hospitality sector', *International Journal of Marketing & Business Communication*, 7 (3): 20–30.
Rather, R. A. and Camilleri, M. A. (2019). 'The effects of service quality and consumer-brand value congruity on hospitality brand loyalty', *Anatolia*, 30 (4). DOI: 10.1080/13032917.2019.1650289.
Rather, R. A. and Hollebeek, L. D. (2019). 'Exploring and validating social identification and social exchange-based drivers of hospitality customer loyalty', *International Journal of Contemporary Hospitality Management*, 31 (3): 1432–1451.
Rather, R. A., Hollebeek, L. D. and Islam, J. U. (2019a). 'Tourism-based customer engagement: The construct, antecedents, and consequences', *The Service Industries Journal*, 39 (7–8): 519–540.
Rather, R. A. and Sharma, J. (2017). 'The effects of customer satisfaction and commitment on customer loyalty: Evidence from the hotel industry', *Journal of Hospitality Application and Research*, 12 (2): 41–60.
Rather, R. A. and Sharma, J. (2019). 'Dimensionality and consequences of customer engagement: A social exchange perspective', *Vision*, 23 (3): 255–266.
Rather, R. A., Tehseen, S., Itoo, M. H. and Parrey, S. H. (2019b). 'Customer brand identification, affective commitment, customer satisfaction, and brand trust as antecedents of customer behavioral intention of loyalty: An empirical study in the hospitality sector', *Journal of Global Scholars of Marketing Science*, 29 (2): 196–217.
Rather, R. A., Tehseen, S. and Parrey, S. H. (2018). 'Promoting customer brand engagement and brand loyalty through customer brand identification and value congruity', *Spanish Journal of Marketing- ESIC*, 22 (3): 319–337.
Risitano, M., Romano, R., Sorrentino, A. and Quintano, M. (2017). 'The impact of consumer-brand engagement on brand experience and behavioural intentions: An Italian empirical study', *British Food Journal*, 119 (8): 1884–1896.
Schmitt, B. (1999a). 'Experiential marketing', *Journal of Marketing Management*, 15 (1-3): 53–67.
Schmitt, B. (1999b). *Experiential Marketing: How to Get Customers to Sense, Feel, Think, Act, and Relate to Your Company and Brands*. New York: Free Press.
Schneider, M. (2004). 'Exploring the world of experience travel', Retrieved May 8, 2004, from http://industry.travelalberta.com/details.cfm?ID=565&subsection=
Sharma, J. and Rather, R. A. (2016). 'The role of customer engagement in ensuring sustainable development in hospitality sector', *International Journal of Hospitality and Tourism Systems*, 9 (1): 33–43.
Sprott, D., Czellar, S. and Spangenberg, E. (2009). 'The importance of a general measure of brand engagement on market behavior: Development and validation of a scale', *Journal of Marketing Research*, 46 (1): 92–104.
Stokburger-Sauer, N. (2011). 'Brand community: Drivers and outcomes', *Psychology & Marketing*, 27 (4): 347–368.
Tsaur, S. H., Chiu, Y. T. and Wang, C. H. (2007). 'The visitors behavioral consequences of experiential marketing: An empirical study on Taipei Zoo', *Journal of Travel & Tourism Marketing*, 21 (1): 47–64.
Tynan, C. and McKechnie, S. (2009). 'Experience marketing: A review and reassessment', *Journal of Marketing Management*, 25 (5–6): 501–517.
United Nations World Tourism Organization. (2014). 'Tourism highlights', Retrieved from http://dtxtq4w60xqpw.cloudfront.net/sites/all/files/docpdf/unwtohighlights12enhr_1.pdf [Accessed September 10, 2014].
Zeithaml, V. A., Berry, L. L. and Parasuraman, A. (1996). 'The behavioral consequences of service quality', *Journal of Marketing*, 60 (2): 31–46.

25

Relationship marketing

A consumer experience perspective

Christy Yen Nee Ng and Brian Kee Mun Wong

Introduction

The increasingly competitive global marketplace has compelled organizations to transform themselves in the way they conceptualize and conduct business. As one of the largest industries in the service sector, hospitality and tourism firms must gain a broader understanding of the essentials that are required if they are to thrive and survive in the rapidly evolving and competitive global market. Service organizations are shifting their focus from "transactional exchange" to "relational exchange" for expanding an equally satisfying relationship with customers (Jain and Jain 2005). In transactional marketing, competition and self-interest are the drivers of value conception. Within a competitive situation, buyers can be presented with a choice, and this choice of sellers/suppliers encourages the marketers to generate a higher value offering to maintain their market position. Independence of choice among marketers creates a more efficient system for distributing marketing value in transactional marketing. This process provides more freedom to the marketers to choose their transactional partners on the basis of protecting their own interests at each decision point. This results in lower cost purchases through bargaining and bidding. On the other hand, when producers and consumers directly deal with each other, there is a greater potential for emotional bonding that transcends economic exchange. In that process, producers and customers can understand each other's needs and develop a willingness to cooperate with each other; therefore, the process becomes more relationship oriented.

Relationship marketing is a form of marketing developed from direct response marketing campaigns conducted in the 1970s and 1980s, which emphasizes customer retention and satisfaction rather than a dominant focus on point-of-sale transactions (Kotler 1999). Relationship marketing is one of the oldest approaches to marketing, yet one of the least understood (Zineldin and Philipson 2007). The concept of relationship marketing was introduced by Berry (1983), who defined relationship marketing, as "attracting, maintaining, and enhancing customer relationship" (p. 25). Another definition was proposed by Gronroos (1991), which further included "establishing relationships with customers and other parties at a profit by mutual exchange and fulfilment of promises" (p. 8). Morgan and Hunt (1994) define relationship marketing as "… all marketing activities directed toward establishing, developing, and maintaining successful relational exchange … adequately conceptualizing relationship marketing requires a definition that accommodates all forms of relational exchange". Relationship marketing differs from other forms of marketing in that it recognizes the long-term value to the firm of keeping customers, as opposed to direct marketing, which focuses upon the acquisition of new clients by targeting key demographics based upon prospective client lists.

Many scholars provide their own definitions (e.g., Berry 1983; Buttle 1996; Grönroos 1994, 2004; Morgan and Hunt 1994) of relationship marketing. Regrettably, there is no common definition of relationship marketing. Table 25.1 shows the various definitions of relationship marketing. Among all the definitions, that of Grönroos (2004) is one of the most cited.

Previous research (Crosby, Evans, and Cowles 1990; Peterson 1995) in relationship marketing has focused on the benefit of customer loyalty, but mostly from the perspective of the firm. Morgan and Hunt (1994) continued to explore when, how, and why to engage in relational exchanges from the firm's perspective as well. Therefore, the benefits to service providers of having developed a strong relationship with their customers, and as a result of a loyal customer base, are well documented. Loyal customers can lead to increased revenue, sales, and profit for the firm (Aaker 1992; Reichheld 1993), and loyal customers are more likely to purchase additional goods or services from the firm in the future. Although the positive outcome for the firm is clear, in order to create a long-term relationship, both the firm and the customer must benefit. Customers who have developed a relationship with a service provider expect to receive satisfactory delivery of the core service. However, customers likely receive other benefits, separate from the core service performance, as a result of the development of a long-term relationship with a service provider. Those benefits, labeled as relational benefits, motivate customers to engage in relationship

Table 25.1 Definition of relationship marketing.

Authors	*Definitions*
Berry (1983)	Attracting, maintaining, and – in multi-service organizations – enhancing customer relationships.
Grönroos (1990)	Marketing is to establish, maintain, and enhance relationships with customers and other partners, at a profit, so that the objectives of the parties involved are met. This is achieved by a mutual exchange and fulfilment of promises.
Berry and Parasuraman (1991)	Relationship marketing concerns attracting, developing, and retaining customer relationships.
Gummesson (1994: 2)	Relationship marketing emphasizes a long-term interactive relationship between the provider and the customer, and long-term profitability.
Sheth and Parvatiyar (1994)	The understanding, explanation, and management of the ongoing collaborative business relationship between suppliers and customers.
Evans and Laskin (1994)	… the process whereby a firm builds long-term alliances with both prospective and current customers so that both buyer and seller work toward a common set of specified goals.
Morgan and Hunt (1994)	All marketing activities directed toward establishing, developing, and maintaining successful relational exchanges.
Sheth and Parvatiyar (1995)	Attempts to involve and integrate customers, suppliers, and other infrastructural partners into a firm's developmental and marketing activities.
Buttle (1996)	Relationship marketing is concerned with the development and maintenance of mutually beneficial relationships with strategically significant markets.

marketing with a service provider. Also, by being a "regular customer" and developing a relationship with a service provider, it allows the provider to better understand the customer's tastes and preferences; ensure better treatment; and encourages more interest in the consumer's satisfaction (Zeithaml 1981).

The relationship development process

Relationship marketing can also be viewed as the marriage of buyer and seller. Levitt (1983: 111) stated that

> ... the sale merely consummates the courtship. Then the marriage begins. How good the marriage is depends on how well the relationship is managed by the seller.

Levitt's (1983) marriage analogy is fitting in the relationship of management to the seller (only one "spouse"). In fact, research analyzing the interpersonal attractions and the interdependence relationships between husbands and wives provides an appropriate framework for describing the evolution of buyer–seller relations. The benefits of companionship, intimacy, parenting, personal growth, shared household maintenance, and social support will not function efficiently without the husband–wife relationship. The buyer–seller relationship involves similar benefits and costs. The former include reduced uncertainty, managed dependence, exchange efficiency and social satisfaction from the association (Dwyer, Schurr, and Oh 1987).

In order to have a strong relationship between the firm and the customer, it must be mutually beneficial (Czepiel 1990). Frequent, honest, and open communication will help to improve and create a strong relationship as well. Relationship marketing is built on the foundation of trust (Crosby et al. 1990; Morgan and Hunt 1994; Parasuraman, Berry, and Zeithaml 1990). Trust is "a willingness to rely on an exchange partner in whom one has confidence" (Moorman, Deshpande, and Zaltman 1993). Without trust, the interdependence between two parties cannot be accomplished.

According to Dwyer et al. (1987), a relationship develops through five general phases, identified as (1) awareness, (2) exploration, (3) expansion, (4) commitment, and (5) dissolution. The author described each phase as:

1. *Awareness* – Party A's recognition that party B is a feasible exchange partner. Situational proximity between the parties facilitates awareness. Interaction between parties has not transpired in phase 1.
2. *Exploration* – The search and trial phase in relational exchange. Potential exchange partners first consider obligations, benefits and burdens, and the possibility of exchange. The exploration phase may be very brief or it may include an extended period of testing and evaluation. The exploration phase is conceptualized in five subprocesses: (1) attraction, (2) communication and bargaining, (3) development and exercise of power, (4) norm development, and (5) expectation development. Attraction results from the degree to which the buyer and seller achieve bilateral communication, based on their interaction with each other. Communication and bargaining is the process where buyers and sellers rearrange their mutual distribution of obligation, benefits, and burdens. Power is considered as the ability to achieve intended effects or goals (Dahl 1957). The one that has the power usually is in control. According to Lipset (1975), norms are "expected patterns of behavior". By adopting norms and establishing standards of conduct, it helps the buyers and sellers to set the ground rules for future exchange. During expectation

development, trust is an important concept in understanding expectations for cooperation and planning in a relational contract.

3. *Expansion* – The continual increase in benefits obtained by exchange partners adds to their increasing interdependency. The five subprocesses introduced in the exploration phase also operate in the expansion phase. The critical distinction is that the rudiments of trust and joint satisfaction established in the exploration stage now lead to increased risk-taking within the dyad. As a result, the range of depth of mutual dependency increases.
4. *Commitment* – An implicit or explicit pledge of relational continuity between exchange partners. At this most advanced phase of buyer–seller interdependency, the exchange partners have achieved a level of satisfaction from the exchange process that virtually precludes other primary exchange partners who could provide similar benefits. Customer loyalty is achieved at this stage. Three measurable criteria of commitments are inputs, durability, and consistency.
5. *Dissolution* – At this phase, the possibility of withdrawal or disengagement has been implicit throughout the relationship development framework.

The benefits of relationship marketing

Relationship marketing benefits the customer as well as the firm. Sheth and Parvatiyar (1995: 256) propose:

> That consumers engage in relational market behavior to achieve greater efficiency in their decision making, to reduce the task of information processing, to achieve more cognitive consistency in their decisions, and to reduce the perceived risks associated with future choice.

Consumers will only engage in relational exchange with trustworthy partners. According to Hunt, Arnett, and Madhavaram (2006), customers who engage in relational exchange experiences decrease their search cost, and receive risk-reducing benefits with a given supplier. Relationship marketing also allows service providers to become more familiar with the customer's requirements and needs. Familiarity with the customers, combined with social bonding built over a series of service encounters, helps in the customization which results in better satisfying the customer's needs and preferences.

From the firm's perspective, engaging in relational exchange can help to decrease the customer-defection rate. Reichheld and Sasser's (1990) study showed that a firm could improve profits from 25% to 85% by reducing customer defections by just 5%. Compared to acquiring new customers, costs associated with maintaining existing customers are generally lower (Rust, Moorman, and Dickson 2002). In addition, loyal customers help to generate revenue for many more years.

Relational benefits and their application in tourism

In the relational benefit approach, it is assumed that both parties in a relationship must benefit for it to continue in the long term. Relational benefits can be referred to as the benefits that customers expect to receive as a result of having developed a long-term relationship with a service provider above and beyond the core services provided (Gutek et al. 1999; Gwinner, Gremler, and Bitner 1998). Three types of relational benefits include confidence benefits, social benefits, and special treatment benefits.

Confidence benefits can be described as a "feeling of reduced anxiety, trust, and confidence in the provider" (Gwinner et al. 1998). According to Morgan and Hunt (1994) and Bitner (1995), confidence benefits are derived from trust in the reliability and integrity of the service provider in delivering the core services. Gwinner et al. (1998) discovered that confidence benefits are the most important relational benefits in numerous types of services. Confidence benefits result in a reduction of uncertainty in the transaction and therefore increase the realistic expectations from the service provider (Yen and Gwinner 2003).

Social benefits focus more on a relationship than on the outcome of the transaction (Goodwin 1994; Goodwin and Gremler 1996). Repeat purchases on a regular basis will allow the service provider to get to know their customer's preferences and needs, and they then start to develop a friendship. Social benefits also comprise a sense of belonging, recognition, feeling of familiarity, friendship, and social support (Berry 1983; Czepiel 1990; Price and Arnould 1999).

Special treatment benefits are associated with economic and customization advantages for customers (Gwinner et al. 1998). Economic benefits involve economic considerations that relate to price discounts for those customers who have developed a long-term relationship with the service provider (Gwinner et al. 1998). According to Parvatiyar and Sheth (2000), customers must recognize receiving value at a lower cost in order to engage in a marketing relationship. Gwinner et al. (1998) also identified a non-monetary benefit, time-saving benefits, as a result of developing a relationship with the service provider.

At the end of the 1990s, competition increased and the product differentiation rate decreased in the tourist market, influencing the necessity for applying relationship marketing in tourism (Rodasavljevic and Borisavljevic 2014). The process of building up relations between different market participants represents basic tourism activity. Applying the relationship marketing model in tourist business implies the development and maintenance of long-term relations with customers and other stakeholders (suppliers, competitors, state, distributors, employees, and customers). All these are relevant to companies' success in tourism. Research studies (Kim, Han, and Lee 2012; Lombard and Steyn 2008; Palmer and Mayer 1996; Sin et al. 2006) have confirmed that an important contribution of all business participants in the tourist offer chain relates to the provision of customer satisfaction and loyalty.

Different approaches can be found to building up relations with customers and providing value in tourism literature. Hyatt introduced a Golden Passport system, based on giving opportunities to guests to gain membership and realize different benefits (e.g. free-of-charge overnight stays on weekends, staying in a better hotel room). One approach provides financial benefits such as free-of-charge overnight stays and comfortable and better quality rooms for repeat guests, which influences their decision to patronize and prefer one hotel brand. Another approach offers social benefits to the customer by means of tourist services personalization. For instance, a waiter or waitress in a restaurant recognizes and addresses guests by their names; therefore he/she develops closer contact and a social connection with the guests. The third approach in building up customer relations implies the introduction of structural changes in service provision which are hard for competitors to copy. For example, airline companies will provide rooms for their first-class customers and hotels will offer their best customers flexible times for checking in and out. The Intercontinental Hotel Group provides special benefits for its permanent guests, including a special reserve regime, welcoming and seeing off guests, a special accommodation process during the weekend, and the delivery of free daily newspapers to the hotel room.

The concerns of relationship marketing

From value-in-exchange to value-in-use

According to Prahalad and Ramaswamy (2000), there has been a progressive shift over the past two decades from considering customers as a "passive audience" to appreciating them as "active players". The perception of value shift from "value creation through exchange" to "value creation through use" possesses an important implication for relationship marketing. Changes in the market environment signify substantial opportunities to build and enhance customer relationships through co-creation. Suppliers can seek to build closer relationships through opportunities provided by technological breakthroughs, changes in industry logics and changes in customer preferences and lifestyles (Payne, Storbacka, and Frow 2008). With all these changes, companies need to change from a unidirectional perspective of viewing customers as largely passive receivers of value to one where they can more actively engage in mutual value co-creation. According to Vargo and Lusch (2008), the contemporary view, which is now becoming more widely acknowledged, is that "value-in-use" is co-created. Therefore, co-creating value-in-use in a "joint sphere of responsibility" (Grönroos and Voima 2013) becomes the concern of both customers and companies and is one that is critical to strengthening customer–company relationships. Value-in-use seeks the active participation of the customer in resource-sharing and contributing to and enhancing relational outcomes.

Therefore, relationship marketing research should focus on understanding how to create ongoing value-in-use as the key priority, rather than considering just how to build a relationship. Within business markets, a broader focus is needed than focusing on managing relationships with direct customers. It is also necessary to consider the customer's customer. This is especially important in intermediated markets. Relationship marketing initiatives will be substantially enhanced where the supplier company places greater emphasis on helping their customer's customer create value-in-use.

"Dark side" behaviors and relationship marketing

Historically, relationship marketing has been proposed as a "good thing". Much of the relationship marketing literature extols the benefit of adopting a relational approach to customers. However, less attention has been directed at instances of poor application of relationship marketing and deliberate or unintentional behavior that damages customer relationships. This topic is substantively under-researched, and there is evidence that dark side behavior is widespread and may even be growing.

a) Information misuse and customer confusion

In order to serve customers better in the future, assembling detailed information about them is needed in CRM and relationship marketing. But this information may be used in ways that customers disapprove of. Service providers are able to collect and integrate information from a variety of sources including the web and in-store tracking and supplement this with additional information purchased from data brokers. Turow, Feldman, and Meltzer (2005) mentioned that information about customers is often sold on to other firms to use without the customer's permission. This can weaken the position of the customer: "because marketers can now survey and analyze consumer behavior in such a detailed way that they achieve what has been unachievable heretofore: turning the consumer's interior inside out" (Zwick and Dholakia 2004: 31).

Misleading or confusing information and the hiding of relevant information from customers can encourage them to make a decision that disadvantages them. McGovern and Moon (2007: 80) call attention to the many companies that infuriate customers by deliberately "binding them with contracts, bleeding them with fees, and confusing them with fine print". Sad to mention, mobile phone providers, car rental companies, banks, health clubs, video stores, and credit card companies are especially prone to these dark side practices. Frequent price and rate changes add more confusion and customers do not have enough time to fully adapt to the new tariffs.

b) Dishonesty and privacy invasion

There are a few categories that fall under the heading of dishonesty. CRM systems frequently place pressure on staff to up-sell and cross-sell. Management pressure, performance measurement systems and employee reward structures may result in customers being sold products or services that they do not need. There are also instances of fraudulent activity. For example, car servicing centers charge for replacement parts and repairs that are not needed and doctors charge for "blanket" medical screening when it is not called for. According to Varzhabedian (2007), bars and restaurants are notorious for permitting staff to illicitly overcharge customers.

Dark side behavior can arise as a result of the flow of information from customers to service providers. Companies can access information about customers from many sources, like transaction records and observations of customer behavior. Providers can learn more about customers than they wish through this and it leads to complaints. For example, hotels keep records of clients' expenditure and may even check a guest's room and waste bins for clues as to their likes and dislikes (Jaipuria 2006). This is done in the name of better serving the customer in the future, but collecting information about such behavior may exceed the information that some customers are comfortable with the provider knowing. Dark side behavior in relationship marketing represents an area requiring critical attention by researchers and much-needed attention by policymakers and regulators. This negative behavior may have a greater impact on close relationships than positive behaviors (Baumeister et al. 2001).

Case study 25.1: Walt Disney Company

Considered a modern-day marketing marvel, The Walt Disney Company has a long-established and loyal consumer base. Beginning as Disney Brothers Studio in 1923, the Walt Disney Company has diversified into five business components – Media Networks, Parks and Resorts, Studio Entertainment, Consumer Products, and Interactive. With a constant emphasis on creating timeless family entertainment, founder Walt Disney passed on to the company the commitment to maintaining the fantasy by which Disney and its theme parks have come to be known. If there is one brand that knows a thing or two about keeping customers happy, it is Disney. This is reflected in the increasing demand for the Disney experience. In 2016, its Parks & Resorts accounted for 31% of the company's total revenues, with its Media Networks generating 43%. More than 13 million people a year visit the newly opened Shanghai Disney Resort, as well as there being greater attendance levels in parks in the United States.

Delivering a great customer experience might seem like an easy task for the "happiest place on earth", but Disney uses much more to delight visitors than Mickey Mouse pancakes. Let us have a look at how Disney's approach can create an unforgettable customer experience.

Making the mundane magical

Disney's brand promise has been the same since Walt Disney opened the first theme park in 1955. It is to make a magical experience come alive and to create happiness via these experiences. Anyone who has visited a theme park before knows that "magic" is often suppressed by the mundane. This means basic tasks like buying a ticket, queuing for rides, and even getting to the resort in the first place if you are an international visitor.

Disney aims to deliver its brand promise by making even mundane details magical and focusing on the unexpected ways it might bring happiness to customers. For example, since recognizing that children often queue up for rides only find out that they are not tall enough, it now hands out special passes to enable disappointed kids to skip to the front on their next ride.

The Disney brand promise is also a natural extension of its internal company culture, with Disney cast members or employees embodying values such as openness, communication, and courtesy. One interesting example of this is how cast members can become "language certified", which means that they can wear pins which indicate what languages they speak, in turn making it clear to guests that they can assist them if needed.

Unexpected moments of personalization

With thousands of people visiting its parks every day, creating a personal connection with visitors is a huge part of Disney's CX strategy. Technology plays a critical part in delivering personalization, specifically tools that help to streamline and elevate the park and hotel experience.

The MyMagic+ vacation planning system is one of the most notable examples, allowing visitors to plan and access information and perks such as advance ride booking and restaurant reservations. With the MagicBand acting as a room key, park ticket, and optional payment method, the brand creates a more seamless experience. Meanwhile, My Magic+ enables Disney to increase levels of personalization – not only in the messages it sends via the mobile app but also at particular touchpoints within the park and resorts. For example, MyMagic+ members might see their name appear on a screen as they walk by, or a photo taken on a ride might unexpectedly appear on the app.

The MyMagic+ system is continuously evolving with reports suggesting that slimmer bands are in the works, as well as the possibility of the technology being transferred to smart devices at some point in the near future.

Immersion

Disney is about much more than just the rides – and in turn, visitors want to entirely immerse themselves in its world. The company extends this immersion even further through its hotels, which allows consumers to enjoy Disney from the moment they wake up to the moment they go to sleep. Skift reported that U.S. hotels contributed $2.8 billion in revenue in 2016, while occupancy rates reached 89%. At the D23 event in 2018, it was announced that Disney is to open a "100% immersive" Star Wars hotel, which allows visitors to watch or participate in themed stories throughout their stay. While it is being pointed out that many hotels also have immersive qualities, Disney seems intent on raising the bar. A 3D scale model of the park shows the intended attention to detail, with guests even seeing a depiction of outer space outside their windows instead of the real world.

Listen, listen and listen

Disney uses "listening posts" to evaluate the customer experience and identify which areas need to be improved. Unlike data which enables the brand to better target and engage customers, listening posts allow Disney to understand the expectations, needs, and wants of visitors. In turn, this technology can be used to address gaps in customer service or areas where the aforementioned "magic" might be lacking.

Just like the previous example of children being given free queue passes, Disney continuously introduces features to improve the customer experience based on this feedback. For instance, it created subtle "Special Assistance" passes for disabled guests, to take away the need for any potentially intrusive questions. Also, after discovering that visitors would often ask about the location of characters, it introduced the CHIP (Character Hotline and Information Program) to let visitors find out where they are located at any given time. These might be fairly small but significant details that visitors might not even think about. Without them, Disney and its customer experience might not be quite so magical.

Conclusion

Although the field of relationship marketing has been a topic of academic research for over two decades, the field appears even more relevant within the highly complex and multifaceted relationships that exist currently. Interest in relationship marketing continues to grow strongly as evidenced by a continuing stream of scholarly work. However, relationships are not always characterized by harmony. Relationships have become an asset that can be evaluated indirectly through company stock values or directly as sellable customer information or contact points. The focus on "relationship" in relationship marketing got relegated to "marketing". Relationship marketing aims to generate a long-term profitable relationship between partners, through customer experiences, satisfaction, trust and commitment, and therefore, a better understanding of overcoming dark side behavior in CRM will definitely improve relationship marketing in the future.

References

Aaker, D. A. (1992). 'The value of brand equity', *Journal of Business Strategy*, 13 (July/August): 27–32. DOI: 10.1108/eb039503.

Baumeister, R. F., Bratslavsky, E., Finkenauer, C. and Vohs, K. D. (2001). 'Bad is stronger than good', *Review of General Psychology*, 5 (4): 323–370. DOI: 10.1037/1089-2680.5.4.323.

Berry, L. L. (1983). 'Relationship marketing'. In L. L. Berry, G. L. Shostack and G. Upah (Eds.), *Emerging perspectives on services marketing* (pp. 25–28). Chicago, IL: American Marketing Association.

Berry, L. L. and Parasuraman, A. (1991). 'Marketing services', *The Free Press*, New York.

Bitner, M. J. (1995). 'Building service relationships: It's all about promises', *Journal of the Academy of Marketing Science*, 23 (4): 246–251. DOI: 10.1177/009207039502300403.

Buttle, F. (1996). 'Relationship marketing'. In F. Buttle (Ed.), *Relationship marketing theory and practice* (pp. 1–16). London: Paul Chapman Publishing Ltd.

Crosby, L. A., Evans, K. R. and Cowles, D. (1990). 'Relationship quality in services selling: An interpersonal influence perspective', *The Journal of Marketing*, 54 (3): 68–81. DOI: 10.1177/002224299005400306.

Czepiel, J. A. (1990). 'Service encounters and service relationships: Implications for research', *Journal of Business Research*, 20 (1): 13–21. DOI: 10.1016/0148-2963(90)90038-F.

Dahl, R. A. (1957). 'The concept of power', *Behavioral Science*, (July): 201–215. DOI: 10.1002/bs.3830020303.

Dwyer, F. R., Schurr, P. H. and Oh, S. (1987). 'Developing buyer-seller relationships', *Journal of Marketing*, 51 (2): 11–27. DOI: 10.2307/1251126.

Evans, J. R. and Laskin, R. L. (1994). 'The relationship marketing process: A conceptualization and application', *Industrial Marketing Management*, 23 (2): 439–452. DOI: 10.1016/0019-8501(94)90007-8.

Goodwin, C. (1994). 'Between friendship and business: Communal relationships in service exchanges (working paper)', Winnipeg, Canada: University of Manitoba.

Goodwin, C. and Gremler, D. D. (1996). 'Friendship over the counter: How social aspects of service encounters influence consumer service loyalty', *Advances in Services Marketing and Management*, 5: 247–282. DOI: 10.1016/S1067-5671(96)05059-7.

Grönroos, C. (1990). 'Service management: A management focus for service competition', *International Journal of Service Industry Management*, 1 (1): 6–14. DOI: 10.1108/09564239010139125.

Gronroos, C. (1991). 'The marketing strategy continuum: Towards a marketing concept for the 1990s', *Management Decision*, 29 (1): 7–13. DOI: 10.1108/00251749110139106.

Grönroos, C. (1994). 'From marketing mix to relationship marketing: Towards a paradigm shift in marketing', *Management Decision*, 32 (2): 4–20. DOI: 10.1016/S1320-1646(94)70275-6.

Grönroos, C. (2004). 'The relationship marketing process: Communication, interaction, dialogue, value', *The Journal of Business and Industrial Marketing*, 19 (2): 99–113. DOI: 10.1108/08858620410523981.

Grönroos, C. and Voima, P. (2013). 'Critical service logic: Making sense of value creation and co-creation', *Journal of the Academy of Marketing Science*, 41 (2): 133–150. DOI: 10.1007/s11747-012-0308-3.

Gummesson, E. (1994). 'Marking relationship marketing operational', *International Journal of Service Industry Management*, 5 (5): 5–20. DOI: 10.1108/09564239410074349.

Gutek, B. A., Anita, D. B., Matthew, A. and Bennett, C. (1999). 'Distinguishing between service relationship and encounters', *Journal of Applied Psychology*, 84 (2): 218–233. DOI: 10.1037/0021-9010.84.2.218.

Gwinner, K. P., Gremler, D. D. and Bitner, M. K. (1998). 'Relational benefits in services industries: The customer's perspective', *Journal of the Academy of Marketing Science*, 26 (2): 101–114. DOI: 10.1177/0092070398262002.

Hunt, S. D., Arnett, D. B. and Madhavaram, S. (2006). 'The explanatory foundations of relationship marketing theory', *Journal of Business and Industrial Marketing*, 21 (1): 72–87. DOI: 10.1108/10610420610651296.

Jain, R. and Jain, S. (2005). 'Towards relational exchange in services marketing: Insights from hospitality industry', *Journal of Services Research*, 5 (2): 139–150. Available at https://search.proquest.com/docview/195575039?pq-origsite=gscholar [Accessed: November 12, 2019].

Jaipuria, A. (2006). 'The dark side of customer relationship management in the luxury segment of the hotel industry', MA in Management thesis, University of Nottingham, Nottingham.

Kim, W., Han, J. and Lee, E. (2012). 'Effects of RM on repeat purchase and WOM', *Journal of Hospitality and Tourism Research*, 25 (3): 272–288. DOI: 10.1177/109634800102500303.

Kotler, P., Armstrong, G., Saunders, J. and Wong, V. (1999). *Principles of marketing: European edition* (2nd ed.). NJ: Prentice Hall.

Levitt, T. (1983). *The marketing imagination*. New York: The Free Press. Available at https://pdfs.semanticscholar.org/1620/9ef8b881ae04fa4baf906af4a180338edb2d.pdf [Accessed November 12, 2019].

Lipset, S. M. (1975). 'Social structure and social change'. In P. M. Blau (Ed.), *Approaches to the study of social structure* (pp. 172–209). New York: The Free Press.

Lombard, R. and Steyn, T. (2008). 'Relationship marketing practices of travel agencies in the Western Cape Province', *South African Journal of Business Management*, 39 (4): 15–25. DOI: 10.4102/sajbm.v39i4.568.

McGovern, G. and Moon, Y. (2007). 'Companies and the customers who hate them', *Harvard Business Review*, 85 (6): 78–84. Available at https://hbr.org/2007/06/companies-and-the-customers-who-hate-them [Accessed: November 12, 2019].

Moorman, C., Deshpande, R. and Zaltman, G. (1993). 'Factors affecting trust in market research relationships', *Journal of Marketing*, 57 (1): 81–101. DOI: 10.1177/002224299305700106.

Morgan, R. M. and Hunt, S. D. (1994). 'The commitment-trust theory of relationship marketing', *Journal of Marketing*, 58 (3): 20–38. DOI: 10.1177/002224299405800302.

Palmer, A. and Mayer, R. (1996). 'A conceptual evaluation of the multiple dimensions of relationship marketing', *Journal of Strategic Marketing*, 4 (4): 207–220. DOI: 10.1080/09652549600000005.

Parasuraman, A., Berry, L. L. and Zeithaml, V. A. (1990). *An empirical examination of relationships in an extended service quality model.* Cambridge, MA: Marketing Sci- ence Institute.

Parvatiyar, A. and Sheth, J. N. (2000). 'The domain and conceptual foundations of relationships marketing'. In J. N. Sheth and A. Parvatiyar (Eds.), *Handbook of relationship marketing* (pp. 3–38). Thousand Oaks, CA: Sage.

Payne, A., Storbacka, K. and Frow, P. (2008). 'Managing the co-creation of value', *Journal of the Academy of Marketing Science*, 36 (1): 83–96. DOI: 10.1007/s11747-007-0070-0.

Peterson, R. A. (1995). 'Relationship marketing and the consumer', *Journal of the Academy of Marketing Science*, 23 (Fall): 278–281. DOI: 10.1177/009207039502300407.

Prahalad, C. K. and Ramaswamy, V. (2000). 'Co-opting customer competence', *Harvard Business Review*, 78: 79–90. Available at https://hbr.org/2000/01/co-opting-customer-competence [Accessed: November 12, 2019].

Price, L. L. and Arnould, E. (1999). 'Commercial friendships: Service provider–client relationships in social context', *Journal of Marketing*, 63 (4): 38–56. DOI: 10.1177/002224299906300405.

Reichheld, F. F. (1993). 'Loyalty-based management', *Harvard Business Review*, 71 (March/April): 64–73. Avaliable at https://hbr.org/1993/03/loyalty-based-management [Accessed: November 12, 2019].

Reichheld, F. F. and Sasser, W. E. (1990) 'Zero defects: Quality comes to services', *Harvard Business Review*, 68 (5): 105–111. Available at http://matlesiouxx.free.fr/Cours/HKU/Courses/CSIS0404/Lecture%201/Module1_3_Zero_Defections_-Quality_Comes_to_Services.pdf [Accessed: November 12, 2019].

Rodasavljevic, G. and Borisavljevic, K. (2014) 'Applying relationship marketing models in tourism', *Actual Problems of Economics*, 156 (6): 143–152. Available at www.irbis-nbuv.gov.ua/cgi-bin/irbis_nbuv/cgiirbis_64.exe?I21DBN=LINK&P21DBN=UJRN&Z21ID=&S21REF=10&S21CNR=20&S21STN=1&S21FMT=ASP_meta&C21COM=S&2_S21P03=FILA=&2_S21STR=ape_2014_6_17 [Accessed November 12, 2019].

Rust, R. T., Moorman, C. and Dickson, C. R. (2002). 'Getting return on quality: Revenue expansion, cost reduction, or both?', *Journal of Marketing*, 66: 7–24. DOI: 10.1509/jmkg.66.4.7.18515.

Sheth, J. N. and Parvatiyar, A. (Eds.). (1994). *Relationship marketing: Theory, methods and applications.* Atlanta: Emory University.

Sheth, J. N. and Parvatiyar, A. (1995). 'Relationship marketing in consumer markets: Antecedents and consequences', *Journal of the Academy of Marketing Science*, 23 (4): 255–271. DOI: 10.1177/009207039502300405.

Sin, L. Y. M., Tse, A. C. B., Chan, H. and Yim, F. H. K. (2006). 'The effects of relationship marketing orientation on business performance in the hotel industry', *Journal of Hospitality and Tourism Research*, 30 (4): 407–426. DOI: 10.1177/1096348006287863.

Turow, J., Feldman, L. and Meltzer, K. (2005). 'Open to exploitation: American shoppers online and offline', Philadelphia, PA: Annenberg Public Policy Center, University of Pennsylvania.

Vargo, S. L. and Lusch, R. F. (2008). 'Service-dominant logic: Continuing the evolution', *Journal of the Academy of Marketing Science*, 36 (1): 1–10. DOI: 10.1007/s11747-007-0069-6.

Varzhabedian, R. (2007). 'Status quo: The restaurant owner's guide to building profits', Arizona, USA: Wheatmark.

Yen, H. J. and Gwinner, K. P. (2003). 'Internet retail customer loyalty: The mediating role of relational benefits', *International Journal of Service Industry Management*, 14 (5): 483–500. DOI: 10.1108/09564230310500183.

Zeithaml, V. A. (1981). 'How consumer evaluation processes differ between goods and services'. In J. H. Donnelly and W. R. George (Eds.), *Marketing of services* (pp. 186–190). Chicago, IL: American Marketing Association.

Zineldin, M. and Philipson, S. (2007). 'Kotler and Borden are not dead: Myth of relationshipmarketing and truth of the 4Ps', *Journal of Consumer Marketing*, 24 (4): 229–241. DOI: 10.1108/07363760710756011.

Zwick, D. and Dholakia, N. (2004). 'Whose identity is it anyway? Consumer representation in the age of database marketing', *Journal of Macromarketing*, 24 (1): 31–43. DOI: 10.1177/0276146704263920.

26

Interpreting relationships among Oktoberfest tourists' experiences and perceived value, overall satisfaction and loyalty behaviors for better marketing strategy decisions

Robert J. Harrington, Michael C. Ottenbacher, Burkhard von Freyberg, Alexandros Paraskevas and Laura Schmidt

Introduction

At its core strategic management and marketing is nothing more than "sensing the need for action, deciding upon an action and executing it ..." (Ansoff 1987: 508). In order for tourism and hospitality firms to sense a need for action, some form of information processing and interpretation must take place. This process can be very formal or informal and is defined in this study as an organization's process of sharing and organizing information into a usable form that provides a basis for desired resource allocations and strategic marketing.

Traditional marketing strategies have focused on the 7 Ps and assessed outcomes such as satisfaction and loyalty (Harrington et al. 2017a). While this type of marketing strategic thought process can be applied to the experiential nature of tourism experiences, many researchers and practitioners have questioned the validity of these traditional concepts for tourism and other service experiences (Chandler and Lusch 2015). For instance, more recent concepts look at the entire customer journey (Lemon and Verhoef 2016) and its impact, assessing service/experience design (Dasu and Chase 2013; Pine and Gilmore 2011) as well as shifting from goods-dominant logic to memory-dominant logic, (Harrington et al. 2019) as a more meaningful or impactful approach to address consumer behavior questions related to service experiences.

Figure 26.1 provides a framework tying together the customer journey, service experience design and memory-dominant logic literature to facilitate the discussion within this chapter. The part contained within the dotted lines of the framework relates to areas tested in the context of Oktoberfest, with results used to tease out implications for tourism management and marketing as well as implications for future research and practical application.

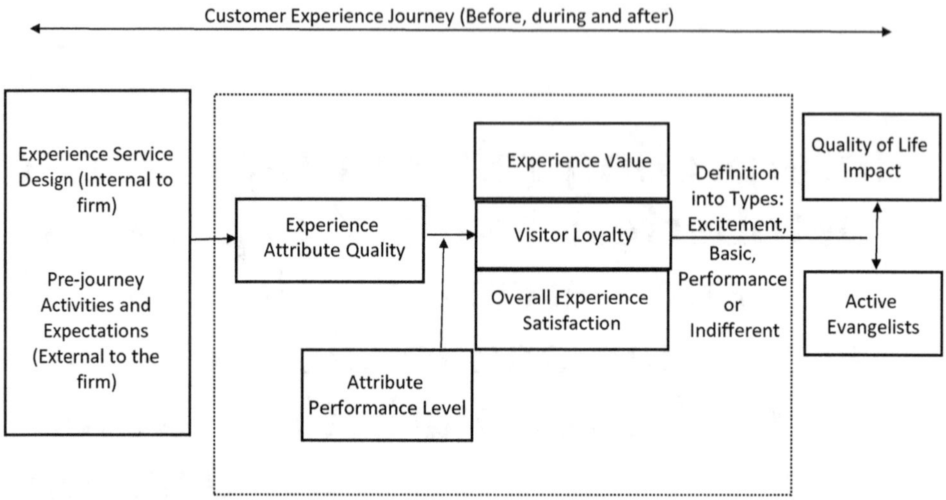

Figure 26.1 Elements of the Tourism Customer Experience

In Figure 26.1, key aspects of the tourism customer experience are shown, starting to the right (before the experience) and leading to the left (during and after the experience). In the box on the far right, experience service design is an aspect of tourism management that occurs before the actual experience and service design is a function internal to the firm (although design decisions are likely to be impacted by customer feedback, trends and the competitive environment). Service experience design includes aspects such as the level of customer control and choice, how these aspects might be sequenced, the management of the actual and perceptual experience duration and the level of personalization/co-creation of the experience delivery (Dasu and Chase 2013). To enhance the attribution of positive experiences and deflect negative ones, this internal experience design revolves around an organizational-driven process and culture to facilitate emotional responses, trust and a feeling of control (Harrington et al. 2019; Lemon and Verhoef 2016). Also, within this box in Figure 26.1, there are customer-driven pre-journey activities that are likely to impact on the actual experience as well as impact expectations prior to the experience that can lead to satisfaction or dissatisfaction depending on whether expectations are exceeded or not met for the consumer (Ottenbacher et al. 2016). Therefore, Figure 26.1 represents an integration of earlier frameworks of the customer journey (Lemon and Verhoef 2016), service experience (Dasu and Chase 2013), attribute classification (Lin et al. 2010), the Oktoberfest experience (Harrington et al. 2017b) and memory-dominant logic (Harrington et al. 2019) to form a basis for empirical tests and discussion used in this chapter in the context of Oktoberfest.

As highlighted above, this chapter focuses primarily on issues associated with the assessment of the quality of attributes as part of the Oktoberfest experience, the level of quality/performance, multiple indicators of success (value, loyalty and satisfaction) and the definition of attributes into four possible types. Past studies have focused primarily on satisfaction levels and/or loyalty intentions as criteria for marketing strategy and resource allocations. In today's competitive environment and information accessibility, we argue that this type of marketing strategy and its impact on experiential service design is necessary but insufficient. As shown in the boxes to the far left in Figure 26.1, newer theoretical models indicating outcomes such as perceived wellbeing or increased quality of life are important indicators of successful consumer experiences. Also, when experiences become memorable consumers are more likely to become active evangelists in the form of recommendations, sharing of experiences on social media or electronic word-of-mouth (Harrington et al. 2019).

The upcoming sections of this chapter focus on the central portions of Figure 26.1 (within the dotted lines) using a modified empirical approach and the context of Oktoberfest to 1) assess the impact of 16 beer tent attributes on attendees' perceptions of value, overall satisfaction, word-of-mouth and return visits, 2) articulate the implications for marketing strategy decisions and the impact on experiential service design, and 3) discuss implications for 21st century-driven service design decision-making, expanded notions of desired consumer behaviors and methods that provide insights for marketing strategy decisions in travel and tourism.

The case of the Oktoberfest beer tent experience allows us to highlight an enhanced understanding of attribute categories using moderated regression and multiple dependent variables. This chapter's approach provides an enhanced mathematical structure of Kano's theory to minimize weaknesses in earlier studies using methods that were either too complex or simplistic, and likely to provide biased results. This case will categorize key Oktoberfest beer tent attributes and how (or if) these are associated with overall satisfaction, visitor loyalty and perceived value. Based on these findings, a discussion will highlight important similarities and differences with using multiple consumer behavior DVs for classification and decision-making.

Specifically, the use of the Oktoberfest experience as a basis for this chapter will allow for better articulation and understanding of 1) how attributes impact consumer behaviors and experiential service design, 2) how this approach can be useful for researchers and practitioners for marketing strategy insights, and 3) how this approach can be expanded on for reliable 21st century strategic decision-making approaches.

Background

The Oktoberfest tourist experience

Each year Oktoberfest provides a positive impact for the tourism sector in Munich and the surrounding areas. The total number of arrivals in the Bavarian capital continues to grow with numbers increasing from 3.74 million in 2000 to 6.59 million in 2014. While the distribution of overnight stays during Oktoberfest is still dominated by travelers from Germany and Europe, it has become more balanced over the years with nearly 50% of all overnight stays in Munich currently taken by foreign travelers.

As usual, the Munich Oktoberfest opens in traditional fashion with a tapping of the first keg by Munich's mayor. Oktoberfest in Munich covers 85 acres with 16 festival beer tents. The overall seating capacity in the beer tents is about 119,000 visitors. The festival is organized by the City of Munich and employs a total of about 13,000 staff. It features about 175 rides and shows with traditional dishes featuring roast chicken, radishes, Obatzda (Bavarian cheese spread), sausages, grilled fish, roasted ox and Wiesnbrezn (large Bavarian pretzel). About 6 million people visit the Munich Oktoberfest each year and consume over 7 million liters of beer. The largest contingent of travelers is from Munich, Bavaria and other parts of Germany, while other primary nationalities include other Europeans, those from the USA and elsewhere. The largest age group is generally 30 years of age and older (more than 50%) with close to a 50/50 split of men and women. This diversity makes it a very international tourism event serving a range of age groups and nationalities (Papke 2015).

Given the previously limited access to researchers for Oktoberfest, many questions remain unanswered with regards to primary elements that drive satisfaction, loyalty, perceived value and positive word-of-mouth for beer festival attendees. Given the importance of festivals as tourism attractions, economic drivers and image builders for local communities, as well as the growth of interest in artisan beer consumption, a better understanding of these relationships is an important outcome of this chapter.

Refining attribute models and measurement

The notion of attribute types, importance and performance having asymmetric effects on customer satisfaction and other behavioral intentions has a long history in the broader consumer behavior literature (Brechan 2006; Gustafsson et al. 2005), in services marketing (Bolton and Lemon 1999; Lin et al. 2010;) and more recently in hospitality and tourism businesses (Knutson et al. 2006; Mathe-Soulek et al. 2015). The value proposed by the classification of attributes by type is the enhanced ability for practitioners to maximize the effectiveness of resource allocation to attributes that have an impact on the underlying satisfaction-profit chain (Harrington et al. 2017b).

To assess the relationships shown in the central portion of Figure 26.1, we followed earlier studies using the classification of attributes from Kanos' attribute–satisfaction model (Kano et al. 1984). While this classification approach has received strong theoretical support, many researchers have criticized it as being too complex and suggest that most early attempts to simplify the classification process are replete with shortcomings (Mikulić and Prebežac 2016).

In this chapter, a test is provided of attribute classification in the context of Oktoberfest using a validated approach of moderated regression (Harrington et al. 2017b; Lin et al. 2010). Oktoberfest, attracting thousands of international and domestic travelers to Munich, Germany each year, has a long history and substantial brand recognition in the festival sector. This sector has become increasingly competitive and represents an important economic segment of tourism (Manthiou et al. 2014). The study sheds light on the classification of attributes with a variety of customer perceptions and behavioral intentions associated with the Oktoberfest experience. These outcomes include overall satisfaction, a composite of customer loyalty and perceptions of value. The chapter's contribution builds on earlier research and, by using multiple dependent variables, checks the validity of attribute classification with implications for managing the Oktoberfest and similar tourism experiences.

While numerous studies have been conducted to assess and determine attribute categories and the measurement of the impact of these attribute categories on customer satisfaction and behavior (Harrington et al. 2017b), Kano's quality attribute-satisfaction model (Kano et al. 1984) addressed the concept of a non-linear (asymmetric) relationship among attribute types and overall customer satisfaction. In essence, the researchers proposed that not all attributes are equal in a consumer's mind and, therefore, an understanding of the inequities of attribute performance and their impact on customer satisfaction would be the key to managing the consumer's experience.

Kano's model proposed five attribute types that have been described using a variety of terms: attractive (excitement/wow factors), one-dimensional (performance/satisfiers), basic (must-be/dis-satisfiers), indifferent and reverse. Attractive or excitement quality factors are defined as unexpected attributes where consumers will likely be more satisfied if the attribute is provided but will accept the product/service if not present with no dissatisfaction. One-dimensional (performance) quality factors follow a relatively straight relationship between attribute satisfaction-dissatisfaction and overall satisfaction-dissatisfaction. Basic (must-be) quality factors are those that create dissatisfaction when not provided but satisfaction increases at a decreasing rate with quality increases in basic attributes. Indifferent factors have no impact on satisfaction or dissatisfaction when present. The reverse quality factors cause dissatisfaction when present and satisfaction when absent. As Chen (2012) pointed out, since a reverse factor does not represent a real-world consumer need, there is no compelling need to include an assessment of reverse quality factors in real-world business situations. An example in the tourism world might be going through security at the airport; while the time and effort it takes is likely to increase dissatisfaction in the travel experience, the process is required and efforts should be made to minimize the duration (perceived or actual) as part of the tourism experience.

Prior to this model, relationships between attributes and overall satisfaction were assumed to be one-dimensional (improvement in any attribute quality would increase customer satisfaction). The realization of non-linear relationships among attributes and customer satisfaction or other outcomes is an important consideration in prioritizing resource allocations by managers (Lin et al. 2010).

While Kano's model and proposed methods have been partially supported, the original process has been criticized as too complex and difficult to implement. Several researchers have proposed alternative methods to classify attributes using a dummy coding method in regression (i.e. penalty-reward contrast analysis [PRCA]) (Witell and Lofgren 2007) or importance-performance grid (IPA) (Vavra 1997). Lin et al. (2010) and Chen (2012) demonstrated the weaknesses of classification of attributes using dummy coding (PRCA) compared to other regression methods that are capable of more closely modeling linear and non-linear relationships in Kano's model while being far easier to implement than the original. Mathe-Soulek et al. (2015) used mixed methods to classify the relationship between Quick Service Restaurants' (QSR) drive-through attributes and overall customer satisfaction. The study attempted to address earlier weaknesses using a combination of qualitative, PRCA and IPA methods. Overall, the findings appeared more valid using this triangulation approach but, as the authors pointed out, interaction effects should be explored in the future as these were one limitation in their study. Mikulić and Prebežac (2016) pointed out that tourism research applying Kano's model has generally lacked sufficiently rich information for assessment, resulting in questionable reliability and validity.

In this chapter, several weaknesses are addressed to more fully understand attribute performance relationships. While these relationships have been considered in other contexts and methods – restaurants (Mathe-Soulek et al. 2015), hotels (Slevitch and Oh 2010) and other services (Lin et al. 2010) – this chapter considers these relationships in the context of a unique festival experience to increase our understanding of classification methods and implications of marketing mix strategies that drive beer festival success. A regression approach that integrates the interaction effects of attribute performance and the level of performance was used to more closely model non-linear relationships and address earlier critiques of overly simplistic methods (Harrington et al.

2017b). Finally, the chapter results move beyond the sole consideration of overall satisfaction as the dependent variable in assessing attribute categories. This approach uses multiple customer behaviors as dependent variables to assess consistencies or inconsistencies between attribute performance on overall satisfaction, customer loyalty and perceptions of value.

Pre-journey activities

The notion of the customer journey tied to the entire experience (and shown in Figure 26.1) includes activities before the purchase decision (Lemon and Verhoef 2016). This process seems particularly relevant when tied to a tourism experience. In the context of Oktoberfest, this pre-journey might include time spent researching the upcoming visit (Google search, travel sites, etc.), time spent in preparation for the visit once travel plans have been finalized (booking rooms, shopping for the trip, finalizing plans, etc.). While these activities are beyond the traditional classification of attributes, this time spent is likely to impact perceptions of the experience and the level of expectations developed prior to the actual visit. When surveyed, 43% of Oktoberfest attendees followed or visited one or more social media sites/programs: 31% the Oktoberfest webpage, 6.5% Facebook, 3% Instagram, 6.9% used the Oktoberfest App and 3.7% used some other type of social media prior to attending. On average, attendees spent about 75 minutes researching before attending and somewhere between one to two hours in visit preparation activities.

Oktoberfest attributes and dependent variables

In order to classify attributes of the Oktoberfest experience into four possible types (excitement, basic, performance or indifferent), Oktoberfest attribute quality and the interacting effect of attribute performance level was assessed on three different dependent variables prior to attribute classification. A description of each dependent variable and the regression model is discussed in the following sections.

Satisfaction

While customer satisfaction has long been recognized as an important consideration for commercial enterprise success, its definition may vary, with at least nine theories established in the literature. The common denominator of most definitions is that customer satisfaction is the outcome or end-state resulting from a consumption experience (Pizam and Ellis 1999). Kotler (1994) defines customer satisfaction as "the level of a person's felt state resulting from comparing a product's perceived performance or outcome with their own expectations" (p.40). The present study adopted Oliver's (2011) definition which states that customer satisfaction is the result of a cognitive and affective evaluation, where some comparison standard is compared to the actually perceived performance. In general, the conceptualization of customer satisfaction is not thought to be simply the arithmetical sum of attribute impressions/satisfaction but instead it varies by type and importance (e.g., Kano et al. 1984). Therefore, the present study aimed to assess the relationship among key Oktoberfest attributes' performance and the linear and non-linear relationships with visitor overall satisfaction.

Loyalty

In contrast to customer satisfaction, customer loyalty appears more difficult to define. It is thought to be multi-faceted and includes both an attitudinal and a behavioral dimension toward the provider of the product or service (Kandampully and Suhartanto 2000). Attitudinal loyalty is expressed as the likelihood to recommend and repurchase (or revisit in the context of a destination) and is normally measured as intentions through self-administered surveys (Tzetzis et al. 2014). Behavioral loyalty is expressed with consumer action such as continued patronage and the act of recommendation. It is measured with a number of different metrics that normally include retention (or defection/churn), usage, lifetime duration, cross buying and share of wallet (Bolton et al. 2004). The integration of attitudinal and behavioral loyalty is often termed "composite" loyalty and is expressed as a behavioral intention which includes revisit intention and intention to recommend (Chen and Tsai 2007).

A significant body of studies (e.g., Homburg and Furst 2005; Kandampully and Suhartanto 2000) indicate that customer satisfaction has a direct, linear impact on customer loyalty. However, there are also many researchers who maintain that this relationship is non-linear (e.g., Baumann et al. 2012; Bowen and Chen 2001). The present study assumes the latter; this conceptualization of customer loyalty underscores the value in using it as an additional dependent variable to explore the impact of attribute performance on customer loyalty behaviors. This relationship appears likely to enhance the understanding of linear or non-linear relationships of attribute performance – and establish if these relationships are similar or different compared to overall satisfaction and the resulting attribute categorization.

With regards to the operationalization of the loyalty construct, this study uses a composite measure of customer loyalty that includes intentions to return, recommendation behaviors and sharing behaviors on social media of the Oktoberfest beer tent experience.

Value

Perceived value is the consumer's trade-off between the various benefits they receive from a consumer experience and the different costs they incur (Sánchez-Fernández et al. 2009). It has been shown to be a subjective and dynamic construct that varies by 1) consumer, location and context, 2) types of value (functional, social, emotional, etc.) and 3) its temporal nature (prior to purchase, at the point of purchase, during use, etc.) (Sanchez et al. 2006). Several studies (Chen and Tsai 2008; Yuksel 2007) have identified perceived value as the critical construct for forging relationships with customers and it is seen as a predictor for both attitudinal and behavioral loyalty. Therefore, this study assumed that the visitor's perceived value of the Oktoberfest beer tent experience was likely to be tied to attribute performance, and used it as a dependent variable in order to provide additional means to assess the relationships between attribute performance and perceived value.

In summary, the chapter was designed to further the understanding of relationships among Oktoberfest tent experience attributes and a number of additional dependent variables to assess consistencies or contrasts in attribute categorization based on consumer behavior outcomes. Figure 26.1 provides an outline of the study's constructs and relationships; from left to right, the direct effect of Oktoberfest attribute-satisfaction was assessed with its relationship with each dependent variable. Additionally, the interaction with performance level was also tested to separate attribute category. This direct and interactive approach provides for an assessment of attribute category that models Kano's four constructs of performance (significant direct effects only), excitement (significant interaction and coefficient > 0), basic (significant interaction and coefficient < 0) and indifferent (non-significant) (Lin et al. 2010). As with earlier studies, the reverse category was not included in the analysis as this type is not relevant in a commercial endeavor situation (Chen 2012).

Methods

A survey was carried out in the Armbrustschützenzelt, a beer tent opened in 1895. The survey was presented in both English and German; it consisted of the performance of 16 attributes, overall satisfaction, customer loyalty questions, price/value of the experience, number of times attending Oktoberfest and demographics. Items included in the survey instrument were based on 1) recommendations by a research faculty with expertise in tourism and festival research, 2) a review of the literature and 3) feedback and desired items suggested by industry representatives. Thus, the Oktoberfest attributes used in this study reflect the physical, social and cultural aspects of the experience as defined by a review of earlier research and a synthesis of experts in Oktoberfest events and tourism in general.

A team of 16 students and two faculty members carried out the survey over the course of 21 shifts directly in the tent, from noon until early evening. The sampling strategy employed was plain convenience sampling, commonly used in this type of research (e.g., Sung et al. 2016). The surveys were completed by people sitting at beer tables in all areas of the tent: a large main open area in the middle, different VIP boxes bordering the main area and a second-floor balcony area overlooking the main area. In total, 1,323 usable surveys were completed during the 2015 Oktoberfest from September 20 through October 3, 2015. Of these, 58.3 were male (41.7% female); 59.7% were from somewhere in Germany, 20.6% from other parts of Europe and 19.4% were from outside of Europe. Respondents were divided into five age groups with 11.3% aged 16–21, 57.7% aged 22–35, 13.9% aged 36–46, 11.4% aged 47–60 and 5.6% older than 60 years.

Analysis

Following Kano's model of quality attribute types (Kano et al. 1984), the chapter used a moderated regression approach to sort out relationships among attributes that should be categorized as basic (or must-be) elements, performance (one-dimensional) elements, excitement (attractive) elements and indifferent elements, as part of the Oktoberfest tent experience. While dummy coding (PRCA) and importance-performance analysis (IPA) are commonly used in tourism research to group attributes into these main types, the common practice of grouping attribute variables into high and low performance categories has been criticized in earlier research (Mikulić and Prebežac 2011) and does not fit with Kano's model (Harrington et al. 2017b). Earlier research supported the need to consider moderated regression (Lin et al. 2010), ridge regression (Chen 2012) or other approaches (Mathe-Soulek et al. 2015) to minimize the potential for misclassification of attributes.

Thus, a moderated regression approach was used to enhance the validity of attribute categorization and minimize the flaws of dummy coding methods. Specifically, the moderated approach assesses the moderating effect of perceived attribute quality level, lessens the misclassification of attributes, simplifies Kano's data collection process, minimizes the effects of biased data (the majority of responses being above average in performance) and includes the average (common) attribute performance level, thus minimizing skewed results due to excluded responses (Lin et al. 2010).

Regression model definitions

Specifics on the regression model used can be found in Lin et al. (2010) and Harrington et al. (2017b). The basics of this moderated regression approach and how it was used are outlined below:

$$DV_i = \alpha + \text{ß}^*{}_{1j}X_{ij} + \text{ß}^*{}_{2j}X_{ij} \times Z_{ij,} \text{ where } Z_{ij,=} \begin{cases} X_{ij} < m \\ X_{ij} = m \\ X_{ij} > m \end{cases}$$

DV_i = Because the current study looked at several dependent variables of interest, regression models were run for four dependent variables (using a 5-point Likert-type scale ranging from 1 [very low] to 5 [very high]):

- CS_i represents the overall satisfaction of the *ith* customer
- CL_i represents a composite measure of customer loyalty of the *ith* customer
- VR_i represents the overall perception of value received of the *ith* customer

α = this constant is the average of all reference groups with regard to the dependent variable of interest

X_{ij} = the performance of the Oktoberfest attribute *j* as perceived by the *i*th customer. Performance level was based on a five-point scale ranging from 1 (not at all satisfied) to 5 (extremely satisfied).

Z_{ij} = the moderator variable was created to verify the moderated effect of performance of the *j*th attribute from the *i*th customer. The symbol *m* is the average (or in this case moderately satisfied [=3 on 5-point scale]). Therefore, if the ith customer's performance rating for the jth attribute was 1 or 2 on the five-point scale, the value was coded as 1. If the rating was 3 (of 5), it was coded as 2 and if a rating of 4 or 5 it was coded as 3, resulting in three performance levels.

ß^*_{1j} = is the impact of the *j*th attribute on the dependent variable of interest.

ß^*_{2j} = the coefficient representing the interaction effect of attribute performance and perceived performance level.

Once regression models were run, the sign and significance of the regression coefficients were used to classify each attribute as a basic, excitement, performance or indifferent element (Lin et al. 2010). Following the recommendations of Hair et al. (2006) and Lin et al. (2010) two steps were used to avoid collinearity issues.

Measures

The measures used included 16 attributes as independent variables, performance level as part of an interaction term and three dependent variables.

Attributes

Table 26.1 provides a description of the 16 attributes used in the study and the coding used in Tables 26.2–26.5. For each attribute performance measure, the survey used a 5-point scale with 1 = not at all satisfied and 5 = extremely satisfied. The far right-hand column provides the means and standard deviations for all independent and dependent variables. Staff friendliness, fun group experience and atmosphere had the highest means of the 16 attributes.

Performance interaction

The moderator variable was described in the earlier section on the regression model. Specifically, the attribute performance was coded into three possible levels (1, 2 or 3) and this performance level was multiplied times the attribute performance rating to assess the moderated effect of performance on each dependent variable (i.e., Lin et al. 2010).

Dependent variables

The wording used for the dependent variables of overall satisfaction, composite measure of customer loyalty (intention to return, recommendations and social media recommendations) and price/value perceptions is also provided in Table 26.1. Overall satisfaction used a five-point scale ranging from very low (1) to very high (5) overall satisfaction. The composite score for customer loyalty was calculated as a mean of three items; each having a five-point scale from very low to very high. The three items included intention to return, the likelihood to recommend the visit to friends or family and the likelihood of sharing experience through social media. Finally, a dependent variable representing perceived value was used and respondents were asked to indicate their perception of the experience quality relative to overall prices. Here again, this used a five-point scale ranging from very inexpensive to very expensive.

Results

The analysis of Oktoberfest attribute categorization results is divided into tables based on the dependent variable. This separation allows additional understanding of attribute categories and the impact on consumer perceptions.

Overall satisfaction

Table 26.2 provides results using overall satisfaction as the dependent variable, and each attribute-satisfaction and the interaction with level of performance as predictor variables. With overall satisfaction as the dependent variable, the 16 attributes were categorized into two categories. Thirteen were categorized as one-dimensional (performance) (Atmosp, Tradition, PartyEnv, MusicQual, NumPeople, Dirndl, NewPeople, FunGrp, TntAccess, FoodQual, BevQual, Shopping and Staff) and three as excitement (BavCult, Flirting and Beer).

Table 26.1 Survey Items and Coding

Attribute Coding	*Complete Item Wording*	*Mean (SD)*
BavCult	Finding Bavarian culture on the Oktoberfest	3.62 (1.06)
Atmosp	Enjoying the overall atmosphere of the Oktoberfest	4.00 (.94)
Tradition	Getting in touch with Oktoberfest traditions in the tent (e.g. Wiesn classic hits, drinking Mass, standing on benches)	3.77 (.99)
PartyEnv	Party environment in the tent	3.82 (.97)
MusicQual	Quality of live music in the tent	3.62 (1.09)
NumPeople	High amount of people in the tent	3.55 (1.02)
Dirndl	Finding lots of people in Dirndl or Lederhose	3.86 (1.05)
NewPeople	Meeting new people	3.56 (1.13)
FunGrp	Having an excellent and fun group experience	4.02 (1.00)
Flirting	Flirting and getting in touch with women or men	3.29 (1.18)
Beer	Drinking a lot of beer without being exposed	3.70 (1.14)
TntAccess	The tent is easy to reach/fast access to the tent	3.78 (1.15)
FoodQual	Food quality in the tent	3.64 (1.05)
BevQual	Beverage quality in the tent	3.89 (.99)
Shopping	Shopping and special souvenirs in the tent	3.10 (1.16)
Staff	How satisfied were you with friendliness and competence of staff in the tent?	4.10 (1.01)
OverallSat	How would you rate your overall satisfaction with the Oktoberfest tent experience?	4.07 (.87)
CompLoyalty	Intnt2Retn: How would you rate your intention to return to an Oktoberfest beer tent based on the quality of your experience? Recommend: Based on your experience, would you recommend an Oktoberfest beer tent visit to friends and family? SocialMedia: Based on the quality of your Oktoberfest beer tent experience, how likely is it that you will share through social media (blog, Twitter, Facebook, etc.)?	3.88 (.89)
PriceValue	Based on the quality of your experience in the beer tent, how would you rate the value received for overall prices?	3.90 (.97)

Customer loyalty

Table 26.3 provides results using a composite score of customer loyalty as the dependent variable with the IV and interaction as stated above. The composite was calculated as the mean of intention to return, recommendations to friends and family and sharing through social media.

With the composite as the dependent variable, the 16 attributes were categorized into three categories. Ten were categorized as performance (Atmosp, Tradition, PartyEnv, NumPeople, Dirndl, NewPeople, FunGrp, Beer, BevQual and Staff), five as excitement (BavCult, MusicQual, Flirting, TntAccess and Shopping) and one as basic (FoodQual).

Table 26.2 Overall Satisfaction Level Based on Satisfaction of Each Attribute

		Moderated Regression			
Attribute	R^2	ΔR^2	$ß^*_{1j}$	$ß^*_{2j}$	*Quality category*
BavCult	.15	.003*	.14	.25	E
Atmosp	.23	.001	.62	-.15	P
Tradition	.15	.000	.30	.10	P
PartyEnv	.17	.002	.62	-.21	P
MusicQual	.14	.001	.20	.17	P
NumPeople	.13	.000	.33	.02	P
Dirndl	.12	.000	.46	-.12	P
NewPeople	.08	.000	.28	.01	P
FunGrp	.14	.001	.25	.13	P
Flirting	.06	.005**	-.09	.33	E
Beer	.08	.006**	-.13	.41	E
TntAccess	.08	.000	.26	.02	P
FoodQual	.10	.001	.20	.12	P
BevQual	.15	.001	.26	.13	P
Shopping	.06	.001	.12	.13	P
Staff	.22	.000	.37	.10	P

Table 26.3 Composite Customer Loyalty Level Based on Satisfaction of Each Attribute

		Moderated Regression			
Attribute	R^2	ΔR^2	$ß^*_{1j}$	$ß^*_{2j}$	*Quality category*
BavCult	.10	.004**	.003	.32	E
Atmosp	.16	.000	.47	-.08	P
Tradition	.14	.001	.24	.14	P
PartyEnv	.15	.000	.43	-.05	P
MusicQual	.10	.005**	-.07	.38	E
NumPeople	.08	.000	.27	.01	P
Dirndl	.08	.000	.25	.04	P
NewPeople	.09	.001	.14	.15	P
FunGrp	.10	.01	.31	.01	P
Flirting	.06	.005**	-.07	.33	E
Beer	.09	.001	.15	.16	P
TntAccess	.04	.004**	-.15	.35	E
FoodQual	.06	.002+	.49	-.25	B
BevQual	.10	.001	.20	.13	P
Shopping	.04	.004*	-.09	.28	E
Staff	.05	.000	.21	.01	P

Perceived value

Table 26.4 provides attribute categories using perceived value as a dependent variable. This relationship considered if there was a relationship between the satisfaction of attributes and perceptions of price/value. The results in these models resulted in vastly different relationships; due to non-significant relationships, ten of the attributes could be considered in the indifferent category (Atmosp, Tradition, PartyEnv, MusicQual, NumPeople, Dirndl, NewPeople, Flirting, TntAccess and FoodQual). Four were categorized as basic (BavCult, FunGrp, BevQual and Staff), one as performance (Beer), and one as excitement (Shopping).

Discussion

The use of multiple dependent variables and the interaction term with attribute performance level provided some interesting insights into the impact of various Oktoberfest attributes on customer satisfaction, loyalty behavior and perceptions of value. Table 26.5 provides a summary of the categorization of each attribute based on individual DVs. The right-hand column provides the mode or most common categorization across all DVs.

Overall, the DV used had an impact on how each attribute was categorized. Thus, if researchers or practitioners use the standard DV of overall satisfaction vs. other desired behaviors, the attributes are likely to vary, in many cases, from performance to excitement, basic or indifferent depending on the consumer behaviors used in the assessment. Assuming a cause and effect relationship between attribute performance and consumer behaviors, this finding has implications for research and practical implications for resource allocations to beer tent attributes starting with the question – what consumer behaviors do we want to impact?

Table 26.4 Perceived Value Level Based on Satisfaction of Each Attribute

		Moderated Regression			
Attribute	R^2	ΔR^2	$ß^*_{1j}$	$ß^*_{2j}$	*Quality category*
BavCult	.01	.003*	.22	-.27	B
Atmosp	.003	.001	-.22	.18	I
Tradition	.000	.000	-.04	.05	I
PartyEnv	.000	.000	-.07	.07	I
MusicQual	.000	.000	.06	-.05	I
NumPeople	.000	.000	-.06	.05	I
Dirndl	.000	.000	.08	-.07	I
NewPeople	.000	.000	.04	-.02	I
FunGrp	.003	$.002^+$	.27	-.26	B
Flirting	.001	.000	-.04	.07	I
Beer	.01	.000	.04	-.11	P
TntAccess	.000	.000	.10	-.10	I
FoodQual	.001	.001	.12	-.14	I
BevQual	.004	.004*	.32	-.33	B
Shopping	.004	.003*	-.26	.25	E
Staff	.01	.01***	.49	-.49	B

Table 26.5 Attribute Category Based on DV Behavior

Attribute	*Overall Satisfaction*	*Comp Loyalty*	*Value*	*Mode*
BavCult	E	E	B	**E**
Atmosp	P	P	I	**P**
Tradition	P	P	I	**P**
PartyEnv	P	P	I	**P**
MusicQual	P	E	I	**P/E/I**
NumPeople	P	P	I	**P**
Dirndl	P	P	I	**P**
NewPeople	P	P	I	**P**
FunGrp	P	P	B	**P**
Flirting	E	E	I	**E**
Beer	E	P	P	**P**
TntAccess	P	E	I	**P/E/I**
FoodQual	P	B	I	**P/B/I**
BevQual	P	P	B	**P**
Shopping	P	E	E	**E**
Staff	P	P	B	**P**

With overall satisfaction as the dependent variable, 13 of the 16 attributes appear to be one-dimensional in nature. This relationship indicates these 13 attributes are anticipated by Oktoberfest visitors and as performance/quality increases (decreases) so does satisfaction (dissatisfaction) with the overall Oktoberfest tent experience. Additionally, Bavarian culture, flirting and drinking without being exposed were unanticipated, were perceived as excitement factors and created higher satisfaction when present as part of the experience.

The composite measure of customer loyalty provides some support to the satisfaction DV findings but also some contrasts. These contrasts appear to be associated with differences in the impact of the interaction term associated with attribute performance level. These relationships appear to support a non-linear and sometimes asymmetrical relationship between attribute-satisfaction and loyalty with an impact of critical thresholds in performance level perceptions (Bowen and Chen 2001). In the case of composite customer loyalty as the dependent variable, ten attributes were categorized as performance; five as excitement (Bavarian culture, music quality, flirting, tent access and shopping) and one as basic (food quality). Therefore, to drive customer loyalty behavior (return visits, recommendations and social media sharing), the music, shopping and tent access were perceived as excitement elements that increase the likelihood of positive word-of-mouth and returns to Oktoberfest.

The relationship among perceptions of experience quality relative to a price/value connection to the attributes lowered the numbers of performance elements and increased indifferent elements substantially. With perceived value as the dependent variable, ten attributes were categorized as indifferent, four as basic (Bavarian culture, fun group experience, beverage quality and staff friendliness), one as performance (drinking without being exposed) and one as excitement (shopping). Therefore, if using the perception of value to categorize tourism attributes, results and interpretation will be greatly impacted.

Conclusions

Using the mode across all dependent variables as an indicator of attribute categorization, ten of the attributes appeared to have clear designations of types. Bavarian culture, flirting opportunities and beer tent shopping were most frequently categorized as excitement factors. These attributes are unanticipated and when present enhance satisfaction, loyalty behavior and value perceptions.

Clear performance attributes including atmosphere, traditions, the number of people in the tent, seeing attendees in Dirndl or Lederhose, meeting new people, a party environment, drinking without being exposed, beverage quality and a fun group experience can be generally described as anticipated and areas where higher quality increases the likelihood of satisfaction, loyalty and perceptions of value. Given the promoted visuals of Oktoberfest, this finding seems to mirror the expectations of prior attendees and first-timers. Strategic investments to facilitate these experiences for attendees appear likely to positively impact multiple types of consumer outcomes.

Based on the mode categorization, there were no attributes that could be defined clearly as basic or indifferent. Three attributes had split modal relationships. Music quality, tent access and food quality varied as a performance, excitement, basic or indifferent category based on DV. Therefore, the implication is that attendees appear to have a level of expectation in these attribute areas. But these attributes represent an opportunity to further assess their importance to attendees and if unanticipated the levels of quality would result in more consistent categorization as performance or excitement factors. Questions regarding how these service areas are managed in the beer tents require further analysis. For instance, is the inconsistent finding for food quality a result of inconsistent control/quality or a lack of choice? Can the tent access issue be better managed to reduce the perception of wait duration or sequencing of the service process? Can music quality be enhanced to create a stronger emotional response from the attendees? Etc.

Overall, this chapter contributed to a better understanding of attribute categories based on Kano's model, the use of moderated regression and multiple dependent variables. The original methods of Kano's model were too complex; the moderated regression more closely follows the mathematical structure of Kano's theory and addresses the shortcomings of dummy coding methods and distorted classification due to skewed samples (e.g., extreme performance levels and overlooking average/common performance ratings using dummy coding exclusively). The use of multiple dependent variables provided a better understanding of how attributes impact consumer perceptions and behaviors – both when consistent and contrasting.

It should be acknowledged that more recent thoughts on tourism outcomes go well below notions or assessments of satisfaction or loyalty. For instance, several tourism research streams have considered the role or impact of tourism experiences on emotional feelings such as higher-order outcomes like memories, wellbeing, quality of life and happiness (Harrington et al. 2019). In a follow-up survey of Oktoberfest attendees, we found that on a 5-point scale (1 = strongly disagree to 5 = strongly agree) the mean was generally within the "agree" range for the Oktoberfest experience, positively impacting attendees' "sense of wellbeing" (3.87/5) and "life satisfaction" (3.86/5). While these results are preliminary and not empirically tested in categorizing attributes here, the business logic or desire to impact these visitor feelings as part of an outcome of a tourist experience has implications for future research and practitioners' innovative implementation of tourism experiences.

Limitations, implications and future research

While the methods in this chapter address shortcomings of more simplistic methods, the approach is not without criticism. First, other research streams have suggested differing hierarchical models of attribute factors than Kano's framework (Brechan 2006). Thus, future research could assess other models in the context of hospitality or tourism. Second, this research stream lacks validity tests for the method used in this study as well as other methods to enhance attribute classification power (Chen 2012). While this chapter contributes to our understanding when classifying attributes using moderated regression and multiple dependent variables, additional studies should test these relationships in different contexts, industries and geographic locations, as well as how they are impacted by the entire customer journey (e.g. pre-activities through post-experience feelings, memories and behaviors).

The impact of repeat customers on attribute classification could also be considered in future studies as well as considering the shift of attribute categories over time, customer segments and unanticipated vs. anticipated attribute-satisfaction relationships.

Kano's classification of quality attribute factors provided a framework for practitioners to more thoughtfully allocate limited resources to sustain competitive advantage. But the original framework was too complex to allow for meaningful articulation of these relations. The approach in this chapter enhanced this ability in terms of both statistical calculation and with regards to the impact on varying consumer perceptions and behaviors. In general, the findings indicate a relationship between the majority of attribute types (12 out 16 attributes) and overall satisfaction and a composite of loyalty behaviors. With that said, it appears that basing attribute categorization and alignment of firm resources on loyalty behaviors rather than overall satisfaction is likely to be beneficial. While research has long supported the empirical relationship between loyalty and substantially higher profits and lower costs per customer (Bowen and Chen 2001; Reichheld and Sasser 1990), managers investing in attributes most closely associated with positive customer loyalty behaviors are likely to accrue benefits in profits and marketing impact (positive-word-of-mouth and more believed recommendations).

References

Ansoff H I (1987). The emerging paradigm of strategic behavior. *Strategic Management Journal* 8(3): 501–515. DOI: 10.1002/smj.4250080602.

Baumann C, Elliott G and Burton S (2012). Modeling customer satisfaction and loyalty: Survey data versus data mining. *Journal of Services Marketing* 26(3): 148–157. DOI: 10.1108/08876041211223951.

Bolton R N, Lemon K and Verhoef P C (2004). The theoretical underpinnings of customer asset management: A framework and propositions for future research. *Academy of Marketing Science* 32(3): 271–292. DOI: 10.1177/0092070304263341.

Bolton, R N, Lemon, K N. (1999). A dynamic model of customers' usage of services: Usage as an antecedent and consequence of satisfaction. *Journal of Marketing Research*, 36(2), 171–186.

Bowen J T and Chen S-L (2001). The relationship between customer loyalty and customer Satisfaction. *International Journal of Contemporary Hospitality Management* 13(5): 213–217. DOI: 10.1108/09596110110395893.

Brechan I (2006). The different effect of primary and secondary product attributes on customer satisfaction. *Journal of Economic Psychology* 27: 441–458. DOI: 10.1016/j.joep.2005.10.003.

Chen C F and Tsai D C (2007). How destination image and evaluative factors affect behavioral intentions? *Tourism Management* 28(4): 1115–1122. DOI: 10.1016/j.tourman.2006.07.007.

Chen C F and Tsai M H (2008). Perceived value, satisfaction, and loyalty of TV travel product shopping: Involvement as a moderator. *Tourism Management* 29(6): 1166–1171. DOI: 10.1016/j.tourman.2008.02.019.

Chen L F (2012). A novel approach to regression analysis for the classification of quality attributes in the Kano model: An empirical test in the food and beverage industry. *Omega* 40(5): 651–659. DOI: 10.1016/j.omega.2011.12.004.

Chandler J D and Lusch R F (2015). Service system: A broadened framework and research agenda on value propositions, engagement, and service experience. *Journal of Service Research* 18(1): 6–22. DOI: 10.1177/1094670514537709.

Dasu S and Chase R (2013). *The customer service solution: Managing emotions, trust, and control to win your customer's business.* New York: McGaw Hill Education.

Gustafsson A, Johnson M D and Ross I (2005). The effects of customer satisfaction, relationship commitment dimensions, and triggers on customer retention. *Journal of Marketing* 69(October): 210–218. DOI: 10.1509/jmkg.2005.69.4.210.

Hair, J F Jr, Black, W C, Babin, B J, Anderson, R E, and Tahtam, R L (2006). *Multivariate data analysis.* Upper Saddle River, NJ: Pearson.

Harrington R J, Fauser S and Ottenbacher M C (2017a). QSR brand value strategies: Assessing perceived relative differences in marketing mix dimensions among brands, loyal customers and non-customers. *International Journal of Contemporary Hospitality Management* 29(1): 551–570. DOI: 10.1108/IHR-06-2018-0002.

Harrington R J, Hammond R, Ottenbacher M C, Chathoth P and Marlowe B (2019). From goods-service logic to a memory-dominant logic: Demand-side strategy research and application in hospitality. *International Journal of Hospitality Management* 76(A): 252–260. DOI: 10.1016/j.ijhm.2018.05.014.

Harrington R J, von Freyberg B, Ottenbacher M C and Schmidt L (2017b) The different effects of dis-satisfier, satisfier and delighter attributes: Implications for Oktoberfest and beer festivals. *Tourism Management Perspectives* 24: 166–176. DOI: 10.1016/j.tmp.2017.09.003.

Homburg C and Furst A (2005). How organizational complaint handling drives customer loyalty: An analysis of the mechanistic and the organic approach. *Journal of Marketing* 69(July): 95–114. DOI: 10.1509/jmkg.69.3.95.66367.

Kandampully J and Suhartanto D (2000). Customer loyalty in the hotel industry: The role of customer satisfaction and image. *International Journal of Contemporary Hospitality Management* 12(6): 346–351. DOI: 10.1108/09596110010342559.

Kano N, Seraku N, Takahashi F and Tsuji S (1984). Attractive quality and must-be quality. *Hinshitsu (Quality, Journal of Japanese Society for Quality Control)* 14: 39–48. Retrieved from www.jsqc.org/en/.

Knutson B, Beck J, Kim S and Cha J (2006). Identifying the dimensions of the experience constructs. *Journal of Hospitality and Leisure Marketing* 15(3): 31–47. DOI: 10.1300/J150v15n03_03.

Kotler P (1994). *Marketing management: Analysis, planning, implementation, and control,* (8th edition). Englewood Cliff, NJ: Prentice Hall, Inc.

Lemon K N and Verhoef P C (2016). Understanding customer experience throughout the customer journey. *Journal of Marketing* 80: 69–96. DOI: 10.1509/jm.15.0420.

Lin S-P, Yang C-L, Chan Y-H and Sheu C (2010). Refining Kano's 'quality attributes-satisfaction' model: A moderated regression approach. *International Journal of Production Economics* 126: 255–263. DOI: 10.1007/s10901-018-9627-0.

Manthiou A, Lee S, Tang L and Chiang L (2014). The experience economy approach to festival marketing: Vivid memory and attendee loyalty. *Journal of Services Marketing* 28(10): 22–35. DOI: 10.1108/JSM-06-2012-0105.

Mathe-Soulek K, Slevitch L and Dallinger I (2015). Applying mixed methods to identify what drives quick service restaurant's customer satisfaction at the unit-level. *International Journal of Hospitality Management* 50: 46–54. DOI: 10.1016/j.ijhm.2015.07.007.

Mikulić J and Prebežac D (2016). The Kano model in tourism research: A critical note. *Annals of Tourism Research* 61: 25–27. DOI: 10.1016/j.jdmm.2015.12.003.

Mikulić, J and Prebežac, D (2011). A critical review of techniques for classifying quality attributes in the Kano model. *Managing Service Quality* 21(1): 46–66.

Oktoberfest.info (2016, January 27) *Anstich – Opening ceremony.* Retrieved from www.oktoberfest.info/en/events/opening-ceremony/. [Accessed: March 26th, 2017].

Oliver R L (2011). Customer satisfaction/dissatisfaction. In J N Sheth and N Malhotra (Eds.), *Wiley international encyclopedia of marketing.* Chichester, UK: John Wiley.

Ottenbacher M C, Harrington R J, Fauser S and Löwenhagen N (2016). Should culinary and hospitality service attributes be defined as primary tourism drivers? An expectancy-fulfillment grid approach. *Journal of Foodservice Business Research* 19(5): 425–440. DOI: 10.1080/15378020.2016.1185879.

Pine B J and Gilmore J H (2011). *The experience economy, updated edition*. Boston, MA: Harvard Business Review Press.

Pizam A and Ellis T (1999). Customer satisfaction and its measurement in hospitality enterprises. *International Journal of Contemporary Hospitality Management* 11(7): 326–339. DOI: 10.1108/09596119910293231.

Papke, G (2015). Vorlaeufiger Oktoberfest-Schlussbericht 2015: Wiesn auf gut Münchnerisch. Press release from Landeshauptstadt München Referat für Arbeit und Wirtschaft. Retrieved Oktober 4, 2015, from http://Downloads/W45VorlaeufigerWiesnSchussbericht2015%20(1).pdf

Reichheld F and Sasser W E (1990). Zero defections: Quality comes to services. *Harvard Business Review* 68((September/October): 105–111. Retrieved from https://hbr.org/1990/09/zero-defections-quality-comes-to-services. [Accessed: November 22nd, 2017].

Sanchez J, Callarisa L, Rodrıguez R M and Moliner M A (2006). Perceived value of the purchase of a tourism product. *Tourism Management* 27: 394–409. DOI: 10.1016/j.tourman.2004.11.007.

Sánchez-Fernández R, Iniesta-Bonillo M A and Holbrook M B (2009). The conceptualisation and measurement of consumer value in services. *International Journal of Market Research* 51(1): 93–113. DOI: 10.1177/147078530905100108.

Slevitch L and Oh H (2010). Asymmetric relationship between attribute performance and customer satisfaction: A new perspective. *International Journal of Hospitality Management* 29: 559–569. DOI: 10.1016/j.ijhm.2009.09.004.

Sung Y-K, Su C-S and Chang W-C (2016). The quality and value of Hualien's amis harvest festival. *Annals of Tourism Research* 56: 128–163. DOI: 10.1016/j.annals.2015.10.001.

Tzetzis G, Alexandris K and Kapsampeli S (2014). Predicting visitors' satisfaction and behavioural intentions from service quality in the context of a small-scale outdoor sport event. *International Journal of Festival and Event Management* 5(1): 4–21. DOI: 10.1108/IJEFM-04-2013-0006.

Vavra T G (1997). *Improving your measurement of customer satisfaction: A guide to creating, conducting, analyzing, and reporting customer satisfaction measurement program*. Milwaukee, WI: ASQ Quality Press.

Witell L and Lofgren M (2007). Classification of quality attributes. *Managing Service Quality* 17(1): 54–73. DOI: 10.1108/09604520710720674.

Yuksel A (2007). Tourist shopping habitat: Effects on emotions, shopping value and behaviors. *Tourism Management* 28(1): 58–69. DOI: 10.1016/j.tourman.2005.07.017.

27

Delivering quality and memorable tourism experience

Anestis K. Fotiadis and Anastasia Ef. Spyridou

Introduction

Service quality has long been debated in different industrial settings, including tourism, as a means to meet customer needs, develop customer loyalty, and induce trust (Fotiadis and Vassiliadis 2016). However, world markets have registered a significant shift in customer behavior, where the primary trend is to offer options for variety and customization. Tourists and visitors are no exception. Discerning customers are not only looking for outstanding service quality but also the opportunity to develop quality and memorable experiences. With the globalization of disparate economies, growing competition has arisen in the tourism industry. Thus, a significant challenge facing managers in the tourism industry is to meet, and even exceed, visitors' expectations while catering to diverse cultural backgrounds, needs, and expectations. Therefore, successful tourism businesses are those capable of developing and managing service quality as well as maintaining facilities that allow tourists to have high quality and memorable experiences. The delivery of quality service encounters and the creation of memorable experiences will depend primarily on how well employers and their staff are trained to deliver these promises.

In this chapter, critical characteristics of services are presented and distinguished between service and product dimensions. This chapter also introduces the service quality model commonly called SERVQUAL, the most widely used service quality model (Fotiadis and Vassiliadis 2010). Given the widespread use of the SERVQUAL model, we suggest consideration before implementing this model because its dimensions cannot satisfy the countless service features that exist in the context of tourism experiences. We also emphasize the need for managers to anticipate visitors' expectations and potential experiences to develop a good-fit strategy for service quality management. Additionally, the notion of memorable tourism experiences, its fundamentals, and perceived importance is also discussed. It further outlines the role of managers as facilitators and providers of means that help create memorable experiences through the process of co-creation and customization. The chapter concludes by highlighting the strategic importance of memorable tourism experiences.

Service quality

The tourism industry is one of the most complex service sectors. The globalization of economies has brought a significant shift in terms of competitiveness within the different components of the tourism industry around the world. While different components (e.g., attractions, adventure tourism and recreation, accommodation, etc.) of the tourism industry become continually more competitive, tourists and visitors have become more experienced, knowledgeable and demanding than ever before (Swarbrooke 2001). Therefore, the most successful tourism businesses are those capable of developing and implementing outstanding service quality that satisfies the needs of visitors/tourists, and meets or exceeds their expectations. Note that we mention visitors/tourists, not customers. The reason we emphasize the term visitors/tourists is due to the particularity of services offered by tourism businesses and how individuals consume these services. First, tourism-related services are not consumed on a regular basis. Second, people who consume tourism services are adventurers who are less likely to consume the same services over and over, though we can expect a revisit in the future. Consequently, we use the word "visitors/tourists" throughout this chapter.

Prior to defining or understanding what service quality is, managers must identify or anticipate the needs and expectations of visitors well in advance. More importantly, they need to develop service quality management strategies that help to meet visitors' expectations, and possibly exceed them. By meeting or exceeding the expectations of visitors/tourists, managers are more likely to satisfy their needs. Furthermore, managers need to understand that the competitiveness of their tourist business can only be assured over the long term through the offer of intrinsic higher service quality compared to the competition. To do so, they need to understand what constitutes service quality and how it applies to the tourism business.

Service quality is a concept that managers, customers, and even stakeholders emphasize while transacting across different service industries. Service quality is a familiar concept to almost everyone. The challenge, however, rests on how to define service quality. It is important to consider that there is not one single satisfactory definition for "service quality". The problem of defining service quality stems from the highly abstract nature of the concept "service". However, it is critical to understand the distinctive features of services to avoid service marketing problems by applying tools and techniques that are designed for tangible products. Distinct from products, services are characterized by their intangibility, heterogeneity, and by the fact that their production, distribution and consumption are simultaneously processed, in addition to the fact that they cannot be kept in stock (Grönroos 1990).

Over the years, many authors have attempted to define the concept of "service". In this chapter, we just mention the two most cited ones. For Grönroos (1990),

> A service is an activity or series of activities of more or less intangible nature that normally, not necessarily, take place in interactions between the customer and service employees and/or physical resources or goods and/or systems of the service provider, which are provided as a solution to customer problems.

Later, Philip Kotler (1997) defines a service "as an action or an activity which can be offered by a party to another party, which is basically intangible and cannot affect any ownership. Its production may or may not be tied to a physical product". Similar to the concept of service, many authors have similarly attempted to define *service quality*. But keep in mind that quality means *innate excellence*. For Yarimoglu (2014), *quality* is "a mark of uncompromising standards and high achievement, universally recognizable, and realized only through experience". Instead of providing several definitions for service quality, we consider the definition of Zeithaml and Bitner (2003) purposely because of the integration of dimensions of service into their definition. According to these authors, "Service quality is a focused evaluation that reflects the customer's perception of specific dimensions of service, namely reliability, responsiveness, assurance, empathy, and tangibles" (2003).

Fundamentally, visitors are the lens through which service quality must be assessed rather than the personal judgment of managers. Thus, service quality should measure how service(s) meet(s) the expectations of the visitors/tourists. The measurement of tourist-related services is all about perception because of the intangible nature of services. Therefore, we often refer to the term "perceived service quality", which is a result of the comparison of perceptions (perceptions of visitors/tourists) about the service delivery process and the actual outcome of the service (Grönroos 1984; Lovelock and Wirtz 2011; Dixit 2017).

Service quality models and dimensions

There are several service quality models that managers can tap into. But, for the purpose of simplicity, we only introduce the two most cited ones. Before introducing the service quality model that we purposely selected, there are many factors that you need to take into account when evaluating the quality of the service you offer to your visitors. For example, security, consistency, attitude, completeness, condition, availability, and training of service providers are factors that affect how visitors perceive the level of quality of services (Sasser et al. 1978). Other factors that may affect visitors' perceptions of the quality of your service include *corporate image* (quality related to all aspects of your organization), *physical quality*, and *interactive quality*, which is the quality of interaction between your staff and visitors (Lehtinen and Lehtinen 1982). By improving quality factors, tourism businesses may enjoy a more positive corporate image (Mombeuil 2018).

The first service quality model was developed by Grönroos (1984). By developing the first model, Grönroos was able to provide guidelines in measuring perceived service quality using *technical quality, functional quality, and corporate image* as the dimensions of total service quality. From a tourism management perspective, technical quality refers to how visitors evaluate your service, whereas functional quality refers to how visitors consume or experience the service. Corporate image, or how a business presents itself to the public, also influences the perceptions of visitors. In the model proposed by Grönroos, six criteria serve to determine the quality of services. These are: adequate professionalism and qualifications of company employees; attitude and behavior; accessibility of the location of the company and flexibility of the schedule; safety (confidence that the service is accomplished according to the provisions of the company); ability of the company to solve an unexpected problem; and reputation and credibility of the company. In addition to this model, Parasuraman et al. (1988) developed the widely used service quality model referred to as SERVQUAL, possessing five dimensions; namely, reliability, assurance, tangibles, empathy, and responsiveness.

Continuous improvement of service quality is compulsory for the survival of a tourism business. With an improvement in service quality, you are likely to satisfy your visitors. Since people visit and travel for various reasons, managers need to anticipate the needs and expectations of visitors adequately. More importantly, managers need to ensure that services are delivered as promised. In today's information savvy society where visitors have ample access to information, they are becoming more and more demanding. As such, managers cannot rely on the adage of excellent service quality and visitor satisfaction as the sole means to ensure competitive advantage. One of the many challenges known to managers is the fact that visitors' expectations are subject to change over time. Therefore, developing the capability to anticipate their future expectations and plan accordingly becomes a must, aside from all the constraints related to the constant changes in visitors' expectations. As such, doing systematic marketing surveys may yield significant insights about visitor expectations and their reaction to the service they consume, and the likelihood that they will revisit and/ or recommend a particular tourism business to others. Sincere feedback and suggestions collected from visitors through surveys are important sources of information that may help to improve service quality and anticipate future expectations. Also, comments, reviews, posts from social media, and travel agency websites (community of travelers) are sources of information that managers need to investigate. Importantly, these websites may provide insights on what works, or not, and how well the competition is doing.

A more reliable way to assess service quality is by directly assessing your customers' perception of the performance of quality attributes (Oriadea and Schofield 2019). It is well supported that visitor perceptions of attribute performance positively influence their perceptions of service quality (Tian-Cole et al. 2002), and thus their level of satisfaction. Therefore, this encourages managers to investigate what attributes customers consider as essential and why. This leads to the possibility of identifying attributes that need to be improved for an overall improvement of service quality. Some attributes that managers may consider during their investigation include but are not limited to: percentage of satisfied visitors versus unsatisfied visitors; number of visitor complaints received; number of visitor complaints that were effectively resolved; number of new orders/visits from visitors' referrals; number of visitors' calls for help or information; percentage of on-time deliveries to visitors; average time to deliver services to visitors; average time to resolve a visitor concern; average time that visitors waited for a response after calling, and others. For continuous service quality management, these attributes need to be an integrated part of a tourism business marketing plan and must be presented into the scheme of SMART service quality improvement objectives (Fotiadis et al. 2018). While improvement in service quality is a critical aspect of competitiveness, we encourage service providers not to limit their strategy to providing visitors with excellent service quality. The reason is that services tend to increasingly become commoditized; thus, the importance for managers to focus on delivering meaningful experiences to visitors in order to add value to their offerings (Berry et al. 2002).

Memorable tourism experience

Having emphasized service quality, customer experiences have become the competitive battleground for marketing (Klaus and Maklan 2013). Therefore, it is crucial to understand the value of memorable tourism experience and its dimensions (Fernandes and Cruz 2016). Hence, a memorable tourism experience is defined as "a tourism experience remembered and recalled after the event has occurred" (Kim et al. 2010: 2). In fact, a meaningful tourism experience tends to stick longer in the memory (Tung and Ritchie 2011). Essentially, long-lasting vivid memories are a highly desired result of a tourism experience (Neuhofer et al. 2012; Campos et al. 2016). More importantly, visitors are more in search of experiences that are compelling and memorable (Gilmore and Pine 2002), rather than having services that meet or exceed their expectations. One essential aspect a manager needs to consider is that a primary goal of each journey in a visitors' life is an experience with extraordinary souvenirs. The main challenge here is that there is no set list of techniques that may help develop a memorable experience. Moreover, visitors or tourists possess different needs and expectations that often seek personalized services, in addition to the variety of services that are offered within the tourism industry. However, memorable experiences may be established through the process of co-creation in a particular service setting. That is, the experiences are created through a collaboration between hosts who provide the tourist experience and visitors/tourists consuming particular services (Lončarić et al. 2018).

We need to acknowledge that there is not a set of one-size-fits-all dimensions that can help you create memorable experiences for your visitors. To identify the best drivers for memorable tourism experiences in a given market, managers need to understand the needs, expectations, or reasons people do visit their destination. For instance, summer vacations are major events that help visitors/tourists take a break from their daily routines (Decrop and Snelders 2004). Also, the search for pleasure and enjoyment are the primary motives that influence countless people to leave the confinement of the workplace (Kim 2014). Therefore, managers are expected to demonstrate the distinctiveness between the tourism experience and everyday life. This is because there are many psychological needs and wants (e.g., relaxation) that cannot be satisfied in a tourist's daily routine (Kim and Ritchie 2014).

Many authors (e.g., Kim et al. 2012; Kim and Ritchie 2014) have agreed upon seven dimensions linking to the development of a memorable tourism experience. These are hedonism, novelty, local culture, refreshment, meaningfulness, involvement, and knowledge. The hedonism dimension of tourist experiences is just the "pleasure of sensuous gratifications for oneself" that can contribute to the overall visitor satisfaction (Triantafillidou and Petala 2016). As such, activities involving hedonic consumption, pleasure-seeking, and positive emotions may generate lasting memories (Scott et al. 2009; Ma et al. 2013) that visitors are likely to share with friends directly or via social media networks. Similarly, the novelty dimension of services may also develop memorable tourism experiences. It is important to remember that visitors/tourists generally set out to see and experience things that are new and out of the ordinary. This is one of the several factors behind their motivation for traveling. More importantly, those who travel constantly tend to be novelty-seekers stimulated by new adventures and surprise (Tung and Ritchie 2011). Therefore, tourism services offered can be of greater significance if accompanied by adventurous activity. Consistently, direct contact with residents of the local community may not only help create a memorable tourism experience but is also an important motivational factor for a number of visitors (Tung and Ritchie 2011; Kim and Ritchie 2014). Consequently, the extent to which tourism service providers facilitate visitors to interact with local culture may result in a unique and unforgettable experience (Lončarić et al. 2018).

Another important dimension that facilitates the creation of memorable tourism experiences is the "knowledge" that visitors acquire during their period of enjoyment with a tourism organization. This may be achieved through helping visitors to obtain new insights, skills, and information through different activities during their stay, such as "learning the history, local culture, way-of-life, natural physiography, and language of the destination" (Tung and Ritchie 2011:1379). Also, activities aimed at enhancing social relationships, self-discovery, intellectual development, and overcoming physical challenges are contributing factors to the development of tourism experience (Tung and Ritchie 2011) that managers may tap into, given their education outcomes. We have mentioned several times the need for managers to understand the needs, expectations, and interests of visitors to ensure quality service is delivered accordingly. This also applies for the provision of memorable tourism experiences; thus, the importance for managers to dig into the interests of their visitors rather than relying solely on an exhaustive list of dimensions linked to the creation of memorable tourism experiences. Here, we do not mean that these dimensions are not significant. But having an indication about visitors' interests is more likely to provide managers with leverage that heightens memorable experiences. It is generally acknowledged that visitors are more likely to remember an experience which is personally relevant and closely related to their interests, more than other experiences (Kim 2014; Kim and Ritchie 2014).

Importance of memorable tourism experience

Striving for positive and memorable experiences must be fundamental to the marketing purpose of every tourism business (LaSalle and Britton 2003) because special experiences and unforgettable memories are what businesses need to achieve for sustainable competitiveness in a highly competitive tourism industry. Thus, a memorable tourism experience has the potential to provide tourism organizations with substantial competitive advantages and sustainability (Pizam 2010; Tung and Ritchie 2011; Chen and Rahman 2018) by developing destination loyalty (Chen and Rahman 2018a). Being able to deliver memorable tourism experiences creates distinctiveness and competitive advantage for tourism businesses (Morgan and Xu 2009; Kim et al. 2012). Memorable tourism experiences encourage revisit, loyalty, and positive word-of-mouth recommendations (Berry et al. 2002; Chen and Rahman 2018). For example, vivid recollection of experiences, spaces, encounters, and destinations are factors that contribute to future consumption (Marschall 2012).

Case study 27.1: Port wineries tours

Wine tourism, a sector of the tourism industry, is a composite of events that includes visits to vineyards, wineries, cellar doors, wine festivals and shows, and accommodation and wine tasting. These events have become a direct or indirect source of revenue for winery businesses. Given the competitive nature of the wine industry, wineries are expanding their offerings in order to attract visitors by developing lodging facilities and/or restaurants with the expectation of establishing a customer relationship and building a stronger brand image through the provision of outstanding service quality and the development of a memorable tourism experience. Port (Porto in Portuguese), the second-largest city in Portugal, is well-known for hosting major wine tourism activities and events because it represents the bastion of wine production in the Douro valley. Because of the reputation of Porto, wine-producing there bears the label of Port wine (Porto means Port in English). Taylor's Vintage Port 1985, Warre's Vintage Port 2000, Quinta Do Noval Nacional Vintage Port 1985, and Graham's Vintage Port 2003 are the most popular Port wineries. In search of building long-lasting relationships with consumers and visitors, these Port wineries often organize cellar door visits as well as wine tasting and showcasing of the winery's products.

Given the growing interest of Port wineries, Fernandes and Cruz (2016) conducted research that aimed at developing and testing a higher-order model of quality experience Port wine cellars, which highlighted relevant dimensions and outcomes. The authors have chosen visitors of three Port wineries that offer extensive cellar door service as the population of their study. The authors acknowledge wine tourism as being a vast area of interest for tourism in Portugal. Port wine cellar visits are particularly significant within the Northern region, in the region of the

Douro Valley. This region has become a popular worldwide wine tourism destination, with several accommodation facilities, vineyards, and wineries to attract visitors, and it has an international reputation for its wines bearing the label of Port wine.

Fernandes and Cruz developed a questionnaire survey in which customers' demographics, experience evaluation, and subsequent behavioral intentions were evaluated. The authors randomly collected data from visitors of three Port wine cellars in the Douro Valley. They retrieved a total of 290 usable responses. The authors surveyed the respondents before the final wine tasting but just after the visit to the cellars was completed, to minimize bias associated with retrospective recall. They collected the data over a two-week period during the peak season, with the assistance of cellar staff to whom they gave explicit instructions on how to brief the respondents without revealing the specific objectives of their research. Besides demographic data, the questionnaire developed by Fernandes and Cruz comprised 30 questions regarding experience quality dimensions, satisfaction, advocacy, and loyalty intentions. The authors found the majority of the respondents (51%) were male, aged between 20 and 40 years old (59%), mainly French (22%) or Spanish (20%) tourists, and who had never visited a Port wine cellar before (69%). They duly noted that they divided the respondents equally among the three cellars (with approximately one third of the respondents each). Consistent with observations mentioned earlier in this chapter, Fernandes and Cruz found that a pleasant tourist experience has a positive impact on satisfaction and intention to revisit and to recommend.

In addition to the results of Fernandes and Cruz (2016) regarding the Port wine cellars, Porto welcomes 1.6 million tourists annually, the equivalent of eight visitors per local according to Xinhua News (2018). While visitors are expected to visit the Port wine cellars, they could also possibly discover places such as: 1) Ribeira, highly frequented by locals and tourists as a relaxation venue, where people can enjoy views of the estuary and also have an enjoyable time at seafood restaurants and taverns; 2) Clerigos Tower, the 18th-century tower of Church of the Clergy, a Porto landmark and a can't-miss attraction; and 3) Livraria Lello, with an inventory of more than 60,000 books. An exuberant mix of Art Deco and Gothic Revival architectural styles makes Livraria Lello one of the most famous bookstores in Europe. People believe that Livraria Lello inspired author J.K. Rowling to write the Harry Potter books.

Conclusion

The multi-faceted nature of the tourism experience makes it impossible to have a pre-established list of dimensions that satisfy the interests of each visitor. Also, the differences in people's backgrounds, beliefs, attitudes, and values determine how they interact with their environment, which causes variations in tourism experiences among individuals (Knutson et al. 2007). Therefore, managers of tourism destinations must have a holistic view of the process of creating a memorable tourism experience. For example, education, aesthetics, escapism, entertainment (Stamboulis and Skayannis 2003), environment, benefit, accessibility, convenience, utility, incentive, and trust (Knutson et al. 2007; Fernandes and Cruz 2016) are other dimensions that may facilitate memorable tourism experiences. The fundamental remit of managers within the realm of the tourism industry is to facilitate the development of the destination artifacts that help visitors to develop their own memorable experiences (Tung and Ritchie 2011). In other words, managers should create circumstances and a design service environment in which tourism experience is formed (Slak Valek 2015, 2018; Zatoria et al. 2018). However, it may be challenging for managers to develop a suitable set of activities that can bring memorable experiences to a particular tourism business because needs, experiences, and interests among visitors are not similar. Given this challenge, managers can tap into the process of co-creation in experiences for better memorable experience development. In the context of tourism, co-creation is defined as the active participation and interaction of visitors/tourists during an experience (Bertella 2014; Tan et al. 2014). To a greater extent, active participation and interaction facilitate the creation of more memorable experiences (Kuhl and Chun 2014). It is essential to know that co-creation of experience is quite relevant to the process of "customization", which serves the purpose of adequately meeting visitors' expectations. In particular, the process of customization is directed by the visitor and may be implemented when they are emotionally, mentally, and physically ready (Binkhorst and Dekker 2009). Using the process of customization, managers are able to meet visitors' expectations which in turn may lead to enjoyable, meaningful, and memorable outcomes

References

Berry, L., Carbone, P., and Haeckel, H. (2002). Managing the total customer experience. *MIT Sloan Management Review*, 43(3): 85–89.

Bertella, G. (2014). The co-creation of animal-based tourism experience. *Tourism Recreation Research*, 39 (1): 115–125. DOI: 10.1080/02508281.2014.11081330.

Binkhorst, E., and Dekker, T. D. (2009). Agenda for co-creation tourism experience research. *Journal of Hospitality Marketing & Management*, 18(2): 311–327. DOI: 10.1080/19368620802594193.

Campos, A. C., Mendes, J., do Valle, P. O., and Scott, N. (2016). Co-creation experiences: Attention and memorability. *Journal of Travel & Tourism Marketing*, 33(9): 1309–1336. DOI: 10.1080/10548408.2015.1118424.

Chen, H., and Rahman, I. (2018). Cultural tourism: An analysis of engagement, cultural contact, memorable tourism experience and destination loyalty. *Tourism Management Perspectives*, 26: 153–163. DOI: 10.1016/j.tmp.2017.10.006

Decrop, A., and Snelders, D. (2004). Planning the summer vacation: An adaptable process. *Annals of Tourism Research*, 31(4): 1008–1030. DOI: 10.1016/j.annals.2004.03.004.

Dixit, S. K. (Ed.). (2017). *The Routledge handbook of consumer behaviour in hospitality and tourism*. London: Routledge. DOI: 10.4324/9781315659657.

Fernandes, T., and Cruz, M. (2016). Dimensions and outcomes of experience quality in tourism: The case of Port wine cellars. *Journal of Retailing and Consumer Services*, 31: 371–379. DOI: 10.1016/j.jretconser.2016.05.002.

Fotiadis, A., Mombeuil, C., and Valek, N. S. (2018). Designing and implementing a marketing plan. In Sotiriadis, M. (Ed.), *The Emerald handbook of entrepreneurship in tourism, travel and hospitality: Skills for successful ventures* (pp. 297–311). Bingley, UK: Emerald Publishing Limited.

Fotiadis, A., and Vassiliadis, C. (2010). Rural tourism service quality in Greece. *e-Review of Tourism Research (eRTR)*, 8(4): 69–84.

Fotiadis, A., and Vassiliadis, C. (2016). Service quality at theme parks. *Journal of Quality Assurance in Hospitality & Tourism*, 17(2): 178–190. DOI: 10.1080/1528008X.2015.1115247.

Gilmore, J. H., and Pine, B. J. (2002). Customer experience places: The new offering frontier. *Strategy & Leadership*, 30(4): 4–11. DOI: 10.1108/10878570210435306.

Grönroos, C. (1984). A service quality model and its marketing implications. *European Journal of Marketing*, 18(4): 36–44. DOI: 10.1108/EUM0000000004784.

Grönroos, C. (1990). *Service management and marketing: Managing the moments of truth in service competition*. Massachusetts: Lexington Books.

Kim, J. H. (2014). The antecedents of memorable tourism experiences: The development of a scale to measure the destination attributes associated with memorable experiences. *Tourism Management*, 44: 34–45. DOI: 10.1016/j.tourman.2014.02.007

Kim, J. H., and Ritchie, J. B. (2014). Cross-cultural validation of a memorable tourism experience scale (MTES). *Journal of Travel Research*, 53(3): 323–335. DOI: 10.1177/0047287513496468.

Kim, J.-H., Ritchie, J. R. B., and McCormick, B. (2010). Development of a scale to measure memorable tourism experiences. *Journal of Travel Research*, 51(1), 12–25. DOI: 10.1177/0047287510385467

Kim, J. H., Ritchie, J. R. B., and McCormick, B. (2012). Development of a scale to measure memorable tourism experiences. *Journal of Travel Research*, 51(1): 12–25. DOI: 10.1177/0047287510385467.

Klaus, P., and Maklan, S. (2013). Towards a better measure of customer experience. *International Journal of Market Research*, 55(2): 227–246. DOI: 10.2501/IJMR-2013-021.

Knutson, B. J., Beck, J. A., Kim, S. H., and Cha, J. (2007). Identifying the dimensions of the experience construct. *Journal of Hospitality & Leisure Marketing*, 15(3): 31–47. DOI: 10.1300/J150v15n03_03.

Kotler, P. (1997). *Marketing management: Analysis, planning, implementation, and control* (9th ed.). Upper Saddle River: Prentice Hall.

Kuhl, B. A., and Chun, M. M. (2014). Memory and attention. In C. C. Nobre, and S. Kastner (Eds.), *The Oxford Handbook of Attention* (pp. 806–836). Oxford: Oxford University Press.

LaSalle, D., and Britton, T. A. (2003). *Priceless: Turning ordinary products into extraordinary experiences.* Boston: Harvard Business School Press.

Lehtinen, U., and Lehtinen, J. R. (1982). *Service quality: A study of quality dimensions.* Helskinki: Service Management Institute.

Lončarić, D., Dlačić, J., and Prodan, M. P. (2018). What makes summer vacation experience memorable? An empirical study from Croatia. *Zbornik Veleučilišta u Rijeci*, 18(1): 67–80. DOI: 10.31784/zvr.6.1.2.

Lovelock, C., and Wirtz, J. (2011). *Services marketing.* Upper Saddle River, NJ: Pearson.

Ma, J., Gao, J., Scott, N., and Ding, P. (2013). Customer delight from theme park experiences: The antecedents of delight based on cognitive appraisal theory. *Annals of Tourism Research*, 42: 359–381. DOI: 10.1016/j.annals.2013.02.018

Marschall, S. (2012). Personal memory tourism' and a wider exploration of the tourism−memory nexus. *Journal of Tourism and Cultural Change*, 10(4): 321–335. DOI: 10.1080/14766825.2012.742094.

Mombeuil, C. (2018). Interaction from tourism development in Port-au-Prince, Haiti. [MPRA Paper]. *Journal of Tourism, Heritage & Services Marketing*, 4(1): 8–14. DOI: 10.5281/zenodo.1247526.

Morgan, M., and Xu, F. (2009). Student travel experiences: Memories and dreams. *Journal of Hospitality Marketing & Management*, 18(2–3): 216–236. DOI: 10.1080/19368620802591967.

Neuhofer, B., Buhalis, D., and Ladkin, A. (2012). Conceptualising technology enhanced destination experiences. *Journal of Destination Marketing and Management*, 1(1–2): 36–46. DOI: 10.1016/j.jdmm.2012.08.001.

Oriadea, A., and Schofield, P. (2019). An examination of the role of service quality and perceived value in visitorattraction experience. *Journal of Destination Marketing & Management*, 11: 1–9. DOI: 10.1016/j.jdmm.2018.10.002

Parasuraman, A., Zeithaml, V. A., and Berry, L. L. (1988). Servqual. *Journal of Retailing*, 64(1): 12–40.

Pizam, A. (2010). Creating memorable experiences. *International Journal of Hospitality Management*, 29(3): 343. DOI: 10.1016/j.ijhm.2010.04.003.

Sasser, W. E., Olsen, R. P., and Wyckoff, D. D. (1978). *Management of service operations: Text, cases, and readings.* Boston: Allyn and Bacon.

Scott, N., Laws, E., and Boksberger, P. (2009). The marketing of hospitality and leisure experiences. *Journal of Hospitality Marketing & Management*, 18(2-3): 99–110. DOI: 10.1080/19368620802590126.

Slak Valek, N. (2015). Tourism expenditure according to mode of transportation: A comparative study between 2009 and 2012. *Journal of Tourism, Heritage & Services Marketing*, 1(1): 3–9. DOI: 10.5281/zenodo.376250.

Slak Valek, N. (2018). Word-of-art: Contribution of artists-in-residence to a creative tourism destination. *Journal of Tourism and Cultural Change*, 1–15. DOI: 10.1080/14766825.2018.1467920.

Stamboulis, Y., and Skayannis, P. (2003). Innovation strategies and technology for experience-based tourism. *Tourism Management*, 24(1): 35–43. DOI: 10.1016/S0261-5177(02)00047-X.

Swarbrooke, J. (2001). Key challenges for visitor attraction managers in the UK. [journal article]. *Journal of Retail and Leisure Property*, 1(4): 318–336. DOI: 10.1057/palgrave.rlp.5090130.

Tan, S.-K., Luh, D.-B., and Kung, S.-F. (2014). A taxonomy of creative tourists in creative tourism. *Tourism Management*, 42: 248–259. DOI: 10.1016/j.tourman.2013.11.008

Tian-Cole, S., Crompton, J. L., and Willson, V. L. (2002). An empirical investigation of the relationships between service quality, satisfaction and behavioral intentions among visitors to a wildlife refuge. *Journal of Leisure Research*, 34(1): 1–24. DOI: 10.1080/00222216.2002.11949957.

Triantafillidou, A., and Petala, Z. (2016). The role of sea-based adventure experiences in tourists' satisfaction and behavioral intentions. *Journal of Travel & Tourism Marketing*, 33: 67–87. DOI: 10.1080/10548408.2015.1008667

Tung, V. W. S., and Ritchie, J. B. (2011). Exploring the essence of memorable tourism experiences. *Annals of Tourism Research*, 38(4): 1367–1386. DOI: 10.1016/j.annals.2011.03.009.

Xinhua News. (2018). *Lisbon, Porto have more tourists per local than Barcelona, Prague: Report.* www.xinhuanet.com/english/2018-04/05/c_137088807.htm.

Yarimoglu, E. K. (2014). A review on dimensions of service quality models. *Journal of Marketing Management*, 2(2): 79–93.

Zatoria, A., Smith, M. K., and Puczko, L. (2018). Experience-involvement, memorability and authenticity: The serviceprovider's effect on tourist experience. *Tourism Management*, 67: 111–126. DOI: 10.1016/j.tourman.2017.12.013

Zeithaml, V. A., and Bitner, M. J. (2003). *Services marketing: Integrating customer focus across the firm* (3rd ed.). New York: Irwin McGraw-Hill.

28

Marketing gastronomic tourism experiences

Saurabh Kumar Dixit

Introduction

The onset of the experience economy has resulted in differentiated patterns of consumption, especially in hospitality and tourism. Tourists now travel more in search of novel experiences, validation of their personality and realization of their self-worth. Tourism contributes significantly to the contemporary experience economy, in which gastronomic delicacies/attractions/events play a pivotal role in tourists' decision-making. The food, besides satisfying hunger and promoting growth and energy to the body, also enhances friendliness and social warmth (Dixit 2019a). It acts as a stimulus in performing rituals and advancing hospitality. In hospitality, the pleasure from food is not only based on the taste and flavor but it also extends to sharing pleasant meal experiences with others (Dixit 2019b). Long (2004) defined gastronomic tourism as participation, including but not limited to the consumption and presentation of food or cuisine in special or unique food experiences related to a tourist destination.

Food is therefore regarded as much more than physical sustenance; rather it epitomizes a major construct of global intangible cultural heritage for tourists. Travelers' experiences of local gastronomy and dining culture are acknowledged as destination experiences, as part of travelers' daily routines and as a major part of a traveler's travel expenditures (Hjalager and Corigliano 2000). The growing trend amongst tourists to seek authenticity and a novel narrative through the local gastronomy and dining pattern has become a significant factor for tourists looking to co-create an extraordinary experience at a destination. Local gastronomic offerings, dining experiences, tastings, cooking classes and many other forms and opportunities of experiencing local gastronomy are being gradually introduced as differentiation themes at global destinations. The bonds between gastronomy and tourism also provide avenues for local economic development, which can be further stimulated by the use of gastronomic experiences for branding and marketing destinations.

Travelers' gastronomic experiences are diverse and persuasive on many levels and they comprise varied characteristics that require careful attention. Björk and Kauppinen-Räisänen (2016) grouped gastronomic experiences into five key dimensions as *the food, the social dimension, place including the dimensions of the external environment, the service place* and *the time dimension.* Another attempt to discover the characteristics of gastronomic experiences is found in the study of Hansen et al. (2005), who focused on restaurant customers' experiences. They have found five dimensions of diners' experience: *the food, the interior of the restaurant, the interaction between the customer and employees/other customers, the company they were among* and *the restaurant atmosphere*. Hence, gastronomic tourism experiences not only assist in place branding, but also help to encourage sustainable tourism through protecting culture, empowering and nurturing pride amid societies, and augmenting intercultural understanding (Dixit 2019a). Gastronomic travelers regularly seek out gastronomic experiences that may involve attending a cooking class, relishing local delicacies, attending gastronomic events or shopping in supermarkets/local shops to take local condiments back home.

Gastronomic tourism experience marketing

Gastronomic tourism inspires tourists to visit food producers, food and wine festivals, specialty restaurants and specific destinations, where experiencing gastronomic delicacies is the primary motivating factor for travel. Food consumption and tourists' experience is manifested not merely as a basic need, but also in its various forms provides a number of benefits that augment the overall travel experience. As gastronomic tourism is gaining importance, food is now considered a key element in formulating destination-marketing strategies. It can therefore be argued that there is a huge potential for the development and augmentation of gastronomic tourism by embracing suitable marketing strategies to target and engage the wider market. Tourism destinations, therefore, have now started to integrate local gastronomy and tourism products to lure tourists and offer a plethora of gastronomic prospects to differentiate themselves from others.

Experiential marketing is a relatively new marketing orientation and provides a contrast to traditional marketing. Whereas traditional marketing frameworks view consumers as rational decision-makers focused on the functional features and benefits of products, experiential marketing views consumers as emotional beings, focused on achieving pleasurable experiences (Williams 2006). Experiential marketing for gastronomic tourism in the present scenario is quite complex because of the availability of the varied range of marketing tools to the consumers. Therefore, marketers need to devise marketing strategies targeting three kinds of consumers: i.e. *Traditional consumers*, who make their decision based on traditional marketing tools such as newspaper/television advertisements, leaflets and peer-to-peer discussion; *Digital consumers*, who rely on a mix of traditional and online marketing tools to make their decisions; and *Connected consumers*, also called generation Y, who explore the marketing information from the most advanced modes such as social media, mobile apps, smartphones and website marketing channels.

In an attempt to enhance the competitiveness of a destination, many destinations focus on building a favorable brand and a distinguishable gastronomic image and making it attractive for tourists (Henderson 2009). In recent years, image has been increasingly applied to gastronomy as an indicator of destination branding. Destination branding researchers have begun to investigate food and cuisine image as a discrete element independent of other image attributes. Berg and Sevón (2014: 289) assert that "food and gastronomy is directly and indirectly affecting the character of the place and its brand-image." As a marketing strategy, branding involves a process of drawing on and managing a product's elements to develop a positive image to attract and retain consumers, including effective strategies for marketing and positioning food as an appealing gastronomic image to target markets. Hong Kong, Singapore and Thailand are popular examples of highly revered gastronomic destinations and are regarded as a gastronomic heaven that gives tourists the opportunity to indulge in a wide variety of their authentic culinary delights.

In order to market gastronomy effectively, one needs to understand the origin of gastronomic experiences, tourism experience building through gastronomy, tourists' food consumption behavior, authenticity in gastronomic tourism experiences and issues relating to memorable gastronomic experiences. They are elaborated in detail in the following sections:

The origin of gastronomic experiences

The growing trend towards emphasizing the increasing role of experiential perspectives has stressed the debate on the origin of experience. Dube and Le Bel (2003) stress that gastronomic experiences initiate from the following four sources: sensory, emotional, social and intellectual (see Figure 28.1).

The sensory origins of gastronomic tourism experiences are rooted in taste, smell, touch, sight and hearing (López-Guzmán and Sánchez-Cañizares 2012). Maier et al. (2007) emphasized the impact of these experiences by concluding that gastronomic sensory experiences may be remembered and considered influential. Kauppinen-Räisänen et al. (2013) also supported this and found that travelers tend to recall in detail the tastes and flavors of their gastronomic experiences.

Holbrook and Hirschman (1982), who stress the stimulation of the senses, imagination and emotions, support the *emotional origin* of experience. Desmet and Schifferstein (2008) found that gastronomic experiences emerge from and are associated with positive and negative emotions. Therefore, experiences may arouse emotions like pleasure and provide

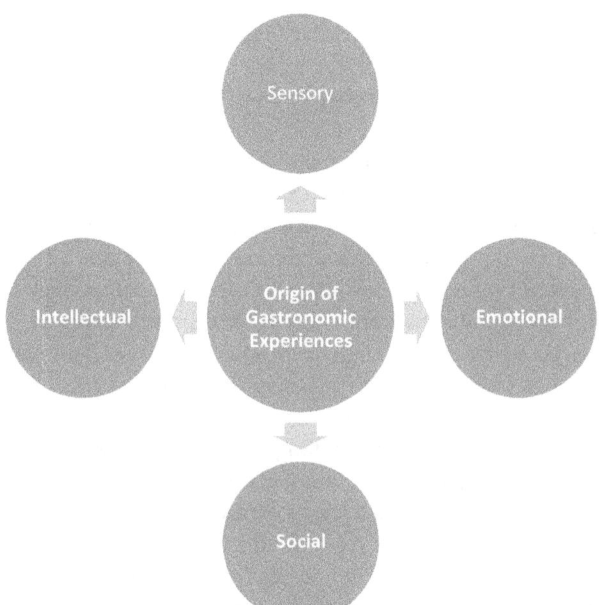

Figure 28.1 The origin of gastronomic experience

Source: Author based on Dube and Le Bel (2003)

hedonic and aesthetic benefits. The research conducted by Kauppinen-Räisänen et al. (2013) also validates that gastronomic experiences recollected from infancy were perceived as positive ones, while experiences evoked from previous vacations involved a spirit of audacity, i.e. the willingness to break out from one's comfort zone.

The *social origin* of experiences is associated with engagement with friends and family (Dube and Le Bel 2003), which is equally applicable to gastronomic experiences. Kauppinen-Räisänen et al. (2013) established that many recalled experiences included elements of symbiotic dining, which is associated with characteristics such as with whom, where and when these experiences were originated. Lupton (1994) too revealed that gastronomic experiences add to shared interactions and may clarify existing routines and habits. Furthermore, the influence of the social origin of experiences supports social functions and is associated with social pleasure and enhances food intake (Lupton 1994; Rozin et al. 2002).

Finally, the *intellectual origin* of gastronomic experiences is related to accomplishment (Dube and Le Bel 2003) and is perhaps more about food production or connoisseurship than routine gastronomic consumption. Kauppinen-Räisänen et al. (2013) support the impact of this experiential origin, as they established that experiences involving engagement and food production together lead to memorable experiences that are remembered and shared.

Tourism experience co-creation through gastronomy

Numerous studies have portrayed the importance of food, wine, and dining as a key contributor to tourist experience co-creation (Correia et al. 2007; Gross et al. 2008; Hall and Sharples 2003; Kivela and Crotts 2006). Dining on local delicacies is described as the most frequent leisure activity of travelers and represents the second largest item per head of daily expenditure (Hall and Sharples 2003).

Tourism destinations have now realized the importance of gastronomy in their marketing strategies, to achieve the competitive differentiation of their products (Dixit 2017). The core of this differentiation is based on a sense of place (constructed via local gastronomy) and represents the bundle of products and services that make up a tourist experience. This bundle concept has been described as "*touristic terroir*" (Hall et al. 2003: 34). The concept of *touristic terroir* for hospitality products and services has also been described in culinary terms as the gastronomic identity. In tasting the typical gastronomic products and dishes, and by acquiring the mentioned gastronomic experiences, one may expand one's knowledge of the destination visited (Dixit 2019a: 20). As a result, many destinations have put together gastronomic tourism initiatives to augment and endorse gastronomic tourism as part of their tourism marketing strategy.

Destination management organizations, specialized associations and national governments are striving hard to promote local gastronomic tourism experiences. Thai Fest (Thai Government), Restaurant Australia (Australian Government), Full of passion strategy (Croatian Government), Hawker Centers (Singapore Government), etc. are some global government efforts to bring gastronomy, tourism, culture and rural and agricultural sectors together. In addition to this, associations such as the World Food Travel Association, the Ontario Culinary Tourism Alliance, the World Gastronomy Institute, etc., are also attempting to position and market gastronomic tourism offerings.

The FoodTreX ("Food Travel Experience") event is a brand that was introduced by the World Food Travel Association in 2018 to increase the level of quality and innovation in trade events in this industry. FoodTreX Regional Summits bring discussion of food and beverage tourism issues and trends to stakeholders in local areas. So far, FoodTreX events have been held in the UK, Spain, Greece, Australia, Nepal and online. The World Tourism Organization (UNWTO) and the Basque Culinary Center (BCC) co-organize the UNWTO World Forum on Gastronomy Tourism every year with the aim of promoting the exchange of experiences between experts in tourism and gastronomy, in order to identify good practices and to promote gastronomy tourism as a factor in the development of countries.

Quest for authenticity in gastronomic tourism experiences

Food producers, chefs, foodies and travelers do have one thing in common: the quest for authenticity in "traditional" and "local" gastronomy. Gastronomy is often described as being authentic if it is specific to a region, is prepared in a certain way, makes use of traditional preparation methods, combines and/or recombines new with traditional ingredients and is made by a specific person. Authenticity has now become the common and most important baseline for developing and sustaining travelers' food experiences. The construct of authenticity is also examined, as is its application to gastronomic tourism as a means of securing a competitive advantage in the contemporary market place. According to Wang (1999: 351), "the complex nature of authenticity in tourism is exhibited in the fact that it can be classified into objective, constructive, and existential authenticity".

Objective authenticity hangs on an act of (re-) production and it is the past that must be adopted as a model of the original. With regard to gastronomic tourism, objective authentic gastronomy would need to be made in a region, according to the traditional recipes and ingredients of that region. MacCannel (2004) has introduced *constructive authenticity*, more popularly known as "*staged authenticity*", into the discussion of authenticity in tourism. He argues that host communities do not allow travelers to live an authentic experience, as they protect their culture, their customs and traditions through a double identity: there exists a frontstage interpretation of their culture to satisfy the tourists' needs and expectations, and a backstage culture that is protected by and shared only among the members of the community. For gastronomy, staged authenticity can hardly be achieved, since it does not usually allow for double identity. However, the adjustment of original flavors and ingredients (downgrading spicing, substituting certain meats, etc.) to better suit the appreciations of tourists may fall into that category.

Existential authenticity is not centered on associations with an object but is based on the search for the "(inner-) self". Existential authenticity thus describes the way in which tourists, by participating in holiday activities, can construct their identity to experience a more authentic sense of self. It denotes a potential existential state of being that is initiated by tourism activities, "and can have nothing to do with the authenticity of toured objects" (Wang 1999). It has become an essential driver in shaping prospects for the construction of authentic gastronomic offerings in relation to travel.

According to Cohen (2002), contemporary tourists seek both *objective* and *existential* authenticity in their holidays because while some tourists are spending more, traveling further and experiencing more discomfort in order to experience encounters with "untouched" environments and cultures, others are happy to simply relax, have a good time and experience the existential authenticity that comes from "being themselves".

Tourist food consumption behavior

Gastronomic tourism can be described as tourism where an opportunity for memorable food and drink experiences contributes significantly to travel motivation and behavior (Wolf 2006). Destination-marketing organizations stress that there is nowadays a rising appeal for genuine gastronomic tourism experiences. In order to successfully market the gastronomic resources of a destination, the marketers need to identify gastronomic motivation, involvement and place attachment and interpret consumers' ongoing gastronomic behavior. The gastronomic delicacies/events constitute lived experiences that, when stored in consumers' memories, may become the key issue for understanding consumers' consumption behavior, and can be drawn upon to meet their needs for future experiences as well.

The researches have also shown that gastronomic experiences affect travel satisfaction (Björk and Kauppinen-Räisänen 2014, 2017b), holiday well-being (Björk and Kauppinen-Räisänen 2017a) and even intentions to revisit a destination (Kim et al. 2011; Quan and Wang 2004). An engaging gastronomic experience can potentially lead to much more than satisfaction and can extend itself into the realm of loyalty. Hence, an experiential analysis of consumers' behavior appreciates hedonism and aesthetics and the involvement of fantasy, feelings and fun (Holbrook and Hirschman 1982). Knowledge regarding gastronomic tourists' behavior will allow gastronomic tourism stakeholders to effectively target and develop markets, thereby intervening in the decision-making process and persuading the consumer to purchase local and regional food products and services (Mitchell and Hall 2003).

Motivation has been recognized as a multi-dimensional construct exerting significant influence on tourist food consumption behavior (bMak et al. 2012b). Motivations for food consumption can be seen as an oscillation between fulfilling the "obligatory" and the "symbolic" facets in the encounter of food in foreign destinations. The "obligatory" facet reflects the essentiality of food consumption in tourism, and the "symbolic" facet signifies the symbolic meanings of food consumption to tourists, such as acquiring cultural capital, exploring the local culture and seeking an authentic experience (Mak et al. 2012a).

McIntosh et al. (1995) indicated four categories of motivation for gastronomic tourism, i.e. *physical motivators*, *cultural motivators*, *interpersonal motivators* and *status and prestige motivators*. They are considered as basic driving forces to make food and drink appealing, which lead to the inclusion of food and drink as part of the tourism experience. Boniface (2003) addressed the motivations of gastronomic tourism by summarizing them into five different dimensions: (i) anxiety – pleasure through food and drink on a vacation and to escape through traditional and non-routine foods with nostalgia; (ii) the need to show distinction, affluence, and individualism; (iii) curiosity and the wish for acquaintance and discovery to display education and superiority to others; (iv) a desire to feel grounded amid globalization – to visit one's roots and the countryside and to see how food is derived; (v) a desire for sensory and tactile pleasure and for experience.

A destination's gastronomic appeal may lure those tourists who reconnoiter for extraordinary dining experiences and those who search for authenticity and locality. These are labeled as *gastronomic tourists*. Therefore, for these tourists eating local delicacies in an authentic dining environment enables them to gain pleasure, entertainment and fuel the senses and provides a means to understand local philosophies, customs and traditions. For such tourists, gastronomy goes beyond being a daily routine, but rather it leads to emotional motivation and is a leading element of travelers' holistic travel experiences. The *second category* of travelers comprises those who view gastronomic experiences as vital but not significant in deciding the destination. They wish to enjoy gastronomic experiences, but instead of having a push effect for deciding a destination gastronomic experiences have a pull effect (Pesonen et al. 2011). Therefore, these travelers take a more casual attitude towards gastronomic appeal, although it is perceived as a pleasurable pastime activity (Henderson 2009). According to Tikkanen (2007), the *third category* of travelers is those who consider food a requirement only to fulfill physiological needs.

Other researches have recognized various other kinds of travelers due to their commitment to gastronomy and dining experiences (Hjalager 2004). Hjalager (2004) has put forward four types of travelers who vary in their food-related lifestyle and on-site food-related behavior (existential, experimental, diversionary and recreational gastronomy travelers). *Existential* gastronomy travelers aim at learning about the cuisine, having low requirements on a sophistication level, in contrast to *experimental* gastronomy travelers, who specifically search for sophisticated, peak experiences. Familiarity is preferred by both *diversionary* and *recreational* gastronomy travelers, the former being less committed and passive while the latter value local dining and actively search for such experiences.

Memorable gastronomic experience

Kim et al. (2012) define a memorable tourism experience (MTE) as a tourism experience that is positively remembered and recalled after the event has occurred. So if we explore a destination, our pleasant unique experiences are usually retained in our memory, which guides us in our future decision-making. Therefore, the service operators that are able to render memorable experiences usually find themselves differentiated from their competitors. Memorable experiences epitomize a novel benchmark that destination managers and tourism businesses strive to deliver (Kim et al. 2012), and they are central to becoming competitive in the marketplace (Kim and Ritchie 2014). Tourists usually rely more on their authentic experiences in decision-making than being convinced by any other marketing efforts. Therefore, in the contemporary hospitality business scenario there is an increasing shift amongst the service providers to deliver services to create memorable experiences.

Relishing local foods during a trip is a mark of an authentic experience that most visitors aspire to engage in. Studies indicate that food consumption can be categorized into three categories: the *individual, the food*, and *the environment* (Meiselman et al. 1999). The *food* deals with sensory features such as flavor, aroma, texture and appearance; the *environment* presents cultural, social, economic and physical influences. Individual, sociocultural, psychological and physiological factors are recognized as exerting direct or indirect effects on food consumption behavior. Quan and Wang (2004) proposed a travel experience model to analyze the role of gastronomy in their travel experiences. They also concluded that the levels of memorability govern whether tourists' dining behavior during travel becomes a peak tourist experience rather than a supporting consumer experience – hence, if the dining behavior is a peak tourist experience, it leads to an unforgettable memory. In addition, the consumption of local food provokes specific emotional responses, including enjoyment, sensory stimulation and fulfillment (Mak et al. 2012b).

Sthapit (2017) has proposed a model of tourists' memorable gastronomic experience based on the review of the existing literature. The results of the analysis based on grounded theory approach identified a number of key components such as *local specialities and food attributes (taste), authenticity, novelty, togetherness and social interaction, hospitality*, and *servicescape (including food souvenirs)* (as illustrated in Figure 28.2). A closer look at the components reveals that tourists enjoy eating local food and search for locality, newness and authenticity when consuming food at a tourist destination. This study represents an initial underpinning, and it has developed an understanding of what contributes to tourists' memorability of local food experiences.

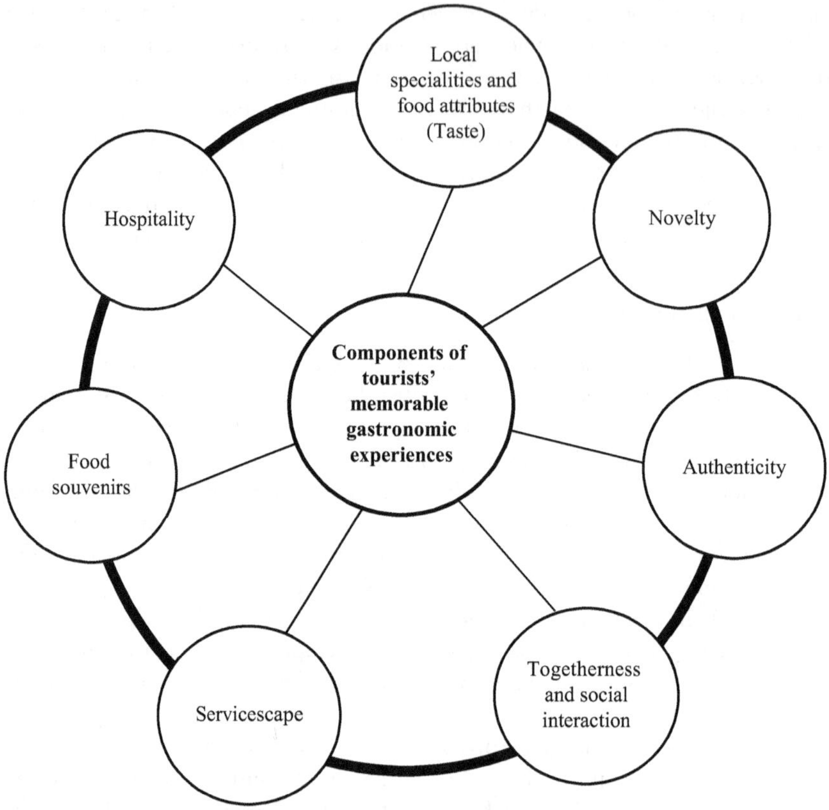

Figure 28.2 The conceptual framework of tourists' memorable tourist experiences
(*Source:* Sthapit, E. 2017)

Case study 28.1: Tea tourism in Meghalaya (India)

As an agricultural product with strong social and cultural connections, the production and consumption of tea has a high potential to be promoted as a means of persuading visitors to choose a related travel experience. Tea is used widely in India as a consumable product and has been changed and adapted in tourism. The social changes within society with regard to tea are widespread and today are experienced as tourism. The available literature has acknowledged evident linkages between multiple components of the tea industry and those of the tourism sector. When incorporating tea production and related cultural practices and gastronomic experiences, tea tourism can supply a variety of income generating opportunities and is a development tool that can be used to promote the tea industry, revitalize local culture and reinforce the local identity of rural areas (Jolliffe and Aslam 2009).

Although its fertile, acidic soil and abundant rainfall attracted and motivated potential planters, tea cultivation in Meghalaya is still in its infancy. It was a visit in 1974 by Tea Board of India scientists that marked the beginnings of organized tea cultivation in the state. Inresponse to the Tea Board of India report nurseries were established from 1977 onwards, and suitable varieties were brought in from neighboring Assam and Darjeeling. The seedlings were subsequently transplanted in experimental gardens at Umsning (in Ri Bhoi District), Tebronggre (in the West Garo Hills) and also in Riangdo (in the West Khasi Hills) in 1978. Meghalaya tea cultivation has spread to all corners of the state, and the teas it produces are rapidly earning a reputation in India and abroad as superior in quality, rich, fragrant in aroma and powerful in liquor. The prominent tea brands of Meghalaya are presented in the following table (see Table 28.1):

Table 28.1 Prominent Tea Brands of Meghalaya

SN	*Tea Brands*	*Produced at*
1.	Meghalaya Tea (CTC)	Arenga Tea Industries, West Garo Hills District
2.	Anderson Tea (CTC and Orthodox)	Anderson Tea Co., Ri Bhoi District
3.	Slalung Tea (CTC)	Slalung Tea Enterprise, Ri Bhoi District
4.	Lakyrsiew Tea (Premium Orthodox Tea)	Tara Tea Estate, East Khasi Hills District
5.	Durama Tea (CTC)	Durama Tea Factory, West Garo Hills District
6.	Urlong Tea (Orthodox Tea)	Mawlyngot, East Khasi Hills District

Meghalaya's Department of Agriculture had launched a brand of certified organic tea under the name "MEG" produced from its stateowned tea farms at Umsning and Upper Shillong (see Figure 28.3).

Visiting tea estates and sale outlets (see also Figures 28.4 and 28.5) has enormous tourism potential if carefully promoted in the region. Many tea estate owners of Meghalaya (such as Anderson Tea Co. in Ri Bhoi District) and Assam are developing cottages to lure tourists to experience the tea tourism of Meghalaya. In addition, the GTAC Tea Lounge has been set up at the premises of the Guwahati Tea Auction Centre in the neighboring state of Assam. The idea behind setting up this Tea Lounge is to promote the tea produced in the North Eastern region of India to the tourists who wish to relish, purchase and explore the variety of teas produced in the region.

Figure 28.3 Tea product range of Anderson Tea, Meghalaya
(*Source:* Saurabh Kumar Dixit)

Figure 28.4 Tea product range of Urlong Tea, Meghalaya
(*Source:* Saurabh Kumar Dixit)

Figure 28.5 Certified organic tea "MEG"
(*Source:* Saurabh Kumar Dixit)

Conclusion

There has been a rising interest in gastronomic tourism within the social sciences, exploring the impact of food and drink on the tourist experience. Gastronomy is getting its due prominence in contemporary tourists' travel decision-making. Tourists' motivations and cravings for local gastronomy as a vacation activity are centered around novelty, change, pleasure, something special and something different – escaping the mundane and attaining acquaintance and cultural capital through consuming traditional and local foods. Local gastronomic delights and cooking/dining culture and events usually leave eternal lived experiences in the memory of the visitors. Local gastronomy also imparts significant marketing potential to the destination by means of branding the destination through the authentic image of the local foods/cuisine. Previous researches have noted the impact of gastronomy on the destination choice and on-site experiences and also recommend that the discourse of a holistic gastronomic tourist experience on a destination has a spillover effect (Swanson 1994). By examining travelers' gastronomic experiences, the chapter not only encourages exploration associated with destinations' strategic marketing efforts, but also endorses the potential of applying an experiential perspective to understanding gastronomic travelers.

References

Berg, P. O., and Sevón, G. (2014). Food-branding places – A sensory perspective. *Place Branding and Public Diplomacy*, 10(S4): 289–304. DOI: 10.1057/pb.2014.29.

Björk, P., and Kauppinen-Räisänen, H. (2014). Culinary tourism: A search for local food experiences. *Food Sciences and Nutrition*, 44(4): 294–309. DOI: 10.1108/NFS-12-2013-0142.

Björk, P., and Kauppinen-Räisänen, H. (2016). Exploring the multidimensionality of travelers' culinary-gastronomic experiences. *Current Issues in Tourism*, 19(12): 1260–1280. DOI: 10.1080/13683500.2013.868412.

Björk, P., and Kauppinen-Räisänen, H. (2017a). A destination's gastronomy for holiday wellbeing. *British Food Journal*, 119(7): 1578–1591. DOI: 10.1108/BFJ-09-2016-0394.

Björk, P., and Kauppinen-Räisänen, H. (2017b). Interested in eating and drinking? How food affects travel satisfaction and the overall holiday experience. *Scandinavian Journal of Hospitality and Tourism*, 17 (1): 9–26. DOI: 10.1080/15022250.2016.1215871.

Boniface, P. (2003). *Tasting tourism: Travelling for food and drink*. Burlington, NJ: Ashgate. DOI: 10.4324/9781315241777.

Cohen, E. (2002). Authenticity, equity and sustainability in tourism. *Journal of Sustainable Tourism*, 10: 267–276. DOI: 10.1080/09669580208667167.

Correia, A., Oom do Valle, P., and Moço, C. (2007). Why people travel to exotic places. *International Journal of Culture, Tourism and Hospitality Research*, 1(1): 45–61. DOI: 10.1108/17506180710729600.

Desmet, P. M. A., and Schifferstein, H. N. J. (2008). Sources of positive and negative emotions in food experience. *Appetite*, 50(2–3): 290–301.

Dixit, S. K (2019a). Gastronomic tourism: A theoretical construct. In Dixit, S. K. (Ed.), *The Routledge handbook of gastronomic tourism* (pp. 13–23). London: Routledge. DOI: 10.4324/9781315147628.

Dixit, S. K. (2019b). Management and marketing of wine tourism business: Theory, practice, and cases. *Information Technology and Tourism*, 21(1): 135–137. DOI: 10.1007/s40558-018-0137-4.

Dixit, S. K. (2017). Introduction. In Dixit, S. K. (Ed.), *The Routledge handbook of consumer behaviour in hospitality and tourism* (pp. 1–3). London: Routledge. DOI: 10.4324/9781315659657.

Dube, L., and Le Bel, J. (2003). The content and structure of laypeople's concept of pleasure. *Cognition and Emotion*, 17(2): 263–295. DOI: 10.1080/02699930302295.

Gross, M. J., Brien, C., and Brown, G. (2008). Examining the dimensions of a lifestyle tourism destination. *International Journal of Culture, Tourism and Hospitality Research*, 2(1): 44–66. DOI: 10.1108/17506180810856130.

Hall, C. M., Mitchell, R., and Sharples, L. (2003). Consuming places: The role of food, wine and tourism in regional development. In C. M. Hall, L. Sharples, R. Mitchell, N. Macionis, and B. Cambourne (Eds.), *Food tourism around the world* (pp. 25–59). Oxford: Elsevier Butterworth-Heinemann. DOI: 10.1016/B978-0-7506-5503-3.50005-1.

Hall, M., and Sharples, L. (2003). The consumption of experiences or the experience of consumption? An introduction to the tourism of taste. In M. Hall, L. Sharples, R. Mitchell, N. Macionis, and B. Cambourne (Eds.), *Food tourism around the world: Development, management and markets*. Oxford: Butterworth-Heinemann.

Hansen, K. V., Jensen, Ø., and Gustafsson, I.-B. (2005). The meal experiences of á la carte restaurant customers. *Scandinavian Journal of Hospitality and Tourism*, 5(2): 135–151. DOI: 10.1080/15022250510014417.

Henderson, J. C. (2009). Food tourism reviewed. *British Food Journal*, 111(4): 317–326. DOI: 10.1108/00070700910951470.

Hjalager, A. (2004). What do tourists eat and why? Towards sociology of gastronomy and tourism. *Journal of Tourism*, 52(2): 195–201.

Hjalager, A., and Corigliano, M. (2000). *Tourism and gastronomy*. London: Routledge.

Holbrook, M. B., and Hirschman, E. C. (1982). The experiential aspects of consumption: Consumer fantasies, feelings, and fun. *Journal of Consumer Research*, 9(2): 132–140. DOI: 10.1086/208906.

Jolliffe, L., and Aslam, M. S. M. (2009). Tea heritage tourism: Evidence from Sri Lanka. *Journal of Heritage Tourism*, 4(4): 331–344. DOI: 10.1080/17438730903186607.

Kauppinen-Räisänen, H., Gummerus, J., and Lehtola, K. (2013). Remembered eating experiences described by self, food, place, time, and scape. *British Food Journal*, 115(5): 666–685. DOI: 10.1108/00070701311331571.

Kim, J., and Ritchie, J. (2014). Cross-cultural validation of a memorable tourism experience scale (MTES). *Journal of Travel Research*, 53: 323–335. DOI: 10.1177/0047287513496468.

Kim, J.-H., Ritchie, J., and McCormick, B. (2012). Development of a scale to measure memorable tourism experiences. *Journal of Travel Research*, 51(1): 12–25. DOI: 10.1177/0047287510385467.

Kim, Y. H., Kim, M., Goh, K. B., and Antun, M. J. (2011). The role of money: The impact on food tourists' satisfaction and intention to revisit food events. *Journal of Culinary Science & Technology*, 9(2): 85–98. DOI: 10.1080/15428052.2011.580708.

Kivela, J., and Crotts, J. C. (2006). Tourism and gastronomy: Gastronomy's influence in how tourists experience a destination. *Journal of Hospitality & Tourism Research*, 30(3): 354–377. DOI: 10.1177/1096348006286797.

López-Guzmán, T., and Sánchez-Cañizares, S. (2012). Culinary tourism in Córdoba (Spain). *British Food Journal*, 114(2): 168–179. DOI: 10.1108/00070701211202368.

Long, L. M. (2004). *Culinary tourism*. Lexington: University Press of Kentucky. DOI: 10.7202/014992ar.

Lupton, D. (1994). Food, memory and meaning: The symbolic and social nature of food events. *Sociological Review*, 42(4): 665–685. DOI: 10.1111/j.1467-954X.1994.tb00105.x.

MacCannel, D. (2004). Staged authenticity: Arrangements of social space in tourist settings. In S. Williams (Ed.), *Tourism: The experience of tourism* (pp. 195–209). London: Routledge.

Maier, A., Chabanet, C., Schaal, B., Leathwood, P., and Issanchou, S. (2007). Food-related sensory experience from birth through weaning: Contrasted patterns in two nearby European regions. *Appetite*, 49(2): 429–440. DOI: 10.1016/j.appet.2007.02.007.

Mak, A. H. N., Lumbers, M., and Eves, A. (2012a). Globalization and food consumption in tourism. *Annals of Tourism Research*, 39(1): 171–196. DOI: 10.1016/j.annals.2011.05.010.

Mak, A. H. N., Lumbers, M., Eves, A., and Chang, R. C. Y. (2012b). Factors influencing tourist food consumption. *International Journal of Hospitality Management*, 31(3): 928–936. DOI: 10.1016/j.ijhm.2011.10.012.

McIntosh, R. W., Goeldner, C. R., and Ritchie, J. R. (1995). *Tourism: Principles, practices, philosophies* (7th ed.). New York: John Wiley.

Meiselman, H. L., Mastroianni, G., Buller, M., and Edwards, J. (1999). Longitudinal measurement of three eating behavior scales during a period of change. *Food Quality and Preference*, 10: 1–8. DOI: 10.1016/S0950-3293(98)00013-5.

Mitchell, M. A., and Hall, C. M. (2003). Consuming tourists: Food tourism consumer behaviour. In C. M. Hall, L. Sharples, R. Mitchell, N. Macionis, and B. Cambourne (Eds.), *Food tourism around the world* (pp. 60–80). Oxford: Butterworth-Heinemann.

Pesonen, P., Komppula, R., Kronenberg, C., and Peters, M. (2011). Understanding the relationship between push and pull motivations in rural tourism. *Tourism Review*, 66(3): 32–49. DOI: 10.1108/16605371111175311.

Quan, S., and Wang, N. (2004). Towards a structural model of the tourist experience: An illustration from food experiences in tourism. *Tourism Management*, 25: 297–305. DOI: 10.1016/S0261-5177(03)00130-4.

Rozin, P., Kurzer, N., and Cohen, A. B. (2002). Free associations to 'food': The effects of gender, generation, and culture. *Journal of Research in Personality*, 36(5): 419–441. DOI: 10.1016/S0092-6566(02)00002-8.

Sthapit, E. (2017). Exploring tourists' memorable food experiences: A study of visitors to Santa's official hometown. *Anatolia*, 28(3): 404–421. DOI: 10.1080/13032917.2017.1328607.

Swanson, K. K. (1994). *Souvenir marketing in tourism retailing: Shopper and retailer perceptions* (Dissertation). Texas Tech University, Lubbock, TX.

Tikkanen, I. (2007). Maslow's hierarchy and food tourism in Finland: Five cases. *British Food Journal*, 109 (9): 721–734. DOI: 10.1108/00070700710780698.
Wang, N. (1999). Rethinking authenticity in tourism experience. *Annals of Tourism Research*, 26(2): 349–370. DOI: 10.1016/S0160-7383(98)00103-0.
Williams, A. (2006). Tourism and hospitality marketing: Fantasy, feeling and fun. *International Journal of Contemporary Hospitality Management*, 18(6): 482–495. DOI: 10.1108/09596110610681520.
Wolf, E. (2006). *Culinary tourism: The hidden harvest.* Dubuque, IA: Kendall/Hunt Publishing Company.

29

Managing brand experience and reputation in the hotel business

Hanım Kader Şanlıöz–Özgen and Metin Kozak

Introduction

Brand experience and reputation management are two interrelated concepts developed in the late 1990s as an extension of new thinking around the distinct features of service concept and subjective aspects leading to the increasing value of experience-related issues. Businesses have thus discovered another fountain head to be utilized in the competition arena. Businesses have recognized, as an important part of their reputation, the key role of customers' subjective responses generated by experiences promised and delivered by specific brands (Figure 29.1).

In this regard, this chapter attempts to explicate these two interrelated concepts within the scope demonstrated in Figure 29.1. The first section deals with fundamental orientations, which have paved the way to the development of the brand experience concept and is followed by reflections on the hotel industry, which is described as the "business of brands". In pursuit of this, the concept of reputation and its management is clarified by recalling the role of social media (SM) and examining the ways in which hotel businesses implement this. Finally, the case of Hilton Worldwide presents how managing brand and reputation is applicable in the hotel industry, examining specific aspects in the light of concepts and relationships explained in previous sections.

Brand experience

The term brand experience stems from the emergence of the experience concept in marketing literature. This new thinking challenged the basic attributes of the service concept and underlined essential attributes such as intangibility and co-creation (Vargo and Lusch 2004), as well as the role of sensations, feelings and emotions (Pine and Gilmore 1999) and multi-dimensionality, including cognitive, sensual, affective, relational and behavioral (lifestyle) aspects (Schmitt 1999). This orientation has paved the way to conceptualize the brand on an experience-based construct, leading to the definition of brand experience as:

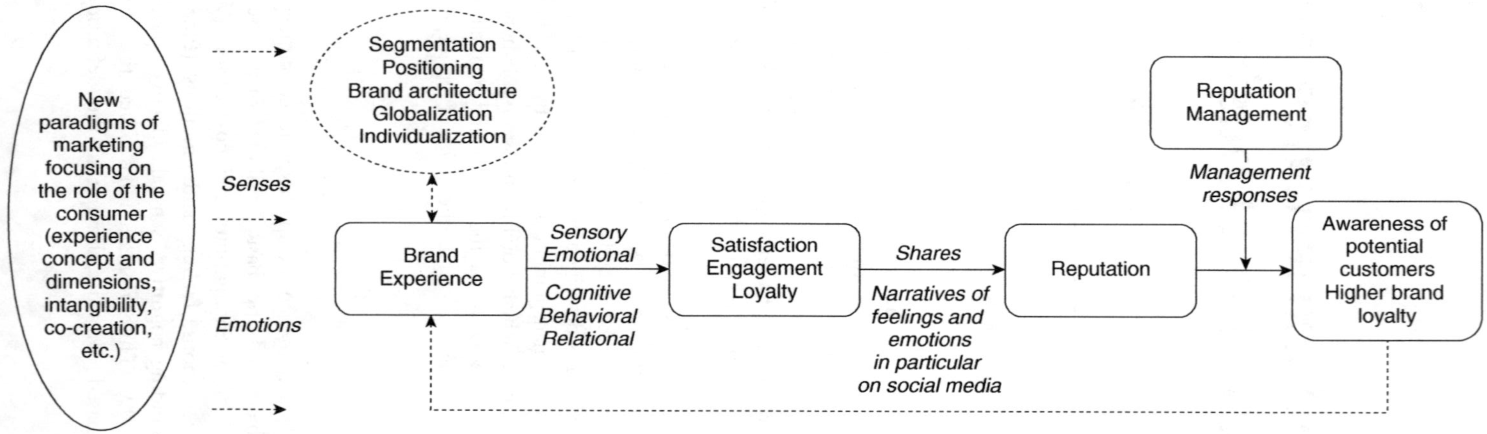

Figure 29.1 Brand Experience and Reputation Management – Relationships and Concepts

> subjective, internal consumer responses (sensations, feelings, and cognitions) and behavioral responses evoked by brand-related stimuli that are part of a brand's design and identity, packaging, communications, and environments.
>
> *(Brakus, Schmitt and Zarantonello 2009: 53)*

This definition highlights the potential and strength of brands to create customer experiences as a result of other drivers beyond utilitarian attributes (Brakus, Schmitt and Zarantonello 2009; Morrison and Crane 2007), leading to much positive impact for businesses, such as customer satisfaction, engagement and loyalty (Morrison and Crane 2007; Nysveen, Pedersen and Skard 2013; Santini et al. 2018; Zarantonello and Schmitt 2010). Furthermore, especially in the case of luxury brands, experiential aspects form the essence through various levels of involvement (participation) and intensity (feelings), together with personalization and interaction (Atwal and Williams 2009).

One critical aspect of the brand experience concept is its roots in various dimensions, namely sensory, emotional, cognitive, behavioral (Brakus, Schmitt and Zarantonello 2009) and relational (Nysveen, Pedersen and Skard 2013) experiences. These dimensions can be evoked in different ways by various brands (Brakus, Schmitt and Zarantonello 2009), leading to the creation of various groups with differing experiential appeals (Zarantonello and Schmitt 2010). Segmentation therefore becomes a crucial action to identify common features (i.e. values and needs) of target groups to communicate brand experiences in an optimal way to create a positive response (Morrison and Crane 2007).

Segmentation recalls the need for a clear understanding of brand architecture in terms of designing multiple brands within a brand portfolio around differing experiences promised to specific target groups. Brand architecture is an organization of structure to specify the scope, roles and interrelations of master brands (points of reference), sub-brands (connected to master brands) and endorsed brands (independent brands) included in a brand portfolio, so that creating a new brand or extending existing ones is relatively uncomplicated (Aaker 2004; Aaker and Joachimsthaler 2000). Moreover, businesses can clarify levels of awareness for each brand and improve brand image through brand extensions based on line (existing brand) or category (new brand) development (Keller 2013). In particular, the house of brands, a brand architecture strategy, supports positioning in regard to functional benefits and appealing to niche markets (Aaker and Joachimsthaler 2000). Thus, businesses can enlarge their market coverage by addressing different target groups regarding prices, channels, geographies and other features with multiple brands in their portfolios (Keller 2013).

Finally, globalization is also critical to the brand experience as businesses expand their operations on a wider geographic scope across national boundaries through international brand architecture (Douglas, Craig and Nijssen 2001). Global businesses seek a balance between the need for consistency around the world and for adaptation to local features so that their brands can form strong relationships across various cultures with a great brand experience, which is the key action for global success (Hollis 2010). For success, businesses need to consider the worldwide needs or interests of target markets (i.e. luxury products), the extent to which locality is applicable, and competitive market environment, including both global and local players (Douglas, Craig and Nijssen 2001).

Brand experience in the hotel business

Being a "business of brands" (Dev 2009: 1), the hotel business is highly fragmented and competitive, with no one dominant brand (Dev 2009). Although the introduction of hotel brands dates back to the 1800s, the most important period for brand development was the 1980s, with the emergence of brand extensions, new brands and brands with numerous tiers. This orientation highlights the hotel industry as a fruitful domain to develop brand categories aiming to meet various needs of customers through differentiated offerings in terms of distinct amenities and services at certain price levels (Dev, Buschman and Bowen 2010). Even the needs of a single customer may change according to their purpose of travel and hotel groups take this as an advantage, retaining the same individuals through a variety of brands in various cases (O'Neill and Mattila 2010). In particular, sub-brands under master brands are dedicated to differing attributes to create niche opportunities around differing levels of price and service parameters (Olsen et al. 2005). For example, Marriott International manages a brand portfolio including various brand categories (e.g. iconic luxury, luxury, lifestyle/collection) and classifies the hotels in this respect (the Ritz-Carlton, JW Marriott, Edition respectively). All these brands are formed as part of a formal brand architecture strategy, which is also announced on the corporate website to prospective guests (Keller 2013).

While establishing these brands at the expansion stage in various countries, groups need to choose on an axis somewhere between globalization (home-country orientation) and localization (host-country orientation) (Cunnill 2006). On the other hand, hotel groups may need to choose a point between a harder brand with strict consistency and a softer brand with nearly no consistency as they realize that the brand experience is so difficult to render consistent through various brands targeting individuals with differing needs (Connell 1992). This decision is also dependent on willingness to implement some established and globally known standards for more effective positioning (hard branding) or to limit a partnership by extending sales and marketing channels, keeping control of management (soft branding) (Holverson and Revaz 2006).

Another continuum affecting hotel brand strategies relates to decisions about standardization and individualization. In parallel with globalization, standardization is effective to implement quality management practices and interrelationships between hotels or departments whereas individualization is more prevalent for integration of feedback and managing complaints (Bouncken, Pick and Hipp 2008). This consideration also recalls the need for questioning consistency, which may, in some cases, be flexible according to the various needs of customers in both practical and emotional aspects, especially with service-related intangible attributes (Dev and Withiam 2012).

Once branding studies emerged in the 1990s in the tourism domain, the first definitions of the concept included the experience focus on pleasure and memorability (Kim 2008). Accordingly, the hotel industry was inevitably influenced by the emerging importance of experience as a new basis of all offerings (Pine and Gilmore 1999). Tourists always tend to look for something new, such as experiences, destinations, places and accommodation facilities associated with diverse competitors (Cunnill 2006). Branding is therefore critical for the hotel business given its principal share of the tourism industry, which is, in essence, based on experiences (Wang 2000; Westwood 2007). One reason why the emphasis on experience has recently increased is that customers may have difficulty distinguishing among sub-brands under master brands or those belonging to other master brands. Hotels may thus prefer to create more individual brands reflecting more unique offerings, bundling tangible and intangible attributes to create value for customers (Olsen et al. 2005). This orientation is even applicable for sub-brands in the same class (i.e. luxury, upscale) of one master brand. These brands promise outstanding levels of memorable experiences in different styles (O'Neill and Mattila 2010).

Brand experience in hotels can be constructed in relation to three interrelated aspects: theme, physical environment reflecting this theme and facilities for social interactions. These aspects can be integrated so as to comprise basic elements – functional (tangible assets rationally affecting product performance), emotional and symbolic (including intangible and emotionally assessed features) and psycho-social (social interactions to gratify belonging needs and to establish self-identity) (Zhang, Cai and Kavanaugh 2008).

Indeed, the experiential essence of the hotel business offers the opportunity to design the offering as a brand based on emotional attributes rather than a product which is heavily dependent on rational aspects (Cai and Hobson 2004) in order to create a pleasant experience through various sensory levels (O'Neill and Mattila 2010). This mindset takes the essence of an offering and presents it in a set of tangible, physical and interactive components so that its impact will be reinforced by creating some particular feelings (Atwal and Williams 2009). This is because the image of a hotel brand is highly related to the meanings of the brand for the customer through experiences and emotions evoked. The distinction of a brand therefore shows up in a memorable, personal and sensational manner, leading to sustainable competitive advantage for hotel brands (Cai and Hobson 2004). Similarly, coherent, multi-sensory brand experiences trigger memorable experiences, leading to increased added value in customer perception, especially for luxury brands, where a substantial orientation in this direction is notable (Wiedmann et al. 2018).

From the experiential point of view, hotel groups may implement a branding strategy incorporating both hotel-driven (rational) product differentiation and customer-driven (emotional) market segmentation in the long run. However, this branding strategy will be successful in terms of higher added value, resulting in higher loyalty, profits and more sustainable performance only if emotional bonds are established with the target markets (Cai and Hobson 2004).

Despite the increasing emphasis on brand experience in the hotel industry, owners and managers may not be aware of its total scope given the limits of their understanding of the offering through tangible attributes and neglect of the experiential mindset (Zhang, Cai and Kavanaugh 2008). Moreover, they sometimes fail to articulate the experiences they offer or use their authenticity (Dev and Withiam 2012). If such constraints are overcome, brand experience offers a variety of opportunities for hotel businesses to raise competitive advantage by, for example, learning from customers who keep seeking unique and personalized experiences, going beyond segmentation to more personalized offerings thanks to technological innovations, and developing brands through distinct experiences (Richard 2017).

Reputation management

Reputation is a concept with a close link to the brand experience and both concepts are critical for building a strategy in business. For many hotels, branding is a critical decision relating to the best fit, including reputation and positioning (Holverson and Revaz 2006). As an intangible asset of a business, reputation is mainly a concept of values, communication and public relations (Davies and Miles 1998). This term incorporates the representation and overall impression of a business, including all its actions in the eyes of a wide network of stakeholders (customers, partners, suppliers, retailers, shareholders, government, financial institutions) in terms of their opinions, feelings and responses. Management therefore focuses on the promotion and protection of brands (Griffin 2008) by creating a favorable impression of business attributes (Gray and Balmer 1998). This impression is mainly based on previous actions and performances of stakeholders or business interactions with them and perceptions or awareness constructed around those actions, which can sometimes be shown as placement in rankings (Rindova and Martins 2012). In particular, a favorable reputation in the eyes of customers, created through communication tools referring to their identity, will support sales, resulting in higher competitive advantage (Gray and Balmer 1998). Reputation management is especially relevant and functional in service businesses in terms of branding to position businesses among competitors, given the interactive role of customers (Davies and Miles 1998).

Following studies which developed reputation management theory in relation to corporate identity, image and branding concepts in the late 1990s, recent work has concentrated much more on managing reputation on online media, particularly SM. SM is the most influential product from Web 2.0 technologies and has enabled individuals to share their views, especially by expressing unfavorable feelings from dissatisfactory experiences, remaining anonymous, to a wider audience including potential customers who can do this at a reduced cost (van Noort and Willemsen 2012). Moreover, as the most effective medium to make a brand more visible and accessible, SM reflects overall customer experience as part of brand building (Gavurova et al. 2018). In addition, given the importance of feedback to strengthen corporate identity and reputation (Gray and Balmer 1998), SM plays a critical role in that sense as the current platform of responses and interactions.

This platform, including shares of customers engaged with a business, is positively related to corporate reputation, especially among non-customers (Dijkmans, Kerkhof and Beukeboom 2015). This is because SM is a stage of hedonic value based on feelings of enjoyment and sociability offered by brands (Barcelos, Dantas and Sénécal 2018), and is a collection of stories in the form of experience narratives appealing to people's feelings and emotions. Prospective customers therefore perceive brands in a positive way and may even be willing to pay more for their experience (Lundqvist et al. 2013). As a reputation channel where brand image is strengthened, SM thus reinforces the competitiveness of brands through positive impact based on experiences and recommendations to potential customers (Gavurova et al. 2018). Positive brand experiences create higher satisfaction, resulting in higher sharing using word of mouth (Santini et al. 2018). For these reasons, businesses make the utmost effort to monitor all relevant SM channels and to undertake reactive or proactive actions, particularly to overcome the pitfalls of unfavorable shares, responding to negative reviews to help more favorable evaluations of brands (van Noort and Willemsen 2012). This action represents an opportunity to contact customers in a more personal way (Barcelos, Dantas and Sénécal 2018).

From another point of view, brand experience, especially with its emotional touch, is an opportunity for businesses to engage customers at the pre-purchase, purchase and post-purchase stages (Morrison and Crane 2007). As a web tool including an environment where experience occurs indirectly and consumers are exposed to brand communications (Brakus, Schmitt and Zarantonello 2009), SM is highly integrated with all these stages, with extensive text (reviews), scores and visual content. Creating a positive brand experience through these stages is therefore strategically required for businesses (Gavurova et al. 2018; Morrison and Crane 2007). Contents created or shared by businesses on SM have a positive impact on brand experience, particularly in the cognitive dimension given the informative function of this channel (Beig and Khan 2018).

Reputation management in the hotel business

Reputation is a concept linked to brand experience in various aspects, including via its favorable effects on positive perceptions of brand experience and the influence of both brand experience and reputation on behavioral responses (Ahn and Back 2018), including higher brand loyalty (Christou 2003). As a part of reputation, the importance of SM is significant for hotel businesses in terms of converting the customer–brand relationship into a continuous interaction. The final purchase may even actualize as a different brand from that previously considered if a consistent experience at all points of brand contact is missing (Dev and Withiam 2012). Both standardization (distribution of concepts) and individualization (more focus on creativity and learning) strategies tend to create positive word of mouth as part of reputation (Bouncken, Pick and Hipp 2008).

With reference to this significant role of reputation and SM for hotel businesses, its management has also become equally critical, particularly in terms of continuous monitoring of the current situation on review sites and responding to reviews. Mainly posted by high-class hotels (Liu, Schuckert and Law 2015), these responses can occur in various forms, such as thanking customers, apologizing for problems, inviting customers for a return visit, ensuring actions have been taken, giving details about such actions, avoidance of recurring problems, and keeping interactions (via email correspondence) beyond reviews and responses (Zhang and Vasquez 2014). Zhang and Vasquez (2014) also draw attention to the remarkably higher share of non-specific and standardized responses, which may be a corporate policy.

Management responses usually address negative reviews and this action may have favorable or unfavorable effects on reputation. On one hand, these responses may enhance reputation by mitigating the adverse effects of negative reviews which stem from controllable factors (Rose and Bodgett 2016). On the other hand, responding may not have a favorable effect on behavioral responses (i.e. purchasing intentions), given their inability to correspond with customers' concerns (Mauri and Minazzi 2013; Xie, Zhang and Zhang 2014).

Finally, concerning ways of how to respond to reviews to improve reputation management, in addition to encouraging positive reviews by customers, hotel businesses can apply various strategies to increase the positive effects of reviews on reputation – for example, posting responses in accordance with review ratings (Xie, Zhang and Zhang 2014) and implementing a targeted response management (Liu, Schuckert and Law 2015).

Conclusion

Despite the recent emergence of the brand experience concept, branding and experience have long been practically in the focus of hotel businesses in, for example, their design of various offerings targeting different customer segments. However, hotel businesses (and academia) have recently noticed and understood the real essence of experience, with its potential to offer new horizons for guest satisfaction, leading to higher levels of favorable impact in terms of reputation, loyalty and other behavioral responses. Thanks to SM, one of the most influential technological revolutions in today's world, the impact of experience has multiplied, with customer shares requiring a specific management process to reap the benefits generated by these platforms. As seen in the case study, these critical interconnections between experience, brand and reputation are intensively analyzed and relevant actions are taken, especially by international hotel groups, in order to maintain competitive advantage through distinguishing brands from rivals.

Case study 29.1: Hilton Worldwide

One of the most recognizable hotel brands (Balmer and Thomson 2009; Olsen et al. 2005), "Hilton" is considered as a synonym of the word "hotel" (Cunnill 2006; Hilton 2019a). As the first global hotel company, now celebrating its 100th anniversary, the history of Hilton Worldwide shows the dynamic change in the hotel industry, extending beyond family businesses to management by other professionals (Vallen and Vallen 2013), and pioneering innovations in terms of implementing new management systems (Balmer and Thomson 2009).

The very first accommodation, a family-operated rooming house, was opened by Conrad Hilton in 1919, while the family name was used for a hotel a few years later in 1925. After establishing and acquiring hotels (Stevens and Palmer House) in the years which followed, Hilton became the first coast-to-coast hotel in the US in 1943, and Hilton Hotels Corporation came to life in 1946. Another milestone was in 1999, with the acquisition of DoubleTree Hotels (seven brands). The first international Hilton hotel was opened in Spain in 1953, followed by Turkey in 1955, representing the very first international hotels in those countries.

Following these openings, Hilton International became a separate brand. The ownership changed twice and returned to Hilton in 2005. Several purchases, acquisitions and other financial cases occurred in the history of the corporation and, finally, the Hilton family split from the operations and ownership in 2007. Moreover, Hilton Worldwide introduced the hotel industry to the first examples of initiations, such as the launch of the Hilton Honors Loyalty Program in 1987 and the 100% satisfaction guarantee (friendly service, clean rooms, and comfortable surroundings) in the Hampton Hotels in 1989. To celebrate its 100th anniversary, the corporation introduced a new concept – "The Hilton Effect" – in order to underline the significant impact of the brand for guests (3 billion), team members (10 million), communities and economies ($1 trillion) (Hilton 2018a).

Branding in hotel groups necessitates adopting certain specifications for various market segments in a consistent way through established standards in all the properties of a brand (Cunnill 2006). These standards can be distinct even in the sub-brands of the same class in a hotel group – for instance, Waldorf Astoria and New York Hilton promise different experiences from each other although they are both first-class properties (O'Neill and Mattila 2010). These kinds of strategies are highly applicable in Hilton Worldwide properties, with promising experiences for specific target groups and, in some cases, multiple brands in the same category (Appendix 29.1).

Today the group operates nearly 6,000 properties under 17 brands (Figure 29.2) in 117 countries, covering a spectrum ranging from economy to super deluxe/high-end luxury properties and opening more than one hotel a day (Hilton 2019a). According to data from 2017 (Hotelsmag 2018), the group is ranked as the second biggest hotel group by number of rooms in the world (nearly 1.2 million rooms in 2017), while Hampton by Hilton is the third biggest brand with approximately 2,500 hotels including more than 250,000 rooms. This expansion has been possible thanks to various strategies implemented by the corporation such as cost leadership with Hampton by Hilton and Home2 Suites, differentiation, with the Hilton brand dedicated to high-class hotels in big business cities, and focus strategy based on certain market segments through multiple brands (Cunnill 2006 and please also see Appendix 29.1).

Figure 29.2 Hilton Worldwide Brands

Cunnill (2006) recalled the necessity of incorporating subjective attributes as part of differentiation strategies for branding. This experiential orientation is evident in the Hilton Worldwide brands (Appendix 29.1), particularly in the upper midscale and superior classes (i.e. authentic, friendly experiences, passionate and independent-minded travelers, little touches of care, personalized attention). Even some brands, such as Canopy by Hilton, Hilton Hotels & Resorts, the Curio Collection, and the Tapestry Collection by Hilton, underline the distinct experiences promised.

Like all businesses, Hilton Worldwide also benefits from the advantages of SM for many purposes, including corporate/group-, brand- and property-level marketing, human resources and career opportunities, loyalty programs and news. The engagement of the corporation with SM is evident in the top management position of Vice President of Social Media Marketing (Hilton 2018b). In this respect, Hilton Worldwide is a remarkable example of integrating reputation management with brand experience. The most recent example is the launching of the corporation's new brand, Signia Hilton (for details please see Appendix 29.1). As part of the company's marketing activities, one of the first steps taken involved creating SM pages and developing content, including the attributes of the brand, so that potential customers become engaged with experiences to be offered by the new brand (Hilton 2019b). As a part of reputation management, the group has also developed an instant messaging service for specific questions, to be replied to within a few hours on Facebook. On the other hand, Hilton Worldwide appears to be active in posting individual responses to both positive and negative reviews on both Facebook and TripAdvisor.

References

Aaker, D. A. (2004). *Brand Portfolio Strategy: Creating Relevance, Differentiation, Energy, Leverage, and Clarity*. New York: Free Press.

Aaker, D. A. and Joachimsthaler, E. (2000). 'The brand relationship spectrum: The key to the brand architecture challenge', *California Management Review*, 42(4): 8–23. DOI: 10.1177/000812560004200401.

Ahn, J. and Back, K. (2018). 'Beyond gambling: Mediating roles of brand experience and attitude', *International Journal of Contemporary Hospitality Management*, 30(10): 3026–3039. DOI: 10.1108/IJCHM-07-2017-0473.

Atwal, G. and Williams, A. (2009). 'Luxury brand marketing – The experiences is everything!', *Brand Management*, 16(5/6): 338–346. DOI: 10.1057/bm.2008.48.

Balmer, J. M. T. and Thomson, I. (2009). 'The shared management and ownership of corporate brands: The case of Hilton', *Journal of General Management*, 34(4): 15–37. DOI: 10.1177/030630700903400402.

Barcelos, R. H., Dantas, D. C. and Sénécal, S. (2018). 'Watch your tone: How a brand's tone of voice on social media influences consumer responses', *Journal of Interactive Marketing*, 41: 60–80. DOI: 10.1016/j.intmar.2017.10.001.

Beig, F. A. and Khan, M. F. (2018). 'Impact of social media marketing on brand experience: A study of select apparel brands of Facebook', *Vision*, 22(3): 264–275. DOI: 10.1177/0972262918785962.

Bouncken, R. B., Pick, C. and Hipp, C. (2008). 'Standardization and individualization strategies of hotel brands: Matching strategy to quality management instruments and marketing in Germany', *Journal of Hospitality & Leisure Marketing*, 13(3/4): 29–51. DOI: 10.1300/J150v13n03_03.

Brakus, J. J., Schmitt, B. H. and Zarantonello, L. (2009). 'Brand experience: What is it? How is it measured? Does it affect loyalty?', *Journal of Marketing*, 73(3): 52–68. DOI: 10.1509/jmkg.73.3.052.

Cai, L. A. and Hobson, J. S. P. (2004). 'Making hotel brands work in a competitive environment', *Journal of Vacation Marketing*, 10(3): 197–208. DOI: 10.1177/135676670401000301.

Christou, E. (2003). 'Guest loyalty likelihood in relation to hotels' corporate image and reputation: A study of three countries in Europe', *Journal of Hospitality & Leisure Marketing*, 10(3–4): 85–99. DOI: 10.1300/J150v10n03_05.

Connell, J. (1992). 'Branding hotel portfolios', *International Journal of Contemporary Hospitality Management*, 4(1): 26–32. DOI: 10.1108/09596119210007965.

Cunnill, O. M. (2006). *The Growth Strategies of Hotel Chains: Best Business Practices by Leading Companies*. New York: The Haworth Hospitality Press. DOI: 10.4324/9780203820810.

Davies, G. and Miles, L. (1998). 'Reputation management: Theory versus practice', *Corporate Reputation Review*, 2(1): 16–27. DOI: 10.1057/palgrave.crr.1540064.

Dev, C. (2009). 'Branding hotels', Available at: https://cayugahospitality.com/chekitan-dev/branding-hotels-cheki-dev/ [Accessed: February 25, 2019].

Dev, C. S., Buschman, J. D. and Bowen, J. T. (2010). 'Hospitality marketing: A retrospective analysis (1960–2010) and predictions (2010–2020)', *Cornell Hospitality Quarterly*, 51(4): 459–469. DOI: 10.1177/1938965510376353.

Dev, S. C. and Withiam, G. (2012). 'Branding hospitality: Challenges, opportunities and best practices', *Cornell Hospitality Roundtable Proceedings*, 4(5): 6–14.

Dijkmans, C., Kerkhof, P. and Beukeboom, C. J. (2015). 'A stage to engage: Social media use and corporate reputation', *Tourism Management*, 47: 58–67. DOI: 10.1016/j.tourman.2014.09.005.

Douglas, S. P., Craig, C. S. and Nijssen, E. J. (2001). 'Integrating branding strategy across markets: Building international brand architecture', *Journal of International Marketing*, 9(2): 97–114. DOI: 10.1509/jimk.9.2.97.19882.

Gavurova, B., Bacik, R., Fedorko, R. and Nastisin, L. (2018). 'The customer's brand experience in the light of selected performance indicators in the social media environment', *Journal of Competitiveness*, 10 (2): 72–84. DOI: 10.7441/joc.2018.02.05.

Gray, E. R. and Balmer, J. M. T. (1998). 'Managing corporate image and corporate reputation', *Long Range Planning*, 31(5): 695–702. DOI: 10.1016/S0024-6301(98)00074-0.

Griffin, A. (2008). *New Strategies for Reputation Management: Gaining Control of Issues, Crises, and Corporate Social Responsibility*. London: Kogan Page Ltd.

Hilton. (2018a). 'The Hilton Effect: As Hilton nears 100-year milestone, new research uncovers world-changing impact', Available at: https://newsroom.hilton.com/brand-communications/news/33560 [Accessed: February 18, 2019].

Hilton. (2018b). 'Hilton appoints Noha Abdalla as vice president of social media marketing', Available at: https://newsroom.hilton.com/corporate/news/hilton-appoints-noha-abdalla-as-vice-president-of-social-media-marketing [Accessed: March 15, 2019].

Hilton. (2019a). 'Welcome to Hilton', Available at: www.hilton.com/en/corporate/ [Accessed February 25, 2019].

Hilton. (2019b). 'Hilton launches new brand, Signia Hilton, delivering sophisticated travel while reimagining meetings and events', Available at: https://newsroom.hilton.com/corporate/news/33924 [Accessed: February 25, 2019].

Hollis, N. (2010). *The Global Brand: How to Create and Develop Lasting Brand Value in the World Market*. New York: Palgrave Macmillan.

Holverson, S. and Revaz, F. (2006). 'Perceptions of European independent hoteliers: Hard and soft branding choices', *International Journal of Contemporary Hospitality Management*, 18(5): 398–413. DOI: 10.1108/09596110610673538.

Hotelsmag. (2018). '325 hotels' (Ed. J. Weinstein), Available at: hotelsmag.com [Accessed November 27, 2018].

Keller, L. K. (2013). *Strategic Brand Management*. Upper Saddle River, NJ: Pearson Education Inc.

Kim, W. G. (2008). 'Branding, brand equity and brand extensions.' In H. Oh and A. Pizam (Eds.), *Handbook of Hospitality Marketing Management* (pp. 87–118). Oxford: Butterworth-Heinemann. DOI: 10.4324/9780080569437.

Liu, X., Schuckert, M. and Law, R. (2015). 'Can response management benefit hotels? Evidence from Hong Kong hotels', *Journal of Travel and Tourism Marketing*, 32(8): 1069–1080. DOI: 10.1080/10548408.2014.944253.

Lundqvist, A., Liljander, V., Gummerus, J. and van Riel, A. (2013). 'The impact of storytelling on the consumer brand experience: The case of a firm-originated story', *Journal of Brand Management*, 20(4): 283–297. DOI: 10.1057/bm.2012.15.

Mauri, A. G. and Minazzi, R. (2013). 'Web reviews influence on expectations and purchasing intentions of hotel potential customers', *International Journal of Hospitality Management*, 34: 99–107. DOI: 10.1016/j.ijhm.2013.02.012.

Morrison, S. and Crane, F. G. (2007). 'Building the service brand by creating and managing an emotional brand experience', *Brand Management*, 14(5): 410–421. DOI: 10.1057/palgrave.bm.2550080.

Nysveen, H., Pedersen, P. E. and Skard, S. (2013). 'Brand experiences in service organizations: Exploring the individual effects of brand experience dimensions', *Journal of Brand Management*, 20(5): 404–423. DOI: 10.1057/bm.2012.31.

O'Neill, J. W. and Mattila, A. S. (2010). 'Hotel brand strategy', *Cornell Hospitality Quarterly*, 51(1): 27–34. DOI: 10.1177/1938965509352286.

Olsen, M. D., Chung, Y., Graf, N., Lee, K. and Madanoglu, M. (2005). 'Branding: Myth and reality in the hotel industry', *Journal of Retail and Leisure Property*, 4(2): 146–162. DOI: 10.1057/palgrave.rlp.5090205.

Pine, B. J. and Gilmore, J. (1999). *The Experience Economy*. Boston: Harvard Business School Press.

Richard, B. (2017). 'Hotel chains: Survival strategies for a dynamic future', *Journal of Tourism Futures*, 3(1): 56–65. DOI: 10.1108/JTF-06-2016-0018.

Rindova, V. and Martins, L. M. (2012). 'Show me the money: A multidimensional perspective on reputation as an intangible asset'. In T. G. Pollock and M. L. Barnett (Eds.), *The Oxford Handbook of Corporate Reputation* (pp. 18–32). Oxford: Oxford University Press. DOI: 10.1093/oxfordhb/9780199596706.013.0002.

Rose, M. and Bodgett, J. G. (2016). 'Should hotels respond to negative online reviews?', *Cornell Hospitality Quarterly*, 57(4): 396–410. DOI: 10.1177/1938965516632610.

Santini, F. O., Ladeira, W. J., Sampaio, C. H. and Pinto, D. C. (2018). 'The brand experience extended model: A meta-analysis', *Journal of Brand Management*, 25(6): 519–535. DOI: 10.1057/s41262-018-0104-6.

Schmitt, B. (1999). *Experiential Marketing*. New York: Free Press.

STR Global. (2018). 'STR chain scales', Available at: www.strglobal.com/Media/Default/Documents/STR-ChainScales2018.xlsx [Accessed February 25, 2019].

Vallen, G. K. and Vallen, J. J. (2013). *Check-In Check-Out: Managing Hotel Operations*. Upper Saddle River, NJ: Pearson Eduation Inc.

van Noort, G. and Willemsen, L. M. (2012). 'Online damage control: The effects of proactive versus reactive webcare interventions in customer-generated and brand-generated platforms', *Journal of Interactive Marketing*, 26(3): 131–140. DOI: 10.1016/j.intmar.2011.07.001.

Vargo, S. L. and Lusch, R. F. (2004). 'The four marketing myths: Remnants of a goods-based, manufacturing model', *Journal of Services Marketing*, 6(4): 324–335. DOI: 10.1177/1094670503262946.

Wang, N. (2000). *Tourism and Modernity: A Sociological Analysis* (Ed. J. Jafari). Oxford: Emerald.

Westwood, S. (2007). 'What lies beneath? Using creative, projective and participatory techniques in qualitative tourism inquiry'. In I. Ateljevic, A. Pritchard and N. Morgan (Eds.), *The Critical Turn in Tourism Studies: Innovative Research Methodologies* (pp. 293–316). Oxford: Elsevier.

Wiedmann, K., Labenz, F., Haase, J. and Hennigs, N. (2018). 'The power of experiential marketing: Exploring the causal relationships among multisensory marketing, brand experience, customer perceived value and brand strength', *Journal of Brand Management*, 25(2): 101–118. DOI: 10.1057/s41262-017-0061-5.

Xie, K. L., Zhang, Z. and Zhang, Z. (2014). 'The business value of online consumer reviews and management response to hotel performance', *International Journal of Hospitality Management*, 43: 1–12. DOI: 10.1016/j.ijhm.2014.07.007.

Zarantonello, L. and Schmitt, B. H. (2010). 'Using the brand experience scale to profile consumers and predict consumer behavior', *Brand Management*, 17(7): 532–540. DOI: 10.1057/bm.2010.4.

Zhang, J., Cai, L. A. and Kavanaugh, R. R. (2008). 'Dimensions in building brand experience for economy hotels: A case of emerging market', *Journal of China Tourism Research*, 4(1): 61–77. DOI: 10.1080/19388160802099782.

Zhang, Y. and Vasquez, C. (2014). 'Hotels' responses to online reviews: Managing consumer dissatisfaction', *Discourse, Context and Media*, 6: 54–64. DOI: 10.1016/j.dcm.2014.08.004.

Appendix 29.1
Hilton Worldwide Brands

Brand name	STR class*	Launch	# CoO	# (C)	# (P)	For whom	Distinct features
Waldorf Astoria Hotels & Resorts	Luxury	1949 (purch.)	14	31	19	Discerning and sophisticated travelers	Iconic upper-luxury brand Landmark destinations Modern design of rich and authentic legacy Personalized attention for a singular experience
LXR Hotels & Resorts	N/A; listed as luxury by HW	2018	1	1		People craving unrivaled accommodation with authentic connection	Located in the world's most alluring places Minimum brand standards with a unique identity
Conrad Hotels & Resorts	Luxury	1982	22	33	16	New generation of smart luxury travelers for whom life, business and pleasure seamlessly intersect	Located in the world's most exciting places Luxury properties in the world's major business and tourism capitals Smart luxury
Canopy by Hilton	Upper upscale	2014	4	8	14	People seeking an energizing experience and who want to get the most out of travel and experience the best of a great neighborhood	Positive stay with a simple, guest-directed service, local choices, thoughtful details

(Continued)

(Cont.)

Brand name	STR class*	Launch	# CoO	# (C)	# (P)	For whom	Distinct features
Signia Hilton	N/A; listed as upper upscale by HW	2019**				Meeting professionals, sophisticated business travelers	Minimum 500 rooms Modern architecture with innovative (state-of -the-art) technology Destination bar with a view A signature restaurant with a chef-driven concept Premium wellness Large meeting spaces with smart design
Hilton Hotels & Resorts	Upper upscale	1925	94	586	179	Travelers seeking unique, memorable and sharable experiences	Destination-inspired design Activated bars Elevated restaurants Leading-edge meetings and events space
Curio Collection	Upper upscale	2014	22	68	59	Passionate and independent-minded travelers seeking local discovery and authentic experiences	Independent and remarkable hotels with glorious individuality Urban landmark properties or grand beach resorts
Doubletree by Hilton	Upscale	1969	45	559	213	Today's business and leisure travelers	Located in metropolitan areas or vacation destinations Comfortable stay with little touches of care Upscale accommodation
Tapestry Collection by Hilton	Upscale	2016	1	18	40	Travelers who do not want to have exactly the same experience twice	Upscale independent hotels with unique style and connection to the city
Embassy Suites by Hilton	Upper upscale	1984	6	253	43	Business travelers and families	All-suite hotel category Space and service Combination of full-service benefits and focused-service staffing model
Motto by Hilton	N/A; but listed as upscale by HW	2018				Today's modern global community of travelers	Micro-hotel with an urban vibe in prime global locations Centrally located, reasonably priced, less traditional lodging (one-of-a-kind experience)

(*Continued*)

(Cont.)

Brand name	*STR class**	*Launch*	*# CoO*	*# (C)*	*# (P)*	*For whom*	*Distinct features*
Hilton Garden Inn	Upscale	1990	40	815	155	People wishing to sleep deep, work smart, eat well, stay fit and treat themselves every time	Mid-market brand Focus on providing only those features travelers truly value
Hampton by Hilton	Upper midscale	1984	25	2.433	320	Business and leisure travelers mainly from local market	Friendly and consistent experience at a competitive price Free hot breakfast, free WiFi Minimal but flexible meeting spaces
Tru by Hilton	Midscale	2016		53		Cross-generational consumers with a zest for life mindset seeking a fresh, consistent, affordable, midscale hotel	Lobby as a huge space for social activation Focus on bed, shower, in-room entertainment, desired technology Engaging and lively public spaces to connect, eat, work or play
Homewood Suites	Upscale	1989	3	482	125	People wishing for longer stays in their routines	Upscale, all-suite, residential-style, extended stay properties
Home2 Suites	Upper midscale	2011	2	290	400	Value-conscious extended stay travelers	Midscale, all-suite Dedicated to extended stays (5+ nights)
Hilton Grand Vacations	Upper upscale	1992	4	51		Club members enjoying spacious and luxurious accommodation offering home comfort	Resorts in spectacular settings, spacious and comfortable elegance with innovative vacation ownership program

CoO: Number of countries where operated; # (C): Number of current properties; # (P): Number of properties in the pipeline.

* Source: STR Global 2018.

** as an innovative product at the 100th birthday of the corporation.

Source: Authors' production with reference to detailed information about brands (Hilton 2019a).

30

Managing service failure and recoveries

Kawon Kim and Melissa A. Baker

Introduction

Due to the intangible and heterogeneous nature of services, providing zero-defect service is almost impossible. Therefore, service failure is inevitable, especially within hospitality and tourism organizations (Baker 2017). The tourism industry is no exception, where interactions between service providers and customers happen in multiphases including pre-trip, during-trip and post-trip stages (Baker 2016) with a number of different parties involved during the process (Gretzel et al. 2006). For example, one of the worst service failures that happened during the pre-trip phases involved United Airlines. This involved airport security forcibly dragging a customer off an overbooked flight to provide seats for crew from another flight, hurting and threatening the customer in the process. The incident was recorded and the video posted online. Far from providing appropriate service recovery, the company made matters worse with an insensitive and ineffective response to the incident, calling the passenger "disruptive" and "belligerent" (Matousek 2018). Such inappropriate service failure recovery added fuel to the fire of the public's anger toward the service firm. After the incident, many customers boycotted the firm, resulting in an approximate one billion dollar stock drop. This example demonstrates that service failure can have a disastrous effect on a firm as it results in customer dissatisfaction, negative word-of-mouth, and decreased sales and profits. Therefore, it is critical to understand how best to handle a service failure by examining the way in which customers react to service failure and how they respond to different types of service recovery strategies.

To better understand service failures and recoveries in tourism experiences, this chapter discusses service failure, customer complaining behavior, and service recovery. Specifically, it details the types of service failures, the buffering effects, and other customer service failures. Second, the chapter discusses customer complaining behavior including the key categories, motivations, and illegitimate customer complaints. Finally, the chapter discusses how to recover from service failures and complaints by examining types of service recoveries, consequences of service recovery, and the role of other customers in service recovery processes. The chapter seeks to educate readers on the critical management of service failures and recoveries in managing tourist experiences.

Service failure

Typologies of service failure

A service failure occurs when the consumers' perceptions of the service delivery do not meet their expectations of adequate service (Kelley et al. 1993). In the 1990s, researchers tried to develop a typology of service failure. For instance, Bitner (1990) suggested two types of service failure, namely outcome service failure when customers receive service that largely fails to meet their needs and wants, and process service failure when the service delivery is not done properly. Furthermore, since service failures are determined by the customer's point of view, Bitner et al. (1990) have identified and classified service failure into three broad service failure categories; 1) failure of employee response to service delivery, 2) failure of employee response to customer needs and requests, and 3) unprompted and unsolicited employee actions. Namkung and Jang (2010) viewed service failure as a series of events and adopted a time-based service stage perspective, as service failure can trigger various customer responses in different service stages: 1) greeting and seating, 2) order taking and delivery, 3) consumption, and 4) payment and exiting. Yang and Mattila (2012) categorized service failure into core service failure and interactional service failure. Core service failure refers to tangible outcomes that customers receive from the service, and interactional service failure reflects intangible elements of the service.

Buffering factors of service failure

The negative consequences of service failure include decreased satisfaction, decreased revisit intention, and increased negative word-of-mouth (Baker 2017). One of the significant factors that can mitigate the negative effects is the relationship with the service provider or firm. The greater the strength of the relationship with the service provider or brand, the stronger the buffering effect on negative outcomes through the trust (Hui and Au 2001). Similarly, individuals who have high affective commitment are more likely to return to the establishment after a service failure and recovery (Mattila 2004). Brand relationships are helpful in shielding a hotel organization from the negative effects of service failures as customers with a high level of brand relationship feel reluctant to cause damage to a valued exchange partner with whom they feel psychologically connected (Xie and Heung 2012). Lastly, the negative effect of stability attribute and controllability attribute caused by service failure on service evaluation is mitigated by globality attribute (the same type of failure occurs in different organizations as well) (Baker 2017). However, all of these moderators are applicable only when the service failure is not severe.

Other customer service failure

In addition to failures that are caused by the firm or an employee's actions, a new type of failure involves the failure being the fault of the other customer (Baker and Kim 2018). Other customer failure refers to actions taken by another customer, whether intentional or unintentional, that disrupt one's service experience. Examples may include drunk customers, tourists being late for their group tour, unruly behavior, or rudeness (Baker and Kim 2018; Kim and Baker 2019). Other customer failure negatively impacts on company value, focal customer satisfaction, loyalty, and revisit intention (Kim and Baker 2019). According to attribution theory (Kelley 1971), when faced with unexpected or negative events, consumers search for explanations for the cause and make attributions based on that. When customers experience other customer generated service failures, although they acknowledge that the failure is the other customer's fault, they still believe that the firm is responsible for a recovery (Baker and Kim 2018). Therefore, it is important to consider not only failures that are the fault of the firm, but failures that derive from other customers.

Customer complaining behavior

Categories of complaining behavior

Consumer complaining behavior refers to responses to dissatisfaction (Singh 1988). Customer complaints can lead to negative outcomes such as negative word-of-mouth and revisit intention. However, as Bill Gates once said: "Your most unhappy customers are your greatest sources of learning." As such, listening and responding to customer complaints can increase efficiency and minimize issues with the service delivery. When customers experience service failure, they respond to a service failure in various ways. These mostly include voicing the complaint and not voicing the complaint. After experiencing a service failure, customers become dissatisfied and experience negative emotions. Following this, they can either voice their complaint or not voice their complaint. The second-level decision concerns whether to take private action such as engaging in negative word-of-mouth or public action such as complaining to public or private agencies. Stated differently, consumers engage in one of three sets of responses when they are dissatisfied with the service: voice responses, private responses, and third-party responses. Voice complaining behavior is aimed at objects that are directly involved in the dissatisfying exchange, such as directly complaining to the service firm for compensation. Private responses are directed at objects that are external to the consumer but not directly involved in the dissatisfying experience such as negative word-of-mouth communication with friends and families. And third-party responses refer to actions directed at formal external parties, such as taking legal action against the service firm.

Initial dissatisfaction caused by service failure leads to cognitive appraisal, which in turn determines the consumer's coping strategies (customer complaining behavior), which have been categorized as remaining silent, complaining to third parties, spreading negative WOM, and voice (Kim et al. 2010). The worst of these complaining actions is for the customer to remain silent. In this case, the employee and the firm do not know the customer is dissatisfied and therefore can do nothing to fix the issue. Conversely, voicing one's dissatisfaction directly to the service provider can result in recovery opportunity where the employee can fix the issue immediately and make the customer happy. It is also less desirable from a management perspective for the customer to voice their complaints via a third-party site such as TripAdvisor or Yelp (Baker and Kim 2019). This is because the customer has already left the firm so a recovery cannot be made immediately in the guest's presence. Second, the reach of the online audience is significantly larger and can therefore more negatively impact the firm (Baker 2017). It is important to note that guests who experience information inadequacy as a result of a service failure are likely to seek a richer form of communication to deliver their complaint such as face-to-face with management (Susskind 2015).

Motivations of customer complaining behavior

A number of motivational factors influence customer complaining behavior. These include individual differences such as personality traits (i.e., guilt prone, aggressiveness), personal attitude toward the situation, culture (a customer with collectivist culture traits is less likely to complain compared to one with individualist traits), propensity to seeking redress, and demographic differences (Gronhaug and Zaltman 1981; Yuksel et al. 2006). In addition, anger is one of the most significant factors that leads to a customer's desire to complain (Mattila and Ro 2008) and customers who have high levels of affective commitment to a service firm exhibit both a strong motivation to help a firm improve its business and a great need to maintain positive relationships with the firm's frontline employees (Liu and Mattila 2015). Consequently, affectively committed customers are more willing to help the firm by lodging complaints when such feedback is not used for employee evaluation purposes. According to McQuilken et al. (2013), perceived failure stability has the greatest influence on customer complaining behavior.

Illegitimate complaining behavior

Although the general principle of customer complaining behavior is that complaints are genuine customer complaints resulting from dissatisfaction, increasingly customers complain without facing service failure or dissatisfaction, deliberately fabricating service failure and fraudulently complaining (Baker et al. 2012). Such behavior refers to illegitimate complaining behavior during which customers purposely act in a dysfunctional manner by knowingly making inauthentic complains (Harris and Reynolds 2003). Illegitimate customer complaining behaviour takes the form of customers voicing fictitious complaints to service firms or service providers with the goal of receiving compensation for their make-believe service failure (Baker et al. 2012). In the tourism industry, illegitimate complaints occur in the form of voicing inflated and fake claims exploiting service recovery policies, and cheating on service guarantees (Wirtz and Kum 2004). Companies such as REI and LLBean have had to change their 100% satisfaction guarantee policies as a result of customer abuse in returning items where there is no actual flaw in the product. Similarly, estimates find that approximately 33% of reviews on TripAdvisor are fake or contain fake elements (Baker and Kim 2019), further demonstrating the significant issue of illegitimate customer complaints.

Case study 30.1

Read the following scenario and answer the following questions.

You and your friend visit a restaurant in a city while on vacation. You take your place at one of the tables next to a group of customers. Once you are seated, you see the sign showing that the restaurant has a 100% satisfaction guarantee policy whereby any customer that complains gets a full refund for their meal. Throughout your stay, you receive good and attentive service from your server. Toward the end of your meal, you overhear the other customers next to you yelling loudly and aggressively at the server, complaining that their dishes tasted terrible. You strongly believe the customer is creating a fictitious complaint just to receive the full refund as you notice that they ate their entire meal. The server shows a high amount of concern for the customers, apologizes for the inconvenience, and takes the food off the bill.

1. What is your perception towards the other customers who are complaining?
2. Discuss the appropriate service recovery strategy to deal with illegitimate complaining customers.

Service recovery

Service recovery refers to the actions taken by an organization in response to a service failure, including all the activities and efforts employed to rectify, amend, and restore the loss incurred after the failure (Gronroos 1988). Benefits of a successful service recovery include improved satisfaction, greater loyalty, increased profits, and positive word-of-mouth. Furthermore, a customer's post-failure satisfaction can exceed pre-failure satisfaction when customers receive high recovery performance, which is known as "service recovery paradox" (Baker 2017; McCollough 2009). On the other hand, if service recovery strategies are inappropriately handled, it may aggravate the original dissatisfaction leading to a "double deviation effect" (Joireman et al. 2013). Studies support the idea that a service recovery paradox is most likely to occur when the service failure is not severe, no prior failure occurred with the firm, the service provider has no control, less critical consumption and so on. However, in a longitudinal study (Maxham and Netemeyer 2002), after a customer experienced a second failure with the same company their satisfaction and revisit intention were considerably lower than after the first recovery even though both recoveries were evaluated as acceptable by the customer. This contradicts the service recovery paradox and supports double deviation.

Types of service recovery

One key managerial action is to identify different types of service recovery strategies by examining the impact of service recovery performance and its effectiveness. There is not always a one-size-fits-all response to how best to address a complaining customer. It is therefore critical for management to determine which recovery action can be most effective, including 1) apology (confessions of responsibility for negative events which include some expression of remorse), 2) problem solving (correcting the mistake and problem to achieve justice), 3) courteous behavior (showing politeness, respect and friendliness when interacting with the customers), 4) explanation (explaining the cause of service failure to customers), 5) prompt handling (service provider's quick response to customer complaint), and 6) compensation (monetary or non-monetary compensation). Stated differently, factors influencing customer evaluations following service failure can include five types of service recovery performance (apology, problem solving, courteous behavior, explanation, prompt handling) which positively influence customer satisfaction, leading to customer repurchase intention (Liao 2007). While the traditional definition of service recovery is limited to the activities and efforts employed by the firm or service provider, recently managers are beginning to examine the customer's participation in the service recovery process. In other words, researchers have started to understand the role of co-creation in the service recovery process. For instance, Dong et al. (2008) reveal that when customers participate in the service recovery process they are more satisfied with the service recovery, which leads to higher intention toward future co-creation since they believe they have the ability to shape outcomes (perceived control, empowerment). Furthermore, Xu et al. (2014) show that when co-recovery is initiated by an employee, they perceive higher justice and greater satisfaction than when it is initiated by themselves. Similarly, Roggeveen et al. (2012) show that when customers experience service delay, co-creating the recovery improves the evaluation of service recovery.

Factors affecting service recovery

Key factors that affect service recovery include perceptions of justice and attribution. A great deal of evidence shows that effective service recovery can mitigate the negative consequences of service failure and identifies a number of key factors that explain the effectiveness of a service recovery effort. The major mediators that link the relationship between a service recovery effort and positive consequences are 1) perceived fairness, which is based on fairness theory (Adams 1965) and 2) attribution, based on attribution theory (Kelley 1971). First, justice theory provides a theoretical framework that explains fairness perceptions regarding recovery efforts from three different perspectives: distributive (perceived outcomes of service failure), procedural (perceived fairness of policies, procedures, and criteria used by decision makers), and interactional justice (interpersonal treatment people receive during the enactment of the procedure) (Baker 2017). In conjunction with justice theory, the research shows that perceived justice mediates the relationship between service recovery performance and satisfaction, trust and commitment (Tax et al. 1998). For instance, Gelbrich and Roschk (2011) examined the whole causal chain of post-complaint behavior by testing the model of the effect of organizational response (compensation, favorable behavior, organizational procedures) on satisfaction and behavioral intention through justice perception. They found that distributive justice was the salient driver of service recovery only for transaction-specific satisfaction, while cumulative satisfaction depends more on interactional justice. Liao (2007) showed that apology, problem solving, courteous behavior, explanation, and prompt handling positively influence customer satisfaction and then customer repurchase intention through perceived fairness. Furthermore, satisfaction with the service recovery leads to customer community behavior and loyalty.

Second, according to attribution theory, when faced with unexpected events customers search for meaningful explanations of their causes. When service failure occurs, customers attempt to determine the cause and make attributions. Weiner (1979) identifies three common dimensions of causal attributions: locus of causality (responsibility for the failure), controllability (cause is perceived to be under the firm's control), and stability (failure is relatively temporal or permanent). In other words, when customers obtain compensation and fast recovery, then they will perceive that the cause is unstable and controllable. Also, repeated failure leads customers to infer that failures are due to stable, inherent problems of the service company, and such stable attribution makes customers feel negative toward the company which makes it harder to please them with recovery efforts. Huang (2010) showed that when a customer perceives a lack of employee effort the compensation is not viewed as sincere, which in turn results in even greater customer dissatisfaction.

In addition, moderating factors can influence the effect of service recovery performance on customer outcomes. Maxham and Netemeyer (2002) argue that if the customer has experienced similar service failure with the company in the past, even a high service recovery effort in addressing the current complaint may not be enough to generate customer satisfaction. This is because repeated failure leads customers to infer that failures are due to stable and inherent problems of the service provider (Russell 1982). Therefore, such stable attribution makes a customer feel heightened dissatisfaction with the company, making it harder to satisfy customers with recovery efforts. On the other hand, customer–organization relationships help to enhance the firm's recovery efforts. Hess et al. (2003) show that customers with higher expectations of relationship continuity have lower service recovery expectations after a service failure and attribute that failure to a less stable cause, which leads to greater satisfaction with service recovery performance. Similarly, corporate social responsibility (CSR) and customer identification with the firm (tie strength) can enhance the recovery efforts.

The role of other customers in the service recovery process

The influence of interaction with other customers on the victim customer's emotions and behavior has been examined in the previous literature. Wu et al. (2014) examined the impact of other customers' territorial behaviors (refusing to share a table with another customer) on observing customers. When observing a customer's territorial tendency (individual factor), they found that the similarity of the rejected intruder (individual factor) and crowding (situational factor) jointly impact on observing a customer's negative emotions and fairness perception, which subsequently influence satisfaction. They found that when in a crowded setting, customers with low territorial tendency had a higher level of negative emotion and fairness perception when the rejected intruder is similar, compared to high territorial tendency.

According to deontic justice theory, people react to a perceived wrongdoing not because of their own self-interest but due to an inherited predisposition to be sensitized to unfair treatment. People judge fairness based on their a priori standards, regardless of their involvement with the other party. Related to third-party justice perception, the observation of other customers being treated unfairly negatively affects perceived fairness. Mattila et al. (2014) examine the effect of observing service recovery aimed another customer (good vs. bad) on emotional responses and perceived fairness, moderated by the valence of own experience and the processing frame of the focal customer (experiential vs. rational). They found that when service recovery is bad, they have negative emotional responses. Furthermore, even when they have positive own experience, their innate sense of fairness causes negative emotions. When primed to process the situation emotionally, they experienced a higher level of negative emotions when the victim customer was treated badly. Wan et al. (2011) examine the role of similarity and customers' regulatory focus on service evaluation when observing servicer recovery aimed at other customers. When there is a personal similarity with the other customer, and the observer customer has prevention focused regulation strategies, they blame the company more and have a poorer perception of service quality.

Conclusion

In conclusion, even with the most meticulous preparation and execution, service failures are inevitable in hospitality and tourism settings. As such, it is critical for firms to properly manage these service failures, as improper management can lead to significant negative consequences such as decreased satisfaction, lack of return intention, and increased negative word-of-mouth. Once a failure occurs, customers can either remain silent and potentially never visit the firm again or voice their complaints. Finally, firms and managers need to consider how to most effectively manage the recovery actions in order to satisfy the dissatisfied customers.

References

Adams, J. (1965). 'Inequity in social exchange', *Advances in Experimental Social Psychology*, 2: 267–299. DOI: 10.1016/S0065-2601(08)60108-2.

Baker, M. (2016). 'Managing customer experiences in hotel chains'. In M. Ivanva and V. Magnini (eds.), *The Routledge Handbook of Hotel Chain Management* (pp. 240–250). Milton Park, Oxfordshire, UK: Routledge. DOI: 10.4324/9781315752532.

Baker, M. (2017). 'Service failures and recovery: Theories and models'. In E. Koc (ed.), *Service Failures and Recovery in Tourism and Hospitality: A Practical Manual* (pp. 27–41). Oxford: Cabi.

Baker, M. and Kim, K. (2018). 'Other customer service failures: Emotions, impacts, and attributions', *Journal of Hospitality and Tourism Research*, 42(7): 1067–1085. DOI: 10.1177/1096348016671394.

Baker, M. and Kim, K. (2019). 'Value destruction in exaggerated online reviews: The effects of emotion, language, and trustworthiness', *International Journal of Contemporary Hospitality Management*, 31(4): 1956–1976. DOI: 10.1108/IJCHM-03-2018-0247 [Accessed: April 29th, 2019].

Baker, M., Magnini, V. and Perdue, R. (2012). 'Opportunistic customer complaining: Causes, consequences, and managerial alternatives', *International Journal of Hospitality Management*, 31(1): 295–303. DOI: 10.1016/j.ijhm.2011.06.004.

Bitner, M. (1990). 'Evaluating service encounters: The effects of physical surroundings and employee responses', *Journal of Marketing*, 54(2): 69–82. DOI: 10.1177/002224299005400206.

Bitner, M., Booms, B. and Tetreault, M. (1990). 'The service encounter: Diagnosing favorable and unfavorable incidents', *Journal of Marketing*, 54(2): 71–84. DOI: 10.1177/002224299005400105.

Dong, B., Evans, K. and Zou, S. (2008). 'The effects of customer participation in co-created service recovery', *Journal of the Academy of Marketing Science*, 36(1): 123–137. DOI: 10.1007/s11747-007-0059-8.

Gelbrich, K. and Roschk, H. (2011). 'A meta-analysis of organizational complaint handling and customer responses', *Journal of Service Research*, 14(1): 24–43. DOI: 10.1177/1094670510387914.

Gretzel, U., Fesenmaier, D., Formica, S. and O'Leary, J. T. (2006). 'Searching for the future: Challenges faced by destination marketing organizations', *Journal of Travel Research*, 45(2): 116–126. DOI: 10.1177/0047287506291598.

Gronhaug, K. and Zaltman, G. (1981). 'Complainers and noncomplainers revisited: Another look at the data', *Journal of Economic Psychology*, 1(2): 121–134. DOI: 10.1016/0167-4870(81)90034-9.

Gronroos, C. (1988). 'Service quality: The six criteria of good perceived service', *Review of Business*, 9(3): 10–19.

Harris, L. and Reynolds, K. (2003). 'The consequences of dysfunctional customer behavior', *Journal of Service Research*, 6(2): 144–161. DOI: 10.1177/1094670503257044.

Hess, R., Ganesan, S. and Klein, N. (2003). 'Service failure and recovery: The impact of relationship factors on customer satisfaction', *Journal of the Academy of Marketing Science*, 31(2): 127–145. DOI: 10.1177/0092070302250898.

Huang, W. (2010). 'Other-customer failure: Effects of perceived employee effort and compensation on complainer and non-complainer service evaluations', *Journal of Service Management*, 21(2): 191–211. DOI: 10.1108/09564231011039286.

Hui, M. and Au, K. (2001). 'Justice perceptions of complaint-handling: A cross-cultural comparison between PRC and Canadian customers', *Journal of Business Research*, 52: 161–173. DOI: 10.1016/S0148-2963(99)00068-5.

Joireman, J., Grégoire, Y., Devezer, B. and Tripp, T. (2013). 'When do customers offer firms a "second chance" following a double deviation? The impact of inferred firm motives on customer revenge and reconciliation', *Journal of Retailing*, 89(3): 315–337. DOI: 10.1016/j.jretai.2013.03.002.

Kelley, H. (1971). *Attributions in Social Interactions*. Morristown. General Learning Press.

Kelley, S., Hoffman, K. and Davis, M. (1993). 'A typology of retail failures and recoveries', *Journal of Retailing*, 69(4): 429–452. DOI: 10.1016/0022-4359(93)90016-C.

Kim, M., Wang, C. and Mattila, A. (2010). 'The relationship between consumer complaining behavior and service recovery: An integrative review', *International Journal of Contemporary Hospitality Management*, 22(7): 975–991. DOI: 10.1108/09596111011066635.

Kim, Y. and Baker, M. (2019). 'Observer reactions to other customer incivility: Emotional labor, gratitude, loyalty to employee and tipping intention', *International Journal of Contemporary Hospitality Management*, 31(3): 1292–1308. DOI: emeraldinsight.com/doi/abs/10.1108/IJCHM-03-2018-0262 [Accessed: April 29th, 2019].

Liao, H. (2007). 'Do it right this time: The role of employee service recovery performance in customer-perceived justice and customer loyalty after service failures', *Journal of Applied Psychology*, 92 (2): 475–489. DOI: 10.1037/0021-9010.92.2.475.

Liu, S. and Mattila, A. (2015). '"I want to help" versus "I am just mad" how affective commitment influences customer feedback decisions', *Cornell Hospitality Quarterly*, 56(2): 213–222. DOI: 10.1177/1938965515570939.

Matousek, M. (March, 2018). *United Airlines has a long history of infuriating customers — Here are its worst customer service incidents*. Available from www.businessinsider.com/united-airlines-worst-customer-service-incidents-2018-3 [Accessed: April 29th, 2019].

Mattila, A. (2004). 'The impact of service failures on customer loyalty: The moderating role of affective commitment', *International Journal of Service Industry Management*, 15(2): 134–149. DOI: 10.1108/09564230410532475.

Mattila, A., Hanks, L. and Wang, C. (2014). 'Others service experiences: Emotions, perceived justice, and behavior', *European Journal of Marketing*, 48(3/4): 552–571. DOI: 10.1108/EJM-04-2012-0201.

Mattila, A. and Ro, H. (2008). 'Discrete negative emotions and customer dissatisfaction responses in a casual restaurant setting', *Journal of Hospitality and Tourism Research*, 32(1): 89–107. DOI: 10.1177/1096348007309570.

Maxham, III, J. and Netemeyer, R. (2002). 'A longitudinal study of complaining customers' evaluations of multiple service failures and recovery efforts', *Journal of Marketing*, 66(4): 57–71. DOI: 10.1509/jmkg.66.4.57.18512.

McCollough, M. (2009). 'The recovery paradox: The effect of recovery performance and service failure severity on post-recovery customer satisfaction', *Academy of Marketing Studies Journal*, 13(1): 89–104.

McQuilken, L., McDonald, H. and Vocino, A. (2013). 'Is guarantee compensation enough? The important role of fix and employee effort in restoring justice', *International Journal of Hospitality Management*, 33: 41–50. DOI: 10.1016/j.ijhm.2013.01.005.

Namkung, Y. and Jang, S. (2010). 'Service failures in restaurants: Which stage of service failure is the most critical?', *Cornell Hospitality Quarterly*, 51(3): 323–343. DOI: 10.1177/1938965510364488.

Roggeveen, A., Tsiros, M. and Grewal, D. (2012). 'Understanding the co-creation effect: When does collaborating with customers provide a lift to service recovery?', *Journal of the Academy of Marketing Science*, 40(6): 771–790. DOI: 10.1007/s11747-011-0274-1.

Russell, D. (1982). 'The causal dimension scale: A measure of how individuals perceive causes', *Journal of Personality and Social Psychology*, 42(6), 1137–1145. DOI: 10.1037/0022-3514.42.6.1137

Singh, J. (1988). 'Consumer complaint intentions and behavior: Definitional and taxonomical issues', *Journal of Marketing*, 52 (January): 93–107. DOI: 10.1177/002224298805200108.

Susskind, A. (2015). 'Communication richness: Why some guest complaints go right to the top—and others do not', *Cornell Hospitality Quarterly*, 56(3): 320–331. DOI: 10.1177/1938965514560161.

Tax, S., Brown, S. and Chandrashekaran, M. (1998). 'Customer evaluations of service complaint experiences: Implications for relationship marketing', *Journal of Marketing*, 62(2): 60–76. DOI: 10.1177/002224299806200205.

Wan, L., Chan, E. and Su, L. (2011). 'When will customers care about service failures that happened to strangers? The role of personal similarity and regulatory focus and its implication on service evaluation', *International Journal of Hospitality Management*, 30(1): 213–220. DOI: 10.1016/j.ijhm.2010.07.004.

Weiner, B. (1979). 'A theory of motivation for some classroom experiences', *Journal of Educational Psychology*, 71(1): 3–25.

Wirtz, J. and Kum, D. (2004). 'Consumer cheating on service guarantees', *Journal of the Academy of Marketing Science*, 32(2): 159–175. DOI: 10.1177/0092070303261416.

Wu, L., Mattila, A. and Han, J. (2014). 'Territoriality revisited: Other customer's perspective', *International Journal of Hospitality Management*, 38: 48–56. DOI: 10.1016/j.ijhm.2013.12.001.

Xie, D. and Heung, V. (2012). 'The effects of brand relationship quality on responses to service failure of hotel consumers', *International Journal of Hospitality Management*, 31(3): 735–744. DOI: 10.1016/j.ijhm.2011.09.010.

Xu, Y., Tronvoll, B. and Edvardsson, B. (2014). 'Recovering service failure through resource integration', *The Service Industries Journal*, 34(16), 1253–1271. DOI: 10.1080/02642069.2014.942652

Yang, W. and Mattila, A. (2012). 'The role of tie strength on consumer dissatisfaction responses', *International Journal of Hospitality Management*, 31(2): 399–404. DOI: 10.1016/j.ijhm.2011.06.015.

Yuksel, A., Kilinc, U. and Yuksel, F. (2006). 'Cross-national analysis of hotel customers' attitudes toward complaining and their complaining behaviours', *Tourism Management*, 27: 11–24. DOI: 10.1016/j.tourman.2004.07.007.

31

Dark tourism

Negative experiences as part of modern tourism

Dinesh Kumar and Punam Gupta

Introduction: dark tourism

Dark places – associated with stories of death, ghosts and tragedy – have always fascinated human beings. Such places generate feelings of revulsion and fear among many people, but an increasing number of tourists are seeking out such places in search of extreme experiences. Sometimes the motive is to learn about events that happened in history, such as wars, murders and tragedies. At other times, people just want to experience what others went through in the past, such as the imprisonment of Nelson Mandela or victims of the slave trade. Such places and experiences are now increasingly being packaged and marketed just like traditional tourist destinations.

Experiencing negative incidents has seen a rise in recent years. Kumar (2012) points to the numerous books written about and movies made around death, torture, kidnappings, vampires, ghosts and destruction, which have become hugely popular in recent years. Dark marketing – described as "the application or adaptation of marketing principles and practices to domains of death, destruction and the ostensibly reprehensible" – is becoming widespread indeed. Foley and Lennon (1996) write that dark tourism, that is, visits to places of death, disaster and atrocities, increasingly form part of the tourism landscape. They write that dark tourism and its marketing simply implies "the presentation and consumption of real and commodified death and disaster sites."

Dark tourism must be distinguished from historical or heritage tourism, which involves visiting places of historical interest. Most people for instance, visit the Taj Mahal in India for its architecture and beauty, but it is essentially a tomb. A visit to the Tower of London is of historical interest but one may well discover dungeons or torture chambers there. Similarly, a visit to Jallianwala Bagh in India, where General Dyer opened fire on a peaceful gathering in 1919, is undertaken more for historical reasons. Would such visits be termed dark tourism?

The answer lies in the motivation of such visits. If they are undertaken to learn about history or heritage and not primarily for negative experiences, they do not qualify as dark tourism. Dark tourism refers to visits that are undertaken to places that are primarily linked to death and evil and not as a subset of heritage tourism.

Types of dark tourism

Known by many names, dark tourism is also referred to as black tourism or grief tourism. Another term associated with dark tourism is thanatourism. Light (2017) explains that while dark tourism refers to visits related to "death, suffering, atrocity, tragedy or crime," thanatourism refers to travel "motivated by a specific desire for an encounter with death." Brown et al. (2012) explain that "thanatourism" has been around for a long time and has attracted visitors for a very long time. Some of the sites that are associated with death and tragedy globally can be described as:

- Sites associated with war and conflict, war cemeteries and sites of mass burials
- Sites associated with the Holocaust, concentration camps, transit camps
- Prisons, dungeons, detention centers and prisoner of war camps
- Disaster sites such as Ground Zero, New York and Chernobyl in Russia
- Genocide sites
- Places associated with the slave trade
- Sites of natural disasters
- Cemeteries, burial sites and exhibits of bones and skulls at various sites; sites of individual or mass murder
- Ghost tours, walks and experiences
- Entertainment-based dark tourism sites
- Staying in a prison or in a haunted house; experiencing fright or torture

There is indeed a kind of fascination for such places. A growing number of people want to visit places that are associated with anything unpleasant such as death, suffering and the macabre. It may be termed as "dark travel" or travel that leads to traumatizing, disturbing or unsettling experiences, as described by Clark (2014). The primary reason for undertaking the visit is not for sightseeing or history, but an encounter with the negative. Korstanje and George (2015) write that such an experience "entails fascination with death as a primary reason of attraction."

Ashworth (2008) defines dark tourism as when the tourist experiences "dark" emotions such as pain, death, horror or sadness, which are not usually associated with an entertainment experience.

Several types of dark tourism have been identified (see Table 31.1). The motivation and interest varies according to the visitors' fascination with death. Nevertheless, places attract visitors too in which the original intent was not to attract visitors, but they are still sought after by tourists. Stone (2006) visualizes the types of dark tourism in a spectrum from "darkest" to "lightest." The scale showing its primary interest is shown as a continuum with serious experiences of death on one side, elements of fun on the other (Figure 31.1) and various shades of gray in between. The intention to visit these sites varies from experience and remembrance at one end to fun on the other.

Several types of dark tourism have been identified within this spectrum:

Dark fun factories: This category consists of tours that primarily provide entertainment. Such places are highly commercialized with scare tunnels, skeletons and macabre exhibits. Set up by entertainment firms, they provide family fun and entertainment and a sanitized fright experience.

Dark exhibitions: Exhibitions intended to educate and inform about past events, usually with a commemorative and reflective message. The intent of these exhibits is serious, not

Table 31.1 Types of dark tourism

penal/prison tourism	Places of confinement, such as Alcatraz and Robben Island
fright tourism	Fright as an experience: haunted houses, dark fun factories
genocide tourism	Sites of mass killings: Holocaust, Cambodia's Killing Fields
grief tourism	Places where people have lost their ancestors
disaster tourism	Natural and man-made disasters
pagan tourism	Tourism to old churches and temples, outdated religions
suicide tourism	Places associated with famous suicides
conflict heritage tourism	Places of past conflicts and struggles
dystopian dark tourism	Representing uncertainty about the future and an increased insecurity about death
poverty tourism	Experience of extreme poverty

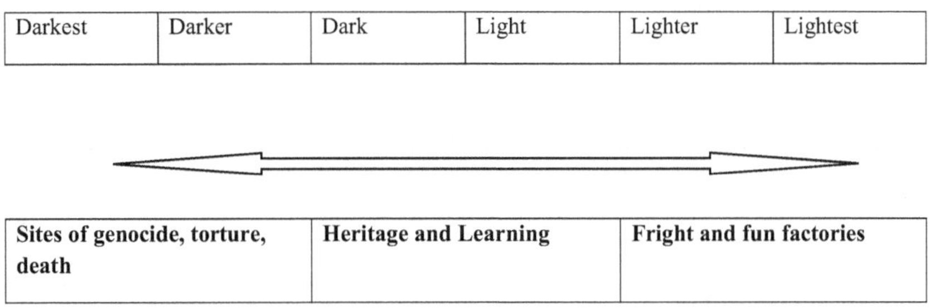

Figure 31.1 Shades of dark tourism

entertainment. Exhibitions of the 9/11 attacks, or the collection of bones of people killed, would be termed dark exhibitions.

Dungeons: Prisons and dungeons have become popular in recent times. Commercialized to a great extent, they are a mixture of dark experiences and entertainment. The National Justice Museum in Nottingham, UK, for instance, offers "inspirational curriculum-linked educational visits … to help pupils gain a practical understanding of the law and justice system." It offers visits to old courthouses and shows exhibits of how prisoners were held, tortured and killed. It is the place where you can "Feel the Fear." Housed in former prisons, it is the "only site in the country where you could be arrested, sentenced and executed." The site allows the visitor to "become part of the dramatic history" and therefore provides both education and entertainment.

Cemeteries and graves: Gravestones and tombs have a fascination for many visitors. Some sites have been developed into parks and museums. Places of death of leaders and famous people also attract interest.

Dark conflict sites: Places that have seen wars are also marketed as tourism sites. These sites are intended to show past happenings and provide an opportunity to learn about history. They are also referred to as heritage sites.

Genocide sites: Sites where mass murders have taken place, such as the killing fields of Cambodia, Rwanda and World War II, stand as stark reminders of how people were led to murdering others because of some belief or other. Such places serve as "a cry of despair and a warning to humanity," as the memorial at the Auschwitz-Birkenau camp reads.

While such sites attract tourists for education and remembrance purposes, in recent times they are being marketed and positioned as tourist places. The next section shows the marketing of dark tourism and its related criticisms.

Dark marketing: sick sells

Dark marketing refers to the marketing of tours and places of death and suffering. Such dark places are packaged, promoted and priced just like any other product or service. Brown et al. (2012) mention that just like the glossy marketing of tourist destinations, sites of death or suffering are marketed with tools like positioning and packaging. Dark sites are sold and branded like any tourist destination, with merchandising and take-away products. Death and tragedy are thus marketed as unique tourism experiences. A British company, Historical Trips, for instance, organizes a tour of concentration camps and torture chambers of Nazi Germany called the "Face of Evil" tour. Its website explains: "Our tour takes in the camp huts, gas chambers and crematoria, as well as the memorials to those who died."

Here, one can witness a particularly evil period of human history, in which millions of people were sent to their death under the Nazi campaign of ethnic cleansing (see case study). These sites bring the stark reality of German death camps to life.

The authors' visit to the Holocaust museum in Washington was a sober and humbling experience. People wept to see the exhibits showing how the Jews were isolated and systematically sent to their deaths. Most visitors were shocked into silence by the visuals.

Criticism arises when such visits are undertaken for less serious purposes. The Face of Evil tour has been condemned for its grotesque sensationalism and bad taste and has been called a "perverse pilgrimage." Such perverse pilgrimages are organized around the drugs trade, the sex trade, blood diamonds, counterfeit goods, funerals and natural and man-made disasters. A whole industry has been built around negative experiences, including the entertainment industry, the gaming industry and even the fashion industry.

Since there is a segment of tourists who want to experience fear for the thrills, we can surmise the reasons for seeking such experiences: apart from learning about history, people want others to see where they have been, in what may be termed "Facebook tourism." People posting pictures and boasting about where they have been has helped tourism the world over and dark tourism is no exception.

Perhaps that is the reason that perversity sells. Television and movies have promoted evil themes liberally in recent years, complete with forced incarcerations, vampires, dark plots and super-villains. The success of the *Twilight* series, *Hunger Games*, *Fifty Shades of Grey* and *Game of Thrones* shows that dark experiences do indeed sell. Movies today depict images of violence and destruction in increasingly graphic ways, thanks to modern technology. The places shown in these movies and TV shows add to further attractions for the dark tourist. Computer games go a step further – they provide the experience of killing and destruction in a world that is increasingly looking less virtual and more real.

Although concerns are often expressed about the impact of violent or disturbing images, such experiences are marketed widely, proving that sick indeed sells. Apart from movies and television and gaming, product advertising too uses such images frequently. *Business Insider* (2012) reports about 13 brands that use sex to sell products. Among them, an advertisement by Dolce & Gabbana is mentioned that shows half-naked men standing around a lone woman, who's lying on the ground in a position suggesting rape. Other brands too have experimented with negative images, sometimes inviting consumer backlash. However, such messages and images are increasingly becoming popular in popular media. While the earlier horror films existed as a separate genre, today dark experiences are found in mainstream entertainment and media.

Case study 31.1

For visitors to the former Nazi death camps, the brutality and the evil that existed there still sends shock waves down their spines. It is one thing to read about the war or see movies about it, but it is quite another to actually visit the sites where the unspeakable atrocities were committed.

Tourism to Europe's sites of German atrocities – concentration camps, death factories and the systematic extermination of a very large population – stands monument to human depravity and is the largest example of dark tourism.

Today these sites attract visitors and tourists from all over the world. Some have lost families in the Holocaust, others have come to learn about history. People come from all over the world to experience and understand what happened during those dark years in World War II (see Figure 31.2). Though Auschwitz was the largest of the death camps, many others were built by the Germans. Poland is most visited by Holocaust tourists because of the number of death camps in the country.

Figure 31.2 Visitors at the Auschwitz camp. Photograph by Adam Jones (Wikimedia commons)

Source: https://commons.wikimedia.org/wiki/File:Auschwitz_I_Camp_with_Arbeit_Macht_Frei_Gate_-_Oswiecim_-_Poland.jpg

Accounts by visitors show that this is no ordinary tourism, but brings immeasurable pain and disbelief that such evil actually existed. An account by Alexander Schwabe (2005) of a visit to the Auschwitz death camp, where more than one million people were killed in death factories, describes how the tour has a chilling effect on the visitors.

Here, visitors see the gas chambers where people, arriving in cattle cars from different parts of Europe, were gassed to death. Their bodies were incinerated or burnt in huge bonfires and the ashes scattered in fields or thrown into the river. "There is not one spot of ground where one can walk without stepping on the ashes of cremated prisoners," he writes.

The tour shows up more horror and cruelty. Visitors can see torture chambers, which were so narrow that prisoners could not sit down. They were forced to spend the night here, and then were sent to hard labor during the day. "Many were kept so long in the cells that they simply died," writes Schwabe. Visitors also see the chambers where the prisoners had to undress before they were shot. Many prisoners simply suffocated to death in the narrow and dark basement chambers.

More shocks await. An exhibit of empty canisters of the insecticide used to gas people to death, another of a two-ton pile of human hair taken from the victims, which was used to make blankets and clothes, and collections of personal belongings of the victims show the stark reality of what was done at the death camps. A pile of prisoners' glasses, another of prosthetic arms and legs and crutches, a room filled with suitcases still with the names, addresses and birthdays on them – such exhibits show how unsuspecting victims were transported and simply sent to their deaths. Among the victims were some 200,000 children, which makes the crimes unforgivable. A pile of thousands of children's shoes and another of adult shoes brings home the stark reality that people were killed deliberately and cruelly here.

The tours and the exhibits are a humbling experience. "... no one is talking. The visitors keep their heads down. No one makes eye contact. Each person has to find a way of coming to terms with the reality behind the glass," writes Shwabe. Many visitors break down.

Another account is given by Patrick Regan, who visited sites in Europe, including Anne Frank's house in Amsterdam. Anne and seven others had hidden here during the German occupation, made famous by her diary. "I felt the history, just as I had hoped. It was somber, informative and thrilling all at once," he writes. For example, the author could peek out of the window and steal a look at the world outside in just the way that Anne Frank would have done. Such things provide visitors with a strong connection to history and they can feel the immense fear of hiding from the Nazis.

These experiences leave a lasting mark on visitors. "Leaving the memorial, I was drained. It was then that I doubted my itinerary. Over eight days, I needed several breaks from my trip's gloomy focus," explains Regan.

Such tours help people to experience the horror that people went through before being sent to their deaths. Visitors can also ponder the question, "What would you have done under the Nazi rule?" through interactive exhibits. Visitors are forced to think of their own beliefs and assess the risks before answering whether they would collaborate with the Nazis, resist or do nothing.

Historical Trips organizes "The Face of Evil" tours that take visitors to Hitler's home and retreat, the infamous "Eagle's Nest," Dachau, the first Nazi concentration camp, and other sites associated with the war. "To make this journey is to truly understand modern history's greatest tragedy," says its website.

In the last two decades, Jewish heritage travels have become popular. Auschwitz was the largest of the Nazi death camps and thus attracts the maximum interest. In Israel, the March of the Living (MOTL) organizes Holocaust tours, sending young people to Poland and Israel. Poland is the most common destination of Holocaust tourists, where over three million Jews or 90% of their population was exterminated.

Though the tours bring visitors to experience the horror of the war, they are not without critics. People wonder about the morality of turning sites of mass murder into tourist sites. Others have asked whether tour operators should use the Holocaust sites and sites related to other crimes to make profits. Taking selfies or posting photoshopped images among dead bodies is distasteful and insults the memory of those who were killed there.

Whatever the criticism, it cannot be denied that such dark tourism sites serve the purpose of bringing history to life. Apart from experiencing what people went through during the Holocaust, they bring ethical questions starkly before visitors: at what point do they discard humanity in the face of nationalist or religious fervor? What would it take for people to stand up against atrocities?

The authors had visited the Holocaust museum in Washington some years back. We were moved beyond words to see exhibits showing how people were denied basic survival during the war. "Think about what you saw," said a notice at the exit gate. For many visitors, including the authors, the experience left permanent memories.

References

Schwabe, A. (2005). "Holocaust Tourism: Visiting Auschwitz, the Factory of Death". *Spiegel Online*. Available at: http://www.spiegel.de/international/holocaust-tourism-visiting-auschwitz-the-factory-of-death-a-338815.html. [Accessed March 24, 2019].

Historical Trips website: https://www.historicaltrips.com/tour/17/Germany/The-Face-of-Evil.html. [Accessed September 10, 2019].

Regan, Patrick (2018). "Dark tourism: 3-city trip through Europe's WWII history brings painful past to life". *Chicago Tribune*. https://www.chicagotribune.com/lifestyles/travel/ct-trav-dark-tourism-europe-0318-20180302-story.html [Accessed September 12, 2019].

Motivations of visiting dark places

Why should anyone visit unpleasant places? Why should people subject themselves to the inconvenience and horror of being incarcerated or witness the torture inflicted on others? There are no easy answers but the motivations could be related to remembrance, education or entertainment.

Ashworth and Hartmann (2005) describe three principal motives for visiting sites of tragedy:

- Curiosity
- Attraction to horror, which may border on the voyeuristic
- Empathy or identification with the victims

However, there are other motives for undertaking dark tourism. People may often visit these sites out of curiosity, to learn about what happened there. People with a genuine interest in history and culture want to visit such places and learn about the past. It may serve as a remembrance for the victims or those close to them: a pilgrimage of sorts if one has lost family members in such events. The motivations for visiting dark tourism spots thus relate to the visitors' identity and expectations. Tinson et al. (2015) explain, "dark tourism experiences can be used to (re)affirm individual roles, enhance feelings of national identity and co-create a self as well as a national identity."

Very often, such tours are undertaken out of a sense of moral duty: to educate future generations so that past atrocities never take place again. The Holocaust museums in several cities, for instance, stand as monuments to past evil, educating visitors about the dangers of a repeat of circumstances. For instance, one can buy merchandise at such sites saying, "Never Again!"

Dark tours may also be taken as a pilgrimage for inner purification, to know and learn about the baser instincts of human beings and to seek redemption. If one has lost family members in such disasters, the pilgrimage aspect is strong, as is the interest in personal genealogy and family history. Quite often, the journey is undertaken to connect with one's roots, when today's generation wishes to find out about the lives of their ancestors and undertake such visits. Slavery sites, for example, attract young people who wish to learn about how their ancestors were transported to distant lands.

It cannot be denied that there is a fascination among human beings with evil as well. Followers of overthrown regimes and religions may visit places of atrocities for nostalgia and hope for their revival. Others may just look for thrills and titillation, and still others may undertake dark tours out of a morbid curiosity.

Another explanation is offered with regard to the world in which we live, which is a world in which everything is a commodity to be marketed. In such a world, people search for "authentic" places and experiences. Places that have been notorious offer experiences that people actually lived through and suffered.

Finally, there is the motivation of seeking experiences just as a normal tourist. People want to tell their friends about the new places they have been to, posting pictures on their walls on social networking sites. Such tourism is often criticized, because it seems incongruous to be taking selfies in front of torture or death chambers. Yet, many tourists do so.

Political ideology and marketing of dark tourism

Many dark tourism sites have been marketed for political influence, which has led to their commercialization. Strange and Kempa (2003) study the political influence upon two former prisons: the US prison of Alcatraz, home to well-known prisoners, and Robben Island in South Africa, where many political prisoners including Nelson Mandela were once held. The former has been made famous by Hollywood movies and is known as "the Rock." Both prisons are presented through the prism of the political and cultural interpretation. "Memory managers" seek to interpret the pasts of the two sites, which are guided by political considerations.

Other sites across the globe are also interpreted through the prism of nationalist identities and presented as such. Movies, even portraying imaginary characters, have been the target of ire as was shown in widespread threats and violence about a recent release, *Padmavat*, a story about a queen and other women who commit suicide by jumping into a large bonfire. Such politically motivated reinterpretation, and its subsequent marketing, revives interest in places historically associated with violence and destruction and leads to their commodification.

Highly emotional and politically charged history is easy to market and political parties seek gains through dark tourism marketing of historical sites. In Stone's (2006) spectrum, the darkest sites have higher political ideology, while the lightest ones have little or no political undertones. However, attaching political motives and reinterpretation attracts charges of trivialization and historical inaccuracy.

The dark experience

In this section, we will find out about the feelings and emotions that visitors go through at such stark places. Cooke (2012) writes that people who undertake tours to dark places experience complex reactions. Buda (2015) mentions that dark places generate feelings of danger and fear mixed with excitement. But we may add that it is experiencing the fear safely. People know that their experience of staying in a prison, for instance, is artificial and short, as they are not going to be imprisoned. They know they will return to the safe environment of their homes without being harmed. It may be added that such an experience helps people to learn about history and past events.

Nawijn et al. (2018) write that dark tourism sites generate a wide range of emotions, reactions and feelings: the most common feelings are those of anger, sorrow, disgust, shock, horror and fear. Kidron (2013) observes that visiting them may also produce personal and social catharsis, while Tinson et al (2015) write that many people seek out historical tragedy sites out of a concern to learn about and better understand what actually happened. This could be out of a need to affirm their identity as educated and responsible people.

Other studies, such as one by Buda and McIntosh (2013), have pointed to the psychoanalytical origins of voyeurism. An inborn but subconscious fascination with death and suffering has also been mentioned.

Ethical question

If marketing is said to be the fulfillment of human needs, dark marketing simply caters to the hidden or latent needs of people. These needs have existed since time immemorial – human history is a ledger of deeds meant to inflict misery on others, as testified by wars, dungeons and torture and death chambers found across the world. Horror movies have also been popular.

The ethical question arises about how places associated with death and suffering are presented to visitors. One problem is when such places are shown simply as sanitized tourist destinations: do they disrespect the memory of victims who were tortured or murdered there? Other ethical problems are related to associating dark places with distorted or misrepresented historical events: the events are rewritten from the point of view of a particular nation, religion or identity. Another problem arises when heritage sites prioritize entertainment over education and historical accuracy. With a focus on entertainment, "dark edutainment" is created when the educational role of places of death is compromised by creating a spectacle and entertainment out of serious past events.

Present day leaders and governments often try to build a nationalist fervor and distort facts to present history that is favorable to them or their parties. Very often visitors are treated to "bogus history" and heritage sites are presenting them with a version of history that creates nostalgia and reassurance. Nationalist parties often distort historical facts and present dark events as nationalist victories. Ethical concerns arise when such places are used for broader political agendas.

Finally, there is a concern that dark tourism commodifies death for tourist consumption. Questions have been raised about making profits out of places associated with suffering and tragedy. The marketing of such destinations attract tourists who have no interest in the sites but to post their selfies. It is extremely distasteful to find tourists posting selfies and "liking" mass murder sites like Auschwitz, or expressing fun near Chernobyl, which people had to leave hurriedly during one of the worst nuclear disasters of our age.

Writing in *National Geographic*, Robert Reid says that "the problem lies not with the choice of destination, but with the intention behind the choice." The ethical question is thus rooted in the motivations of tourists when they visit such places.

Conclusion

Dark tourism is a phenomenon that is rising over the years. It has become popular because of many motivations and results in a range of emotions for the visitors. Attempts have been made to understand why people undertake journeys that belong to unpleasant times in history, especially if these are not related to family history. Do they appeal to a certain kind of tourist? Do they satisfy hidden but base personalities? Since dark tourism is becoming popular across wide sections of the population, these questions cannot be answered on the basis of personality only.

The answer may lie in Jung's concept of the collective unconscious, the legitimate sublimation of unconscious drives. History is part of our collective unconscious, consisting of our primordial images or what we have inherited. The evils of the past give us a guilty conscience – how could society or humanity have been so cruel? Dark tourism results from humanity's desire to answer this question because dark sites challenge our very ideas of culture, nation and humanity.

Walter (2009) surmises that the unquiet dead – people who were killed needlessly, not from accidents or illnesses – haunt us and our societies. Society is guilty of slavery and racism and tries to find answers about how they flourished. We want to find answers to why so many people died in vain and for meaningless causes, and these deaths challenge our understanding of civilization, nationalism and modernity. There is also an unstated desire to prevent such things from happening again. The Holocaust, for instance, challenges our very notions of nation, progress and rationality. Many visitors at the Holocaust sites try to wrestle with the question of how this could have happened in the twentieth century.

Indeed, the power of such places is precisely that they stir the imagination: Anne Frank is just not an individual but represents all the children who were sent to their deaths, and the families whose lives were destroyed by war and racism. Dark tourism sites confront people with mortality and suffering.

Whatever the motivations of dark tourism, it cannot be denied that dark places of tragedy attract human beings. It is a natural instinct, such as slowing down our cars when we see an accident on the highway or watching TV for every single detail of disasters and crime. The factors range from cultural and personal to socio-psychological. Dark tourism has a deeper purpose as well – it connects people through shared emotions and educates generations about past events. If history teaches us a lesson that such events should not be repeated, the purpose of dark tourism has been firmly established. Convery, Corsane, and Davis (2014) write in the introduction of their book, *Displaced Heritage: Responses to Disaster, Trauma, and Loss*: "Disasters can also make us better informed about the heritage we have lost, or are losing, leading to attempts to better preserve what remains."

References

Ashworth, G.J. (2008). The memorialisation of violence and tragedy: human trauma as heritage. In B. Graham and P. Howard (Eds.), *The Ashgate companion to heritage and identity* (pp. 231–244). Hampshire, UK: Ashgate.

Ashworth, G.J. and Hartmann, R. (2005). *Horror and Human Tragedy Revisited: The Management of Sites of Atrocities for Tourism*. New York: Cognizant.

Brown, S., McDonagh, P., and Shultz, C. (2012). 'Dark marketing: ghost in the machine or skeleton in the cupboard?' *European Business Review*, 24(3): 196–215. DOI: 10.1108/09555341211222477l.

Buda, D.M. (2015). *Affective Tourism: Dark routes in conflict*. London: Routledge. DOI: 10.4324/9781315742786.

Buda, D.M. and McIntosh, A.J. (2013). 'Dark tourism and voyeurism: tourist arrested for "spying" in Iran' *International Journal of Culture, Tourism and Hospitality Research*, 7(3): 214–226. DOI: 10.1108/IJCTHR-07-2012-0059.

Clark, L.B. (2014). Ethical spaces: ethics and propriety in trauma tourism. In B. Sion (Ed.), *Death tourism: Disaster sites as recreational landscape* (pp. 9–35). London: Seagull.

Convery, I., Corsane, G., and Davis, P. (2014). *Displaced Heritage: Responses to Disaster, Trauma, and Loss*. Woodridge, UK: The Boydell Press.

Cooke, S. (2012). Sebald's ghosts: traveling among the dead in the rings of Saturn. In J. Skinner (Ed.), *Writing the dark side of travel* (pp. 47–62). Oxford: Berghahn.

Foley, M. and Lennon, J.J. (1996). 'JFK and dark tourism: A fascination with assassination' *International Journal of Heritage Studies*, 2(4): 198–211. DOI: 10.1080/13527259608722175.

Historical Trips website: www.historicaltrips.com/faceofevil.html. [Accessed: March 20, 2019].

Kidron, C. (2013). 'Being there together: dark family tourism and the emotive experience of co-presence in the holocaust past'. *Annals of Tourism Research*, 41: (175–194). DOI: 10.1016/j.annals.2012.12.009.

Korstanje, M. and George, B. (2015). 'Dark tourism: revisiting some philosophical issues'. *e-Review of Tourism Research*, 12(1/2): 127–136. Available at: 3ws1wk1wkqsk36zmd6ocne81.wpengine.netdna-cdn.com/files/2015/08/4.-eRTR_ARN_Vol.12-No1.2_Korstanje-and-George.pdf.www.historicaltrips.com/faceofevil.html. [Accessed: March 2, 2019]:.

Kumar, D. (2012). *Consumer Behaviour*. New Delhi: Oxford University Press.

Light, D. (2017). 'Dark tourism and thanatourism research: an uneasy relationship with heritage tourism'. *Tourism Management*, 61. DOI: 10.1016/j.tourman.2017.01.011.

National Justice Museum website: www.nationaljusticemuseum.org.uk/[Accessed March 15, 2019].

Nawijn, J., Isaac, R.K., Gridnevskiy, K., and A. van Liempt, A.V. (2018). 'Holocaust concentration camp memorial sites: an exploratory study into expected emotional response' *Current Issues in Tourism*, 21(2): 175–190. DOI: 10.1080/13683500.2015.1058343.

Rogers, A. (2012). Brands that use sex to sell their products. *Business Insider*, February 3, 2012. Available at: www.businessinsider.com/13-brands-that-use-sex-to-sell-their-products-2012-2?op=1#ixzz2gXEbn3BZ [Accessed: March 27, 2019].

Schwabe, A. (2005). Holocaust tourism: Visiting Auschwitz, the factory of death. *Spiegel Online*, Available at: http://www.spiegel.de/international/holocaust-tourism-visiting-auschwitz-the-factory-of-death-a-338815.html [Accessed March 24, 2019].

Stone, P.R. (2006). 'A dark tourism spectrum: towards a typology of death and macabre related tourist sites, attractions and exhibitions'. *Tourism*, 54(2): 145–160. Available at: http://clok.uclan.ac.uk/27720/1/27720%20fulltext_stamped.pdf. [Accessed: March 10, 2019].

Strange, C. and Kempa, M. (2003). 'Shades of dark tourism: alcatraz and Robben Island' *Annals of Tourism Research*, 30(2): 386–405. DOI: 10.1016/S0160-7383(02)00102-0.

Tinson, J.S., Saren, M.A.J., and Roth, B.E. (2015). 'Exploring the role of dark tourism in the creation of national identity of young Americans' *Journal of Marketing Management*, 31(7/8): 856–880. DOI: 10.1080/0267257X.2014.995207.

Walter, T. (2009). Dark tourism: mediating between the dead and the living. In R. Sharpley and P. R. Stone (Eds.), *The darker side of travel: The theory and practice of dark tourism* (pp. 39–55). Bristol: Channel View.

Part IV

Technology enabled tourism experiences

32

Sharing economy and tourism experience

Adam Pawlicz

Introduction

The sharing economy virtually revolutionized tourism, especially in the hospitality market. The model of exchanging services among peers using innovative platforms facilitated entry to hospitality markets for many non-professional property owners who had idle apartments at their disposal. As a result, the supply of available beds in major city destinations more than doubled, creating a completely new social and economic environment. One of the key factors driving sharing economy growth was the specific feature of customer experience connected with the social aspects of staying among the local population and dealing with non-professional sellers. Shaping customer experience became an important issue, not only for platforms and accommodation providers who are interested in propelling the short-term rental market but also for destination marketing organizations. Accommodation experience constitutes an important part of the overall tourism destination experience and may potentially lessen or enhance it. Hence, it is important to understand the specific features of the sharing economy experience.

The notion of the sharing economy

Although the sharing economy emerges from the idea that people can earn additional income from using their underused resources it is now viewed as a model of peer-to-peer (P2P) exchange that is intermediated online by electronic platforms. At first, the term was associated with non-profit activities, then later with the usage of idle resources that was supposed to benefit the local community and the natural environment and finally it became a multimillion-dollar industry that completely changed the nature of transportation and hospitality markets. The sharing economy may be found in various markets that cater to tourism such as transportation (e.g. BlaBlaCar), guiding (Trip4Real), equipment rental (Boatsetter) or even dining (EatWith). All of these services might affect the tourism experience but only the hospitality sector with sharing economy platforms like Airbnb or HomeAway really changed the nature of the industry and therefore the hospitality part of the sharing economy will be examined closely in this chapter.

Discussing the origins of this term one should bear in mind that it belongs to those issues that are most difficult to define in both academic and popular tourism literature. The sharing economy has been referred to as a market, a phenomenon, an exchange model, a concept and many other things. Very often the sharing economy is associated with the part of the market that is intermediated by giant platforms, especially those companies that can be called iconic: Airbnb and Uber (Mikhalkina and Cabantous 2015). The fact that data for scientific research is almost entirely based on the platforms' activity renders previous definitions almost obsolete, as research is just carried out on given platform users.

Sharing economy platforms in the value chain

Although P2P markets are as old as the economy itself, the formation of enterprises which were able to obtain significant economies of scale and scope clearly made the individual production of most goods inefficient. This process was particularly visible in markets for tangible goods, whereas a majority of services, especially those which are labor-intensive, can still be produced by micro-enterprises. A vital characteristic of P2P markets relates to the fact that each peer might be either a seller or a customer in different transactions, so the clear division between professional entrepreneur and non-professional customer is blurred.

The hospitality industry provides an exemplary model of this process. The first hotel chains were developed in the mid-20th century and, especially in Europe, a majority of hospitality services are still provided by enterprises which operate with just one hotel (called independent hotels). The main value a hotel chain provides to its customers consists of standardized service and a brand which promises the same quality regardless of location. The Holiday Inn's famous commercial 'The best surprise is no surprise', from 1975, represents this concept in its purest form. A brand is, therefore, a tool which effectively ameliorates the endemic hospitality problem of information asymmetry between tourists and provider. As such a brand is a substitute for the classification of accommodation establishments provided either by public bodies or hospitality associations. A relatively homogenous product renders a comparison between different offers easy for potential tourists. A hotel chain's other competitive advantage was always an efficient reservation system, which allowed tourists to quickly complete a transaction. Central reservation offices, toll-free telephone lines, the administration of the chain's inventory by Global Distribution Systems, corporate websites, loyalty systems and finally mobile applications allowed tourists to significantly decrease transaction costs (O'Connor 2005; O'Gorman 2005, 2010; Dixit 2017, Comp. Andrews 2007).

The share of chain hotels in the total number of hotels is, however, different in various tourist destinations. Regions where the tourism product is based on natural attractions tend to have fewer hotel chains. Hospitality accommodation usually provides tourists with various leisure activities, which are meant to provide a unique experience, whereas a hotel chain provides a rather predictable product. Unlike a typical business trip and city break the average length of stay is much longer, which renders information asymmetry not an issue any more as tourists are prepared to invest more time in searching for information about a potential holiday resort than in the case of a 1–2 nights' business stay, when tourists' needs are much more limited. In the case of international leisure tourism, especially when a destination is reached only by air, information asymmetry is addressed mostly by powerful tour operators (Law, Leung, and Wong 2004; Zhang, Guillet, and Kucukusta 2015).

Leisure destinations that are visited by domestic tourists usually provide a wide range of available accommodation. The hospitality supply of hotels, pensions and bed and breakfast is complemented by the short-term rental of apartments, rooms or even beds. Traditionally these services catered for the budget market segment and were distributed either directly by customer word-of-mouth, straight encounters in local transportation hubs or via specialized intermediaries which, however, never had similar market power to tour operators and hotels.

The emergence of sharing economy platforms changed this traditional dichotomy and enabled the rapid growth of P2P hospitality markets also in cities. Sharing economy platforms introduced innovative markets which not only provided a secure payment platform but, most importantly, created trust between transaction parties. Similarly to platforms that intermediated on P2P markets for tangible goods (eBay-like), they introduced the review process which effectively transformed individual transactions into recurring ones. This innovation increased the costs of opportunistic behavior. Moreover, sharing economy platforms created an input system which converted heterogeneous products into a list of homogeneous offers. The basic classification of a hospitality product in the sharing economy is related to the presence of the owner in the flat during service delivery. The hospitality product can be classified into (1) whole apartment, (2) a room in a flat or (3) a bed in the owner's room. The other innovation was entry conditions which included ID verification, photos and an availability calendar and finally most platforms also provide insurance for both sides. By introducing those trust mechanisms, platforms were able to balance entry costs with the need for security.

To be successful, sharing economy platforms need to reach a critical mass of both providers and customers, as networking effects (which work in a very similar way to the telephone market) on this market are strong. The more peers that join the platform the more value it has for all users. Creating a platform is a simple technical task, but reaching potential customers even in niche markets is much more daunting. Sharing economy travel blog service sharetraveller.com identified 500 different platforms, out of which more than 160 were devoted to the hospitality market. It is not possible to check their market share because of a lack of appropriate data availability, but an analysis of organic search values which basically refer to traffic on that website revealed that there is just one leader – Airbnb – and only six other websites have reached more than 1% of its value (Pawlicz 2019). This may confirm the oligopolistic nature of competition between sharing economy platforms but increasingly OTAs (online travel agencies) are entering the short-term rental market (e.g. Booking.com).

Case study 32.1: Misterb&b

Misterb&b is a sharing economy platform which, similarly to its main competitors Airbnb and HomeAway, creates a short-term rental marketplace where customers are connected with apartment owners. The reservation process and the outlook of the website are very similar to other platforms but its size is considerably smaller (e.g. organic search traffic on Airbnb.com is 12.1M, homeaway.com is 994.7K while on misterbandb.com/ it is just 36.1K, according to the SEMrush service accessed on 05.12.2019 (SEMrush 2019)). Still, Misterb&b, unlike many other small platforms, has not been merged into one of the major market players (e.g. according to Crunchbase.com service, Airbnb has made 21 acquisitions (almost half of them between the years 2017 and 2019) and according to Wikipedia they operate 13 subsidiaries). This French start-up founded as late as 2013 is able to compete with other platforms due to its niche market strategy.

Misterb&b is advertising itself as 'the largest gay hotelier in the world with hosts in 135 countries and guests from all over the world'. The platform that meets the needs of the gay community emerged as its founder, Jost Matthieu, had experienced an uncomfortable welcome in the apartment he rented with his partner in Barcelona. Apart from security, similarly to non-profit sharing economy platforms, Misterb&b stresses social benefits as well. Providers are supposed to be tolerant and have knowledge about guy-oriented tourist attractions, so the platform's main competitive advantage is the harassment-free experience. Moreover, the website promotes gay events, provides guides to major tourist destinations and offers general safe traveling tips. First-hand advice from a host is stressed by website founders as more reliable. Similarly to Airbnb, Misterb&b collects tourism tax in various cities. Misterb&b now plans to be a platform for the whole LGBTQ community.

People identifying themselves as lesbian, gay or bisexual represent around 2% of the population in the UK, according to the Office for National Statistics (Bulman 2019) and in the USA the figure is supposed to be ca. 4% (Gates 2014). These numbers are higher among the younger part of the population. As this community is subject to the 'coming out' conundrum and often meets harassment or anti-gay sentiment, it is likely that their members represent a substantial market group to which major sharing economy platforms are also not indifferent. Almost all platforms have strict policies that ban any form of discrimination and threaten to remove from the platform those peers that fail to abide by these principles. The opposite strategy was undertaken in 2017 by Russian portal MyLinker.com, an alternative to Russian property owners unwilling to host gay people. From October 2016 the major player, Airbnb, bans users from turning away guests because of gender or sexuality. In 2015 Airbnb even launched a video campaign, 'Host with Pride', to promote LGBT travel. Still, gender discrimination and a huge number of listings cannot fully eliminate the inconveniences connected with meeting gay unfriendly guests.

Founders of Misterb&b chose to cater to a niche market and, more importantly, were able to correctly identify niche market benefits for the LGBT community. Those user benefits offset the fact that a smaller number of peers cannot ensure the same market efficiency (e.g. occupancy for hosts and value for money for guests) that the major players provide.

Reasons behind participation in the sharing economy

Very important for understanding the relationship between tourist experience and the sharing economy are motives for participation in the sharing economy. Regardless of the good that is needed, every customer is facing a fundamental choice: rent or buy. The decision is influenced by several factors:

1. Frequency of using a good. The more often a particular good is utilized the more likely it will be purchased. When a good is already bought, it may be later rented to others. A common example is a property in a resort, which is often used by the owner for just a few weeks and is re-rented in the remaining period.
2. Transaction costs of renting vs. costs of maintenance. One of the very first examples of the sharing economy given by Rachel Botsman was a drill, which is used on average between 12 and 15 minutes a year. Still, a drill is cheap to buy and does not take much space, whereas renting a drill means not only costs which include a platform fee, but also a time of delivery. This explains why sharing economy platforms intermediating on the markets for e.g. workshop tools have ceased to exist.

Sharing economy platforms are developed in the markets where a customer needs a different product each time they use the platform. In the case of renting services such as cleaning, teaching or walking a dog there is little value that platform may offer once an appropriate transaction partner is found. This can also explain why sharing economy platforms are more developed in cities rather than in areas with natural attractions. Repeated visits are much more frequent in resorts than in the case of city breaks and, unlike tour operators, sharing economy platforms are not offering packages which include tour escorts and are lower in price in comparison to purchasing via direct distribution channels.

As motives and experiences are strongly linked (Aşan and Emeksiz 2018) it is important to understand the reasons for joining the sharing economy from both the providers' and the customers' point of view. The motives of the decision whether to rent a good via a sharing economy platform rather than purchase services from traditional enterprises can be classified into two main groups (Balck and Cracau 2015; Tussyadiah and Pesonen 2016). The first one is connected with costs: tourists purchase on the P2P markets simply because (1) it is cheaper than on the traditional market or (2) it is more convenient, i.e. sharing economy platforms are able to reduce transaction costs and information asymmetry. The other group of motives can be classified as either social or related to ecology. Various studies clearly showed that economic-related motives are dominant regardless of country, type of goods traded on the market, age and method (Balck and Cracau 2015; Lampinen and Cheshire 2013; Teodoro et al. 2014; Ikkala and Lampinen 2015; Lee et al. 2015; Shaheen, Chan, and Gaynor 2016). Some studies showed, however, that the difference between economic and ecological motives is much higher in hospitality than in the transportation market (Balck and Cracau 2015; Möhlmann 2015). According to Belk, social motives can be predominant only in the case of non-profit platforms (Belk 2014). Tussyadiah (2016) showed that motives in the hospitality market can be determined by the type of product. For those renting whole apartments economic motives were most important, while for those renting just a room or a bed social aspects played a major role.

Impact of the sharing economy on tourist destinations

The proliferation of the sharing economy in tourism meant that supply in the hospitality market almost doubled in major tourist city destinations. This was possible because many flats, especially in core tourism areas, changed their primary use from apartments occupied by permanent residents and rented on a long-term basis to short-term rentals. Many new flats are bought and sold (and even advertised) as an investment for short-term rentals which is facilitated by the emergence of new intermediaries that act between owners and sharing economy platforms. Their services cover designing, renovation, cleaning, maintaining profiles on the sharing economy platform, checking guests in and out and clearing. Of course, some citizens make use of their idle resources and rent spare rooms and beds but according to AirDNA (*AirDNA* 2018) – a company that provides information about Airbnb listings – this represents just a fraction of supply in the major sharing economy platform. The supply is also much more elastic than before, as in case of major events many new offers are available. The growth of short-term rentals causes several externalities, which are neither internalized by transaction parties nor by intermediaries. First of all, tourists stay in living quarters, create additional noise and have less incentive to care about common areas etc., which in effect decreases the quality of life for permanent residents (Widener 2015; Edelman and Geradin 2016; Oskam and Boswijk 2016). These arguments constitute a basic ground for a call for sharing economy regulations. On the other hand, the sharing economy might contribute to the development of social capital in a destination (Deng 2016).

Characteristics of the sharing economy experience

Total tourist destination experience is a complex phenomenon which is based on tourists' experiences from various parts of the destination product. It can be argued that a destination experience is not just a sum of accommodation, attractions, guiding, transportation, infrastructure, etc., but rather a product, while the components of a destination are factors. Assuming a destination experience is a product, then if one of the components produces a zero experience the result is also zero. The main contribution of the sharing economy to total tourist experience is connected with its alleged social contribution to local areas. Social aspects are also featured in advertising materials produced by sharing economy platforms. It must be underlined that the concept of Like-a-local has not been introduced by sharing economy platforms but was known before and was even used in DMOs' marketing campaigns (Binkhorst and Den Dekker 2009).

When discussing the impact of the sharing economy on tourism experience, three distinct types of hospitality product must be differentiated:

1. Services obtained via non-profit sharing economy platforms
2. Rooms or beds rented via for-profit platforms
3. Whole apartments

Non-profit sharing economy platforms are usually a place where peers offer fee-free accommodation opportunities. Tourists using this type of accommodation usually get a separate room, a bed or just a floor where they can put their sleeping bag. When choosing a location, or on the other hand accepting a guest location, with conditions playing a minor role, the most important features are characteristics of peers, their interests, nationality and previous reviews (Jung et al. 2016). Social interactions and the development of language skills are the most important motives (Cova and White 2010). Peers spend evenings talking, cooking together and even bring presents. In this case, experience from sharing economy accommodation is decisive for the overall experience and the personal characteristics of the service provider (not a quality of the service itself) are decisive when choosing a place to stay.

In the case of monetary-based accommodation the social relationship is much weaker than in the first group. Although renting a room or a bed means that there is social interaction between peers, the author's research showed that this type of accommodation can be either sold by enterprises (hostels etc.) or by peers who have converted their houses into small hostels. Still, this was an exception (Pawlicz 2019).

Renting whole apartments to tourists rarely offers a distinct experience to a traditional hotel service. The authors' research showed that a majority of apartment owners actually never see their guests because they use intermediaries – so-called Host Management Companies – or simply leave their keys in a post box or (if available) at the reception. Others limit their interaction with guests to a welcome coffee. A major value of renting the whole apartment for tourists was based on lower costs (especially for families, small groups and for longer stays – perhaps 1-3 months), anonymity, convenience, a casual atmosphere and availability of a kitchen (Pawlicz 2019). In this case, reviews concentrate more on the place and its features than the host. If anything, a special experience may be created by the fact that by using any sharing economy services a customer accepts either a very limited range of additional services or none at all. It can therefore be seen as a 'second generation of experiences' which are based on co-creation (Boswijk, Thijssen, and Peelen 2007; Binkhorst and Den Dekker 2009). The possibility of cooking and accepting guests can be particularly attractive from both a social and an economic point of view for two groups of travelers: medium-term rentals (1–3 months) and families.

Scholars often highlight differences between the sharing economy experience (referred to as the Airbnb experience) and a traditional hotel experience, which is also underlined by platforms as its competitive advantage. That is, meaningful social encounters and a unique local experience. This authentic and unique experience can also generate new tourism products and enhance destination attributes (Heo 2016; Tussyadiah 2016; Tussyadiah and Pesonen 2016; Lin 2018). Still, as stated above, without direct contact between hosts and guests the local experience is even less than in the case of a hotel, where at least there is e.g. a possibility of a chat with the receptionist.

Another important aspect of experiencing hospitality services via sharing economy platforms is the element of risk which is endemic in all P2P markets. Although platforms were able to significantly reduce it for all transaction partners, the perceived risk is still higher than in the case of traditional hospitality service (Comp. Mittendorf 2017). Hospitality organizations use this argument as a rationale for public quality certification for sharing economy providers, partly because this potential risk deters the overall tourism experience. On the other hand, the involvement of risk when purchasing accommodation via a sharing economy platform might potentially enhance the total tourism experience as tourists, similarly to all other human beings, look for a certain level of excitement. This elated state of mind cannot be too low or high to maximize the experience. Still, assessment of the optimal level of risk depends greatly on the individual situation (Andersson 2007; Csikszentmihalyi 2014). The availability of sharing economy accommodation along with traditional accommodation establishments gives a potential tourist a choice of the accepted level of risk.

The main question which needs to be answered from a destination point of view consists of a relationship between the presence of the sharing economy and overall tourists' experience. Sharing economy providers, especially those renting just rooms or beds (also via a non-profit platform), are the first contact for tourists in a destination. As such they, similarly to tour guides, may influence the perception of a destination with their knowledge and attitude toward the destination. Local DMOs should first and foremost identify this group of providers, which can be a tedious task, as part of them are operating in the gray economy (Comp. Miller 2016). Ideally, DMOs should create an institutional cooperation platform and provide them with basic knowledge about the tourism product of a destination. DMOs might also organize familiarization trips, similar to those organized for tour operators, and training. The creation of incentives to participate in these undertakings for providers would constitute a challenge though. The cooperation with providers who offer whole apartments should follow the support scheme that is granted to other hospitality establishments.

Conclusions

The sharing economy is a new phenomenon which certainly revolutionized city tourism and brought destinations a wide range of opportunities and problems. Apart from enlarging the supply base, its growth constitutes an opportunity to provide tourists with a unique experience provided by the local community. This value is contributed mostly by those sharing economy providers who rent rooms or beds and stay at the property, whereas providers that rent whole apartments provide an experience similar to traditional hospitality companies.

References

AirDNA (2018). Available at: www.airdna.co/.[Accessed: April 04, 2018].

Andersson, T. D. (2007) 'The tourist in the experience economy', *Scandinavian Journal of Hospitality and Tourism*, 7(1): 46–58. DOI: 10.1080/15022250701224035.

Andrews, S. (2007) *Introduction to Tourism and Hospitality Industry*. New Delhi: Tata McGraw-Hill.

Aşan, K. and Emeksiz, M. (2018) 'Outdoor recreation participants' motivations, experiences and vacation activity preferences', *Journal of Vacation Marketing*, 24(1): 3–15. DOI: 10.1177/1356766716676300.

Balck, B. and Cracau, D. (2015) *Empirical Analysis of Customer Motives in the Shareconomy: A Cross-sectoral Comparison, Working Paper Series*. Otto von Guericke University Magdeburg.

Belk, R. (2014) 'Sharing versus pseudo-sharing in web 2.0', *Anthropologist*, 18(1): 7–23.

Binkhorst, E. and Den Dekker, T. (2009) 'Agenda for co-creation tourism experience research', *Journal of Hospitality Marketing & Management*, 18(2–3): 311–327. DOI: 10.1080/19368620802594193.

Boswijk, A., Thijssen, T., and Peelen, E. (2007) *The Experience Economy: A New Perspective*. Amsterdam: Pearson Education.

Bulman, M. (2019). Number of people identifying as lesbian, gay or bisexual hits record high, figures show. *Independent*.

Cova, B. and White, T. (2010) 'Counter-brand and alter-brand communities: The impact of web 2.0 on tribal marketing approaches', *Journal of Marketing Management*, 26(3–4): 256–270. DOI: 10.1080/02672570903566276.

Csikszentmihalyi, M. (2014) 'Toward a psychology of optimal experience', In *Flow and the Foundations of Positive Psychology* (pp. 209–226). Dordrecht: Springer Netherlands,. DOI: 10.1007/978-94-017-9088-8_14.

Deng, F. (2016) *The Sharing Economy and Urban Property Rights*. Chongqing. Available at: https://papers.ssrn.com/sol3/papers.cfm?abstract_id=2861152 [Accessed: December 19, 2016].

Dixit, S. K. (Ed). (2017). *The Routledge Handbook Of Consumer Behaviour In Hospitality And Tourism*. London: Routledge. DOI: 10.4324/9781315659657.

Edelman, B. and Geradin, D. (2016) 'Efficiencies and regulatory shortcuts: How should we regulate companies like Airbnb and Uber?', *Stanford Technology Law Review*, 19: 293–328. Available at: http://papers.ssrn.com/sol3/papers.cfm?abstract_id=2658603 [Accessed: December 27, 2016].

Gates, G. J. (2014). *LGBT Demographics: Comparisons among Population-based Surveys*. Available at: www.escholarship.org/uc/item/0kr784fx [Accessed December 27, 2018].

Heo, C. Y. (2016) 'Sharing economy and prospects in tourism research', *Annals of Tourism Research*, 58: 166–170. DOI: 10.1016/j.annals.2016.02.002.

Ikkala, T. and Lampinen, A. (2015) 'Monetizing network hospitality', in *Proceedings of the 18th ACM Conference on Computer Supported Cooperative Work & Social Computing – CSCW '15*. New York, USA: ACM, pp. 1033–1044. DOI: 10.1145/2675133.2675274.

Jung, J. et al. (2016) 'Social or Financial Goals?', In *Proceedings of the 2016 CHI Conference Extended Abstracts on Human Factors in Computing Systems – CHI EA '16* (pp. 2857–2863). New York, New York, USA: ACM. DOI: 10.1145/2851581.2892328.

Lampinen, A. and Cheshire, C. (2013) 'Hosting via Airbnb: Motivations and financial assurances in monetized network hospitality', in *Proceedings of the 2016 CHI Conference on Human Factors in Computing Systems*. San Jose: ACM (pp. 1669–1680). Available at: http://dl.acm.org/citation.cfm?id=2858092 [Accessed: December 7, 2017].

Law, R., Leung, K., and Wong, R. (2004) 'The impact of the Internet on travel agencies', *International Journal of Contemporary Hospitality Management*, 16(2): 100–107. DOI: 10.1108/09596110410519982.

Lee, M. et al. (2015) 'Working with machines: The impact of algorithmic and data-driven management on human workers', in *Proceedings of the 33rd Annual ACM Conference on Human Factors in Computing Systems*. ACM, pp. 1603–1612. Available at: http://dl.acm.org/citation.cfm?id=2702548 [Accessed: December 7, 2017].

Lin, P. M. C. (2018) 'Is Airbnb a good choice for family travel?', *Journal of China Tourism Research*, 1–18. DOI: 10.1080/19388160.2018.1551167.

Mikhalkina, T. and Cabantous, L. (2015) 'Business model innovation: How iconic business models emerge', *Business Models and Modelling*, 33: 59–95. DOI: 10.1108/S0742-332220150000033024.

Miller, S. R. (2016) 'First principles for regulating the sharing economy', *Harvard Journal on Legislation*, 53(1): 147–202. DOI: 10.2139/ssrn.2568016.

Mittendorf, C. (2017) 'The implications of trust in the sharing economy–An empirical analysis of Uber', in *Proceedings of the 50th Hawaii International Conference on System Sciences*, pp. 5837–5846. Available at: http://scholarspace.manoa.hawaii.edu/handle/10125/41866 [Accessed: January 5, 2017].

Möhlmann, M. (2015) 'Collaborative consumption: Determinants of satisfaction and the likelihood of using a sharing economy option again', *Journal of Consumer Behaviour*, 14(3): 193–207. DOI: 10.1002/cb.1512.

O'Connor, D. (2005) 'Towards a new interpretation of "hospitality"', *International Journal of Contemporary Hospitality Management*, 17(3): 267–271. DOI: 10.1108/09596110510591954.

O'Gorman, K. (2010) 'Introduction to the origins of hospitality and tourism', In *The Origins of Hospitality and Tourism* (pp. 1–4). Available at: https://strathprints.strath.ac.uk/id/eprint/15100 [Accessed: April 29, 2019].

O'Gorman, K. D. (2005) 'Modern hospitality: Lessons from the past', *Journal of Hospitality & Tourism Management*, 12(2): 141–151. Available at: https://strathprints.strath.ac.uk/id/eprint/5843 [Accessed: April 29, 2019].

Oskam, J. and Boswijk, A. (2016) 'Airbnb: The future of networked hospitality businesses', *Journal of Tourism Futures*, Emerald Group Publishing Limited, 2(1): 22–42. DOI: 10.1108/JTF-11-2015-0048.

Pawlicz, A. (2019) *Ekonomia Współdzielenia. Niedoskonałości – Regulacje – Pośrednicy*. Szczecin: Wydawnictwo Naukowe Uniwersytetu Szczecińskiego.

SEMrush. (2019). Available at https://www.semrush.com/ [Accessed: December 5, 2019].

Shaheen, S. A., Chan, N. D., and Gaynor, T. (2016) 'Casual carpooling in the San Francisco Bay area: Understanding user characteristics, behaviors, and motivations', *Transport Policy*, 51: 165–173. DOI: 10.1016/j.tranpol.2016.01.003.

Teodoro, R. et al. (2014) 'The motivations and experiences of the on-demand mobile workforce', in *Proceedings of the 17th ACM conference on Computer supported cooperative work & social computing*. ACM, 236–247. Available at: http://dl.acm.org/citation.cfm?id=2531680 [Accessed: December 7, 2017].

Tussyadiah, I. P. (2016) 'Factors of satisfaction and intention to use peer-to-peer accommodation', *International Journal of Hospitality Management*, 55: 70–80. DOI: 10.1016/j.ijhm.2016.03.005.

Tussyadiah, I. P. and Pesonen, J. (2016) 'Impacts of Peer-to-Peer accommodation use on travel patterns', *Journal of Travel Research*, 55(8): 1022–1040. DOI: 10.1177/0047287515608505.

Widener, M. (2015) 'Shared spatial regulating in sharing-economy districts', *Seton Hall Law Review*, 46(1): 111–187. DOI: 10.1525/sp.2007.54.1.23.

Zhang, L., Guillet, B. D., and Kucukusta, D. (2015) 'Online travel agents—hotels' Foe or friend? A case study of Mainland China', *Journal of China Tourism Research*, 11(4): 349–370. DOI: 10.1080/19388160.2015.1111823.

33

An exploration of experiential travel behavior during the traveler journey

Roberta Minazzi

Introduction

The increasing attention given to the topic of travel experience is the result of more complex journeys based on multi-channel and multi-media systems of interaction between customers and companies (Minazzi and Mauri 2015). Moreover, the advancements in information technology and the development of social media and of mobile devices stimulate peer-to-peer communication among customers over the various steps of the travel behavior process (pre-trip, during-trip and post-trip) (Parra-Lòpez et al. 2012; Xiang et al. 2014; Zeng and Gerritsen 2014). Therefore, travel experience is the result of a co-creation process increasingly affected by peer-to-peer relationships (Neuhofer et al. 2014). During their experience, travelers create context-related information (Buhalis and Foerste 2013) that they can share in real-time with others (Litvin et al. 2008), influencing the behavior of further tourists, who are at the destination at that moment or who will visit it in the future or are searching for information about the destination.

Recent trends affecting the "traveler journey" show a gradual convergence of travel behavior activities from the pre-trip and the post-trip stages to the during-trip stage (Jun et al. 2007; Munar and Jacobsen 2014; Xiang et al. 2014; Minazzi and Mauri 2015). This means that travelers increasingly delay some kinds of travel activities to the during-trip stage (i.e., the choice of the activities to do at the destination, transportation, etc.) and anticipate other kinds of actions from the post-trip stage (i.e. sharing, WOM, etc.). This transformation has important implications for travel companies. It could be a risk for companies that generally offer the services in advance, before the customer leaves, but it could also represent an opportunity for local companies able to develop appropriate customer services and marketing activities (i.e., hotels, restaurants, attractions, etc.).

In the light of these assumptions, the first part of the chapter provides a conceptualization of the "traveler journey" and explores travelers' experience at the different stages of their behavior (pre, during and post-trip). In the second part, recent trends affecting the travel behavior process are examined. Finally, the implications of travel companies are discussed.

The concept of travel experience

The topic of "consumer experience" has become increasingly important in the travel sector. In developing countries, providing experiences is a way for creating a competitive advantage (Pine and Gilmore (1999) and remaining competitive in a global market more and more affected by information technology (Schmitt 1999).

Travel experience can be defined as "an individual's subjective evaluation and undergoing (i.e. affective, cognitive, and behavioral) of events related to his/her tourist activities that begins before (i.e. planning and preparation), during (i.e. at the destination) and after the trip (i.e. recollection)" (Tung and Ritchie 2011). According to Pine and Gilmore (1999: 11), an experience is created when "a company intentionally uses services as the stage, and goods as props, to engage individual customers in a way that creates a memorable event". A person is engaged with an event on an emotional, physical, spiritual or intellectual level.

What emerges from the literature is the highly subjective nature of the tourism experience (Holbrook and Hirschman 1982; Pine and Gilmore 1999; Kotler et al. 2001): specific individuals involved in a specific context interpret experiences in different ways (Urry 1990; Jennings 2006). Moreover, people can desire different kinds of tourist experiences according to their specific features (Cohen 1979) and the final evaluation of quality comes from a comparison between expectations and perceptions. According to Parasuraman et al. (1985) and Zeithaml et al. (1993) expectations are influenced by personal needs, previous experience, word-of-mouth, explicit service promises (i.e. advertising), implicit service promises (i.e. price, tangibles), transitory service intensifiers (i.e. emergencies, service problems) and situational factors (i.e. bad weather, a strike, etc.), while perceptions are influenced by the result of the service delivery and external communication to the consumers. However, expectations and perceptions are dynamic: this means that during the service delivery they continue to change due to a multitude of interactions. Moreover, they are influenced by the competitive environment (Fournier and Mick 1999; Seth et al. 2005). Therefore, the evaluation of the travel experience cannot be linked to a temporal dimension (Jennings and Weiler 2006) but it should consider the systemic nature of the company and the global perspective of analysis of the tourism experience (Mauri et al. 2013). The tourism experience is highly complex because it concerns a process that involves multiple decisions about the various services that compose the overall experience (Hyde and Lawson 2003; Decrop and Snelders 2004). Travelers create their personal experiences by subjectively combining the different tourism services provided by the travel operators (Uriely 2005) during the whole travel process, co-creating the experience with all the actors who participate in the process (Prahalad and Ramaswamy 2004). According to Pine and Gilmore (1999) people have different experiences because they are the result of the interactions (moments of truth) with the various service providers. Experiences are influenced by the emotional, physical, spiritual and/or intellectual state of mind of each individual, by other customers present during the service supply (Lehtinen and Lehtinen 1991; Lin et al. 2001; Orsingher 2003; Ekinci and Dawes 2009) and by the physical environment and personnel (Nickerson 2006; Mossberg 2007). All of these actors and features are considered mediators of the tourism experience: other tourists, tourist providers, local governments and communities are personal mediators while design, signage, aesthetics and settings are non-personal mediators (Jennings and Weiler 2006). Thus, the overall satisfaction judgment of a tourism experience comes from a subjective and emotional response to the various aspects of all the services provided (Otto and Richie 1995) and to all the interactions that occurred during the process.

With the development of technology, new mediators influence the travel experience: smartphones, digital cameras, tablets and wearable devices allow people to share real-time their experience with others. The content posted on the internet and new devices based on artificial intelligence allow people far from a specific destination to practice tourism within everyday life and immerse themselves in a travel experience (Jansson 2002; Uriely 2005; Tussyadiah and Fesenmaier 2009; Wang et al. 2012). As a result, tourism experience, which in the past was conceptualized as something in contrast with everyday life (Cohen 1979; Uriely 2005), enters in our routine (Neuhofer et al. 2014). The blurred boundary between everyday life and tourism experiences can influence the nature of each step of the trip (Wang et al. 2014, 2016). Therefore, travel experience is not only something that happens in a specific destination, but a more complex process composed of various steps with an increasingly virtual component (Tung and Ritchie 2011). The next paragraph describes the process of travel decision-making integrating two established approaches: decision-making studies and travel planning theory.

Consumer behavior in tourism

Tourism services are generally regarded as experience products (Cohen 1979; Uriely 2005; Morgan et al. 2010; Moscardo 2010; Sundbo and Sørensen 2013). Like other services, they are mainly intangible (Murray and Schlacter 1990; Gremler et al. 1994) and their quality could hardly be evaluated prior to the consumption (Dye 2000; Rosen 2000, 2009; Zeithaml et al. 2000; Dixit 2017). For these reasons, travelers are generally highly involved in the buying decision process and perceive a high risk that frequently produces a longer and more complex consumer behavior process (Murray and Schlacter 1990; Laroche et al. 2004).

Tourists' behavior may be analyzed according to different theories. Decision-making studies divide the traveler buying process into five stages: need recognition, information search, evaluation of alternatives, purchase decision and post-purchase behavior (Kotler et al. 2014; Zeithaml et al. 2000). However, this approach is generally related to single transactions (Jun et al. 2007). A more specific approach focused on the tourists' planning process (travel planning theory) identifies a temporal and linear perspective based on various phases (Jennings and Oppermann 1997, 2006; Jennings et al. 2009): pre-trip, during-trip, and post-trip (Engel et al. 1990); the anticipatory phase, the experiential phase, and the reflective phase (Craig-Smith and French 1994); planning, outward travel, on-site activities, return travel and recollection (Clawson 1963). Travel planning theory postulates that travel planning is more complex in nature than other transactions as it implies interrelated actions in respect of a combination of many services at the same time to achieve multiple goals (Stewart and Vogt 1999; Dellaert et al. 2014). For this reason, the temporal perspective of the travel process has been differently represented by scholars. Killion (1992), for example, adapts the linear model by Clawson (1963) in a circular way.

Other scholars have focused on other dimensions of the tourist experience such as activity, experiences, meanings, benefits and satisfaction (Jennings et al. 2009). Starting from the "tourist gaze" of Urry (1990) some scholars try to understand the meaning of the tourist experience (Beeton et al. 2006; Jennings and Weiler 2006).

In this chapter, the temporal perspective has been adopted integrating the two previously cited approaches (decision-making studies and travel planning theory). As a result, the stages of need recognition, information search, evaluation of alternatives and booking/purchase could be considered as part of the pre-trip phase (anticipatory), consumption as part of the during-trip stage (experiential) and post consumption as part of the post-trip (reflective) stage (see Figure 33.1).

In the pre-trip phase, the first step of travel planning is the need recognition that can be produced by internal (previous experience) and external stimuli (suggestions of others,

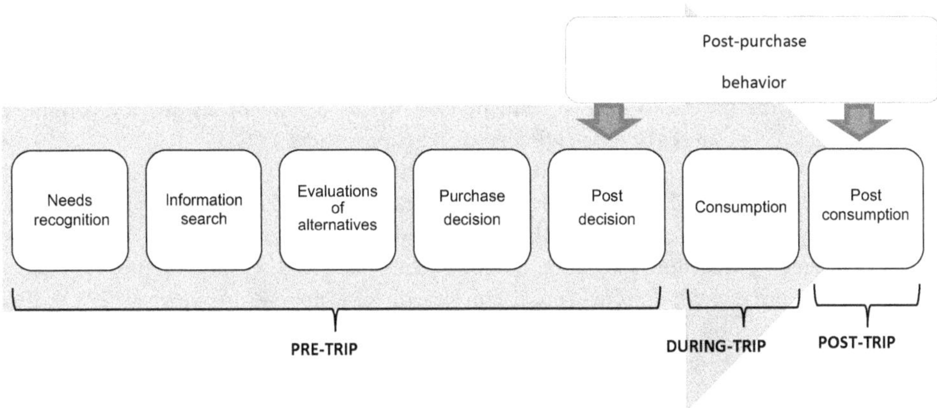

Figure 33.1 The travel behavior process
Source: Adapted from Minazzi R. 2015

commercials or other marketing stimuli) (Kotler et al. 2010). After that, travelers gather information about goods and services able to satisfy the specific need. Consumers generally employ both personal and non-personal sources of information (Zeithaml et al. 2000): personal information coming from word-of-mouth of family, friends, neighbors, colleagues, etc., and non-personal online and offline information coming from commercial sources (corporate website, advertising, etc.) and public/third parties sources (official classifications, customers' reviews, ratings, rankings, awards, etc.) (Kotler et al. 2010).

Once the traveler has gathered information, he or she identifies a set of alternatives that represents acceptable options to satisfy the need. The evaluation of alternatives is influenced by subjective factors; this means that each traveler gives different importance to various attributes of the alternatives identified (Kotler et al. 2010). At the end of this process, the customer proceeds to the buying intention that is then translated into an actual decision. Between intention and decision other factors, generally not under the control of the company, could intervene and influence the final decision (i.e., attitudes of others, unexpected situational factors). For travel services, it is important to notice that the decision and the purchase are not the same. The decision sometimes is only a reservation that does not imply a binding decision. Therefore, we have for tourism a step that can be called "post-decision", in which customers could change something about the travel service. For example, in the case of unexpected situations or cognitive dissonance, especially when the reservation presents the option "free cancelation", travelers could decide to cancel and change the hotel or the destination or simply add new services. This is confirmed by a study by Fotis et al. (2012), who found that a large majority of interviewees made some sort of changes to the original travel plan after having consulted online user-generated content. Moreover, due to the high perception of risk and the significant amount of time existing generally between the moment of the reservation and the actual consumption, in the post-decision step customers may try to find cues to reassure themselves about the decision taken (Minazzi 2015).

At the end of the pre-trip, the traveler leaves toward the destination/s. Travel consumption (during-trip) is a complex process composed of various activities and relationships with different actors that mediate the overall experience (Wang et al. 2012): companies (transportation companies, hotels, travel agents, local attractions, etc.), different employees of the companies according to the kind of service experience (check-in, food & beverage, etc.), and other people (customers, residents, etc.). The interactions that occur between the consumers and the company staff are defined by Normann (1984) as "moments of truth" because they are moments in which the staff can really demonstrate the service quality. Nevertheless, even the interactions with other customers affect the travel experience (Zeithaml et al. 2000). The company and the consumers are partners that co-produce the service (Prahalad and Ramaswamy 2004).

After the consumption step, the consumer evaluates the service quality of the overall experience. As mentioned in the previous paragraphs, travelers compare expectations and perceptions in a progressive process that comprehends all the steps of the process. The overall customer satisfaction judgment combines the perceived quality of the experience with the price/prices paid, personal factors and situational factors (Zeithalm et al. 2012). The result of this evaluation (satisfaction/dissatisfaction) will guide the subsequent actions of the traveler. In the case of disconfirmation, the traveler could decide whether to stop buying the product (exit) or give voice to dissatisfaction (voice) (Hirschman 1970), taking actions in order to reduce dissonance (Oliver 1980, 1993) (i.e., complaints, offline or online negative word-of-mouth). For example, travelers may share their experiences with other people to recollect their memories about the journey. The act of sharing in the post-trip step, both in the case of positive and negative valence, is confirmed as improving the quality evaluation of the experience (Kim and Fesenmaier 2017).

However, the concept of "quality tourism experience" presents a multiplicity of interpretations due to the personalized nature of experiences (Jennings et al. 2009). As mentioned in the previous paragraph, expectations and perceptions are dynamic concepts that continue to change during the consumption, due to many interactions and the actions of competitors. Therefore, the overall quality of the experience is the result of a multitude of individual emotional and subjective evaluations taking place during the travel behavior process.

Recent trends affecting the travel behavior process

The most recent trends that affect consumer behavior in tourism are determined by the development of information and communication technologies: the proliferation of social media and mobile devices along with widespread connectivity changes travelers' habits and the ways in which they make decisions (Parra-Lòpez et al. 2012; Xiang et al. 2014; Zeng and Gerritsen 2014). Tourists are increasingly "multi-device" consumers who use more than one device during different kinds of activities (Minazzi and Mauri 2015). All the steps of the travel behavior process are involved (Schindler and Bickart 2005; Christou and Nella 2012; Law et al. 2019) to a different extent and with diverse objectives according to the specific stage (Fotis et al. 2012). It is important to notice that cultural and age differences could affect the propensity of people to use social media in the various stages of travel planning (Cox et al. 2009; Fotis et al. 2012; Wilson et al. 2012).

As already mentioned before, the travel experience is the result of a co-creation process among the various stakeholders involved (Prahalad and Ramaswamy 2004). With the development of new technologies, co-creation is increasingly connected with the relationship among consumers rather than between the company and consumers (Grönroos 2008). Especially during the experience, people create context-related information (Buhalis and Foerste 2013) that they then share online with others (Litvin et al. 2008). The sharing activity of user-generated content (UGC) may influence the behavior of other tourists who are at the destination and that of other people who are in another place but linked through social media (Litvin et al. 2008; Qualman 2012). Other tourists at the destination are searching for information about what to do (i.e., attractions, restaurants) (Fotis et al. 2012;

Xiang et al. 2014) and UGC could deeply influence their choices. People linked through social media can be involved at different levels. Considering the grade of involvement and the social intensity of the connection, Neuhofer et al. (2014) identified various dimensions of the co-creation activity: a light contact based on messaging (social intercommunication), an intense dialogue in which both sides create and exchange meaning (social interaction), real-time sharing of the tourism experience (co-participation) and living the experience together through the eyes of the tourist (co-living).

Nevertheless, UGC may affect all the steps of the travel process, not only the consumption stage (Buhalis and Law 2008). In the pre-trip phase, UGC could stimulate travelers' imagination (i.e., a video of a specific destination) (Tussyadiah and Fesenmaier 2009) generating new ideas (White 2010; Fotis et al. 2012) or helping travelers in the stages of information search, evaluation of alternatives and decision-making. According to Koo et al. (2016) travelers' exposure to mass media and social media contributes to the development of a positive image of the destination, increasing the intention to visit it. Another study by Narangajavana et al. (2017) found that tourists' social media usage affects expectations about a destination.

The literature confirms the importance of UGC also during the step of travel planning (Gretzel 2007; Anderson 2012; Xiang et al. 2014), as a source of information for tourists (Pan et al. 2007; Murphy and Chen 2016). Concerning the stage of evaluation of alternatives, new intermediaries of the travel distribution system allow customers to compare various services, decreasing the perception of risk. Online travel agencies (OTAs) and meta-search websites like Trivago or Skyscanner offer reviews, rating and rankings of the various travel services, helping tourists in making decisions (Kracht and Wang 2010; Ert and Fleischer 2016; Minazzi 2017). These operators are particularly important also at the post-decision stage because travelers continue to search for information that is able to reassure them about the choice taken. Negative WOM about one or more services booked could lead to changes in the reservation. In general, the amount of content published online by peers gives the opportunity to confirm/disconfirm the decision made.

Online information consulted during the decision-making process can be generated by customers (actual customers and prospects) that post content on their social media profiles, on those of other people (friends or bloggers), on the corporate website, blog or social media page, as well as on third parties' websites (i.e. OTAs, meta-search websites, travel review websites and travel blogs). This last kind of information is in part authentic content published by the customers (reviews, scores, photos, videos, etc.) and in part an elaboration of this information realized by the operator by means of a specific algorithm (ranking, overall score, etc.). Furthermore, travel companies can generate online information on their websites, blogs, social media pages (social networks and content communities) as well as on third parties' websites. Finally, third party organizations of associations, travel clubs and tourist guides, that generally exist even offline, offer information about travel companies (official ratings, hotel information, destination descriptions and attractions, etc.).

At the post-trip stage, UGC could provide the opportunity to re-experience the past (Tussyadiah and Fesenmaier 2009) with tourists going on to share their travel experience with other people on social media. A study by Murphy et al. (2010) found that some travelers usually share UGC about the trip on social media (Facebook) in the post-trip stage. Moreover, tourists could post a review that reflects the level of satisfaction about the experience. Online travel agencies and travel review websites (i.e., Tripadvisor) allow travelers to share their experience online combining text, photos and videos.

Considering the entire travelers' behavior process, IT developments generate a progressive moving of some activities from the pre-trip and the post-trip stages to the during-trip phase (Jun et al. 2007; Munar and Jacobsen 2014; Xiang et al. 2014; Minazzi and Mauri 2015). This is true especially for high experience customers or in the case of well-known destinations (Jun et al. 2007). Thanks to mobile phone tracking (GPS technologies), wider connectivity (i.e., free roaming), and a wide range of apps that offer specific location-based services, tourists can easily make real-time decisions about various services directly at the destination. For example, during the trip they can search for transportation from the airport to the hotel, find a restaurant, identify the attractions to visit, etc. Moreover, travelers can manage unexpected situations and complete travel activities more efficiently and effectively (Wang et al. 2012). The choice is easier when you can have a direct experience at the destination as opposed to taking decisions at home without knowledge of the context. In the past, these activities were part of the pre-trip step. People used to print maps, verify the timetables of transportation, and identify the main attractions before leaving.

In the same way, some activities that generally were part of the post-trip step have moved to the during-trip stage. In particular, people can now share their experience online in real-time with other people directly when they are at the destination. This means instant messaging, social media and online reviews that generate electronic word-of-mouth (eWOM) (Mauri and Minazzi 2013). In the past people came back from holidays and told the story of the trip, sharing the experience with the family, friends and colleagues. Now people continue to tell their stories offline and online to recollect their experience, but in a more structured way (i.e., creating travel diaries).

Given the importance of user-generated content, it is interesting to investigate which are the motivations that push the traveler to share. Previous studies on the topic identified two main types of motivation for sharing travel experience (Hennig-Thurau et al. 2004; Munar and Jacobsen 2014): community-related and self-centered. In the first case, the purpose is adding value to the community. On the one hand, the intention of travelers is to help other customers (altruism) (Sundaram et al. 1998; Cheng et al. 2006; Yoo and Gretzel 2011), and on the other hand the travelers' intention is "to give something in return" to the company for the favorable experience (Sundaram et al. 1998; Cheung and Lee 2012). In this group of motivations, consumers could search for social integration and the feeling of belonging to a community (Qu and Lee 2011; Cheung and Lee 2012; McWilliam 2012).

In the second case, travelers post a comment to gain informal (private or public online conversations) or formal (ranking systems and scoring for reviewers) approval of the usefulness of their feedback. The reasons for this behavior could be a self-enhancement motivation, based on the need to gain a reputation on a consumer-opinion platform (Lampel and Bhalla 2007; Gretzel and Yoo 2008; Munar 2010; Viglia et al. 2016), or the intention to obtain a reward from the operator that manages the platform (i.e., an economic benefit).

Managerial implications for travel companies

Progress in information and communication technologies (ICTs) and the synergy between connectivity, the use of mobile devices, and of social media has been deeply influencing the tourism sector, from the point of view of both the demand and the supply (Law et al. 2019). The travel distribution system has changed due to the development of new intermediaries and travelers' behavior is constantly influenced by the opportunity offered by innovation. Thus, to create, manage, and attempt to control travelers' experience in the face of an increasingly competitive sector has become demanding but pivotal for tourism companies.

Starting from the previous analysis, in this paragraph we try to identify the main managerial implications for travel companies. First, the increasing trend to delay some kinds of travel decisions/purchases to the during-trip stage (especially for the "activities to do" at the destination), thanks to the development of mobile technologies, could be both an opportunity and a threat for travel and hospitality companies. For local companies (small hotels, restaurants, etc.), that do not belong to large partnerships and that sell services through OTAs, this could be a great opportunity. It is difficult for them to catch the attention of the tourists during the planning step, when they are searching for information online. However, the propensity to move some of the information search activities to the during-trip stage allows local companies to intercept the customers at the destination. These operators should be able to develop appropriate technologies (i.e., iBeacon) in order to promote their presence on main location-based apps (i.e. Google maps). On the other hand, for large and well-known companies, this trend could be a risk. Travelers could change their plans during the travel experience (Stewart and Vogt 1999; Jun et al. 2007). For example, a tourist could have the intention of renting a car from a specific company directly at the airport, but the same service offered by local competitors, maybe at a more convenient price, could affect the final choice. This could happen for other travel services purchased during the trip. To avoid this risk, travel companies could try to convince customers to purchase the services in the pre-trip stage by means of partnerships with other travel operators. For example, some airlines or metasearch websites, before concluding the purchase process, give tourists the chance to book other services such as accommodation, car rental, tickets to attractions, etc. In this way, they try to optimize sales at the pre-trip stage, reducing the risk of losing customers during the process due to the competition of local operators.

A second implication concerns the opportunity offered by travel reviews about the experience. They represent an additional instrument for travel operators to monitor customer satisfaction, integrating this kind of information with that coming from other more traditional methods. This is true for hotels, restaurants, attractions and destinations. Sometimes, travel reviews are considered a threat rather than an opportunity by operators due to the possible manipulation of feedback (fake reviews) on anonymous websites (i.e., Tripadvisor). Academic literature investigates the concern about the level of trustworthiness of online reviews (Filieri et al. 2014; Filieri 2016). However, nowadays other websites like OTAs and metasearch engines verify the reviews, allowing only actual customers to publish a comment about an experience. Given that we know from research that travelers have a high propensity to share their experience online, a high number of passive customers (few reviews) could mean that travelers prefer to give their opinions on social media like Facebook and Instagram, a kind of eWOM that can hardly be controlled by the company. In this context, a great opportunity for travel companies is represented by the active use of social media to interact and engage with customers. The development of appropriate online interactions on social media with customers allows companies to manage real-time service recovery and to stimulate long-term relationships, customer satisfaction and loyalty. Moreover, this kind of relationship could give the company the opportunity to sell the service directly without paying commissions to OTAs.

References

Anderson, C.K. (2012). 'The impact of social media on lodging performance', *Cornell Hospitality Report*, 12(5):4–12.

Beeton, S., Bown, H.E., and Santos, C.A. (2006). 'State of knowledge: Mass media and its relationship to perceptions of quality', In G.R. Jennings, and N. Nickerson (eds.), *Quality tourism experiences* (pp. 25–37). Burlington, MA: Elsevier.

Buhalis, D., and Foerste, M.K. (2013). 'SoCoMo marketing for travel and tourism', In Z. Xiang, and I. Tussyadiah (eds.), *Information and communication technologies in tourism 2014* (pp. 175–185). Switzerland: Springer.

Buhalis, D., and Law, R. (2008). 'Progress in information technology and tourism management: 20 years on and 10 years after the Internet—the state of eTourism research', *Tourism Management*, 29(4):609–623. DOI: 10.1016/j.tourman.2008.01.005.

Cheng, S., Lam, T., and Hsu, C.H. (2006). 'Negative word-of-mouth communication intention: An application of the theory of planned behavior', *Journal of Hospitality & Tourism Research*, 30(1): 95–116. DOI: 10.1177/1096348005284269.

Cheung, C.M., and Lee, M.K. (2012). 'What drives consumers to spread electronic word of mouth in online consumer-opinion platforms', *Decision Support Systems*, 53(1):218–225. DOI: 10.1016/j.dss.2012.01.015.

Christou, E., and Nella, A. (2012). 'Web 2.0 and pricing transparency in hotel services', In M. Sigala, E. Christou, and U. Gretzel (eds.), *Social media in travel, tourism and hospitality. Theory, practice and cases* (pp. 11–24). Surrey: Ashgate.

Clawson, M. (1963). *Land and water for recreation: Opportunities, problems and policies*. Chicago, IL: Rand McNally.

Cohen, E. (1979). 'A phenomenology of tourist experiences', *Sociology*, 13:179–201. DOI: 10.1177/003803857901300203.

Cox, C., Burgess, S., Sellitto, C., and Buultjens, J. (2009). 'The role of user-generated content in tourists' travel planning behaviour', *Journal of Hospitality Marketing & Management*, 18(8):743–764. DOI: 10.1080/19368620903235753.

Craig-Smith, S., and French, C. (1994). *Learning to live with tourism*. Melbourne, Australia: Pitman.

Decrop, A., and Snelders, H. (2004). 'Planning the summer vacation: An adaptable process', *Annals of Tourism Research*, 31(4):1008–1030. DOI: 10.1016/j.annals.2004.03.004.

Dellaert, B.G., Arentze, T.A., and Horeni, O. (2014). 'Tourists' mental representations of complex travel decision problems', *Journal of Travel Research*, 53(1):3–11. DOI: 10.1177/0047287513506297.

Dixit, S.K. (ed.). (2017). *The Routledge handbook of consumer behaviour in hospitality and tourism*. London: Routledge. DOI: 10.4324/9781315659657.

Dye, R. (2000). 'The buzz on buzz', *Harvard Business Review*, 78(6):139–146.

Ekinci, Y., and Dawes, P.L. (2009). 'Consumer perceptions of frontline service employee personality traits, interaction quality, and consumer satisfaction', *The Service Industries Journal*, 29(4):503–521. DOI: 10.1080/02642060802283113.

Engel, J.F., Blackwell, R.D., and Miniard, P.W. (1990). *Consumer behaviour* (6th ed.) Hinsdale, IL: Dryden Press.

Ert, E., and Fleischer, A. (2016). 'Mere position effect in booking hotels online', *Journal of Travel Research*, 55(3):311–321. DOI: 10.1177/0047287514559035.

Filieri, R. (2016). 'What makes an online consumer review trustworthy?', *Annals of Tourism Research*, 58:46–64. DOI: 10.1016/j.annals.2015.12.019.

Filieri, R., and McLeay, F. (2014). 'E-WOM and accommodation: An analysis of the factors that influence travellers' adoption of information from online reviews', *Journal of Travel Research*, 53(1):44–57. DOI: 10.1177/0047287513481274.

Fotis, J., Buhalis, D., and Rossides, N. (2012). 'Social media use and impact during the holiday travel planning process', In M. Fuchs, F. Ricci, and L. Cantoni (eds.), *Information and communication technologies in tourism 2012* (pp. 13–24). Austria: Springer-Verlag.

Fournier, S., and Mick, D.G. (1999). 'Rediscovering satisfaction', *Journal of Marketing*, 63(4):5–23. DOI: 10.1177/002224299906300403.

Gremler, D.D., Bitner, M.J., and Evans, K.R. (1994). 'The internal service encounter', *International Journal of Service Industry Management*, 5(2): 34–56. DOI: 10.1108/09564239410057672.

Gretzel, U. (2007). *Online travel review study: Role and impact of online travel reviews*. Texas A&M University. Laboratory for Intelligent Systems in Tourism, College Station.

Gretzel, U., and Yoo, K.H. (2008). 'Use and impact of online travel reviews', In U. Gretzel, and K. H. Yoo (eds.), *Information and communication technologies in tourism 2008* (pp. 35–46). Austria: Springer-Verlag.

Grönroos, C. (2008). 'Service logic revisited: Who creates value? And who co-creates?', *European Business Review*, 20(4):298–314. DOI: 10.1108/09555340810886585.

Hennig-Thurau, T., Gwinner, K.P., Walsh, G., and Gremler, D.D. (2004). 'Electronic Word-of-mouth via consumer-opinion platforms: What motivates consumers to articulate themselves on the Internet?', *Journal of Interactive Marketing*, 18(1):38–52. DOI: 10.1002/dir.10073.

Hirschman, A.O. (1970). *Exit, voice and loyalty*. Cambridge: Harvard University Press.

Holbrook, M.B., and Hirschman, E.C. (1982). 'The experiential aspects of consumption: Consumer fantasies, feelings and fun', *Journal of Consumer Research*, 9(2):132–140. DOI: 10.1086/208906.

Hyde, K., and Lawson, R. (2003). 'The nature of independent travel', *Journal of Travel Research*, 42(1):13–23. DOI: 10.1177/0047287503253944.

Jansson, A. (2002). 'Spatial phantasmagoria: The mediatization of tourism experience', *European Journal of Communication*, 17(4):429–443. DOI: 10.1177/02673231020170040201.

Jennings, G., Lee, Y.S., Ayling, A., Lunny, B., Cater, C., and Ollenburg, C. (2009). 'Quality tourism experiences: Reviews, reflections, research agendas', *Journal of Hospitality Marketing & Management*, 18(2-3):294–310. DOI: 10.1080/19368620802594169.

Jennings, G.R. (2006). 'Perceptions on quality tourism experience', In G. Jennings and N. Nickerson (eds.), *Quality tourism experiences* (pp. 1–22). Oxford: Elsevier Butterworth-Heinemann.

Jennings, G.R., and Oppermann, M. (1997). 'The travel experience of cruisers', In M. Oppermann (ed.), *Pacific Rim 2000: Issues, interrelations, inhibitors* (pp. 94–105). London: CAB International.

Jennings, G.R., and Weiler, B. (2006). 'Mediating meaning: Perspectives on brokering quality tourist experiences', In G. Jennings and N. Nickerson (eds.), *Quality tourism experiences* (pp. 57–78). Oxford: Elsevier Butterworth-Heinemann.

Jun, S.H., Vogt, C.A., and MacKay, K.J. (2007). 'Relationships between travel information search and travel product purchase in pre-trip contexts', *Journal of Travel Research*, 45(3):266–274. DOI: 10.1177/0047287506295945.

Killion, K.L. (1992). *Understanding tourism, study guide*. Rockhampton, Australia: Central Queensland University.

Kim, J., and Fesenmaier, D.R. (2017). 'Sharing tourism experiences: The post-trip experience', *Journal of Travel Research*, 56(1):28–40. DOI: 10.1177/0047287515620491.

Koo, C., Joun, Y., Han, H., and Chung, N. (2016). 'A structural model for destination travel intention as a media exposure: Belief-desire-intention model perspective', *International Journal of Contemporary Hospitality Management*, 28(7):1338–1360. DOI: 10.1108/IJCHM-07-2014-0354.

Kotler, P., Adam, S., Brown, L., and Armstrong, G. (2001). *Principles of marketing*. Frenchs Forest, Australia: Pearson–Prentice Hall.

Kotler, P., Keller, K.L., Ancarani, F., and Costabile, M. (2014). *Marketing management*, 14th ed. Milan: Pearson.

Kracht, J., and Wang, Y. (2010). 'Examining the tourism distribution channel: Evolution and transformation', *International Journal of Contemporary Hospitality Management*, 22(5):736–757. DOI: 10.1108/09596111011053837.

Lampel, J., and Bhalla, A. (2007). 'The role of status seeking in online communities: Giving the gift of experience', *Journal of Computer-Mediated Communication*, 12(2):434–455. DOI: 10.1111/j.1083-6101.2007.00332.x.

Laroche, M., McDougall, G.H.G., Bergeron, J., and Yang, Z. (2004). 'Exploring how intangibility affects perceived risk', *Journal of Service Research*, 6(4):373–389. DOI: 10.1177/1094670503262955.

Law, R., Leung, D., and Chan, I.C.C. (2019). 'Progression and development of information and communication technology research in hospitality and tourism: A state-of-the-art review', *International Journal of Contemporary Hospitality Management*. ahead of printing. DOI: 10.1108/IJCHM-07-2018-0586.

Lehtinen, U., and Lehtinen, J.R. (1991). 'Two approaches to service quality dimensions', *The Service Industries Journal*, 11(3):287–303. DOI: 10.1080/02642069100000047.

Lin, N.P., Chiu, H.C., and Hsieh, Y.C. (2001). 'Investigating the relationship between service providers' personality and customers' perceptions of service quality across gender', *Total Quality Management*, 12 (1):57–67. DOI: 10.1080/09544120020010093.

Litvin, S.W., Goldsmith, R.E., and Pan, B. (2008). 'Electronic word-of-mouth in hospitality and tourism management', *Tourism Management*, 29(3):458–468. DOI: 10.1016/j.tourman.2007.05.011.

Mauri, A.G., and Minazzi, R. (2013). 'Web reviews influence on expectations and purchasing intentions of hotel potential customers', *International Journal of Hospitality Management*, 34(5):99–107. DOI: 10.1016/j.ijhm.2013.02.012.

Mauri, A.G., Minazzi, R., and Muccio, S. (2013). 'A review of literature on the gaps model on service quality: A 3-decades period: 1985–2013', *International Business Research*, 6(12):134–144. DOI: 10.1016/j.ijhm.2013.02.012.

McWilliam, G. (2012). 'Building stronger brands through online communities', *Sloan Management Review*, 41(3):43–54.

Minazzi, R. (2015). *Social media marketing in tourism and hospitality*. Switzerland: Springer International Publishing.

Minazzi, R. (2017). 'The travel distribution system and information and communication technologies (ICT)', In Dixit, S.K. (ed.), *The Routledge handbook of consumer behaviour in hospitality and tourism* (pp.377–386). London and New York: Routledge.

Minazzi, R., and Mauri, A.G. (2015). 'Mobile technologies effects on travel behaviours and experiences: A preliminary analysis', In A. Inversini and I. Tussyadiah (eds.), *Information and communication technologies in tourism* (pp. 507–521). Switzerland: Springer.

Morgan, M., Lugosi, P., and Ritchie, J.R.B. (eds.). (2010). *The tourism and leisure experience. Consumer and managerial perspectives*. Bristol: Channel View Publications.

Moscardo, G.M. (2010). 'The shaping of tourist experience. The importance of stories and themes', In M. Morgan, P. Lugosi, and J.R.B. Ritchie (eds.), *The tourism and leisure experience. Consumer and managerial perspectives* (pp. 43–58). Bristol: Channel View Publications.

Mossberg, L. (2007). 'A marketing approach to the tourist experience', *Scandinavian Journal of Hospitality and Tourism*, 7(1):59–74. DOI: 10.1080/15022250701231915.

Munar, A.M. (2010). 'Tourist-created content: Rethinking destination branding', *International Journal of Culture, Tourism and Hospitality Research*, 5(3):291–305. DOI: 10.1108/17506181111156989.

Munar, A.M., and Jacobsen, J.K.S. (2014). 'Motivations for sharing tourism experiences through social media', *Tourism Management*, 43:46–54. DOI: 10.1016/j.tourman.2014.01.012.

Murphy, H.C., and Chen, M.M. (2016). 'Online information sources used in hotel bookings: Examining relevance and recall', *Journal of Travel Research*, 55(4):523–536. DOI: 10.1177/0047287514559033.

Murphy, H.C., Gil, E.A.C., and Schegg, R. (2010). 'An Investigation of motivation to share online content by young travellers—Why and where', In R.L.U. Gretzel, and M. Fuchs (eds.), *Information and communication technologies in tourism* (pp. 467–478)). Austria: Springer.

Murray, K.B., and Schlacter, J.L. (1990). 'The impact of services versus goods on consumers' assessment of perceived risk and variability', *Journal of the Academy of Marketing Science*, 18(1):51–65. DOI: 10.1007/BF02729762.

Narangajavana, Y., Fiol, L.J.C., Tena, M.Á.M., Artola, R.M.R., and García, J.S. (2017). 'The influence of social media in creating expectations. An empirical study for a tourist destination', *Annals of Tourism Research*, 65: 60–70. DOI: 10.1016/j.annals.2017.05.002.

Neuhofer, B., Buhalis, D., and Ladkin, A. (2014). 'A typology of technology-enhanced tourism experiences', *International Journal of Tourism Research*, 16(4):340–350. DOI: 10.1002/jtr.1958.

Nickerson, N.P. (2006). 'Some reflections on quality tourism experiences', In G. Jennings, and N. Nickerson (eds.), *Quality tourism experiences* (pp. 227–235). Oxford: Elsevier Butterworth-Heinemann.

Normann, R. (1984). *Service management*. New York: John Wiley.

Oliver, R.L. (1980). ''A cognitive model of the antecedents and consequences of satisfaction decision', *Journal of Marketing Research*, 17(4):460–469. DOI: 10.1177/002224378001700405.

Oliver, R.L. (1993). 'A conceptual model of service quality and service satisfaction: Compatible goals, different concepts', In T.A. Swartz, D.E. Bowen, and S.W. Brown (eds.), *Advances in services marketing and management: Research and practice* (pp. 65–85). Greenwich: JAI Press.

Orsingher, C. (2003). 'Un'analisi della soddisfazione del cliente nei servizi ad elevata partecipazione', *Micro & Macro Marketing*, 12(1):3–20. DOI: 10.1431/8580.

Otto, J., and Richie, J. (1995). 'Exploring the quality of the service experience. A theoretical and empirical analysis', *Advances in Services Marketing and Management*, 4(1):37–61.

Pan, B., MacLaurin, T., and Crotts, J.C. (2007). 'Travel blogs and the implications for destination marketing', *Journal of Travel Research*, 46(1):35–45. DOI: 10.1177/0047287507302378.

Parasuraman, A., Zeithaml, V.A., and Berry, L.L. (1985). 'A conceptual model of service quality and its implications for future research', *Journal of Marketing*, 49(4), 41–50. DOI: 10.2307/1251430.

Parra-Lòpez, E., Gutiérrez-Tano, D., Diaz-Armas, R.J., and Bulchand-Gidumal, J. (2012). 'Travellers 2.0: Motivation, opportunity and ability to use social media', In M. Sigala, E. Christou, and U. Gretzel (eds.), *Social media in travel, tourism and hospitality. Theory, practice and cases* (pp. 171–187). Surrey: Ashgate.

Pine, B.J., and Gilmore, J.H. (1999). *The experience economy: Work is theatre & every business a stage*. Boston, MA: Harvard Business School Press.

Prahalad, C.K., and Ramaswamy, V. (2004). 'Co-creation experiences: The next practice in value creation', *Journal of Interactive Marketing*, 18(3):5–14. DOI: 10.1002/dir.20015.

Qu, H., and Lee, H. (2011). 'Travellers' social identification and membership behaviours in online travel community', *Tourism Management*, 32(6):1262–1270. DOI: 10.1016/j.tourman.2010.12.002.

Qualman, E. (2012). *Socialnomics: How social media transforms the way we live and do business*. Hoboken, NJ: John Wiley.

Rosen, E. (2000). *The anatomy of buzz: How to create word-of-mouth marketing*. New York: Double-day.

Rosen, E. (2009). *The anatomy of buzz revisited: Real-life lessons in word-of-mouth marketing*. New York: Doubleday.

Schindler, R., and Bickart, B. (2005). 'Published "word of mouth": Referable, consumer-generated information on the Internet', In C. Hauvgedt, K. Machleit, and R. Yalch (eds.), *Online consumer psychology: Understanding and influencing consumer behaviour in the virtual world* (pp. 35–61). Mahwah, NJ: Lawrence Erlbaum Associates.

Schmitt, B. (1999). *Experiential marketing: How to get customers to sense, feel, think, act and relate to your company and brands*. New York: The Free Press.

Seth, N., Deshmukh, S.G., and Vrat, P. (2005). 'Service quality models: A review', *International Journal of Quality and Reliability Management*, 22(9):913–949. DOI: 10.1108/02656710510625211.

Stewart, S.I., and Vogt, C.A. (1999). 'A case-based approach to understanding vacation planning', *Leisure Sciences*, 21(2):79–95. DOI: 10.1080/014904099273165.

Sundaram, D.S., Mitra, K., and Webster, C. (1998). 'Word-of-mouth communications: A motivational analysis', In J.W. Alba, and J.W. Hutchinson (eds.), *NA – Advances in consumer research* (pp. 527–531). Provo, UT: Association for Consumer Research.

Sundbo, J., and Sørensen, F. (2013). *Handbook on the experience economy*. Cheltenham: Edward Elgar Publishing.

Tung, V.W.S., and Ritchie, J.R. (2011). 'Exploring the essence of memorable tourism experiences', *Annals of Tourism Research*, 38(4):1367–1386. DOI: 10.1016/j.annals.2011.03.009.

Tussyadiah, I.P., and Fesenmaier, D.R. (2009). 'Mediating tourist experiences: Access to places via shared videos', *Annals of Tourism Research*, 36(1):24–40. DOI: 10.1016/j.annals.2008.10.001.

Uriely, N. (2005). 'The tourist experience: Conceptual developments', *Annals of Tourism Research*, 32:199–216. DOI: 10.1016/j.annals.2004.07.008.

Urry, J. (1990). *The tourist gaze: Leisure and travel in contemporary societies, theory, culture & society*. London: Sage.

Viglia, G., Minazzi, R., and Buhalis, D. (2016). 'The influence of e-word-of-mouth on hotel occupancy rate', *International Journal of Contemporary Hospitality Management*, 28(9):2035–2051. DOI: 10.1108/IJCHM-05-2015-0238.

Wang, D., Park, S., and Fesenmaier, D.R. (2012). 'The role of smartphones in mediating the touristic experience', *Journal of Travel Research*, 51(4):371–387. DOI: 10.1177/0047287511426341.

Wang, D., Xiang, Z., and Fesenmaier, D.R. (2014). 'Adapting to the mobile world: A model of smartphone use', *Annals of Tourism Research*, 48:11–26. DOI: 10.1016/j.annals.2014.04.008.

Wang, D., Xiang, Z., and Fesenmaier, D.R. (2016). 'Smartphone use in everyday life and travel', *Journal of Travel Research*, 55(1):52–63. DOI: 10.1177/0047287514535847.

White, L. (2010). Facebook, friends and photos: A snapshot into social networking for generating travel ideas. In *Tourism informatics: Visual travel recommender systems, social communities, and user interface design* (pp. 115–129). Hershey: IGI Global.

Wilson, A., Murphy, H., and Fierro, J.C. (2012). 'Hospitality and travel. The nature and implications of user-generated content', *Cornell Hospitality Quarterly*, 53(3):220–228. DOI: 10.1177/1938965512449317.

Xiang, Z., Wang, D., O'Leary, J.T., and Fesenmaier, D.R. (2014). 'Adapting to the internet: Trends in travelers' use of the web for trip planning', *Journal of Travel Research*, 54(4):511–527. DOI: 10.1177/0047287514522883.

Yoo, K.H., and Gretzel, U. (2011). 'Influence of personality on travel-related consumer-generated media creation', *Computers in Human Behavior*, 27(2): 609–621. DOI: 10.1016/j.chb.2010.05.002.

Zeithaml, V.A., Berry, L.L., and Parasuraman, A. (1993). 'The nature and determinants of customer expectations of service', *Journal of the Academy of Marketing Science*, 21(1):1–12. DOI: 10.1177/0092070393211001.

Zeithaml, V.A., Bitner, M.J., Gremler, D.D., and Pandit, A. (2000). *Services marketing: Integrating customer focus across the firm*. Madison: McGraw-Hill.

Zeng, B., and Gerritsen, R. (2014). 'What do we know about social media in tourism? A review', *Tourism Management Perspectives*, 10:27–36. DOI: 10.1016/j.tmp.2014.01.001.

34

Co-construction of the tourist experience via social networking sites

Sabrina Seeler and Heike Schänzel

Introduction

Driven by the development of mobile technologies and digital media, recent decades have witnessed drastic changes in the online behavior of tourists throughout the different stages of a trip (pre-trip, during-trip and post-trip). Most research has concentrated on the managerial implications and applications of social media in overall marketing strategies, with an emphasis on how social media influences pre-trip information searching and booking behavior (Huertas and Marine-Roig 2018). More recently, scholars' attention has shifted toward the motivations of tourists to consume and create travel-related content on social networking sites (SNSs). This chapter introduces the emergence and growth of digital media and the role of digital narratives as tools to co-construct tourist experiences. Using Germany and New Zealand as case studies, the chapter demonstrates how the co-construction of experiences influences the process of experience accumulation and what this means for the online behavior of tourists. It provides an overview of tourists' intentions and motivations to share experiences by exploring the sociology of co-constructed tourist experiences. With special emphasis on visuals, audiovisuals and written narratives, the chapter explores how tourist experiences are co-constructed online and how tourism practitioners are leveraging on these new developments of co-constructed experiences and first-person voice narratives.

The emergence and growth of social networking sites

Technological advances have progressively contributed to a shift from Web 1.0 to Web 2.0, the emergence of SNSs and the growing importance of user-generated content (UGC; Bilgihan et al. 2016). SNSs facilitate interaction between different stakeholders in a two-way communication, including Customer-to-Customer (C2C) communication. Marketing is no longer a unidirectional stream of information that equals broadcasting but is participatory and a form of multidirectional communication between different players who build their own dynamic ecosystems (Audrezet, de Kerviler and Guidry Moulard 2018). The growth of social media gradually democratizes the user, leading to changes in information asymmetries and the growth of crowd wisdom, and providing the consumer with more bargaining power and empowerment (Leung et al. 2013).

Advances in mobile technologies and SNSs have provided tourists with quicker and more personally relevant access to travel-related information. This in turn has revolutionized the inspiration and planning stages of travel and added a new dynamic to tourism marketing. As tourists become editors and narrators of their experiences and multidirectional interaction in the form of C2C communication is established, the top-down approach is weakened, and the bottom-up approach strengthened (Huertas and Marine-Roig 2018; Kang and Schuett 2013). Individuals can immediately co-construct their lived experiences and narrate, share and store their personal memories through online communities (e.g., Facebook, Instagram), crowd-sourced reviews (e.g., TripAdvisor), media sharing sites (e.g., YouTube) and (micro-) blogs or social knowledge sharing sites (e.g., Wikitravel; Leung et al. 2013). The digital narratives enable tourists to organize experiences and re-engage with memories so they can transform them into meaningful moments. However, the opportunity to narrate and share experiences online supports more than personal interests of recollecting and storing memories: UGC also establishes new ways of electronic word-of-mouth (eWOM) that reach wider circles beyond those of immediate acquaintances in social networks (Leung et al. 2013). In this vein, Qualman (2013) postulates that the global dissemination of information and increased transparency has led to an upgrade from eWOM to *world of mouth*. However, developments in mobile technologies and SNSs have not only contributed to a shift in travel behavior, they have also resulted in a growing segment of *experienced tourists* (Seeler 2018; Seeler, Lück and Schänzel 2019).

Case study 34.1: Today's experienced tourists and their quest for co-constructed experiences

Recent years have shown incremental, yet decisive, changes in tourism demand. While tourists in the past were mostly understood as passive gazers and spectators, they are becoming more active participants in the creation and consumption of the tourism product. At the same time, they are no longer primarily driven by hedonic needs, but are also motivated by eudaemonic reasons. Consequently, they seek experiences that can act as catalysts for personal growth and contribute to shaping self-identity. This desire for self-development and self-expression, together with improved global mobility and access to information, has led to the emergence of a new type of traveler: the *experienced tourist* (Seeler 2018). With the aim of exploring the dimensions of an experienced tourist from a supply-side perspective, 18 experts from 15 Destination Marketing Organizations (DMOs) in Germany and New Zealand were asked to define an experienced tourist (Seeler 2018). Some key results from this research are presented in the following case study.

Tourism marketers in both countries agreed that today's tourists are increasingly experienced. The transparency and availability of information encourages tourists to compare offers and obtain more knowledge prior to their trip. This leads to optimal experiencing and positively contributes to the desired personal growth. As experienced tourists aim to maintain the optimal flow of experiencing when they are not traveling, they actively search for new travel inspirations and dream about where their next trip will be. Not only does the tourism industry gain from tourists' desire to travel frequently, as reflected in the increasing number of tourism arrivals worldwide, tourists themselves continue accumulating experiences, become more experienced and, in many cases, change their travel behavior toward more responsible and sustainable ways of traveling. The ability to co-construct tourist experiences in SNSs, and to actively engage in sharing one's own experiences through digital narratives, adds to the memorability of experiences. These shared and digitally stored narratives encourage sense-making through reflections and redefinitions of lived experiences. Thus, the technological advances that have created the ability to co-construct experiences online meet experienced tourists' evolving demands.

In addition to recording their reflections, those sharing experiences online gain feedback both from within their immediate network and beyond their closer social circles, which further supports the reinterpretation of their experiences. Actively engaging in SNSs also entails the consumption of other people's co-constructed experiences, which further contributes to knowledge gathering and experience accumulation (Seeler, Lück and Schänzel 2019). One tourism expert in the study summarized this as follows: "In terms of past travel experience it has to be distinguished between what was built up through actual, lived through travel behavior and what has been added virtually through the media behavior." Through the asynchronous and passive consumption of co-constructed experiences, tourists accumulate secondary experiences in the form of other people's narratives. This was also noted by Kim and Fesenmaier (2017: 28), who argue that social media creates opportunities for people to "experience the traveler's footsteps without physically being there."

The co-construction of tourist experiences becomes a process with its own internal dynamics. Against the background that today's experienced tourists are continuously seeking opportunities to further accumulate experiences that shape their personal identity and contribute to personal growth, tourism marketers are facing a challenging and somewhat contradictory situation. Tourism marketers want to encourage their guests to share their experiences and become editors and distributors themselves. This supports the guest in maintaining positive memories of the trip, which also means enhancing the memorability of the destination and providing inspiration for future tourists (Seeler, Lück and Schänzel 2019). However, the consumption of digitally co-constructed experiences introduces the risk that tourists will feel that they no longer need to physically travel to a place they have already consumed online. These new dynamics and challenges will need to be more thoroughly addressed in future tourism marketing.

The sociology of co-constructed tourist experiences

Although most tourists seek and consume information online and do not actively engage in sharing their own experiences, the practice of co-constructing tourist experiences online is growing in significance (Huertas and Marine-Roig 2018). Prahalad and Ramaswamy (2004) suggest that co-creation encompasses the active involvement of customers with the primary aim of co-constructing an experience that is personalized and leads to the most fulfilling and satisfying outcome for the individual customer, while also adding value to the business. Thus, customers are steadily empowered and seek experiences that have the potential to provide the space for co-constructed value. Co-constructing experiences in tourism is multi-phased. The three phases of experiencing are spatially and temporally separated, yet interconnected, and all are influenced by the use of SNSs (Kim and Fesenmaier 2017). Particularly the post-trip co-construction of experience is of huge significance, as tourists newly interpret and re-define their experiences and frequently share these new interpretations online (Tussyadiah and Fesenmaier 2009). Sharing stories and photographs after the trip has always been important as it enables tourists to relive their experience with its associated mental rewards. While this is not a new phenomenon, the channels and the pace of diffusion have changed. Travel experiences in the pre-internet era were judged by the quality of the images and memories analogously delivered. The worth of a trip in today's dynamic and digital environment is defined by online posts and the numbers of "likes" and comments the posts receive.

Similarly, Jansson (2007) explains that the *classical style* of tourism, in which stories were told on postcards and then later with photos and stories that were shared once the social, cultural and geographical distance had been overcome, has shifted with the emergence and growth of digital, networked communication. Digital photography can be reworked and edited, can be accompanied by audiovisual recordings and can be shared instantly as well as consumed by the audience in their home environment. This has the potential to intensify tourist experiences and fulfill the encapsulation process of liminality and temporary spatial separateness. Jansson (2007: 8) describes encapsulation as "being immersed in a liminal realm" and decapsulation as "being hooked up to the world of routines, duties and ordinariness." Liminality refers to a state of distancing the self from societal structures and everyday routines in order to achieve emotional freedom and support the development of the authentic self. However, White and White (2007: 101) postulate that through the use of mobile technologies and digital media "the liminal experience is transformed into a continuing engagement with established relationships." This was echoed by Jansson (2007: 19), who argues that the "mediatization can break this magic, abolishing the boundaries between tourism and everyday life, leading to decapsulation." In this regard, the phenomenon of the *touristification* of everyday life was addressed by scholars, who found that digitization has contributed to an inseparability of life domains and a blurring of boundaries. Eventually, this touristification of everyday life leads to a dissolution of classic tourist rituals that further influences the co-construction of tourist experiences (MacKay and Vogt 2012; Urry and Larsen 2011).

Paradoxically, contemporary tourists who actively engage in online sharing frequently describe escape from everyday routines as a major motivation; however, by immediately sharing experiences online and maintaining relationships with home, traditional notions of distance and mobility are challenged (MacKay and Vogt 2012). Despite the contradiction between pursuing distance while staying close, studies confirm the increasing use of SNSs and active engagement in co-constructing tourist experiences online.

Social media as a tool to co-construct tourist experiences

A possible explanation for why tourists continue to share their lived experiences online might be related to McCabe and Foster's (2006: 194) assumption that "stories provide members of a society with a mechanism of escape" and that these stories of escape contribute to shaping self-identity. This is echoed by Bosangit, Hibbert and McCabe (2015: 4) who suggest that "[W]riting stories about their experiences is an important activity through which people act upon their experience to connect them with the self." This explanation is further underpinned by tourists' increasing quest for eudemonia and for memorable experiences that go beyond object-related authenticity to meet the demand for existential authenticity and, eventually, authentic self-branding (Wang and Alasuutari 2017). Digital immediacy is facilitated through mobile technologies, particularly smartphones, which become tools of production, as well as SNSs that act as distributors and communication channels. This enables tourists to co-construct and narrate their experiences *in situ*, which adds to the demonstration of being there in the moment and at the same time reduces the spatial and temporal liminality to home.

Mobile technologies and SNSs facilitate not only the digital storage of personal memories that are primarily self-oriented, but also enable tourists to share their meaningful stories and thereby outwardly brand themselves. Bearing in mind the significant role of eWOM in the inspiration and decision-making stages, tourism marketers have more recently realized the power of co-constructed marketing and informal branding through first-person narratives. Consequently, they have progressively integrated digital narratives and influencer marketing in their overall marketing strategies (Lund, Cohen and Scarles 2017). Thus, co-constructed experiences and digital narratives in the form of text, visuals and audiovisuals take on a dual role. Firstly, they serve subjective aims and are directed toward the development of self-identity and self-branding. Secondly, they are of managerial interest as they are becoming vitally important in the tourist's inspirational and decision-making phases.

Enhanced memorability, self-development and self-branding

The digital co-construction of tourist experience varies throughout the stages of a trip and is deployed to different degrees and with different overall aims by the individual tourist. For some it is mostly about maintaining the magic and keeping trip memories alive over a prolonged period (Jansson 2007); for others it is about self-realization and chasing one's own personal dreams of self-ownership (van Nuenen 2016). Moreover, for some the presentation of self through strategic self-promotion and self-branding is at the fore, and impression management is consciously or subconsciously leveraged to create a "beneficial image of self in the presence of others" (Lo and McKercher 2015: 106). Through sharing experiences with others and receiving feedback, the experiences are redefined, become more memorable and more strongly contribute to shaping self-identity. Comments and endorsements are provided faster in the online context, which means that tourists can engage in reflection through feedback immediately. Central to the visual and textual co-construction of tourist experiences are travel photography, selfie-taking and audiovisuals/videos, as well as travel narratives in the form of blogs. Munar and Jacobsen (2014) postulate that sharing visuals and audiovisuals is more related to sharing experiences, while narratives and textual content aim for information sharing.

Co-constructed tourist experiences through travel photography, selfie-taking and videography

Travel photography as a form of self-expression is of great significance for the tourist's recollection of past trips and strengthens the memory of experiences. Jansson (2007: 13) acknowledges photographs as "signifiers of the tourist's liminal experiences" and Urry and Larsen (2011: 169) describe the act of taking a photograph as a "power/knowledge relationship," as the visual consumption eventually becomes owning the moment. Consequently, tourists actively engage in travel photography to co-construct the experience and to tell something about the self. Dinhopl and Gretzel (2016a) highlight the interconnectedness of the tourist gaze and tourist photography and refer to the objectification of places and sights, while Lo and McKercher (2015) bring attention to the ideal framing of these places and sights and the potential of manipulating the reality to ensure that one's own needs and wishes are fulfilled. This manipulation of reality includes staging the perfect picture. This might bring harm to some touristic spots that have experienced an influx of tourists due to picture-perfect images shared online. One such example is Roy's Peak in the Wanaka region of New Zealand. Roy's Peak is one of the most Instagrammed spots in New Zealand with a constantly growing number of hikers who queue for a chance to take the perfect Instagram picture (Figure 34.1).

Figure 34.1 Creating the perfect Instagram moment, Roy's Peak, November 2017 (author's own images)

Not only does the queuing interfere with the message delivered in the *perfect* Instagram picture that depicts emptiness, space and individuality, but the growing number of tourists visiting every day also causes harm to both the track providing access to the site and to the surrounding natural habitat (New Zealand Herald 2018). Engaging in travel photography also means freezing moments that are otherwise gone as the experience is fully consumed once the photo is taken. Thus, travel photography is "predominantly an emotional, nostalgic experience of time, place, self, and others" (Lo and McKercher 2015: 105). Physical/digital storage is required to ensure that the memory of the moment is preserved. These stored memories in the form of photographs can encourage later reflection on the co-constructed tourist experience that eventually leads to enrichment and positively contributes to shaping personal identity (Bosangit, Dulnuan and Mena 2012).

More recently, studies have drawn attention to the growing significance of selfie-taking in travel photography documentation, where tourist experiences are co-constructed through a reversed lens (Cardell and Douglas 2018; Dinhopl and Gretzel 2016a). A self-directed tourist gaze is established that objectifies the self and contributes to an "othering of the self" (Dinhopl and Gretzel 2016a: 132) and "idealised versions of the self" (Cardell and Douglas 2018: 104). This also means that the tourist gaze directly reflects on the tourist and is centered inwardly. The tourist might strategically choose a touristic landmark as the background for the selfie to increase the impression of outwardness; however, the site or attraction itself moves into the background. As selfies are not limited to an individual, but often encompass others in the picture, they become communicative tools to maintain social connections. Selfie-taking is also often referred to as neoliberal (even narcissistic) self-branding and a presentation of the self as superior and extraordinary in comparison to others (Gómez Cruz and Thornham 2015). Tourists who engage in selfie-taking are not necessarily driven by a desire to learn about the self, but are rather aiming to present the self in an idealized way and to demonstrate that the self takes center stage in the experience and its respective narrated co-construction.

Another method used to co-construct tourist experiences is through audiovisuals and videos. Beyond the development of mobile technologies that facilitate photo-taking and immediate uploading, recent advances in wearable cameras (e.g., GoPro, Spectacles) that allow for automated recording and photography as well as the development of commercial aerial drones bring new dynamics to the co-construction of tourist experiences (Dinhopl and Gretzel 2016b; Stankov et al. 2019). Generally, it is assumed that videography creates stronger immersion, enables more continuity and richness in storytelling, is less mediated and is considered more real as it goes beyond the appeal of the visual senses (Dinhopl and Gretzel 2016c). After analyzing snowboarders' use of wearable cameras, Dinhopl and Gretzel (2016b) challenged the authenticity of the recorded experiences as they were performed and celebrated silently. A major distinction between the co-construction of tourist experiences through photography and videography lies in the immediacy of sharing. The immediacy of videos is compromised by their high-profile editing (Dinhopl and Gretzel 2016c). However, this digital distance is being progressively closed with the continuous development of mobile technologies and apps (e.g., Go Pro Quick) that simplify editing and encourage more immediate sharing and thus co-construction of the experiences.

By adding a short text, quotes, and/or location tags to uploaded content, individuals create digital narratives that become shorter versions of personal diaries. Particularly the use of hashtags on Instagram goes beyond the inwardly oriented memorability of the co-constructed tourist experiences, as connections are not always reciprocal but asymmetrical and a wider scope can be achieved using specific hashtags (e.g., #travelphotography, used more than 69.3 million times, as of 17 March 2019). Thus, the use of hashtags is often strategic as they function both as a broadcasting tool and a means of supporting increased visibility (Erz, Marder and Osadchaya 2018). Tourists who engage in co-constructing their experiences within online communities mostly aim at immediate short messages and showcasing the most compelling images or videos. The use of blogs to co-construct tourist experiences through storytelling provides an additional tool that aims at creating longer and more in-depth narratives that replicate offline personal diaries.

Co-constructed tourist experiences through digital storytelling and blogging

Travel blogging is constantly gaining in importance as a way for tourists to co-construct their experiences (Bosangit, Dulnuan and Mena 2012; van Nuenen 2016). Travel writing goes beyond the narration of travel experiences. It encompasses a reflection of the self and thus contributes to the shaping of personal identity (Bosangit, Hibbert and McCabe 2015). A general distinction is made between interaction-oriented blogging behavior, such as writing for self-expression or maintaining social relationships, and travel-documenting and information-oriented blogging behaviors (Chun-Yao et al. 2007). Co-constructed experiences in the form of narratives on travel blogs entail a wider range of emotions, have richer descriptions and are more personalized. Blogging entails the subjective interpretation of a person's own experiences in a reversed chronological order. As the narrator subjectively chooses what to include in the travel diary, it supports the conscious or subconscious identification of memorable experiences. Experiences that are meaning-making in the long term have always been incremental for tourists; however, contemporary tourists more consciously engage with their emotions and are more explicitly seeking experiences that support them in expressing themselves and growing as individuals.

While blogging is mostly associated with the post-consumption of tourist experiences, the development of mobile technologies and specific apps (e.g., WordPress) has encouraged *moblogging*, which means that bloggers can tweak and share selected content immediately (Tussyadiah and Fesenmaier 2009). This immediacy is often strongly associated with authenticity, as its spontaneity indicates "realness" and seemingly unmediated experiences as the narrator has less chance to manipulate or alter the event after the trip (McNeill 2003).

Regardless of whether online blogs are written during or after the trip, the temporal distance between production and consumption is minimized compared to offline travel diaries. The narratives circulate faster, reach a larger audience and communities can be formed through social interaction. Bloggers' intentions to narrate autobiographical stories are never isolated and purely subjective acts. It is more about self-presentation and interactive community building in a spiraling way between production and consumption, as there are always two sides to a story: the narrator and the listener (Karlsson 2006). However, when looking at the number of tourists who actively engage in writing travel blogs, it becomes evident that tourists are mostly consumers and not necessarily narrators (Huertas and Marine-Roig 2018; Seeler, Lück and Schänzel 2019). At the same time, the segment of professional bloggers is constantly growing. These individuals actively engage in blogging to support their living and become highly influential in eWOM communication (Duffy 2017).

Scholars have addressed the paradoxes of professional blogging and refer to the balancing act between *becoming* and *selling* themselves, which van Nuenen (2016: 208) describes as "trajectories of existential authenticity and self-branding." This act of self-branding was also addressed by Erz, Marder and Osadchaya (2018: 57) who summarize that "recent years have specifically experienced a growing number of influencers and microcelebrities, who use social media to connect, yet not for the sake of community but for the sake of broadcasting themselves." While travel bloggers use the rhetoric of existential authenticity to persuade readers that their lifestyle enables them to distance themselves from capitalist society and live the desired freedom, they need to ensure that they can secure an income with this lifestyle.

In this vein, Karlsson (2006: 305) argues that "the tales of travel are highly crafted and written with a specific audience in mind." This means that the used rhetoric of existential authenticity and seemingly purely intrinsically motivated behavior might be performed to some degree, and that bloggers' motivation is both self-centered and altruistically community-oriented to gain recognition while maintaining relationships (Munar and Jacobsen 2014). Consequently, the realness of co-constructed tourist experiences in professional travel blogs is vulnerable to becoming exploited for commercial purposes, which would lead to a loss of credibility and perceived trustworthiness (Huertas and Marine-Roig 2018; Kang and Schuett 2013). Most bloggers ensure authenticity by strategically creating a self-brand that is topical, such as adventure travel, solo-female travel or family travel. They usually provide background information on their humble lives before travel to express ordinariness and to encourage the reader to believe that the blogger's nomadic life can be replicated by anyone (van Nuenen 2016). An example is Matt Kepnes (better known as "Nomadic Matt"), who is the person behind today's most-read and successful travel blog (www.nomadicmatt.com) with an estimated income of 50,000 USD/month. On his "About" page, Matt welcomes his readers with "Welcome Home Fellow Traveler!" and introduces himself as a cubicle worker with no overseas travel until he was 23 years old. However, he does not spend much time rambling about his own experiences but turns the lens to the reader. He promises to help them realize their dreams and finishes the introduction page with: "We're going to make you a smarter traveler." Topical blogs and those that succeed in creating emotional bonds through authentic narrators have the potential to become highly influential in the inspiration and decision-making processes of others (Huertas and Marine-Roig 2018; MacKay and Vogt 2012). The potential of first-person voices as inspiration for future tourists has recently been more understood by tourism marketers, who have engaged more actively in integrating social media influencers (SMIs) in their overall marketing strategies.

First-person voices as inspiration for future tourists

Generally, it is expected that first-person voice narratives that are non-commercially shared online in the form of storytelling, images or videos are more real (Leung et al. 2013). They are believed to be more emotional and subjective as they are perceived as having greater freedom to not conform and to reveal their real feelings about an experience. Greater emotional bonding can be achieved, which is especially important for intangible products and services like tourism products (McCabe and Foster 2006). Thus, a significant role is performed by travel bloggers and other SMIs who co-construct their experiences online and thereby become the inspiration for future tourists.

In the past, DMOs feared the integration of SMIs into their overall marketing strategies, due to the potential risk that traditional marketing would further lose significance and the pressure to demonstrate the legitimacy of a DMO's work could further increase (Seeler 2018). However, more recently, DMOs have understood the value of influencer marketing and have felt the urgency to keep up with trends related to online behavior and the growing desire to co-construct tourist experiences online. Consequently, SMIs are increasingly approached by tourism marketers to endorse and promote their products and brands, and influencer marketing has become a vital part of overall marketing strategies (Audrezet, de Kerviler and Guidry Moulard 2018). Due to the lack of credibility of paid influencers and commercial bloggers, who are known to be paid to write positive reviews, micro influencers (with less than 100,000 followers) are gaining momentum. This also means that anyone can become an influencer by co-constructing their own experiences in the public domain. Through *world of mouth*, tourists have more access points through which to consume these co-constructed experiences and have become more aware of digital information. With this new awareness comes an improved ability to identify authentic and real experiences that are created outside of the commercial bubble, and thus the integration of SMIs into tourism marketing strategies will be further challenged. In the future, tourism marketing will probably require constant adjustment between the use of material supplied by lifestyle social influencers and that of micro influencers who do it for the love of traveling rather than for monetary gain. These new dynamics will need to be balanced in order to leverage on the potential that the consumption and production of co-constructed tourist experiences brings to both tourists and tourism businesses.

Conclusion

Tourists progressively leave their spectator-modes and passenger seats, get behind the steering wheel and actively engage in experience co-construction. This chapter demonstrates that co-construction of experiences in an increasingly digital society goes beyond the physical world. Technological advances and the emergence of Web 2.0 facilitate a digital immediacy that encourages tourists to co-construct their experiences throughout the different stages of a trip by means of different social media channels and media formats. Motivations vary; they can be inward-oriented and mostly for reasons of personal memory storage and prolonging the magic of the moment or outward-oriented as a form of impression management. Irrespective of the mechanisms behind experience co-construction via social media, messages spread more quickly and go beyond immediate contacts. This uplifts the ordinary social media user to a powerful social media influencer and stimulates the shift toward more consumer-centric and experiential forms of marketing that appreciate more improvised and multidirectional communication. Today's experienced tourists consult crowd-sourced and peer-validated information for travel inspiration and planning while actively engaging in experience-sharing themselves. Thus, a duality of *being influenced* and *becoming an influencer* emerges. *Being influenced* through other people's narratives needs to be seen beyond a managerial and marketing lens; it encapsulates a sociological perspective that can be interpreted as a curse or a blessing. The consumption of secondary experiences might result in the impression that one has seen a place already and no longer needs to physically travel there. This can be a blessing for destinations that suffer from overtourism, but a curse for those fighting undertourism. Tourism businesses need to reflect on their roles and responsibilities in tourists' processes of experience co-construction more actively to ensure that it brings added value to all stakeholders involved.

References

Audrezet, A., de Kerviler, G. and Guidry Moulard, J. (2018). 'Authenticity under threat: when social media influencers need to go beyond self-presentation', *Journal of Business Research*. DOI: 10.1016/j.jbusres.2018.07.008.

Bilgihan, A., et al. (2016). 'Consumer perception of knowledge-sharing in travel-related online social networks', *Tourism Management*, 52: 287–296. DOI: 10.1016/j.tourman.2015.07.002.

Bosangit, C., Dulnuan, J. and Mena, M. (2012). 'Using travel blogs to examine the postconsumption behavior of tourists', *Journal of Vacation Marketing*, 18(3): 207–219, DOI: 10.1177/1356766712449367.

Bosangit, C., Hibbert, S. and McCabe, S. (2015). '"If I was going to die I should at least be having fun": travel blogs, meaning and tourist experience', *Annals of Tourism Research*, 55: 1–14. DOI: 10.1016/j.annals.2015.08.001.

Cardell, K. and Douglas, K. (2018). 'Visualising lives: "the selfie" as travel writing', *Studies in Travel Writing*, 22(1): 104–117, DOI: 10.1080/13645145.2018.1463843.

Chun-Yao, H., et al. (2007). 'Bloggers' motivations and behaviors: a model', *Journal of Advertising Research*, 47(4): 472–484.

Dinhopl, A. and Gretzel, U. (2016a). 'Selfie-taking as touristic looking', *Annals of Tourism Research*, 57: 126–139. DOI: 10.1016/j.annals.2015.12.015.

Dinhopl, A. and Gretzel, U. (2016b). 'GoPro panopticon: performing in the surveyed leisure experience'. In S. Carnicelli, D. McGillivray and G. McPherson (eds.), *Digital Leisure Cultures: Critical Perspectives* (pp. 66–79). London: Routledge.

Dinhopl, A. and Gretzel, U. (2016c). 'Conceptualizing tourist videography', *Information Technology and Tourism*, 15(4): 395–410, DOI: 10.1007/s40558-015-0039-7.

Duffy, A. (2017). '"The locals are friendly!" An empirical analysis of mobility and mooring in travel blogs', *Information, Communication and Society*, 20(3): 444–459, DOI: 10.1080/1369118X.2016.1187192.

Erz, A., Marder, B. and Osadchaya, E. (2018). 'Hashtags: motivational drivers, their use, and differences between influencers and followers', *Computers in Human Behavior*, 89: 48–60. DOI: 10.1016/j.chb.2018.07.030.

Gómez Cruz, E. and Thornham, H. (2015). 'Selfies beyond self-representation: the (theoretical) f(r)ictions of a practice', *Journal of Aesthetics and Culture*, 7. DOI: 10.3402/jac.v7.28073.

Huertas, A. and Marine-Roig, E. (2018). 'Searching and sharing of information in social networks during the different stages of a trip', *Cuadernos de Turismo*, 42: 185–212. DOI: 10.6018/turismo.42.08.

Jansson, A. (2007). 'A sense of tourism: new media and the dialectic of encapsulation/decapsulation', *Tourist Studies*, 7(1): 5–24, DOI: 10.1177/1468797607079799.

Kang, M. and Schuett, M. A. (2013). 'Determinants of sharing travel experiences in social media', *Journal of Travel and Tourism Marketing*, 30(1–2): 93–107, DOI: 10.1080/10548408.2013.751237.

Karlsson, L. (2006). 'The diary weblog and the travelling tales of diasporic tourists', *Journal of Intercultural Studies*, 27(3): 299–312, DOI: 10.1080/07256860600779303.

Kim, J. and Fesenmaier, D. R. (2017). 'Sharing tourism experiences: the post-trip experience', *Journal of Travel Research*, 56(1): 28–40, DOI: 10.1177/0047287515620491.

Leung, D., et al. (2013). 'Social media in tourism and hospitality: a literature review', *Journal of Travel and Tourism Marketing*, 30(1–2): 3–22. DOI: 10.1080/10548408.2013.750919.

Lo, I. S. and McKercher, B. (2015). 'Ideal image in process: online tourist photography and impression management', *Annals of Tourism Research*, 52: 104–116. DOI: 10.1016/j.annals.2015.02.019.

Lund, N. F., Cohen, S. and Scarles, C. (2017). 'The power of social media storytelling in destination branding', *Journal of Destination Marketing and Management*, 8: 271–280. DOI: 10.1016/j.jdmm.2017.05.003.

MacKay, K. and Vogt, C. (2012). 'Information technology in everyday and vacation contexts', *Annals of Tourism Research*, 39(3): 1380–1401, DOI: 10.1016/j.annals.2012.02.001.

McCabe, S. and Foster, C. (2006). 'The role and function of narrative in tourist interaction', *Journal of Tourism and Cultural Change*, 4(3): 194–215, DOI: 10.2167/jtcc071.0.

McNeill, L. (2003). 'Teaching an old genre new tricks: the diary on the Internet', *Biography*, 26(1): 24–47.

Munar, A. M. and Jacobsen, J. K. S. (2014). 'Motivations for sharing tourism experiences through social media', *Tourism Management*, 43: 46–54. DOI: 10.1016/j.tourman.2014.01.012.

New Zealand Herald. (2018, November 27). 'Roy's Peak: photo shows the reality behind "New Zealand's most Instagrammed hike"', [online]. Available at: www.nzherald.co.nz/travel/news/article.cfm?c_id=7&objectid=12166932 [Accessed November 28, 2018].

Prahalad, C. K. and Ramaswamy, V. (2004). 'Co-creation experiences: the next practice in value creation', *Journal of Interactive Marketing*, 18(3): 5–14, DOI: 10.1002/dir.20015.

Qualman, E. (2013). *Socialnomics: How Social Media Transforms the Way We Live and Do Business*. 2nd Ed. Hoboken, NJ: John Wiley.

Seeler, S. (2018). *Continuum of an Experienced Tourist's Multidimensionality – Explorations of the Experience Levels of German and New Zealand Tourists*. Unpublished PhD dissertation, Auckland University of Technology, Auckland, New Zealand.

Seeler, S., Lück, M. and Schänzel, H. A. (2019). 'Exploring the drivers behind experience accumulation – the role of secondary experiences consumed through the eyes of social media influencers', *Journal of Hospitality and Tourism Management*, 41: 80–89. DOI: 10.1016/j.jhtm.2019.09.009.

Stankov, U., et al. (2019). 'The view from above: the relevance of shared aerial drone videos for destination marketing', *Journal of Travel and Tourism Marketing*. DOI: 10.1080/10548408.2019.1575787.

Tussyadiah, I. P. and Fesenmaier, D. R. (2009). 'Mediating tourist experiences: access to places via shared videos', *Annals of Tourism Research*, 36(1): 24–40, DOI: 10.1016/j.annals.2008.10.001.

Urry, J. and Larsen, J. (2011). *The Tourist Gaze 3.0*. London: SAGE.

van Nuenen, T. (2016). 'Here I am: authenticity and self-branding on travel blogs', *Tourist Studies*, 16(2): 192–212, DOI: 10.1177/1468797615594748.

Wang, L. and Alasuutari, P. (2017). 'Co-construction of the tourist experience in social networking sites: two forms of authenticity intertwined', *Tourist Studies*, 17(4): 388–405, DOI: 10.1177/1468797616687559.

White, N. R. and White, P. B. (2007). 'Home and away: tourists in a connected world', *Annals of Tourism Research*, 34(1): 88–104, DOI: 10.1016/j.annals.2006.07.001.

35

Technologies for enhancing wine tourism experience

Roberta Garibaldi and Fabiola Sfodera

Introduction

Technology is pervasive in people's daily lives. It has changed their way of living and relating to others. However, not all of the technology has been accepted and adopted by the market, but only that which is able to create value. Technology is pervasive even throughout the tourist customer journey and concerns all tourist activities, from the choice of destination to the purchase of an event ticket to the choice of the vinery and evaluation of the wine being tasted. ICT has enabled the economic and technical feasibility of digital innovation and has fulfilled more sophisticated and relevant functions, such as becoming an enabler and spreader of experiences and a driver in the co-creation of value. Since the 1980s, studies have focused on technology support functions to improve service quality and cost control (Buhalis and Law 2008). In the last few decades, however, attention has focused on the ability of ICT, and specifically digital technologies to support consumer centricity in the purchasing process and creating and consuming experiences (Navío-Marco et al. 2018).

This is particularly relevant compared to those tourist activities and those attractors whose efficacity requires information and knowledge for visitors to fully appreciate their meaning and value in relation to wine tourism. In recent years, wine and food tourism have assumed significant dimensions both in the old world (e.g. France and Italy) and in the new (e.g. Australia, USA and New Zealand). Wine is a sophisticated product and the consumer experience refers to functional, symbolic, aesthetic and hedonistic (e.g. Charters and Ali-Knight 2002) dimensions, whose symbolic values (conviviality and socialization) and emotional values (knowledge, wine destination, discovery and tasting) determine the main motivations for the wine tourist journey. Similarly, wine tourism and winery experiences are a multi-dimensional experience articulated on several levels and characterized by the presence of a narrative and a multi-sensory dimension (World Tourism Organization and Basque Culinary Center 2019). In this chapter, we will analyze how technology is able to activate the narrative dimension of wine and contribute to creating memorable experiences.

Technology enabling consumer experiences

Technology is one of the most studied topics of the last decades, and is relevant to many disciplines. The technology literature review, applied to the consumer experience, shows a broad and extremely fragmented picture of studies. The role of technology in the consumption experience is fragmented and conditioned by the lack of a theoretical framework. According to the S-D logic approach, technology enables the integration of resources processes and the value of co-creation within the service ecosystem (Vargo and Lusch 2004; Lusch and Vargo 2006a, 2006b; Vargo and Akaka 2012; Akaka and Vargo 2014; Sfodera et al. 2018). Lusch and Nambisan (2015) have theorized how technology, according to the S-D Logic perspective, is a lever for service innovation that facilitates exchanges between ecosystem actors; that improves the efficiency and effectiveness of the exchange in service platforms and the dissemination of knowledge; and, finally, that activates the processes of value co-creation.

With tourist empowerment moving clients toward more dynamic behavior in experience creation (Kim et al. 2016), technology contributes to their transformation, supporting tourists at all stages of the experience, pre, on-site and post-journey (Sfodera 2011). Technology, therefore, enables and facilitates the value creation process through co-created, dynamic and personalized experiences (Neuhofer et al. 2014; Zhang et al. 2018), overcoming a specific planning formalism of the Pine and Gilmore framework (1998), according to which the company plans the experiential offer. The customer experience is, therefore, a multi-dimensional construct that enhances subjective, personal and dynamic dimensions (Lemon and Verhoef 2016). The responses of consumers, according to cognitive, emotional, sensorial, behavioral and social dimensions, to the stimuli deriving from the offer of companies, create the customer experience. Tourism is, for several reasons, the ideal domain to study the potential of technology to enhance the customer experience: it is a domain of services (servicescape), characterized by a hedonic consumption (e.g. Holbrook and Hirschman 1982; Dhar and Wertenbroch 2000) as well as being intrinsically intangible and experiential (Tussyadiah et al. 2018). Tourism is also the industry where technologies like virtual reality, augmented reality and mixed reality have expressed great potential, starting from cultural heritage to wine tourism (e.g. Flavián et al. 2018).

Wine tourism and technology

Wine tourism experience

According to Getz (1998), wine tourism is 'a form of consumer behavior, where wine lovers or those interested in wine regions travel to preferred destinations'. From the earliest studies, it was found that the winery visitor, especially if she/he is a tourist, is not a conventional consumer (Hall and Mitchell 2000). Although among scholars the wine tourism and wine experience definitions are accepted, the theoretical and empirical corpus of the phenomenon is still fragmented (Quintal et al. 2015). According to Quadri-Felitti and Fiore (2012) and Thanh and Kirova (2018), wine tourism studies appeared from 1990–2000 and involved both destinations, wine actors and the analysis of wine tourist consumer behavior. The wine tourist's experience is a multi-dimensional experience and the result of the interaction of several variables such as natural setting, wine, food, cultural aspects and local heritage and quality of staff and service (Charters 2006). However, there is still a significant cognitive gap in wine tourism motivations (Ravenscroft and Van Westering 2001) and on the wine experience (Quadri-Felitti and Fiore 2012; Bruwer and Rueger-Muck 2018). If this has led to considering wine tourism as a specific market, on the other hand it has not allowed the definition of an overall cognitive framework of the phenomenon itself.

Winery experience

There is a specific gap, deriving from the copious literature on wine tourism in the 'new world', which is Australia and America in the first place, which allowed the definition of theoretical frameworks and measurement scales relating to these specific contexts (e.g. Quintal et al. 2015). Due to the relevance of the wine sensorial nature of wine tourism, whether related for example to tastings rather than walks in the vineyards, the wine tourism researcher has paid attention to the experience concepts, according to Pine and Gilmore's model (Pine and Gilmore 1998) (e.g. Quadri-Felitti and Fiore 2012; Thanh and Kirova 2018). Winery experience research shows an inevitable delay regarding the evolution of tourist experience studies. The thematic literature review only identifies three study areas on the wineries experience: *learning, hedonic experience and the winescape.*

The cognitive and perceptive stimulation opportunities through the winery visit are in themselves an engaging experience (Hughson and Boakes 2001). Learning concerns the world of wine, from the winemaking process to the history of the winery to the tasting and attributes of the wine (e.g. Cohen and Ben-Nun 2009; Clemente-Ricolfe et al. 2012). Tasting is one of the most practiced activities by visitors (Alant and Bruwer 2010) and is a trigger of the sensory aspects, allowing visitors to experiment and test the product, which influences their overall satisfaction (Charters and O'Neill 1999; Jago and Issaverdis 2001). Moreover, the tasting links the product with the company and the destination and the experience evaluation is the result of several variables (for example, tasting different wines, tested wines' reputation, presence of local products that accompany the tasting, information provided, the overall atmosphere, lighting, interaction with the staff, equipment and cleaning) (e.g. Gill et al. 2007; Charters et al. 2009; McNamara and Cassidy 2015). Learning depends on tourist involvement regarding the wine (Velikova et al. 2015). Learning develops through different touchpoints before and during the visit and through some activities such as the tasting and the information provided, the tour of the vineyard and in the cellar, participation in wine pressing with feet, meeting with the enologist and the cellar staff and the visit to the wine museum (e.g. Ballester et al. 2008; Velikova et al. 2015).

The hedonic dimension of the winery visit is that of entertainment and pleasure defined by Holbrook and Hirschman (1982: 92) as 'characterized by an affective and sensory experience of aesthetic or sensual pleasure, fantasy, and fun'. This dimension is an intrinsic part of the wine experiences consumption (Pikkemaat et al. 2009) and is inherent in several activities and services such as events and festivals, wine shops, wine and food demonstrations, wine museums and heritage, recreational areas, spas and wellness services and walks in the vineyards.

The winescape, deriving from the Bitner (1992) servicescape, is 'a winsome combination of vineyards, wineries and supporting activities necessary for modern production' (Peters 2018), to which Johnson and Bruwer (2007) have added 'natural landscape and environment, people and heritage, city and their architecture and artifacts within them'. The winescape, and specifically the staff and the complementary products, influences the behavioral intentions of tourists according to the framework for 'new world wineries' by Quintal et al. (2015). It influences the visit or revisit intentions to a winery or a wine destination (Byrd et al. 2016) and its positive evaluation has a positive influence on satisfaction and on the willingness to recommend or revisit the winery (Thomas et al. 2010, 2018).

It seems, however, that even today there has been no definition of the experience as a multi-dimensional construct and the winery visitors' behavior according to the customer journey. Such a gap also has repercussions in studies concerning the use of technology in the wine experience, which are few and extremely fragmented. The theoretical framework is still under development, as mentioned. However, the empirical evidence of two relative cases – the use of technology to enhance and implement pre and on-site visits and to enable a dynamic co-creation process – allows us to understand the potential of technology in the wine and winery experience.

Technology as a means to enhance pre- and on-site visits

Technology is gaining attention in the food and wine tourism field as a means to enhance visitors' experience, making it distinct and valuable. The foodservice sector provides a number of examples, with interactive touch screens and a multitouch table that is currently used to create a digitalized dining experience, where customers have access to a wide variety of contents and functions. This digitalized dining experience takes place, for instance, in the Inamo Restaurant in London, which pioneered the introduction of new technologies to regenerate the traditional meal experience. The E-Table™ uses a combination of table touchpads that allows customers, e.g. to see the food and drinks menu projected onto the table surface, change tablecloths accordingly to their mood and preferences, manage their order and watch their food being prepared in the kitchen through a webcam in real time (Neuhofer et al. 2014). A different approach to modernizing the dining experience is that used by the Ultraviolet restaurant in Shanghai. Multi-sensory technology was adopted to recreate an immersive setting and provide a unique meal experience. A 20-courses avant-garde menu is served in a dining room fully covered with video screens, as well as lights, scents, atmosphere and projections that change with each dish (Ultraviolet by Paul Pairet 2019).

The use of technology to implement pre and on-site visits has been recently developed also in wineries, in order to offer new and distinctive experiences that exploit traditional visits and tastings and meets new customers' needs (Getz 2000; Jaffe and Pasternak 2004; Garibaldi and Pozzi 2018). Among types of technology, the use of virtual reality technology has spread thanks to the variety of possible applications, e.g. as a learning tool and as a means to provide public access to sites and attractions from which everyone can potentially benefit but where visiting is limited due to physical barriers (Guttentag 2010). Bodegas Ramòn Bilbao in Spain and Brancott Estate in New Zealand provide examples of the possible application of VR.

Located in the renowned Spanish wine region of La Rioja, Bodegas Ramòn Bilbao has been committed to wine tourism development and innovation for years. Visitors to Haro wine cellar – one of the two wineries of Bodegas Ramòn Bilbao – are today offered the opportunity to enjoy the 'Experiencia Oculus' during the guided tour. This VR-based experience allows visitors to discover the entire winemaking process from 'the birth of the grape to the glass of wine' and to be introduced to the range of products that can be appreciated in the tasting room. Content played through VR headsets has been specifically created to be linked to the visual universe of this Rioja label, which is based on the balance between tradition and innovation (Bodegas Ramón Bilbao 2019). By using this technology, the winery offers a new and more engaging experience that overcomes physical barriers; in other words, VR recreates the 'journey of the wine' and provides access to the winescape during all seasons. Additionally, it functions as an educational tool that permits visitors to gain knowledge about the history and the values of the winery, the winemaking process and the range of products available.

Brancott Estate is a New Zealand-based winery with over 40 years of history of crafting wines, and it has been committed to winemaking innovation and sustainability for years (Brancott Estate 2019). In 2016 the Brancott Estate winery commissioned the creation of the

'Red Shed Experience', a virtual reality-based experience to be also used during company presentation events. Through VR headsets users can navigate the virtual replica of its vineyards and cellar and view it from a different perspective, switching between different periods to discover the maturation of the wines and see the entire winemaking process. To offer users a more realistic taste of the winery (and its wines), specific devices release scents of the grapes and fruits in order to immerse them in the experience of what it would be like to be hanging out in New Zealand, enjoying a glass of wine (Bradley 2017; futurewineexpo.com; 2019). In doing so, potential customers can enjoy the Brancott Estate winery even though they cannot travel to New Zealand; or they can have a pre-experience of a place that is going to be visited afterwards. From this perspective, the 'Red Shed Experience' represents a marketing tool that may encourage future visits, rather than just a way to enhance the real visit.

Technology as an enabler of a dynamic co-creation process

Examples previously mentioned succeeded in taking advantage of technologies to create enhanced experiences where customers actively participate in and interact with virtual contents and places. In this scenario technology mainly plays a complementary role, as it supports the tourism experience. The provision of the winery products still remains the core function, however. New technologies can also empower and become an integral part of the tourism experience, enabling a dynamic co-creation process; at this level, technology has a crucial role and needs to exist for the experience to happen (Neuhofer et al. 2014). The technology role appears particularly relevant to the new generation of wine museums, which are characterized by a multifaceted presentation of wine culture, an experiential approach and the extensive use of multimedia technologies (Bellini and Resnick 2018). La Cité du Vin, in the Bordeaux region of France, represents one of the most innovative examples of this genre.

This cultural facility was opened on the 1 June 2016 and it has rapidly become one of the most visited attractions in the area, reaching the milestone of 1 million visitors in August 2018 (La Cité du Vin 2019). The commitment of the founders to promote and share wine culture led to a new museum concept, where the history of winemaking and its place in cultures around the world is transmitted interactively by encouraging visitors to engage with the contents on exhibition and to dynamically (co-) create their own experience. The permanent tour was therefore conceived as a modular journey with 20 themed spaces explaining the culture of wine through history, geography, geology, enology and the arts, each of them characterized by the extensive use of digital and interactive technologies. There are no standard paths to take, and each visitor can discover the permanent tour according to their own wishes and tastes.

Interactive touch screens and a multitouch table are displayed in themed spaces, which allow visitors to access a variety of videos, images and sounds that make them more informed about the wines of the world, their history and manufacturing methods. For instance, 'The terroir table' re-creates a journey through the most important wine regions, allowing visitors to listen to 50 winemakers from ten wine regions talking about the history of the vineyards and wineries, simply by interacting with a tactile station. The 'Private meeting with experts' consists of a series of interactive PODs where visitors can have a person to person virtual interview with wine experts, e.g. on tips to choose the wine, how to preserve it and serve in the best conditions, etc. Audiovisual systems reproduce HD content on giant screens so that the audience is surrounded by images and sounds and is thus fully immersed in a wine learning experience. In the 'World tour of the vineyards', for example, visitors can enjoy an immersive film experience that plays out over three screens, presenting the different global landscapes where vines are cultivated, from high mountain ranges to sandy coastal plains. In the 'All aboard!' a 180 degree film projection get visitors immersed in a journey across the tides of history, following the galleys and barges of generations of wine merchants. Smell diffusion systems are used to educate visitors to distinguish different aromas as well as to create a more immersive setting. The 'Metamorphoses of wine' provides an example, as this experience allows visitors to acquire information on harvests, fermentation, maturation and the aging process through the use of images, videos, smells and sounds.

In order to enable visitors to discover the permanent tour according to their wishes and tastes, a digital 'travel companion' is provided. This tool was specifically developed for La Cité du Vin and detects and activates the different multimedia animations along the path. For example, in the 'E-vine' experience the 'travel companion' provide access to specific games developed for children (Vigneaud 2016).

Conclusions

Technology has always been important in the tourism sector, from the first function of disintermediation to the subsequent one of sharing information, for example through reviews. But, in recent years it has taken on more sophisticated and relevant functions by becoming both an enhancer and an enabler of experiences in co-creation value. Although its potential uses and benefits are considerable, little attention has been devoted to the use of technology in the wine tourism experience.

Wine tourism is a growing segment according to recent studies (e.g. Garibaldi et al. 2016; Garibaldi 2018, 2019) but has been under-investigated in terms of technological applications, although there appears to be evidence of a number of examples. Case studies examined in this chapter highlight that the use of technology can not only facilitate the development of enhanced experiences where customers actively participate and interact with virtual content and places (such in the cases of VR-based experience in the Bodegas Ramòn Bilbao and Brancott Estate wineries) but also enable a dynamic co-creation process in which visitors are allowed to create their own experience (as currently happens in the themed museum La Cité du Vin). Along with the number of technologies used (virtual reality, touchpoints, multitouch tables, 360 degree and 180 degree projections, etc.), there appears to be evidence that such tools can have a direct impact on the main three elements constituting the wine experience, namely learning, hedonism and the winescape. Digital and interactive technologies installed in the permanent tour of La Cité du Vin, for instance, enable visitors to gain knowledge about the history of winemaking and its place in different cultures in a more interactive way; the 'Red Shed Experience' reinforces the hedonic dimension of the visit by combining virtual reality with the use of specific devices that release scents of the grapes and fruits; 'Experiencia Oculus' re-creates the 'journey of the wine' and provides visitors with virtual access to the winescape during all seasons.

The use of technology in the wine experience certainly provides new opportunities, but it also involves a number of challenges. One of these opportunities is for the wine tourism industry, which has the opportunity to make itself more attractive to a wide range of tourists. By implementing traditional activities (e.g. visits and tasting) and providing co-creation opportunities, technology-based wine experiences can increase their appeal both to primary-motivated tourists and other travelers (such as non-wine drinker tourists). This suggests a number of implications. First, it implies a shift in thinking away from the product or service toward a more holistic approach to the entire experience that also considers how technology can transform tourism service ecosystems. Secondly, as technology has the potential to enable a dynamic co-creation process, wine tourism operators should be stimulated to incorporate travelers into the process of experience design and development.

References

Akaka, M. A., and Vargo, S. L. (2014). 'Technology as an operant resource in service (eco) systems', *Information Systems and e-Business Management*, 12(3): 367–384. DOI: 10.1007/s10257-013-0220-5

Alant, K., and Bruwer, J. (2010). 'Winery visitation sets: Intra-regional spatial movements of wine tourists in branded regions', *International Journal of Wine Business Research*, 22(2): 191–210. DOI: 10.1108/17511061011061748

Ballester, J., Patris, B., Symoneaux, R., and Valentin, D. (2008). 'Conceptual vs. perceptual wine spaces: Does expertise matter?', *Food Quality and Preference*, 19(3): 267–276. DOI: 10.1016/j.foodqual.2007.08.001

Bellini, N., and Resnick, E. (2018). 'The luxury turn in wine tourism. Still good for local development?', In Bellini, N., Clergeau, C., and Etcheverria, O. (eds) *Gastronomy and local development the quality of products, places and experiences*, 1st edition (pp. 214–228). London: Routledge.

Bitner, M. J. (1992). 'Servicescapes: The impact of physical surroundings on customers and employees', *Journal of Marketing*, 56(2): 57–71. DOI: 10.2307/1252042

Bradley, M. (2017). 'Merchants are utilizing virtual reality', *veer.tv*. Retrieved from: https://veer.tv/blog/6-ways-top-winemakers-and-merchants-are-utilizing-virtual-reality/ [Accessed: May 21, 2019].

Bruwer, J., and Rueger-Muck, E. (2018). 'Wine tourism and hedonic experience: A motivation-based experiential view', *Tourism and Hospitality Research*, 19(4): 488–502. DOI: 10.1177/1467358418781444

Buhalis, D., and Law, R. (2008). 'Progress in information technology and tourism management: 20 years on and 10 years after the internet—The state of eTourism research', *Tourism Management*, 29(4): 609–623. DOI: 10.1016/j.tourman.2008.01.005

Byrd, E. T., Canziani, B., Hsieh, Y. C. J., Debbage, K., and Sonmez, S. (2016). 'Wine tourism: Motivating visitors through core and supplementary services', *Tourism Management*, 52: 19–29. DOI: 10.1016/j.tourman.2015.06.009

Charters, S. (2006). *Wine and society*. London: Routledge. DOI: 10.4324/9780080458038

Charters, S., and Ali-Knight, J. (2002). 'Who is the wine tourist?', *Tourism Management*, 23(3): 311–319. DOI: 10.1016/S0261-5177(01)00079-6

Charters, S., Fountain, J., and Fish, N. (2009). '"You felt like lingering…" Experiencing "real" service at the winery tasting room', *Journal of Travel Research*, 48(1): 122–134. DOI: 10.1177%2F0047287508326508

Charters, S., and O'Neill, M. (1999). 'Service quality at the cellar door—Some implications for Western Australia's developing wine industry', Paper presented to the 1999 *Australian Wine Tourism Conference*, Rutherglen, August 1999.

Clemente-Ricolfe, J. S., Escribá-Pérez, C., Rodriguez-Barrio, J. E., and Buitrago-Vera, J. M. (2012). 'The potential wine tourist market: The case of Valencia (Spain)', *Journal of Wine Research*, 23(2): 185–202. DOI: 10.1080/09571264.2012.678934

Cohen, E., and Ben-Nun, L. (2009). 'The important dimensions of wine tourism experience from potential visitors' perception', *Tourism and Hospitality Research*, 9(1): 20–31. DOI: 10.1057%2Fthr.2008.42

Dhar, R., and Wertenbroch, K. (2000). 'Consumer choice between hedonic and utilitarian goods', *Journal of Marketing Research*, 37(1): 60–71. DOI: 10.1509%2Fjmkr.37.1.60.18718

Flavián, C., Ibáñez-Sánchez, S., and Orús, C. (2018). 'The impact of virtual, augmented and mixed reality technologies on the customer experience', *Journal of Business Research*, 100: 547–560. DOI: 10.1016/j.jbusres.2018.10.050

Future Wine Expo. (2019). 'innovative uses for virtual reality in the wine industry, futurewineexpo.com', Retrieved from: https://futurewineexpo.com/en/blog/insights-64/innovative-uses-for-virtual-reality-in-the-wine-industry-252.htm. [Accessed: May 21, 2019].

Garibaldi, R. (2018). *Primo Rapporto sul Turismo Enogastronomico Italiano 2018 (First report on gastronomy tourism in Italy 2018)*. Bergamo: Centro Editoriale Librario Studium Bergomense – CELSB.

Garibaldi, R. (2019). *Rapporto sul Turismo Enogastronomico Italiano 2019 (Report on gastronomy tourism in Italy 2019).*

Garibaldi, R., and Pozzi, A. (2018). 'Creating tourism experiences combining food and culture: An analysis among Italian producers', *Tourism Review*, 73(2): 230–241. DOI: 10.1108/TR-06-2017-0097

Garibaldi, R., Stone, M. J., and Migacz, S. (2016). *2016 Food travel monitor.* Portland, OR: World Food Travel Association.

Getz, D. (1998). 'Wine tourism: Global overview and perspectives on its development', *Wine Tourism Perfect Partners: Proceedings of the First Australian Wine Tourism Conference*, Canberra: Bureau of Tourism Research.

Getz, D. (2000). *Explore wine tourism: Management, development, destinations.* New York: Cognizant.

Gill, D., Byslma, B., and Ouschan, R. (2007). 'Customer perceived value in a cellar door visit: The impact on behavioural intentions', *International Journal of Wine Business Research*, 19(4): 257–275. DOI: 10.1108/17511060710837418

Guttentag, D. A. (2010). 'Virtual reality: Applications and implications for tourism', *Tourism Management*, 31: 637–651. DOI: 10.1016/j.tourman.2009.07.003

Hall, C. M., and Mitchell, R. (2000). 'Wine tourism in the mediterranean: A tool for restructuring and development', *Thunderbird International Business Review*, 42(4): 445–465. DOI: 10.1002/1520-6874(200007/08)42:4%3C445::AID-TIE6%3E3.0.CO;2-H

Holbrook, M. B., and Hirschman, E. C. (1982). 'The experiential aspects of consumption: Consumer fantasies, feelings, and fun', *Journal of Consumer Research*, 9(2): 132–140. DOI: 10.1086/208906

Hughson, A. L., and Boakes, R. A. (2001). 'Perceptual and cognitive aspects of wine expertise', *Australian Journal of Psychology*, 53(2): 103–108. DOI: 10.1080/00049530108255130

Jaffe, E., and Pasternak, H. (2004). 'Developing wine trails as a tourist attraction in Israel', *International Journal of Tourism Research*, 6(4): 237–249. DOI: 10.1002/jtr.485

Jago, L. K., and Issaverdis, J. P. (2001). 'Building cellar door business through accreditation—Victoria's perspective', *International Journal of Wine Marketing*, 13(3): 18–31. DOI: 10.1108/eb008724

Johnson, R., and Bruwer, J. (2007). 'Regional brand image and perceived wine quality: The consumer perspective', *International Journal of Wine Business Research*, 19(4): 276–297. DOI: 10.1108/17511060710837427

Kim, S. R., Youn, S. H., Um, S. H., and Lee, T. J. (2016). 'The mediation of information technology on visitors' experience at a cultural heritage site', *Asia Pacific Journal of Tourism Research*, 21(10): 1126–1141. DOI: 10.1080/10941665.2015.1093513

Lemon, K. N., and Verhoef, P. C. (2016). 'Understanding customer experience throughout the customer journey', *Journal of Marketing*, 80(6): 69–96. DOI: 10.1509%2Fjm.15.0420

Lusch, R. F., and Nambisan, S. (2015). 'Service innovation: A service-dominant logic perspective', *MIS Quarterly*, 39(1): 155–175. DOI: 10.25300/MISQ%2F2015%2F39.1.07

Lusch, R. F., and Vargo, S. L. (2006a). 'Service-dominant logic: Reactions, reflections and refinements', *Marketing Theory*, 6(3): 281–288. DOI: 10.1177/1470593106066781

Lusch, R. F., and Vargo, S. L. (2006b). 'Service-dominant logic as a foundation for a general theory', In Lusch, R. F., and Vargo, S. L. (eds.), *The service-dominant logic of marketing: Dialog, debate, and directions*, 406–420. New York: M. E. Sharpe.

McNamara, N., and Cassidy, F. (2015). 'Wine tasting: To charge or not to charge?', *International Journal of Hospitality Management*, 49: 8–16. DOI: 10.1016/j.ijhm.2015.05.004

Navío-Marco, J., Ruiz-Gómez, L. M., and Sevilla-Sevilla, C. (2018). 'Progress in information technology and tourism management: 30 years on and 20 years after the internet-revisiting Buhalis and Law's landmark study about eTourism', *Tourism Management*, 69: 460–470. DOI: 10.1016/j.tourman.2018.06.002

Neuhofer, B., Buhalis, D., and Ladkin, A. (2014). 'A typology of technology-enhanced tourism experiences', *International Journal of Tourism Research*, 16(4): 340–350. DOI: 10.1002/jtr.1958

Peters, G. L. (2018). *American winescapes: The cultural landscapes of America's wine country.* New York: Routledge.

Pikkemaat, B., Peters, M., Boksberger, P., and Secco, M. (2009). 'The staging of experiences in wine tourism', *Journal of Hospitality Marketing and Management*, 18(2–3): 237–253. DOI: 10.1080/19368620802594110

Pine, B. J., and Gilmore, J. H. (1998). 'Welcome to the experience economy', *Harvard Business Review*, 76: 97–105.

Quadri-Felitti, D. L., and Fiore, A. M. (2012). 'Experience economy constructs as a framework for understanding wine tourism', *Journal of Vacation Marketing*, 18(1): 3–15. DOI: 10.1177%2F1356766711432222

Quintal, V. A., Thomas, B., and Phau, I. (2015). 'Incorporating the winescape into the theory of planned behaviour: Examining 'new world' wineries', *Tourism Management*, 46: 596–609. DOI: 10.1016/j.tourman.2014.08.013

Ravenscroft, N., and Van Westering, J. (2001). 'Wine tourism, culture and the everyday: A theoretical note', *Tourism and Hospitality Research*, 3(2): 149–162. DOI: 10.1177%2F146735840100300206

Sfodera, F. (2011). *Turismi, destinazioni e internet: La rilevazione della consumer experience nei portali turistici (Tourisms, destinations and internet: The consumer experience recognition on tourism web portal).* Milano: Franco Angeli.

Sfodera, F., Mingo, I., Mattiacci, A., and Colurcio, M. (2018), 'Technology as operant resources for the cultural heritage reconstruction experience', *IRSSM-9 Symposium proceedings*. ISBN:978-961-240-342-3 vol. 9.

Thanh, T. V., and Kirova, V. (2018). 'Wine tourism experience', *A Netnography Study' Journal of Business Research*, 83: 30–37. DOI: 10.1016/j.jbusres.2017.10.008

Thomas, B., Quintal, V., and Phau, I. (2010). 'Predictors of attitudes and intention to revisit a winescape', *Proceedings of Australian and New Zealand Marketing Academy conference*. Australian and New Zealand Marketing Academy.

Thomas, B., Quintal, V. A., and Phau, I. (2018). 'Wine tourist engagement with the winescape: Scale development and validation', *Journal of Hospitality and Tourism Research*, 42(5): 793–828. DOI: 10.1177/1096348016640583

Tussyadiah, I. P., Wang, D., Jung, T. H., and Tom Dieck, M. C. (2018). 'Virtual reality, presence, and attitude change: Empirical evidence from tourism', *Tourism Management*, 66: 140–154. DOI: 10.1016/j.tourman.2017.12.003

World Tourism Organization and Basque Culinary Center (2019). *Guidelines for the development of gastronomy tourism*. Madrid: UNWTO. DOI: 10.18111/9789284420957.

Vargo, S. L., and Akaka, M. A. (2012). 'Value cocreation and service systems (re) formation: A service ecosystems view', *Service Science*, 4(3): 207–217. DOI: 10.1287/serv.1120.0019

Vargo, S. L., and Lusch, R. F. (2004). 'Evolving to a new dominant logic for marketing', *Journal of Marketing*, 68(1): 1–17. DOI: 10.1509%2Fjmkg.68.1.1.24036

Velikova, N., Howell, R. D., and Dodd, T. (2015). 'The development of an objective wine knowledge scale: The item response theory approach', *International Journal of Wine Business Research*, 27(2): 103–124. DOI: 10.1108/IJWBR-11-2014-0052

Vigneaud, J. P. (2016). *La Cité du vin,un monde de cultures*. Bordeaux: Editions Sud-Ouest.

Zhang, H., Gordon, S., Buhalis, D., and Ding, X. (2018). 'Experience value cocreation on destination online platforms', *Journal of Travel Research*, 57(8): 1093–1107. DOI: 10.1177%2F0047287517733557

Sitography

Bodegas Ramón Bilbao (2019). *Ramón Bilbao, the first wine cellar to offer a virtual reality journey*. Retrieved, from www.bodegasramonbilbao.es/en/press/ramon-bilbao-the-first-wine-cellar-to-offer-a-virtual-reality-journey/ [Accessed: May 21, 2019].

Brancott Estate (2019). *Brancott Estate Visit our vineyards*. Retrieved from www.brancottestate.com/en/visit-our-vineyard [Accessed: May 21, 2019].

La Cité du Vin (2019). *La Cité du Vin. Un monde de cultures*. Retrieved from www.laciteduvin.com/fr [Accessed: May 21, 2019].

Ultraviolet by Paul Pairet (2019). Retrieved from https://uvbypp.cc [Accessed: May 21, 2019].

36

Pictures, colors and emotions

Shaping the UK's e-tourism experience

Anke Schneider, Wilhelm Loibl and Ann Hindley

Introduction

Rising competition between tourism destinations has led to an increase in the research about destination competitiveness (Abreu Novais, Ruhanen and Arcodia 2018). To be competitive a destination has to create a positive perception (Dwyer and Kim 2003), meaning that potential tourists need to become inspired well in advance of their visit. Therefore, digital channels have become increasingly important for destination marketing (Hays, Page and Buhalis 2013; Mariani, Di Felice and Mura 2016; Uşaklı, Koç and Sönmez 2017). The aim is to provide a positive image that has a positive impact on the consumer's buying decision. After all, customers no longer buy only "products" but the experiences and emotions associated with them (Gilmore and Pine 2002). In particular, visual experiences play a decisive role (Tuch, Bargas-Avila, Opwis and Wilhelm 2009). In this online environment consumers rapidly scan website pages, but do not take in the full content. Pictures draw significant amounts of attention, making them important in website design (Pieters and Wedel 2004). It is already known from psychology that pictures bring about emotions (Bradley, Greenwald, Petry and Lang 1992; Ou, Luo, Woodcock and Wright 2004; Valdez and Mehrabian 1994). Pictures can affect people cognitively and emotionally very quickly. Unlike textual building blocks, pictures can be processed and stored faster. Therefore, it is of significant importance for marketers to optimally integrate visual experiences into their marketing or advertising strategy (Kroeber-Riel and Esch 2011). Certain visual stimuli can highlight important key information and control the viewer's attention.

A central communication element here is color (Küthe and Venn 1996), as cultural identity and sense of place are uniquely reflected through color (Lenclos and Lenclos 2004). However, the pictures used in tourism are not objective, but are intended to evoke in the viewer the desire to see and perceive the picture in real life. By today's technology, the embellishment of a picture, such as the brightening of a sandy beach, is common practice (Urry and Larsen 2011). *"Even though advertisers are aware of the various attributes of a picture, they usually choose their stimulatory material based on their good feeling rather than on reasoned analyses"* (Labrecque and Milne 2012). Too much distortion of reality is thus easily possible, which of course can have negative effects.

Due to the importance of website pictures and their color characteristics, this chapter will use an automated approach to explore the color content of pictures on UK websites. This will determine the positive, neutral or negative emotions conveyed to the viewer. The ability to awaken positive emotions in a viewer would help build a competitive advantage for a destination. As a consequence, potential visitors would likely expect a more positive tourism experience when visiting this destination.

Pictures in e-tourism settings

Since the late 1950s pictures have played a crucial part in advertising, conveying information about products and brands (Baker 1961). Especially in the service industry, pictures are often used to reduce the inherent risk associated with the intangibility (Cutler and Javalgi 1993). With the rise of digital marketing channels, visual content like pictures and videos has also become more important (Hanlon 2019: 107). Within an online promotion context in hospitality and tourism, Han and Mills (2006) argue that the use of pictures and colors is an important element of any website's aesthetics.

Destination brands are formed of different meanings and experiences that promise value, these being used as marketing tools to differentiate from the competition (Jiménez Barreto, Rubio and Campo Martínez 2019). To provide a destination with a tangible picture, extensive levels of visuals are required (Buhalis and Law 2008), with stereotypical visuals used in promotional materials to build or create the unique image of the destination (Page, Steele and Connell 2006). Although for Paris this might include the Eiffel Tower, the Champs Elysées or Disneyland Paris, a review of 10,000 pictures suggested, "the top-scoring elements that make Paris look like Paris are the doors, balconies, windows with railings, street signs and special Parisian lampposts" (Doersch, Singh, Gupta, Sivic and Efros 2015: 106). However, color plays its part. Destination Management Organisation (DMO) websites are vital promotional tools, because they express the potential tourist experience and through color they evoke positive feelings and sensations that influence attitudes and intentions (Jiménez Barreto, Rubio and Campo Martínez 2019). In a study of color in hotels, it was determined that although high Chroma blue and green hues were less noticeable by participants, customers could be influenced by the right color choices by tourism organizations (Hunjet, Glogar, Osterman and Kozina 2016). In reviewing promotional materials from the Malta Tourism Association, Francesconi (2011) identified varying degrees of light strength covering sunset, night and dawn, but mainly focusing on strong daylight; strong chromatic tints providing coherence; and reoccurring yellows, blues and reds creating a positive mood.

Composition, lighting and color

Imagery in the large part is made up of pictures and in the destination context these are formed of often standardized content such as palm trees, bedrooms with a sea-view and the colors of exotic plants (Urry 1992). Picture composition, lighting and color are used to select, organize and beautify the objects (Kress and van Leeuwen 1996). Nevertheless, the manipulation of photographic pictures by digital means through the removal, addition or enhancement of objects and color can present an idealized picture that no longer reflects reality (Park 2018). Picture composition in tourism photography involves the ideological inclusion of some elements but the exclusion of others from the tourist gaze, such as social conflict, poverty and waste (Urry 2002). The composition has three interrelated systems: information value, salience and framing (Kress and van Leeuwen 2006). Information value relates to the meaningful location of elements in a picture (Francesconi 2011). Salience considers the different ways in which pictures are depicted to capture attention: the presence of cultural symbols or human figures; size and sharpness of focus; tonal and color contrasts; hard or soft lighting (Kress and van Leeuwen 2006). Framing is achieved through the thickness of frame lines, discontinuities in shape or color and the use of space or connectedness between the elements (Kress and van Leeuwen 2006). To analyze the interpersonal function within pictures modality is used to represent the credibility and reliability of the picture, demonstrated through color saturation, color differentiation and color modulation (Francesconi 2011). Color saturation is the degree to which a hue is present (strong saturation-high pigment; low saturation-low pigment); hues in the coloration range are determined by the dominant wavelength (red, blue or yellow); and lightness, brightness and luminance determine the light or dark characteristics of the color (Lichtlé 2007).

A picture's color characteristics are critical in maintaining interest through emotional connections (Bellizzi and Hite 1992; Lee and Andrade 2010; Solli and Lenz 2011; Valdez and Mehrabian 1994). These color characteristics concern hue, saturation and luminance (Gonzales and Woods 2008; Zimbardo and Gerrig 2004), which create a positive or negative emotional response in the prospective user. To evoke different emotions within pictures, colors can be manipulated (Ou et al. 2006) or they can be substituted if the emotional response is negative.

Emotions and online experience initiated through colors

Pictures are often used in marketing to trigger and transport emotions. Emotionally arousing materials like pictures can also be used to improve memory performance, attention and recall (Bradley, Greenwald, Petry and Lang 1992; Childers and Houston 1984). Affectively loaded pictures are also observed longer (Bradley and Lang 1999) and advertisements have a varying emotional response when they contain both positive and negative pictures (Chowdhury, Olsen and Pracejus 2008). Generally, it can be said that emotions are related to the four goals of media messages: attracting attention, being remembered, entertaining and persuading (Bolls, Lang and Potter 2001; Ravaja 2004).

Although few studies have been undertaken on the use of color within tourism pictures (Park 2018), colors are known to evoke emotions (Ou, Luo, Woodcock and Wright 2004). In a review of color-emotion literature by Lee, Denizci Guillet and Law (2018) yellow is joyous and cheerful; green is relaxing; blue is cool and calm, although it is also linked to depression; orange can be upsetting or can encourage socialization; and red can be positive (excitement) or negative (anger). Further, the energetic and bold oranges, yellows and reds are warm colors, while the comforting colors of green and blue are cool colors (Lee, Denizci Guillet and Law 2018). It is the emotional connection with a product or brand that can stimulate buying interest, intentions and buying decisions (Jiménez Barreto, Rubio and Campo Martínez 2019). Emotion, mood and atmosphere can be created through the color-mix, color-brightness and clarity of color (Hench and Van Pelt 2009; Ou, Luo, Woodcock and Wright 2004). In a study by Ou, Luo, Woodcock and Wright (2004), color-emotion factors were developed using color weight, color heat and color activity. Color emotions based on appearance were heavy–light, active–passive, warm–cool and hard–soft (Ou, Luo, Woodcock and Wright 2004). These could be used as an objective measure, as they were culture independent in the study regions: Hong Kong, Thailand, China, Japan and Britain (Ou, Luo, Woodcock and Wright 2004). Research has shown that warm colors, or colors with longer wavelengths (e.g., red, orange and yellow), increase the level of arousal. By contrast, cool colors, or colors with shorter wavelengths (e.g., green, blue and violet), are emotionally and physically relaxing, calming and pleasant (Bellizzi, Crowley and Hasty 1983; Bellizzi and Hite 1992). Furthermore, research on hue, saturation and luminance has shown that these low-level features influence emotional responses under certain circumstances (Ring and Schneider 2012). In addition, it is empirically confirmed that ads with higher saturated colors and a higher level of luminance increase the likeability of the respective ad (Gorn, Chattopadhyay, Yi and Dahl 1997).

Unique destination image through color

A study by Lenclos and Lenclos (2004), considered that destinations uniquely feature colors that are specific to them, being extracted from the land and reflecting a sense of place or culture. Marrakech is known as the "red city", because of its red, blush or ochre hue buildings and walls (VisitMorocco n.d.). However, uniqueness is not always a factor of extraction. Following a 2018 decree, downtown buildings in Mombasa were required to be painted blue and white to enhance the touristic appeal, resulting in the city becoming the most photographed in Africa (Jasiński 2019). Natural features provide visitor attractions at destinations, such as the seven-colored earth of Chamarel, a volcanic phenomenon in Mauritius (Mauritius, n.d.) and the Rainbow Mountain in Peru, which started gaining popularity after *National Geographic* magazine listed it as a place to visit (Garvey n.d.). In the case of Kentucky, US, the "bluegrass state" is so-called because of the spring buds, which give a bluish tint to the landscape. New England, US, is noted for the display of autumn colors when leaves turn to an array of reds, yellows and purples (Discover New England, n. d.). Australia's Blue Mountains appear to have a blue haze when viewed from Sydney, because of the eucalyptus trees (Ennion n.d.), while the Emerald Isle is a poetic name for Ireland, reflecting the rich green landscape and the cultural traditions of St. Patrick's Day, with the wearing of green clothing and the shamrock.

Labrecque and Milne (2012) substantiated that colors have a strong influence on consumers' brand perceptions, familiarity and likeability, as well as on purchase intention. They explicitly highlight that it is not only the color but also saturation and brightness (luminance) that matters when it comes to emotional judgments on brands. Therefore, marketers have to choose and use colors strategically. But is there any evidence that DMO marketers optimize pictures used on destination websites to achieve a positive emotional reaction from potential visitors? The following research analyzes the low-level features of pictures taken from 22 UK destinations (counties, cities and national parks). This will provide an overview of colors used, but will also provide an emotional landscape of UK destinations.

Study design and method

Data collection

The process of discovering potentially useful, and previously unknown information, from web sources is commonly referred to as web mining (Johnson and Gupta 2012). Depending on how structured the website data is, different approaches can be used. Web pages are built using the standardized Hypertext Markup Language (HTML), which specifies elements for pictures (https://html.spec.whatwg.org/). Therefore, it is logical to assume that a simple page content mining approach would be sufficient to extract this structured data from target web pages (Johnson and Gupta 2012). Extracting data from web pages can be done in two ways. One approach would be to create a purpose-built software tool using web scraping libraries for programming languages like Python (Jarmul and Lawson 2017) or Java (Alag 2009). While such software would be very flexible, implementation effort would likely be too high for a simple task like picture extraction. For such tasks generic tools like curl (https://curl.haxx.se/) and wget (www.gnu.org/software/wget/) can be used. Ultimately wget was chosen as it offers useful options like defining how many levels of the web site wget crawls, accepting regular expressions and limiting downloads to specific file types. Testing individual options revealed that wget does not work equally well on all chosen websites and results varied strongly. Ultimately, a more forceful approach was taken downloading the first three levels of the website using wget and then deleting all files not containing "jpeg", "jpg" or "png" in their filename. One needs to be careful with such an approach as it takes a lot of time to gather the data and may create a lot of traffic on the respective web servers. Therefore, the use of timeouts is advised. Another area to consider is copyright law. In the UK it is stated explicitly that limited extracts of work can be copied when the use is non-commercial research (www.gov.uk/guidance/exceptions-to-copy right#non-commercial-research-and-private-study).

Extracting and analyzing pictures

All pictures were sourced from UK destination websites. The sample consisted of DMO websites of all English counties, major UK cities and two major national parks. Unfortunately, wget was not able to extract pictures from all these sources, which resulted in a reduced set of 22 DMO websites (see Table 36.1 for a complete list of analyzed sites). The pictures mined from these sites were analyzed using the PIXEM software tool developed by Schneider (2017). PIXEM calculates the HSL color histogram for each picture providing the H-categories "red", "blue", "green", "cyan", "yellow" and "magenta" and the saturation and lightness in percentage terms, which are used to define arousal and valence values. Based on these, PIXEM can allocate general emotional categories ("neutral", "positive", "negative") but also more salient emotional categories ("gloomy", "distressing", "unpleasant", "sleepy", "arousing", "neutral", "relaxing", "exciting", "pleasant") to individual pictures.

Findings

In order to shape the UK's e-tourism experience, we collected data from 22 UK websites (see Table 36.1). Collectively 42,945 pictures were returned. Most of them were from the website of Gloucestershire (10,513), whereas only 23 pictures were from the website of Northamptonshire. Several websites (e.g. London) did not return any pictures.

The used software tool PIXEM returns an HSL-Histogram of each picture. Figure 36.1 shows the mean value of the hues, the median saturation and the median lightness over all website pictures. The average lightness of the pictures is 44%, whereas saturation of 26% is very low. The dominant hue is red, with 32%, followed by yellow, also with 32%.

Based on PIXEM's emotional categorization 11,263 (27%) are negative, 28,989 (65%) are neutral, and only 3,930 (8%) are positive.

Colors and emotions of the UK

As Table 36.1 shows, the colors red and yellow dominate on almost all websites. The values range from 22.9% (Cornwall) to 40.5% (Country Durham) in red and from 15.21% (Derbyshire) to 40.27% (Peak District) in yellow. The solitary exceptions are the pictures of Derbyshire. Here Hblue is highest with 34.08%. Green and magenta are hardly used on all of the websites. Saturation is lowest in Northampton with only 16% and highest with 32%

Table 36.1 Collected data – Colors and emotions of the UK

			Colors of the UK								Emotions of the UK		
	Website	*#*	*Hyellow*	*Hgreen*	*Hcyan*	*Hblue*	*Hmagenta*	*Hred*	*S*	*L*	*positive*	*neutral*	*negative*
Kent	www.visitkent.co.uk/	1,202	31.36	5.07	12.93	14.27	2.98	33.39	30	46	8.15	72.30	19.55
Aberdeenshire	www.visitabdn.com/	263	30.07	3.44	12.33	16.39	2.95	34.81	22	51	7.60	50.19	42.21
Birmingham	visitbirmingham.com/	1,589	27.28	4.12	10.18	16.01	4.64	37.77	27	40	8.43	65.89	25.68
Cornwall	www.visitcornwall.com/	6,260	37.24	5.66	18.73	13.22	2.24	**22.92**	25	49	6.57	69.14	24.30
County Durham	www.visitcountydurham.org/	269	26.06	2.47	13.70	14.43	2.84	**40.50**	26	**53**	4.83	58.36	36.80
Derbyshire	www.visitderbyshire.co.uk/	195	20.80	5.58	8.13	**34.08**	6.49	24.93	24	46	10.26	65.64	24.10
Edinburgh	edinburgh.org/	2,029	26.34	3.83	14.45	14.41	3.70	37.26	27	40	8.92	53.72	37.36
Glasgow	visit-glasgow.info/	61	34.41	4.98	5.48	11.07	5.35	38.72	22	**36**	**3.28**	72.13	24.59
Gloucestershire	www.exploregloucestershire.co.uk/	10,513	29.56	6.95	10.82	11.84	3.04	37.78	26	43	11.77	63.55	24.68
Hertfordshire	www.visitherts.co.uk/	710	28.94	7.90	10.71	13.34	4.60	34.50	30	43	13.24	58.45	28.31
Lake District	www.lakedistrict.gov.uk/	1,106	35.73	6.86	14.44	13.77	4.22	24.98	22	46	8.86	64.29	26.85
Lincolnshire	www.visitlincolnshire.com/	280	38.39	4.58	15.55	14.23	2.24	25.01	30	48	7.50	75.00	17.50
Liverpool	www.visitliverpool.com/	1,460	23.71	4.01	15.13	17.22	4.15	35.77	29	41	6.51	69.73	23.77
Manchester	www.marketingmanchester.com/	127	15.21	2.44	20.92	27.80	3.04	30.59	23	46	6.30	58.27	35.43
Northamptonshire	ww.visitnorthamptonshire.co.uk/	23	29.05	7.85	16.76	9.11	6.36	30.86	**16**	48	8.70	**47.83**	**43.48**
Oxfordshire	www.experienceoxfordshire.org/	5,848	33.03	4.59	11.26	15.19	3.70	32.24	29	42	7.51	67.97	24.52
Peak District	www.visitpeakdistrict.com/	2,857	**40.27**	5.94	11.73	11.77	2.79	27.49	26	40	7.77	64.65	27.58
Shropshire	www.shropshiretourism.co.uk/	1,659	37.07	7.44	12.19	13.34	2.85	27.10	26	42	8.98	67.57	23.45
Sussex	www.sussextopattractions.co.uk	581	37.66	8.38	10.02	12.35	4.52	27.07	**32**	40	**15.49**	**72.46**	12.05
Worcestershire	www.visitworcestershire.org/	1,084	31.61	5.30	12.00	14.89	2.61	33.58	27	42	6.00	71.31	22.69
Yorkshire	www.yorkshire.com/	4,829	34.66	4.99	13.16	14.94	2.68	29.57	27	44	8.53	65.25	26.22
Cheshire	www.visitcheshire.com/	1,237	33.82	5.94	11.42	13.94	3.96	30.92	26	42	9.62	63.54	26.84

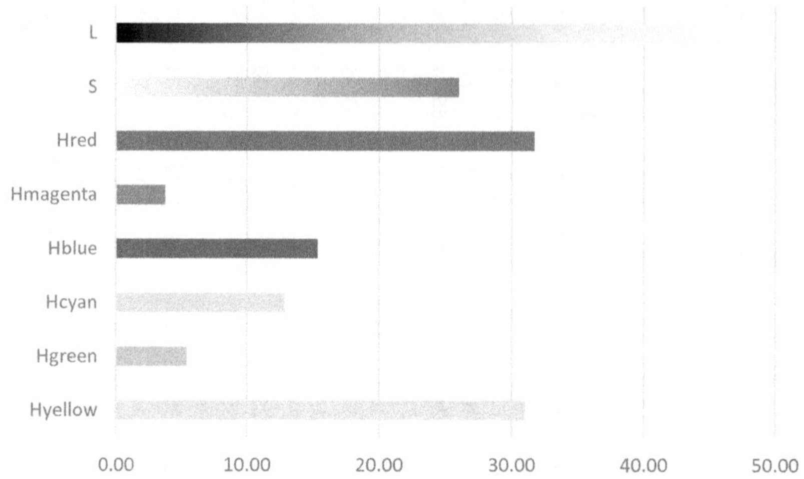

Figure 36.1 Colors of the UK (in %)

in Sussex. Glasgow impresses with its dark pictures (36%), while the lightness of the pictures on the County Durham website is very high (53%). The Manchester website has pictures that are more balanced in their color values. There is no prominent dominant color. Yellow, cyan, blue and red are almost equally occupied.

Most pictures are assigned by PIXEM to the neutral category (64.42%). The negative category (27.18%) is the second most frequently used. Only a few pictures are assigned to the positive category (8.4%). Pictures on the website of Sussex are an exception: there are more positive than negative pictures on their site (see Table 36.1). Aberdeenshire (42.21%) and Northamptonshire (43.48%) show a lot of negative pictures on their websites, while Glasgow's website has the fewest positive categorized pictures (3.28%).

Conclusion

As can be seen in the small study above, most of the pictures have been assigned to the neutral category. That is perfectly acceptable as not all pictures have to be positive, but the number of negative pictures should be reduced. It is noticeable that some outstanding websites only have a few pictures on their website. These pictures are either very well chosen, as on the website of Derbyshire, or have some color design deficits, so that they are perceived as negative. Most websites use pictures with warm colors. Red and yellow are clearly the dominant colors in the pictures studied. As already known from the literature, warm colors increase the level of arousal (Bellizzi, Crowley and Hasty 1983; Bellizzi and Hite 1992). Similarly, the color red is not only positive but can also trigger negative emotions (Lee, Denizci Guillet and Law 2018). This is exactly what is found in the cited study. The greater the proportion of red and the more neglected the blue and green content, the more negatively the pictures are perceived. If, on the other hand, the proportion of blue increases, they are clearly perceived positively (e.g. Derbyshire).

As noted by Gorn et al. (1997), it has also been seen in the present study that high saturation has a positive effect on the emotional perception of a picture (e.g. Sussex) and low saturation on the other hand has a negative effect (e.g. Northamptonshire). Low luminance also reduces the amount of positively perceived pictures (e.g., Glasgow). It is not possible to recognize a unique color design for each DMO. So, you cannot say county "XY" is the green county or city "AB" is the blue city. For this purpose, further studies on user-generated content, for example, might be more useful. Finally, the gut feeling of the marketer cannot be considered bad, but it is in need of improvement. The overall visual e-tourism experience can be improved with only small post-processing changes, which are so minimal that there is no danger of reality distortion. For example, increasing saturation by just 2% can bring a more positive result. These small changes concerning the pictures' colors would make it possible to captivate the potential tourist with a unique visual experience in advance and consequently would help build a competitive advantage for a destination.

References

Abreu Novais, M., Ruhanen, L. and Arcodia, C. (2018). 'Destination competitiveness: A phenomenographic study', *Tourism Management*, 64: 324–334. DOI: 10.1016/j.tourman.2017.08.014.

Alag, S. (2009). *Collective intelligence in action*. New York: Manning.

Baker, S. (1961). *Visual persuasion*. New York: McGraw-Hill.

Bellizzi, J. A., Crowley, A. E. and Hasty, R. W. (1983). 'The effects of color in store design', *Journal of Retailing*, 59(1): 21–45.

Bellizzi, J. A. and Hite, R. E. (1992). 'Environmental color, consumer feelings, and purchase likelihood', *Psychology & Marketing*, 9(5): 347–363. DOI: 10.1002/mar.4220090502

Bolls, P. D., Lang, A. and Potter, R. F. (2001). 'The effects of message valence and listener arousal on attention, memory, and facial muscular responses to radio advertisements', *Communication Research*, 28 (5): 627–651. DOI: 10.1177/009365001028005003.

Bradley, M. M., Greenwald, M. K., Petry, M. C. and Lang, P. J. (1992). 'Remembering pictures: Pleasure and arousal in memory', *Journal of Experimental Psychology*, 18(2): 379–390. DOI: 10.1037/0278-7393.18.2.379.

Bradley, M. M. and Lang, P. J. (1999). 'Fearfulness and affective evaluations of pictures', *Motivation and Emotion*, 23(1): 1–13. DOI: 10.1023/A:1021375216854.

Buhalis, D. and Law, R. (2008). 'Progress in information technology and tourism management: 20 years on and 10 years after the Internet—The state of eTourism research', *Tourism Management*, 29(4): 609–623. DOI: 10.1016/j.tourman.2008.01.005

Childers, T. L. and Houston, M. J. (1984). 'Conditions for a picture-superiority effect on consumer memory', *Journal of Consumer Research*, 11(2): 643–654. DOI: 10.1086/209001

Chowdhury, R. M., Olsen, G. D. and Pracejus, J. W. (2008). 'Affective responses to images in print advertising: Affect integration in a simultaneous presentation context', *Journal of Advertising*, 37(3): 7–18. DOI: 10.2753/JOA0091-3367370301

Cutler, B. D. and Javalgi, R. G. (1993). 'Analysis of print ad features: Services versus products', *Journal of Advertising Research*, 33(2): 62–69.

Discover New England. (n.d.). 'Fall foliage maps', from: www.discovernewengland.org/fall-foliage-maps [Accessed May 30, 2019].

Doersch, C., Singh, S., Gupta, A., Sivic, J. and Efros, A. (2015). 'What makes Paris look like Paris?', *Communications of the ACM*, 58(12): 103–110. DOI: 10.1145/2830541

Dwyer, L. and Kim, C. (2003). 'Destination competitiveness: Determinants and indicators', *Current Issues in Tourism*, 6(5): 369–414. DOI: 10.1080/13683500308667962.

Ennion, J. (n.d.). 'Guide to the blue mountains', Australia Inc. from: www.australia.com/en-gb/places/sydney-and-surrounds/guide-to-the-blue-mountains.html [Accessed May 30, 2019].

Francesconi, S. (2011). 'Images and writing in tourist brochures', *Journal of Tourism and Cultural Change*, 9 (4): 341–356, DOI: 10.1080/14766825.2011.634914

Garvey, C. (n.d.). Top 5 places to visit in Peru. from: www.nationalgeographic.com/travel/destinations/south-america/peru/places-visit-peru-trip/ [Accessed May 30, 2019].

Gilmore, J. H. and Pine, B. J. (2002). *The experience is the marketing*. Louisville: BrownHerron Publishing.

Gonzales, R. C. and Woods, R. E. (2008). *Digital image processing*. Upper Saddle River, NJ: Pearson Education.

Gorn, G. J., Chattopadhyay, A., Yi, T. and Dahl, D. W. (1997). 'Effects of color as an executional cue in advertising: They're in the shade', *Management Science*, 43(10): 1,387–1,400. DOI: 10.1287/mnsc.43.10.1,387

Han, J.-H. and Mills, J. E. (2006). 'Zero acquaintance benchmarking at travel destination websites: What is the first impression that national tourism organizations try to make?', *International Journal of Tourism Research*, 8(6): 405–430. DOI: 10.1002/jtr.581

Hanlon, A. (2019). *Digital marketing. Strategic planning & integration*. London: SAGE.

Hays, S., Page, S. J. and Buhalis, D. (2013). 'Social media as a destination marketing tool: Its use by national tourism organisations', *Current Issues in Tourism*, 16(3): 211–239. DOI: 10.1080/13683500.2012.662215

Hench, J. and Van Pelt, P. (2009). *Designing disney (A walt disney imagineering book)*. New York: Disney Editions.

Hunjet, A., Glogar, M. I., Osterman, D. P. and Kozina, G. (2016). 'The importance of color as a marketing tool in tourism'. (Eds.) I. Vrankic and D. Tomic, *Economic and social development: Barcelona, 14–16 April 2016: Book of proceedings*: 382–388 (Varazdin Development and Entrepreneurship Agency, Varazdin: Croatia University North, Koprivnica, Croatia).

Jarmul, K. and Lawson, R. (2017). *Python web scraping-: Hands-on data scraping and crawling using PyQT, Selnium, HTML and Python*. Birmingham: Packt Publishing.

Jasiński, A. (2019). 'Colour as a tool in shaping the city image – Based on the case of Mombasa', *Technical Transactions, Vol.* 4/2019: 23–36. from: www.ejournals.eu/Czasopismo-Techniczne/2019/Volume-4/art/14087 [Accessed May 30, 2019].

Jiménez Barreto, J., Rubio, N. and Campo Martínez, S. (2019). 'The online destination brand experience: Development of a sensorial–cognitive–conative model', *International Journal of Tourism Research*, 21(2): 245–258. DOI: 10.1002/jtr.2258.

Johnson, F. and Gupta, S. K., (2012). 'Web content mining techniques: A survey', *International Journal of Computer Applications*, 47(11): 44–50. DOI: 10.5120/7236-0266.

Kress, G. R., and van Leeuwen, T. (1996). *Reading images: The grammar of visual design*. London & New York: Routledge.

Kress, G. and van Leeuwen, T. (2006). *Reading images: The grammar of visual design*. 2nd Edn. London: Routledge.

Kroeber-Riel, W., and Esch, F. R. (2011). *Strategie und Technik der Werbung-Verhaltens-und neurowissenschaftliche Erkenntnisse*. Hrsg: Hermann Diller und Richard Köhler (7. Aufl., Verlag W. Kohlhammer, Stuttgart, S. 21f, 35).

Küthe, E., and Venn, A. (1996). *Marketing mit Farben*. Köln: Dumont Verlag.

Labrecque, L. I. and Milne, G. R. (2012). 'Exciting red and competent blue: The importance of color in marketing', *Journal of the Academy of Marketing Science*, 40(5): 711–727. DOI: 10.1007/s11747-010-0245-y

Lee, A. H., Denizci Guillet, B. and Law, R. (2018). 'Tourists' emotional wellness and hotel room colour', *Current Issues in Tourism*, 21(8): 856–862. DOI: 10.1080/13683500.2016.1217830

Lee, C. J. and Andrade, E. (2010). 'The effect of emotion on color preferences', *Advances in Consumer Research*, 37: 846–847.

Lenclos, J.-P. and Lenclos, D. (2004). *Colors of the world: A geography of color*. New York: W.W. Norton.

Lichtlé, M.-C. (2007). 'The effect of an advertisement's colour on emotions evoked by attitude towards the ad', *International Journal of Advertising*, 26(1): 37–62, DOI: 10.1080/02650487.2007.11072995.

Mariani, M. M., Di Felice, M. and Mura, M. (2016). 'Facebook as a destination marketing tool: Evidence from Italian regional Destination Management Organizations', *Tourism Management*, 54: 321–343. DOI: 10.1016/j.tourman.2015.12.008

Mauritius. (n.d.). *The 7-coloured earth of Chamarel*. from: www.tourism-mauritius.mu/en-int/7-the-7-coloured-earth-chamarel [Accessed May 30, 2019].

Ou, L. C., and Luo, M. R. (2006). 'A colour harmony model for two-colour combinations', *Color Research & Application*, 31(3), 191–204.

Ou, L. C., Luo, M. R., Woodcock, A. and Wright, A. (2004). 'A study of colour emotion and colour preference. Part I: Colour emotions for single colours', *Color Research & Application*, 29(3): 232–240. DOI: 10.1002/col.20010

Page, S. J., Steele, W. and Connell, J. (2006). 'Analysing the promotion of adventure tourism: A case study of Scotland', *Journal of Sport & Tourism*, 11(1): 51–76. DOI: 10.1080/14775080600985358.

Park, E. (2018). 'Histogram-based colour image analysis on tourism photography'. (Ed.) R. Nunkoo, *Handbook of research methods for tourism and hospitality management* (pp. 278–291). Cheltenham: Edward Elgar Publishing.

Pieters, R. and Wedel, M. (2004). 'Attention capture and transfer in advertising: Brand, pictorial, and text-size effects', *Journal of Marketing*, 68(2): 36–50. DOI: 10.1509/jmkg.68.2.36.27794

Ravaja, N. (2004). 'Contributions of psychophysiology to media research: Review and recommendations', *Media Psychology*, 6(2): 193–235. DOI: 10.1207/s1532785xmep0602_4

Ring, A. and Schneider, A. (2012). Pictures as emotional stimuli. Going beyond content to trigger emotional response. *Marketing to Citizens: Going beyond Customers and Consumers*: 1–5.

Schneider, A. (2017). 'Color influence factors for the emotional impact of a picture and their relevance for the retrieval of tourism pictures', *Wissenschaftliche Schriftenreihe Dissertationen der Medieninformatik*, (6). from: http://nbn-resolving.de/urn:nbn:de:bsz:ch1-qucosa-209553

Solli, M. and Lenz, R. (2011). 'Color emotions for multi-colored images', *Color Research & Application*, 36 (3): 210–221. DOI: 10.1002/col.20604

Tuch, A. N., Bargas-Avila, J. A., Opwis, K. and Wilhelm, F. H. (2009). 'Visual complexity of websites: Effects on users' experience, physiology, performance, and memory', *International Journal of Human-computer Studies*, 67(9): 703–715. DOI: 10.1016/j.ijhcs.2009.04.002.

Urry, J. (1992). 'The tourist gaze "revisited"', *American Behavioral Scientist*, 36(2): 172–186. DOI: 10.1177/0002764292036002005

Urry, J. (2002). *The tourist gaze*. London: Sage.

Urry, J., and Larsen, J. (2011). *The tourist gaze 3.0*. London: Sage.

Uşaklı, A., Koç, B. and Sönmez, S. (2017). 'How 'social' are destinations? Examining European DMO social media usage', *Journal of Destination Marketing & Management*, 6(2): 136–149. DOI: 10.1016/j.jdmm.2017.02.001

Valdez, P. and Mehrabian, A. (1994). 'Effects of colour on emotions', *Journal of Experimental Psychology: General*, 123(4): 394–409. DOI: 10.1037/0096-3445.123.4.394

Visit Morrocco. (n.d.) 'Marrakech eternal, Unmissable', from: www.visitmorocco.com/en/travel/marrakech [Accessed May 30, 2019].

Zimbardo, P. G. and Gerrig, R. J. (2004). *Psychologie*. München: Pearson Studium.

37

Managing tourist experience through social media

Evidence from South Korea

Yeonjung (Alice) Kang and Tiffany S. Legendre

Introduction

Social media have become increasingly important in the tourism industry (Fotis, Buhalis and Rossides 2012; Munar and Jacobsen 2014) and significant research endeavors are being made to examine the role of social media in tourists' decision-making process (Litvin, Goldsmith and Pan 2008). Sigala (2016) postulates that social media act as a value co-creation venue between tourists and destinations that influences tourist experience. Therefore, examining the role of social media provides meaningful assistance to many industry practitioners, as the platform has a great effect on creating positive tourist experience (Fuchs, Scholochov and Höpken 2009).

Despite its functional importance and the body of existing literature, the lack of understanding about the role of social media in value co-creation and tourist experience prohibits scholars and industry practitioners from implementing effective strategies (Berger and Schwartz 2011; Bronner and De Hoog 2011; Yoo and Gretzel 2011). Thus, there is a need for examining the influences of social media on tourist experience by understanding the value co-creation process. This chapter begins with explaining the meaning and the functions of social media in tourist experience. Specifically, it introduces a value co-creation model to explain how destinations and tourists' interactions contribute to a positive experience. To illustrate this process, this chapter adopts a case from Korea Tourism Organization (KTO) and provides practical suggestions for industry practitioners.

The role of social media in tourist experiences

Social media have reshaped tourist experience (Neuhofer, Buhalis and Ladkin 2012; Sigala, Christou and Gretzel 2012). Tourists can influence each other by sharing information on social media, which makes them become the "media" themselves (Li and Wang 2011; Thevenot 2007). This type of social media provides tourists with various platforms on which to share their experiences and also lets destinations disseminate marketing information, which in turn co-creates value and tourist experience at all stages of travel (Sigala 2016).

With the proliferation of social media, tourists increasingly utilize the platforms to search for and share information (Xiang and Gretzel 2010). Because of this, social media fulfills multiple functions. First, it helps travelers with the decision-making process. Previous studies show that more than one third of tourists make decisions based on the reviews on social media such as TripAdvisor and Facebook (Koumelis 2011). Moreover, Wang and Fesenmaier (2004) assert that social media can be an effective place to attract travelers because these venues can increase travelers' engagement and interactions. Consequently, many tourism industry practitioners are exploring ways to exploit social media (Fuchs, Scholochov and Höpken 2009).

The second function of social media is that unlike the pre-internet era tourists no longer act as passive receivers, but instead they become "prosumers", which combines the characteristics of an information producer and a consumer (Silver 2009). Prosumers, together with social media, are key to value co-creation and tourist experience, which positively influences the industry (Sigala 2016). With the ease of communication on social media platforms, travelers are more empowered than ever before and are actively participating in the value co-creation process (Sigala 2016). As a result, tourist experience has expanded its scope to encompass the pre, during and post-trip stages – the broadened scope of travel explaining the greater importance of value co-creation (Sigala 2016).

Justification for the case selection

In order to explain how social media are used to assist destination–tourist co-creation of value, this chapter introduces a case from the Korea Tourism Organization (KTO). There are several reasons why the KTO is a suitable case for the analysis. First, the organization actively implements social media to effectively communicate with tourists, which comprises all of the elements in the value co-creation framework. Second, KTO has been actively producing interaction opportunities for their tourists to be involved in the co-creation process, which has yielded positive outcomes. Examining the KTO's case through a value co-creation framework provides practical insights into the development of effective travel strategies.

Case studies using a value co-creation framework

As tourists become more active and participative, the research stream has moved from just considering one party (e.g., destinations or tourists, not both) to value co-creation. Ranjan and Read (2016) propose a conceptual framework of consumer–provider value co-creation and identify dimensionality and measurement tools. This endeavor clarifies the theoretical foundation of value co-creation, addressing two main dimensions and three elements under each dimension. The two dimensions are co-production and value-in-use. Knowledge, equity and experience are three elements under the co-production dimension, while experience, personalization and relationship are included in value-in-use. As social media allows tourists to become more important in value co-creation, investigating the framework and its conceptual elements in order to understand the impact of social media on value co-creation is necessary. Therefore, the chapter adopts a tourist–provider value co-creation framework to analyze the destinations–tourists value co-creation process by using an exemplary case. Figure 37.1 shows the value co-creation framework adopted from Ranjan and Read (2016).

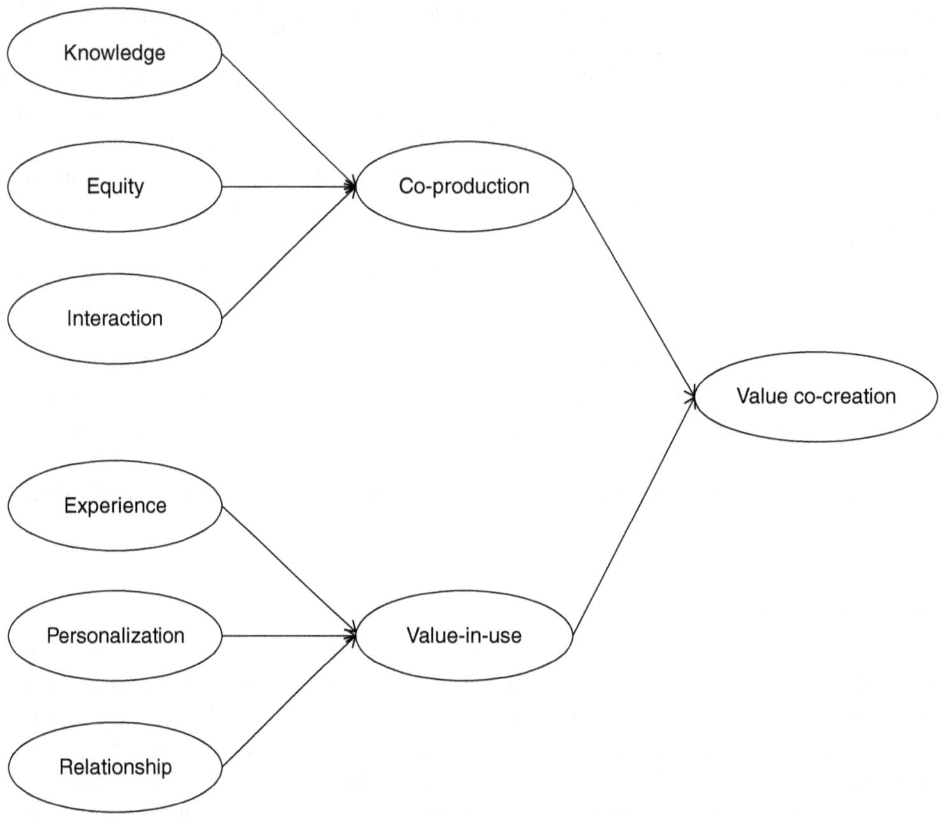

Figure 37.1 Value Co-creation Framework Adapted from Ranjan and Read (2016)

Knowledge

Knowledge refers to sharing tourists' knowledge and ideas in the expression of their needs (Ranjan and Read 2016; Zhang and Chen 2008). It includes sharing ideas and creativity between tourists and destinations, which can co-create value (Maglio and Spohrer 2008). Many studies maintain that sharing creates a better outcome compared to working independently because this activity covers various tourists' perspectives, resulting in better reconciliation and expression (Enz and Lambert 2012; Fisher and Smith 2011; Grover and Kohli 2012; Powell and Swart 2010; Ramirez 1999). Example measurement items for knowledge are "The party was open to my ideas and suggestions about its existing products or towards developing a new product" and "The party provided sufficient illustrations and information to me" (Ranjan and Read 2016: 301).

Case study 37.1

KTO has utilized social media to provide tourists with knowledge that shares ideas based on their needs. For example, they conducted a survey on the organization's Facebook page to identify the needs of Japanese tourists in order to increase the destination's attractiveness. The survey targeted

Japanese tourists and asked them which was their most preferred item sold at Korean traditional markets, among over 70 items. The survey collected 495 responses and the result showed that Japanese tourists preferred to purchase traditional socks (10.3%), socks (9.3%), flavor enhancer (6.1%), etc. Moreover, the survey asked why participants chose specific items, in order to understand the tourists' needs and concerns. After analyzing the results, KTO produced shopping-related content and distributed it through different social media channels such as YouTube, Facebook and Twitter. The content aimed to attract Japanese tourists by showing them diverse purchase options. This is based on the premise that Japanese tourists prefer purchasing various goods from traditional markets. As such, KTO stated that conducting the survey helped them to devise a range of products that could contribute to a development of the traditional market industry. The case shows that KTO developed marketing content by adopting tourists' ideas and suggestions, which contributes to knowledge-related value co-creation.

KTO also conducted different surveys via their social media accounts such as Facebook, Weibo, etc. The survey aimed to identify "the most unique Korean food I would like to try". A total of 944 responses were collected and the result showed that tourists were willing to try live octopus (26%), soy sauce crab (14.6%), traditional sausage (14.2%), etc. The purpose of conducting the survey was to advertise some unique Korean food that is not as popular as other traditional food, such as Bulgogi. Thus, based on the survey results, KTO produced and posted a video via YouTube that provided detailed information about the unique Korean food, characteristics of the food and places to eat the food. The organization expected tourists to share this information via the social network, which can increase knowledge-related value co-creation.

Equity

Equity means a company's willingness to share organizational control and provide tourist empowerment in order to fulfill tourists' desire to participate in the co-creation process (Gebauer, Johnson and Enquist 2010). Equity is derived from a company's tourist centrism with the aim of achieving congruence in interests and goals, which leads to value actualization and integration (Ranjan and Read 2016). "The party had an easy access to information about my preferences" and "The party considered my role to be as important as its own in the process" are some examples of equity measurement in the value co-creation framework (Ranjan and Read 2016: 301).

Case study 37.2

To provide empowerment to tourists, KTO has conducted many campaigns and analyses. For example, they implemented a social media-based incentive travel campaign called "15 seconds of success" which aimed to create a positive perception of incentive travel in Korea. First, they produced an advertising video that addressed the needs of target companies. Also, the video was designed to help tourists understand what incentive travel is. After this, KTO organized a video contest in order to provide specific information to target tourists as well as to share empowerment with them. This activity created value actualization which resulted in tourists' active participation in the form of 211 videos in three weeks. Moreover, after conducting the campaign, they conducted an online survey to examine whether the preference for incentive travel in Korea had increased. The result indicated that the preference for incentive travel in

Korea had increased by 6.6% for normal tourists and 15.1% for campaign participants, compared to the previous year of the campaign.

Furthermore, KTO utilized big data from social media in order to identify tourists' preferences of destinations, as a means of developing their marketing strategy. For example, they initiated a marketing project for Pyeongchang where the Winter Olympics were held in 2018. In their marketing analysis, they collected the data from Google Trends, Instagram and TripAdvisor to identify tourists' needs. For instance, they analyzed Instagram data by using "hashtag pyeongchang (#pyeongchang)" and gathered 119,205 items of data. With the data, they categorized the keywords that were frequently used by tourists and developed marketing strategies based on those keywords, showing tourists' preferences for a destination. Therefore, their marketing strategies considering the role of tourists led the organization to share control and to provide empowerment to tourists.

Interaction

Interaction is defined as "the primary interface between parties undertaking co-production" (Ranjan and Read 2016: 293). Studies have addressed the importance of tourists' interactions, engagement, and participation (Edvardsson, Ng, Zhi Min, Firth and Yi 2011; Grönroos and Ravald 2011; Ordanini, Miceli, Pizzetti and Parasuraman 2011). During this exchange process, tourists actively suggest solutions about the product and/or service, which yields positive outcomes such as increased involvement and response (Archpru Akaka and Chandler 2011; Chan, Yim and Lam 2010; Ramani and Kumar 2008; Verbeke, Belschak, Bakker and Dietz 2008). As the interaction produces social exchange, it becomes a source of value (Nambisan and Baron 2007; Vallaster and von Wallpach 2013). Sample measurement questions of interaction are: "During the process I could conveniently express my specific requirements" and "The party conveyed to its consumers the relevant information related to the process" (Ranjan and Read 2016: 301).

Case study 37.3

KTO has provided their tourists with interaction opportunities via different social media channels. For instance, the organization's Facebook page actively communicates with tourists. Moreover, the organization has responded to users' comments from various channels such as YouTube and Facebook. This interaction accentuates tourists' involvement and boosts their responses, resulting in an increase in the number of likes on their Facebook page from 97,238 to 665,137 between 2013 and 2019. The organization normally posts tour information and tourists share these posts with their friends. Also, tourists ask questions related to the posts and KTO responds to the questions by providing valuable information that helps to resolve their inquiries or concerns.

In addition, KTO has held numerous contests in order to enhance tourist engagement and participation. For example, the organization held a video contest about "tips for traveling in Korea" so that the tourists became involved in active interaction to co-create value. The organization let the contest attendees select the video content by themselves, which helped tourists express their specific preferences or requirements. The winning videos were posted on social media such as YouTube, to attract more tourists. Therefore, KTO implemented value co-creation through the interaction between the organization and tourists by considering tourists' own preferences.

Experience

Experience is related to a memorable and emotional interaction as a result of products and services, and it is also associated with tourists' physical and cognitive dimensions (Ballantyne and Varey 2008; Bolton 2004; Edvardsson, Gustafsson and Roos 2005; Lusch and Vargo 2006). Therefore, experience leads tourists to co-create value-in-use through intensive participation and actions (Edvardsson, Enquist and Johnston 2010; Lemke, Clark and Wilson 2011; O'Neill and Palmer 2003). It is a form of self-transformation where individuals experience emotional, cognitive and hedonic growth, which ultimately creates value (Maclaran and Brown 2005). Example measurement questions for an experience are "It was a memorable experience for me (i.e., the memory of the process lasted for quite a while)" and "Depending upon the nature of my own participation, my experiences in the process might be different from other consumers" (Ranjan and Read 2016: 301).

Case study 37.4

KTO considers experience as one of the most valued elements for value co-creation. For example, they have provided many experience opportunities to tourists such as themed tours and seasonal festivals. Their social media channels such as Facebook and Twitter have introduced different festivals depending on the season. Since Korea has four seasons, there are numerous seasonal festivals. In spring, they post content about the Jinhae Gunhangje festival, which is one of the biggest cherry blossom festivals in Korea. During the summer, they post information about the Boryeong mud festival, an internationally recognized summer festival. In the fall, the channels provide information about the Icheon rice festival, which celebrates the year of good harvest in October. During the winter, they post content about the Pyeongchang trout festival, which has different programs related to snow or fishing activities.

Also, with regard to themed travel, KTO provides information about medical tourism, demilitarized zone (DMZ) tours, temple-stays, etc. One of the most popular themed tours, the Changdeokgung moonlight tour, is also posted on social media in order to create a memorable experience for tourists. By providing information that can create a memorable and emotional interaction, KTO achieves intensive participation among tourists that can enhance value co-creation.

Personalization

Personalization means "a unique consumption value through the enjoyment of doing, or an idiosyncratic use process" (Ranjan and Read 2016: 293). Because of the uniqueness of the process, it can affect cultural reinforcement. This means that tourist value can be extended or refigured by a personalized proposition that includes the boundaries of value-in-use (Cova, Dalli and Zwick 2011). Questions for measuring personalization include "The benefit, value, or fun from the process (or, the product) depended on the user and the usage condition" and "Different consumers, depending on their taste, choice, or knowledge, involve themselves differently in the process (or, with the product)" (Ranjan and Read 2016: 301).

Case study 37.5

KTO offers diverse programs that personalize tourists' consumption value. One of the examples is the offer of a specially-themed monitoring program called the "traditional culture tour". The organization invites users of social media to experience the specially-themed tour. During the program, it creates a unique tourist experience such as tasting Korean traditional cuisine and role-playing at a traditional wedding. After attending the program, the participants post a record of their tour experience on social media, which can increase awareness of traditional culture in Korea. This activity offers a unique consumption value that influences tourists' personalized value proposition.

Furthermore, KTO used big data gathered from multiple social media channels for developing a foreign individual tourists (FIT) course. The course was particularly designed for foreign travelers in order to provide a personalized experience. One of the popular courses, named "Busan to Gyeongju one-day tour", was developed by analyzing the data from social media in order to identify preferred destinations for foreign travelers. The organization also globally advertised the course to attract more foreign tourists. For example, they posted the course advertisement video on YouTube. In addition, they invited influencers from social media. Therefore, the examples indicate that KTO aims to provide uniqueness in value co-creation.

Relationship

Tourists establish a solid relationship with providers through active interaction that enhances tourists' participation and empowerment, which helps in developing solutions and value-in-use (Bonsu and Darmody 2008). Example measurement questions for the relationship are "There was usually a group, a community, or a network of consumers who are a fan of the party" and "The party was renowned because its consumers usually spread positive word about it in their social networks" (Ranjan and Read 2016: 301).

Case study 37.6

Building a relationship between tourists and destinations is critical in the value co-creation process. KTO also puts effort into establishing a strong relationship with tourists by using social media. For instance, the organization has recruited student reporters who advertise Korea tourism on social media. The recruitment program offers different seminars such as tips for posting on social media in a way that can establish a robust relationship between reporters and the organization. As a result, the program has run for 11 years. Furthermore, it was nominated as the best organization program among college students in Korea.

Moreover, the organization recruited social media reporters called "supporters" in order to promote tourism content on social media. For example, 125 reporters were recruited from China, Taiwan and Hong Kong. These reporters posted 2,923 items of content after receiving an opportunity to travel to over 40 cities in Korea. Also, KTO provided the supporters with a specialized training program showing them how to create a post on social media. In addition, they recruited supporters from Japan (40 people), Middle East Asia (40), and America (25) to extend the program's applicability. Therefore, these examples show that KTO utilizes the impact of social media to build a relationship with tourists, which fortifies tourist empowerment and value co-creation.

Toward a future of managing tourist experience through social media

From the case study of KTO and by using a value co-creation framework, this chapter shows how social media contribute to tourist experience and value co-creation in South Korea. First, in the analysis of knowledge, the study shows that sharing information and understanding tourists' needs plays a pivotal role in value co-creation. Using social media, the organization identifies tourists' preferences. In the example, KTO not only categorized tourists' preferred items but also identified the reason why they preferred the items, which led the organization to develop more useful content on social media that could effectively attract target tourists. This implies that destinations can develop a better product or marketing strategy by understanding co-created knowledge in value co-creation. This suggests that considering tourist preferences using social media can co-create value.

Second, in the analysis of equity, KTO utilized big data from social media when developing marketing strategies for Pyeongchang as well as conducting a "15 seconds of success" campaign for incentive tourism. This provides the insight that destinations should be able to identify tourists' requirements and should also be willing to incorporate their needs. To advance tourists' value co-creation from an equity perspective, the industry needs to adopt big data analysis to develop marketing strategies while considering tourists' needs.

Third, the study examines how KTO implemented interaction through social media. The case demonstrates that the organization puts effort into communicating with different channel users by responding to their questions or comments. They also persuaded tourists to join the video contest through social media, which increased tourists' participation and involvement. This demonstrates that industry practitioners should devise ways to interact with tourists on social media, which becomes a source of value.

Fourth, the study analyzes the experience perspective of KTO that provided a seasonal festival and a themed travel program. By posting information about special events that created a memorable tourist experience on social media, value was co-created between tourists and destinations. For example, the organization utilized examples of national heritage that had creative and unique characteristics, such as Changdeok palace.

Fifth, in the analysis of personalization, the study shows that KTO provided a specialized tour program after investigating tourists' personalized needs through social media. The case offers a suggestion that destinations should develop unique and personalized tour programs by identifying the preferences of target tourists. Having them experience the specialized course can ultimately accentuate value co-creation.

Lastly, the chapter investigates how KTO manages the relationship element in value co-creation. Establishing a strong relationship with specific groups such as student reporters and supporters contributes to the co-creation of value because the strong relationship leads the groups to spread positive word of mouth comments on social media. Therefore, the case demonstrated that destinations enhancing tourist empowerment by forging a solid relationship with tourists can yield positive outcomes.

Although these implications and suggestions seem easy to apply, destinations still need to identify effective ways to take advantage of social media in creating positive tourist experience co-creation. As Sigala (2016) discusses, social media play an important role in value co-creation, so understanding different elements in value co-creation framework and using them to implement a strategic approach creates a positive tourist experience.

References

Archpru Akaka, M. and Chandler, J.D. (2011). 'Roles as resources: A social roles perspective of change in value networks', *Marketing Theory*, 11 (3): 243–260. DOI: 10.1177/1470593111408172.

Ballantyne, D. and Varey, R.J. (2008). 'The service-dominant logic and the future of marketing', *Journal of the Academy of Marketing Science*, 36 (1): 11–14. DOI: 10.1007/s11747-007-0075-8.

Berger, J. and Schwartz, E.M. (2011). 'What drives immediate and ongoing word of mouth?', *Journal of Marketing Research*, 48 (5): 869–880. DOI: 10.1509/jmkr.48.5.869.

Bolton, R.N. (2004). 'Invited commentaries on "evolving to a new dominant logic for marketing"', *Journal of Marketing*, 68 (1): 18–27. DOI: 10.1509/jmkg.68.1.18.24035.

Bonsu, S.K. and Darmody, A. (2008). 'Co-creating second life: Market—consumer cooperation in contemporary economy', *Journal of Macromarketing*, 28 (4): 355–368. DOI: 10.1177/0276146708325396.

Bronner, F. and De Hoog, R. (2011). 'Vacationers and eWOM: Who posts, and why, where and what?', *Journal of Travel Research*, 50 (1): 15–26. DOI: 10.1177/0047287509355324.

Chan, K.W., Yim, C.K. and Lam, S.S. (2010). 'Is customer participation in value creation a double-edged sword? Evidence from professional financial services across cultures', *Journal of Marketing*, 74 (3): 48–64. DOI: 10.1509/jmkg.74.3.048.

Cova, B., Dalli, D. and Zwick, D. (2011). 'Critical perspectives on consumers' role as "producers": Broadening the debate on value co-creation in marketing processes', *Marketing Theory*, 11 (3): 231–241. DOI: 10.1177/1470593111408171.

Edvardsson, B., Enquist, B. and Johnston, R. (2010). 'Design dimensions of experience rooms for service test drives: Case studies in several service contexts', *Managing Service Quality: An International Journal*, 20 (4): 312–327. DOI: 10.1108/09604521011057469.

Edvardsson, B., Gustafsson, A. and Roos, I. (2005). 'Service portraits in service research: A critical review', *International Journal of Service Industry Management*, 16 (1): 107–121. DOI: 10.1108/09564230510587177.

Edvardsson, B., Ng, G., Zhi Min, C., Firth, R. and Yi, D. (2011). 'Does service-dominant design result in a better service system?', *Journal of Service Management*, 22 (4): 540–556. DOI: 10.1108/09564231111155114.

Enz, M.G. and Lambert, D.M. (2012). 'Using cross-functional, cross-firm teams to co-create value: The role of financial measures', *Industrial Marketing Management*, 41 (3): 495–507. DOI: 10.1016/j.indmarman.2011.06.041.

Fisher, D. and Smith, S. (2011). 'Cocreation is chaotic: What it means for marketing when no one has control', *Marketing Theory*, 11 (3): 325–350. DOI: 10.1177/1470593111408179.

Fotis, J.N., Buhalis, D. and Rossides, N. (2012). *Social Media Use and Impact during the Holiday Travel Planning Process* (pp. 13–24). Vienna: Springer Verlag.

Fuchs, M., Scholochov, C. and Höpken, W. (2009). 'E-Business adoption, use, and value creation: An Austrian Hotel Study', *Information Technology & Tourism*, 11 (4): 267–284. DOI: 10.3727/109830510X12670455864168.

Gebauer, H., Johnson, M. and Enquist, B. (2010). 'Value co-creation as a determinant of success in public transport services: A study of the Swiss Federal Railway operator (SBB)', *Managing Service Quality: An International Journal*, 20 (6): 511–530. DOI: 10.1108/09604521011092866.

Grönroos, C. and Ravald, A. (2011). 'Service as business logic: Implications for value creation and marketing', *Journal of Service Management*, 22 (1): 5–22. DOI: 10.1108/09564231111106893.

Grover, V. and Kohli, R. (2012). 'Cocreating IT value: New capabilities and metrics for multifirm environments', *Mis Quarterly*, 36 (1): 225–232. DOI: 10.2307/41410415.

Koumelis, T. (2011). 'Social media continues to play role in travel decisions'. *Travel Daily News*. [online]. Available from: www.traveldailynews.asia/news/article/46174/social-media-continues-to-play. [Accessed: October 30, 2015].

Lemke, F., Clark, M. and Wilson, H. (2011). 'Customer experience quality: An exploration in business and consumer contexts using repertory grid technique', *Journal of the Academy of Marketing Science*, 39 (6): 846–869. DOI: 10.1007/s11747-010-0219-0.

Li, X. and Wang, Y. (2011). 'China in the eyes of western travelers as represented in travel blogs', *Journal of Travel & Tourism Marketing*, 28 (7): 689–719. DOI: 10.1080/10548408.2011.615245.

Litvin, S.W., Goldsmith, R.E. and Pan, B. (2008). 'Electronic word-of-mouth in hospitality and tourism management', *Tourism Management*, 29 (3): 458–468. DOI: 10.1016/j.tourman.2007.05.011.

Lusch, R.F. and Vargo, S.L. (2006). 'Service-dominant logic: Reactions, reflections and refinements', *Marketing Theory*, 6 (3): 281–288. DOI: 10.1177/1470593106066781.

Maclaran, P. and Brown, S. (2005). 'The center cannot hold: Consuming the utopian marketplace', *Journal of Consumer Research*, 32 (2): 311–323. DOI: 10.1086/432240.

Maglio, P.P. and Spohrer, J. (2008). 'Fundamentals of service science', *Journal of the Academy of Marketing Science*, 36 (1): 18–20. DOI: 10.1007/s11747-007-0058-9.

Munar, A.M. and Jacobsen, J.K.S. (2014). 'Motivations for sharing tourism experiences through social media', *Tourism Management*, 43: 46–54. DOI: 10.1016/j.tourman.2014.01.012.

Nambisan, S. and Baron, R.A. (2007). 'Interactions in virtual customer environments: Implications for product support and customer relationship management', *Journal of Interactive Marketing*, 21 (2): 42–62. DOI: 10.1002/dir.20077.

Neuhofer, B., Buhalis, D. and Ladkin, A. (2012). 'Conceptualising technology enhanced destination experiences', *Journal of Destination Marketing & Management*, 1 (1–2): 36–46. DOI: 10.1016/j.jdmm.2012.08.001.

O'Neill, M. and Palmer, A. (2003). 'An exploratory study of the effects of experience on consumer perceptions of the service quality construct', *Managing Service Quality: An International Journal*, 13 (3): 187–196. DOI: 10.1108/09604520310476454.

Ordanini, A., Miceli, L., Pizzetti, M. and Parasuraman, A. (2011). 'Crowd-funding: Transforming customers into investors through innovative service platforms', *Journal of Service Management*, 22 (4): 443–470. DOI: 10.1108/09564231111155079.

Powell, J.H. and Swart, J. (2010). 'Mapping the values in B2B relationships: A systemic, knowledge-based perspective', *Industrial Marketing Management*, 39 (3): 437–449. DOI: 10.1016/j.indmarman.2008.11.011.

Ramani, G. and Kumar, V. (2008). 'Interaction orientation and firm performance', *Journal of Marketing*, 72 (1): 27–45. DOI: 10.1509/jmkg.72.1.027.

Ramirez, R. (1999). 'Value co-production: Intellectual origins and implications for practice and research', *Strategic Management Journal*, 20 (1): 49–65. DOI: 10.1002/(SICI)1097-0266(199901)20:1<49::AID-SMJ20>3.0.CO;2-2.

Ranjan, K.R. and Read, S. (2016). 'Value co-creation: Concept and measurement', *Journal of the Academy of Marketing Science*, 44 (3): 290–315. DOI: 10.1007/s11747-014-0397-2.

Sigala, M. (2016). 'Social media and the co-creation of tourism experiences'. In M. Sotiriadis and D. Gursoy (eds.), *The Handbook of Managing and Marketing Tourism Experiences?* (pp. 85–111). Bingley, UK: Emerald Group Publishing. DOI: 10.1108/978-1-78635-290-320161033.

Sigala, M., Christou, E. and Gretzel, U. (2012). *Social Media in Travel, Tourism and Hospitality: Theory, Practice and Cases. Ashgate Publishing.* [online]. Routledge. Available from: https://books.google.com/books?hl=en&lr=&id=ep2kuk7idmkC&oi=fnd&pg=PR9&dq=Sigala,+M.,+Christou,+E.+and+Gretzel,+U.+(2012)+(eds.),+Social+Media+in+Travel,+Tourism+and+Hospitality:+Theory,+Practice+and+Cases.+Ashgate+Publishing,+Ltd.&ots=8GEVoZuhK7&sig=8JDCONK7OzADPWRJFXufV0_pW3o#v=onepage&q&f=false.[Accessed: March 31, 2019].

Silver, D. (2009). *The Social Networking Business Plan:18 Strategies that Will Create Great Wealth.* John Wiley. [online]. Available from: https://books.google.com/books?hl=en&lr=&id=Pqc2TkgWimlC&oi=fnd&pg=PT7&dq=Silver,+D.+(2009).+The+Social+Networking+Business+Plan:18+Strategies+that+Will+Create+Great+Wealth.+John+Wiley+and+Sons.&ots=YnEn8EDg-S&sig=GXLhaH6qUnyd0fVivDGnCe8vREc#v=onepage&q=Silver%2C%20D.%20(2009).%20The%20Social%20Networking%20Business%20Plan%3A18%20Strategies%20that%20Will%20Create%20Great%20Wealth.%20John%20Wiley%20and%20Sons.&f=false.[Accessed: March 31, 2019].

Thevenot, G. (2007). 'Blogging as a social media', *Tourism and Hospitality Research*, 7 (3–4): 287–289. DOI: 10.1057/palgrave.thr.6050062.

Vallaster, C. and von Wallpach, S. (2013). 'An online discursive inquiry into the social dynamics of multi-stakeholder brand meaning co-creation', *Journal of Business Research*, 66 (9): 1505–1515. DOI: 10.1016/j.jbusres.2012.09.012.

Verbeke, W.J., Belschak, F.D., Bakker, A.B. and Dietz, B. (2008). 'When intelligence is (dys) functional for achieving sales performance', *Journal of Marketing*, 72 (4): 44–57. DOI: 10.1509/jmkg.72.4.044.

Wang, Y. and Fesenmaier, D.R. (2004). 'Modeling participation in an online travel community', *Journal of Travel Research*, 42 (3): 261–270. DOI: 10.1177/0047287503258824.

Xiang, Z. and Gretzel, U. (2010). 'Role of social media in online travel information search', *Tourism Management*, 31 (2): 179–188. DOI: 10.1016/j.tourman.2009.02.016.

Yoo, K.H. and Gretzel, U. (2011). 'Influence of personality on travel-related consumer-generated media creation', *Computers in Human Behavior*, 27 (2): 609–621. DOI: 10.1016/j.chb.2010.05.002.

Zhang, X. and Chen, R. (2008). 'Examining the mechanism of the value co-creation with customers', *International Journal of Production Economics*, 116 (2): 242–250. DOI: 10.1016/j.ijpe.2008.09.004.

38

Virtual reality, augmented reality and tourism experience

Sandra Maria Correia Loureiro

Introduction

Creating tourists' experiences that lead them to look back on outstanding or memorable experiences (e.g., Arnould and Price 1993; Tung and Ritchie 2011; Kim, Ritchie, and McCormick 2012; Loureiro 2014) has become a top priority at destinations, hotels, restaurants, tours, museums, or any cultural or entertainment attraction. Schmitt (1999) and Pine II and Gilmore (1998) are the first scholars to point out the importance of experiences in a new economy that goes beyond products, services, or brand.

Tourists are changing the way they live and intend to live their vacations and travel to other destinations. They are interacting with the managers of destinations, hotels, and other tourist and cultural venues in real time, using technologies such as social media platforms (e.g., Libai et al. 2010; Leeflang et al. 2014), augmented reality, virtual reality and other related technologies (e.g., Loureiro et al. 2018; Loureiro and Guerreiro 2018).

Notwithstanding the importance of the tourism experience theme, to date an overview of the phenomenon is not found in the literature. Thus, this chapter intends to give an overview of tourist experience in the real and virtual worlds (virtual and augmented reality) and how tourists are dealing with these two worlds. To achieve such a goal, the chapter examines existing definitions and conceptualizations of experience (and related constructs), virtual reality (VR) and augmented reality (AR) and delivers a historical perspective of real and virtual experiences within tourism. A comprehensive literature review makes it possible to methodologically capture the concepts of the topic and organize the structure of the chapter giving (i) different experience definitions, (ii) definitions of VR and AR, (iii) the root of experience in tourism (e.g., service quality, satisfaction, relationship marketing, tourist focus, tourist engagement, co-creation), and (iv) the tourist experience journey (stages of tourism experience and its relationship with VR and AR).

The chapter is of interest to both academics and practitioners, since it identifies concepts of tourism experience, aggregates the real and virtual worlds, and gives a view of the management of the tourist experience. Finally, the chapter points out critical areas for future research.

Experience conceptualization

When Pine II and Gilmore (1999) present and spread the idea of experience in an age where intangible activities become important, vis-à-vis customization and going beyond the core product or service, they highlight the staged experience as the new economy. Thus, a customer pays "to spend time enjoying a series of memorable events that a company stages – as in a theatrical play – to engage him in a personal way" (Pine II and Gilmore 1999: 2). Although other academics have mentioned the word experience previously, like Abbott (1955), Hirschman and Holbrook (1982) or Holbrook and Hirschman (1982), yet these authors associated experience with hedonic products/services, whereas Pine II and Gilmore (1999) extend it to all types of products/services and are more effective in communicating the concept.

Pine II and Gilmore (1998, 1999) claim four realms of experience economy – educational, aesthetic, escapist, and entertainment – which vary according to the consumer's or tourist's active or passive participation and or immersion or absorption in the experience. Thus, depending on the level of tourist involvement in creating the experience, Pine II and Gilmore (1998, 1999) mention active or passive participation. Indeed, tourists' passive participation characterizes entertainment (e.g., watching television, attending a concert) and aesthetic (e.g., like a tourist who merely views an art gallery) dimensions, while active participation is connected to educational (e.g., attending a class, taking a ski lesson) and escapist (e.g., playing in an orchestra) experiences. Absorption, in turn, is "occupying customers' attention by bringing the experience into the mind" and immersion is "becoming physically or virtually a part of the experience itself" (Pine II and Gilmore 1999: 31). Absorption is associated with entertainment and educational experiences, whereas immersion is related to aesthetic and escapist dimensions.

Experience has been highly associated with emotions (e.g., Hirschman and Holbrook 1982; Holbrook and Hirschman 1982), which is well expounded in the conceptualization of brand experience expressed by Brakus, Schmitt, and Zarantonello (2009: 53) as subjective, internal consumer responses (sensations, feelings, and cognitions) and behavioral responses evoked by brand-related stimuli that are part of a brand's design: sensory (the impression that the experience of the brand makes on consumers' visual or other senses), affective (an emotional brand that induces feelings and sentiments), intellectual (stimulates tourists' curiosity and problem solving), and behavioral (the brand is action-oriented and tourists can engage in physical actions and behaviors).

De Keyser et al. (2015: 23) regard customer experience as "comprised of the cognitive, emotional, physical, sensorial, spiritual, and social elements that mark the customer's direct or indirect interaction with (an) other market actor(s)". This conceptualization is very aligned with Kim, Ritchie, and McCormick (2012: 13), where the authors argue that a positive memorable tourism experience (MTE) has been operationally defined as "a tourism experience positively remembered and recalled after the event has occurred" and propose seven dimensions, citing hedonism, novelty, local culture, refreshment, meaningfulness, involved, and knowledge.

Experiences can also be seen from a co-creation perspective (e.g., Prahalad and Ramaswamy 2003; Sfandla and Björk 2013; Chandler and Lusch 2015). Co-creation experiences relate to the involvement of customers or tourists in defining and designing the experience with those who are on the side of the offer (e.g., brand managers, firms or destinations, hotel and entertainment managers) (e.g., Prahalad and Ramaswamy 2003; Sfandla and Björk 2013). Co-creation experience puts the emphasis on a network of relationships and interactive processes (Sfandla and Björk 2013).

In summary, the concept of experiences is multidimensional and involves an interaction between the firm and its stakeholders, particularly customers or tourists.

These interactions occur before the purchase of the product/service, during the use of such product/service and, after that, when consumers decide to repurchase and recommend to others or not.

Roots of tourist experience

This section is an overview of the past research that gave insights into and contributed to the emergence of the whole conceptualization of experience: (i) buying behavior models, (ii) satisfaction and loyalty, (iii) service quality, (iv) relationship marketing and engagement process, and (v) virtual and augmented reality (see Figure 38.1).

Buying behavior models arose in the 1960s and 1970s and dealt with the customers'/tourists' process of decision-making. Nicosia's model (1966) was the first to emerge, followed by Engel, Kollat, and Blackwell's (1968) model, but one of the most emblematic models was proposed by Howard and Sheth (1969) and extended by other researchers (Pucinelli et al. 2009; Verhoef et al. 2009). Nicosia's (1966) model sowed the seeds for Howard and Sheth's (1969) model by suggesting four stages. Engel, Kollat, and Blackwell's (1968) model introduced for the first time the motion of the feedback or "search" loop, that is, when customers/tourists use not only the information communicated by the firm, but also their own knowledge about the product/service and the contribution by other consumers to the decision process.

Howard and Sheth's (1969) model presented four main components, beginning with the marketing communication process provided by the organizations and social stimuli and followed by the search for information and the evaluation process made by customers/tourists, leading to a frame of mind and finally the decision and outcomes in terms of behavioral consumption. Finally, in the late 1990s McCarthy, Perreault, and Quester (1997) suggested a model of consumer decision-making – based on the previous models – where a cycle of consumer decision-making is proposed: problem recognition, information search, evaluation and selection of alternatives, decision implementation, and post-purchase evaluation.

Satisfaction and loyalty mean the evaluation of products/services or experiences with standardized scales and tools, which had an important development in the 1970s. Oliver

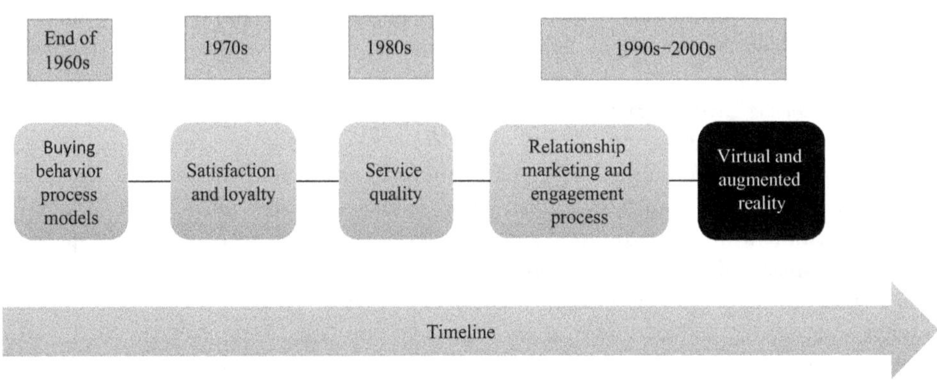

Figure 38.1 Roots of tourist experience

(1980) conceptualize satisfaction as resulting from a comparison of the actual delivered performance with customer expectations (disconfirmation paradigm). Several multiple items have been used to measure customer satisfaction, as well as the non-linear effects of satisfaction and customer delight (e.g., Oliver, Rust, and Varki 1997; Loureiro and Kastenholz 2011; Loureiro, Miranda, and Breazeale 2014). There are also several different ways in which loyalty can be measured, usually associated with behavioral intentions, which are highly associated with the work of Zeithaml, Berry, and Parasuraman (1996). Three major components are usually considered for loyalty, that is, the intention to buy again or revisit, the willingness to recommend and the willingness to pay a price premium (e.g., Loureiro 2014; Loureiro and Roschk 2014; Loureiro and de Araújo 2014).

Service quality has garnered importance as a way to measure service delivery and map the customer journey. One of the most emblematic models comes from the work of Parasuraman, Zeithaml, and Leonard Berry (1988), who propose the SERVQUAL scale. Other researchers have improved and adapted the scale in several contexts, like SERVPERF (e.g., Cronin and Taylor 1992, 1994), or RURALQUAL (Loureiro and Miranda 2009; Loureiro and Kastenholz 2011). Together with service quality, researchers have also incorporated atmospherics and the environment as influences on the customer experience (e.g., Bitner 1990; Roschk, Loureiro, and Breitsohl 2017).

Relationship marketing emerged as a field in the 1990s, giving importance to the establishment and development of strong relationships. Here, we may see constructs like commitment, trust (Morgan and Hunt 1994) and also relationship quality (Verhoef 2003; Palmatier, Gopalakrishna, and Houston 2006). At the beginning of the 21st century more attention has been paid to the emotional aspects of relationships (Verhoef and Lemon 2015), and emerging concepts like love, passion, intimacy, and attachment (e.g., Yim, Tse, and Chan 2008; Bügel, Verhoef, and Buunk 2011; Batra, Ahuvia, and Bagozzi 2012; Loureiro 2014). In the first decade of the 21st century, all conceptualizations of relationships with consumers/tourists have evolved to the idea of optimizing the relationships and creating value for customers/tourists and shareholders (e.g., Venkatesan and Kumar 2004; Shah, Kumar, and Kim 2014).

A different construct, but related to experience, is customer/tourist engagement (e.g., Bowden 2009; van Doorn et al. 2010; Vivek, Beatty, and Morgan 2012; Kumar 2013; Loureiro and Sarmento 2017; Bilro, Loureiro, and Guerreiro 2018). Although we cannot find a consensus regarding the conceptualization of engagement, it is possible to claim that engagement demands that customers/tourists should be active, involved and co-creational in their relationship with firms (e.g., Kumar et al. 2010; Brodie et al. 2011; Hollebeek, Srivastava, and Chen 2019). For instance, Brodie et al. (2011: 260) define consumer engagement as "*a psychological state that occurs by virtue of interactive, co-creative customer experiences with a focal agent/object (e.g., a brand) in focal service relationships*". Hollebeek, Glynn, and Brodie (2014) posit three dimensions of online engagement: cognitive processing (consumer's level of relationship with a brand through processing and elaboration in a particular consumer/brand interaction), affection (degree of positive affective-relation with a brand), and activation (level of energy, effort, and time spent on a brand). Kumar and his colleagues point out four dimensions for customer engagement value, that is, customer purchasing behavior, customer referral behavior, customer influencer behavior, and customer knowledge behavior (e.g., Kumar et al. 2010, 2013; Kumar and Pansari 2016).

Consumer engagement can contribute to the co-creation process of new experiences or improve the existing ones. This process has been increasingly effective in online multiple interactions, using platforms such as crowdsourcing in real time, but also through social media reviews (Bilro, Loureiro, and Ali 2018; Hollebeek, Srivastava, and Chen 2019). In summary, technologies are changing the way consumers and tourists are living experiences and virtual reality (VR) and augmented reality (AR) are playing an important role.

Indeed, virtual reality and augmented reality are transforming and reshaping the way tourists see the world and perceive experiences (Williams and Perry Hobson 1995; Guttentag 2010). A VR environment means a completely synthetic world that could be a totally new, unknown environment or a real-life mimicry. AR is an interactive experience of a real-world environment where tourists can see the objects of the real-world "augmented" by computer-generated perceptual information, sometimes across multiple sensory modalities, like visual, auditory, or olfactory (Milgram et al. 1994; Loureiro et al. 2018). Thus, in AR, tourists see the real world and also virtual objects overlaid on the real world, usually by wearing see-through displays (e.g., Milgram et al. 1994; Bimber and Raskar 2005).

Although the uses of AR date back to 1968, with the first head-mounted display (HMD) developed for research purposes by Ivan Sutherland (Carmigniani, Carmigniani, and Furht 2011), its commercial development began in the 1990s (Javornik 2016). VR and AR have had huge impacts on the tourism marketing industry (Guttentag 2010; Lee, Jeong, and Jeon 2016; Hsu 2018). For instance, through the use of smartphone applications (e.g., Skyline 2018), tourists can learn more about iconic landmarks without a tourist guide, a museum, or an art gallery.

Virtual Reality (VR) technologies allow users to experience immersive simulated reality. VR has been used successfully in several industry sectors (Wang, Yu, and Fesenmaier 2002), such as architecture (Arafa 2017), medicine (Rizzo et al. 2011), military (Manojlovich et al. 2003), or commerce (van Herpen et al. 2016), but it has been increasingly applied in a tourism and hospitality context (e.g., Cheong 1995; Williams and Perry Hobson 1995; Guttentag 2010). Indeed, museums, art galleries, hotels and restaurants are using this technology to promote their experiences, leading to an increasing amount of research being conducted (e.g., Hwang, Yoon, and Bendle 2012; Lee, Jeong, and Jeon 2016; Hsu 2018; Tussyadiah et al. 2018).

Tourist experience journey

The experience journey of a tourist with a destination, hotel or other firm dealing with tourism is an iterative and dynamic process which starts when tourists search for a destination or place they intend to visit, goes through the time at the destination and ends after the visit of the destination, when tourists return to their home country or place of residence (see Figure 38.2) (e.g., Buhalis and Law 2008; O'Connor, Höpken, and Gretzel 2008). To summarize, three major phases emerge: before the visit of the destination, during the visit, and after the visit.

The first phase aggregates all of the interaction with the destination managers and organizations that could give information about the place. Social media is massively used to search for, decide upon, and then purchase the trip, accommodation, and tours or visits that will be made at the destination. Tourists need to recognize a desire and/or a need to visit a destination, search and analyse alternatives, and then decide where to go. VR could perform an important role in promoting destinations, places, or lodgings and then creating an imaginary foretaste of what the tourist could expect at the destination (e.g., Buhalis and Amaranggana 2014, 2015).

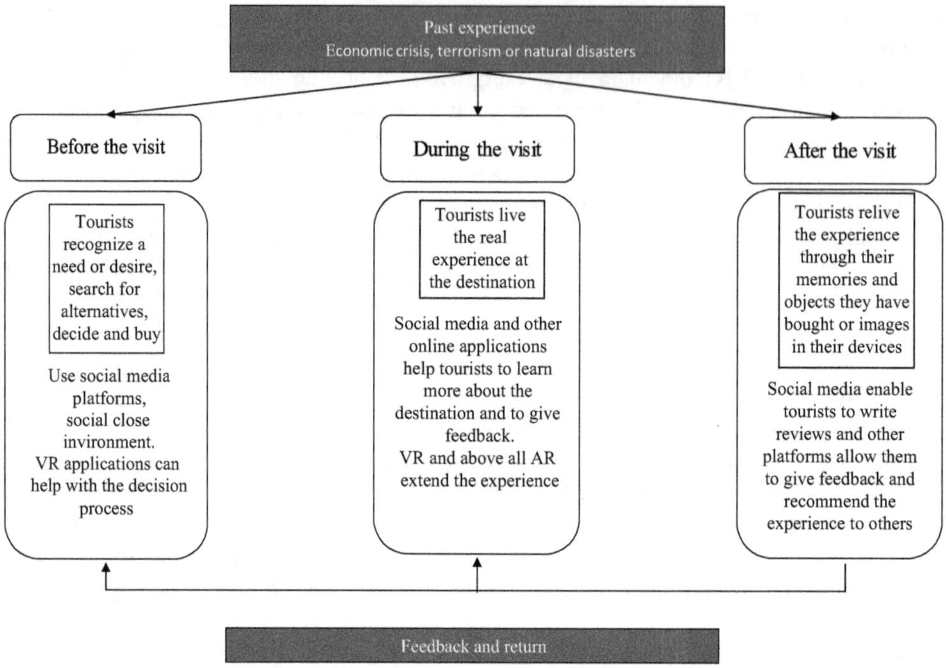

Figure 38.2 Tourist journey experience

The second stage encompasses the real experience at the destination, including the environment, tours, hotels, restaurants, museums, and other attractions (Loureiro 2014; Loureiro and Sarmento 2018). The stimuli of the environment create an image in the tourists' minds which could be positive, leading to favorable emotions and an equally favorable approach to the destination. On the other hand, a negative image and appraisal gives rise to negative emotions and avoidance of the destination. At this stage VR and above all AR can extend the experience by using virtual objects to complement the real world, which can entertain the tourists or even assist them to learn more about the place, for instance, what existed in the past in the same location (e.g., Buhalis and Amaranggana 2015).

The last phase – after the visit – relates to tourists' interactions with others and also with managers of the different places at the destination. These interactions could follow the more traditional lines of communication (e.g., letters, telephone, email) or, above all, the social media platforms (e.g., Twitter, TripAdvisor). Thus, tourists use their memories and destination-related objects (e.g., photos, souvenirs) as a means of recommending/ encouraging others to visit the destination or not, which takes the form of positive/negative reviews and posts (Boes, Buhalis, and Inversini 2016). The feedback should help managers to improve the experience and co-create potential experiences. Tourists may even intend to return to the same destination in the near future (e.g., Xiang, and Gretzel 2010; Benckendorff, Sheldon, and Fesenmaier 2014; Huang et al. 2017).

Past experience may influence the way tourists live their experience in the three stages (Bolton and Lemon 1999). It creates expectations and influences the judgment made by tourists. This cognitive process, with past knowledge being used to evaluate the current situation, can happen before, during, and after the visit. Thus, tourists use the memories and images in their minds of past experiences to analyse the information sought about potential new destinations and places to visit (Tung and Ritchie 2011; Kim, Ritchie, and McCormick 2012; Dixit 2017).

If they find a characteristic or stimulus in the information collected that reminds them of bad situations or experiences, they might decide not to visit such a destination. Past negative memories may be an even more important factor in the decision process than the influence exerted by family and friends or by the reviews on social media platforms (Verhoef et al. 2009; Prayag, Hosany, and Odeh 2013). Yet we may also argue that the stimulus provided by the information about a new destination that cognitively is compared with past situations in a positive way could be a strong factor in persuading tourists to decide in favor of such a destination. Indeed, according to the emotional appraisal theory (Moors et al. 2013), we may claim that the cognitive comparative process between past knowledge and memories with the potential new situation of the next destination to visit – or even the decision to return to a destination visited in the past – combined with the emotions that are provoked by the memories, is a strong mechanism that develops in tourists' minds which could lead to them deciding whether or not to buy the travel to the destination (Verhoef et al. 2009).

During the visit to the destination, tourists continuously compare the present experience with past experience. This cognitive evaluation combined with emotions felt by tourists will determine whether their expectations are fulfilled (tourist satisfaction and positive emotions reinforce the experience) or not fulfilled (tourist dissatisfaction and even very negative emotions may be developed), or the situation where the current experience is perceived to be superior to their expectations (tourists become delighted and the experience could be classified as outstanding) (e.g., Arnould and Price 1993; Oliver, Rust, and Varki 1997; Berry, Carbone, and Haeckel 2002; Loureiro and Kastenholz 2011).

Although past experience is a powerful influencing factor in tourist experience, we may find other factors that have often been referred to, that is, economic crises, terrorism, or natural disasters (e.g., Fornell, Rust, and Dekimpe 2010; Kumar and Pansari 2014). These events may affect how tourists make the decision to select a destination, live the experience at the destination and also behave after the visit (Gijsenberg, van Heerde, and Verhoef 2015). A normal tourist (unless he/she has a particular interest in extreme and radical situations) tends to prefer to travel to a destination where there is security in terms of violence, terrorism, serious illnesses, or environmental disasters that could lead to devastating situations. These aspects will be taken into consideration when searching for information and at the time of making the decision. An economic crisis may create a negative image in tourists' minds. Indeed, for some tourists, an economic crisis could be a negative indicator. Tourists or potential tourists may infer that an economic crisis may reduce internal expenditures that might lead to problems of poorer health care or infrastructure support (e.g., roads, equipment, hospitals, cultural promotion) (e.g., Hunneman, Verhoef, and Sloot 2015).

VR allows tourists to have an experience without being at the destination. This avoids tourists being exposed directly to diseases, terrorism, or natural disasters. Tourists could have an immersive and vivid experience at the destination without having the negative consequences (Boes, Buhalis, and Inversini 2015). An immersive experience could also encourage donations and other support for a trouble-hit destination (e.g., Manojlovich et al. 2003), changing attitudes and promoting more pro-active assistance (Tussyadiah et al. 2018).

However, natural disasters and unexpected situations can also happen, in particular during the tourist's visit to the destination. These events can completely alter tourists' experience at the destination, leading to an earlier return than expected or even placing the tourists in situations of serious danger. In the end, these negative events will influence the decision to revisit or not and how the recommendation is made to others, as well as the comments on social networks. VR can also help citizens with disabilities and other health problems to visit emblematic destinations, museums, and other places in an immersive way (e.g., Rizzo et al. 2011), without needing to travel. These VR tourists may even stay in their own homes.

AR is regarded as a travel technology that uses dynamic packaging in the form of GPS (Global Positioning System) navigation and ubiquitous onsite contextual information to supplement (with more information) or replace the traditional tour guide. This background information on the area of interest for tourists could be commercial (promotional) information, historical information and touristic narrative (Hunter 2014), but it could also provide timely and credible information about diseases, potential terrorism or other calamities (Hunter et al. 2015).

In summary, VR, AR and related technologies allow tourists to acquire different information about tour destinations, receive electronic coupons and make diverse reservation confirmations at intelligent terminals (e.g., Hunter et al. 2015; Li, Hu, Huang, and Duan 2017). Tourists become less restricted by the pre-tour arrangements, because they can change those arrangements at any time and have updated information (for instance through AR) while traveling. During the travel process, tourists can communicate in real time with other tourists and friends (for instance using social networks) (e.g., Arafa 2017; Loureiro et al. 2018).

From the point of view of tour organizations and other national institutions (e.g., public service organizations and information service enterprises) in each destination, they can trace demand and other marketing information and coordinate the dual interactions between tourists and all other stakeholders in the chain of distribution through various media and dissemination channels. Indeed, tourism managers can provide interesting and important information to tourists based on their demands, that is, respond to tourists' requirements and locations (Li et al. 2017).

Conclusions and implications

In the literature very few authors explain how destination managers may operate and deal with the entire tourist journey. Technology (VR, AR, and artificial intelligence) and tourist behavior are changing, however, which demands new and flexible business models.

Regarding technologies we may point out Disney's new approach (in Orlando, USA), which employs technologies throughout all stages of the tourist experience. This is the case of the Magic Bands (bracelets individualized for each tourist) given to each tourist before entering the park that can be used throughout the park. Band sensors enable tourists to swipe the band for entrance onto rides and into their hotel room, to make purchases and allow the organization to find them for a meet-and-greet with a Disney character, to take a photo while the tourist is on a ride, to deliver pre-ordered entrées to a tourist's table, or to email a coupon if a tourist happens to wait too long in line (Disneyworld 2019).

Another example of the use of artificial intelligence is the case of a chain hotel in Japan called Henn-na. In this chain the hotels are run by robots, but it turns out that it all needs a surprising amount of human intervention, except where the tourist would most like it. Behind the reception desk is a motionless but lifelike girl robot wearing a cream jacket and a smirk. Indeed, robots are taking off in Japan and several companies manufacture them for service roles. In April the Tokyo branch of the Bank of Tokyo-Mitsubishi put a humanoid robot on the reception desk; Nestlé is investing in robots to sell coffee makers in stores across the country; and robots guide visitors around the capital's National Museum of Emerging Science and Innovation.

AR and VR are improving and extending the tourist experience by introducing artificial objects and information about the destination, the tours, or the museum in apps that the tourists can use through their mobiles. The immersive VR experience allows simulated real places, like museums or parks that the tourist may visit, to be seen without creating problems such as overcrowding and damaging the actual place. As mentioned previously, VR may also be the first experience during the tourist decision process and could create expectations before the tourist actual travels to the destination. Citizens with some disabilities or who do not like to travel could have a VR experience of the destination at home.

Nevertheless, all of these new technological approaches also have negative aspects. One that is very often discussed is the dramatic changes that could take place in the world of employment and the potential reduction in the number of employees required. Thus, new business models are needed, and people will assist in a reconfiguration of the meaning of work.

In the future more studies are needed on (i) how VR and AR could develop a stronger role after the visit; (ii) how artificial intelligence changes the whole flow of the tourist experience process; (iii) whether or not VR could be a true alternative to a real destination visit in every situation; (iv) how the whole process of selection, purchase, experience, and after experience could be achieved through VR; (v) how the five senses are more effective in the VR environment; (vi) how background music style in the VR atmosphere influences the whole process of purchase decision; (vii) developing a tool to measure VR and AR experiences (and not use adaptations from other contexts). Finally, a full theory on VR and AR experiences is needed and not the mere extension of S-O-R (stimuli, organism, and response) or TPB (theory of planned behavior).

In summary, the current chapter traces the conceptualization, foundation, and roots of the tourist experience, as well as how new technologies are influencing and changing the way tourism is carried out and how tourists behave. The chapter presents an integrated view of tourists' experience across their journey which can help other researchers to know what has been achieved so far in this field and stimulate future research.

References

Abbott, L. (1955). *Quality and Competition*. New York: Columbia University Press.

Arafa, M. N. (2017). Achilles as a Marketing Tool for Virtual Heritage Applications. *International Journal of Architectural Research: ArchNet-IJAR*, 11(3): 109–118. DOI: 10.26687/archnet-ijar.v11i3.1385

Arnould, E. J., and Price, L. L. (1993). River Magic: Extraordinary Experience and the Extended Service Encounter. *Journal of Consumer Research*, 20(June): 24–45. DOI: 10.1086/209331

Batra, R., Ahuvia, A., and Bagozzi, R. P. (2012). Brand Love. *Journal of Marketing*, 76(2): 1–16. DOI: 10.1509/jm.09.0339

Benckendorff, P., Sheldon, P., and Fesenmaier, D. R. (2014). *Tourism Information Technology*. Oxford: CAB International.

Berry, L. L., Carbone, L. P., and Haeckel, S. H. (2002). Managing the Total Customer Experience. *MIT Sloan Management Review*, 43(3): 85–89.

Bilro, R. G., Loureiro, S. M. C., and Ali, F. (2018). The Role of Website Stimuli of Experience on Engagement and Brand Advocacy. *Journal of Hospitality and Tourism Technology*, 9(2): 204–222. DOI: 10.1108/JHTT-12-2017-0136

Bilro, R. G., Loureiro, S. M. C., and Guerreiro, J. (2018). Exploring Online Customer Engagement with Hospitality Products and Its Relationship with Involvement, Emotional States, Experience and Brand

Advocacy. *Journal of Hospitality Marketing & Management*, 28(2): 147–171. DOI: 10.1080/19368623.2018.1506375

Bimber, O., and Raskar, R. (2005). *Spatial Augmented Reality Merging Real and Virtual Worlds*. Cambridge, MA: A K Peters.

Bitner, M. J. (1990). Evaluating Service Encounters: The Effects of Physical Surroundings and Employee Responses. *Journal of Marketing*, 54(April): 69–82. DOI: 10.2307/1251871.

Boes, K., Buhalis, D., and Inversini, A. (2015). Conceptualising Smart Tourism Destination Dimensions. In I. Tussyadiah, and A. Inversini (Eds.), *Information and Communication Technologies in Tourism* (pp. 391–403). Cham: Springer.

Boes, K., Buhalis, D., and Inversini, A. (2016). Smart Tourism Destinations: Ecosystems for Tourism Destination Competitiveness. *International Journal of Tourism Cities*, 2(2): 108–124. DOI: 10.1108/IJTC-12-2015-0032

Bolton, R. N., and Lemon, K. N. (1999). A Dynamic Model of Customers' Usage of Services: Usage as an Antecedent and Consequence of Satisfaction. *Journal of Marketing Research*, 36(2): 171–186.

Bowden, J. (2009). The Process of Customer Engagement: A Conceptual Framework. *Journal of Marketing Theory and Practice*, 17(1): 63–74. DOI: 10.2753/MTP1069-6679170105

Brakus, J. J., Schmitt, B. H., and Zarantonello, L. (2009). Brand Experience: What Is It? How Is It Measured? Does It Affect Loyalty? *Journal of Marketing*, 73(3): 52–68. DOI: 10.1509/jmkg.73.3.52

Brodie, R. J., Hollebeek, L. D., Juric, B., and Ilic, A. (2011). Customer Engagement: Conceptual Domain, Fundamental Propositions, and Implications for Research. *Journal of Service Research*, 14(3): 252–271. DOI: 10.1177/1094670511411703

Bügel, M. S., Verhoef, P. C., and Buunk, A. P. (2011). Customer Intimacy and Commitment to Relationships with Firms in Five Different Sectors: Preliminary Evidence. *Journal of Retailing and Consumer Services*, 18(4): 247–258. DOI: 10.1016/j.jretconser.2010.11.005

Buhalis, D., and Amaranggana, A. (2014). Smart Tourism Destinations. In Z. Xiang, and I. Tussyadiah (Eds.), *Information and Communication Technologies in Tourism* (pp. 553–564). Cham: Springer.

Buhalis, D., and Amaranggana, A. (2015). Smart Tourism Destinations Enhancing Tourism Experience through Personalisation of Services. In I. Tussyadiah, and A. Inversini (Eds.), *Information and Communication Technologies in Tourism* (pp. 377–389). Cham: Springer.

Buhalis, D., and Law, R. (2008). Progress in Information Technology and Tourism Management: 20 Years on and 10 Years after the Internet-the State of eTourism Research. *Tourism Management*, 29(4): 609–623. DOI: 10.1016/j.tourman.2008.01.005

Carmigniani, J., Carmigniani, J., and Furht, B. (2011). *Handbook of Augmented Reality*. New York: Springer.

Chandler, J. D., and Lusch, R. F. (2015). Service Systems: A Broadened Framework and Research Agenda on Value Propositions, Engagement, and Service Experience. *Journal of Service Research*, 18(1): 6–22. DOI: 10.1177/1094670514537709

Cheong, R. (1995). The Virtual Threat to Travel and Tourism. *Tourism Management*, 16(6): 417–422. DOI: 10.1016/0261-5177(95)00049-T

Cronin, J. J., Jr., and Taylor, S. A. (1992). Measuring Service Quality: A Reexamination and Extension. *Journal of Marketing*, 56(July): 55–68. DOI: 10.2307/1252296

Cronin, J. J., Jr., and Taylor, S. A. (1994). SERVPERF versus SERVQUAL: Reconciling Performance-Based and Perceptions-Minus Expectations Measurement of Service Quality. *Journal of Marketing*, 58(January): 125–131. DOI: 10.2307/1252256

De Keyser, A., Lemon, K. N., Klaus, P., and Keiningham, T. (2015). A Framework for Understanding and Managing the Customer Experience. *MSI Working Paper No. 15–121*. Cambridge, MA: Marketing Science Institute.

Disneyworld. (2019). Retrieved from: https://disneyworld.disney.go.com/en-eu/ [Accessed April 15, 2019].

Dixit, S. K. (2017). Introduction. In Dixit, S. K. (Ed.), *The Routledge Handbook of Consumer Behaviour in Hospitality and Tourism* (pp. 1–3). London: Routledge. DOI: 10.4324/9781315659657

Engel, J. F., Kollat, D. T., and Blackwell, R. D. (1968). *Consumer Behavior*. New York: Holt Rinehart and Winston.

Fornell, C., Rust, R. T., and Dekimpe, M. G. (2010). The Effect of Customer Satisfaction on Consumer Spending Growth. *Journal of Marketing Research*, 47(1): 28–35. DOI: 10.1509/jmkr.47.1.28

Gijsenberg, M. J., van Heerde, H. J., and Verhoef, P. C. (2015). Losses Loom Longer than Gains: Modeling the Impact of Service Crisis on Perceived Service Quality over Time. *Journal of Marketing Research*, 52(5): 642–656. DOI: 10.1509/jmr.14.0140

Guttentag, D. A. (2010). Virtual Reality: Applications and Implications for Tourism. *Tourism Management*, 31(5): 637–651. DOI: 10.1016/j.tourman.2009.07.003

Hirschman, E. C., and Holbrook, M. B. (1982). Hedonic Consumption: Emerging Concepts, Methods, and Propositions. *Journal of Marketing*, 46(3): 92–101. DOI: 10.2307/1251707

Holbrook, M. B., and Hirschman, E. C. (1982). The Experiential Aspects of Consumption: Consumer Fantasies, Feelings, and Fun. *Journal of Consumer Research*, 9(2): 132–140. DOI: 10.1086/208906

Hollebeek, L. D., Glynn, M. S., and Brodie, R. J. (2014). Consumer Brand Engagement in Social Media: Conceptualization, Scale Development and Validation. *Journal of Interactive Marketing*, 28(2): 149–165. DOI: 10.1016/j.intmar.2013.12.002

Hollebeek, L. D., Srivastava, R. K., and Chen, T. (2019). S-D Logic-informed Customer Engagement: Integrative Framework, Revised Fundamental Propositions, and Application to CRM. *Journal of the Academy of Marketing Science*, 47(1): 161–185. DOI: 10.1007/s11747-016-0494-5

Howard, J. A., and Sheth, J. N. (1969). *The Theory of Buyer Behaviour*. New York: J. Wiley.

Hsu, C. H. C. (2018). Tourism Education on and beyond the Horizon. *Tourism Management Perspectives*, 25: 181–183. DOI: 10.1016/j.tmp.2017.11.022

Huang, C. D., Goo, J., Nam, K., and Yoo, Ch. W. (2017). Smart Tourism Technologies in Travel Planning: The Role of Exploration and Exploitation. *Information and Management*, 54(6): 757–770. DOI: 10.1016/j.im.2016.11.010

Hunneman, A., Verhoef, P. C., and Sloot, L. M. (2015). The Impact of Consumer Confidence on Store Satisfaction and Share of Wallet Formation. *Journal of Retailing*, 91(3): 516–532. DOI: 10.1016/j.jretai.2015.02.004

Hunter, W. C. (2014). Virtual Reality. In J. Jafari (Ed.), *The Encyclopedia of Tourism* (2nd ed., pp. 34–70). New York: Springer.

Hunter, W. C., Chung, N., Gretzel, U., and Koo, C. (2015). Constructivist Research in Smart Tourism. *Asia Pacific Journal of Information Systems*, 25(1): 105–120. DOI: 10.14329/apjis.2015.25.1.105

Hwang, J., Yoon, S.-Y., and Bendle, L. J. (2012). Desired Privacy and the Impact of Crowding on Customer Emotions and Approach-avoidance Responses: Waiting in a Virtual Reality Restaurant. *International Journal of Contemporary Hospitality Management*, 24(2): 224–250. DOI: 10.1108/09596111211206150

Javornik, A. (2016). Augmented Reality: Research Agenda for Studying the Impact of Its Media Characteristics on Consumer Behaviour. *Journal of Retailing and Consumer Services*, 30(May): 252–261. DOI: 10.1016/j.jretconser.2016.02.004

Kim, J., Ritchie, J., and McCormick, B. (2012). Development of a Scale to Measure Memorable Tourism Experiences. *Journal of Travel Research*, 51(1): 12–25. DOI: 10.1177/0047287510385467

Kumar, V. (2013). *Profitable Customer Engagement: Concept, Metrics and Strategies*. New Delhi: SAGE Publications India.

Kumar, V., Aksoy, L., Donkers, B., Venkatesan, R., Wiesel, T., and Tillmanns, S. (2010). Undervalued or Overvalued Customers: Capturing Total Customer Engagement Value. *Journal of Service Research*, 13(3): 297–310. DOI: 10.1177/1094670510375602

Kumar, V., Bhaskaran, V., Mirchandani, R., and Shah, M. (2013). Creating a Measurable Social Media Marketing Strategy: Increasing the Value and ROI of Intangibles and Tangibles for Hokey Pokey. *Marketing Science*, 32(2): 194–212. DOI: 10.1287/mksc.1120.0768

Kumar, V., and Pansari, A. (2014). The Construct, Measurement, and Impact of Employee Engagement: A Marketing Perspective. *Customer Needs and Solutions*, 1(1): 52–67. DOI: 10.1007/s40547-013-0006-4

Kumar, V., and Pansari, A. (2016). Competitive Advantage through Engagement. *Journal of Marketing Research*, 53(4): 497–516. DOI: 10.1509/jmr.15.0044

Lee, S. A., Jeong, M., and Jeon, M. M. (2016). Effects of Experiential Stimuli on Customers' Responses: An Example of Bed and Breakfast Websites. *Journal of Hospitality and Tourism Technology*, 7(4): 390–404. DOI: 10.1108/JHTT-07-2016-0035

Leeflang, P. S., Verhoef, P. C., Dahlström, P., and Freundt, T. (2014). Challenges and Solutions for Marketing in a Digital Era. *European Management Journal*, 32(1): 1–12. DOI: 10.1016/j.emj.2013.12.001

Li, Y., Hu, C., Huang, C., and Duan, L. (2017). The concept of smart tourism in the context of tourism information services. *Tourism Management*, 58: 93–300. DOI: 10.1016/j.tourman.2016.03.014.

Libai, B., Bolton, R., Bügel, M. S., de Ruyter, K., Götz, O., Risselada, H., and Stephen, A. T. (2010). Customer to Customer Interactions: Broadening the Scope of Word of Mouth Research. *Journal of Service Research*, 13(3): 267–282. DOI: 10.1177/1094670510375600

Loureiro, S. M. C. (2014). The Role of the Rural Tourism Experience Economy in Place Attachment and Behavioral Intentions. *International Journal of Hospitality Management*, 40(5): 1–9. DOI: 10.1016/j.ijhm.2014.02.010

Loureiro, S. M. C., and de Araújo, C. M. B. (2014). Luxury Values and Experience as Drivers for Consumers to Recommend and Pay More. *Journal of Retailing and Consumer Services*, 21(3): 394–400. DOI: 10.1016/j.jretconser.2013.11.007

Loureiro, S. M. C., and Guerreiro, J. (2018). Psychological Behavior of Generation Y: Living between Real and Virtual Reality. In M. Gerhardt, and J. V. E. Peluchette (Eds.), *Millennials: Characteristics, Trends and Perspectives* (pp. 67–90). New York: NOVA science publishers, Inc.

Loureiro, S. M. C., Guerreiro, J., Eloy, S., Langaro, D., and Panchapakesan, P. (2018). Understanding the Use of Virtual Reality in Marketing: A Text Mining-based Review. *Journal of Business Research*. (Available online 9 November 2018). DOI: 10.1016/j.jbusres.2018.10.055

Loureiro, S. M. C., and Kastenholz, E. (2011). Corporate Reputation, Satisfaction, Delight, and Loyalty towards Rural Lodging Units in Portugal. *International Journal of Hospitality Management*, 30(3): 575–583. DOI: 10.1016/j.ijhm.2010.10.007

Loureiro, S. M. C., Miranda, F. J., and Breazeale, M. (2014). Who Needs Delight? the Greater Impact of Value, Trust and Satisfaction in Utilitarian, Frequent-Use Retail. *Journal of Service Management*, 25(1): 101–124. DOI: 10.1108/JOSM-06-2012-0106

Loureiro, S. M. C., and Miranda, G. F. J. (2009). Perceived Quality in Rural Lodgings of Spain and Portugal: RURALQUAL Scale. *European Journal of Management Studies*, 14(1): 33–52. (ISSN: 1647-2225).

Loureiro, S. M. C., and Roschk, H. (2014). Differential Effects of Atmospheric Cues on Emotions and Loyalty Intention with respect to Age under Online/Offline Environment. *Journal of Retailing and Consumer Services*, 21(2): 211–219. DOI: 10.1016/j.jretconser.2013.09.001

Loureiro, S. M. C., and Sarmento, E. M. (2017). Tourist Engagement: The Role of Place Attachment and Place Authenticity. In A. Pizamand, and Y. Wang (Eds.) Conference Book Proceedings of the 4th World Research Summit for Tourism and Hospitality: Innovation, Partnerships and Sharing (accepted), 8–11 of December, Rosen College of Hospitality Management, University of Central Florida, Orlando, Florida, USA.

Loureiro, S. M. C., and Sarmento, E. M. (2018). Engaging Visitors in Cultural and Recreational Experience at Museums. *ANATOLIA: an International Journal of Tourism and Hospitality Research*, 29(4): 581–592. DOI: 10.1080/13032917.2018.1484378

Manojlovich, J., Manojlovich, J. M., Chen, J., and Lewis, M. (2003). UTSAF: A Multi-agent-based Framework for Supporting Military-based Distributed Interactive Simulations in 3D Virtual Environments. In S. Chick, P. J. Sánchez, D. Ferrin, and D. J. Morrice (Eds.), *Proceedings of the International Conference on Machine Learning and Cybernetic* (pp. 960–968). North Bangkok: King Mongkut's University of Technology. DOI: 10.1109/WSC.2003.1261517

McCarthy, J. E., Perreault, W. D., and Quester, P. G. (1997). *Basic Marketing: A Managerial Approach* (1st Australasian ed.). Sydney: Irwin.

Milgram, P., Takemura, H., Utsumi, A., and Kishino, F. (1994). Augmented Reality: A Class of Displays on the Reality-virtuality Continuum. *Systems Research*. In H. Das (Ed.), *Telemanipulator and Telepresence Technologies* (vol. 2351, pp. 282–292). Boston: Photonics for Industrial Applications. DOI: 10.1117/12.197321

Moors, A., Ellsworth, P. C., Scherer, K. R., and Frijda, N. H. (2013). Appraisal Theories of Emotion: State of the Art and Future Development. *Emotion Review*, 5(2): 119–124. DOI: 10.1177/1754073912468165

Morgan, R. M., and Hunt, S. D. (1994). The Commitment–Trust Theory of Relationship Marketing. *Journal of Marketing*, 58(3): 20–38. DOI: 10.2307/1252308

Nicosia, F. M. (1966). *Consumer Decision Processes: Marketing and Advertising Implications*. Englewood Cliffs: Prentice-Hall.

O'Connor, P., Höpken, W., and Gretzel, P. (2008). Information and Communication Technologies in Tourism 2008. *Proceedings of the International Conference in Innsbruck, Austria*, 2008. Innsbruck: Springer.

Oliver, R. L. (1980). A Cognitive Model of the Antecedents and Consequences of Satisfaction Decisions. *Journal of Marketing Research*, 17(4): 460–469. DOI: 10.2307/3150499

Oliver, R. L., Rust, R. T., and Varki, S. (1997). A Customer Delight: Foundations, Findings, and Managerial Insight. *Journal of Retailing*, 73(3): 311–336. DOI: 10.1016/S0022-4359(97)90021-X

Palmatier, R. W., Gopalakrishna, S., and Houston, M. B. (2006). Returns on Business-to-Business Relationship MarketingInvestments: Strategies for Leveraging Profits. *Marketing Science*, 25(5): 477–493. DOI: 10.1287/mksc.1060.0209

Parasuraman, A., Zeithaml, V. A., and Leonard Berry, L. (1988). SERVQUAL: A Multiple-Item Scale for Measuring Consumer Perceptions of Service Quality. *Journal of Retailing*, 64(1): 12–40. file:///C:/Users/Sandra/Downloads/Escala%20Servqual%20-%20Journal%20of%20Retailing%20(1).pdf

Pine, B. J., and Gilmore, J. (1998). Welcome to the Experience Economy. *Harvard Business Review*, 76(4): 97–105.

Pine, B. J., II, and Gilmore, J. H. (1999). *The Experience Economy: Work Is Theater and Every Business a Stage*. Cambridge, MA: Harvard Business School Press.

Prahalad, C. K., and Ramaswamy, V. (2003). The New Frontier of Experience Innovation. *MIT Sloan Management Review*, 44(4): 12–18.

Prayag, G., Hosany, S., and Odeh, K. (2013). The Role of Tourists' Emotional Experiences and Satisfaction in Understanding Behavioral Intentions. *Journal of Destination Marketing and Management*, 2 (2): 118–127. DOI: 10.1016/j.jdmm.2013.05.001

Pucinelli, N. M., Goodstein, R. C., Grewal, D., Price, R., Raghubir, P., and Stewart, D. (2009). Customer Experience Management in Retailing: Understanding the Buying Process. *Journal of Retailing*, 85(1): 15–30. DOI: 10.1016/j.jretai.2008.11.003

Rizzo, A., Requejo, P., Winstein, C. J., Lange, B., Ragusa, G., Merians, A., Patton, J., Banerjee, P., and Aisen, M. (2011). Virtual Reality Applications for Addressing the Needs of Those Aging with Disability. *Studies in Health Technology and Informatics*, 163(Medicine Meets Virtual Reality, 18): 510–516. DOI: 10.3233/978-1-60750-706-2-163

Roschk, H., Loureiro, S. M. C., and Breitsohl, J. (2017). Calibrating 30 Years of Experimental Research: A Meta-analysis of the Atmospheric Effects of Music, Scent, and Color. *Journal of Retailing*, 93(2): 228–240. DOI: 10.1016/j.jretai.2016.10.001

Schmitt, B. H. (1999). *Experiential Marketing*. New York: Free Press.

Sfandla, C., and Björk, P. (2013). Tourism Experience Network: Co-creation of Experiences in Interactive Processes. *International Journal of Tourism Research*, 15(5): 495–506. DOI: 10.1002/jtr.1892

Shah, D., Kumar, V., and Kim, K. H. (2014). Managing Customer Profits: The Power of Habits. *Journal of Marketing*, 51(6): 726–741. DOI: 10.1509/jmr.13.0423

Skyline. (2018). Retrieved from: https://crokepark.ie/ericsson-skyline-app [Accessed February 12, 2019].

Tung, V. W. S., and Ritchie, J. R. B. (2011). Exploring the Essence of Memorable Tourism Experiences. *Annals of Tourism Research*, 38(4): 1367–1386. DOI: 10.1016/j.annals.2011.03.009

Tussyadiah, I. P., Wang, D., Jung, T., and Dieck, M. C. T. (2018). Virtual Reality, Presence, and Attitude Change: Empirical Evidence from Tourism. *Tourism Management*, 66(June): 140–154. DOI: 10.1016/j.tourman.2017.12.003

van Doorn, J., Lemon, K. N., Mittal, V., Nass, S., Pick, D., Pirner, P., and Verhoef, P. C. (2010). Customer Engagement Behavior: Theoretical Foundations and Research Directions. *Journal of Service Research*, 13(3): 253–266. DOI: 10.1177/1094670510375599

van Herpen, E., van den Broek, E., van Trijp, H. C. M., and Yu, T. (2016). Can a Virtual Supermarket Bring Realism into the Lab? Comparing Shopping Behavior Using Virtual and Pictorial Store Representations to Behavior in a Physical Store. *Appetite*, 107: 196–207. DOI: 10.1016/j.appet.2016.07.033

Venkatesan, R., and Kumar, V. (2004). A Customer Lifetime Value Framework for Customer Selection and Resource Allocation Strategy. *Journal of Marketing*, 68(4): 106–125. DOI: 10.1509/jmkg.68.4.106.42728

Verhoef, P. C. (2003). Understanding the Effect of Customer Relationship Management Efforts on Customer Retention and Customer Share Development. *Journal of Marketing*, 67(4): 30–45. DOI: 10.1509/jmkg.67.4.30.18685

Verhoef, P. C., Lemon, K. A., Parasuraman, A., Roggeveen, A., Tsiros, M., and Schlesinger, L. A. (2009). Customer Experience Creation: Determinants, Dynamics and Management Strategies. *Journal of Retailing*, 85(1): 31–41. DOI: 10.1016/j.jretai.2008.11.001

Verhoef, P. C., and Lemon, K. N. (2015). Advances in Customer Value Management. In R. M. Morgan, J. T. Parish, and G. Deitz (Eds.), *Handbook of Research in Relationship Marketing* (pp. 75–103). Northampton: Edward Elgar.

Vivek, S. D., Beatty, S. E., and Morgan, R. M. (2012). Customer Engagement: Exploring Customer Relationships beyond Purchase. *Journal of Marketing Theory and Practice*, 20(2): 122–146. DOI: 10.2753/MTP1069-6679200201

Wang, Y., Yu, Q., and Fesenmaier, D. R. (2002). Defining the Virtual Tourist Community: Implications for Tourism Marketing. *Tourism Management*, 23(4): 407–417. DOI: 10.1016/S0261-5177(01)00093-0

Williams, P., and Perry Hobson, J. S. (1995). Virtual Reality and Tourism: Fact or Fantasy? *Tourism Management*, 16(6): 423–427. DOI: 10.1016/0261-5177(95)00050-X

Xiang, Z., and Gretzel, U. (2010). Role of Social Media in Online Travel Information Search. *Tourism Management*, 31(2): 179–188. DOI: 10.1016/j.tourman.2009.02.016

Yim, C. K., Tse, D. K., and Chan, K. W. (2008). Strengthening Customer Loyalty through Intimacy and Passion: Roles of Customer-Firm Affection and Customer Staff Relationships in Services. *Journal of Marketing Research*, 45(6): 741–756. DOI: 10.1509/jmkr.45.6.741

Zeithaml, V. A., Berry, L. L., and Parasuraman, A. (1996). The Behavioral Consequences of Service Quality. *Journal of Marketing*, 60(2): 31–46. DOI: 10.2307/1251929

Part V
Sustainable tourism experience

39

Sustainable tourism and the visitor experience

Frans Melissen and Bert Smit

Introduction

Realizing sustainable development is one of the most pressing and urgent challenges today's society is faced with. Simultaneously, tourism and sustainable development have so far had a rather complicated relationship. Whereas tourism is oftentimes seen as something that could play a crucial role in the social and economic development of communities, regions and countries, it is also a sector that is responsible for a significant portion of our sustainability problems in the first place. Tourism puts quite a strain on our planet's natural resources and is oftentimes directly or indirectly responsible for significant waste generation and pollution. Moreover, in many regions and countries it has not resulted in everyone actually enjoying the benefits of this potential catalyst for social and economic development and it could even be argued to have contributed to increasing the gap between rich and poor. Therefore, tourism could make a significant contribution to sustainable development if and when this sector could be adapted in such a way that its activities would (contribute to) repair(ing) these damages and prevent(ing) further negative impacts. Simultaneously, this need not be the only, and actually not the most important, contribution that this sector could make. What if tourism could be turned into a catalyst for sustainable development; a catalyst that promotes sustainable behavior and the application of sustainable solutions well beyond the boundaries of the tourism sector itself? This chapter explores this multifaceted relationship between tourism and sustainability by focusing on the concept that is at the core of this sector: the visitor experience. It shows how this concept can explain some of the negative impacts of this sector and the (perceived) difficulties involved with avoiding or mitigating these negative impacts. However, it also shows that deliberately designing visitor experiences to avoid these negative impacts need not be a mission impossible, and that these visitor experiences could even prove the key to transforming tourism into a catalyst for sustainable development.

The visitor experience as the culprit

When our ancestors roamed the plains of Africa, we humans had to focus on fulfilling basic needs such as finding shelter and food. Nowadays, a combination of technological, democratic and economic developments has created a situation in which many of us, especially in so-called developed countries, need not worry about these basic needs; fulfilling them is 'easy'. Most of our lives have become much more peaceful, risk-free and possibly even 'boring'. This might very well explain why so many of us feel the urge to experience strong emotions and look for intense experiences (Le Breton 2002). Whereas our Stone Age forefathers sought security and stability, modern man seems obsessed by novelty, personal development and fulfilling the need to 'move on' (Abrahams 1986). Carù and Cova (2003) suggest that this is part of the explanation for the growth in extreme sports, adventure holidays and the like. However, they also state that not all of us look for such extreme experiences; some of us prefer more relaxing, less dangerous experiences to make us happy and develop ourselves. A nice dinner with friends or reading umpteen books while lying on a tropical beach is exactly what we need to unwind from our busy working lives. All of these types of experiences help us satisfy our psychological and social needs, our self-esteem needs and our self-actualization needs, now that we no longer have to worry about satisfying our physiological and safety needs (Maslow 1943). Many of these experiences are offered to us by the tourism sector.

What's more, the number of people who can afford such experiences, especially in today's Western society, is ever-increasing. High school graduates go on backpacking holidays across the globe to find and develop themselves. Families go on multiple holidays a year to unwind from their busy lives. Weekends are the perfect time to visit zoos and theme parks. Dining out is something many of us can now afford to do on a weekly basis or even more often. Business people travel the world to attract new clients and close deals. Academics from all over the world get together in conference hotels to discuss the latest findings and network. Young couples prefer their next holiday destination to be in the middle of pristine nature on another continent simply because they can afford it and they need new inspiration and stories. In countries like China and India, the new middle class has now developed the necessary resources and desire to discover the world. Tourism is booming, not least because this increasing demand is eagerly met by those 'in power' – local, regional and (inter)national policy-makers, the management of globally operating tour operators and hotel chains, investors, leading international agencies such as the UN's World Travel Organization (UNWTO) and the World Travel and Tourism Council (WTTC) and the like – by vigorously pursuing and stimulating continued growth of this sector (Hall 2019).

Increasing numbers of tourists and increasing numbers of customers consuming the experiences staged for them by the wide variety of experience providers in each of tourism's subsectors – hotels, casinos, theme parks, restaurants, cruise ships and the like – are likely to result in more resource use, more CO_2 emissions, more pollution, more overcrowding, more damage to natural environments and so on. However, growth is not the only problem of tourism from a sustainable development perspective. As Brohman (1996) points out, the most lucrative components of tourism – i.e. marketing and procurement of customers, transportation and food and lodging – are usually handled by vertically integrated and Western-owned companies and networks, thus leaving little room for actual community involvement at destination level and increasing socioeconomic inequities across our globe. Combining this 'inherently imperialistic, dependent nature of tourism production' with 'the characteristics of tourism consumption' creates a situation that is difficult to reconcile with the principles of sustainable development (Sharpley 2000: 14).

This second component of the sustainability problem linked to tourism is directly related to its core product: visitor/guest/customer experiences. To satisfy today's customers' demands, many of the experiences offered by the tourism sector require those same customers to fly across our globe to reach their exotic, authentic and pristine natural destinations. As pointed out by Smit and Melissen (2018: 180), more and more 'hotels and resorts have sprung up in areas that formed the last undisturbed habitats of rare species of animals', with newly constructed roads cutting through wilderness providing access to them for visitors and suppliers. Local resources are (over) used and damaged to build and operate these 'experiencescapes'. Many of these destinations require air travel to reach them, which results in (increased) emissions of CO_2, hydrocarbons, carbon monoxide, nitrogen oxides, sulfur oxides, lead and the like. Meanwhile, meat, bottled water and alcoholic beverages are flown in from yet another continent on our globe to ensure that guests can be served 'the best' food and drinks. The latter is not just commonplace in luxurious resorts and hotels that are the 'habitat' of the 'lucky few'. Your average diner across town is just as likely to serve food and drinks originating from multiple countries and even continents. Hotels in not-so-pristine locations also focus on comfort and luxury, and creating a carefree home away from home. As pointed out by Melissen and Sauer (2018), a typical example of how they hope to do so is through offering copious amounts of bathroom amenities – individually packaged in small plastic bottles – and making sure that the breakfast buffet 'exceeds guests' expectations' both in terms of quantity and quality.

In other words, the fact that tourism today is less sustainable than ever (Hall 2019) is not just the result of an ever-increasing number of people being able to afford the experiences offered by this sector. The situation is aggravated by the characteristics of these experiences. Many experience providers in the tourism industry focus on oversupplying each and every individual visitor/guest/customer. As stated by Melissen and Sauer (2018: 16), 'king-size, luxury, seduction and excitement are not only the principles on which many of these experiences are based but also the terms often used to [promote and] advertise them' to potential customers. Luring today's experience-seeking customer thus has turned into a seduction game that is not only based on offering and providing more experiences but also on offering experiences that come with even more emissions, more damages, more (scarce) resource use and more inequality. Obviously, from a sustainable development perspective, this is a deadly combination – no pun intended!

The visitor experience as part of the solution

If tourism were to become part of the solution instead of part of the problem, the first logical step would be to, at the very least, reduce its excessive use of natural resources and ensure that its activities were not accompanied by negatively impacting people. In fact, this is the least that any business or organization involved with tourism, or any other sector for that matter, could and should do – 'not just from an ethical perspective, but also from a business-economic perspective' (Smit and Melissen 2018: 228). Even though our society might be argued to, so far, have been rather slow in adapting and/or transforming to a sustainable way of living, the seriousness and urgency of the situation is clear to the vast majority of us. Companies that willingly and knowingly use more resources than they would have to, cause pollution that can easily be prevented, harm or mistreat people, and so on, increasingly run the risk of publicly being named and shamed. In today's globalized and technologically advanced world, information travels fast and is accessible to almost anyone – including your competitors, your suppliers and your (potential) customers. Attention to sustainability is clearly here to stay and ignoring this topic altogether is definitely going to make you vulnerable in a marketplace in which sustainable alternatives are on the rise – both in terms of supply and demand. Not paying attention to sustainability is no longer an option, at least not if you want to remain competitive and hang on to your long-term social license to operate.

Moreover, both practitioners and researchers have already developed guidelines, checklists, standards and initiatives, that are available to any organization or business in tourism, to reduce the negative impacts of tourism activities, both from an environmental and a social perspective. For example, books like *Sustainable Event Management* by Meegan Jones (2014) and *Sustainability in the Hospitality Industry: Principles of sustainable operations* by Legrand et al. (2017) provide practitioners in the events and hospitality subsectors of the tourism industry with a wealth of practical tips and guidelines on how to reduce energy use, greenhouse gas emissions and water use, how to build and operate more sustainable physical environments in which to stage experiences for visitors and guests, how to reduce the negative environmental impacts of the food and drinks served to them, and how to ensure that local ecosystems and residents are not negatively impacted. For the specific context of urban tourism, large and influential tourism organizations, such as UNWTO and WTTC, have recently published reports that could support destinations in dealing with some of the sustainability challenges associated with this specific tourism activity. They present a range of strategies that could be used to manage tourism more sustainably as well as sets of parameters that provide insights into how likely a destination is to become overvisited and thus unsustainable, both from an environmental and a social perspective. Practitioners and companies in the tourism sector have also joined forces to exchange information, define ambitions and set standards for sustainability aspects and measures, for instance the International Tourism Partnership (ITP) and the Hotel Carbon Measurement Initiative (HCMI) (Melissen and Sauer 2018).

However, it is important to note that at the core of all of these activities still lies the visitor experience – the core product of tourism. The initial challenge for tourism businesses and organizations, the providers of tourism experiences, could therefore be described as incorporating sustainability in the way they stage these experiences. As elaborated on in previous chapters of this handbook, these experiences consist of tangible and intangible components that together shape touchpoints that, in turn, shape the ultimate customer journey for visitors/guests/customers. Almost all of the tangible components involved require natural resources to create them or to deliver them to customers, and doing so could lead to further negative impacts on ecosystems and people. Therefore, from a sustainability perspective, it would be wise to consider some of the tips, tricks, suggestions, checklists, guidelines and the like, such as the ones mentioned above and many, many more that are already widely available, in reducing the negative impacts of these tangible components. For many of these components, this is not a mission impossible by any means. Opportunities to serve local, seasonal and organic food and drinks are growing by the day, as are the numbers of customers who actually prefer them over produce flown in from across the globe. There is no doubt that hotels and theme parks use a lot of energy but guests and visitors will not enjoy their stay or their ride in a rollercoaster any less if that energy comes from electricity generated by wind turbines and solar panels instead of non-renewable sources.

With respect to the intangible component of visitor experiences – the interaction between staff/employees and visitors – the sustainable solutions that can be incorporated are even more obvious. Unfortunately, companies operating in the tourism sector have a bit of a bad reputation when it comes to how they treat their employees – some of these companies are even described as suffering from 'warm body syndrome' (Kusluvan et al. 2010; Melissen and Sauer 2018; Smit and Melissen 2018). Obviously, changing this situation would already be a step in the right direction from a sustainable development perspective. One of the clear-cut ways of addressing the social component of sustainability in staging visitor experiences would therefore be to pay employees decent salaries, have reasonable working hours, provide safe and healthy working conditions, and simply treat employees fairly and respectfully. However, as pointed out by Melissen and Sauer (2018), treating your employees well is not just the 'right' thing to do, it is also the 'smart' thing to do. Treating your employees well is actually crucial for creating competitive advantage, especially in the tourism sector (Baum 2007; Kusluvan et al. 2010). Time and time again, research has shown that investing in employees is not costly but rather results in more productive, motivated and committed employees. This is especially relevant for companies, such as those in tourism, that rely on their staff to provide their core products to customers, through directly interacting with those same customers. Investing in employees, empowering and developing them, has been shown to result in better employee performance, job satisfaction *and* improved interaction with guests and thus guest satisfaction (Kusluvan et al. 2010). In fact, most companies in this sector would probably also benefit from involving staff in finding and implementing all other sustainability solutions that could be incorporated in staging experiences for visitors/guests/employees. Your employees might very well have interesting perspectives on how to do things more efficiently, how to use fewer natural resources and how to do so in a way that customers would appreciate – after all, they stage these experiences and interact with these customers on a day-to-day basis. Moreover, doing so would create an added benefit because engaging employees in sustainability could not only lead to generating insights that would otherwise be left unexplored but, in turn, could lead to even more engagement and productivity from those same employees, as well as increased wellbeing (Glavas 2012).

Simultaneously, incorporating sustainability aspects in visitor experiences is not always straightforward. In fact, many tourism organizations and businesses are actually scared of changing their core product and their standards, especially if this involves changing how they treat their customers for the sake of the environment (Borden et al. 2017; Font et al. 2017). Some of them feel that providing luxury and oversupplying visitors/guests/customers is crucial to their (continued) success, and that their customers would simply not accept 'more sustainable' visitor experiences (see e.g., Cavagnaro et al. 2018). Moreover, some of the physical components of specific visitor experiences could very well be difficult to adapt or replace. How do you avoid food waste if breakfast service in your hotel is based on a buffet? How do you save water if your experience is based on 'playing in and with water', such as in a water park? In these situations, incorporating sustainability solutions might very well require you to 'go back to the drawing board' and completely redesign your visitor experience. As indicated by Smit and Melissen (2018: 248), incorporating sustainability in visitor experiences might require you to address 'all reference points and complications involved in doing so in the [original] design of these experiences'. One way of doing so would be to add your ambition to reduce your negative impacts, from a sustainable development perspective, as an objective to the list of objectives generated in the first step of the design process described in chapter 11 of this handbook. Obviously, this objective would then also have to be translated into specific requirements in the third step of this design process, which, in turn, will guide generating design alternatives in the fifth step and evaluating those alternatives in the sixth step. One of the added benefits of explicitly addressing sustainability aspects within and by means of a design approach such as the one suggested in chapter 11 is that this supports communicating your objectives and constraints to other parties involved with staging your visitor experience. You might even consider actively involving some of your suppliers in your design process. Some of them might very well be able to suggest or point out possible changes in the design of the experience that would allow them to create or provide you with more sustainable products or services – the products and services that you use to stage the ultimate visitor experience. Finally, using this design process, including all theories, methods, tools and techniques that could be used to shape it in practice, could also prove quite helpful in addressing the social component of sustainability linked to your ultimate visitor experience. Given that the reference point for such a design process is to ensure that the needs, wishes and characteristics of (potential) customers are successfully accounted for in the ultimate design, it could also be used to concurrently design a sustainable 'customer journey' for employees, as well as for other stakeholders impacted by your activities as a visitor experience provider. A full review of all details involved in doing so is beyond the scope of this chapter. However, those interested might benefit from reviewing *Sustainable customer experience design: Co-creating experiences in events, tourism and hospitality* by Smit and Melissen (2018).

The visitor experience as a catalyst

The previous section has shown that any tourism organization or business could and should address sustainability through reducing its negative impacts. This is not a mission impossible but rather serves as a minimum requirement for operating in today's marketplace. However, tourism's contribution to sustainable development should not and cannot stop there. In fact, doing so would only add to one of the main problems associated with tourism from a sustainability perspective: so far, it 'has [had] a largely inward, product-centered perspective' (Sharpley 2000: 14), which is fundamentally incompatible with the principles of sustainable development. Tourism should not and cannot remain stuck in the efficiency stage of organizational change toward sustainability (Benn et al. 2014). As pointed out by Hall (2019: 1), 'there is a need to rethink human-environment relations given the mistaken belief that the exertion of more effort and greater efficiency will alone solve problems of sustainable tourism'.

Case study 39.1: The Eden Project

From the perspective of using visitor experiences to raise awareness about sustainability issues within the context of a fun day out, the Eden Project is an interesting example. This visitor attraction park in Cornwall, UK, is located in a reclaimed kaolinite pit. The main components of the experiencescape that has been created there are formed by two huge biomes. One of these biomes simulates a rainforest environment and the second a Mediterranean environment. Together, these biomes house thousands of plant species. The park also has a botanical garden and it hosts pop concerts and has a visitor center and classrooms, art exhibitions and sculptures, canopy walks, zip wires, climbing frames and more. It resembles a theme park and provides all the fun and excitement that a 'regular' theme park would provide but with a twist. The park has been set up to not only stage a pleasant experience but to also raise awareness about environmental problems and solutions by educating people about the relationship between people and plants. However, the Eden Project also tries to move beyond simply raising awareness, for instance by trying to set up a geothermal power plant. This plant could not only show the public how renewable sources could be used to generate electricity but it will also actually provide electricity to the surrounding community – plans are based on supplying some 7,000 houses near the park with electricity generated by this power plant.

The Eden Project operates as a visitor attraction park but its ambition is to do so by minimizing its negative environmental impact *and* by creating a positive environmental and social impact. It organizes community get-togethers and initiates and participates in numerous social and environmental projects. It is a charity 'disguised' as a tourist attraction. It provides people with the opportunity to experience the beauty and importance of rainforests without having to step into an airplane to Brazil. It shows the crucial importance of water and plants for life on Earth through applying the principles of biophilia but in a way that appeals to people; the park attracts more than a million visitors a year.

Obviously, it is impossible to measure the exact positive impact of the Eden Project. It might very well have had a lasting impact on the behaviors of quite a few of its visitors but whether it has truly managed to stage a transformative experience for all or specific groups of its visitors is difficult to establish with any degree of certainty. Having to rely on subsidies, donations and ticket sales, instead of contributions from commercial investors, has also created problems and has slowed down the realization of some of its plans and ambitions. However, overall, the Eden Project still provides an interesting and inspiring example of how visitor experiences in the tourism domain do not always have to rely solely on luxury, oversupplying visitors, and meeting hedonic goals and values. Sustainability can actually form the basis for and the core of visitor experiences that are appreciated by millions of people.

Reference

Smit, B., and Melissen, F. (2018). *Sustainable customer experience design: Co-creating experiences in events, tourism and hospitality.* New York/London: Routledge.

Sources: www.edenproject.com

One could argue that tourism's core product – visitor experiences – actually offers opportunities to move well beyond that so-called efficiency stage. Staging these experiences revolves around satisfying the needs and wishes of customers through truly understanding them, getting to the bottom of what makes them 'tick' and then using those insights to create satisfaction, excitement, happiness and fond memories. As Smit and Melissen (2018: 186) state, you could not possibly ask for 'better circumstances to address and discuss sustainability'. Moreover, the whole concept of visitor experiences is based on the premise of visiting a place away from your own home and daily routines. This creates an interesting opportunity from a sustainability perspective. What better place to let people discover that sustainable food can be just as tasty as 'unsustainable alternatives' than in a restaurant where that food is prepared by professionals – by people who specialize in making food taste good? Many tourism activities also take place in natural environments. These environments create the perfect opportunity to tap into people's biophilia – we are all endowed, through our genes, with an appreciation of and a desire for natural environments, and immersing us in those environments makes us appreciate them even more and, more importantly, makes us more likely to engage in sustainable behavior (see e.g., Wilson 2006). Why not use visitor experiences to 'awaken' that (hidden) human characteristic to purposely promote sustainable decision-making and behavior? In fact, visitor experiences and experiencescapes could be the perfect time and place to tap into a number of other evolutionary tendencies that explain some of the sustainability challenges our society faces today but that could also be used to promote sustainable behavior (Griskevicius et al. 2012)? Why not use those experiences and experiencescapes to let people discover that sustainability need not be unpleasant, uncomfortable and expensive but can also be fun, exciting and rewarding? As pointed out by Steg et al. (2014), sometimes all that is needed to steer our behavior and routines in a more sustainable direction is to reinforce the normative goals that already favor sustainable behavior that most if not all of us already have. Why could visitor experiences and experiencescapes, staged for us and created for us by tourism providers, not serve as that nudge in the right direction?

What is definitely needed to be able to answer all of these questions with 'Why not – let us do this!' and have tourism live up to its potential to serve as a catalyst for sustainable development is, once again, to carefully and purposely (re)design visitor experiences to make this possible. The design process discussed in the previous section, and presented in more detail in chapter 11 of this handbook, could serve as a reference point for and provide assistance in doing so. Another promising approach to doing so is applying user-centered design methodology to formulating and promoting truly sustainable tourism activities, as suggested and elaborated on by Font et al. (2018). In fact, more and more researchers are now suggesting viable ways of nudging tourists – and customers in other sectors, for that matter – in the direction of 'better' products to stimulate more sustainable consumption patterns (Gunn and Mont 2014). Once again, a full review of all possible and promising approaches to doing so is beyond the scope of this chapter. However, the characteristics of the core product of tourism clearly suggest that turning it into a catalyst for sustainable development need not be a mission impossible.

Conclusion

The previous sections have shown why visitor experiences constitute part of the explanation for tourism's current contribution to unsustainability and tourism's potential to become part of the solution. Simultaneously, turning the tourism sector into a 'sector of hope' instead of a 'sector of destruction' requires much more than pointing out that its core product, especially if and when designed carefully, could come with far less negative impacts and even serve as a catalyst for sustainable behavior beyond the boundaries of this specific sector. It also requires those in power to 'shift [...] from a growth mentality to one that explicitly commits humanity to prospering and travelling within the limits of the ecosystems of which we are a part' (Hall 2019: 13). Realizing this shift is essential for allowing visitor experiences to truly live up to their sustainability potential.

References

Abrahams, R. D. (1986). *Ordinary and extraordinary experience. The anthropology of experience* (pp. 45–72) Urbana: University of Illinois Press.

Baum, T. (2007). 'Human resources in tourism: Still waiting for change', *Tourism Management*, 28: 1383–1399. DOI: 10.1016/j.tourman.2007.04.005

Benn, S., Dunphy, D., and Griffiths, A. (2014). *Organizational change for corporate sustainability* (3rd ed). New York: Routledge.

Borden, D. S., Coles, T., and Shaw, G. (2017). 'Social marketing, sustainable tourism, and small/medium size tourism enterprises: Challenges and opportunities for changing guest behaviour', *Journal of Sustainable Tourism*, 25(7): 903–920. DOI: 10.1080/09669582.2016.1270952

Brohman, J. (1996). 'New directions in tourism for third world development', *Annals of Tourism Research*, 23(1): 48–70. DOI: 10.1016/0160-7383(95)00043-7

Carù, A., and Cova, B. (2003). 'Revisiting consumption experience: A more humble but complete view of the concept', *Marketing Theory*, 3(2): 267–286.

Cavagnaro, E., Düweke, A., and Melissen, F. (2018). 'The host-guest relationship is the key to sustainable hospitality: Lessons learned from a Dutch case study', *Hospitality and Society*, 8(1): 23–44. DOI: 10.1386/hosp.8.1.23_1

Font, X., Elgammal, I., and Lamond, I. (2017). 'Greenhushing: The deliberate under communicating of sustainability practices by tourism businesses', *Journal of Sustainable Tourism*, 25(7): 1007–1023. DOI: 10.1080/09669582.2016.1158829

Font, X., English, R., and Gkritzali, A. (2018). 'Mainstreaming sustainable tourism with user-centred design', *Journal of Sustainable Tourism*, 26(10): 1651–1667. DOI: 10.1080/09669582.2018.1491981

Glavas, A. (2012). 'Employee engagement and sustainability: A model for implementing meaningfulness at and in work', *Journal of Corporate Citizenship*, 46: 13–29. [online] Available from: www.jstor.org/stable/jcorpciti.46.13 [Accessed: December 4, 2019].

Griskevicius, V., Cantú, S., and van Vugt, M. (2012). 'The evolutionary bases for sustainable behaviour: Implications for marketing, policy and social entrepreneurship', *Journal of Public Policy and Marketing*, 31(1): 115–128. DOI: 10.1509/jppm.11.040

Gunn, M., and Mont, O. (2014). 'Choice editing as a retailers' tool for sustainable consumption', *International Journal of Retail and Distribution Management*, 42(6): 464–481. DOI: 10.1108/IJRDM-12-2012-0110

Hall, C. M. (2019). 'Constructing sustainable tourism development: The 2030 agenda and the managerial ecology of sustainable tourism', *Journal of Sustainable Tourism*. DOI: 10.1080/09669582.2018.1560456

Jones, M. (2014). *Sustainable event management: A practical guide* (2nd ed). New York: Routledge.

Kusluvan, S., Kusluvan, Z., Ilhan, I., and Buyruk, L. (2010). 'The human dimension. A review of human resources management issues in the tourism and hospitality industry', *Cornell Hospitality Quarterly*, 51 (2): 171–214. DOI: 10.1177/1938965510362871

Le Breton, D. (2002). *Conduites à risque: Des jeux de mort aux jeux de vivre*. Paris: Puf.

Legrand, W., Sloan, P., and Chen, J. S. (2017). *Sustainability in the hospitality industry: Principles of sustainable operations* (3rd ed). London: Routledge.

Maslow, A. H. (1943). 'A theoy of human motivation', *Psychological Review*, 50(4): 370.

Melissen, F., and Sauer, L. (2018). *Improving sustainability in the hospitality industry*. New York and London: Routledge.

Sharpley, R. (2000). 'Tourism and sustainable development: Exploring the theoretical divide', *Journal of Sustainable Tourism*, 8(1): 1–19. DOI: 10.1080/09669580008667346

Smit, B., and Melissen, F. (2018). *Sustainable customer experience design: Co-creating experiences in events, tourism and hospitality*. New York and London: Routledge.

Steg, L., Bolderdijk, J. W., Keizer, K., and Perlaviviute, G. (2014). 'An integrated framework for encouraging pro-environmental behaviour: The role of values, situational factors and goals', *Journal of Environmental Psychology*, 38: 104–115. DOI: 10.1016/j.jenvp.2014.01.002

Wilson, E. (2006). *The creation: An appeal to save life on earth*. New York: W.W. Notton and Company.

40

Rethinking sustainable tourism through identity decomposition

The stratification of the intensifying factors of the tourism experience

António Sérgio Araújo de Almeida

Introduction

There are many definitions of Identity and its abstract and generic concept hinders the desired operational implementation in tourism activity. However, Identity can provide an operational meaning to tourism because it provides the capacity to illustrate collective facts and events, values and behaviors that differentiate a community that is historically linked to a specific territory. In addition to generic feedbacks associated with the transversal nature of local cultures, the aforementioned differentiated and identitary facts and events, values and collective behaviors are the ones that are valued by tourists in both case studies that we present below: the Ceremonies of Holy Week in Braga and the "Mouraria", a typical neighborhood of Lisbon, both in Portugal.

In these case studies we will see what integral parts of the local identity are seen as experiential attributes: the usufruct of objective authenticity (Wang 1999) and of the community experience in its more intimate side, whether in its religiosity or its popular and profane coexistence. Their deep link to local resources and community involvement positions them as potential elements to promote sustainable tourism products.

Identity is a fundamental factor of sustainable tourism since it consubstantiates to a large extent the local resources. In this sense, it is important to rescue and understand this concept in a logic in which it is assumed as a tool to foster a convergence of mutual advantages between the needs of the tourist and the emancipation of local territories and communities. Understanding the Identity concept involves, in the present-day perspective of interest in sustainable tourism, its multidimensional understanding and, therefore, its decomposition. This process implies the stratification of Identity attributes, seen as Intensifying Factors of Tourism Experience (IFTE). In other words, tourism activities can, under certain circumstances, give rise to approaches and understandings that are different from usual regarding the Identity concept.

The history, the stories, the traditions (including music and gastronomy) and the propensity for interaction and integration in the local collective behaviors are also Intensifying Factors of the Tourist Experience that underline this management opportunity in the process of decomposition of the Identity. In this way, an integration between local communities and tourists is promoted as well as principles of tourism sustainability. Thus, this process leads to cultural valuation as a tourism resource, social involvement and economic opportunities based on differentiation and local Identity assumption.

Identitary decomposition – meaning and touristic reach

Although environmental, cultural and social characteristics are constitutive and differentiating elements of places, it is precisely their Identity that confers originality and uniqueness on places and distinguishes them from others. It is in this context that other fundamental concepts emerge within this theme: authenticity and Identity assumption (Malheiros, Lourenço and Almeida 2016: 72). The literature has a multifaceted and complex understanding of the Identity concept, including it in a wide range of situations, namely in the contexts of philosophy, sociology, history, anthropology, law and biology, among others. Generally speaking, Identity can be pointed out as corresponding to a set of unique and exclusive characteristics in which it is possible to differentiate systems of values, places, people, constructions, animals, plants and all sorts of inanimate or animate objects, including individual and collective behaviors. This complexity nests Identity within the scope of the *Anthopos*, *Oikos*, *Ethos* and *Chronos* points of view. The diagnosis of Identity attributes that can be managed in the tourism experience ends up becoming a kind of Identity decomposition, thus providing an operational aspect to the concept of Identity, which is assumed to be abstract, subjective and complex (Almeida 2018: 432).

While Castells (1999) emphasizes Identity as a kind of source of meaning and experience of a people, Bauman (2005) points to Identity as a continuous process of construction, always being incomplete and adjusting itself to the mechanics of a liquid modernity. In this last perspective, tourism, due to its size and global impacts, is one of the phenomena that contribute the most to this process of local, national and international flux. The intersection of these two works is fundamental to invoke the status quo of world tourism activity, namely in what concerns the power of the icons for the affirmation and differentiation of territories and the ease of communication, learning and integration as a means of acquiring new ideas and values (Almeida 2018: 431).

The direct contact with Identities and the process of learning resulting from this contact are such very important factors of decision-making concerning tourist trips that it will be essential to promote the understanding of this relational process, managing it from the perspective of the common good. The growing appreciation of Identity in tourism raises a better understanding of the concept, a need that drives us toward its multidimensional understanding and to the countless platforms that use it as a tool capable of providing mutual advantages for tourists and local communities.

"By providing tourism-related job opportunities, residents may acquire revenue from tourism. Furthermore, managers should apply recreational opportunity spectrum theory to plan trails for tourists to enable them to appreciate the natural and cultural resources" (Lee and Jan 2019: 378).

Nowadays, tourists are searching for more interactive and fulfilling experiences rather than just being served by the travel industry (Tan, Luh and Kung 2014: 248). The growing appreciation of local specificities, differences and icons has become a global structural reality of tourism systems. In this sense, it is naturally the position of the World Tourism Organization (WTO), which values low density tourism as a sustainable mechanism of response to the seasonality and overtourism that is generally observed in mass tourism. Memorable Tourism Experiences (MTEs) for tourists and collective self-esteem and development for local communities is an interesting and appropriate concept.

Intensifying factors of the tourism experience – building convergence between local communities and tourists

We face an open process that ends up being a complex challenge, as there are innumerable barriers of various kinds within the scope of what is a major challenge in world tourism: the differentiation of the intangible. This has become increasingly complex over the past few years and destinations have been rethinking policies and strategies.

Tourism has assumed pedagogical and emancipatory roles in community contexts associated with non-formal and informal education. Numerous examples illustrate local practices that, in addition to providing their own income, implement the symbolic appropriation of reality and, consequently, promote the collective self-esteem of these same communities.

In the definition of tourism products, the intrinsic qualities of the community, including its mentality, end up being considered at the same level as the resources, as this prior knowledge is needed to adjust the mechanisms of local involvement and the symbolic appropriation of reality. Therefore, in addition to providing MTEs in destination areas, destination managers should successfully transfer the destination attributes of MTEs into effective marketing and advertising strategies. For example, different local tourism programs, a variety of activities, any remarkable tourism superstructures, and the hospitality of local people should be appropriately projected in commercials (Kim 2014: 43).

According to Coelho, Gosling and Almeida (2018: 17), there are at least three processes that contribute to the memorability of the visitor experience, namely: 1) Ambiance, 2) Socialization and 3) Emotion and Reflection. Also, visitors have to (re)signify their experience relative to their life experiences in order to evoke memories and, sometimes, the personal transformation that comes from a MTE. This last process occurs after the trip and can evoke either "signification", when it's the first time it occurs to the tourist, or "resignification", when the tourist already has a previous link to moments, signs and signals that are reconsidered within the tourist experience.

MTEs are not limited to a single type of travel purpose or motivation. Leisure trips, business trips, exchange programs, ordinary daily life experiences, pedagogical or educational trips and health trips can all activate long-lasting memories for tourists. This statement helps in understanding that it is not the type of experience itself that makes a trip memorable or not, but how it represents a meaningful moment in the consumer's life (Coelho, Gosling and Almeida 2018: 18).

The systemic cohesion of destinations leads us to structural and complex issues such as mentalities and the propensity for cooperation and articulation of efforts in the development of tourism products. It is therefore important to take into account social mechanisms that can be triggered in tourism systems, providing new collective attitudes toward an Identity Assumption which can, under certain conditions, result in sustainable tourism and differentiated offers, thus reflecting the concentrated diversity of the territory (Almeida 2010: 40).

The"Mouraria" tourism experience

In Xavier and Almeida (2017) the authors focus on the relationship between local urban rehabilitation policies and the touristic projection of rehabilitated spaces, such as "Mouraria" in Lisbon. The starting point has as its main actors the socio-economic components of this neighborhood: residents, traders, service providers, etc., as well as the institutions and associations that represent them. This research aimed to understand the perspective of the touristic offer and its relationship to the local community and, given the qualitative nature of the information required, this research was performed under comparative and exploratory methods.The "Mouraria" neighborhood is located on the western and northern slopes of the hill on which stands the castle of Lisbon, Portugal, as shown in Figure 40.1. Its history is naturally linked to the history of the city itself and by now it encompasses a vast area marked by a common Identity, whether at the level of architecture or at the level of the people who still inhabit it.

Figure 40.1 Mouraria, Lisbon
Source: António Sérgio Araújo de Almeida

The history of "Mouraria" underlines the importance of maintaining the original local Identity in the specificity of the touristic experience that this type of neighborhood and community promotes. In contrast, another neighborhood in Lisbon, Bairro Alto, as the result of several changes, real estate speculation and abandonment of the area by the local community, gave way to an Identity disorder. It then became a place of evening entertainment, with the local environment being dominated by alcohol and music-based revelry, all of which has nothing to do with the Lisbon of the past.

According to information provided by the City Council of Lisbon and described in this case study, until the eighteenth century "Mouraria" saw the birth of a series of convents, churches and palaces (construction of the Colégios dos Meninos Órfãos e de Sto. Antão o Velho, Mosteiro da Rosa, Ermida de S. Sebastião, Palácio do Marquês de Alegrete, Palácio do Marquês de Tancos e Palácio da Rosa) that demonstrated the commitment of the clergy and nobility to the propogation of religious life. However, the urban network of this area of the city maintains its irregular and spontaneous nature.

Between the end of the nineteenth century and the early twentieth century, Lisbon clearly assumed two axes of communication with the center. On one side, the noble and modern axis of the new avenues linking the Rossio to Campo Grande, and on the other side Avenida Dona Amélia (current Avenida Almirante Reis), linking Areeiro to the Martim Moniz/"Mouraria" entrance to the *saloia* region in Lisbon. "Mouraria" continued to maintain its popular nature, being seen by many as a zone of bohemian marginality, crime and "perdition" and by others as a neighborhood where restaurants, cafes, taverns and botequins were places of conviviality, fundamental to the development of social relations within the neighborhood and the preservation of the collective imaginary of Lisbon – made up of stories of *Fado* and of *Fadistas*, of saints and processions, of games and *Ginjinha*, of fights, of alleys and streets.

Still, according to the City Council, during the twentieth century the most unrefined area of the neighborhood survived, albeit being almost in ruins, in spite of two urbanization projects that proposed the demolition of most of "Mouraria". If, on one hand, this inconsistency of urban rehabilitation plans was dramatic for "Mouraria" and its social fabric, on the other hand it kept it safe from real estate private interests. "Mouraria" "stopped in time", which allowed it to maintain affordable urban lease levels. Other historical neighborhoods, such as "Alfama" and "Bairro Alto" had a different destiny. The suffocating private interests increased the speculative value of space and made it unmanageable for locals and small businesses to maintain their way of life. Therefore a scenario was created for tourists that has lasted to this day, causing residents to increasingly move to other less touristified neighborhoods. Thus, "Mouraria" remained unchanged, far from the adverse speculation of the real estate market, and has been receiving migrants since the nineteenth century.

Until the end of the 1970s, residents of the "Mouraria" neighborhood were rural and Portuguese, but during the last few decades we have witnessed the arrival of many immigrants from all over the world, namely Asians, South Americans, Eastern Europeans and Africans. This phenomenon not only contributed to the cultural enrichment of "Mouraria", making its Rossio (Martim Moniz) a center of multicultural trade, but it also preserved the tradition, saving its residents from the gentrification that could culturally discharacterize the neighborhood. "Mouraria" also remained unharmed after the Portuguese revolution of 25 April 1974, which marked the turning point from the authoritarian regime to democracy. Since then, there have been only one or two episodes of urban rehabilitation of monumental heritage.

This typical neighborhood has thus maintained the essential nature of its Identity, including community practices. During visits made to "Mouraria", there is spontaneous interaction between tourists and residents. Locals are proud to show off their neighborhood and easily offer a little *ginjinha* to tourists who interact with them. The extension of the most intimate family practices to the public highway, such as the popular singing of *Fado*, the smell of fried fish and popular dances, reflects the valorization of these Identity attributes which, under certain conditions and with economic advantages, contributes to the promotion of local self-esteem. This interaction around distinct value systems, respectively of tourists and the local community, promotes an atmosphere of mutual experiential enrichment from the moment the local community feels that its culture and way of being is valued by local policies.

Xavier and Almeida (2017: 328) argued that hospitality itself is increasingly viewed in a transversal and composite way, incorporating a sum of experiences, whether in the field of gastronomy, in the ambit of touristic recreation or in the interaction with the local community. This question inevitably influences the capacity of tourist retention and the very image of destiny.

Identity interaction becomes facilitated from the moment the local community feels its values are protected and valued. In these circumstances there are conditions for the promotion of a collective self-esteem that becomes a stimulus for sharing with tourists who, in turn, are more likely to consume authentically objective services and products (Wang 1999).

An urban rehabilitation project started in 2011 and was, according to the current Prime Minister of Portugal, António Costa, "a project of people and for the people of the neighbourhood". He intended not only to rehabilitate the physical space, but also to revitalize the social fabric and, using tourism as a development tool, to promote "Mouraria" as a neighborhood of "*Lisboetas*", in other words, the people born in Lisbon (Xavier and Almeida 2017: 346).

The solemnities of the holy week in Braga

According to Almeida and Pinto (2017), there are various specific attributes that are successively mentioned by tourists in interviews in the context of the Braga solemnities that deserve due importance. These are the feelings of integration into a community, sharing its transcendent values, the contrast provided by the abandonment of the social lifestyle and the entrance into a kind of a sacred reality and the interaction with secular cultural systems. This was reported by one of the authors, as he had experienced the aforementioned solemnity while he was a professional journalist for over a decade.

Braga, with about 140,000 inhabitants and an area of 183.5 square kilometres, is the oldest city in Portugal, having marked in the year 2000 the same age of existence. It is located in the north of the country. From a tourist-administrative point of view, it is considered the regional Tourism Entity of Oporto and Northern Portugal. "City of the Archbishops" or "Portuguese Rome" are the nicknames of this historical city, reflecting its history, its system of values around religiosity and its status as an archdiocese. The Holy Week celebrations in Braga, demonstrated in Figure 40.2, annually record about 100,000 visitors and according to official data from the Tourism Entity of Oporto and Northern Portugal, 51.6% of the tourists remained, on average, for two to three days in Braga and 23% between four and five days. Since 2014, hotel occupancy rates vary between 87% and 93%. According to the same source, the nationalities who visit the most are Spanish (39.73%), Portuguese (29.99%) and French (7.58%). Regarding the various activities, the religious processions attract the major percentage of tourists (71.3%), followed by the churches with 66.8%, the cultural heritage with 56.7% and the museums with 29.5%.

During Holy Week, everything changes in the historic center of Braga. It is a time marked by a kind of collective theatrical representation, in which the sacred community does not leave anyone indifferent who witnesses it. Almost everyone participates, regardless of their differences and convictions. According to Almeida and Pinto (2017), an intimate identitary side emerges as

Figure 40.2 Holyweek, Braga
Source: Hugo Delgado

an IFTE. The unique aura of solemnity characterized by contrasts is the major attribute highlighted by visitors. Doing justice to its historical and secular nicknames, such as the "Portuguese Rome" or "City of the Archbishops", the fact is that most of Braga's population is openly religious and there is a social involvement, both within the organization of the archdiocese and the various living forces of the city, including the confraternities and the Commercial Association of Braga. In this sense, it is easy to understand the values of authenticity transmitted in the context of the tourist experience of the Holy Week in Braga.

Regarding the solemnities, there is a relational reciprocity between the visitors and the ones who are visited, largely based on mutual respect for the experiences of the two parties which produces equal dividends for both sets of people. Because they enjoy a unique scenario of collective solemnity, they can be assumed to be experiencing a space of existential authenticity (Wang 1999). The local people feel their Identity valued, both in socio-cultural terms and in the context of obtaining economic dividends, as seen in the rates of hotel occupancy and in the areas of consumption as a whole (Almeida and Pinto 2017: 624). The prolonged contact between tourists and the most intimate practices, such as prayers, meditation and the sharing of collective spirituality, is a contribution to a process of integration and learning, regardless of one's religious or spiritual convictions. A convergence of mutual advantages becomes evident, based both on the local Identity assumption and on the search for emotional and/or intellectual, alternative and sacred states by the visitors. A balance is generated based on mutual appreciation of the experience of religiosity and the act of faith and those who witness these same acts are imbued with the solemnity that they then extend to the visited community.

The sacred nature, the local community's predisposition and the organization of the rituality around this event impose a formal rhythm and agenda, materialized mostly in a solemn and intensely lived program, also by the "Bracarense" society. This fact leads to a conditioning of the global experience of those who visit Braga during the solemnities.

The present case study can help to determine that the touristic valuation of attributes such as silence and darkness preserved by the absence of artificial lights during the sacred moments are important in order to derive a feeling of sharing and spiritual integration.

As stated by Almeida and Pinto (2017: 617), a strategic logic of convergence of mutual advantages is reached, bringing together local communities and tourists and providing economic and social benefits and experiences that are unique to everyone. The tourist agents face the double challenge that consists in the diagnosis of community mechanisms of involvement and in the diagnosis of IFTE, so that they can maintain the primacy of the tourist experience.

If, on one hand, there is an imaginary, a pre-experience and a touristic predisposition within the solemnity, it is verified that, during the experience itself, there are emotional impacts arising from the collective solemnity that change those feelings and the tourists begin to assume other meanings of inner growth, until the moment of realization that is not foreseen (Almeida and Pinto 2017: 623).

Conclusions

In an international system marked by transnationalism and cosmopolitanism, Identity has more than ever come to be seen as a dynamic and open concept. This work assumed that global tourism trends affect the observation of Identity while it remains a strategic concept. The diagnosis and management of Identity attributes, while an IFTE, reinforces the metamorphic nature of Identity, which is becoming more dynamic than ever before. This point of view, under certain conditions of observation of principles of cultural, social, economic and environmental appreciation, can contribute to the operationalization of Identity as a sustainability tool. Investigations carried out in the context of the typical Lisbon neighborhood of "Mouraria" and the solemnities of the Holy Week in Braga enhance the usufruct of objective authenticity (Wang 1999) the history and stories, popular music (such as *Fado*), traditions and popular knowledge, community dynamics and local aptitudes for involvement, interaction, sharing Identity and integration as IFTE.

By participating in the solemnities of the Holy Week in Braga, tourists are allowed to incorporate themselves into what is seen as the soul of the ancient and sacred Portuguese Identity. The interaction and development of a more intimate relationship with the community promotes an understanding and an Identity integration. The management of the factors that promote this integration is thus crucial. The destination faces a challenge as it has to use its local and international positioning to reinforce the emotional impact felt by tourists and then encourage their integration by introducing them to a certain way of learning, by way of doing something different.

Tourists are initially attracted by a destination's attributes, which leads them to engage in tourism-related activities and/or travel within the destination. The attributes then become experiential components, which affect the formation of visitors' experiences (Kim 2014: 34).

The concept of sustainability of the tourist destination, observing the cultural, social, economic and ecological/spatial realities, desirably provokes an appropriation of sustainable practices by the communities, which will tend to perpetuate them if advantageous results are obtained. On the other hand, and according to Nunkoo and Gursoy (2012: 264), researchers should also understand that support for tourism is not only influenced by residents' attitudes to tourism impacts, but also by more complex factors such as their Identities. Findings suggest that Identities serve as a guide for behaviors, which serve to verify and maintain the self-meanings associated with these same Identities.

A tourist product should enhance identity attributes that facilitate its enjoyment and understanding. If the tastes and smells in wine tourism are the attributes that are underlined, reinforcing multisensory perception, in the case of religious tourism it is important to emphasize the contemplation and integration of the sacred into the experience. Each product should be managed in a thorough manner.

As a subject for future investigations, it will be interesting to understand how the will for Identity involvement, provided by tourism, could become a changing agent that works on the Identity itself. Thus, Identity and tourism complement each other as agents of change.

References

Almeida, ASA. (2010). 'The inter-community competition as a factor for sustainability and differentiation of tourism product – the case of National Park of Peneda Gerês', *European Journal of Tourism, Hospitality and Recreation*, 1(1): 39–50.

Almeida, ASA. (2018). 'Building tourism experience through Identity decomposition – between local emancipation and international integration', *Lusophone Journal of Cultural Studies*, 5(2): 409–425.

Almeida, ASA. and Pinto, R. (2017). 'Religiosity and tourism: the primacy of experience', *Boletim Do Museu Paraense Emílio Goeldi. Ciências Humanas*, 12(2): 615–627. DOI: 10.1590/1981.81222017000200019

Bauman, Z. (2005). *Identidade*. Rio de Janeiro: Jorge Zahaer.

Castells, M. (1999). *O poder da identidade*. São Paulo: Paz e Terra.

Coelho, M., Gosling, M., and Almeida, ASA. (2018). 'Tourism experiences: core processes of memorable trips', *Journal of Hospitality and Tourism Management*, 37: 11–22. DOI: 10.1016/j.jhtm.2018.08.004

Kim, J-H. (2014). 'The antecedents of memorable tourism experiences: the development of a scale to measure the destination attributes associated with memorable experiences', *Tourism Management*, 44: 34–45. DOI: 10.1016/j.tourman.2014.02.007

Lee, T-H. and Jan, F-H. (2019). 'Can community-based tourism contribute to sustainable development? Evidence from residents' perceptions of the sustainability', *Tourism Management*, 70: 368–380. DOI: 10.1016/j.tourman.2018.09.003

Malheiros, A., Lourenço, PA., and Almeida, ASA. (2016). 'Some contributions to the assumptive identity of Portugal's Jewish heritage – the case of Belmonte', *Revista Portuguesa de Estudos Regionais*, 43: 69–88.

Nunkoo, R. and Gursoy, D. (2012). 'Resident'ssupport for tourism - anIdentityperspective', *Annals of Tourism Research*, 39(1): 243–268. DOI: 10.1016/j.annals.2011.05.006

Tan, SK., Luh, D-B., and Kung, S-F. (2014). 'A taxonomyofcreativetourists in creativetourism', *Tourism Management*, 42: 248–259. DOI: 10.1016/j.tourman.2013.11.008

Wang, N. (1999). 'Rethinking authenticity in tourism experience', *Annals of Tourism Research*, 26(2): 349–370. DOI: 10.1016/S0160-7383(98)00103-0

Xavier, A. and Almeida, ASA. (2017). 'Políticas locais e impactos na conceção da experiência turística em espaços urbanos – os casos do Bairro Alto e da "Mouraria", Lisboa', *Revista Brasileira de Gestão e Desenvolvimento Regional*, 13(2): 322–350.

41

Developing rural tourism through co-creation of sustainable tourist experiences

Jennifer Chan Kim Lian

Introduction

The chapter contributes to the understanding of rural tourism (RT) in Kota Belud, Sabah, Malaysia. First, it reviews rural tourism experiences as experiential products; then it illuminates the importance of co-creation experiences to achieve sustainable tourism experiences (STEs) within the rural tourism context; and finally, it relates co-creation experiences to the four realms of experiences by Pine and Gilmore (1999). Next, the meaning and attractions of rural tourism are explored and the co-creation of rural tourism experiences within the rural tourism experience network (RTEN) is proposed. It suggests sustainable rural tourism experiences (SRTE) can be achieved through co-creation experiences by focusing on active participation and a high level of interaction between the tourist and host communities/physical environment. The co-creation of rural tourism experiences adds value to RT products and services and embraces innovative and creative ways to provide meaningful and memorable experiences. Hence, rural communities and hosts can add value and introduce new ways of enabling themselves to provide unique, authentic and memorable rural tourism experiences (RTEs) to tourists while ensuring sustainable rural development. The chapter concludes with the importance of delivering distinctive, authentic and unforgettable RTEs through the co-creation of STEs to ensure sustainable rural development, and examines issues faced in sustainable rural tourism development in Kota Belud, Sabah.

Rural tourism in Sabah, Malaysia

Sabah, Malaysia is located on the island of Borneo and the majority of Sabah territory can be categorized as rural areas (Figure 41.1). Sabah is blessed with an attractive geographical landscape and many rural sites with diverse natural resources. Along with its natural beauty, Sabah is also home to various cultural heritage attractions and hospitable and friendly people. Indeed, many rural sites are equipped with these rural tourism elements and these sites have great potential to be promoted as unique, attractive and authentic tourism products and places.

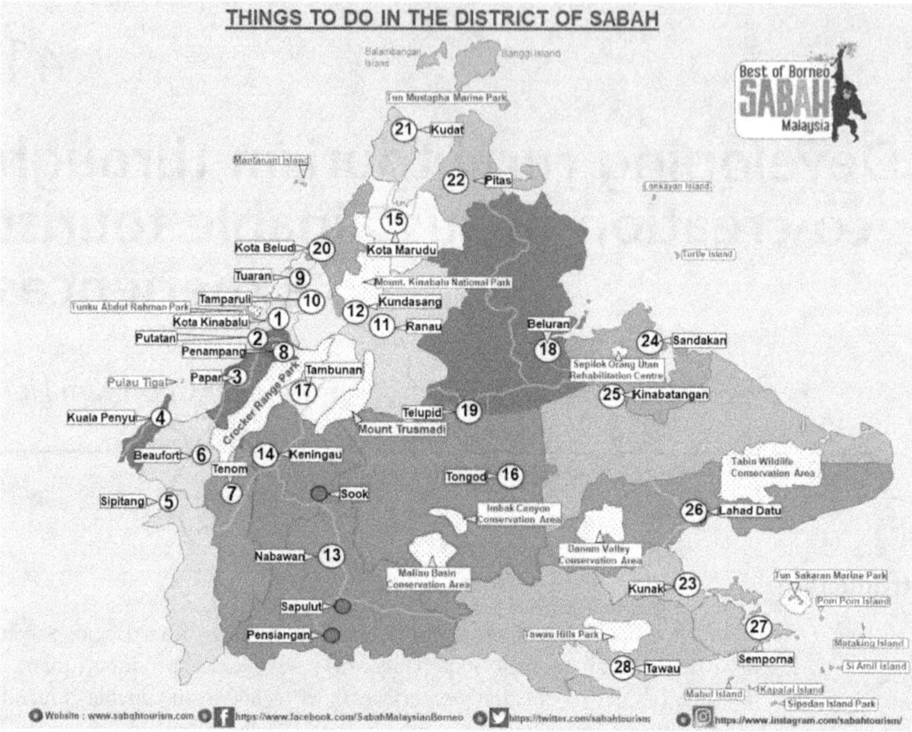

Figure 41.1 Rural tourism sites in Sabah
Source: Sabah Tourism Board

Figure 41.1 presents 28 key rural sites that can be developed as rural tourism destinations. Each has their own unique resources ranging from rural landscapes including mountains, forests and rivers to agriculture, rich cultural heritage, food, home stays and farm stays that provide authentic and memorable rural experiences. Kiulu in the Tuaran parliamentary (No 10) constituency and Kadamaian in the Kota Belud parliamentary constituency (No 20) (Figure 41.1) were then chosen by Sabah Tourism Board management as two pioneer districts for RT development. The efforts made to introduce and develop rural tourism have significantly impacted on the influx of both local and international tourist arrivals and thus generated substantial incomes and employment for the villagers. This has motivated villagers from other districts such as Tambunan (No 17), Kota Marudu (No 15), Tenom (No 7), Keningau (No 14), Kudat (No 21) and Ranau (No 11) to follow in their footsteps.

As RT is becoming a vital and fast-growing sector in Sabah, creating unique rural tourism experiences and generating benefits to the local community is crucial to ensure competitiveness and sustainable growth. Sustainable rural tourism accounts for the abundant resources and unique experiences gained from the interaction or encounters with the hosts and the environment. A greater focus on RTE and the co-creation of sustainable experiences in rural destinations is essential but progress is still limited (Kastenholz, João Carneiro, Peixeira Marques and Correia Loureiro 2018). Although RT has been widely researched from different perspectives, little is known about rural tourism's meaning and SRTE through co-creation, especially in Sabah, Malaysia.

An overview of tourism experiences

Experience has been long seen as central to the tourism product and services. Indeed, experiences are intangible and immaterial and considered to be of great value to tourists as they are subjective, individualistic and memorable. Accordingly, experience could be described as "the subjective mental state felt by participants during a service encounter" (Otto and Ritchie 1996: 166). Similarly, tourism experience can be termed as a psychological occurrence with processes that integrate into different stages and involve the elements of expectations, perceptions and memory (Larsen 2007). Schmitt (1999) states that tourist experiences can be structured into five values: SENSE (sensory), FEEL (emotional), THINK (cognitive), ACT (behavioral) and RELATE (relational), which offer sensory, emotional, cognitive, behavioral and relational values that are reflected in the experience sequence. The experience is thus formed based on perception through the cognitive and emotive aspect of the tourist during a specific encounter, event or engagement. Arguably, experiences in the context of tourism should be regarded as crucial components of prosperous, innovative and competitive tourism supply (Ellis and Rossman 2008). Rural tourists consume a series of experiences derived from the tangible and intangible elements present in the tourism experiences environment and network using their five senses. They seek fascinating, novel and remarkable experiences in destination contexts. There is a necessity to develop fascinating and unique offerings for a challenging and heterogeneous rural tourist market to thrive in the competitive international context (Lane 2009). Rural sites have unique resources, which, if properly managed, can create various attractive, authentic, novel and notable experiences for tourists (Lane and Kastenholz 2015).

Rural tourism as an experiential product

RT provides an array of experiences and services to tourists. The pursuit of distinctive and unforgettable experience settings is the driving force of RT. Such unique and memorable experiences are ascribed to the rural site in respect of its rural heritage and life, activities and territory as presented in Figure 41.2 below. It is an experience-based tourism that is related to a range of rural resources located in rural settings or areas – such as visiting villages, seeing local architecture, hiking, fishing, horse-riding and participating in local festivals and traditions – and territories including mountains, rivers and other natural scenery. Rural sites offer rich and unique resources that provide multiple appealing, "authentic," and memorable experiences (Lane and Kastenholz 2015). Rural activities to some are quite complex and multifaceted (Clemenson and Lane 1997). Therefore, a wide range of on-site rural experiences are "experiential" products enabling feelings, emotions and knowledge for visitors; these experiences are seen as useful ways to add value and make RT more competitive. The value or attractiveness of its products or services is enhanced through a specific experience, and creating memorable experiences is essential. This implies that RTEs are both multifaceted and co-created between the host or the environment and visitors/tourists who have engaged themselves in a specific service transaction. RT suppliers or the hosts become "experience stagers" and offer an experience that involves tourists in a personal and memorable way (Pine and Gilmore 1999: 3–4).

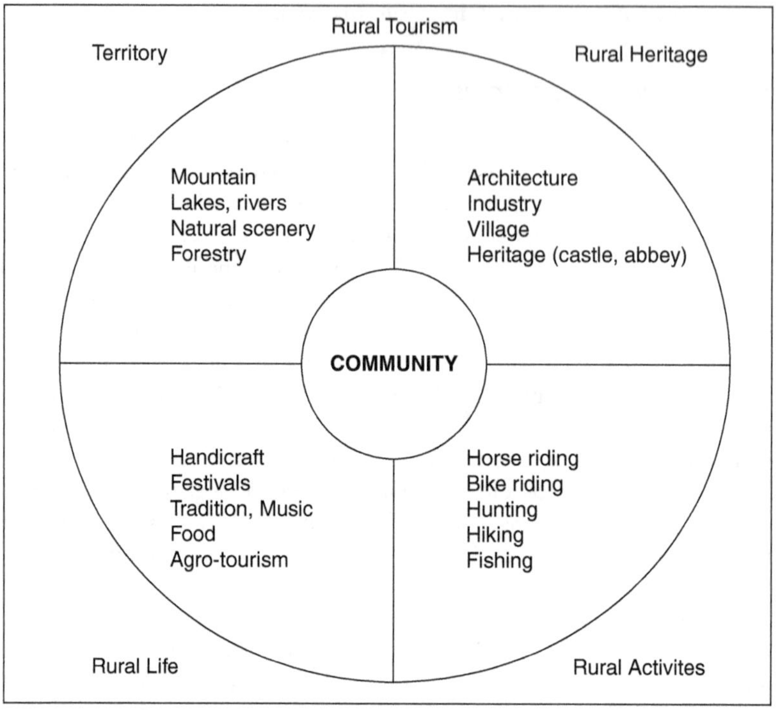

Figure 41.2 Rural tourism resources

Importance of sustainable rural tourism experiences

Sustainable tourism implies achieving an equilibrium between environmental, economic and socio-cultural factors, as stated in the triple bottom line (Elkington 1998). It includes activities, supervision and growth of tourism that maintain natural, economic and social integrity and safeguard natural, environmental and cultural resources while ensuring tourists' satisfaction through meaningful experiences and promoting sustainable tourism practices (Niedziolka 2012). The sustainable tourism model by Hall (1998) indicates the three important pillars of sustainability: economic, environmental and social; and this model serves as a useful guide to understand the co-creation of sustainable rural tourism (SRT). The close connection and interrelation of the three pillars of sustainability constructed on the local community integrate economy and environment and conservation with equity for the social sphere in achieving sustainable development. This model is relevant to the RTEs that are attributed to the economic, environmental and social aspects of the rural community; and is pertinent to the rural tourism experiences in developing quality experiences, adding value to rural tourism and enhancing the rural community's wellbeing.

RT products and experiences are associated with environmental, natural, cultural and social parameters within that site. Rural tourism attractions are presented in Figure 41.2 above. These attractions create a unique rural tourism experience. Arguably, making rural tourism sustainable requires considering environmental, economic and social aspects of the rural community, and these are related to the triple bottom sustainability model (Figure 41.3). For example, the unspoilt and pristine natural environment acts as a unique attraction and contributes to the developing authenticity of the rural setting. It is vital to preserve and conserve landscapes, natural resources and the environment. Concerning the economy, numerous micro-businesses such as home stays, farmhouse stays and handicrafts/local food and leisure activities generate employment for these communities. RT promotes and stimulates the presence of small businesses, thus providing opportunities to generate extra income while maintaining the authenticity of resources and preserving the rural lifestyle, values and traditions. Subsequently, it enhances the wellbeing and wealth of these communities. In the social aspect, rural tourism experiences comprise rural heritage and activities (visiting villages and local architecture, hiking, fishing and participating in local festivals and traditional events). These activities contribute to a unique RTE. Sustainable RT can significantly contribute to local economic development, revitalize rural communities, reduce migration to cities, create entrepreneurial opportunities, positively affect the preservation of local culture, tradition and heritage and increase demand for local products. Hence, economic, environmental, political and social pillars form key components of RT development and influence the RTEs. Similarly, RT offers an alternative way of preserving nature, landscapes and environment and improving the wealth and wellbeing of local communities. The three pillars (economy, environment and social) are part of rural tourism network experiences that are simultaneously co-created within RTEs and which lead to value addition and making rural sites or tourism businesses more competitive. Co-creation among the three pillars and tourists and hosts in developing sustainable rural tourism experiences cannot be ignored. Accordingly, "Sustainable tourism is the only type of tourism that can offer authentic countryside life-style, where they can relax themselves and enjoy nature and country lifestyle" (Sanagustín Fons, Fierro and Patiño 2011: 557). This implies the importance of SRT and the need for further research to obtain an insightful comprehension of how sustainable rural tourism experiences through co-creation can be realized.

Review of rural tourism experience network and co-creation of rural tourist experiences

RT by nature is a service and experience-oriented sector; and is based on people, space and products; and the rural setting, space, people and products co-exist and form important elements of rural tourism (Dragulanescu and Drutu 2012). It requires the presence of both hosts and tourists to interact, engage, connect and participate within the RTEN. The tourists interact, participate, connect and engage with local hosts and rural resources and gain experiences. This experience network is one in which a variety of stakeholders co-create tourism experiences that serve as a holistic rural tourism of stakeholders connected in an experiences environment where everyone operates in a different time-spatial context. This network enables tourists to immerse and surround themselves with the tangible and intangible elements within the environment. RTE is attributed to co-creation by tourists and hosts within that RTEN, who comprise the various key players that are interlinked (Figure 41.3). Such an experience network is referred to as a "co-creation network," and consists of all stakeholders and things that surround tourists within an experience environment; and involves stakeholders in the creation of tourism experiences (Binkhorst 2005: 3)

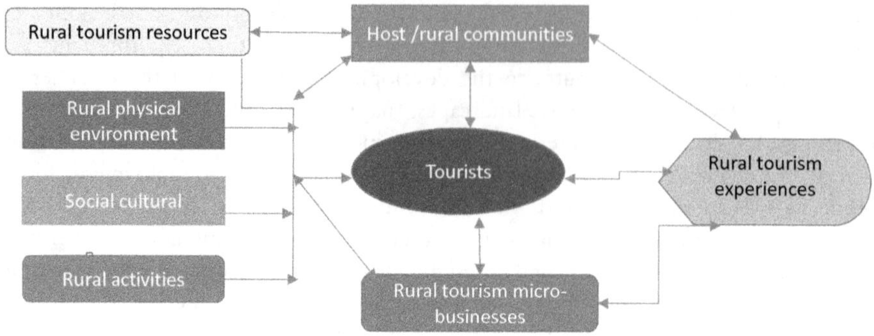

Figure 41.3 Proposed rural tourism experience network

Based on Figure 41.3, RT is an experience involving interaction and co-creation between host and tourists and other rural tourism resources/facilities within RTEN. It is well acknowledged that tourists are central to the experience context, and the space, time and involvement of the individual tourists influences and shapes their respective experiences. Evidently, the rural experience network consists of various stakeholders who co-create and participate in tourism and add value to both hosts and tourists; and thus, this creates the distinctiveness and genuineness of the tourism destination in rural sites. The host environment/rural tourism suppliers and tourists are key players who co-create the experiences in the process of developing a rural tourism experiences environment. RT suppliers or hosts can innovate experience environments for co-creating new experiences and adding values. Hence, the concept of co-creation brings benefits of unique, authentic, memorable experiences, adding value to rural tourism. Likewise, there is a growing interest in co-creation in the tourism context (Campos, Mendes, Oom do Valle and Scott 2015).

Above all, it is postulated that co-creation experiences become the foundation for value creation. The co-creation of experiences encourages high-quality interactions that allow an individual to co-create inimitable experiences, thus offer a competitive advantage to the RT suppliers as both the suppliers and tourists jointly create the value. The need for economic growth has driven RT development, which enhances the wellbeing of local communities, leading to poverty alleviation. This way, value creation and sustainability of the natural resources and environment become critical concerns for RT development.

The co-creation of RT experiences is derived from the individual tourist when participating in activities and interacting with subjects over the duration of their trip in a particular experience environment (Larsen 2007). "A co-creation tourism experience is the sum of psychological events a tourist goes through when contributing actively through physical and or mental participation in activities and interacting with other subjects in the experience environment." (Campos, Mendes, Oom do Valle and Scott 2015: 23). RTE co-creation breaks down the process through which tourists create their own interactions within the experiences network. Tourists in rural sites make a connection and achieve interaction with the rural community and resources to seek experiences. There are four different types of rural tourism experiences co-created within the rural tourism experience network, which can be linked to the four realms of experiences by Pine and Gilmore (1999), namely education, escapism, aesthetics and entertainment; in accordance with these needs, different levels of participation and absorption versus the immersion of tourists is required, thus separating tourists from tourism suppliers. Active participation and interaction in co-creation experiences and on-site experiences is vital as tourists connect in diverse manners through emotional, cognitive, physical and social means and different levels of intensity. In this context, experiences formed and gained are attributed to the presence of hosts, tourists and the environment within RTEN, involving dimensions such as the physical aspects of the environment, social actors, participants, host dynamic and features of the services, which ultimately affects tourists' lives or on-site experiences (Campos, Mendes, Oom do Valle and Scott 2015).

Similarly, the involvement of tourists in rural activities to meet their expectations is discovered to be significant for co-creation (Andrades and Dimanche 2014), as the natural rural environment is considered as a space to stage the unique experiences in different processes in on-site experiences (Mathisen 2013). Thus, the system to engage in co-creation must be in place to enable tourists to partake in the design, production and consumption of experiences or collaborate in the creation of new rural tourism products and services (Eraqi 2011). This implies that tourists must play a part in certain aspects, phases or the whole tourism experience and actively participate in on-site activities as activities are part of tourism experiences (Morgan 2007), and experiences arise from activities that involve their active participation. Such active participation may be in the form of physical, emotional or mental engagement (Bertella 2014). Besides, encouraging tourists' interactions among themselves during on-site experiences is vital, as on-site co-creation experiences require engaging the individual tourist physically, emotionally and intellectually. Accordingly, interactions during travel are sources of experiences (Minkiewicz et al. 2013) and impact tourists' level of engagement in experiences. Active participation and interaction in co-creation experiences are important elements in on-site experiences. Co-creation is a consumer's experiences of specific activities that require active participation and interactive experiences (Cabiddu 2013); and experiences co-created form the basis of a unique value for each individual. Elsewhere, Prahalad and Ramaswamy (2004: 14) state that "co-creation experiences" form the basis for value addition and future innovation. The importance of co-creation in enhancing innovation in the business environment can become an innovative perspective in tourism (Binkhorst 2005).

The co-creation of RTEs can be guided and explained by the four realms of experiences (Pine and Gilmore 1998), resulting from the two dimensions of perception, namely level of participation – active and passive – and four types of experiences, namely (i) entertainment experiences – with passive participation, while absorbed by the environment; (ii) aesthetic experiences – passive participation but being immersed in the environment; (iii) educational experiences – one actively participates in activities being observed; and (iv) escapist experiences – characterised by active participation and immersion in the environment. Based on these experiences and the level of participation, the co-creation experiences' value is created, resulting from the interaction and connection of an individual at a specific place and time. The suppliers/hosts can then focus on that specific context using the environment network to develop co-creation experience value.

The following section presents a finding that is part of an extensive study on rural tourism in Sabah

A study was conducted in Kota Belud District, comprising 23 interviews with rural tourists and eight hosts located at two villages (Nahandang and Tegudon) from December 2018 to March 2019. It aimed to explore the meaning of rural tourism, rural tourism attractions and sustainable rural tourism experiences.

Findings on meaning and attractions of rural tourism in Kota Belud, Sabah, Malaysia

The responses from the rural community reveal four meanings associated with rural tourism, which can be categorized as place, source of income, collaboration and activities (Table 41.1).

Interestingly, rural community relates to RT with a place in the village with natural and local heritage, which offers relaxation, recreation and activities. RT takes place in a countryside or rural area where nature and cultural heritage can be experienced for recreational purposes. This implies that rural tourism is considered a unique tourist destination that offers a range of experiences to tourists away from the urban area. Rural resources are deemed important to the rural community in providing RTEs. Others related rural tourism to a source of income and a collaboration opportunity with others to enhance their living standards. In this context, RT is considered as an alternative means for creating extra income and helping the rural community to increase their living standards, which is consistent with Xue and Kerstetter's (2018) and Dragulanescu and Drutu's (2012) studies. Besides, RT is seen as a form of activity provided to tourists in rural areas, which are administered and owned by the locals.

In short, the meaning of RT to the rural community is consistent with the literature review. However, there is no mention of experience or issues related to sustainability. RT is generally seen as a tool to reduce poverty or create an alternative way to increase income. The locals tend to focus on their village or rural site as an important place of attraction, as well as the activities provided.

RT provides an array of experiences and is widely acknowledged as experience-based tourism that comprises a range of rural resources located in the rural settings, as evidenced from tourists' responses that the meaning of RT can be categorized into experience, place and nature. Significant numbers of tourists interpret rural tourism as "going to a place away from the city, a place to relax, be peaceful, enjoy nature" (R2,5,7,9,10,14,19,200). Other tourists associate RT with experiences and nature (R1,8,11,13,16,170). It seems that the elements of rural place, experience and nature are vital for rural tourism in Kota Belud. The findings widely support the

Table 41.1 Meaning of rural tourism from rural communities' responses

	Response and respondent
1 Place	A place for the release of tension (R1) A place for recreation and gathering (R1) Rural tourism is defined as having a place located in the village that accommodates guests. The place itself displays local heritage and natural resources. (R3) Rural tourism is a place located in the sub-urban countryside that allows visitors to experience nature and the cultural heritage of the said place (R5) Tourism activities that operate in the rural area (R8)
2 Source of income	The source of income for the local community (R1) A continuous collaboration intended to increase the living standards of the local community (R2) Increase the income of the community (R6) Gives chances to the local community to have a source of income (R7)
3. Collaboration	A continuous collaboration intended to increase the living standards of the local community (R2)
4. Activities	Tourism activities that operate in the rural area (R8) Tourism activities owned by the local community (R8)

conclusion that rural sites/places and experiences are important in the context of RT. The place should provide rich and unique resources and numerous alluring, authentic and unforgettable experiences to the tourists (Lane and Kastenholz 2015). In short, it is interesting to note that both rural communities and tourists relate rural tourism to a place; however, the importance of experience was highlighted in tourists' interpretation but not in that of the rural communities. This indicates the gap in the understanding of RT and that rural communities need to consider that the experience dimension is equally important to the place in RT. Simply, RT includes rural attractions and activities, the rural environment/scenery/landscapes, culture and heritage/ lifestyles and leisure facilities; and tourist experiences are derived from the participation in and involvement with these elements. As RT is targeted toward the development of the rural site, a positive attitude of hosts and tourists toward visiting rural sites is vital.

Concerning RT attraction, both the rural community and tourists' responses reveal that a RT site offers a wide range of attractions, namely the landscape in the form of rivers and other scenic features, adventurous jungle trekking, the unique local architectural setting of home stays, and the location of campsites along the rivers, which create a new environment; and they influence and shape the rural tourism experiences. These attractions are categorised and presented in Table 41.2 below.

Both hosts and tourists acknowledge that landscapes and cultural heritage are key attractions that produce unique experiences. The rural sites have a unique landscape feature, which is at the foot of Mt Kinabalu in Kota Belud, along the rivers that offer pristine scenic views. Traditional arts and handicrafts in rural areas and local fabrics, artifacts and other natural products are an additional bonus that boosts RT and changes the economic scenario of a rural settlement.

Concerning the quality of RT, both the rural community and tourists agreed that satisfactory experiences were gained through facilities and the natural environment, which influenced the quality of RT. Many rural communities acknowledge that experience plays an important part in creating quality RT. For the local communities, they perceived the quality of RT to be good when visitors were satisfied with their products and services. To them, giving the right information and providing excellent services and facilities to tourists is an integral part of providing a quality RT experience.

The responses from the rural communities with regards to ways to create SRTEs are shown in Table 41.3 below. Interestingly, teamwork, vision and mission, facilities, community involvement and continuous improvement are ways to sustain rural tourism experiences. Many of the respondents mention facilities and continuous improvement, which – according to them – will enable them to sustain their rural tourism, as these are linked to quality rural experiences. Moreover, RT requires community involvement, teamwork and collaboration among the rural community, besides having a clear vision and mission to determine the direction of rural tourism in their villages. However, the respondents did not consider other relevant key stakeholders and the experience environment within the rural tourism network (Figure 41.3).

Table 41.2 Rural tourism attractions from rural communities' and tourists' responses

Rural territory/landscape	Mountains, rivers, natural scenery and forestry, campsite
Rural life	Handicrafts, festivals, traditions, music, food and agro-tourism
Rural heritage	Architecture, home stays, villages, heritage
Rural activities	Fishing, jungle trekking, bike riding, hunting, hiking, fishing

Table 41.3 Rural communities' responses to create sustainable rural tourism experiences.

Item	*Responses/Respondents*
1 Facilities	Accommodated, they will surely be satisfied with their visit to our village (R5) Good accommodation for the guests – for their satisfaction (R3) Cleanliness of the surroundings – it gives the best impact + feedback if the surroundings are clean (R3) Friendly staff (R4) Be hospitable to the guests – if the guest is well accommodated, they will surely be satisfied with their visit to our village (R5) Good facilities such as room and toilets (R8)
2 Continuous improvement	Give suggestions, feedbacks and comments to further improve this area (R7) Always make improvements and changes without damaging nature (R8) Feedback from visitors/tourists/customers (R2)
3 Community involvement	Unity of the community – if they are united, success will come (R5) Teamwork and unity among tour operators (R7)
4 Teamwork/collaboration/unity	The cooperation from the community must be continuous (R1, R3)
5 Clear vision and mission	The management must be in order (R1, R5) A clear future (R1, R3)

Conclusions and implications

The paper reviews RT and co-creation experiences as an important approach within the context of RT to ensure sustainability. It reports the meaning of RT and attractions, the quality of RT and ways to create SRTEs from the rural community's perspective.

RT in Sabah is growing rapidly and becoming an emerging attractive tourist destination as it offers a diverse range of RTEs. It lets people obtain a serene and calming experience and be close to nature. It is an excellent way of preserving nature, landscape and environment besides enriching human wellbeing in rural areas. However, there are several issues facing SRT development from the study. Human resources in rural areas are generally not well trained to provide quality RT products and services, and most individuals have problems related to an inferiority complex and a lack of confidence in providing tourism services. The study has found that the rural community has limited knowledge and understanding of RT, in particular RTEs and ways to co-create SRTEs. Such understanding is pertinent to RT development as it involves the economy, the environment and social aspects of the rural community, and these pillars together with hosts and tourists co-create quality RTEs. It is important to develop and achieve authentic and memorable experiences and add value and creativity to the rural tourism products; the rural community can't afford to ignore this issue. Likewise, the concept of responsible tourism is closely connected with sustainable development, as it minimizes negative social and environmental impacts. Responsible tourism undertakes to protect the nature, culture, history and heritage of the community.

Acknowledgments

The chapter is part of the research collaboration with Almacrest International College under the research code: GLS0009-2018.

References

Andrades, L., and Dimanche, F. (2014). Co-creation of experience value: A tourist behaviour approach. In M. Chen and J. Uysal (Eds.), *Creating experience value in tourism* (pp. 95–112). London: CABI. DOI: 10.1079/9781780643489.0095

Bertella, G. (2014). The co-creation of animal-based tourism experience. *Tourism Recreation Research*, 39 (1): 115–125. DOI: 10.1080/02508281.2014.11081330

Binkhorst, E. (2005). The co-creation tourism experience. Whitepaper Co-creations, Sitges [online]. Available from: www.esade.edu/cedit2006/pdfs2006/papers/esther_binkhorst_paper_esade_may_06.pdf [Accessed: March 6, 2019].

Cabiddu, F., Lui, T.-W., and Piccoli, G. (2013). Managing value co-creation in the tourism industry. *Annals of Tourism Research*, 42: 86–107. DOI: 10.1016/j.annals.2013.01.001

Campos, A. C., Mendes, J., Oom do Valle, P., and Scott, N. (2015). Co-creation of tourist experiences: A literature review. *Current Issues in Tourism*, 21(4): 1–32. DOI: 10.1080/13683500.2015.1081158

Clemenson, H. A., and Lane, B. (1997). Niche markets, niche marketing and rural employment. In R. D. Bollman and J. M. Bryden (Eds.), *Rural employment: An international perspective* (pp. 410–426). Wallingford: CAB International.

Dragulanescu, I., and Drutu, M. (2012). Rural tourism for local economic development. *International Journal of Academic Research in Accounting, Finance and Management Science*, 2(1), Special issue: 196–203.

Elkington, J. (1998). Partnerships from cannibals with forks: The triple bottom line of 21st-century business. *Environmental Quality Management*, 8(1): 37–51.

Ellis, G. D., and Rossman, J. R. (2008). Creating value for participants through experience staging: Parks, recreation, and tourism in the experience industry. *Journal of Park and Recreation Administration*, 26(4): 1–20.

Eraqi, M. I. (2011). Co-creation and the new marketing mix as an innovative approach for enhancing tourism industry competitiveness in Egypt. *International Journal of Services and Operations Management*, 8 (1): 76–91. DOI: 10.1504/IJSOM.2011.037441

Hall, C. M. (1998). Historical antecedents of sustainable development and ecotourism: New labels in old bottles. In C. M. Hall and A. A. Lew (Eds.), *Sustainable tourism a geographical perspective* (pp. 13–24). London: Longman.

Kastenholz, E., João Carneiro, M., Peixeira Marques, C., and Maria Correia Loureiro, S. (2018). The dimensions of rural tourism experience: Impacts on arousal, memory, and satisfaction. *Journal of Travel and Tourism Marketing*, 35(2): 189–201. DOI: 10.1080/10548408.2017.1350617

Lane, B. (2009). Rural tourism: An overview. In T. Jamal and M. Robinson (Eds.), *The SAGE handbook of tourism studies* (pp. 354–370). London: Sage.

Lane, B., and Kastenholz, E. (2015). Rural tourism: The evolution of practice and research approaches – towards a new generation concept? *Journal of Sustainable Tourism*, 23(8–9): 1133–1156. DOI: doi.org/10.1080/09669582.2015.1083997

Larsen, S. (2007). Aspects of a psychology of the tourist experience. *Scandinavian Journal of Hospitality and Tourism*, 7(1): 7–18. DOI: 10.1080/15022250701226014

Mathisen, L. (2013). Staging natural environments: A performance perspective. *Advances in Hospitality and Leisure*, 9: 163–183. DOI: 10.1108/S1745-3542(2013)0000009012

Minkiewicz, J., Evans, J., and Bridson, K. (2013). How do consumers co-create their experiences? An exploration in the heritage sector. *Journal of Marketing Management*, 30(1–2): 1–30. DOI: 10.1080/0267257X.2013.800899

Morgan, M. (2007). Festival spaces and the visitor experience. In J. Casado-Diaz, M. Everett, and S. Wilson (Eds.), *Social and cultural change: Making space(s) for leisure and tourism* (pp. 113–130). Eastbourne: Leisure Studies Association.

Niedziolka, I. (2012). Sustainable tourism development. *Regional Formation and Development Studies*, 8(3): 157–166.

Otto, J. E., and Ritchie, J. R. B. (1996). The service experience in tourism. *Tourism Management*, 17(3): 165–174. DOI: 10.1016/0261-5177(96)00003-9

Pine, B. J., II, and Gilmore, J. H. (1999). *The experience economy: Work is theatre and every business a stage*. Boston: Harvard Business School Press.

Pine, J. B., II, and Gilmore, J. H. (1998). Welcome to the experience economy. *Harvard Business Review*, 76(4): 97–105.

Prahalad, C. K., and Ramaswamy, V. (2004). Co-creation experiences: The next practice in value creation. *Journal of Interactive Marketing*, 18(3): 5–14. DOI: 10.1002/dir.20015

Sanagustín Fons, M. V., Fierro, J. A., and Patiño, M. G. (2011). Rural tourism: A sustainable alternative. *Applied Energy*, 88(2): 551–557. DOI: 10.1016/j.apenergy.2010.08.031

Schmitt, B. (1999). Experiential marketing. *Journal of Marketing Management*, 15(1–3): 53–67. DOI: 10.1362/026725799784870496

Xue, L., and Kerstetter, D. (2018). Rural tourism and livelihood change: An Emic perspective. *Journal of Hospitality and Tourism Research*, 43(3): 416–437. DOI: 10.1177/1096348018807289

42

Towards green loyalty

Driving from green experiential quality, green emotional attachment, green image, green switching experience, and green experiential satisfaction

Hung-Che Wu and Guo-Wei Chen

Introduction

Lorenzini (1994: 119) defines green restaurants as "new or renovated structures designed, constructed, operated, and demolished in an environmentally-friendly and energy-efficient manner". A study by Vieregge, Scanlon, and Huss (2007) demonstrates that more than 67 percent of participants prefer restaurants that offer locally grown food, and are willing to pay 10 percent more for local products. Within the hospitality and tourism industry, 75 percent of travelers have classified themselves as environmentally minded consumers, of whom 54 percent would choose to stay in environmentally friendly hotels over other lodging options (Watkins 1994). In addition, 80 percent of Americans consider themselves as environmentally concerned patrons of restaurants – when they dine out, they want to feel as if they have done something good for the environment rather than just having a nice meal (Parker 2011). It is reported that 79 percent of consumers prefer to dine at a certified green restaurant (Dewald, Bruin, and Jang 2014). The findings show that the market growth potential of green restaurants is tremendous (Chen, Cheng, and Hsu 2015). Despite the important trend, however, none of the green restaurants are certified in Taiwan. Although none of the "green certified restaurants" are in existence, there are numerous restaurants engaging in green practices, such as recycling, developing their energy and water efficiency, providing sustainable and organic food, and preventing pollution (Teng, Wu, and Huang 2014).

In the restaurant industry, quality plays a critical role in creating satisfaction, which ultimately generates loyalty (Anderson, Fornell, and Lehmann 1994; Kim, Ng, and Kim 2009). Chen and Chen (2010) and Wu and Li (2017) propose that experiential quality has been considered to be an antecedent of satisfaction. Abdullah, Wasiuzzaman, and Musa (2014) indicate that quality has been found to have an impact on emotional attachment. Dinç (2007) argues that emotional attachment might decrease or increase with the level of satisfaction regarding functional and technical criteria. Customers' emotional attachment to an organization might predict their commitment to the organization (e.g., organizational loyalty) and their willingness to make financial sacrifices in order to obtain it (e.g., pay a price premium) (Park, MacInnis, Priester, Eisingerich, and Iacobucci 2010). Several studies (Clemes, Shu, and Gan 2014; Clemes, Wu, Hu, and Gan 2009; Wu, Lin, and Hsu 2011) indicate that perceived quality plays a key role in increasing image, which in turn results in experiential satisfaction and loyalty in the restaurant industry. Bloemer and Ruyter (1998) show that image can be a significant factor affecting satisfaction and loyalty. Matzler, Strobl, Thurner, and Füller (2015) argue that switching experience negatively influences satisfaction and loyalty respectively. In addition, they argue that switching experience negatively influences satisfaction for two reasons: first, customers have less time to develop relationships to accumulate benefits, and to perceive a provider as unique. Second, having more experience with diverse providers implies that more experienced customers might be more demanding and difficult to satisfy, as they have greater knowledge about how a product or service should perform. Moreover, they expect satisfaction to mediate the influence of switching experience on loyalty. However, to the best of current knowledge, no prior research has studied the relationships among green experiential quality, green emotional attachment, green image, green switching experience, green experiential satisfaction, and green loyalty for the green restaurant industry.

Several studies (e.g., Abdullah et al. 2014; Choi and Kim 2013; Gan, Clemes, Wei, and Kao 2011; Lu, Zhang, and Wang 2009; Shonk and Chelladurai 2008; Wu, Li, and Li 2016) point out that image is a moderator that influences the relationship between experiential quality and interaction quality, peer-to-peer quality, physical environment quality, outcome quality, enjoyment quality, and management quality. A poor customer service experience may be ignored because a good restaurant image is established in the customer's mind (Lu et al. 2009; Wu and Mohi 2015). However, very few studies attempt to explore whether green image moderates the relationship between green experiential quality and its dimensions for green restaurants.

Several researchers (Parasuraman, Zeithaml, and Berry 1985; Reichheld and Sasser 1990; Zeithaml and Bitner 1996) indicate that providing quality has been considered to be a key determinant of reaching success and survival in today's competitive environment. Under this criterion, the primary emphasis of both theoretical and managerial efforts has focused on determining what dimensions of quality customers are concerned about and developing strategies to satisfy customers' expectations. The dimensions of experiential quality perceived by customers have been applied in a variety of sectors using multi-dimensional modeling as a theoretical framework (e.g., Wu and Cheng 2018; Wu, Cheng, and Ai 2018a; Wu and Li 2015, 2017; Wu, Li, and Li 2018b). However, little research focuses on assessing customers' perception of green experiential quality using a multi-dimensional model. Therefore, this study attempts to examine green restaurant customers' overall evaluations of the dimensions of green experiential quality. The degree of important dimensions of green experiential quality is also explored in this study.

The first purpose of this study is to explore the characteristics of green loyalty using a multi-dimensional approach and to show that the path formed by the green loyalty construct can be used to identify the dimensions of green experiential quality and their relative importance based on green restaurant customers' perceptions. The second purpose of this study is to examine whether the relationship between green experiential quality and its six dimensions is moderated by green image, as perceived by green restaurant customers. The third purpose of this study is to examine the relationship between the green loyalty of customers and the other higher order constructs: green experiential quality, green emotional attachment, green image, green switching experience, and green experiential satisfaction as perceived by green restaurant customers.

Literature review

Service quality

Parasuraman, Zeithaml, and Berry (1988) define service quality as the ability of the organization to meet or exceed customers' expectations. The most popular measure of service quality is SERVQUAL, an instrument developed by Zeithaml, Parasuraman, and Berry (1985), Zeithaml, Berry, and Parasuraman (1988) which compares the expectations and perceptions of customers regarding a particular service. Notwithstanding the popularity and widespread application of SERVQUAL, this scale has been subjected to criticisms on both theoretical and procedural grounds (Cronin and Taylor 1992, 1994). For example, major criticisms of this scale include its length, the validity of its five generic service quality dimensions, and its predictive power in respect of subsequent consumer purchases (Carman 1990; Finn and Lamb 1991). Saleh and Ryan (1991) have found that the SERVQUAL dimensions cannot be used to accurately measure service quality.

The performance-based service quality scale SERVPERF is identified by Cronin and Taylor (1992). Cronin and Taylor (1992) propose SERVPERF, which is a more concise performance-based scale, as an alternative to the SERVQUAL scale. Numerous researchers (e.g., Babakus and Boller 1992; Brady, Cronin, and Brand 2002; Brown, Churchill, and Peter 1993; Cronin and Taylor 1992) have supported the view that SERVPERF is a better alternative than SERVQUAL. However, Nadiri and Hussain (2005) fail to confirm the five SERVPERF dimensions for the service industry. Robledo (2001) proposes that SERVPERF is not an effective measurement scale in terms of its validity and reliability.

As mentioned above, several studies (e.g., Babakus and Boller 1992; Cronin and Taylor 1992, 1994; Nadiri and Hussain 2005) have identified that the existing measurement of service quality using the SERVQUAL and SERVPERF scales is insufficiently comprehensive to capture the service quality construct for green restaurants.

Green experiential quality

The concept of experiential quality encompasses customers' affective responses to their desired psychological advantages from a visiting experience (Chan and Baum 2007). Experiential quality has been considered to be the outstanding or advantageous perception and judgment of customer experience (Lemke, Clark, and Wilson 2011) in terms of measuring customers' perceived service quality (Lai and Chang 2013). Crompton and Love (1995) define experiential quality as involving "not only the attributes provided by a supplier, but also the attributes brought to the opportunity by the customer" (p. 12). According to Chen and Chen (2010) and MacKay and Crompton (1988), service quality focuses on service performance at the attribute level whereas experiential quality concentrates on customer participation in hospitality activities, which results in the psychological outcome. However, none of the research focuses on the issue of experiential quality in the context of green restaurants.

This study proposes a novel concept – green experiential quality, and defines the term as a subjective measure of a customer's experiences with a green restaurant (Kilkki 2008). Green experiential quality is applied to measure the extent to which green restaurant management considers the specific needs of the customers in delivering environmentally-friendly products or services, as perceived by customers. Therefore, green experiential quality is a subjective measure from green restaurant customers' perspectives of the overall value of the product or service provided, and thus does not replace but augments end-to-end green experiential quality by providing the quantitative link to the customers' perception (Xu and Chan 2010).

Numerous studies have focused on the components of experiential quality. For example, Lemke et al. (2011) reveal that experiential quality is based on hedonism of product category, involvement, product complexity, and relationality. Pareigis, Edvardsson, and Enquist (2011) show that experiential quality consists of customer processes, other customers, physical environment, contact personnel, provider processes, and the wider environment. Jin, Lee, and Lee (2015) argue that experiential quality comprises immersion, surprise, participation, and fun. De Rojas and Camarero (2008) propose that experiential quality should be made up of interaction quality, physical environment quality, and outcome quality. Wu and Li (2017) and Wu et al. (2018b) show that experiential quality comprises interaction quality, physical environment quality, outcome quality, and access quality. Wu et al. (2016) propose that interaction quality, physical environment quality, outcome quality, administration quality, and perceived enjoyment constitute experiential quality. Therefore, several researchers (Jin et al. 2015; Kao, Huang, and Wu 2008; Wu and Li 2017) recommend that perceived experiential quality has been found to be a multi-dimensional model. However, none of the research focuses on measuring the perceptions of the dimensions of experiential quality for green restaurants. Accordingly, in order to accurately measure customers' perceived dining experiences in green restaurants, a multi-dimensional model is applied to measure the dimensions of green experiential quality in this study.

Green emotional attachment

Thomson, MacInnis, and Park (2005) define emotional attachment to organizations as an emotion-laden bond between a person and an organization, characterized by deep feelings of connection, affection, and passion. Customers feeling a strong attachment to an organization will be more loyal and less price sensitive (Dunn and Hoegg 2014). Emotional attachment is very different when comparing material possession bonding with organizational bonding as they differ in irreplaceability or post-acquisition use (Kleine and Baker 2004). However, this concept is entirely new in measuring the intensity of customers in green restaurants. This study creates a novel concept – green emotional attachment, which is defined as the bond that connects a customer with a specific environmentally-friendly restaurant by involving feelings of affection, passion, and connection (Malär, Krohmer, Hoyer, and Nyffenegger 2011). Customers' emotional attachment to an environmentally-friendly restaurant might predict their commitment to the green restaurant (e.g., green loyalty) and their willingness to make financial sacrifices in order to obtain it (e.g., to pay a price premium) (Park et al. 2010).

Green image

Bravo, Montaner, and Pina (2009) define image as the "overall impression" left in the customers' minds as a result of accumulative feelings, ideas, attitudes and experiences with the organization, stored in the memory, transformed into a positive/negative meaning, retrieved to reconstruct an image, and recalled when the name of the organization is heard or brought to mind. Therefore, image is a result of a communication process in which organizations create and spread a specific message that constitutes their strategic intent; mission, vision, goals, and identity, which reflects the core values that they cherish (Bravo et al. 2009; Leuthesser and Kohli 1997). This study uses the definition of Chen (2010), where a novel green image is referred to as the perceptions developed from the interaction among green restaurants, their service providers, and customers that are linked to environmental commitments and concerns, and is also related to the association with a green restaurant's ability to position itself in the customer's mind (Tan and Yeap 2012).

Green switching experience

A few studies in the satisfaction literature have examined the impact of customers' switching experience on satisfaction, wherein switching experience is defined "as the extent to which the consumer has switched between providers in the past" (Burnham, Frels, and Mahajan 2003: 114). Increased switching reduces switching costs by increasing the consumer's familiarity with the process of both switching and learning to use new providers (Nilssen 1992). This study creates a novel concept – green switching experience, indicating a reduced duration with the incumbent environmentally-friendly restaurant, which means that there has been less time to accumulate benefits that might be lost in switching. Furthermore, customers who have switched more have had less time with their current service provider to develop brand and personal relationship bonds, and they are less likely to perceive their service provider relationship as unique (Bhattacharya, Rao, and Glynn 1995).

Green experiential satisfaction

Satisfaction has been increasingly gaining attention from researchers and practitioners as a recognized field of scholarly study and is a fundamental tool used by financial institutions for enhancing loyalty and ultimately organizational performance and profitability (Mohsan, Nawaz, Khan, Shaukat, and Aslam 2011). Experiential satisfaction is conceived as the basis of the concept of service satisfaction, even though it extends beyond service satisfaction in that it focuses on customers' overall evaluations of their experiences after consumption is fulfilled (Kao et al. 2008). Since green restaurants are subjected to a growing pressure to become responsible and green, several stakeholders press their organizations to reduce their negative impacts on society and natural environments (Barnett 2007). With the rise of the environmental movement, not only are customers willing to purchase products that produce minimum impact, but also society is concerned about the environment. Therefore, this study creates a novel concept – green experiential satisfaction, which is defined as customers' overall evaluation of environmentally-friendly content based on their experience in a green restaurant (Tan and Yeap 2012; Wu et al. 2018b).

Green loyalty

Oliver (1999: 34) defines loyalty as "a deeply held commitment to rebuy or repatronize a preferred product/service consistently in the future, thereby causing repetitive same-brand or same brand-set purchasing, despite situational influences, and marketing efforts having the potential to cause switching behavior." According to Backman and Crompton (1991), loyalty involves a positive attitude towards the product or service of an organization, followed by favorable behavior that results in purchasing it and recommending it to others. Based on the aforementioned statement, this study proposes a novel construct – green loyalty, which has relevance to environmental commitments and concerns (Chang and Fong 2010). Green loyalty is defined as the customer desiring to maintain a relationship with a restaurant which involves environmental or green concerns, and their commitment to repurchase or repatronize a preferred product consistently in the future. It is a customer's intention to repurchase, and their desire to recommend, show tolerance for a higher price, and purchase other products with (from) a green restaurant (Chang and Fong 2010).

Research model and hypothesis development

The conceptual research model shown in Figure 42.1 was developed using a multi-dimensional structure based on the framework adapted from several studies (e.g., Abdullah et al. 2014; Brady and Cronin 2001; Choi and Kim 2013; Dabholkar, Thorpe, and Rentz 1996; Wu and Li 2015, 2017). In order to form an overall green experiential quality perception, the multi-dimensional model reflects the proposition that customers form their perceptions of each of six dimensions: green interaction quality, green peer-to-peer quality, green physical environment quality, green outcome quality, green enjoyment quality, and green management quality. Green experiential quality is assumed to influence green emotional attachment and green image respectively. Green image is assumed to moderate the relationship between green experiential quality and its six dimensions. Green experiential quality, green image, and green switching experience are postulated to influence green experiential satisfaction, which, in turn, results in green emotional attachment. The perceptions of green emotional attachment, green experiential satisfaction, green image, and green switching experience are expected to influence green loyalty.

In this study, there are six dimensions used to form green experiential quality. The first dimension, green interaction quality, explores how the environmentally-friendly service is provided (Brady and Cronin 2001; Czepiel, Solomon, and Suprenant 1985). Ko and Pastore (2005) and Wu (2013, 2017) indicate that the interaction between employees and customers in the restaurant is the subjective rating of how the environmentally-friendly service is

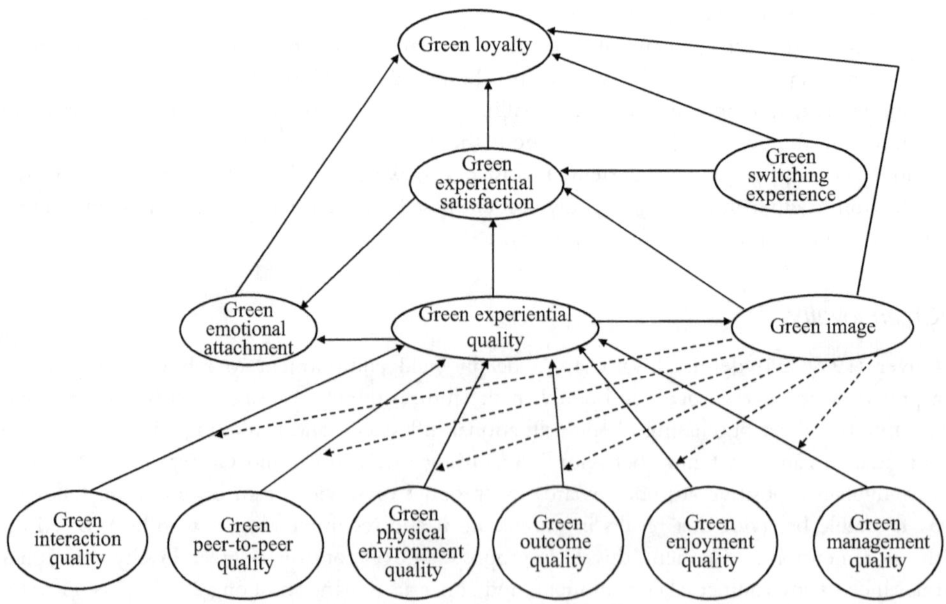

Figure 42.1 A conceptual research model

delivered through the interaction. The second dimension, green peer-to-peer quality, refers to the perceived judgment of the superiority of customers' eco-friendly interaction with others. According to this notion, a customer's helpful (distracting) behavior may influence other customers' service experience positively (negatively) (Choi and Kim 2013). The third dimension, green physical environment quality, is based on the equipment provided when an environmentally-friendly service is given to a customer (Bitner 1992). This dimension is considered to be one of the most important aspects of a green experiential quality evaluation (Wu and Li 2017; Wu et al. 2018b). The fourth dimension, green outcome quality, stresses the results of the environmentally-friendly service and the concept of quality (Ko and Pastore 2005). Green outcome quality reflects the customer's perception of the superiority of an environmentally-friendly service experience (Brady and Cronin 2001; Grönroos 1984; Kang and James 2004). The fifth dimension, green enjoyment quality, is defined as the extent to which a quality eco-friendly *experience* enables potential customers to feel relaxed, joyful, pleased and even wonderful (Wu et al. 2016). This dimension is an intrinsic motivation that emphasizes the eco-friendly usage process and reflects the pleasure and enjoyment associated with using the system (Praveena and Thomas 2014). The last dimension, green management quality, focuses on the elements in service quality including assurance, employees' *environmentally-friendly* knowledge and courtesy, and their *environmentally-friendly* ability to inspire trust and confidence in delivering an outcome desired by the customers (Abdullah et al. 2014). The customers seek environmentally-friendly empathy, responsiveness, and assurance during the development process of service quality which then enables the restaurant to gain extra potential to compete in the market. These six dimensions are based on previous studies.

Summary and implications

This study contributes to the theoretical advancement of the field of attraction management by empirically testing the interrelationships between experiential quality, green emotional attachment, green image, green switching experience, green experiential satisfaction, and green experiential loyalty. In addition, the comprehensive multi-level model developed in this study provides a valuable framework for future studies which focus on examining the relationships between the constructs in the green restaurant industry. Moreover, the multi-level modeling approach will address some of the weaknesses of traditional measurement methods (such as SERVQUAL and SERVPERF) and thus provide a more accurate method of assessing perceived experiential quality in the context of the green restaurant. From a practical perspective, this study can provide further insights for green restaurant management to better understand green experiential loyalty using the proposed model, as well as providing useful information to marketing managers who are designing the eco-friendly products and services in the green restaurant. The study's findings will provide green restaurant management with a clear understanding of relevant experiential quality dimensions because they derive from the customer's perspectives, so that management will know exactly how to make a choice about the best way to allocate scarce resources.

The multi-dimensional factor structure of green experiential quality identified in this study provides green restaurant management with insights into how green restaurant customers form their perceptions of green experiential quality. From a managerial perspective, the measurement scale for green experiential quality developed in this study provides green restaurant management with a method to evaluate customers' perceptions of service delivery using several indicators of green experiential quality. Based on customers' different needs, therefore, green restaurants can measure perceived green experiential quality at a global level, at a dimensional level, or at both levels. For example, green restaurant managers who are interested in customers' general attitudes towards their experience in a green restaurant can use the two global measures to determine customers' overall perceptions of green experiential quality.

References

Abdullah, A. H., Wasiuzzaman, S., and Musa, R. (2014). 'The effects of university quality on emotional attachment: a case from a private higher education institution', *Procedia-Social and Behavioral Sciences*, 130: 282–292. DOI: 10.1016/j.sbspro.2014.04.033

Anderson, E. W., Fornell, C., and Lehmann, D. R. (1994). 'Customer satisfaction, market share, and profitability: findings from Sweden', *Journal of Marketing*, 58 (3): 53–66. DOI: 10.1177/002224299405800304

Babakus, E. and Boller, G. (1992). 'An empirical assessment of the SERVQUAL scale', *Journal of Business Research*, 24 (3): 253–268. DOI: 10.1016/0148-2963(92)90022-4

Backman, S. J. and Crompton, J. L. (1991). 'The usefulness of selected variables for predicting activity loyalty', *Leisure Sciences*, 13 (3): 205–220. DOI: 10.1080/01490409109513138

Barnett, M. L. (2007). 'Stakeholder influence capacity and the variability of financial returns to corporate social responsibility', *Academy of Management Review*, 32 (3): 794–816. DOI: 10.5465/amr.2007.25275520

Bhattacharya, C. B., Rao, H., and Glynn, M. A. (1995). 'Understanding the bond of identification: an investigation of its correlates among art museum members', *Journal of Marketing*, 59 (4): 46–57. DOI: 10.1177/002224299505900404

Bitner, M. J. (1992). 'Servicescapes: the impact of physical surroundings on customers and employees', *Journal of Marketing*, 56 (2): 57–71. DOI: 10.1177/002224299205600205

Bloemer, J. and Ruyter, K. (1998). 'On the relationship between store image, store satisfaction and store loyalty', *European Journal of Marketing*, 32 (5/6): 499–513. DOI: 10.1108/03090569810216118

Brady, M. K. and Cronin, J. J. (2001). 'Some new thoughts on conceptualizing perceived service quality: a hierarchical approach', *Journal of Marketing*, 65 (3): 34–49. DOI: 10.1509/jmkg.65.3.34.18334

Brady, M. K., Cronin, J. J., and Brand, R. R. (2002). 'Performance-only measurement of service quality: a replication and extension', *Journal of Business Research*, 55 (1): 17–31. DOI: 10.1016/S0148-2963(00)00171-5

Bravo, R., Montaner, T., and Pina, J. M. (2009). 'The role of bank image for customers versus non-customers', *International Journal of Bank Marketing*, 27 (4): 315–334. DOI: 10.1108/02652320910968377

Brown, T. J., Churchill, G. A., and Peter, J. P. (1993). 'Improving the measurement of service quality', *Journal of Retailing*, 69 (1): 127–139. DOI: 10.1016/S0022-4359(05)80006-5

Burnham, T. A., Frels, J. K., and Mahajan, V. (2003). 'Consumer switching costs: a typology, antecedents, and consequences', *Journal of the Academy of Marketing Science*, 31 (2): 109–126. DOI: 10.1177/0092070302250897

Carman, J. M. (1990). 'Consumer perceptions of service quality: an assessment of the SERVQUAL dimensions', *Journal of Retailing*, 66 (1): 33–55.

Chan, J. K. L. and Baum, T. (2007). 'Eco-tourists' perception of ecotourism experience in lower Kinabatangan, Sabah, Malaysia', *Journal of Sustainable Tourism*, 15 (5): 574–590. DOI: 10.2167/jost679.0

Chang, N. J. and Fong, C. M. (2010). 'Green product quality, green corporate image, green customer satisfaction, and green customer loyalty', *African Journal of Business Management*, 4 (13): 2836–2844. DOI: DD16E2424448

Chen, C. F. and Chen, F. S. (2010). 'Experience quality, perceived value, satisfaction and behavioral intentions for heritage tourists', *Tourism Management*, 31 (1): 29–35. DOI: 10.1016/j.tourman.2009.02.008

Chen, C. T., Cheng, C. C., and Hsu, F. S. (2015). 'GRSERV scale: an effective tool for measuring consumer perceptions of service quality in green restaurants', *Total Quality Management and Business Excellence*, 26 (3/4): 355–367. DOI: 10.1080/14783363.2013.832478

Chen, Y. S. (2010). 'The drivers of green brand equity: green brand image, green satisfaction, and green trust', *Journal of Business Ethics*, 93 (2): 307–319. DOI: 10.1007/s10551-009-0223-9

Choi, B. J. and Kim, H. S. (2013). 'The impact of outcome quality, interaction quality, and peer-to-peer quality on customer satisfaction with a hospital service', *Journal of Service Theory and Practice*, 23 (3): 188–204. DOI: 10.1108/09604521311312228

Clemes, M. D., Shu, X., and Gan, C. (2014). 'Mobile communications: a comprehensive hierarchical modeling approach', *Asia Pacific Journal of Marketing and Logistics*, 26 (1): 114–146. DOI: 10.1108/APJML-04-2013-0040

Clemes, M. D., Wu, J. H. C., Hu, B. D., and Gan, C. (2009). 'An empirical study of behavioral intentions in the Taiwan hotel industry', *Innovative Marketing*, 5 (3): 30–50.

Crompton, J. L. and Love, L. L. (1995). 'The predictive validity of alternative approaches to evaluating quality of a festival', *Journal of Travel Research*, 34 (1): 11–24. DOI: 10.1177/004728759503400102

Cronin, J. J. and Taylor, S. A. (1992). 'Measuring service quality: a re-examination and extension', *Journal of Marketing*, 56 (3): 55–68. DOI: 10.1177/002224299205600304

Cronin, J. J. and Taylor, S. A. (1994). 'SERVPERF versus SERVQUAL: reconciling performance-based and perceptions-minus-expectations measurement of service quality', *Journal of Marketing*, 58 (1): 125–131. DOI: 10.1177/002224299405800110

Czepiel, J. A., Solomon, M. R., and Suprenant, C. E. (1985). *The Service Encounter*. Lexington, MA: Lexington Books.

Dabholkar, P. A., Thorpe, D. I., and Rentz, J. O. (1996). 'A measure of service quality for retail stores: scale development and validation', *Journal of the Academy of Marketing Science*, 24 (1): 3–16. DOI: 10.1007/BF02893933

De Rojas, C. and Camarero, C. (2008). 'Visitors' experience, mood and satisfaction in a heritage context: evidence from an interpretation center', *Tourism Management*, 29 (3): 525–537. DOI: 10.1016/j.tourman.2007.06.004

Dewald, B., Bruin, B. J., and Jang, Y. J. (2014). 'US consumer attitudes towards 'green' restaurants', *Anatolia*, 25 (2): 171–180. DOI: 10.1080/13032917.2013.839457

Dinç, P. (2007). 'Spatial and behavioral variables that affect "emotional attachment" of users: a multi-dimensional approach for private offices', *Gazi University Journal of Science*, 20 (2): 41–50.

Dunn, L. and Hoegg, J. (2014). 'The impact of fear on emotional brand attachment', *Journal of Consumer Research*, 41 (1): 152–168. DOI: 10.1086/675377

Finn, D. W. and Lamb, C. W. (1991). 'An evaluation of the SERVQAUL scales in a retail setting', In R. H. Holman and M. R. Solomon (eds.), *Advances in Consumer Research* (vol. 18, pp. 483–490). Provo, UT: Association for Consumer Research.

Gan, C., Clemes, M., Wei, J., and Kao, B. (2011). 'An empirical analysis of New Zealand bank customers' satisfaction', *Banks and Bank Systems*, 6 (3): 63–77.

Grönroos, C. (1984). 'A service quality model and its marketing implications', *European Journal of Marketing*, 18 (4): 36–44. DOI: 10.1108/EUM0000000004784

Jin, N. P., Lee, S., and Lee, H. (2015). 'The effect of experience quality on perceived value, satisfaction, image and behavioral intention of water park patrons: new versus repeat visitors', *International Journal of Tourism Research*, 17 (1): 82–95. DOI: 10.1002/jtr.1968

Kang, G. D. and James, J. (2004). 'Service quality: an examination of Grönroos's service quality model', *Journal of Service Theory and Practice*, 14 (4): 266–277. DOI: 10.1108/09604520410546806

Kao, Y. F., Huang, L. S., and Wu, C. H. (2008). 'Effects of theatrical elements on experiential quality and loyalty intentions for theme parks', *Asia Pacific Journal of Tourism Research*, 13 (2): 163–174. DOI: 10.1080/10941660802048480

Kilkki, K. (2008). 'Quality of experience in communications ecosystem', *Journal of Universal Computer Science*, 14 (5): 615–624.

Kim, W. G., Ng, C. Y. N., and Kim, Y. (2009). 'Influence of institutional DENESERV on customer satisfaction, return intention, and word-of-mouth', *International Journal of Hospitality Management*, 28 (1): 10–17. DOI: 10.1016/j.ijhm.2008.03.005

Kleine, S. S. and Baker, S. M. (2004). 'An integrative review of material possession attachment', *Academy of Marketing Science Review*, 1: 1–35.

Ko, Y. J. and Pastore, D. L. (2005). 'A hierarchical model of service quality for the recreational sport industry', *Sport Marketing Quarterly*, 14 (2): 84–97. aA.

Lai, Y. L. and Chang, S. C. (2013). 'How improving the customer experience quality and business performance? A case study by mystery shopper practices', *International Journal of Marketing Studies*, 5 (6): 52–61. DOI: 10.5539/ijms.v5n6p52

Lemke, F., Clark, M., and Wilson, H. (2011). 'Customer experience quality: an exploration in business and consumer contexts using repertory grid technique', *Journal of the Academy Marketing Science*, 39 (6): 846–869. DOI: 10.1007/s11747-010-0219-0

Leuthesser, L. and Kohli, C. (1997). 'Corporate identity: the role of mission statements', *Business Horizons*, 40 (3): 59–66.
Lorenzini, B. (1994). 'The green restaurant, part II: systems and service', *Restaurant and Institutions*, 104 (11): 119–136.
Lu, Y., Zhang, L., and Wang, B. (2009). 'A multi-dimensional and hierarchical model of mobile service quality', *Electronic Commerce Research and Applications*, 8 (5): 228–240. DOI: 10.1016/j.elerap.2009.04.002
MacKay, K. J. and Crompton, J. L. (1988). 'A conceptual model of consumer evaluation of recreation service quality', *Leisure Studies*, 7 (1): 41–49. DOI: 10.1080/02614368800390041
Malär, L., Krohmer, H., Hoyer, W. D., and Nyffenegger, B. (2011). 'Emotional brand attachment and brand personality: the relative importance of the actual and the ideal self', *Journal of Marketing*, 75 (4): 35–52. DOI: 10.1509/jmkg.75.4.35
Matzler, K., Strobl, A., Thurner, N., and Füller, J. (2015). 'Switching experience, customer satisfaction, and switching costs in the ICT industry', *Journal of Service Management*, 26 (1): 117–136. DOI: 10.1108/JOSM-04-2014-0101
Mohsan, F., Nawaz, M. M., Khan, M. S., Shaukat, Z., and Aslam, N. (2011). 'Impact of customer satisfaction on customer loyalty and intentions to switch: evidence from banking sector of Pakistan', *International Journal of Business and Social Science*, 2 (16): 263–270.
Nadiri, H. and Hussain, K. (2005). 'Perceptions of service quality in North Cyprus hotels', *International Journal of Contemporary Hospitality Management*, 17 (6): 469–480. DOI: 10.1108/09596110510612112
Nilssen, T. (1992). 'Two kinds of consumer switching costs', *RAND Journal of Economics*, 23 (4): 579–589. DOI: 10.2307/2555907
Oliver, R. L. (1999). 'Whence customer loyalty?', *Journal of Marketing*, 63: 33–44. DOI: 10.1177/00222429990634s105
Parasuraman, A., Zeithaml, V. A., and Berry, L. L. (1985). 'A conceptual model of service quality and its implications for future research', *Journal of Marketing*, 49 (4): 41–50. DOI: 10.1177/002224298504900403
Parasuraman, A., Zeithaml, V. A., and Berry, L. L. (1988). 'SERVQUAL: a multiple-item scale for measuring consumer perceptions of service quality', *Journal of Retailing*, 64 (1): 12–40.
Pareigis, J., Edvardsson, B., and Enquist, B. (2011). 'Exploring the role of the service environment in forming customer's service experience', *International Journal of Quality and Service Sciences*, 3 (1): 110–124. DOI: 10.1108/17566691111115117
Park, W. C., MacInnis, D. J., Priester, J. R., Eisingerich, A. B., and Iacobucci, D. (2010). 'Brand attachment and brand attitude strength: conceptual and empirical differentiation of two critical brand equity drivers', *Journal of Marketing*, 74 (6): 1–17. DOI: 10.1509/jmkg.74.6.1
Parker, L. D. (2011). 'Twenty-one years of social and environmental accountability research: a coming of age', *Accounting Forum*, 35 (1): 1–10. DOI: 10.1016/j.accfor.2010.11.001
Praveena, K. and Thomas, S. (2014). 'Continuance intention to use Facebook: a study of perceived enjoyment and TAM', *Bonfring International Journal of Industrial Engineering and Management Science*, 4 (1): 24–29. DOI: 10.9756/BIJIEMS.4794
Reichheld, F. F. and Sasser, W. E. (1990). 'Zero defections: quality comes to services', *Harvard Business Review*, 68 (5): 105–111.
Robledo, M. A. (2001). 'Measuring and managing service quality: integrating customer expectations', *Journal of Service Theory and Practice*, 11 (1): 22–31. DOI: 10.1108/09604520110379472
Saleh, F. and Ryan, C. (1991). 'Analyzing service quality in the hospitality industry using SERVQUAL', *The Service Industries Journal*, 11 (3): 324–346. DOI: 10.1080/02642069100000049
Shonk, D. J. and Chelladurai, P. (2008). 'Service quality, satisfaction, and intent to return in event sport tourism', *Journal of Sport Management*, 22 (5): 587–602. DOI: 10.1123/jsm.22.5.587
Tan, B. C., and Yeap, P. F. (2012). 'What drives green restaurant patronage intention?', *International Journal of Business and Management*, 7 (2): 215–223. DOI: 10.5539/ijbm.v7n2p215
Teng, Y. M., Wu, K. S., and Huang, D. M. (2014). 'The influence of green restaurant decision formation using the VAB Model: the effect of environmental concerns upon intent to visit', *Sustainability*, 6 (12): 8736–8755. DOI: 10.3390/su6128736
Thomson, M., MacInnis, D. J., and Park, C. W. (2005). 'The ties that bind: measuring the strength of consumers' emotional attachments to brands', *Journal of Consumer Psychology*, 15 (1): 77–91. DOI: 10.1207/s15327663jcp1501_10
Vieregge, M., Scanlon, N., and Huss, J. (2007). 'Marketing locally grown food products in branded restaurants: do customers care?', *Journal of Foodservice Business Research*, 10 (2): 67–82. DOI: 10.1300/J369v10n02_05

Watkins, E. (1994). 'Do guests want green hotels?', *Lodging Hospitality*, 50 (4): 70–72.

Wu, H. C. (2013). 'An empirical study of the effects of service quality, perceived value, corporate image and customer satisfaction on behavioral intentions in the Taiwan quick service restaurant industry', *Journal of Quality Assurance in Hospitality and Tourism*, 14 (4): 364–390. DOI: 10.1080/1528008X.2013.802581

Wu, H. C. (2017). 'What drives experiential loyalty? A case study of Starbucks coffee chain in Taiwan', *British Food Journal*, 119 (3): 468–496. DOI: 10.1108/BFJ-08-2016-0349

Wu, H. C. and Cheng, C. C. (2018). 'What drives spectators' experiential loyalty: a case study of the 2016 summer Olympics', *Asia Pacific Journal of Marketing and Logistics*, 30 (4): 837–866. DOI: 10.1108/APJML-08-2017-0174

Wu, H. C., Cheng, C. C., and Ai, C. H. (2018a). 'A study of experiential quality, experiential value, trust, corporate reputation, experiential satisfaction and behavioral intentions for cruise tourists: the case of Hong Kong', *Tourism Management*, 66: 200–220. DOI: 10.1016/j.tourman.2017.12.011

Wu, H. C., Li, M. M., and Li, T. (2018b). 'A study of experiential quality, experiential value, experiential satisfaction, theme park image and revisit intention', *Journal of Hospitality and Tourism Research*, 42 (1): 26–73. DOI: 10.1177/1096348014563396

Wu, H. C. and Li, T. (2015). 'An empirical study of the effects of service quality, visitor satisfaction, and emotions on behavioral intentions of visitors to the museums of Macau', *Journal of Quality Assurance in Hospitality and Tourism*, 16 (1): 80–102. DOI: 10.1080/1528008X.2015.966298

Wu, H. C. and Li, T. (2017). 'A study of experiential quality, perceived value, heritage image, experiential satisfaction, and behavioral intentions for heritage tourists', *Journal of Hospitality and Tourism Research*, 41 (8): 904–944. DOI: 10.1177/1096348014525638

Wu, H. C., Li, T., and Li, M. Y. (2016). 'A study of behavioral intentions, patient satisfaction, perceived value, patient trust and experiential quality for medical tourists', *Journal of Quality Assurance in Hospitality and Tourism*, 17 (2): 114–150. DOI: 10.1080/1528008X.2015.1042621

Wu, H. C. and Mohi, Z. (2015). 'Assessment of service quality in the fast-food restaurant', *Journal of Foodservice Business Research*, 18 (4): 358–388. DOI: 10.1080/15378020.2015.1068673

Wu, H. C. J., Lin, Y. C., and Hsu, F. S. (2011). 'An empirical analysis of synthesizing the effects of service quality, perceived value, corporate image and customer satisfaction on behavioral intentions in the transport industry: a case of Taiwan high-speed rail', *Innovative Marketing*, 7 (3): 83–100.

Xu, J. B. and Chan, A. (2010). 'A conceptual framework of hotel experience and customer-based brand equity: some research questions and implications', *International Journal of Contemporary Hospitality Management*, 22 (2): 174–193. DOI: 10.1108/09596111011018179

Zeithaml, V. A., Berry, L. L., and Parasuraman, A. (1988). 'Communication and control processes in the delivery of service quality', *Journal of Marketing*, 52 (2): 35–48. DOI: 10.1177/002224298805200203

Zeithaml, V. A. and Bitner, M. J. (1996). *Services Marketing*. New York, NY: McGraw-Hill.

Zeithaml, V. A., Parasuraman, A., and Berry, L. L. (1985). 'Problems and strategies in services marketing', *Journal of Marketing*, 49 (2): 33–46. DOI: 10.1177/002224298504900203

43

Guest and host relationship in tourism experience building

Rising issue of sustainability

Rashmi Ranjan Choudhury and Saurabh Kumar Dixit

Introduction

Tourism experience is the reflection of a complete process of actions such as the planning of actions, actually undertaking the events and the remembrance of those events. According to Hinch and Higham (2004) tourist experience can be understood as the length of time spent at the destination, activities pursued and tourist behavior. The time and money invested in the tour are relatively influential in experience creation with respect to the value in return. In general, these three courses of action can be termed as expectation, perception and memories, which influence the tourism experience (Larsen 2007). Tourism experiences may also be influenced by the tourism network in which various factors play a considerable part, such as the destination, locals, tour operators, government and NGOs and events, as well as the internet and technology. In the present transformation economy, the consumer not only experiences but also perceives the experience. The tourism industry is driven by diverse experiences; the better the experience the better the perception of tourists as well as others associated with the experience. Tourists' quality experience can influence revisit intention, WOM and lifelong memories whereas the hosts' quality experience can improve their perception towards tourists, boost their involvement and build a better relationship with tourists. In tourism, numerous players are involved who are regarded as stakeholders of the industry. In general, there is the meeting of two different populations, namely: a stable population called hosts/residents and a variable one called guests/tourists. When two different groups interact with each other, an exchange of opinions, attitudes and values occurs which influences the experience as well as the lifestyles of both (Salvatore and Punzo 2016).

This chapter places emphasis on tourism experience building from the perspectives of tourism stakeholders, specifically the guest and host perspective. Categorized into three sections, the chapter focuses on: different aspects of tourism experience (guests' and hosts') elaborated in the foremost section, followed by stakeholders associated in tourism and the relationship between guest and host in tourism experience building. The third and final section focuses on guest–host responsibilities and sustainability in providing a quality tourism experience. Sustainability stands as a major concern in the tourism industry and is the most deliberated issue in contemporary times. The chapter examines 'a case of Mawlynnong: the cleanest village of Asia' to exhibit a clear picture of guest–host relations in tourism and the sustainability issues of tourism at the destination.

Aspects of tourism experience

Tourism experience varies from person to person even at the same destination or event or even in the case of the same tourism activities. A tourism experience turns out to be authentic when both the tourist's and the host's experience of each other are true to their nature and cultural ways. According to Steiner and Reisinger (2004), '*authentic travel experience happens when neither host nor tourist is trying to be someone else, when neither is trying to modify their behavior to fit some cultural stereotype nor show acceptance of something alien and perhaps even objectionable*'. The experience can be arbitrated from the guest and host communities at the destination with various aspects of experience.

Aspects of guests' experience

Quality tourism experience is mostly understood as the tourist or guest experience and several literatures focus on addressing the topic from the tourist's perspective. Tourist experience can be viewed as a function of the individual psychological process of reminiscence relating to the actions or activities conducted during the tour or event undertaken.

> A tourist experience is a past personal travel related event strong enough to have entered long-term memory.
>
> *(Larsen 2007)*

The aspects of guests' experience can be studied under the following heads:

Authenticity

The term authenticity takes a wide number of meanings depending on the various contexts of its usage. In a summary of definitions 'authentic' is described as whether any individual or groups, object, event or communication products are what they appear to be or are claimed to be, and therefore worth the price that is asked for them (Trilling 1972; Wang 1999). In tourism, 'authenticity' can be segregated into two different contexts like toured objects (destination, culture, tradition, etc.) and tourist experience (Wang 1999). Authenticity can also be divided into tangible and intangible aspects with a significant stance in tourism and is further classified into *objective authenticity* and *constructive authenticity*. The originality and genuineness of tourism objects and sites separates objective authenticity from constructive authenticity, in which constructive measures are undertaken to build up tourism attractions and offer services having economic value (Ram et al. 2016). In the context of tourism, constructive authenticity can be exceptional and referential. Exceptional authenticity refers to something that people think has been done exceptionally well and is mostly shaped to someone's individual preferences whereas in referential authenticity people find the authenticity with reference to some other context like destination history, civilization and shared memories (Molleda 2010). Tourists' perception regarding authenticity is mostly dependent on the place where the attraction is located and also by several other factors associated with a destination (Ram et al. 2016).

Uniqueness of resources

The uniqueness of resources in tourism has a significant role in guests' experience building. Uniqueness provides any organization or destination with a certain value and offers competency among other things. A set of both tangible and intangible assets configures the unique resources in tourism which in turn attracts the tourists. The tangible resources providing uniqueness include the destination, the physical resources, the infrastructure and the equipment engaged, etc., whereas the culture, heritage, services, technology and brand value come under the intangible uniqueness of the tourism industry (Erislan 2016). The uniqueness of resources is a major asset in tourism experience building and the uniqueness is central to strategy and implementation.

Entertainment

From ancient times to the modern era, entertainment has been a major concern in the tourism industry. The way of entertainment might have changed over the years but the purpose remains the same; to become more sociable and to derive pleasure from the activities. In the context of tourism, any revenue-generating activity (whether artificial or structured) centered on a core audience of tourists is considered as a tourism entertainment activity (Loi 2009). '*Entertainment is predominantly receptive, but occasionally participative*' (Loi 2009). Hence in tourism, entertainment becomes a part of memory and has a great effect on guests' experience building. Entertainment has various forms and differs from one type to another depending on individuals, places and time, as well as in the form of activities. For example, younger people prefer adventure or recreational type entertainment while older people mostly seek out activities associated with cultural learning and pilgrimage. Likewise, some specific destinations attract a defined category of tourists based on their entertainment preference. In ancient times Greeks and Romans found entertainment in watching gladiator fights inside the Roman amphitheaters while in present times the amphitheaters stand as heritage monuments, entertaining visitors with different preferences. In the modern era many countries open their borders to provide the entertainment facilities available in their destinations for tourists from other countries.

Safety and security

Safety and security have an unavoidable place in any industry; similarly, in the tourism industry they have a significant effect on tourism experience building. Since the last century, safety and security issues have arisen in the tourism industry due to the progression of mass tourism and rapid development at tourism destinations. Tourism is a global industry, hence the safety and security issue is not only limited to the tourism destinations but it has a worldwide impact in the aftermath of any accident or uncertainty. The safety and security aspect has multidimensional notions, like: individual and public safety, socio-political security and environmental concerns (Istvan and Krisztina 2011). In the technology-driven world, the safety and security issue has a greater impact on tourism as it affects the tourist experience. The perception of tourists about any destination is also influenced by their concern for their personal safety and security as well as their belongings and hence this specific topic cannot be overlooked.

Aspects of host experience

As tourists experience tourism products or services the host experiences the benefits and perils of tourism, which affects the host experience from tourism. Hosts' attitudes towards tourism vary and are influenced by various factors like economic dependency, social lifestyle, contact with tourists, proximity to tourism, etc. Unlike the tourist experience the host experience is the after-effect of the tourism process. The hosts are subjected to the effects of tourism for a longer period of time as they are the residents of that destination, rather than the visitors who spend a much shorter period of time at a destination. Various aspects of host experience are highlighted below.

Economic dependency

Undoubtedly, tourism is an economic driver for the country or region where it is practiced, and has a similar effect on the host communities who participate in the process. Tourism may bring economic development to a region or it may well trigger inflation and change aspects of the hosts' economic dependency. Tourism stimulates trade, income and entrepreneurship in and around the tourism destinations. The service industry generates employment for a number of small and medium businesses, providing the local community with an opportunity on their doorstep. As a result of tourism development, the flow of funds increases and accelerates development in the education, transportation and communication sectors. The cascading or multiplier effect of tourism spreads out the tourist's spending and provides opportunities to the several dependents either directly, indirectly or in an induced manner.

Improved socio-cultural life

Economic prosperity through tourism provides long term changes in the social life of the host community. Social values, beliefs and cultural practices are also improved with tourism demand and the authenticity and ethnicity of the community are maintained through tourism practices. Tourism contributes to socio-cultural changes in the host community's lifestyle in terms of the standard of living, social relationships and the moral conduct of the host community, all of which are well appreciated. However, a few social aspects are less welcome, as they can have an impact on the host community, like changes in values, political intervention, individual behavior and traditional changes. All of these factors are directly or indirectly associated with the host community, hence any alteration in them can affect the hosts' experience of tourism. Unless and until tourism provides a sustainable livelihood, the host community may not have a quality tourism experience.

Resource management

Resources are the most important facet of tourism experience for tourists as well as for the host community. Tourism influences the quality, availability and accessibility of resources for the local communities residing in nearby tourism destinations. Whether it is a tangible or an intangible resource, the hosts have all the rights over it. Locals' access to daily resources like land and water is restricted as a result of tourism development. In many cases the rural environs are taken over when urban and sub-urban areas are developed, which means that the locals are forced to change their occupations as natural resources are converted towards tourism purposes (Boley and Green 2015). The hosts' cultural resources provide unique value and identity to the community and stand as a significant resource in tourism. However, over-commercialization of the cultural resources also degrades their value with time. Hence culture preservation is an important aspect for the host communities affected by tourism. In short, tourism resource management needs every stakeholder's attention in conservation and responsible consumption.

Experience and value co-creation in tourism

Value co-creation is a process of creating or maintaining the value of the product or services with the participation of both the consumer and the producer. Value co-creation is said to take place when the consumer plays an active part in the consumption and production of value and deals with the designing of the experience (Prebensen and Foss 2011). The 'first generation experience', described in the late 1990s, was merely based on fun and entertainment, whereas at the present time the 'second generation experience' mostly depends on co-creation (Binkhorst 2005). Experience co-creation is highly dependent on individuals and it varies from person to person according to services, destinations and other associated factors in tourism (Prahalad and Ramaswamy 2004). Products and services hold the value but it's the market which co-creates the value and influences the experience of the customers. In tourism, tourist participation, stakeholder (or service provider) involvement and destination setting are similarly significant in value co-creation of experience. Consumers, i.e. tourists in the tourism industry want to interact with producers or service providers to have a better experience during their tourism-related activities. Grönroos (2006) illustrates that it is not only the tourists who engage in the process with service providers but it is also the service providers' responsibility to create opportunities for the tourists to engage in the value creation process.

Stakeholders' role in tourism experience building

Stakeholders associated with tourism

The term stakeholder has a long history in academic literatures; the Stanford Research Institute (SRI) generalized the term in 1963 and defined it as 'those groups without whose support the organization would cease to exist'. Further down the years, different scholars have defined stakeholders in a different manner. For instance, the term was associated by some with a broad definition (Freeman 1984), whereas others were to define the term 'stakeholder' in a narrow perspective, aimed towards an organizational purpose. Freeman (1984), a pioneer in 'stakeholder theory', refined the stakeholder's definition of SRI by stating that a '*stakeholder can be any group or individual who can affect or is affected by the achievement of the organization's objectives*'. During later years, the study of stakeholder theory became popular among an increasing number of scholars (Clarkson 1995; Donaldson and Preston 1995; Sautter and Leisen 1999) and several studies were carried out on the concept, either in support of it or as critiques. A group can be considered as a stakeholder when the said group has a genuine interest in an organization's activities and objectives. '*The interests of all stakeholders are of intrinsic value, i.e., each group of stakeholders merits consideration for its own sake and not merely because of its ability to further the interests of some other group, such as the shareowner*' (Donaldson and Preston 1995). Each stakeholder needs to be treated as an end in the organization without giving priority to some specific stakeholders, which may compromise the benefit of others' ends. Management should have a stronger voice than stakeholders in order to have control over them and need to provide equal attention to each stakeholder group regardless of its power (Sautter and Leisen 1999). Failure to maintain the participation of even a single stakeholder group can lead to failure of the organization (Clarkson 1995). Understanding three major concepts like identification of the stakeholders, the organizational relationship among them and negotiation among stakeholders is significant for the smooth management of any industry (Freeman 1984).

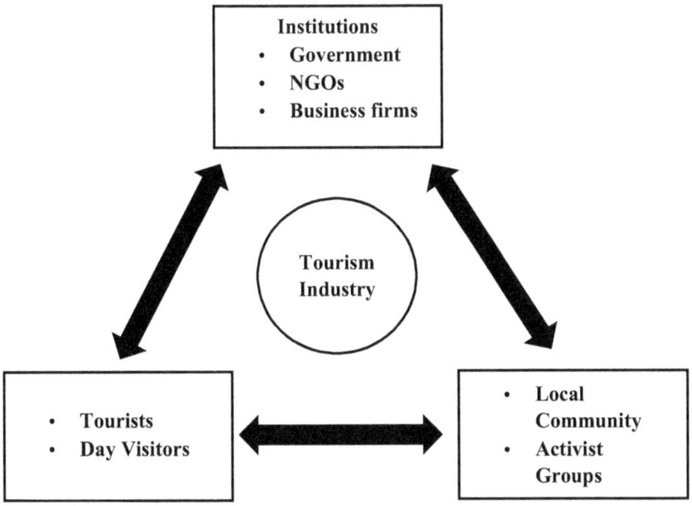

Figure 43.1 Stakeholders associated with tourism
(Adopted from Freeman 1984)

In tourism the stakeholders (See Figure 43.1) are government, tourism planners, Destination Management Organizations (DMOs), travel entities, employees, NGOs, residents and tourists, etc. Each and every one has a specific role in the better management of tourism at any destination.

Stakeholders in experience building

Tourism is a complex service industry accompanied by various stakeholders' groups with different interests and levels of involvement in tourism development. The fundamental concern of stakeholders is directed towards tourism development, which takes different forms according to goals, values, interests, beliefs, individual perception and power concerns in the society. Stakeholders' knowledge and experience, participation in planning and development and long-term community involvement all help in developing the relationship between stakeholders in tourism. Each group of stakeholders plays a significant role in tourism to provide a better tourism experience as well as achieving their goals. Government is more concerned about the infrastructure and policies whereas the tourism business providers are concerned about the factors affecting their business, like marketing and quality of service. The locals or host community are concerned about the effect of tourism on their community, quality of life and sustainability issues while tourists are more focused on the quality and accessibility of the tourism products. In short, every stakeholder's participation and mutual benefits are significant in overall tourism experience building.

Guest and host relationship in tourism

For the sustainable development of tourism, the linkage between various elements is necessary in which the local community (*hosts*) and tourists (*guests*) are the key elements (Zhang et al. 2006). Traditionally, both parties have been at opposite ends of the transaction, as hosts

Table 43.1 Factors affecting the guest–host relationship

Host expectation	*Guest expectation*
Income	Resource and attraction
Employment	Safety, security
Consumer goods	Value for money
Respect and dignity	Transport and infrastructure
Social interaction	Curiosity and novelty
Tourism consciousness	Cultural interaction
Living within a community	

Source: Adapted from Ravindran et al. 2018

are engaged in production or service provision and guests are at the consumption end of the product (Nash and Smith 1991). The guest–host relationship can be considered to be at the heart of tourism, i.e. without a smooth and purposeful relationship between both elements, tourism developments cannot be imagined. Studies suggest that the closer the level of interaction with the host the greater the experience of the tourist, wherein the interaction may not be authentic all of the time (Dixit 2014; Cronauer 2012; Maccannell 1973; Zhang et al. 2006). The interaction between the two parties is transitory as the tourists spend a limited time with the host community and experience a concise version of their lifestyle, culture and tradition (UNESCO 1976). On the other hand, when the number of tourists visiting a destination increases it puts pressure on the local community, resulting in behavioral changes that can be identified as further affecting the host–guest relationship (Valerio 2017). For long-term tourism development, the local community's (host) behavior towards tourists is important as it influences tourist satisfaction, overall experience and attitude towards the destination. The relationship can be strengthened between the two parties if and when tourism brings more benefits rather than existing at a cost to the host community and the tourist gets an enriched experience.

Several factors affect the behavioral aspect of both host and guest in tourism (see Table 43.1) as both parties have certain expectations from tourism.

The sensitive issues of the guest–host relationship need to be researched before any policy implementation at any tourism destination, to avoid conflicts and tension between the two parties.

Sustainability in the development of a quality tourism experience

Quality tourism experience is complex to understand and can be interpreted differently from various perspectives. As many factors are associated with tourism, a quality tourism experience differs accordingly. A tourist's quality experience depends on high value and authentic products or services, which act as a link between the expectation and satisfaction felt by the tourist (Jennings et al. 2009). More often tourism experience is considered as the tourist's experience, placing less emphasis on the host's experience and that of other stakeholders. Quality tourism experience covers a wide range of perspectives like; service quality, the relation between quality and satisfaction, environmental issues, reputation, brand value, guest–host relationship and, more importantly, sustainability issues. The primary concern for the tourism business is to provide memorable experiences to everyone associated with the industry. Tourism is an industry of resources and these resources are inseparable from the activities that are associated with it. In other words, tourism resources cannot be isolated from tourism activities. Hence, the growing trend of quality experience needs special attention given to the overall sustainability of tourism products and resources. Sustainability is applicable to all forms of tourism; from mass tourism to niche tourism sustainable measures should be practiced on a priority basis at tourism destinations. The identification and careful evaluation of tourism resources are the basic features of developing or expanding tourism at a particular destination.

Tangible and intangible sustainability

In general, 'tangible' can be defined as 'capable of being perceived especially by the sense of touch, precisely identified or realized by the mind' (Scholtz and Slabbert 2016). Tangible sustainability is to identify the impacts of tourism development on resources which can be touched or quantified in nature. The impacts on stakeholders' tourism experience can be either positive or negative in nature. The positive impacts that can be identified at tourism destinations include infrastructure and communication development and employment creation, whereas the negative impacts include such factors as resource degradation, excessive use of services, economic leakages, population growth and seasonality.

On the other hand, 'intangible' is defined as 'something that cannot physically be touched, but rather can be just experienced'. Similar to tangible products, intangible products also impart both positive and negative impacts in tourism development. The positive impacts such as social cohesiveness, pride and respect of community and cross-cultural understanding happen because of tourism, while a few negative impacts cannot be avoided, such as: commodification of culture and tradition, cultural degradation, perception and behavioral changes, conflicts between host communities, etc.

As tourism includes both tangible and intangible resources, the industry experiences both positive as well as negative impacts. Each and every stakeholder gets affected by the impacts, but the host community that resides at a destination experiences the most impacts. These impacts need special attention before the destination experience starts to decline in the tourism area life cycle, pointing towards the importance of resource sustainability. Sustainable practices should be undertaken at popular tourism destinations as well as at the declining destinations, to make sure that tourism thrives for the long term through overall sustainable tourism development. The principles of sustainability are as follows:

a) Concerned about the future: Development should be done while keeping in view the positive and negative impacts. The availability of natural resources to future generations should not be prejudiced by the present generation, hence the conservation of resources is significant while development is taking place.
b) Equality in participation: Every stakeholder involved in tourism needs to be given an equal opportunity of participation, despite the social and political influences. The local community's views and opinions should be included in the decision-making process, along with those of the policy makers.
c) Evaluate pressure on the environment: The environment and the ecosystem should be taken into consideration so that the negative impact on the environment can be limited. Before planning any developmental activity an environmental impact assessment must be done at the destination and its peripheral areas.

d) Importance to the local community: The development should be planned to provide maximum benefit to the local community. The community's values, pride, culture and traditions should not be affected by tourism development. Economic leakage should be suppressed to ensure that maximum economic benefit is directed towards locals.

Guest and host responsibilities towards sustainability for a better tourism experience

Sustainability in tourism allows the tourist to enjoy a rich experience with better opportunities for the hosts as well. Hence it is the responsibility of both parties to maintain mutual coordination for long-term development. The hosts need to provide authentic values in exchange for tourist income. Tourism as a service industry mostly delivers experience as an end product and involves extensive interaction between host communities and the guests. Therefore the host community must be aware of the products and services they provide to the guests and the guests, in turn, need to be aware of the practices, culture and traditions of the hosts. Awareness and interaction can influence a positive attitude and increase the concern for sustainability, not only for the tourists but also for the host community (UNWTO 2005). A close social relationship between hosts and guests influences a positive change in the attitude of tourists towards hosts and the destination, providing an opportunity for sustainable tourism development. At a tourism destination where the host community is empathetic towards the tourists, a positive interaction builds a strong social relationship between the hosts and tourists. Cultural differences between guests and hosts are a major concern in tourism. The interaction between international guests and hosts raises concerns over cultural intolerance at destinations. In the case of intolerance, a mutual understanding between different cultures can enhance consciousness and tolerance, reducing conflicts, tension and prejudice between the guest and the host (Zhang et al. 2006).

Case study 43.1: Mawlynnong, the cleanest village of Asia

Mawlynnong is a small village situated in the forest-covered East Khasi Hills District of Meghalaya, India, around 80 kilometers from the state capital Shillong. The pristine village consists of 102 households with a population of around 550 people, who belong to a tribal community indigenous to the state (Lyngdoh 2017). Members of the tribal community of '*Khasi*' are considered as warriors and worshipers of nature. The century-old village is in the limelight because of its natural beauty and more importantly because of the practice of environmental cleanliness and hygiene by the residents as an aesthetic way of living. The pristine village situated in the lap of 'The Abode of the Clouds' (Meghalaya) is best viewed in the monsoon season, when its lush greenery, streams with active waterfalls and flowering colorful orchids add beauty to the village. The traditional tree houses and living root bridges signify the generations-old architectural skills of the local communities of the village.

Prior to 2003, though the village was well known in the North-East region of India, it was not highlighted globally. Despite such natural attractiveness, the connectivity and communication issues back then at the village were so great that the village was isolated and marked as remote on the tourism map. The year 2003 was a landmark year for the village when *Discover India*

Figure 43.2 Living root bridge at Mawlynnong
(*Image*: Saurabh Kumar Dixit)

magazine declared the hamlet as the 'cleanest village of Asia'. The prestigious tag and extensive narration in the global magazine about the village and its aesthetic practices gave the village worldwide prominence. In the aftermath of gaining its new identity, connectivity to the village improved as a result of increased tourist visits. The inflow of tourists in the peak season was around 250 daily in 2005 but this has increased to 1,200 to 1,500 in the time span of a decade. A rising number of tourists in such a short time span impacts any tourism destination hence the issues and effects at Mawlynnong cannot be overlooked. The unique living root bridge located in the village has started deteriorating due to excessive tourist inflow in the village (See Figure 43.2).

On the positive side, tourism provides a better livelihood option to the residents of Mawlynnong as the residents engage in tourism-related activities like providing bed and breakfast and souvenir outlets, all the while maintaining their surroundings. Tourism has increased the revenue of the locals and the basic infrastructure of the village has also been improved: footpaths, community halls, roads, electricity and more importantly communication facilities have all benefitted from tourism. On the other hand, the massive tourist inflow into the small village creates pressure on the locality and resources and even on the residents' lifestyle. As tourism grows issues of noise pollution, vehicle parking, interference with residents' privacy and resource management stand out as major challenges at Mawlynnong village. A community-based tourism model has been adopted at the village and the community people are engaged in the process of tourism, resource management and conservation.

Tourism at Mawlynnong village may be considered as village tourism, ecotourism, responsible tourism or even sustainable tourism, where an equal level of responsibility is required from both hosts and guests. Hosts are quite aware of their responsibility to provide an authentic and unique experience to the visitor. In the same way the guests need to follow the guidelines of the village community and avoid spoiling the community culture and exploiting the locals. Natural resources like the orchids, the tree house and the living root bridge need to be conserved by the locals by making the visitors aware of their importance and significance. Finally, the carrying capacity of the village needs to be identified with respect to the number of tourists and the government needs to push for a sustainable tourism development at the 'cleanest village of Asia'.

Conclusion

Tourism experience is a psychological aspect of stakeholders associated with the industry, either directly or indirectly. In the tourism industry production and consumption run simultaneously which further influences experience building. Though several stakeholders are associated with tourism, guests (*tourists*) and hosts (*local community*) are the major ones. In many studies, tourism experience is studied from the tourists' point of view while giving less preference to the other stakeholders. Guests and hosts can be considered as two separate sides of the same coin in the tourism industry, hence equal attention is needed when it comes to tourism experience. Both parties experience tourism in different ways; tourists search for authenticity, uniqueness, entertainment and most significantly safety and security, whereas the aspects that matter most to the hosts include the economy and employment generation, improved socio-cultural life and resource management. Individual satisfaction regarding the said aspects leads to a quality tourism experience for both parties. A better tourism experience gives the tourists a good perception of the destination, creating an intent to revisit and recommend the service or destination to others, while positive behavioral changes are reflected among the hosts. A superior tourism experience also builds a healthier relationship between the guests and the hosts. Guest–host relationships vary according to place, impacts, consumption pattern, host perception, experiences and destination image. In order to build a quality tourism experience a strong relationship is necessary, where both parties need to benefit; guests get value for money and hosts get economic assistance with less impact on their lifestyle.

In providing a quality tourism experience, the impacts of tourism cannot be overlooked. As different resources are associated with tourism, a three-dimensional impact can be observed; *Economic*, *Socio-cultural* and *Environmental*. Tourism development needs to address the benefits to the stakeholders with less negative impacts from each dimension. Hence the sustainability of resources becomes more significant and sustainable tourism development may be considered as a solution. The guest–host relationship becomes an important factor in sustainable tourism development where both parties have equal responsibilities towards sustainability. Tourists should be aware of and respect the culture and tradition of the host community and need to be restrained from the activities which may exploit or affect (in a negative manner) the host. The host community needs to be aware of the uniqueness and authenticity of its culture, tradition and other tangible resources. Mutual understanding, an improved relationship and a common mindset between guest and host can achieve sustainable tourism goals to make the tourism survive for the long term. To conclude, tourism experience is not only the experience of tourists but also the other stakeholders' experiences, opinions and perceptions, in the process prioritizing the development of a sustainable tourism industry.

References

Binkhorst, E. (2005). [Online] Available at: https://scholar.google.com/scholar_lookup?hl=en&publication_year=2005b&author=E.+Binkhorst&title=+The+co-creation+tourism+experience+ [Accessed: May 5, 2020].

Boley, B. B. and Green, G. T. (2015). 'Ecotourism and Natural Resource Conservation: The "Potential" for a Sustainable Symbiotic Relationship', *Journal of Ecotourism*, 15(1): 36–50. DOI: 10.1080/14724049.2015.1094080

Clarkson, M. B. E. (1995). 'A Stakeholder Framework for Analysing and Evaluating Corporate Social Performance', *The Academy of Management Review*, 20(1): 92–117. DOI: 10.5465/amr.1995.9503271994

Cronauer, D. (2012). *Host-Guest Relationships in Non-Commercial Tourism Settings: Wwoofing in New Zealand.* Wellington: Victoria University of Wellington. Available at: http://hdl.handle.net/10063/2309

Dixit, S. K. (2014). 'Community Involvement and Tourism Development in Meghalaya, India', *South Asian Journal of Tourism and Heritage*, 7(2): 25–32.

Donaldson, T. and Preston, L. E. (1995). 'The Stakeholder Theory of the Corporation: Concepts, Evidence, and Implications', *The Academy of Management Review*, 20(1): 65–91. DOI: 10.5465/amr.1995.9503271992

Erislan. (2016). 'Tourist Attraction and the Uniqueness of Resources on Tourist Destination in West Java, Indonesia', *Review of Integrative Business and Economics Research*, 5(1): 251–266. Available at: https://ssrn.com/abstract=2864086

Freeman, R. E. (1984). *Strategic Management: A Stakeholder Approach.* Boston: Pitman.

Grönroos, C. (2006). 'Adopting a Service Logic for Marketing', *Marketing Theory*, 6(3): 317–333. DOI: 10.1177/1470593106066794

Hinch, T. and Higham, J. (2004). *Sport Tourism Development.* Clevedon: Channel View Publication.

Istvan, K. and Krisztina, Z. (2011). 'Safety and Security in the Age of Global Tourism', *Applied Studies in Agribusiness and Commerce–APSTRACT*, 5(3–4): 59–61. Available at: https://ageconsearch.umn.edu/record/104672

Jennings, G., Lee, Y. S., Ayling, A., Lunny, B., Cater, C. and Ollenburg, C. (2009). 'Quality Tourism Experiences: Reviews, Reflections, Research Agendas', *Journal of Hospitality Marketing & Management*, 18(2–3): 294–310. DOI: 10.1080/19368620802594169

Larsen, S. (2007). 'Aspects of a Psychology of the Tourist Experience', *Scandinavian Journal of Hospitality and Tourism*, 7(1): 7–18. DOI: 10.1080/15022250701226014

Loi, K. L. (2009). *Entertainment as a Tourism Development Tool in Macao.* s.l: James Cook University.

Lyngdoh, B. (2017). *Mawlynnong: Issues through a Dramatic Lens.* [Online] Available at: http://theshillongtimes.com/2017/08/08/mawlynnong-issues-through-a-dramatic-lens/ [Accessed: September 9, 2019].

Maccannell, D. (1973). 'Staged Authenticity: Arrangements of Social Place in Tourist Setting', *American Journal of Sociology*, 79(3): 589–603. DOI: www.journals.uchicago.edu/doi/pdfplus/10.1086/225585

Molleda, J. C. (2010). 'Authenticity and the Construct's Dimensions in Public Relations and Communication Research', *Journal of Communication Management*, 14(3): 223–236. DOI: 10.1108/13632541011064508

Nash, D. and Smith, V. L. (1991). 'Anthropology and Tourism', *Annals of Tourism Research*, 18(1): 12–25. DOI: 10.1016/0160-7383(91)90036-B

Prahalad, C. K. and Ramaswamy, V. (2004). 'Co-creating Unique Value with Customers', *Strategy and Leadership*, 32(3): 4–9. DOI: 10.1108/10878570410699249

Prebensen, N. K. and Foss, L. (2011). 'Coping and Co-creating in Tourist Experiences', *International Journal of Tourism Research*, 13: 54–67. DOI: 10.1002/jtr.799

Ram, Y., Bjork, P. and Weidenfeld, A. (2016). 'Authenticity and Place Attachment of Major Visitor Attractions', *Tourism Management*, 52: 110–122. DOI: 10.1016/j.tourman.2015.06.010

Ravindran, G., Kumar, K. and Rao, N. (2018). *Understanding Tourists and Hosts.* Delhi: IGNOU. Available at: http://egyankosh.ac.in//handle/123456789/42469

Salvatore, B. and Punzo, L. F. (2016). 'Tourist Development and Host–guest Interaction: An Economic Exchange Theory', *Annals of Tourism Research*, 58: 128–139. DOI: 10.1016/j.tourman.2015.06.010

Sautter, E. T. and Leisen, B. (1999). 'Managing Stakeholders: A Tourism Planning Model', *Annals of Tourism Research*, 26(2): 312–328. DOI: 10.1016/S0160-7383(98)00097-8

Scholtz, M. and Slabbert, E. (2016). 'The Relevance of the Tangible and Intangible Social Impacts of Tourism on Selected South African Communities', *Journal of Tourism and Cultural Change*, 14(2): 107–128. DOI: 10.1080/14766825.2015.1031796

Steiner, C. J. and Reisinger, Y. (2004). 'Enriching the Tourist and Host Intercultural Experience by Reconceptualising Communication', *Journal of Tourism and Cultural Change*, 2(2): 118–137. DOI: 10.1080/14766820408668172

Trilling, L. (1972). *Sincerity and Authenticity*. London: Oxford University Press.

UNWTO. (2005). *Making Tourism More Sustainable a Guide for Police Makers*. Madrid: United Nations World Tourism Organisation, pp. 9 and 39. DOI: www.e-unwto.org/doi/book/10.18111/9789284408214

UNESCO. (1976). 'The Effects of Tourism on Socio-cultural Values', *Annals of Tourism Research*, 4(2): 74–105. DOI: 10.1016/0160-7383(76)90100-6

Valerio, S. (2017). Hosts and Guests. In L. L. Lowry (ed.) *The SAGE International Encyclopedia of Travel and Tourism* (pp. 618–621). London: Sage. DOI: 10.4135/9781483368924.n228

Wang, N. (1999). 'Rethinking Authenticity in Tourism Experience', *Annals of Tourism Research*, 26(2): 349–370. DOI: 10.1016/S0160-7383(98)00103-0

Zhang, J., Inbakaran, R. J. and Jackson, M. (2006). 'Understanding Community Attitudes towards Tourism and Host—Guest Interaction in the Urban—Rural Border Region', *Tourism Geographies*, 8 (2): 182–204. DOI: 10.1080/14616680600585455

Part VI

Emerging avenues of tourism experience and co-creation

44

Co-creation of tourism experiences

A conceptual framework

Mine Inanc and Metin Kozak

Introduction

In recent years, social, economic, political and technological changes have led to the birth of marketing-oriented paradigms that are referred to as "consumerism" and rely more on consumers. On the supply side, businesses have started to produce and design products in line with the demands and expectations of consumers (Vargo and Lusch 2004). On the demand side, consumers have become more dynamic and mobile and have begun to give more importance to the subject of experience itself. Being experience-oriented leads consumers to engage in more interactive communication with both service providers and other consumers (Cutler and Carmichael 2010; Tung and Ritchie 2011). The distance between service providers and consumers in the traditional marketing model is about to decline, and various forms of production models with the two parties exist, e.g. co-creation.

Co-creation is a new term defined as a value which is produced as a result of the interaction between two parties; consumers and service providers (Frow, Nenonen, Payne, and Storbacka 2015; Jaakkola, Helkkula, and Aarikka-Stenroos 2015). While the former discover the satisfaction of consumers by differentiating the product, the latter strive to create a difference in their experience. This type of experience directly affects the product value perceived by consumers. From this perspective, the concepts of experience economy and co-creation have become important tools in terms of developing effective marketing strategies by revisiting the marketing mix (Dev and Schultz 2005; Gordon 2012; Grönroos 1994).

Although the subject of co-creation has been introduced in the field of tourism recently, it has gained particular attention by being explored by empirical studies (e.g., Cabiddu, Lui, and Piccoli 2013; Lin, Chen, and Filieri 2017; Lugosi 2014). However, there has been little attempt to indicate how the system is conceptualized and works in practice. Thus, as a conceptual study, this chapter aims to develop a model with a specific reference to how the co-creation system works. The model indicates details of the elements, e.g. inputs (interaction among consumers and between consumers and service providers, etc.), outputs (consumer satisfaction, behavioral intentions, etc.), and instruments (environment, experience, etc.). Finally, the study also provides detailed information about the co-creation model from a holistic perspective, to be followed up in empirical studies of the future.

Literature review

The new millennium has become central to understanding the value of knowledge creation and gaining the utmost benefit from it for the sake of people's lives. This has pushed people to seek a more frequent and intense travel and tourism experience and has created a desire to feel a sense of being respected and belonging to a certain tribe within the society. The types of service offerings have also been modified by putting consumers in the middle of the system, because there has been a kind of shift from production-oriented marketing strategies to consumer-centric marketing, enabling the consumer to become a co-creator of value rather than a co-producer (Lusch and Vargo 2006; Vargo and Lusch 2004). There is also a learning process in this form of marketing when we look at the hierarchical model originally initiated by Maslow in the 1940s (Maslow 1943), but updated early this century with two more needs: learning (or knowledge) and aesthetics. While consumers are interacting with their peers or service providers, they also try to meet their learning needs beyond their respect needs (Rihova, Buhalis, BethGouthro, and Moital 2018).

The changes in consumption habits that began in the new century show themselves in the tourism industry (Pizam 1999). Compared to previous periods, consumers in the tourism industry have become more dynamic and experienced and, as a result, more demanding consumer segments have emerged. It is also increasingly difficult to meet the expectations of such consumer groups. An increase is expected in the number of people who want to travel independently rather than depending on a group. It is also expected that people will take shorter but more frequent vacations rather than long-term vacations or trips. As a result of the facilitation of tourism through intercultural communication, there are some social benefits in terms of providing social harmony and a better understanding of different cultures (Dogan 1989). A person who is looking to do something different in their vacations may want to experience everything that gives them pleasure and happiness. The co-creation experience is also expected to be a significant advantage for tourists traveling in smaller groups for a shorter period and looking for a different sort of vacation experience.

The co-creation experience is inherently interdisciplinary and applies to many areas of the service industry. When the literature is examined, it is evident that co-creation has become a research subject for many fields. These include popular music (Choi and Burnes 2013), body care (Bilgram, Bartl, and Biel 2011), and short video footage (Xie, Tsai, Xu, and Zhang 2017). In the tourism and hospitality industry, more and more attention has been given to studies on the experience of co-creation, resulting from the communication and interaction between consumers themselves or between consumers and service providers (e.g. Buonincontri, Morvillo, Okumus, and van Niekerk 2017; Cabiddu, Lui, and Piccoli 2013).

Co-creation: meaning and characteristics

Co-creation is defined as creating a socio-cultural value for consumers by working together with other consumers and businesses within the same environment (Prahalad and Ramaswamy 2004). It is stated that consumers are collaborators in the business while receiving the goods or services created by the business (Vargo and Lusch 2004). Reay and Seddighi (2012) consider the benefits gained from consumers' opinions and experiences during the production development process. In generic terms, co-creation provides a way of creating a common value by moving personalized experiences to the forefront, based on the cooperation of consumers with their peers, and creating value for consumers by bringing them to the center.

With the occurrence of new paradigms, moving from goods-dominant logic to service-dominant logic (Vargo and Lusch 2004) and customer-dominant logic (Rihova, Buhalis, BethGouthro, and Moital 2018), there has also been a shift in marketing methods and practices. Jaakkola, Helkkula, and Aarikka-Stenroos (2015:193) define the concept by stating "service experience co-creation may encompass lived or imaginary experiences in the past, present or future, and may occur in interaction between the customer and service provider (s), other customers and/or other actors." Service experience co-creation has important implications for the field of marketing, which has a possible influence on promotion, branding, and relationships with stakeholders.

As such, co-creation is based on the inclusion of consumers in all processes of business life, and incorporating them into the value creation process based on their potential creativity and experiences. In this sense, all consumers in all creative forms and processes of businesses want to have a share in the value to be created for themselves. At this point, consumers move from a passive to an active position and are involved in the act of co-creation. The ability to respond to the ever-changing competitive environment of businesses and the effect of co-creation on the fundamental dynamics of innovation processes plays a key role at this point.

The interaction of new technologies has made it difficult to maintain sustainability in competitiveness and it has become more difficult to achieve this with classical marketing approaches. So businesses have turned towards the goal of creating value as a new form of product development and marketing strategy. It is not just about meeting natural needs with consumption, services, or products, it is also a global system of indicators regulated by codes and shapes society as a source of pleasure and prestige based on artificial needs (Baudrillard 1970). In the traditional marketing approach, competition is product-based, while it is consumer-based in the experiential marketing approach. When the historical development of marketing is considered, the commodity, which is important in production-oriented marketing, has currently been replaced by the importance of consumer-oriented experiences (Dev and Schultz 2005; Grönroos 1994).

Different stakeholders can be involved to create a space for co-creation but consumer feedback is an integral part of collaborative creation. Mukhtar et al. (2012) classified the models of co-creation under three categories: DART model, Payne model, and Grönroos model. In particular, as an abbreviation of dialogue, access, risk assessment, and transparency, the DART model is considered as the main model of co-creation. Payne, Storbacka, and Frow (2008) provide a three-part method: consumers, encounter and interactions, and supplier processes. The Grönroos model allows businesses to understand how their value invitations are valuable in the eyes of consumers. To sum up, the examination of all three models indicates the focus of the DART model as a facilitating dialogue and it is important to make sense of consumers' processes with the Payne model. The Grönroos model can be seen as a function of fulfilling value. The DART model initiated and suggested by Payne, Storbacka, and Frow (2008) can be simply considered as revisiting the main characteristics of the co-creation experience.

Dialogue: In terms of co-creation, there appears to be a profound commitment, interaction, and willingness to participate from both sides (Prahalad and Ramaswamy 2004). It is important for businesses to actively communicate with consumers and establish a dialogue based on the equality of the parties. In this way, businesses play an important role within the perspective of consumers by observing their expectations and acting with them to solve the problems that will be experienced. At this point, it is important for businesses to create the communication channels appropriate to their target audiences and to use this environment to create a sustainable dialogue.

Access: It serves many factors in terms of co-creation. It both serves the transparency policy of the model and engages the consumer with new ideas. While a process from consumer to consumer is dominant from a traditional marketing point of view, in new marketing models the consumer does not depend solely on the resources offered by businesses. Therefore, businesses should follow an open policy to share their information with consumers and be accessible, transparent, etc.

Risk assessment: In general terms, it can be expressed as possible damages that may be faced by service providers and consumers (Prahalad and Ramaswamy 2004). Consumers are also involved in the process of becoming a partner in the losses of businesses they co-create. Therefore, it should be recalled that bilateral risks may exist in the process of creating value on both sides. In other words, co-creation is a phenomenon that involves risks as well as benefits. At this point, consumers' trust in the business is an important element because, in the process of co-creation, risk management is based on the fact that it is out of the business's control, which becomes a partner to eliminate these risks to the advantage of its consumers.

Transparency: With the changing profile of consumers and marketing, the consumer wants to have access to all kinds of information. Information that was previously monopolized by the business can now be conveniently accessed by the consumer using advanced technological facilities. Consumers want to gain access to information as much as businesses do. Businesses should adopt a transparent structuring approach to consumers and provide equal access to the information produced and distributed.

From experience economy to co-creation experience

In original terms, co-creation seems to be following similar paths of interactive quality to those initiated by Lehtinen in the 1990s (Lehtinen and Lehtinen 1991). This type of interactive quality refers to the outcomes directly or indirectly produced through the interactions between service providers and consumers or among consumers as peers benefitting from the same service experience or environment. As such, interactive quality can be considered as an initial stage of co-creation, although the primary objective is not to intentionally contribute to any forms of co-creation experience. For instance, tour guides think their tasks require a high level of service orientation, which can be achieved by interacting more with the tour participants. To ensure that the participants are consistently satisfied, the tour guide thinks that the best platform for success is to intensively interact with the tourists he is dealing with. Any outcomes that are produced together can be seen as examples of co-creation.

Subsequently, following the early 2000s, we have become familiar with another new but similar subject – the experience economy. According to Pine and Gilmore (1998), the concept of the experience economy aims to provide a consumer-based experience instead of benefit-oriented goods and services. Experience is the fourth largest economic output by businesses to consumers beyond commodities and services. The authors have argued that these economic outputs will enable businesses to differentiate themselves from their competitors in the future and optimize profit. With the understanding of the experience economy, new marketing strategies have emerged. In this sense, consumer needs and desires should be defined and these expectations should be answered in the most correct way. Consumers do not just buy goods and services but they also purchase experiences covering the entire consumption process. After benefitting from such experiences, consumers tend to buy the same product again, but when they share the experience with environmental factors, they also offer their recommendations to the provider in the form of feedback. At this point, businesses that devote their efforts to consumer satisfaction can gain a competitive advantage.

The experience economy and co-creation concepts, as highlighted above, have become an important tool for the development of effective marketing strategies by reviewing the marketing mix, with production-oriented marketing paradigms becoming consumer-oriented (Dev and Schultz 2005; Gordon 2012; Grönroos 1994). Indeed, as in any other consumer segments, a radical change in tourism demand has also begun in recent years. The vacation period, which previously lasted for two or three weeks, has started to shorten and people prefer to use their annual leave periods on shorter and perhaps more frequent vacations. Empirical studies have also concluded that short-term consumers, particularly those traveling alone, spend more per person per day than long-term vacationers traveling with more than one person, and those who have been to the same region more than once before (Aguiló, Rosselló, and Vila 2017; Alegre, Mateo, and Pou 2011; Alegre and Pou 2006; Barros, Butler, and Correia 2010; Gokovali, Bahar, and Kozak 2007). From there, the importance of the co-creation experience for consumers traveling alone and experiencing a destination for a short time emerges automatically.

The role of co-creation in tourism marketing

In its essence, as a part of the service industry, tourism offers an experience itself. Consumer experiences offer cognitive and emotional values rather than functional values. Tourism experiences are loaded with subjective emotional and symbolic meanings (Holbrook and Hirschman 1982). Centres of tourism with cultural, technological, and recreational factors, gastronomic values, and well-being facilities are suitable areas for creating and being involved in co-creation experiences for both consumers and service providers.

Today, the world is starting to become more individualized. As a consequence, options for this new, emerging individualized segment are increasing day by day. This is the case not only in tourism and travel operations but also in many other manufacturing industries such as automobiles, electronics, etc. It then becomes necessary to make changes in the management and marketing practices of businesses. The scenarios for the future indicate that people usually travel independently as singles and take shorter breaks, and as a consequence they will require more opportunities to be entertained in a social environment. This leads to co-creating experiences by interacting with either other peers or service providers or both parties together.

Businesses that enable consumers to turn into value creators as co-creators may also gain a competitive advantage. This is an indication that the product-based approach has been replaced by service-dominant logic (Dong, Evans, and Zou 2007). With the culture of consumption, commodification has emerged and consumers have sought something different. In order to fulfill this demand, businesses will be forced to design and produce more personalized products. On the other hand, within the framework of meeting the demands of consumers, businesses that use their creativity and create value with consumers are in an advantageous position. Therefore, as a means of keeping up with the variable dynamics of the market, consumers are considered as a tool for innovation and competition for businesses (Prahalad and Ramaswamy 2000) or destinations.

A proposed model of co-creation

Figure 44.1 demonstrates the model of co-creation compiled with two parts: inputs and outputs. Undoubtedly, the experience gained from the consumption of any product or service group has begun to constitute the most important step in the consumer behavior model.

From the service point of view, the outcomes of communication and the interaction of consumers with all elements constituting the service have also begun to be directly proportional to the quality or the quality of service experience. The subject of co-creation, developed in the 2000s and depending on the experience economy, is defined as a process in which a value is produced as a result of the interaction between two parties consisting of consumers and service providers and which benefits both parties (Frow, Nenonen, Payne, and Storbacka 2015; Jaakkola, Helkkula, and Aarikka-Stenroos 2015). When the two parties are brought together, a well-focused service industry may offer an environment where the co-creation experience can easily occur, although additional elements of environmental factors co-exist to influence the experience and its outputs, e.g., technology, economy, culture, politics, etc.

The most important aspect of service marketing, which distinguishes it from other types of marketing, is that a business must create the perception that the service offered has a distinctive feature that is not offered by the types of services produced by its competitors. While it is very difficult to make such a distinction, it is possible that the service is different and even differentiated for each consumer, especially with the small refinements that can be added during the experience stage. Since quality is also a subjective concept, how consumers perceive a service can directly affect its quality. Addressing consumers by name, having a conversation about a subject they like, and giving the feeling that their thoughts are taken into consideration at the experience stage are examples of factors that can contribute to the success of service marketing.

As such, human life has become a world of experience today (Pine and Gilmore 1999, 2008). In short, giving a service refers to the activity of meeting the wishes or expectations of consumers. Service marketing, on the other hand, has an abstract characteristic and is a concept that needs to be consumed as soon as it is produced. In a way, the service reflects non-stored and perishable production. Again, it is a continuous process. Due to the consequences of such characteristics, the role of consumer experiences in service marketing is quite evitable. In summary, as indicated in Figure 44.1, experience is shaped not only by consumers' participation in production, but also by

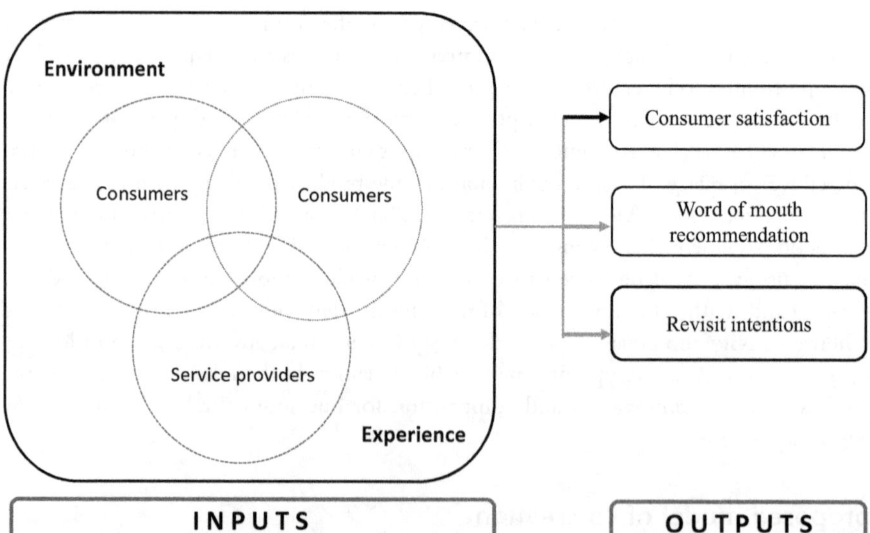

Figure 44.1 The co-creation model
Source: Authors' own elaboration

the output from their interaction with other consumers and service providers (e.g. employees). The final output (e.g. satisfaction or dissatisfaction) from this experience can also affect consumers' future trends (e.g. behavioral intentions and word of mouth recommendations).

Conclusion

Today, due to the rapid changes on the demand side, consumers are likely to become more demanding and responsive. Businesses are (in)directly forced to be ready to comply with consumers' changing needs and wants. As such, the competitive environment in marketing has become more dynamic and complex. It is dynamic because consumers expect a quicker response to their needs and desires. It is complex because maintaining and sustaining consumer satisfaction is difficult, due to a competitive business environment. The service experience is co-created by multiple actors. In particular, the fact that digitalization has been a key medium has led to a series of indicators in which even consumers' reviews have turned into a purchase intention. Consumer reviews that share a consumer's experiences with the service directly affect purchasing behavior. At this point, there is a platform that refers to a co-creation experience including consumers' own participation and contribution to upholding the system.

Having said that, this chapter has suggested the development of a basic and practical model with specific reference to how the co-creation system works. The model also indicates the details of the elements: inputs, outputs, and instruments. The term "inputs" includes the personal interactions among consumers and/or between consumers and service providers whereas "outputs" encompass the side effects of lived experiences with the varying elements of a vacation/destination such as consumer satisfaction and behavioral intentions, etc. "Instruments" refers to the impacts of environmental factors such as technology, economy, politics, etc., and the quality of experience overall within the co-creation system.

References

Aguiló, E., Rosselló, J., and Vila, M. (2017). Length of stay and daily tourist expenditure: A joint analysis. *Tourism Management Perspectives*, 21: 10–17. DOI: 10.1016/j.tmp.2016.10.008

Alegre, J., Mateo, S., and Pou, L. (2011). A latent class approach to tourists' length of stay. *Tourism Management*, 32(3): 555–563. DOI: 10.1016/j.tourman.2010.05.003

Alegre, J., and Pou, L. (2006). The length of stay in the demand for tourism. *Tourism Management*, 27(6): 1343–1355. DOI: 10.1016/j.tourman.2005.06.012

Barros, C. P., Butler, R., and Correia, A. (2010). The length of stay of golf tourism: A survival analysis. *Tourism Management*, 31(1): 13–21. DOI: 10.1016/j.tourman.2009.02.010

Bilgram, V., Bartl, M., and Biel, S. (2011). Successful consumer co-creation: The case of Nivea body care. *Marketing Review St. Gallen*, 28: 34–42.

Buonincontri, P., Morvillo, A., Okumus, F., and van Niekerk, M. (2017). Managing the experience co-creation process in tourism destinations: Empirical findings from Naples. *Tourism Management*, 62: 264–277. DOI: 10.1016/j.tourman.2017.04.014

Cabiddu, F., Lui, T.-W., and Piccoli, G. (2013). Managing value co-creation in the tourism industry. *Annals of Tourism Research*, 42: 86–107. DOI: 10.1016/j.annals.2013.01.001

Choi, H., and Burnes, B. (2013). The internet and value co-creation: The case of the popular music industry. *Prometheus*, 31(1): 35–53. DOI: 10.1080/08109028.2013.774595

Cutler, S. Q., and Carmichael, B. A. (2010). The dimension of the tourist experience. In M. Morgan, P. Lugosi and J. R. B. Ritchie (eds.), *The Tourism and Leisure Experience: Consumer and Managerial Perspectives*, 3–26. Bristol: Channel View Publications.

Dev, C. S., and Schultz, D. E. (2005). A customer focused approach can bring the current marketing mix into the 21st century. *Marketing Management*, 14(1): 18–24.

Dogan, H. Z. (1989). Forms of adjustment: Sociocultural impacts of tourism. *Annals of Tourism Research*, 16(2): 216–236. DOI: 10.1016/0160-7383(89)90069-8

Dong, B., Evans, K. R., and Zou, S. (2007). The effects of customer participation in co-created service recovery. *Journal of the Academy of Marketing Science*, 36(1): 123–137. DOI: 10.1007/s11747-007-0059-8

Frow, P., Nenonen, S., Payne, A., and Storbacka, K. (2015). Managing co-creation design: A strategic approach to innovation. *British Journal of Management*, 26(3): 463–483. DOI: 10.1111/1467-8551.12087

Gokovali, U., Bahar, O., and Kozak, M. (2007). Determinants of length of stay: A practical use of survival analysis. *Tourism Management*, 28(3): 736–746. DOI: 10.1016/j.tourman.2006.05.004

Gordon, R. (2012). Re-thinking and re-tooling the social marketing mix. *Australasian Marketing Journal*, 20(2): 122–126. DOI: 10.1016/j.ausmj.2011.10.005

Grönroos, C. (1994). From marketing mix to relationship marketing: Towards a paradigm shift in marketing. *Management Decision*, 32(2): 4–20. DOI: 10.1016/S1320-1646(94)70275-6

Holbrook, M. B., and Hirschman, E. C. (1982). The experiential aspects of consumption: Consumer fantasies, feelings, and fun. *Journal of Consumer Research*, 9(2): 132–140. DOI: 10.1086/208906

Jaakkola, E., Helkkula, A., and Aarikka-Stenroos, L. (2015). Service experience co-creation: Conceptualization, implications, and future research directions. *Journal of Service Management*, 26(2): 182–205. DOI: 10.1108/JOSM-12-2014-0323

Lehtinen, U., and Lehtinen, J. R. (1991). Two approaches to service quality. *The Service Industries Journal*, 11(3): 287–303. DOI: 10.1080/02642069100000047

Lin, Z., Chen, Y., and Filieri, R. (2017). Resident-tourist value co-creation: The role of residents' perceived tourism impacts and life satisfaction. *Tourism Management*, 61: 436–442. DOI: 10.1016/j.tourman.2017.02.013

Lugosi, P. (2014). Mobilising identity and culture in experience co-creation and venue operation. *Tourism Management*, 40: 165–179. DOI: 10.1016/j.tourman.2013.06.005

Lusch, R. P., and Vargo, S. L. (2006). *The Service Dominant Logic of Marketing: Dialog, Debate and Directions*. New York: ME Sharpe.

Maslow, A. H. (1943). A theory of human motivation. *Psychological Review*, 50(4): 370–396.

Mukhtar, M., Ismail, M. N., and Yahya, Y. (2012). A hierarchical classification of co-creation models and techniques to aid in product or service design. *Computers in Industry*, 63(4): 289–297. DOI: 10.1016/j.compind.2012.02.012

Payne, A. F., Storbacka, K., and Frow, P. (2008). Managing the co-creation of value. *Journal of the Academy of Marketing Science*, 36(1): 83–96.

Pine, B., and Gilmore, J. (2008). The eight principles of strategic authenticity. *Strategy & Leadership*, 36 (3): 35–40. DOI: 10.1108/10878570810870776

Pine, B. J., and Gilmore, J. H. (1998). Welcome to the experience economy. *Harvard Business Review*, July–August: 98–105.

Pine, B. J., and Gilmore, J. H. (1999). *The Experience Economy: Work Is Theatre & Every Business a Stage*. Boston: Harvard University Press.

Pizam, A. (1999). Life and tourism in the year 2050. *International Journal of Hospitality Management*, 18(4): 331–343. DOI: 10.1016/S0278-4319(99)00041-9

Prahalad, C. K., and Ramaswamy, V. (2000). Co-opting customer competence. *Harvard Business Review. Harvard Bussiness Review*, 78(1): 79–87.

Prahalad, C., and Ramaswamy, V. (2004). Co-creating unique value with customers. *Strategy & Leadership*, 32(3): 4–9. DOI: 10.1108/10878570410699249

Reay, P., and Seddighi, H. R. (2012). An empirical evaluation of management and operational capabilities for innovation via co-creation. *European Journal of Innovation Management*, 15(2): 259–275. DOI: 10.1108/14601061211221002

Rihova, I., Buhalis, D., BethGouthro, M., and Moital, M. (2018). Customer-to-customer co-creation practices in tourism: Lessons from customer-dominant logic. *Tourism Management*, 67: 362–375. DOI: 10.1016/j.tourman.2018.02.010

Tung, V. W. S., and Ritchie, J. R. B. (2011). Exploring the essence of memorable tourism experiences. *Annals of Tourism Research*, 38(4): 1367–1378. DOI: 10.1016/j.annals.2011.03.009

Vargo, S., and Lusch, R. (2004). Evolving to a new dominant logic for marketing. *Journal of Marketing*, 68 (1): 1–17. DOI: 10.1509/jmkg.68.1.1.24036

Xie, X.-Z., Tsai, N.-C., Xu, S.-Q., and Zhang, B. Y. (2017). Does customer co-creation value lead to electronic word of mouth? An empirical study on the short-video platform industry. *The Social Science Journal*, 56(3): 401–416. DOI: 10.1016/j.soscij.2018.08.010

45

Co-creation of tourism experience through service-dominant logic

Xiang Ying Mei

Introduction

While experiences have existed as long as human beings have existed, it was Pine and Gilmore (1998) who coined the notion of the experience economy as part of the progress of economic value which moved from commodities, goods and services and then to experiences. Following such a development, there has been an immense focus on creating experiences to gain a competitive advantage for service and experience-based industries such as tourism. The notion of Service-Dominant Logic (SDL) was introduced by Vargo and Lusch (2004) as a shift from the traditional Goods Dominant Logic (GDL). Vargo and Lusch (2004) argue that consumers do not buy tangible products or services, but rather life-enhancing experiences in which the value resides in personalized experiences. Essentially, SDL advocates that value is co-created with the customers, rather than being created by one actor such as the tourism provider and being subsequently delivered to the tourists (Vargo and Lusch 2016). In addition, "intangibility, exchange processes, and relationships are central" in this process (Vargo and Lusch 2004:4).

Developed primarily in the marketing context, SDL is particularly applicable to areas such as the tourism industry and tourist experiences (Beckman and Khare 2018) as it is essentially an experience-based industry. A network and stakeholder approach is also required to provide the complete tourist experience. Tourism destinations are arguably considered as complete tourist experience suppliers that assemble various services and experiences (Ashworth and Voogd 1990, Volo 2006). Hence, a destination can be regarded as an amalgam of all products, services and ultimately experiences provided locally (Buhalis 2000). In addition to the concept of complete tourist experiences, tourism destinations are considered dynamic, evolving and complex systems, which encompass numerous factors and activities that are interdependent and require various stakeholders to work together in non-linear relationships (Baggio 2008). Subsequently, value co-creation through SDL is particularly relevant in the tourism industry (Cabiddu et al. 2013).

Co-creation and SDL

The notion of SDL was introduced by Vargo and Lusch (2004:9) as "a mindset, a lens through which to look at social and economic exchange phenomena so they can potentially be seen more clearly." Moving from the traditional GDL where the manufacturer is the

Table 45.1 Principles of SDL

The essential principles of SDL	*How to achieve it?*
In the value of service provision, customers are the arbiters	This is achieved in interaction with the tourism operator
Operant resources will determine competitive advantage	Such advantage can be achieved as a mutual collaboration between customers and the tourism operator
SDL focuses on the dynamic development of relationships	Such relationships are established by engaging customers as well as value-network partners
Value creation is a phenomenological concept	Such value creation is influenced by the resource integrators

Source: Lusch and Vargo (2006), Vargo and Lusch (2004, 2008), Tronvoll (2012)

creator of the goods and products, SDL is where the customer is the co-creator of value when the organization offers a value proposition (Vargo and Lusch 2008:7). Value co-creation is the foundation of SDL (Vargo and Lusch 2004). Hence, the organization together with the customer creates value. Essentially, value co-creation is a function of interaction with customers to create value-in-use (Vargo and Lusch 2008). On the basis of such a notion, it can be argued that value is subjective as it is determined by the customer (Prebensen et al. 2017). The essential principles of the SDL perspective are summarized in the following table (see Table 45.1).

The customers' point of view

As discussed, the role of the customer is emphasized in SDL, contrary to the traditional GDL. To achieve such a situation, providers need to assist customers in their own value-creation processes. In SDL, customers are involved in the context of their own networks rather than being considered as isolated entities. More importantly, customers are regarded as resources rather than passive targets (Vargo and Lusch 2008). Another term which is frequently applied in service and tourist consumption is co-production. It is, however, important to note the differences between value co-creation through SDL and co-production. While co-production also requires a higher degree of involvement by the customer (Payne et al. 2008), it essentially focuses on the provider's own conception of how customer–provider integration should take place, whereas the customer's point of view is largely disregarded (Chathonth et al. 2017, Payne et al. 2008).

Nevertheless, despite a large amount of interest and numerous studies, value co-creation through SDL is not easily defined and studied. Some of the challenges are attributed to the elusiveness of the definition of value. According to Grönroos (2008), value creation is a process in which the result increases the well-being of the user or customer. In a simplified definition, Grönroos (2008:303) explains that value for customers means that after they have been assisted by a self-service process or a full-service process they feel better off than before, in some respect. As discussed, only the customers can determine what type of value has been created for them. For instance, some customers may see value in going on a luxury holiday while others believe that value is spending time with the family on a holiday where the destination and location itself is not as significant. Hence, suppliers cannot fundamentally create value as they can only make value propositions. In order to achieve a competitive advantage, a provider has to develop its capacity to add to a customer's resources or to provide the customer with tools that enable him or her to make better use of their present resources as a way to create value (Payne et al. 2008:86).

Understanding tourist experiences

The understanding of value as discussed bears some similarity to the elusive understanding of an individual's tourist experiences. Likewise, tourist experiences are also very subjective, meaning different things to different individuals. While some experiences may be regarded as unique and memorable for some, they are not as out of the ordinary for others. All depends on the tourists' or individuals' previous experiences and expectations. Just like value, experience cannot be created by the tourism providers as it is considered as a tourists' mental response to the stimulus in the environment (Mei et al. 2018). Thus, tourism providers can neither deliver value nor make the experience for the tourist; they can only facilitate or stage the context so that the best experience or value is gained by the customer as a result (Grönroos 2011, Mossberg 2007). As value may mean different things to different people, the extent to which a tourist wants to participate in order to co-create value such as their own tourist experience also varies. For instance, although all tourist experiences are co-creative by nature, the degree of participation and immersion required will vary depending on the type of experiences and the tourists themselves. In addition, the two concepts are closely linked as customers assign value to experiences, and tourist experiences are the result of something which was co-created by the customers and tourism providers to obtain value (Pine and Gilmore 1998, Ramaswamy 2011, Suntikul and Jachna 2016, Yang and Mattila 2016). Furthermore, value co-creation of tourist experiences may occur before the stay at various levels, on-site at the destination, or afterwards upon returning home (Campos et al. 2018). As it has been argued that value is created in use, the on-site focus seems to be an appropriate stage to investigate as the core of tourism lies in the experiences that tourists consume on-site at the destination.

The justification of value co-creation and SDL in tourist experiences

A key justification of co-creation in tourist experiences is that when tourists themselves actively participate in the production and consumption of the experience, they co-create the experience with the tourism provider, making it unique to each individual versus a "standardized" service experience, which is provided to everyone (Buonincontri et al. 2017). Since the tourists themselves must invest time and effort in active participation to co-create the experiences, they are likely to appreciate them more, especially when they lead to hedonic benefits and ultimately memorability. Such a development is a direct result of intensive market competition in the tourism industry as well as changes in the attitudes and needs of tourists, moving from traditional mass tourism products to unique and memorable experiences (Cracolici and Nijkamp 2009). Hence, the co-creation of unique and life-enhancing tourist experiences is the ultimate goal of the tourism industry, including the tourists, providers, and destinations (Campos et al. 2018). Furthermore, Grönroos (2006) argues that while tourists get the opportunity to immerse and engage themselves in the process, co-creation also advocates and creates opportunities for providers to engage themselves with the tourists in the value-generating process.

As per the definition of SDL, the customer or tourist is always the co-creator of experiences (Blazquez-Resino et al. 2015, Prebensen et al. 2017). This suggests that a close relationship between the tourism provider and tourists is imperative at all stages of experience production and consumption (Shaw et al. 2011). As discussed, SDL recognizes the importance of tourists' active involvement in order to develop the experience or the "product". Hence, when developing tourist offerings and experiences, the tourist becomes central to the value co-creation process (Payne et al. 2008). In addition, based on the stakeholder and network approach, the presence of other tourists and the locals as well as other tourism suppliers in a destination will also influence the experiences and value co-creation process (Lichrou et al. 2008, Poulsson 2014). The tourism industry may be in a unique position, as the tourism destination is not a product owned by a single company. Several small encounters at the business level with a variety of tourism principals such as taxi drivers, hoteliers and waiters, as well as elements of the local attractions including museums, theaters, beaches, theme parks and so on, actually compose the tourists' overall experience (Buhalis 2000). As tourists' experiences are holistic, it is therefore necessary to integrate the destination's stakeholders into coordinated strategies that seek to create "experience environments" (Blazquez-Resino et al. 2015). According to (UNWTO 2005:7), stakeholders in a destination may consist of the following:

- national governments;
- local governments with specific competence in tourism matters;
- tourism establishments and tourism enterprises, including their associations;
- institutions engaged in financing tourism projects;
- tourism employees, tourism professionals and tourism consultants;
- trade unions of tourism employees;
- tourism education and training centers;
- travelers, including business travelers, and visitors to tourism destinations, sites and attractions;
- local populations and host communities at tourism destinations;
- other juridical and natural persons having stakes in tourism development including non-governmental organizations specializing in tourism and directly involved in tourism projects and the supply of tourism services.

As a result, the tourism industry needs to consider the roles of all of these stakeholders in a network approach as part of tourists' value co-creation through the SDL.

Critiques of co-creation and SDL

Just like terms such as innovation, co-creation seems to have become a buzzword that both practitioners and researchers alike are eager to study and explore. As discussed, despite the immense interest and attention given to it in recent years, co-creation, just like tourist experiences, is not easily measured and defined. Furthermore, tourism studies that adopt SDL to understand the co-creation of tourist experiences are currently limited, despite it being particularly relevant to the tourism context (Blazquez-Resino et al. 2015). Most studies also remain at a largely conceptual level, due to the difficulties, inconsistencies and reluctance to adopt the notion in practice (Campos et al. 2018, Langlotz 2016, Prahalad 2004). At the practical level, SDL's limited adoption is sometimes caused by a reluctance by providers to change their business models and move from traditional ideas about the roles of a consumer and company, as well as a fear of its processes breaking down when customers are given more power in the process (Langlotz 2016, Prahalad 2004). Besides the reluctance and difficulties in application, there are some other criticisms of SDL, as some authors argue the novelty of the word "service" (Achrol and Kotler 2006) while Stauss (2005) points out the lack of precision and conceptual clarity, making the notion difficult for practitioners to comprehend and apply in practice.

Challenges in the tourism industry

Similar arguments are also found in the tourism industry, as tourist experiences are also not new. Although the degree of involvement varies according to the type of tourist experiences, experiences in the tourism industry have always required some level of involvement by the consumers. For instance, any tourism consumption would require the presence and participation of the tourists themselves, hence it is co-creative by nature. This is for example characterized as the inseparability between production and consumption, a term coined to reflect the uniqueness of services and the service industry, including the tourism industry (Kandampully 2003). Thus, the idea of co-creation and SDL has arguably always existed in terms of tourist experiences; the question is whether co-creation and the SDL approach in the tourism industry is really something new and revolutionary. On the other hand, it must be noted that tourists' need to actively participate and co-create their experiences varies. Such a degree of involvement was, for instance, conceptualized as adoption, immersion, active participation and passive participation by Pine and Gilmore (1998). Beach and sun tourists may merely choose passive participation in their holiday experiences in comparison to tourists who seek unique farm tourist experiences. Hence, it cannot be assumed that all tourist experiences must be co-created or require a complex system that integrates the notion of SDL in the process. However, in certain types of tourist experiences, which do require tourists' active participation and immersion to gain unique and hedonic experiences, a clear understanding and adoption of co-creation through SDL may lead to unique opportunities and outcomes for both the tourist and the provider. Furthermore, the complexity of terms such SDL, value, experience, value-in-use and co-creation may have contributed to the reluctance among practitioners, as it is often assumed that co-creation would require drastic changes in operation and current business models. Nevertheless, some examples indicate that co-creation of tourist experiences through SDL does not necessarily require significant investment in resources and time. The key is to understand the needs and wants of the tourists and what is regarded as value and a unique memorable experience in a destination (Blazquez-Resino et al. 2015).

Case study 45.1: Snow safari

In the northern hemisphere, snow and snow tourism has always been attractive to many tourists. Traditionally, snow tourism has been associated with winter sports activities such as cross-country skiing, snowboarding and alpine skiing (Cristobal-Fransi et al. 2017). In recent years, however, there are a growing number of tourists who display an interest in snow, but not for the sole purpose of undertaking winter sports activities. Increasingly, destinations are offering snow safaris as unique tourist experiences for tourists who may not have experienced snow previously and by whom snow is regarded as something exotic and unique. As a relatively new type of tourism offering, there is no definite description or definition of snow safaris to date. However, the definition of "safari" is an organized journey to look at, or to hunt, wild animals (Cambridge 2019). Hence, snow safaris can be understood as organized tours in which the snow and its natural surroundings, including animals, are the main backdrop. Snow safaris may include snow as the main attraction as well as animals in their natural surroundings, such as wolf and bear experiences (Arnesen 2015).

Tourists' fascination with the snow cannot be underestimated. In addition to admiring the snow, tourists who participate in a snow safari tour in Northern Norway, for instance, can almost not comprehend that they can drink water straight from a stream (Grønvik 2017). Such experiences are taken for granted by the locals, but the tourists cannot wait to experience this and to share such experiences on social media. This suggests that some unique experiences, which are co-created with the tourists, are unplanned and surprising. Other activities, which require active participation from the tourists themselves, include snow shoveling, which is considered a common household chore in winter for most locals in Norway. Some tourists who seek snow safari experiences have never seen snow previously, let alone shoveled snow. In addition, they delight in using the traditional "spark", a type of sleigh commonly used in Norway, Sweden and Finland, which requires the person to kick backwards to gain speed, as a mode of transport to reach the city area (Figure 45.1).

While these are considered as common winter activities for the locals, they lead to unique experiences and value for the tourists. In order for value co-creation to occur, the tourism provider and other relevant stakeholders, as discussed in the previous section, need to facilitate

Figure 45.1 Norwegian Spark
Source: Author's own image

the process by providing the proper information and equipment as well as educating the tourists about such activities and the environment. This includes providing a user manual on how to use the "spark" for instance, which the local tourist information center has done. In addition, maps should be provided as well as information and guidelines on the necessary clothing and equipment needed when exploring nature in the winter. Also, as the tourists participate in such experiences with their families it provides additional value, as they are spending time with their families to experience something new and exotic together (Strøm 2015).

Case study 45.2: Farm tourist experiences

As illustrated in case study 1, the co-creation of tourist experience does not have to be complicated or require heavy investment in unique products, as the tourism destinations can take advantage of the existing resources. The key is facilitation and cooperation between all of the stakeholders in a network process, as advocated by SDL. A good example of this is farm tourist experiences. Farm tourist experiences can be had when tourists visit farms (both working and non-working) and interact with the farmers or the staff working at the farm. Such interaction and the dynamic between the guest and the host is one of the key elements in creating unique and memorable farm tourist experiences. In many instances, such interactions go beyond interactions between a business and its customers, as they can be more accurately described as personal relationships (Mei et al. 2018). This suggests that the participation and involvement of both parties is important in co-creating value in such farm experiences; hence, value is created in use. In addition, the interaction between the tourists themselves is also an essential element in the co-creation process and the SDL approach (Mei et al. 2018).

As an example from the Inland region, located in the eastern part of Norway, farmers as tourist providers are considered as a part of the product and the experiences cannot be created without the providers' motivation and active involvement in interacting with the tourists. Such interaction includes various activities such as storytelling to convey a particular story or concept of the farm, including the history of the farm, food preparation, where the tourist must actively participate and tractor driving. Other activities include informative sessions in order to educate the tourists about farming and farm life (Mei et al. 2018). As discussed, the relationship and interaction between the tourist and the farmer are the key elements in creating such experiences. In addition, the destination setting, which is conceptualized as the experiencescape by Campos et al. (2018) and Mei et al. (2018), also plays a significant part in the co-creation process, as well as the interaction between the tourists themselves. The setting and the environment facilitate the story or concept that the provider wants to convey. Farmers also work with other stakeholders such as local government, local suppliers and other tourist providers consisting of other farmers, local accommodation providers, restaurants and activity providers in a local network called "Mjøsgårdene" – The Farms at Mjøsa Lake – to provide a holistic tourist experience while tourists are visiting the region of the Inland. Logically, locals living in the region are also important stakeholders in the process, as how tourists interact with the locals will also affect the value and experiences that are created. These elements are the foundation of SDL, where the roles of producers and consumers, as well as the interaction among and between other stakeholders, are not viewed as separate entities and tasks because value is always co-created (Prebensen et al. 2017).

Conclusion

Despite the increasing number of studies focusing on co-creation, many providers find the term difficult to implement and relate to in practice. Co-creation of tourist experiences through SDL in the tourism industry may be more challenging due to the subjectivity of both the terms "experiences" and "value", as well as the notion of the tourism destination, which does not consist of a single product. At the same time, the nature of tourist experiences indicates that value co-creation and the SDL approach is particularly relevant in the tourism industry, as any tourism consumption would require some degree of participation and involvement by the tourists.

As the case studies in this book chapter have illustrated, co-creation of value does not necessarily require heavy investment in resources and equipment. Some unique experiences which lead to value co-creation occur unplanned and even surprise the providers. The importance is the value that the tourists believe they are gaining from the experience as a result. Other experiences would require more systematic planning to create a concept through storytelling and experiencescape, as well as collaboration through networks of tourism providers, local government and other stakeholders.

Importantly, the essence of the discussion on co-creation and SDL is value. As value must be co-created, the presence and participation of the customer must also exist. Since both value and experiences are essentially determined by the consumers themselves, the focal point should be to understand what value means for the tourists who are seeking certain tourist experiences. Based on such understanding, tourism providers would be able to assist tourists, as advocated by SDL, in gaining value and unique experiences as a result.

References

Achrol, R. and Kotler, P. (2006) 'The service-dominant logic for marketing: A critique', in Lusch, R. and Vargo, S. L. (eds.) *The Service-Dominant Logic of Marketing: Dialog, Debate and Directions* (pp. 320–337). Armonk: ME Sharpe.

Arnesen, M. (2015) *Tilbyr ulvesafari i Rendalen*: Østledingen. Available at: www.ostlendingen.no/rovdyr/nyheter/nord-osterdal/tilbyr-ulvesafari-i-rendalen/s/5-69-15955?fbclid=IwAR0XBm5z3pr2L6Mr3umPVGE3KQO_jSA4f9XNcGLo63FBQAuj1u-RsEz4gMc [Accessed: February 11, 2019].

Ashworth, G. J. and Voogd, H. (1990) 'Can places be sold for tourism', in Ashworth, G. J. and Goondall, B. (eds.) *Marketing Tourism Places* (pp. 1–16). London: Belhaven.

Baggio, R. (2008) 'Symptoms of complexity in a tourism system', *Tourism Analysis*, 13(1): 1–20. DOI: 10.3727/108354208784548797.

Beckman, T. and Khare, A. (2018) 'A service-dominant logic and value co-creation approach for online business education', in Khare, A. and Hurst, D. (eds.) *On the Line: Business Education in the Digital Age* (pp. 21–35). Zurich: Springer.

Blazquez-Resino, J. J., Molina, A. and Esteban-Talaya, A. (2015) 'Service-dominant logic in tourism: The way to loyalty', *Current Issues in Tourism*, 18(8): 706–724.

Buhalis, D. (2000) 'Marketing the competitive destination of the future', *Tourism Management*, 21(1): 97–116. DOI: 10.1016/S0261-5177(99)00095-3.

Buonincontri, P., Morvillo, A., Okumus, F. and van Niekerk, M. (2017) 'Managing the experience co-creation process in tourism destinations: Empirical findings from Naples', *Tourism Management*, 62: 264–277. DOI: 10.1016/j. tourman.2017.04.014.

Cabiddu, F., Lui, T.-W. and Piccoli, G. (2013) 'Managing value co-creation in the tourism industry', *Annals of Tourism Research*, 42: 86–107. DOI: 10.1016/j.annals.2013.01.001.

Cambridge. (2019) *Safari Meaning in the Cambridge Dictionary*. Available at: https://dictionary.cambridge.org/dictionary/english/safari [Accessed February 11, 2019].

Campos, A. C., Mendes, J., Valle, P. O. d. and Scott, N. (2018) 'Co-creation of tourist experiences: A literature review', *Current Issues in Tourism*, 21(4): 369–400. DOI: 10.1080/13683500.2015.1081158.

Chathonth, P. K., Ungson, G. R., Harrington, R., Altinay, L., Okumus, F. and Chan, E. S. W. (2017) 'Conceptualization of value co-creation in the tourism context', in Prebensen, N., Chen, J. S. and Uysal, M. S. (eds.) *Creating Experience Value in Tourism* (pp. 48–64). Oxfordshire: CABI.

Cracolici, M. F. and Nijkamp, P. (2009) 'The attractiveness and competitiveness of tourist destinations: A study of Southern Italian regions', *Tourism Management*, 30(3): 336–344. DOI: 10.1016/j.tourman.2008.07.006.

Cristobal-Fransi, E., Daries-Ramon, N., Mariné-Roig, E. and Martin-Fuentes, E. (2017) 'Implementation of Web 2.0 in the snow tourism industry: Analysis of the online presence and e-commerce of ski resorts', *Spanish Journal of Marketing – ESIC*, 21(2): 117–130.

Grönroos, C. (2006) 'Adopting a service logic for marketing', *Marketing Theory*, 6(3): 317–333. DOI: 10.1177/1470593106066794.

Grönroos, C. (2008) 'Service logic revisited: Who creates value? And who co-creates?', *European Business Review*, 20(4): 298–314. DOI: 10.1108/09555340810886585.

Grönroos, C. (2011) 'Value co-creation in service logic: A critical analysis', *Marketing Theory*, 11(3): 279–301. DOI: 10.1177/1470593111408177.

Grønvik, Y. (2017) *Penger å spare og bedre inneklima*: Sør-Varanger Avis. Available at: www.sva.no/nyheter/–penger-aa-spare-og-bedre-inneklima [Accessed: February 11, 2019].

Kandampully, J. (2003) *Services Management: The New Paradigm in Hospitality*. Upper Saddle River: Hospitality Press.

Langlotz, N. (2016) *Co-creation in the Innovative Sharing Economy – An Analysis Using the Example of AirBnB*. Master thesis, Copenhagen Business School (CBS), Copenhagen.

Lichrou, M., O'Malley, L. and Patterson, M. (2008) 'Place-product or place narrative(s)? Perspectives in the marketing of tourism destinations', *Journal of Strategic Marketing*, 16(1): 27–39. DOI: 10.1080/09652540701794429.

Lusch, R. and Vargo, S. L. (2006) 'Service-dominant logic as a foundation for a general theory', in Lusch, R. and Vargo, S. L. (eds.) *The Service-Dominant Logic of Marketing: Dialog, Debate, and Directions* (pp. 43–56). Armonk: ME Sharpe.

Mei, X. Y., Hågensen, A.-M. S. and Kristiansen, H. S. (2018) 'Storytelling through experiencescape: Creating unique stories and extraordinary experiences in farm tourism', *Tourism and Hospitality Research*. DOI: 10. 1177/1467358418813410.

Mossberg, L. (2007) 'A marketing approach to the tourist experience', *Scandinavian Journal of Hospitality and Tourism*, 7(1): 59–74. DOI: 10.1080/15022250701231915.

Payne, A. F., Storbacka, K. and Frow, P. (2008) 'Managing the co-creation of value', *Journal of the Academy of Marketing Science*, 36(1): 83–96. DOI: 10.1007/s11747-007-0070-0.

Pine, B. J. and Gilmore, J. H. (1998) 'Welcome to the experience economy', *Harvard Business Review*, 76: 97–105.

Poulsson, S. H. G. (2014) *On Experiences as Economic Offerings*. Phd Thesis, BI, Oslo.

Prahalad, C. K. (2004) 'Co-creation unique value with customers', *Strategy & Leadership Journal*, 32(3): 4–9. DOI: 10.1108/10878570410699249.

Prebensen, N., Chen, J. S. and Uysal, M. S. (2017) 'Co-creation of tourist experience: Scope, definition and structure', in Prebensen, N., Chen, J. S., and Uysal, M. S. (eds.) *Creating experience Value in Tourism* (pp. 14–24). Oxfordshire: CABI.

Ramaswamy, V. (2011) 'It's about human experiences … and beyond, to co-creation', *Industrial Marketing Management*, 40(2): 195–196. DOI: 10.1016/j.indmarman. 2010.06.030.

Shaw, G., Bailey, A. and Williams, A. (2011) 'Aspects of service-dominant logic and its implications for tourism management: Examples from the hotel industry', *Tourism Management*, 32(2): 207–214. DOI: 10.1016/j.tourman.2010.05.020.

Stauss, B. (2005) 'A Pyrrhic victory: The implications of an unlimited broadening of the concept of services', *Managing Service Quality: An International Journal*, 15(3): 219–229. DOI: 10.1108/09604520510597782.

Strøm, P. (2015) *Turister som familien Michie truer hotellnæringen*: NRK. Available at: www.nrk.no/troms/reiser-fra-sor-afrika-for-a-make-sno-1.12128219 [Accessed February 11, 2019].

Suntikul, W. and Jachna, T. (2016) 'The co-creation/place attachment nexus', *Tourism Management*, 52: 276–286. DOI: 10.1016/j.tourman.2015.06.026.

Tronvoll, B. (2012) 'A dynamic model of customer complaint behaviour from the perspective of service-dominant logic', *European Journal of Marketing*, 46(1/2): 284–305. DOI: 10.1108/03090561211189338.

UNWTO. (2005). *Report on the WTO Survey on the Implementation of the Global Code of Ethics for Tourism*. Madrid: UNTWO.

Vargo, S. L. and Lusch, R. (2004) 'Evolving to a new dominant logic for marketing', *Journal of Marketing*, 68(1): 1–17. DOI: 10.1509/jmkg.68.1.1.24036.

Vargo, S. L. and Lusch, R. F. (2008) 'Service-dominant logic: Continuing the evolution', *Journal of the Academy of Marketing Science*, 36(1): 1–10. DOI: 10.1007/s11747-007-0069-6.

Vargo, S. L. and Lusch, R. F. (2016) 'Institutions and axioms: An extension and update of service-dominant logic', *Journal of the Academy of Marketing Science*, 44(1): 5–23. DOI: 10.1007/s11747-015-0456-3.

Volo, S. (2006) 'A consumer-based measurement of tourism innovation', *Journal of Quality Assurance in Hospitality & Tourism*, 6(3–4): 73–87. DOI: 10.1300/J162v06n03_05.

Yang, W. and Mattila, A. S. (2016) 'Why do we buy luxury experiences?: Measuring value perceptions of luxury hospitality services', *International Journal of Contemporary Hospitality Management*, 28(9): 1848–1867. DOI: 10.1108/IJCHM-11-2014-0579.

46

Place-based business models of value creation in tourism

Case of the Albergo Diffuso

Francesco Raggiotto, Cristiana Compagno and Michela C. Mason

Introduction

In the last couple of years the Italian touristic sector has experienced remarkable growth, providing a key contribution to the national economy by accounting for 13% of national GDP in 2017 (Statista 2019). However, the recent, massive changes the touristic sector has gone through have determined a massive shift in the industry, from the traditional models of touristic development (e.g., mass tourism) towards new models supporting a sustainable approach to tourism. In this sense, changes in environmental, cultural, societal, and political equilibria are increasingly calling for new models of value creation in tourism, satisfying the complex needs of a number of tourism niches, which are gradually substituting, in terms of profitability and customer participation, more traditional touristic models.

Accordingly, new models of touristic development are emerging, which are focused on new goals, business models and new concepts of value, and involve not only tourism-related value *tout court*, but also other kinds of values (e.g., Naidoo and Sharpley 2016). Such new models embed, for instance, values related to the preservation of local communities, heritage, and environment, as well as contrasting the overexploitation of places and resources that characterizes traditional models of touristic development (Loureiro and Gonzalez 2008).

The emergence of *Place-Based Business Models* in tourism can be considered in the light of this evolutionary pattern. *Place-Based Business Models* base their competitive advantage on unique conditions occurring in a specific location, with which they develop strong ties (Di Gregorio 2017).

This chapter presents the case of Albergo Diffuso, an Italian model of hospitality, as a *Place-Based Business Model*, examining how the strong bond place-business models develop with a location have an impact on the model's ability to organize resources (Peteraf 1993), obtain stakeholder support (Donaldson and Preston 2011), and ultimately create value.

Place-based business models

The business model can be defined as a framework that "articulates the logic and provides data and other evidence that demonstrates how a business creates and delivers value to customers. It also outlines the architecture of revenues, costs, and profits associated with the business enterprise delivering that value." (Teece 2010: 173)

The primary functions of the business model are value creation and value capture/appropriation (Osterwalder and Pigneur 2012; Teece 2010; Zott and Amit 2010), and its major components are the value proposition and the revenue model (Morris, Schindehutte, and Allen 2005; Zott, Amit, and Massa 2011). A value proposition concerns value creation, and relates to the "content, structure, and governance of transactions designed so as to create value through the exploitation of business opportunities" (Amit and Zott 2001: 511). It specifies how and for whom the value will be created. The revenue model concerns value capture/appropriation, and specifies the mechanisms allowing value appropriation (i.e., profit generation). Both the value proposition and the revenue model are sustained by the organization's resources, networks, and processes that, ultimately, allow value creation and value appropriation (Osterwalder, Pigneur, and Tucci 2005).

In the case of *Place-Based Business Models*, the processes of value creation and appropriation are strongly tied to location-specific resources. In broader terms, *Place-Based Business Models* create value from and contribute to the local "sense of place" or *genius loci* (Di Gregorio 2017), that is a concept encompassing the diverse emotional relationships that may develop between individuals and a certain place (like: place bonding, identity, and attachment) (Jepson and Sharpley 2015). The concept of "sense of place" may either apply to locals (e.g., concerning an individual and/or collective identity) or to outsiders/tourists (De Blij 2008): in this sense, from a tourism management perspective, sense of place is likely to be a key source of value and competitive advantage for *Place-Based Business Models*.

The value associated with the "sense of place" results from the combination of a number of tangible and intangible elements peculiar to a certain place, like its geography (e.g., climate, landscape, natural heritage) and its cultural and historical heritage, whose combination contributes to the definition of "peculiarity of place" (Trubek 2008). Those elements are the ones on which *Place-Based Business Models* typically rely.

The Albergo Diffuso in Friuli Venezia Giulia (Northeastern Italy): a place-based, sustainable business model for value creation

The Albergo Diffuso

The Albergo Diffuso is a typically Italian form of hospitality (Confalonieri 2011; Paniccia and Leoni 2019), differing from most of the traditional tourist and holiday accommodation formats like hotels, residences, and B&Bs.

Albergo Diffuso gathers a number of different businesses related to tourist accommodation (e.g., lodging, dining, entertainment) which are bonded to each other by an operational integration (i.e., they are part of the same accommodation structure) and denoted by physical dispersion (Confalonieri 2011). Typically, Albergo Diffuso is located in small rural areas, like small towns and villages (in Italian, *borgo* and *villaggio*) characterized by a significant cultural and/or historical heritage (Presenza, Messeni Petruzzelli, and Sheehan 2019). Albergo Diffuso is a model which combines the sustainable use of location-specific resources (like local heritage, landscape, traditions) for the touristic development of the area.

Through such development, Albergo Diffuso aims at providing support for the socioeconomic development of such communities, fostering the reconversion of often isolated and disadvantaged rural areas and promoting the preservation of and the respect for local communities and heritage (Romolini, Fissi, and Gori 2017).

Albergo Diffuso provides tourists with unique opportunities to experience the *genius loci*, immersing them in an authentic local experience that allows them, for instance, to access buildings permeated by cultural and historical significance – that otherwise would be inaccessible for tourists – or to interact with locals in their discovery of the place, thus allowing tourists to engage in multiple co-creation processes of experiential value (Paniccia and Leoni 2019).

Such relevance of the *genius loci* for the Albergo Diffuso is confirmed if the resources needed for its existence are considered (Confalonieri 2011):

- The presence of structures denoted by an historical/cultural value, suitable for touristic accommodation;
- The presence of basic services supporting hospitality and tourism;
- The presence of a location-specific cultural, social, and historical heritage (for instance, traditions, culture, cuisine, and natural landscape);
- A well-developed local consciousness of the relevance of local cultural heritage which stimulates initiatives related to preserving and promoting local heritage;
- A proactive attitude from the side of public administration to favor and gather the support of the many stakeholders involved in the Albergo Diffuso venture (local communities, public and private bodies, and so forth).

Albergo Diffuso Comeglians

The present section considers the case of the application of the Albergo Diffuso model to the small town of Comeglians (500 inhabitants), located in Carnia, a rural area of the Italian Friuli Venezia Giulia region (northeastern Italy), as an example of the *Place-Based Business Model* relying upon the uniqueness of the place to create value for tourists in order to accomplish its primary aim, that is, to create value for the surrounding community. Despite several Italian regions having implemented the Albergo Diffuso model (e.g., Confalonieri 2011; Presenza et al. 2019; Romolini et al. 2017), the case of Comeglians appears relevant as it was there that the conceptual model of Albergo Diffuso was first conceived and introduced, primarily as a system for the reconversion of rural areas in some Italian regions where large depopulation processes were taking place. The evolutionary pattern of the Albergo Diffuso model in Comeglians, described in the reminder of this chapter, shows the pattern the model has gone through, from a purely reconstruction model to a *Place-Based Business Model*, a necessary transformation to accomplish the goals set by the Albergo Diffuso conceptualization.

The harsh economic conditions that Italy experienced at the end of the Second World War triggered massive emigration waves, mostly directed to Europe, the United States, and South America, which particularly involved some Italian rural areas, which were subject to a constant, progressive depopulation that ultimately resulted in the abandonment of entire rural towns and villages. The Friuli Venezia Giulia region experienced remarkable emigration levels, leading to the full depopulation of entire rural and mountain areas (Toson 2014).

The already critical economic situation of such areas was further worsened by a devastating earthquake which hit Friuli in 1976, which caused extensive damage and destroyed entire towns and villages. Destruction brought by the earthquake further boosted depopulation in these areas, which were already experiencing very high emigration and depopulation rates. Despite the economic and working opportunities deriving from the reconstruction and restoration initiatives supported by the national government, such depopulation was seemingly unstoppable.

One of the areas that suffered the most from the emigratory waves and the 1976 earthquake was the Carnia area, a remote, rural area located close to the mountains (Alpi Carniche). These adverse events further worsened an already difficult socioeconomic situation, largely determined by the region's peculiar geographical positioning (close to the mountains and far from the major regional urban centers).

In 1979, in an effort to stop the demographic and economic breakdown, a group of young inhabitants of Comeglians began promoting the development of initiatives aimed at boosting the weak local economy and increasing competitiveness through the restoration of the abandoned villages in the area, with the aim of achieving the economic development of the area through its touristic conversion. From this seminal idea the Albergo Diffuso model was introduced.

In its early form, the model did not in any way consider a tourist industry role; rather, the model was fundamentally conceived to achieve community-related goals, like urban preservation, improvement of local socioeconomic conditions, and the preservation of the local heritage. The conceptual proposals regarding the key components of the Albergo Diffuso model, including the preliminary studies to identify, for instance, its cost structure, the revenue model, and the required resources, were almost entirely developed in Comeglians. The early developmental steps of the model were enthusiastically perceived both by the public and the private sector, and interest extended beyond the Carnia area: the Polytechnic of Zurich considered the urban restoration aims extremely innovative and relevant, to the point of providing close support for the creation of the architectural recovery plan, a necessary asset for the development of the entire Albergo Diffuso model. The early conceptual phase was followed by increasingly operational phases devoted to setting up the entire business model sustaining the Albergo Diffuso model. These conceptual phases were followed by some early experimental applications of the model in the nearby towns of Sauris and Sutrio.

Case study 46.1

Given the exploratory nature of this study and the relative scarcity of previous studies on the Albergo Diffuso phenomenon (e.g., Confalonieri 2011; Presenza et al. 2019), we employed a qualitative approach (Myers 2013). Where possible, we considered multiple data sources to take advantage of possible data triangulations (Jick 1979).

To develop our preliminary knowledge about the Albergo Diffuso model in Carnia we accessed several data sources like specialized magazines, webpages, local publications, multimedia sources (e.g., podcasts, videos, and other materials), and regional regulations, as well as scientific journals.

Then we considered primary data collection (qualitative data) from structured interviews with several individuals involved in the Albergo Diffuso model in various roles, including owners of the buildings that were part of this accommodation system, managers, and founders of the conceptual model.

The structured interviews explored the characteristics of the Albergo Diffuso *Place-Based Business Model*, and considered the mechanisms through which it creates touristic value through the place and how it transmits such value to the place itself.

We decided to continue the interviews until convergence was reached. Accordingly, we did not set a predefined number of interviews, but instead the interviews continued until there was no new information gain (Grace and O'Cass 2002). Following this approach, we concluded the research after 12 interviews.

We conducted in-depth interviews as an open dialogue to encourage participants to describe their ideas and to report their own experiences (mean length=60 minutes). Each interview contained a first part, introducing a set of questions (McCracken 1988) about the participants' personal background. In its second part, the interview turned to the description of individuals' professional and personal experiences regarding the Albergo Diffuso model. In the third part of the interview, we focused on gathering the data required for pursuing the study's objectives. We guided the interviews through questions such as: "What are the goals of the Albergo Diffuso model"; "When did the Albergo Diffuso model emerge as a *Place-Based Business Model*?"; and "What are the key sources characterizing Albergo Diffuso's competitive advantage?"

Analysis and results

We content-analyzed the data using an interpretational analysis to reveal the core themes emerging from such interviews.

Despite the different backgrounds and professional roles within the Albergo Diffuso model, opinions and ideas regarding the origins and key rationale of the model were almost unanimously confirmed by the individuals included in our sample, whose responses suggest that the creation of the Albergo Diffuso model stems from a strong willingness to help Carnia survive, by finding a way to reduce depopulation and its consequent physical and cultural degradation.

> The project of Albergo Diffuso was (and still is) for our Carnia. When we started thinking about it, about forty years ago, we wanted to bring life back to our villages and towns that were almost doomed to extinction, as people (especially young generations) were leaving to find a job abroad, leaving behind a desolating landscape of empty houses.
>
> *Mario, one of the founders of the original model*

> No doubt. To engage in the project of Albergo Diffuso you need to have enthusiasm and to love your community. The Albergo Diffuso is something inspired by the need to support our small beautiful villages.
>
> *Patrizia, owner*

In the Albergo Diffuso model, the concept of value is central, at least because of its inner nature (i.e., creating value for the local community in social and economic terms).

It became thus straightforwardly clear that Albergo Diffuso needed to consider multiple recipients of value and multiple types of value.

> The initial conceptualization of the model was basically focusing on creating value in the form of territorial development, nothing more, nothing less. However, there was the need to understand what would have been the mechanism allowing the creation of such value: we needed a market to give economic value to our strengths: culture, nature, and so forth. That market was the tourism market.
>
> *Giorgio, manager*

> We needed tourists to help our people.
>
> *Laura, manager*

> We considered that the value we were creating was multimodal, in the sense that, we wanted to make our land come back to life, but to maximize the value created for locals, we were aware of the fact that we needed to be able to maximize the value provided to our customers.
>
> *Mario*

It was hence crucial to define a value proposition grounded on the uniqueness of Albergo Diffuso to the touristic market, as the ability of Albergo Diffuso to create value for tourists was likely to depend, to a great extent, on its ability to create value for the local community.

> We considered that, basically, we were the major competitive advantage of Albergo Diffuso: our culture, traditions, nature, history create an original tourist experience, highly sustainable.
>
> *Gigliola, owner*

The nature of the key resources required by Albergo Diffuso reflects its place-based nature. A key challenge in the case of Comeglians regarded the enhancement of many components that were yet present but underestimated or only partially available, to support the touristic reconversion of the area. If, on the one hand, Carnia is denoted by a wide, rich heritage articulating in many tangible and intangible aspects (Battaglia 2007), on the other hand, the culture of touristic accommodation, as well as facilities and services supporting tourism-related activities (e.g., basic facilities like restaurants, services like public transportation, as well as further initiatives to promote local heritage) needed a further (or even full) development. Accordingly, before putting the business model in place, some preliminary steps were needed to develop the missing resources, and empower the existing ones.

> Actually, we were in need of everything. This is why, before actually starting the Albergo Diffuso, we first came up with some proposals concerning key resources allowing Albergo Diffuso to work, like: enhancement of public services, development of commercial spaces, and the empowerment/reactivation of typical Carnia economic activities, like farming and woodworking, which due to the depopulation were drastically reducing but, actually, were – and are – our story, our richness, where we make the difference.
>
> *Antonio, one of the founders of the Albergo Diffuso model*

In this sense, architectural recovery was one of the most relevant challenges, as architectural enhancement of existing buildings and structures represented the backbone of the overall Albergo Diffuso system.

> We knew that what we needed most was finding a way out of that physical degradation taking place in the area following emigration. In this process, we received a great support from the Polytechnic of Zurich, that considered us a relevant case to study, and through extensive research it provided intelligent solutions for the successful recovery of isolated, disadvantaged areas.
>
> *Mario*

The Albergo Diffuso had then to be equipped with the managerial and economic tools satisfying its operational needs. Accordingly, the Albergo Diffuso Comeglians Cooperative was introduced, and put in charge of managing the entire system, an entity epitomizing the willingness of the community to make this project succeed to improve future socioeconomic conditions.

Costs were initially distributed according to the revenues generated by each of the houses involved in the Albergo Diffuso (which remained in the hands of the owners, who only "rented" their houses). The revenue model initially hypothesized embraced all of the economic activities supporting the Albergo Diffuso, from the production of local foods to the cultural initiatives provided for the tourists.

Nowadays, the Albergo Diffuso can still count on a strong competitive advantage, basing on unique, location-specific, and hardly replicable conditions, deriving from the unique mixture of intangible and tangible place-specific components. In this sense, it still represents a business model creatively employing the local resources in a sustainable manner, creating value for the community and translating the value created for the tourists, who in turn are involved in unique processes of value co-creation (Paniccia and Leoni 2019).

> We don't market a product. We create relationships between tourists and our place. We make them feel like they have been always part of this land. We do not provide accommodation: we create personal relationships with our guests, helping them feel part of our community, discover our land, telling them stories, and showing them all the beautiful things our land can offer.
>
> *Patrizia*

Such strong competitive advantage is likely to have relevant marketing implications: consumer loyalty is likely to occur when Albergo Diffuso succeeds in conveying a "sense of place" to the tourists:

> Most of the tourists may come here because of marketing: they see the Albergo Diffuso on Booking.com, TripAdvisor, and so on and so forth. But if they come back, it's not because of marketing, but because they fall in love with this place, its houses, its people, its treasures. Because they feel a bond with this place.
>
> *Gigliola*

However, over the years, the development of Albergo Diffuso Comeglians has been only partial. This is likely due, to a great extent, to the inner characteristics of the business model it adopts. *Place-Based Business Models* are characterized by their strong bond with the place, which represent, at the same time, their key source of value creation as well as their inner *raison d'être*. This strong bond with the external environment may appear in open contrast with the behavior of the majority of organizations, which, according to extant literature, strive to minimize the impact and the dependence on environmental fluctuations of their functioning (Thompson 2017). In the case of *Place-Based Business Models*, such a deep bond with the place is likely to be a double-edged sword, with relevant drawbacks potentially hampering the potential for value creation.

This appears quite clear with respect to the case of Albergo Diffuso Comeglians, which has suffered (and still suffers) from a general lack of support from its local environment, especially with respect to the local administration, a support which was instead guaranteed to other Albergo Diffuso programs located in the same region (in other towns and villages) and in other Italian regions (e.g., Presenza et al. 2019). Over the years, such lack of support has undermined the development and/or the efficacy of the programs of resource development (hampering therefore also the revenue model), and consequently the value creation processes. In this sense, the lack of support from the local public administration reverberates not just in terms of missing financial support, but also in terms of missing key resources and, further, in terms of support from the local community. Such uncertainty is further worsened by the lack of a specific regional regulation concerning Albergo Diffuso, that, however, is partially overcome when local administrations directly support the Albergo Diffuso venture both from a managerial and a financial point of view (as in the case of Sauris, a nearby town whose local administration has devoted many resources to the venture).

> There must be someone that believes in you. In Sauris there has been, from the beginning, support from wise, forward-looking administrations that saw in the Albergo Diffuso an integrated model for local development. This strategy still pays off, as Sauris is one of the few towns in which emigration has almost stopped, economic activities prosper, and many tourists come and go. Young generations find jobs because the Albergo Diffuso has succeeded in stimulating a new wave of community. There, everything works. Resources exist because the local administration has helped people understand the value of Albergo Diffuso.
>
> *Mario*

> You know, there is this road that leads to my house, that is so narrow and steep tourists experience huge difficulties in reaching my place, and to visit the surrounding territory as well – with any means of transport. I've repeatedly asked for the help of the local municipality, but I have received just vague promises. I cannot work in this way.
>
> *Patrizia*

> We need restaurants, places where people can taste local food. So far, I need to send them to the local bar, which is literally not suitable for tourists: no flexible opening times, no ability to welcome tourists, no willingness to take care of them. If they want to eat, I need to send them nearby, not to Comeglians. But they come here for Comeglians.
>
> *Gigliola*

Finally, a recent issue appears to hamper the ability of Albergo Diffuso to create value for customers and the community in the near future. Development of Albergo Diffuso has been sustained, so far, by a public initiative – which began in the mid-2000s – which awarded with non-repayable funding owners of holiday homes located in Albergo Diffuso areas. In order to obtain this partial funding to restore their houses, owners had to express their willingness to restore their houses with the aim of joining the Albergo Diffuso project for a ten-year period, after which they could decide whether to continue the venture or have their houses back. The ten-year period obligation is now close to expiring for most of the houses of which Albergo Diffuso is currently composed: however, instead of continuing the Albergo Diffuso venture, most of the owners are requiring their houses back.

According to our study sample, this problematic situation derives, again, from a lack of a coordinated, holistic policy of regional and local administrations, able to effectively communicate the value creation potential of Albergo Diffuso to the target community.

> If the regional policy does not change, or if we don't find public administrators able to explain that the Albergo Diffuso is important, we will have less and less people willing to give their homes to Albergo Diffuso. This is exactly what is happening in Comeglians: we do not have any support from public institutions, they are not interested in our case, so nothing happens. And people want their houses back.
>
> *Mario*

Concluding remarks

Albergo Diffuso is an example of a *Place-Based Business Model* basing its competitive advantage on strong ties with the specific location in which it is developed and operated. Its value creation processes are based on a competitive advantage determined by the uniqueness of the local resources, which ties the entire business model to the specificities of the place, making Albergo Diffuso a sustainable, original model of hospitality hardly replicable elsewhere. However, the analysis of the Albergo Diffuso Comeglians case study also highlights that such a close bond between the business model and a certain place does not guarantee the success of the entire business venture, as key challenges are created as well. The analysis of Albergo Diffuso Comeglians reveals the strategic challenges *Place-Based Business Models* entail. Accordingly, the actual use of location-specific competitive advantage is determined by the simultaneous presence of location-specific resources, which contribute to value creation processes if, and only if, the model is able to engage all the relevant stakeholders in the place (both public and private). The analysis also suggests that such stakeholder support (which also plays a key role in terms of resources development and management) could be stimulated by leveraging the ability of the model to create value not so much on the side of tourists (Confalonieri 2011) but rather on the side of the local communities, preserving in this way the *genius loci*.

References

Amit, R., and Zott, C. (2001). Value creation in E-business. *Strategic Management Journal, 22*(6–7): 493–520. DOI: 10.1002/smj.187

Battaglia, F. (2007). urismo rurale: l'albergo diffuso per la conservazione del paesaggio. Il caso della Carnia. In C. Donato (Ed.), *Turismo Rurale, Agriturismo Ed Ecoturismo Quali Esperienze Di Un Percorso Sostenibile* (pp. 108–122). Trieste: EUT Edizioni Università di Trieste.

Confalonieri, M. (2011). A typical Italian phenomenon: The "albergo diffuso". *Tourism Management*, 32 (3): 685–687. DOI: 10.1016/j.tourman.2010.05.022

De Blij, H. (2008). *The Power of Place: Geography, Destiny, and Globalization's Rough Landscape*. New York: Oxford University Press.

Di Gregorio, D. (2017). Place-based business models for resilient local economies. *Journal of Enterprising Communities: People and Places in the Global Economy*, 11(1): 113–128. DOI: 10.1108/JEC-02-2015-0016

Donaldson, T., and Preston, L. E. (2011). The stakeholder theory of the corporation: Concepts, evidence, and implications. *Academy of Management Review*, 20(1): 65–91. DOI: 10.5465/amr.1995.9503271992

Grace, D., and O'Cass, A. (2002). Brand associations: Looking through the eye of the beholder. *Qualitative Market Research: An International Journal*, 5(2): 96–111. DOI: 10.1108/13522750210423797

Jepson, D., and Sharpley, R. (2015). More than sense of place? Exploring the emotional dimension of rural tourism experiences. *Journal of Sustainable Tourism*, 23(8–9): 1157–1178. DOI: 10.1080/09669582.2014.953543

Jick, T. D. (1979). Mixing qualitative and quantitative methods: Triangulation in action. *Administrative Science Quarterly*, 24(4): 602–611. DOI: 10.2307/2392366

Loureiro, S. M. C., and Gonzalez, F. J. M. (2008). The importance of quality, satisfaction, trust, and image in relation to rural tourist loyalty. *Journal of Travel and Tourism Marketing*, 25(2): 117–136. DOI: 10.1080/10548400802402321

McCracken, G. (1988). *The Long Interview* (vol. 13). Newbury Park: Sage. DOI: 10.4135/9781412986229

Morris, M., Schindehutte, M., and Allen, J. (2005). The entrepreneur's business model: Toward a unified perspective. *Journal of Business Research*, 58(6): 726–735. DOI: 10.1016/j.jbusres.2003.11.001

Myers, M. D. (2013). *Qualitative Research in Business and Management*. London: Sage.

Naidoo, P., and Sharpley, R. (2016). Local perceptions of the relative contributions of enclave tourism and agritourism to community well-being: The case of Mauritius. *Journal of Destination Marketing & Management*, 5(1): 16–25. DOI: 10.1016/j.jdmm.2015.11.002

Osterwalder, A., Pigneur, Y., and Tucci, C. L. (2005). Clarifying business models: Origins, present, and future of the concept. *Communications of the Association for Information Systems*, 16(1): 1.

Osterwalder, P., and Pigneur, Y. (2012). *Business Model Generation: A Handbook for Visionaries, Game Changes, and Challengers*. New York: John Wiley & Sons.

Paniccia, P. M. A., and Leoni, L. (2019). Co-evolution in tourism: The case of Albergo Diffuso. *Current Issues in Tourism*, 22(10): 1216–1243. DOI: 10.1080/13683500.2017.1367763

Peteraf, M. A. (1993). The cornerstones of competitive advantage: A resource-based view. *Strategic Management Journal*, 14(3): 179–191. www.jstor.org/stable/2486921

Presenza, A., Messeni Petruzzelli, A., and Sheehan, L. (2019). Innovation trough tradition in hospitality. The Italian case of Albergo Diffuso. *Tourism Management*, 72(November 2018): 192–201. DOI: 10.1016/j.tourman.2018.11.020

Romolini, A., Fissi, S., and Gori, E. (2017). Integrating territory regeneration, culture and sustainable tourism. The Italian albergo diffuso model of hospitality. *Tourism Management Perspectives*, 22: 67–72. DOI: 10.1016/j.tmp.2017.02.002

Statista (2019). Total contribution as a share of GDP of the tourism industry in Italy from 2014 to 2028, from www.statista.com/statistics/628849/tourism-total-contribution-to-gdp-italy-share/ [Accessed: April 1, 2019].

Teece, D. J. (2010). Business models, business strategy, and innovation. *Long Range Planning*, 43(2–3): 172–194. DOI: doi:10.1016/j.lrp.2009.07.003

Thompson, J. D. (2017). *Organizations in action: Social science bases of administrative theory*. New York: Routledge.

Toson, C. (2014). *Il racconto dell'Albergo Diffuso: 1978–2014*. Scotts Valley, CA: CreateSpace Independent Publishing Platform.

Trubek, A. B. (2008). *The Taste of Place: A Cultural Journey into Terroir* (vol. 20). Berkeley, CA: University of California Press.

Zott, C., and Amit, R. (2010). Business model design: An activity system perspective. *Long Range Planning*, 43(2): 216–226.

Zott, C., Amit, R., and Massa, L. (2011). The business model: Recent developments and future research. *Journal of Management*, 37(4): 1019–1042. DOI: 10.1177/0149206311406265

47

Chinese meaning of face in Arctic Norway

Cultural co-creation for tourist experiences

Yan Zhang and Young-Sook Lee

Introduction

The Arctic region is fast gaining a position as a unique place to visit for tourists around the world. This relatively new phenomenon is related to macro-level global issues such as climate change, sustainable use of resources and shared concerns over the sustainable future life of humanity on the planet (Lee et al. 2017). Another notable macro-level issue at international level is the potential rise of China/Asia as an influential player in the coming decades. Indeed, the Asian Development Bank (2011) describes the increasing economic and political powers of Asia in its report titled "Realizing the Asian Century". The report gives a convincing argument based on an extensive economic dataset that by 2050 Asia will account for over 50% of the global Gross Domestic Product (GDP). The growing influence of Asia is witnessed in the Arctic tourism sector, one of the last places on earth to be called a "tourist destination" per se.

Given this context, the current chapter presents a case study on Chinese face as part of Arctic tourism experiences. In so doing, the chapter highlights an increasingly significant number of Asian/Chinese tourists in the Arctic region. This illustration further brings out diverse aspects of co-creating tourist experiences in Arctic destinations. Chinese face is an important cultural concept deeply rooted in Chinese society. Face plays a significant role in Chinese peoples' social interactions (Du and Fan 2012; Hu 1944). As the Chinese economy develops further, there are increasing opportunities for the Chinese outbound tourist market.

Employing a qualitative approach, the current chapter presents the findings from the interviews of 26 Chinese tourists traveling to Arctic Norway in groups. Content analysis was used to code the collected data. The analysis identified three themes: a) face-driven motivation and face-earning; b) factors influencing face loss, and c) effective face-saving strategies.

Both theoretical and industry implications of Chinese tourists in the Arctic, particularly regarding co-creating experiences as part of an emerging trend, are drawn from the current study.

Chinese tourists in the Arctic: dealing with Western and Chinese "face" concepts

With the increasing importance of the Chinese tourist market in Arctic destinations (Huang et al. 2017), enquiries into the concept of face are a timely study. The concept of face in Chinese society is one of the quintessential aspects of the society. Indeed, research on "face" demonstrated that 83.33% of the Chinese population feel that face is very important in their social communication (Zhang and Si 2010). It is obvious that face plays a special and significant role in Chinese people's daily life and Chinese culture. The concept of face further extends to the people's value and behavior intentions (Guo and Lin 2015).

The importance of face in Chinese society and culture has attracted scholars' attention for a long time. Therefore, a wide variety of studies on face have been carried out for many years. In the English language, the concept of face was put forward by Hu in 1944 and was subsequently introduced to the West. She argued that face has two meanings. One is "lien (脸)", while the other one is "mien (面)". "Mien (面)" is also equal to "mianzi (面子)" in Chinese, which means "honor" or "prestige". "Mien (面)" represents a good social reputation earned by success, praise and achievement, etc., while "lien (脸)", also known as "lian (脸")", refers to a good moral reputation (Hu 1944).

Western societies have different interpretations of the concept of face from that of the Chinese. Indeed, Goffman (1955: 213) defined face as "positive social value a person effectively claims for himself [*sic*] by the line others assume he has taken during a particular contact". In his research, it was pointed out that face comes into play in certain social interactions. In addition, face is also an image that people want to show to others, and one which they ask from others (Goffman 1955). Comparing Goffman's understanding with Hu's, Goffman emphasized a dynamic association – contact between an individual and other people. Stover (1962) further highlighted the different emphasis placed on conceptualizing face in the West. In the understanding of Western face, he argued that face is a formal and formalized product, thus reflecting a person's status and position in a social system (as cited in Du and Fan 2012). Stover emphasized the connection between an individual and the whole society, where face is an element in placing an individual in a given society. In this way, face is an attribute of a society that can help better conceptualize an individual.

Ho (1976) declared that face is not a standard of behavior, status, dignity, honor or prestige. However, "it is meaningful only when his [*sic*] face is considered in relation to that of others in the social network" (Ho 1976: 882). In other words, a single concept such as a standard of behavior, status, dignity, honor and prestige does not make any sense unless it occurs as part of a social interaction. The concept, face, becomes meaningful when a person gains respect and expectations from contact in a social encounter. Similar to Ho's ideas, Chen defined face as a socialized and interpersonal self-image created when people realize the judgments other people have made of them (Zhang and Si 2010). Self-image comes into being during a social communication. To be more specific, face is an impression aiming to please a social group and win their approval (Du and Fan 2012). The impression, to some degree, is what people want to deliver to others. It shows their value and self-identity, or is a good image that they create and want to share with others. Taking outbound tourism as an example, the main motivation for Chinese females traveling abroad is to obtain knowledge and prestige from others (Li et al. 2011). Traveling abroad impresses people with an image of erudition and uniqueness, which endows the traveler with a sense of prestige or honor.

Differences exist between Western and Chinese perceptions. Western researchers pay more attention to individual and personal characteristics. People are self-focused and responsible solely unto themselves only if their behavior would not influence other people. For example, although Goffman mentions that face only works in an interaction, face is an image that a person wants to perform to others. The person makes a private decision about the public persona that he wishes to project to others in the form of a face performance, which does not depend on social interaction. The Chinese concept of face differs from that of Western observers owing to the interactional difference (Mao 1994). Within tourism studies enquiries, the dynamic patterns of face have been studied. Tracing back the origin of face to Confucian values, Kwek and Lee (2015) demonstrated that while the fast-paced modernization of China is undeniable, the dynamics of face among Chinese corporate travelers to Australia reflect its underlying Confucian interactions.

Face, Chinese consumer behavior and Chinese face-driven consumption in tourism

Chinese consumer behavior and consumer psychology is closely linked to the cultural concept of face. Chinese consumers are concerned about other people's impression of them, thus their consumer behavior is closely aligned with their concept of face. Four characteristics have been captured in Chinese consumer behavior. First, Chinese people pay a great deal of attention to the satisfaction of their own psychological feelings by assuring themselves that they have won face (Hong 2011; Wang and Huang 2010). Tourists would achieve both material value and additional spiritual value, which satisfies them. Material value is equivalent to use value, while additional spiritual value refers to the pleasant sensation of earning face. Thus, psychological satisfaction drives the Chinese consumers to buy. For example, it leads to a loss of face if a Chinese tourist does not follow his or her companions to purchase a product. They decide to buy it because others have done so, even if it is valueless to them. They therefore not only purchase the commodity for its use value but also to maintain face or gain spiritual value amongst their companions. It is worth mentioning that an intangible product also has "material value". Taking a package tour to Norway as an example, Chinese tourists buy the intangible product – the trip – and then they receive an experience, which is a "material value". In addition, they also acquire a spiritual value as they obtain more knowledge of a different culture. The learning process satisfies them and identifies them as prestigious people, which earns them face.

Second, similar to one of the aspects of face, consumer behavior is regarded as a resource exchange – that is, face can be bought with money (Hong 2011). Money is exchanged for products, which demonstrates a consumer's fortune and ability. During the purchasing process, they get approval and admiration from other members of society. As a result, they experience psychological satisfaction and ensure their social status is fixed. These feelings help them maintain and win face. When a tourist buys souvenirs for friends, he or she exchanges money for souvenirs. Friends appreciate gifts, which pleases the tourist, and consequently earns her or him face.

Third, Chinese purchasers pay more attention to the packaging of a product instead of the product's use value (Hong 2011; Wang and Huang 2010). Well-designed packaging easily attracts them. The purpose of this consumer behavior is to *show off* (emphasis added by the authors) and to be praised by others. Specifically, beautiful and unique packaging easily draws a consumer's attention, despite the truth that the commodity is the same and cheaper in conventional packaging. Thus, when most Chinese customers purchase goods with special and attractive packaging it makes them think that it represents their great taste and social status as well as winning them face. Similarly, if a hotel has unique buildings (e.g. an ice hotel) and decoration, it might draw Chinese tourists' attention.

Fourth, some Chinese people take pride in buying more costly products compared to others (Hong 2011; Wang and Huang 2010). Wealth is a symbol that demonstrates one's status and identity, which in turn contributes to gaining face. The desire to earn face results in ignorance of prices. It creates a feeling that their consumer power is higher than other people's when they buy something expensive. As a result, they are proud of earning good face. According to Xie and Li (2009), a growing number of Chinese high-income tourists prefer high quality and serviced tour packages. They flock to buy these packages even though they are not cheap because these tourists identify themselves as high-class people and believe that they earn face by doing so.

With improving living standards, the Chinese consumption domain has increasingly expanded to international travel and tourism activities. This newly found consumption domain also reflects Chinese people's face. Four distinct traits are presented, which are related to the current research.

The first trait might be termed as being a "*show-off*". Studies found that face consciousness is associated with an eagerness to flaunt (Guo and Lin 2015; Wu and Tao 2010; Yuan et al. 2009). Chinese tourists regard showing-off as an appropriate approach to earn face. The power of this desire is too strong to ignore. It is common for Chinese people to buy something luxurious even when the high-priced product is beyond their financial ability (Wu and Tao 2010).

The second trait is to follow a trend (Wu and Tao 2010). This type of consumer behavior follows prevalent choices and trends in order to seek common ground. Chinese tourists would rather sacrifice their own wishes than be different from others, because they do not want to be looked down upon or lose face. For instance, they prefer to follow most people's decision to go to the most popular tourist attractions and cope with traffic jams, even if these places are not their number one choice.

The third trait is trying to be unique and fashionable (Guo and Lin 2015; Wu and Tao 2010). As traveling is gradually becoming part of a normal lifestyle, some tourists choose special trips to highlight their uniqueness and good taste. Such unconventional choices enhance their face, as it easily catches others' eyes and generates praise.

The last trait is to consume blindly (Wu and Tao 2010). Blind tourism consumption indicates that some tourists make irrational decisions to buy something that is useless, unfamiliar to them or unaffordable. Searching for the reason behind such decisions, it is easy to find out that maintaining or gaining face pushes the tourists to shop unwisely.

Adding to the identified traits of face-driven consumption in tourism, Guo (2015) proposed four types of tourism that result in face-earning: culture type, high consumption type, harmonious relationship-building type and special activities type.

Based on the literature reviewed, the current research sets its research question as below:

How do Chinese tourists perceive the "showing-off" element, a feature of the Chinese concept of face, in the decision to travel to Norway?

Method

The current chapter's study design has adopted an interpretive paradigm due to its emphasis on the understanding of people's behavior/social phenomenon (Matthews and Ross 2010).

As such, the Chinese cultural concept of face and Chinese tourists' behavior were interpreted and understood in order to "get inside the minds of subjects and see the world from their point of view" (Veal 2011: 32). Approaching from this premise, non-probability sampling was chosen for an in-depth understanding of the Chinese face concept in relation to the decision to travel to Norway, and was particularly focused on "showing-off".

All informants were selected based on purposive features, as recommended by Patton (2002). The informants came from China, chose Norway as their destination and they all traveled in groups. For an in-depth investigation (Blaxter et al. 2006: 172; Jennings 2005) on the Chinese tourists' perception in relation to face, semi-structured interviews were conducted.

In order to get an access to Chinese tourists, permission was sought and granted at a Chinese restaurant in Northern Norway. In return, the first author helped the restaurant and tour guide with the tourists. When groups arrived at the restaurant, the aims of the research and relevant research protocols were explained. The interviewer (first author) was introduced to the tour group members by the tour guides and was allowed to commence the interviews.

Two main interview questions posed were:

1 Many people think that travel to Norway can show their wealth. Some people think that traveling to Norway (a country with a high level of wealth) can show that they have more ability to enjoy life. What is your opinion?
2 Do you think it is worthwhile spending a great amount of money traveling to Norway to win others' admiration?
    a What can one/you gain from this?

All interviews were conducted in Chinese in order to make sure that interviewees fully understood the questions and were able to communicate freely. Words were carefully chosen because of the sensitive topic: "showing-off". The Chinese are sensitive towards face issues. Therefore, it is not wise to directly ask them if they can earn face by showing-off or spending a great amount of money traveling to Norway. It would be considered impolite and disrespectful to judge someone as a show-off. Although showing one's wealth can earn face, Chinese people are reluctant to be described as someone who is a show-off. To avoid embarrassing Chinese tourists, it was more appropriate to use language in a euphemistic way. For example, the use of commendatory terms, such as "show wealth" and "show ability", instead of mentioning the word "show-off". Further, it was better to talk about the topic in the third person, seeking their opinions about other people's views. Some interviewees were reluctant to be audio-recorded and thus notes were made by the interviewer in those cases. After data collection, audio recordings were transcribed into Chinese, while notes were also compiled into Chinese. The current study obtained ethical approval from the NSD (Data Protection Official for Research of Norway) before the data collection.

Content analysis was used to analyze the transcribed interview data. Content analysis is suitable for all contexts, which includes interpersonal and group messaging (Neuendorf 2002), which made this applicable for the face-to-face interviews conducted. To analyze data, interview transcripts were coded. The first step was to read the transcripts. The second step was to extract meanings, commenting in the margins of pages, which helped to reveal themes. The third step was to compare these comments and code them. The next step was to review these codes. The last step was to compile codes into higher and lower levels.

Findings and discussions

Background information on the informants

Background information on the informants was gathered at the beginning of the interviews as shown in Table 47.1. Following the protocols laid down by the NSD, ensuring the interviewees' anonymity is essential. Therefore, the interviewees are quoted using the informant number in this chapter.

Among the 26 interviewed Chinese tourists, nine were male. As for females, they accounted for around 65%, with 17 people. Their ages ranged from 20 to 70. Most of them were over 40 years old. Many interviewees had reasonably well-paid professions, such as doctor, economist, civil servant and teacher. Nine people worked as teachers – the profession that had the largest proportion of interviewees. There were five informants who were retired, three office clerks, two entrepreneurs and two civil servants. The remaining five people were respectively a doctor, an economist, an accountant, a manager and a freelancer.

Table 47.1 Informants

Informant number	*Gender*	*Age*	*Profession*
1.	Female	60	Doctor
2.	Male	65	Economist
3.	Female	28	Office clerk
4.	Female	29	Kindergarten teacher
5.	Male	49	Entrepreneur
6.	Female	38	Accountant
7.	Male	50	Teacher
8.	Female	51	Teacher
9.	Female	50	Teacher
10.	Male	52	Teacher
11.	Female	30	Manager
12.	Male	70	Teacher
13.	Female	62	Retired
14.	Female	40	Teacher
15.	Male	60	Civil servant
16.	Female	35	Office clerk
17.	Female	59	Retired
18.	Male	52	Retired
19.	Female	59	Freelancer
20.	Female	30	Civil servant
21.	Female	45	Teacher
22.	Female	49	Office clerk
23.	Female	48	Teacher
24.	Male	39	Entrepreneur
25.	Male	58	Retired
26.	Female	62	Retired

Chinese tourists' perceptions of "showing-off" in traveling to Norway

Two main interview questions were put forward to the informants regarding their perception of traveling to Norway and the element of showing-off. The content analysis identified that only a small number of informants, five in total, thought that traveling to Norway does have the showing-off element. The majority of the interviewees considered traveling to Norway is not related to showing-off. Figure 47.1 shows the findings of the analysis.

When asked their opinions regarding whether traveling to Norway demonstrated their wealth, only five informants in total said that traveling to Norway was a way to demonstrate one's wealth. Four people indicated that they felt they earned face when others knew they had spent a lot of money traveling to Norway:

> I agree with the idea (that traveling to Norway leads to showing off wealth). Many Chinese would take photos and post them on social media when they travel to European countries, especially a European country like Norway. They would get "likes" after sharing photos. To some degree, these "likes" earn those visitors face because it is not everyone who could come to Norway due to the high cost. The high price results in the fact that those who can visit Norway are from a well-off family or middle or higher social class.
>
> *(Informant 2)*

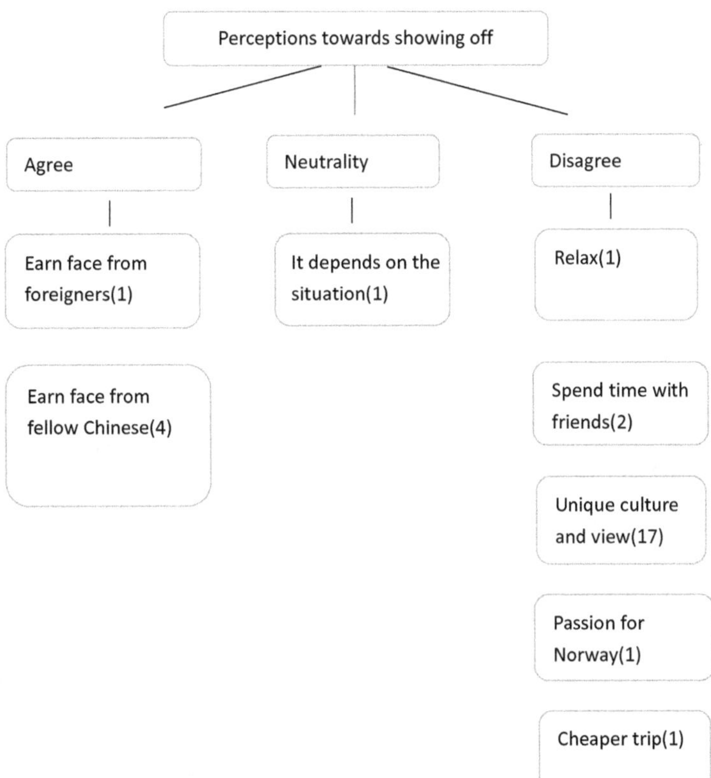

Figure 47.1 Informants' perceptions towards "showing-off" in traveling to Norway

> To some extent it can show wealth since commodity prices are so high in Norway. Therefore, people should have enough money to travel to an expensive country like Norway.
>
> *(Informant 4)*

> I totally agree. A good economic background is the basic requirement for buying the costly package tour to Norway – a place, which is so beautiful but also expensive.
>
> *(Informant 11)*

> I agree with the idea. (It) shows that those visitors have more abilities or money to enjoy life. However, it doesn't mean that you could only enjoy your life when you are a rich or competent person. Sometimes you can still enjoy life even when you spend little money.
>
> *(Informant 14)*

Their answers demonstrate that visitors need a certain level of prosperity to travel to Norway, owing to the high cost of living in Norway as well as the expensive package tour. Therefore, visiting Norway is a way of showing off one's wealth.

One informant pondered some time before expressing her view that it may not always be to show off one's wealth when Chinese people visit Norway, although a certain level of economic means understandably is needed:

> I am uncertain about it. It depends on which angle you are speaking from. The consumer price level in Northern European countries may be higher than other countries. If you want to come to Norway, for example, you have to pay a lot of money and this can't be avoided. Living in a second-tier city in China is completely different from living in a big city like Shanghai. Can we actually say that living in Shanghai is an expression of showing-off? I do not think so. If you want to live in Shanghai, you have to spend more money compared to living in a second-tier city. Therefore, coming to Norway is for the different landscape .... not really to show wealth. On the contrary, if you visit Norway many times, it is probably an approach to show off ...
>
> *(Informant 8)*

Conversely, the rest of the tourists firmly proposed opposite views. The majority (17 interviewees) expressed a willingness and desire to experience a different culture and landscape or spend quality time with friends and family in a pristine natural environment and gave that as the reason for visiting Norway rather than for showing-off. For example, informants said:

> I cannot agree with that. We choose Norway as our destination country purely for the special scenery and exotic culture.
>
> *(Informant 16)*

> Absolutely disagree. It is not important to care about what others would think. We only want to see the world and have fun. Most importantly, our purpose is to see the different view and hear local stories. We love traveling.
>
> *(Informant 10)*

> Traveling to Norway, you need to have money, time and interest. For me, money is not the most significant part and many Chinese are wealthier than before. However, they do not necessarily have the time or interest to come to Norway.
>
> *(Informant 13)*

> The backing of an economic base is necessary to travel abroad. Nowadays, I think many Chinese can afford an outbound trip. Hence, it is not appropriate to say that traveling to Norway can show people have more ability. Besides some people are wealthy but don't have enough time to enjoy a long-distance trip. Regarding traveling to Norway, the desire to experience the culture and views were the most important motivations.
>
> *(Informant 21)*

The answers quoted above indicate that the willingness or interest to experience a special culture and scenery outweighs other factors, such as financial considerations, although they agreed that visiting Norway requires a certain level of wealth.

The passion for Norway is another reason for the disagreement:

> For me, the goal of visiting Norway was not to show my wealth. Instead, the passion for Norway impels me to come. There is a song called "Norwegian forest" in China. Although the song is mainly about expressing feeling, it arises curiosity about Norway and also leads people to imagine what it is like to be in Norway. Many people also regard Northern Europe as the last pure land with unique culture, which had less invasion than other cultures. I am fascinated by it. If I want to show my wealth, I maybe go to America or Britain and go shopping.
>
> *(Informant 6)*

One traveler pointed out that the possibility of a cheap form of travel to Norway contradicts the showing-off element in traveling to Norway:

> I don't agree. For instance, a backpacker could still have a relatively cheaper trip (to Norway), which does not necessarily need a large financial capacity.
>
> *(Informant 15)*

As illustrated, the findings from the current study show that the majority of the interviewed Chinese did not perceive showing-off as the integral element in traveling to Norway.

Face, showing-off and traveling to Norway for the Chinese market

Showing-off is a feature in the concept of face. The Chinese would feel that they had earned face by displaying their wealth, social status, capability and the like (Zhang and Si 2010). The Chinese also tend to perceive that buying expensive products could bring pride and face-earning (Hong 2011; Wang and Huang 2010). From these earlier arguments, it was reasonable to speculate that purchasing costly package tours to Norway, hence showing off their wealth, may be perceived as helping to earn Chinese tourists face. However, only a small number of the interviewed Chinese tourists thought the trip to Norway was a way to show off wealth. They recognized that wealth and showing-off was not the essential driving force for the interviewees. Their reasoning included China's great economic progress, which enabled a higher standard of living and the economic capacity for leisure traveling (informants 7, 13, 20, 21 and 22).

The findings confirm two types of tourism that arguably earn face: the culture type and the activities type, according to Guo (2015). Most of the interviewed Chinese tourists said that the different culture in Norway attracted them and regarded their travel to Norway as a special cultural experience. Many of them were also impressed by the trip to North Cape because it was a unique activity to watch the midnight sun and step into the Arctic Circle.

On the whole, showing off an expensive price tag was not what the Chinese tourists would perceive to have earned them face when traveling to Norway. The interviewed tourists were more focused on unique and exotic cultural and natural experiences. This trend reflects the fact that cultural tourism and special tourism activities are becoming more and more popular in China. These two types of tourism have a common feature – distinctiveness. To be unique is becoming a significant motivation for Chinese tourists to make the decision to travel to Norway. They can gain face through the trip by identifying themselves as unique and gain the appreciation of others through sharing.

Conclusion

This chapter reported a study on Chinese tourists in Arctic Norway. As the Arctic is increasingly perceived as the last place on earth to be accessible to outsiders of the region, a fact that is intendedly and unintendedly associated with the melting of the Arctic ice, arrival numbers are growing from the world's biggest tourist markets: Asia and China. Face in the Chinese cultural context is an important issue to be considered for Arctic tourism practitioners, who are already receiving Chinese travelers and/or preparing for the new market. Literature on face and its adaptation within the context of tourism indicated that visiting Arctic Norway, which is costly and perceived as a world beyond the realities of home in China, may be related to a way of showing off, an element of face. However, the findings indicate that the showing-off element was only marginal in the responses of the informants. The findings highlight the dynamic nature of the world's biggest tourist market with its economic, political, and technological stances constantly interacting. While dynamic and fast-paced, it is important to recognize that the underlying cultural values (i.e. face, Confucian values) play a major role within the dynamic interactions, thus requiring further research.

References

Asian Development Bank. (2011). *Asia 2050: Realizing the Asian Century*. Manila: Asian Development Bank. www.adb.org/sites/default/files/publication/28608/asia2050-executive-summary.pdf.

Blaxter, L., Hughes, C., and Tight, M. (2006). *How to Research*. Maidenhead: Open University Press.

Du, J. and Fan, X. (2012). 服务失败情境下面子丢失对顾客抱怨倾向的影响 The Impact of Face Lost on Complaint Intention under the Service Failure Circumstance. *Management Review*, 24(3): 91–99.

Goffman, E. (1955). On Face-Work: An Analysis of Ritual Elements in Social Interaction. *Psychiatry: Interpersonal and Biological Processes*, 18: 213–231. DOI: 10.1080/00332747.1955.11023008

Guo, X. (2015). 中国旅游者的面子结构与旅游行为– 一项探索性研究 The Structure of Face on Chinese Tourists: An Exploratory Study. *Human Geography*, 30(1): 122–128.

Guo, X. and Lin, D. (2015). 中国本土消费者的面子意识与消费行为研究述评 A Literature Review of Face Consciousness and Local Consumer Behavior in China. *Foreign Economics & Management*, 37(11): 63–71.

Ho, D. Y.-F. (1976). On the Concept of Face. *American Journal of Sociology*, 81(4): 867–884.

Hong, Q. (2011). 中国面子心理与旅游消费的研究 Research on "Mianzi" and Chinese Tourist's Consumption Behavior. *Journal of Wuhan Polytechnic*, 10(3): 92–95.

Hu, H. C. (1944). The Chinese Concepts of 'Face'. *American Anthropologist*, 46(1): 45–64.

Huang, M. F., Tang, C., and Weaver, D. B. (2017). The Arctic Tourism Experience from an Evolving Chinese Perspective. In Y. S. Lee, D. B. Weaver, and N. K. Prebensen (Eds.), *Arctic Tourism Experiences: Production, Consumption and Sustainability* (pp. 89–99). Wallingford: CABI. DOI: 10.1079/9781780648620.0089.

Jennings, G. R. (2005). Interviewing: A Focus on Qualitative Techniques. In B. W. Ritchie, P. Burns, and C. Palmer (Eds.), *Tourism Research Methods: Integrating Theory with Practice* (pp. 99–117). Wallingford: CABI. DOI: 10.1079/9780851999968.0099.

Kwek, A. and Lee, Y.-S. (2015). How 'Face' Matters: Chinese Corporate Tourists in Australia. *Journal of Travel & Tourism Marketing*, 32(1–2): 120–140. DOI: 10.1080/10548408.2014.986016.

Lee, Y.-S., Weaver, D. B., and Prebensen, N. K. (eds) (2017). *Arctic Tourism Experiences. Production, Consumption and Sustainability*. Wallingford: CABI. DOI 10.1079/9781780648620.0000.

Li, M. M., Wen, T., and Leung, A. (2011). An Exploratory Study of the Travel Motivation of Chinese Female Outbound Tourists. *Journal of China Tourism Research*, 7(4): 411–424. DOI: 10.1080/19388160.2011.627020.

Mao, L. R. (1994). Beyond Politeness Theory: 'Face' Revisited and Renewed. *Journal of Pragmatics*, 21(5): 451–486. DOI: 10.1016/0378-2166(94)90025-6.

Matthews, B. and Ross, L. (2010). *Research Methods: A Practical Guide for the Social Sciences*. Harlow: Pearson Longman.

Neuendorf, K. A. (2002). *The Content Analysis Guidebook*. California: Sage Publications.

Patton, M. Q. (2002). *Qualitative Research & Evaluation Methods*. Thousand Oaks, CA: Sage Publications.

Stover, L. E. (1962). 'Face' and Verbal Analogues of Interaction in Chinese Culture: A Theory of Formalized Social Behavior based on Participant-Observation of an Upper-Class Chinese Household, Together with a Biographical Study of the Primary Informant, Unpublished doctoral dissertation, Columbia University, New York.

Veal, A. J. (2011). *Research Methods for Leisure and Tourism: A Practical Guide*. Harlow: Financial Times Prentice Hall.

Wang, R. and Huang, A. M. (2010). 浅析面子文化对中国旅游者消费行为的影响 A Brief Analysis on the Influence of Face Culture on Chinese Tourists' Consumption Behavior. *Journal of Leshan Teachers College*, 25(1): 69–85.

Wu, J. and Tao, D. (2010). 论"面子"文化与旅游消费行为的关系 On the Relationship between "Mianzi" Culture and Tourism Consumption Behavior. *Economic Research Guide*, 96(22): 167–169.

Xie, Y. and Li, M. (2009). Development of China's Outbound Tourism and the Characteristics of Its Tourist Flow 中国出境旅游的发展及旅游流特征 Development of China's Outbound Tourism and the Characteristics of Its Tourist Flow. *Journal of China Tourism Research*, 5(3): 226–242. DOI: 10.1080/19388160903149965.

Yuan, S., Gao, Y., and Zheng, Y. (2009). 面子意识,地位消费倾向与炫耀性消费行为–理论关系模型及实证检验 Face Consciousness, Status Consumption Tendency and Conspicuous Consumption Behavior–Theoretical Relationship Model and Empirical Research. *Collected Essays on Finance and Economics*.

Zhang, Y. and Si, H. (2010). 中国人面子心理的文化解读 An Cultural Interpretation of Chinese Face Psychology. *Theoretic Observation*, 1(61): 52–54.

48

The emotional labor of the co-created tourism experience

Richard Ek, Anne Hardy, Mia Larson and Can-Seng Ooi

Introduction

Demonstrating a cheerful and positive disposition at work has become an expectation in many jobs. This demand is particularly strong for frontline staff members who have to meet customers directly, such as those in the hospitality industry. Maintaining a friendly and helpful demeanor can be difficult, especially when one is having a tough period in other aspects of one's life. Maintaining a charming and happy front is emotional work (Hing 2010, Ooi and Ek 2010).

Emotional labor has for decades been addressed in tourism studies and tourism management. Originally coined by Arlie Hochschild (1979, 1983) the concept has been elaborated upon in relation to tourism and hospitality service work and has predominantly been regarded as a clear-cut social interaction between employee and customer with an employer in the background. Through a managerial point of view, emotional labor has been imagined, grasped and understood in more or less instrumental and functionalistic ways. Emotional labor is something that has to be formalized in a certain way in order to be addressable from a managerial rationality. However, in contemporary capitalism, with the rise of neoliberalism, the experience economy and the sharing economy, the clear-cut interaction mentioned above is unfolded and rearranged. The employee can just as well be his or her own employer, and at the same time find himself or herself regulated by an assemblage of mobile technologies and digital software. The established view on emotional labor as situated within a triangle of employee, customer and employer does not apply in the same way. As a consequence, emotional labor as a societal phenomenon needs to be rethought. To some degree this has been done in tourism studies, but this research is still in its infancy. In particular, there is a lack of more nuanced takes on emotional labor in the tourism experience economy and the sharing economy. This chapter offers such a take on emotional labor, highlighting the emotional labor of being an Airbnb host as a case.

The chapter starts with a short introduction to emotional labor and a literature review on how it has been elaborated upon in tourism and hospitality management studies. The next section outlines a societal contextualization and interpretation of emotional labor in contemporary capitalism. Thereafter, the sharing economy and the case of Airbnb hosting are presented. The chapter finishes off with a conclusive call for further research.

Emotional labor in the contemporary capitalist economy

Emotional labor conjoins two societal trajectories: the rise of a service economy and the apprehended need of self-awareness and individual management of emotions in workplace situations. This management of emotions takes place, metaphorically speaking, on a surface level and on a deep level. The surface level is very much a visual performance in social interaction while the deep level refers to how the individual really "feels". The surface level is often in the service encounter, when the employee is required to follow a script produced by the employer. From the perspective of the employer the surface display of emotions becomes genuine only if there is a correspondence between the employee's level on the surface and on a deeper level. The employee does not only have to act in a service-minded way towards a customer but also feel "deep inside" that s/he wants to be kind and service-oriented in the customer encounter. To Hochschild (1983: 90) if the employee who is expected to display a certain set of emotions does not feel comfortable with their feelings, emotional dissonance may be developed over time. Either the surface acting has to change, or the employee needs to recalibrate their deep feelings. From a managerial perspective this implies that either the service-performance script has to change, or the employee's deep feelings need to be adjusted (thus, becoming genuinely service-minded).

In the current age of seeking rational efficiency and ultimate service, hospitality businesses aim to provide increasingly authentic and warm services. Emotions have become part of the service commodity. The fetishization of experiences and emotion have led to the intensification of emotional labor. There are at least two interrelated reasons for this.

First, emotional work suffers from Baumol's "cost disease" (2012). Labor productivity in many service industries, such as in education, the arts and nursing, does not increase because the nature of the products could not change drastically. The example that Baumol provided is the live orchestral music performance. Orchestral performances that were performed centuries ago are still performed in similar ways, despite the advent of technological advancement. Similarly, warm and friendly service at a restaurant or a hotel is still appreciated by guests despite the pressures to automate back-room processes and innovate service procedures in the hospitality industry. These service products remain the same so the productivity gains have been limited, as compared to other sectors such as manufacturing. While there is limited productivity growth in such work, wages have to be increased to ensure that there are workers to serve customers. But the productivity and cost pressures remain. This often results in stressful circumstances, as these workers have to maintain their composure while serving more and more customers. To Baumol, cost disease is inevitable under current economic circumstances unless and until society accepts that such services are necessarily becoming more expensive, in relation to manufactured and other automated products.

Second, under the current state of capitalism almost everything is being "neoliberalised" (Higgins and Larner 2017). Things are reconfigured via the creation of economic value, including nature, culture and, in the case of emotional work, experiences. The emotion-inducing behavior of service staff members is part of the service commodity. Tourism has become a driver of neoliberalism as it has become an economic platform to monetize landscapes, wild animals, lived cultures, heritage and authentic hospitality (Duffy 2013). "Authentic" warm hospitality has been assumed to be part of mass tourism experiences, whether one is on a cruise ship, in a busy restaurant or in an Airbnb home. Warm hospitality is not a unique but an essential selling proposition. This has led to the creation of incentive structures that encourage good service in some countries, such as in the USA and Canada. Restaurant servers work for tips. Their encounters with customers have become a direct economic exchange. They do not only present a pleasant and warm front but also attempt to serve as many guests as possible. Any service "failures" will render their service economically worthless. The creation of economic value in the current neoliberal regime has been reduced to individualized encounters, in which the market structures are seen to be a neutral platform for exchange rather than structures of emotional and economic exploitation.

These two points assume the commodification of emotional expressions. Neoliberalism is based on a theory of political economic practices that aims to advance human well-being by liberating individual entrepreneurial freedoms and skills (Harvey 2005). The translation of neoliberal idealism into practice has been varied (Duffy 2013; see chapters in Higgins and Larner 2017). Nonetheless, one of the focuses in neoliberalism is the individual's own ability to exercise their freedom and exploit their own capability to further their own well-being (Comunian and Ooi 2016, Harvey 2005, Ooi 2012). The focus on personal choices and responsibility assumes that the social, economic and political structures in which those choices are made are also fair to all. This perspective has implications for emotional labor. The focus on the individual and the responsibility of individuals ignores the employee as a human with multiple concerns and aspirations. Instead, the person compartmentalizes and becomes a worker who has to provide pleasing experiences even during difficult periods in their personal life (Andersson Cederholm 2010, Ooi 2010).

Emotional labor in tourism and hospitality studies

Even if emotional labor as a concept stems back to the late 1970s, it did not find its way into tourism marketing and tourism studies until the early 21st century. After that, however, the number of studies discussing emotional labor has rapidly increased in tourism studies in general and tourism management in particular. The research on emotional labor has striven to understand the emotional labor of the employed from different perspectives. Some research has put the employed in focus, stressing individual psychological characteristics like emotional intelligence or mindfulness (Li, Wong and Kim 2017, Prentice and King 2011), or gender and cultural background (Li, Canziani and Barbieri 2018, Riley 2007, Torland 2011a) in terms of, for instance, the presence of individualism or collectivism (Newnham 2017). Other research, still with a focus on the employed, has investigated the intertwinement of emotional labor with other more articulated individual ambitions, positions and stances, like coping strategies (Kim and Han 2009, Lundberg and Mossberg 2008), job satisfaction (Torland 2011b) and customer orientation (Sohn 2017). In some cases, a variable is understood to have a strong influence on the nature of the emotional labor, like the presence of affect and empathy (Chu, Baker and Murrmann 2016, Heimtun 2016). In other cases the nature and amount of emotional labor are supposed to have specific relevance to, for instance, turnover intention, emotional exhaustion and work-life balance (Xu, Martinez and Lv 2017). There has been a particular interest in the effect of surface acting and deep acting on the conduct of service provision (Sohn and Lee 2012, van Dijk, Smith and Cooper 2011).

Additional research contextualizes the emotional labor of the employed in the wider dynamic framework of the service encounter from a more explicit managerial perspective, which makes it possible to increase the service quality, create customer equity (Choi et al. 2014) and increase managerial support for the employed (Bratton and Watson 2018). Some research efforts have been made to explicitly understand the emotional labor connected to troublesome customers (Hu, Hu and King 2017). Finally, one strand of research has attempted to construct more holistic frameworks in which emotional labor is one variable of many, albeit a crucial one (Shani et al. 2014, Uriely, Maoz and Reichel 2008).

Research on emotional labor in the specific context of the staging and management of tourist experiences is however scarcer. Some work exists on how personnel working at theme parks and zoos have been involved in new sorts of emotional labor as these tourist attractions have been oriented and managed into sites of the tourist experience economy (Baerenholdt and Jensen 2009, Beardsworth and Bryman 2001, Mathisen 2019, Wijeratne et al. 2014). However, this research, like the other research outlined above, often frames emotional labor as something that takes place in a traditional employee–employer relationship. What is thus interesting to further discuss is how emotions and emotional labor can be addressed in the growing tourism and hospitality experience sector, which lacks a traditional employer. We therefore continue with addressing emotional work and emotional labor in the growing tourist- and hospitality-oriented sharing economy.

The rise of sharing economy labor

The vast digital evolution during recent years has given rise to an expanding so-called sharing economy. The terms and definitions around the concept "sharing economy" are numerous and blurred – some call it the collaborative economy (Dredge and Gyimóthy 2015), others the on-demand economy, the gig economy, etc. Mair and Reischauer (2017) distinguish the sharing economy from the traditional economy via its use of various forms of compensation, its focus on the redistribution of resources and access to them, the fact that individuals are transaction partners and not so often organizations and that transactions occur on digital platforms operated by organizations. Airbnb and Uber are iconic examples of platform organizations (Reischauer and Mair 2018) that have emerged, based on social production and consumption, in communities on digital platforms (cf. Dredge and Gyimóthy 2015).

The sharing economy shapes social and economic dynamics and blurs the boundaries between production and consumption, between full employment and casual labor and between private and public life (Reischauer and Mair 2018). The platform companies consider their workers to be independent contractors and not employees subject to benefits. Instead they are encouraging the empowering benefits of labor flexibility and entrepreneuralism (Regan Shade 2018). The sharing economy is, thus, reinforcing the neoliberal paradigm by endorsing individualization and market-based solutions to employment (Cockayne 2016, Martin 2016, Regan Shade 2018). Although the platform companies consider their contractors as independent, it is important to note that the people working for a platform organization are strictly governed by the design and rules of the platform, with very limited possibilities to influence it. So, although the entrepreneurs are their own agents, they are unable to reform the digital business model they work within, and thus the way they interact with customers online.

It can be argued that the sharing economy poses an opportunity for people enjoying flexible work and wanting to have an extra income, and that they can quickly and easily enter the market by using the platforms provided. At the same time, workers in the sharing economy are often people out of the traditional labor market with no other option than the often precarious work the sharing economy offers. Moreover, the platforms operate in discriminatory ways (Schor and Attwood-Charles 2017). For instance, Hannák, Wagner and Garcia (2017) show that gender and race are significantly correlated with worker evaluations. This context has, of course, consequences for the workers involved in the sharing economy and thus it is important to consider it when discussing emotional labor in the sharing economy.

Work for the contractors consists to a large extent of digital communication with customers on the sharing platform, so it is here that substantial parts of the service encounter take place. Platform mediators play a significant role in facilitating trust-building and exercising social control through these platforms (Gyimóthy 2017). How the platform steers the communication in terms of what information is required by the contractors influences the digital service encounter. The contractors must balance competing demands for openness on the platform with their need for integrity, which requires emotional labor with specific characteristics. The rating system on the platform, shared by both the service provider and the customer, is a way of measuring performance. How the rating system is set and what communication is allowed in the comments influences the workers' level of participation in the platform and their emotional response (cf. Banning 2016). The feedback to the service providers made by the customers is sometimes extensive and the distance between the sender and receiver of the communication influences the rhetoric and the degree of harshness. Since the service providers sometimes share private assets, such as their homes, this may be particularly sensitive.

The sharing economy does, as discussed, elevate the complexity of emotional labor in the tourism industry. The often precarious working conditions, the blurring of private and commercial space and the distanced nature of digital social media communication contributes to a challenging emotional situation for workers in the sharing economy. Airbnb, for instance (as will be demonstrated next) celebrates unique hospitable experiences of someone's home. Hosts are encouraged to be genuine and welcoming. The home away from home concept disguises the emotional efforts required in being warm and hospitable.

Case study 48.1: Emotional labor in the sharing economy – the case of Airbnb

From its humble beginnings in 2008, when it was designed as a platform that could make use of under-utilized space, Airbnb has grown into a global behemoth worth US $38 billion (Forbes 2018). Its rapid rise in popularity has resulted in polarized reactions; in the first place, it is loved by many travelers – approximately 115 million guests used Airbnb in 2017 across 81,000 cities and 191 countries (Forbes 2018). This rapid growth is due to the product that it offers: Airbnb markets itself to travelers as offering affordable, authentic accommodation where tourists can "*live like a local*" (Chesky 2014).

But despite its enormous growth, Airbnb has often been criticized by some sectors of the tourism industry, the local community and town planners. The platform challenges traditional regulatory requirements by allowing hosts to instantaneously list their properties with very few requirements. Its rapid rise has been attributed to overtourism and issues related to housing affordability and access (Gutiérrez et al. 2017). However, while much research has been undertaken on the supply side of Airbnb, far fewer studies address the phenomenon of hosting. Moreover, the "work" involved in hosting an Airbnb has rarely been addressed within tourism literature.

There are a variety of very positive aspects to hosting that arguably *reduce* emotional labor for hosts, relative to other platforms. The Airbnb app offers tourists and hosts the opportunity to make bookings via an app that manages payment prior to guests' arrival, thus removing the need for financial discussions between the host and the guest upon their arrival. Once guests arrive, all monies have been transferred, allowing the guests to focus solely on their experience and the host to focus on the act of hosting.

On an individual level, there are further positive benefits that have been reported from hosting. It generates income, can give hosts a sense of pride in sharing their home and city with travelers and can connect hosts with like-minded guests. Another positive outcome of hosting that has been reported within the literature is that it has facilitated the formation of groups. Around the world, hosting groups have emerged, who meet both online and in person. Research has demonstrated that for hosts these groups provide a supportive environment where they can share ideas, vent their frustrations and work together to lobby governments or community groups, when faced with criticism. These hosting groups have been conceptualized as neo-tribes as they can form, reform and dissipate as is characteristic of neo-tribes (Bennett 2010). Hosting groups have been reported as facilitating supportive fellowship, giving hosts a feeling of being part of a greater movement, particularly when faced with criticism (Hardy and Dolnicar 2017a).

However, while individual emotional labor benefits do arise from hosting, negative emotional labor costs are also apparent. The Airbnb model does not appeal to only one market of travelers; research has determined that a variety of hosts exist. Host types have been defined as *capitalists* who have minimal interaction with guests and seek to maximize financial returns from their propertie(s), *befrienders* who spend a great deal of time and effort ensuring their guests have good experiences and are not as driven by money and *ethicists*, who are guided by the principles of sustainability and, as such, choose to host in order to occupy unused space (Hardy and Dolnicar 2017b). Negative emotional labor is arguably more likely to be a phenomenon for befrienders and ethicists, whose drive is more focused on the host experience and who are more likely to interact with their guests. For these hosts, fatigue is particularly apparent towards the end of the tourist season, especially for those who undertake hosting on top of their existing forms of work. Following the departure of guests, work is required to clean and prepare for the next arrival. In addition, the requirement by Airbnb for hosts to respond to enquiries within 24 hours, or risk losing their status, places additional pressure on hosts. This pressure is exacerbated by the fact that many hosts run Airbnb businesses in addition to their normal work. However, while Airbnb imposes rules that require hosts to remain responsive at all times, they also provide mechanisms that can reduce emotional labor. Hosts may choose to "block out" days between guests when they need time to recover from the work involved in hosting, or would like a break from "performing" their hosting activities.

While blocking out dates for recovery between guests can alleviate emotional labor, negative emotional labor can be created when a mismatch occurs between hosts and guests (Hardy and Dolnicar 2017b). Like hosts, a typology of guests has been identified, ranging from those who seek out large amounts of guest interaction through to those who would rather have complete privacy. If a highly social guest who is seeking out interaction with local people stays in an Airbnb that is run by a *capitalist*, the chances are that dissatisfaction may occur. Similarly, if an introverted guest is hosted by a *befriender* host, discontentment could also arise. When this occurs, negative emotional labor is likely to result. It is far more likely that this dissatisfaction will become apparent after departure, when both guests and hosts are requested to review their guest/host experience.

The review process underpins the concept of Airbnb and involves hosts and guests in reviewing their experiences. This can also be a source of negative emotional labor. For hosts who have had to endure rowdy guests, untidy guests or guests whose personalities do not align with their own, a conundrum can exist. They have hosted their guests in their home, or on their property and have had to perform duties as a host – responding to guest requests, maintaining a neat property, maintaining their composure when dealing with difficult guests and ensuring that guests have a satisfactory experience. The emotional labor that these tasks require is magnified for *befriender* and *ethicist* hosts, who are more likely to have guests staying in their own home, and who have high levels of interaction with their guests. At 3 p.m. on the day of departure, hosts are sent a message requesting that they review their guests. A dilemma exists: does the host perform an about-face and criticize the guest five hours after departure? The host must also consider how to respond in the event that the guest provides a poor review before they do – or even worse, a positive one when the host has provided a poor one. The emotional energy this process consumes is not insignificant; host forums for Airbnb are commonly filled with discussions around this issue.

Conclusion

In this chapter we have offered an overview of how emotional labor has been elaborated upon in tourism and hospitality studies. We have concluded that the notion of emotional labor as something that plays out between the employed person and the customer, with an employee in the background, does not hold in an increasingly experience-based sharing economy of companies based on digital platforms and neo-tribalism. This does not imply the diminishment of emotional labor. Rather, emotional labor takes new shapes and becomes even more transgressive as the distinction between work and free time, work and lifestyle, employed and customer and employed and employer continue to bleach out. The demands of the individual connected to a neoliberal societal ecology change and probably also increase as every employee is more and more his or her own entrepreneur. At the same time, digital business models and solutions, at first sight offering freedom to choose, and individual flexibility, in a multitude of ways restrict and structure the service encounter and the emotional labor that takes place.

Nevertheless, it is important to search for a nuanced picture of emotional labor during these contemporary circumstances. However, the approaches, models, and concepts connected to management and marketing theory chiseled-out during an era of Fordist economy only take us to a certain heuristic understanding. Additional societal theories and concepts have to be included to make it possible for experience management and marketing scholars and students to frame the emotional labor taking place in this at least partly new work environment. One useful approach could be biopolitical. Roelofsen and Minca's recent paper (2018) on how the "Superhost" is crystallized through biopolitical processes, including the imagined constructions of communities, self-regulation and self-discipline and the quantification of performative work offers such an avenue. Emotional labor is, after all, connected to the individual and his and her subject positions and care of the self. At the same time, as Foucault stresses, biopolitical self-disciplining at work is in no way only oppressive, but constructive in a multitude of ways. The individual is never completely locked into a restrictive frame of demands and limitations, but free in a substantial way to chisel out a life that is rewarding and nurturing. Dehumanizing emotional labor is in many ways a reality among service personnel today, but never an inevitable outcome, not even in an increasingly demanding neoliberal work ecology.

References

Andersson Cederholm, E. (2010). 'Effective emotions: The enactment of a work ethic in the Swedish meeting industry', *Culture Unbound: Journal of Current Cultural Research*, 2 (3): 381–400. DOI: 10.3384/cu.2000.1525.10222381

Baerenholdt, J. O. and Jensen, H. L. (2009). 'Performative work in tourism', *Scandinavian Journal of Hospitality and Tourism*, 9 (4): 349–365. DOI: 10.1080/15022250902978710

Banning, M. E. (2016). 'Shared entanglements – Web 2.0, info-liberalism & digital sharing', *Information, Communication & Society*, 19 (4): 489–503. DOI: 10.1080/1369118X.2015.1061573

Baumol, W. J. (2012). *The Cost Disease*. New Haven: Yale University Press.

Beardsworth, A. and Bryman, A. (2001). 'The wild animal in late modernity. The case of the Disneyization of zoos', *Tourist Studies*, 1 (1): 83–104. DOI: 10.1177/146879760100100105

Bennett, A. (2010). 'Subcultures or neo-tribes? Rethinking the relationship between youth, style and musical taste'. In M. Pickering (ed.), *Popular Culture, Volume 3* (pp. 145–162). London, UK: Sage.

Bratton, J. and Watson, S. (2018). 'Talent management, emotional labour and the role of line managers in the Scottish hospitality industry: A roundtable discussion', *Worldwide Hospitality and Tourism Themes*, 10 (1): 57–68. DOI: 10.1108/WHATT-10-2017-0063

Chesky, B. (2014). *Belong anywhere*. Airbnb 16.07.2014.07.10.2017.

Choi, C. H. et al. (2014). 'Testing the stress-strain-outcome model of customer-related social stressors in predicting emotional exhaustion, customer orientation and service recovery performance', *International Journal of Hospitality Management*, 36 (1): 272–285. DOI: 10.1016/j.ijhm.2012.09.009

Chu, K. H., Baker, M. A. and Murrmann, S. K. (2016). 'When we are onstage, we smile: The effects of emotional labor on employee work outcomes', *International Journal of Hospitality Management*, 31 (3): 906–915. DOI: 10.1016/j.ijhm.2011.10.009

Cockayne, D. G. (2016). 'Sharing and neoliberal discourise: The economic function of sharing in the digital on-demand economy', *Geoforum*, 77: 73–82. DOI: 10.1016/j.geoforum.2016.10.005

Comunian, R. and Ooi, C.-S. (2016). 'Global aspirations and local talent: The development of creative higher education in Singapore', *International Journal of Cultural Policy*, 22 (1): 58–79. DOI: 10.1080/10286632.2015.1101085

Dredge, D. and Gyimóthy, S. (2015). 'The collaborative economy and tourism: Critical perspectives, questionable claims and silenced voices', *Tourism Recreation Research*, 40 (3): 286–302. DOI: 10.1080/02508281.2015.1086076

Duffy, R. (2013). 'The international political economy of tourism and the neoliberalisation of nature: Challenges posed by selling close interactions with animals', *Review of International Political Economy*, 20 (3): 605–626. DOI: 10.1080/09692290.2012.654443

Forbes. (2018). 'As a rare profitable unicorn, Airbnb appears to be worth at least $38 billion', *Forbes*, May 11, 2018.

Gutiérrez, J. et al. (2017). 'The eruption of Airbnb in tourist cities: Comparing spatial patterns of hotels and peer-to-peer accommodation in Barcelona', *Tourism Management*, 62: 278–291. DOI: 10.1016/j.tourman.2017.05.003

Gyimóthy, S. (2017). 'Business models of the collaborative economy'. In D. Dredge and S. Gyimóthy (eds.), *The Collaborative Economy: Tourism Social Science Perspectives* (pp. 31–39). Cham: Springer.

Hannák, A. et al. (2017).'Discrimination in online freelance marketplaces: Evidence from TaskRabbit and Fiverr', *CSCW 17*, Portland, OR, 25 February–1 March, http://personalization.ccs.neu.edu/papers/hannak-cscw17.pdf.

Hardy, A. and Dolnicar, S. (2017a). 'Networks and hosts – A love hate relationship'. In S. Dolnicar (ed.), *Peer to Peer Accommodation Networks: Pushing the Boundaries* (pp. 182–193). Oxford: Goodfellow Publishers.

Hardy, A. and Dolnicar, S. (2017b). 'Types of network members'. In S. Dolnicar (ed.), *Peer to Peer Accommodation Networks: Pushing the Boundaries* (pp. 170–181). Oxford: Goodfellow Publishers.

Harvey, D. (2005). *A Brief History of Neoliberalism*. Oxford: Oxford University Press.

Heimtun, B. (2016). 'Emotions and affects at work on Northern Lights tours', *Hospitality & Society*, 6 (3): 223–241. DOI: 10.1386/hosp.6.3.223_1

Higgins, V. and Larner, W. (Eds.) (2017). *Assembling Neoliberalism: Expertise, Practices, Subjects*. New York: Palgrave Macmillan.

Hing, A. Y. (2010). 'Service workers: Governmentality and emotion management', *Culture Unbound: Journal of Current Cultural Research*, 2 (3): 311–327.

Hochschild, A. (1979). 'Emotion work, feeling rules, and social structure', *The American Journal of Sociology*, 85 (3): 551–575. DOI: 10.1086/227049

Hochschild, A. (1983). *The Managed Heart: Commercialization of Human Feelings*. Berkeley: University of California Press.

Hu, H.-H., Hu, H.-Y. and King, B. (2017). 'Impacts of misbehaving air passengers on frontline employees: Role stress and emotional labor', *International Journal of Contemporary Hospitality Management*, 29 (7): 1793–1813. DOI: 10.1108/IJCHM-09-2015-0457

Kim, M. J. and Han, S. Y. (2009). 'Relationship between emotional labor consequences and employees' coping strategy', *Asia Pacific Journal of Tourism Research*, 14 (3): 225–239. DOI: 10.1080/10941660903023929

Li, J., Canziani, B. F. and Barbieri, C. (2018). 'Emotional labor in hospitality: Positive affective displays in service encounters', *Tourism and Hospitality Research*, 18 (2): 242–253. DOI: 10.1177/1467358416637253

Li, J., Wong, I. W. and Kim, W. G. (2017). 'Does mindfulness reduce emotional exhaustion? A multilevel analysis of emotional labor among casino employees', *International Journal of Hospitality Management*, 64: 21–30. DOI: 10.1016/j.ijhm.2017.03.008

Lundberg, C. and Mossberg, L. (2008). 'Learning by sharing: Waiter' and bartenders' experienses of service encounters', *Journal of Foodservice*, 19: 44–52. DOI: 10.1111/j.1745-4506.2007.00079.x

Mair, J. and Reischauer, G. (2017). 'Capturing the dynamics of the sharing economy: Institutional research on the plural forms and practices of sharing economy organizations', *Technological Forecasting & Social Change*, 125: 11–20. DOI: 10.1016/j.techfore.2017.05.023

Martin, C. J. (2016). 'The sharing economy: A pathway to sustainability or a nightmarish form of neoliberal capitalism?', *Ecological Economics*, 121: 149–159. DOI: 10.1016/j.ecolecon.2015.11.027

Mathisen, L. (2019). 'Storytelling: A way for winter adventure guides to manage emotional labour', *Scandinavian Journal of Hospitality and Tourism*, 19 (1): 66–81. DOI: 10.1080/15022250.2017.1411827

Newnham, M. P. (2017). 'A comparison of the enactment and consequences of emotional labor between frontline hotel workers in two contrasting societal cultures', *Journal of Human Resources in Hospitality & Tourism*, 16 (2): 192–214. DOI: 10.1080/15332845.2016.1202729

Ooi, C.-S. (2010). 'Cacophony of voices and emotions: Dialogic of buying and selling art', *Culture Unbound*, 2: 347–364.

Ooi, C.-S. (2012). 'Education and becoming an artist: Experiences from Singapore'. In C. Mathieu (ed.), *Careers in Creaive Industries* (pp. 254–269). New York: Routlege.

Ooi, C.-S. and Ek, R. (2010). 'Culture, work and emotion', *Culture Unbound: Journal of Current Cultural Research*, 2 (3): 303–310.

Prentice, C. and King, B. (2011). 'The influence of emotional intelligence on the service performance of casino frontline employees', *Tourism and Hospitality Research*, 11 (1): 49–66. DOI: 10.1057/thr.2010.21

Regan Shade, L. (2018). 'Hop to it in the gig economy: The sharing economy and neo-liberal feminism', *International Journal of Media and Cultural Politics*, 14 (1): 35–54. DOI: 10.1386/macp.14.1.35_1

Reischauer, G. and Mair, J. (2018). 'Platform organizing in the new digital economy: Revisiting online communities and strategic responses', *Research in the Sociology of Organizations*, 57: 113–135. DOI: 10.1108/S0733-558X20180000057005

Riley, M. (2007). 'Role interpretation during service encounters: A critical view of modern approaches to service quality management', *Hospitality Management*, 26 (2): 409–420. DOI: 10.1016/j.ijhm.2006.03.005

Roelofsen, M. and Minca, C. (2018). 'The superhost. Biopolitics, home and community in the Airbnb dream-world of global hospitality', *Geoforum*, 91: 170–181. DOI: 10.1016/j.geoforum.2018.02.021

Schor, J. B. and Attwood-Charles, W. (2017). 'The "sharing" economy: Labour, inequality, and social connection on for-profit platforms', *Sociology Compass*, 11: 8. DOI: 10.1111/soc4.12493

Shani, A. et al. (2014). 'Emotional labor in the hospitality industry: The influence of contextual factors', *International Journal of Hospitality Management*, 37: 150–158. DOI: 10.1016/j.ijhm.2013.11.009

Sohn, H. K. and Lee, T.J. (2012). 'Relationship between HEXACO personality factors and emotional labour of service providers in the tourism industry', *Tourism Management*, 33 (1): 116–125. DOI: 10.1016/j.tourman.2011.02.010

Sohn, H.-K. (2017). 'Verifying the moderating effects of personality factors on the relationship between emotional labour and customer orientation', *International Journal of Tourism Sciences*, 17 (1): 1–14. DOI: 10.1080/15980634.2016.1274158

Torland, M. (2011a). 'Emotional labour and job satisfaction of adventure tour leaders: Does gender matter?', *Annals of Leisure Research*, 14 (4): 369–389. DOI: 10.1080/11745398.2011.639419

Torland, M. (2011b). 'Effects of emotional labor on adventure tour leaders' job satisfaction', *Tourism Review International*, 14 (2–3): 129–142. DOI: 10.3727/154427211X13044361606450

Uriely, N., Maoz, D. and Reichel, A. (2008). 'Israeli guests and Egyptian hosts in Sinai: A bubble of serenity', *Journal of Travel Research*, 47 (4): 508–522. DOI: 10.1177/0047287508326651

van Dijk, P. A., Smith, L. D. G. and Cooper, B. K. (2011). 'Are you for real? An evaluation of the relationship between emotional labour and visitor outcomes', *Tourism Management*, 32 (1): 39–45. DOI: 10.1016/j.tourman.2009.11.001

Wijeratne, A. J. C. et al. (2014). 'Rules of engagement: The role of emotional display rules in delivering conservation interpretation in a zoo-based tourism context', *Tourism Management*, 42: 149–156. DOI: 10.1016/j.tourman.2013.11.012

Xu, S., Martinez, L. R. and Lv, Q. (2017). 'Explaining the link between emotional labor and turnover intentions: The role of in-depth communication', *International Journal of Hospitality & Tourism Administration*, 18 (3): 288–306. DOI: 10.1080/15256480.2016.1276003

49

Tourism experience and destination competitiveness

Abijith Abraham and Saurabh Kumar Dixit

Introduction

A tourism destination could be considered as an amalgam of the individual products and opportunities for experience creation which integrate to form a sum of experience at the areas visited (Murphy et al. 2000). All tourism products are considered to be experiential in nature and while visiting a place traveler engagement, extensively, shapes the visitor experience formation. When people travel to a place, they tend to form an understanding of the attractions/activities at the destination visited, i.e. a 'sense of place'. This perception, the 'sense of place', is quite important in tourism experience management because a strong sense of place leads to an engaging destination experience. Even though the reasons for travel may vary from one tourist to another, vacation trips are commonly taken voluntarily to meet either an individual's personal needs or their hedonic needs, or both. These trips are undertaken not because the tourists have to, but because they want to (Prebensen et al. 2014), and there will be those elements at the destination that act as stimuli for their enjoyment. These stimuli, in turn, will also act as the deciding factor for the time, money and effort spent by tourists at a destination. Due to their important role in the overall tourism experience, quality standards need to be studied and partnered, along with the various reasons for tourists to be at a particular destination (Prebensen et al. 2014). The whole experiential component of the tourism environment is to be considered as a super set of all the stakeholders involved. This is an approach where people, who are at the center of attention, try to engage in tourism experience with other people while tourism itself is treated as a network of experiences, co-created by all of the stakeholders involved (Binkhorst and Dekker 2009). In order to attain a competitive advantage, the destination must make sure that the overall appeal as well as the experience provided is of the highest quality and ranks superior to that of an alternative destination catering to potential visitors (Dwyer and Chulwon 2003).

Competitiveness in tourism may also be defined as the ability to increase tourism expenditure; that is, to draw a large number of visitors while at the same time ensuring a memorable experience for them in a profitable manner. The aforementioned is done with the intention of adding to the well-being of the host community and the preservation of the natural resources of the destination for the coming generations (Ritchie and Crouch 2012; Vilić and Dujaković 2016).

Tourism destination competitiveness

Evolution of the concept

Scholastic interest in the topic became plainly evident from the late 1990s, with academicians citing the importance of studying the competitiveness aspect associated with a destination (Buhalis 2000). Although it has proven to be quite a popular subject, recent research into tourism destination competitiveness has shown the topic to be associated with a lot of confusion and controversy (Novais et al. 2017). The causes of all this bitter discourse are mainly due to its intricate nature (Novais et al. 2017); for example, the complexity of the concept (Cracolici and Nijkamp 2009), the difference in the perspectives applied to understanding the topic (Echtner and Ritchie 1993) and the more recent holistic multi-layered approach (Crouch and Ritchie 1999).

The early stages of tourism are synonymous with mass tourism; beginning in the mid-20th century and continuing to the current 21st century (Ritchie and Crouch 2003). Past studies have shown that the components that make up the competitive dimension have themselves evolved over the course of time. The very basics of competition have been transformed into something quite different from what they were in the last century (Tyson 1997).

The phenomenon of tourism has been the root cause of the significant changes in the manner in which destinations are managed. Kozak and Rimmington (1998) outlined the four components that define the attractiveness of a destination, namely: attractions; facilities and services; infrastructure; and hospitality. There has been a considerable change in the amount of time, money and effort diverted by the destinations (whether regions, cities or countries) towards the travel trade in order to attract and serve visitors as well as manage the aforementioned components (Ritchie and Crouch 2003). As the competition for the share of the tourism market escalated, destinations were found to embrace a marketing orientation. Places started being marketed like any other products in order to woo more investments, industry, residents and, obviously, tourism (Kotler et al. 1993). The most common of the references that are made to the elements of competitiveness is the ability of a tourism destination to attract visitors. This so-called ability also refers to the capacity of a destination to fulfill certain goals (Novais et al. 2017), mainly: economic and improving the well-being of the people associated with the destination (Ritchie and Crouch 2003); attractiveness and attraction (Enright and Newton 2004); and finally, sustainability (Hassan 2000).

Subsequently, the destinations were found to take a number of steps forward, starting with the setting up of national tourism offices (at regional as well as state level) and corporations, in order to monitor the situations and take action accordingly. Furthermore, the local authorities, along with the support of departments of commerce and various industrial associations, have been found to direct their keen attention towards the tourism sector (Ritchie and Crouch 2003). This step has proven to be of massive importance as the involvement of the interdepartmental committees has seen progress in the form of basic infrastructural development (transportation, public health and education); such developmental initiatives are taken as positive signs for potential investment and they also help in creating a thriving environment for the local tourism industry. The committees also try and incorporate sustainable practices through the integration of beneficial policies for local communities into their developmental plans. The scenario that ensued as a result of all the new changes saw some destinations managing to weather the heightened competition fairly well, albeit the others labored.

Destination competitiveness determinants: Over the years, a wide array of variables has been found to be associated with destination competitiveness (see Table 49.1). These include the

Table 49.1 Academic works associated with the concept of tourism destination competitiveness

S. N.	*Publications*	*Themes discussed*
1.	Ritchie and Crouch (1993) Crouch and Ritchie (1995) Crouch and Ritchie (1999) Ritchie and Crouch (2000)	Developed a model that was applicable to tourism destinations in contexts spanning from companies and products, national industries and national economies and competitiveness related to service industries.
2.	Murphy et al. (2000)	Explained destination as an amalgam of all the individual products and opportunities for experience creation which integrate to form a sum of experience at the areas visited.
3.	Hassan (2000)	Identified the four determinants of market competitiveness: a. *comparative advantage* b. *demand orientation* c. *industry orientation* d. *environment orientation.*
4.	Heath (2003) Goffi and Cucculelli (2014) Prebensen et al. (2014) Vilić and Dujaković (2016)	Explained the importance of studying the quality of standards due to their impact on tourism experience and tourism experience as a crucial element of sustained destination competitiveness.
5.	Heath (2003)	Developed a model that showcased the importance of factors such as the people themselves as key success drivers and the crucial linkages (e.g.: communication and information management) in developing a comprehensive structure of sustainable destination competitiveness.
6.	Dwyer and Chulwon (2003), Ritchie and Crouch (2003), Gândara and Chim-Miki (2016)	Explained the importance of destination competitiveness in the regional and national economy.
7.	Dwyer and Chulwon (2003)	Proposed a model recognizing the importance of the demand factor as an important determinant of destination competitiveness.
8.	Trauer and Ryan (2005) Gao (2017)	Explained how experiences can be shaped by the attitude of the host community towards the tourism activities in a particular region.
9.	Oh et al. (2007)	Explained how the destination's distinct attributes prompt the creation of experiences of a certain nature and scope.
10.	Binkhorst and Dekker (2009)	Explained tourism as a network of experience co-created by all of the stakeholders.

variables that are objectively measured as well as those which are subjectively measured. The objectively measured variables include the likes of market share, visitor numbers, employment and tourist expenditure, as well as the value added by the tourism industry. The subjectively measured variables include the quality of tourism experience, the affluence of culture and heritage, etc. (d'Harteserre 2000; Dwyer and Chulwon 2003).

According to Hassan (2000), there are four determinants of market competitiveness. They are, namely, *comparative advantage* (concerned with the factors associated with both micro- and macro-economic environments); *demand orientation* (a destination's ability to remain dynamic according to market demands); *industry orientation* (the presence or absence of an established tourism-related industry); and *environment orientation* (the commitment of a destination towards the environment) (Dwyer and Chulwon 2003). Dwyer and Chulwon (2003) proposed a model recognizing the importance of the demand factor as an important determinant of destination competitiveness. The works by Crouch and Ritchie (1995, 1999) and Ritchie and Crouch (1993, 2000) were focused on developing a model that was applicable to a tourism destination in the context spanning *companies and products, national industries and national economies and competitiveness related to service industries* (Dwyer and Chulwon 2003). The works of Ernie Heath (2003) resulted in a model that showcased the importance of factors such as the people themselves as key success drivers and the crucial linkages (e.g.: communication and information management) in developing a comprehensive structure of sustainable destination competitiveness (Dwyer and Chulwon 2003; Heath 2003).

Travel and tourism competitiveness index (TTCI)

The Travel and Tourism Competitiveness Index, or TTCI, is a tool used to measure the competitiveness of economies (see also Figure 49.1). The TTCI aims to provide a comprehensive strategic means for measuring the assortment of factors and policies which enable the development of the travel and tourism sector in a sustainable manner. The TTCI, in turn, will be helpful in furthering the development and competitiveness of a country. The results are found to be beneficial for other stakeholders involved as well, as they help in tracking a country's progress over various areas measured, boosting the industry's competitiveness.

The Travel and Tourism Competitiveness report, first published by The World Economic Forum in the year 2007, is a report based on the data obtained from the major economies of the world. The initial report contained data pertaining to 124 economies (World Economic Forum 2007), while the more recent 2017 report dealt with the statistics of 136 economies. In this report, the economies are globally ranked by means of the 90 indicators, 14 pillars and four sub-indexes included in the TTCI (World Economic Forum 2017). The four sub-indexes being: *Enabling Environment (e.g. health and hygiene, safety and security, etc.)*; *Travel and Tourism Policy and Enabling Conditions (e.g. international openness, price competitiveness, etc.)*; *Infrastructure (e.g. air transport infrastructure, tourist service infrastructure, etc.)*; and, *Natural and Cultural Resources* (World Economic Forum 2017).

Tourism experience

Experience, in the context of a concept, is often regarded as relatively old. Activities such as leisure, tourism, cultural activities, internet services, marketing, etc., have been both

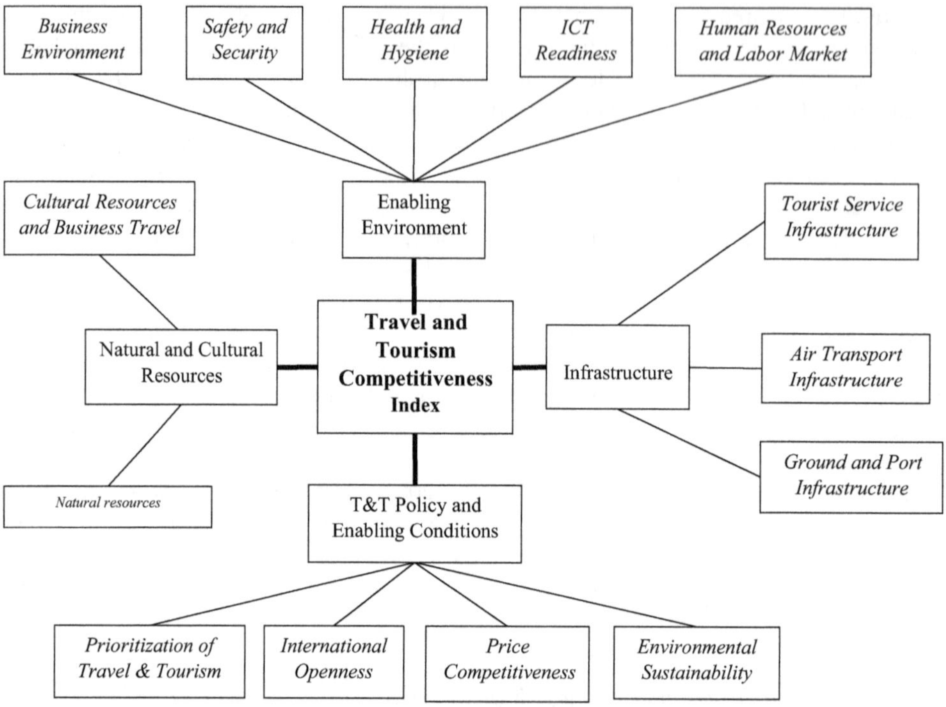

Figure 49.1 The Travel and Tourism Competitiveness Index framework
(Adapted from The Travel and Tourism Competitiveness Index framework published by the World Economic Forum 2017)

discussed and analyzed for many years (Darmer and Sundbo 2008). Initial studies conducted on tourism experience tended to revolve around the examination of the trips as units of the analysis and it was widely considered that all trips resulted in the same type of experience. These earlier studies employed sociological theories on travel in order to give justification to why the tourist wanted to travel in the first place (Scott and Le 2017). By the 1990s, the researchers started employing a more refined approach towards the topic, along with a more dynamic view of tourism experiences (Cutler and Carmichael 2010). The studies, in accordance with the immediate approach (Csikszentmihalyi 1990), were oriented around the lived experience rather than post hoc summative evaluations that are implicit in theories of, for example, satisfaction or value (Scott and Le 2017). The late 1900s saw the introduction of theories related to the experience economy and the idea that experiences have dimensions (Pine and Gilmore 1999). Four dimensions were described: affect, expectations, consequential reality and recollection; these essential components of a memorable tourism experience were later identified by researchers Tung and Ritchie (2011) and at around the same time Kim et al. (2010) established seven such components. Lastly, the early 2000s saw a cognitive psychological approach being implemented by researchers (Larsen 2007) and this approach underlined the importance of the individual, perceptual and mental characteristics of tourists in ascertaining the fact that experience covers both psychological and cognitive science theories (Scott and Le 2017).

Tourism experience and destination competitiveness

Tourism experience symbolizes the fundamental process for the successful management of a destination's competitiveness (Vilić and Dujaković 2016). It has been a growing notion that actual tourism experience can be substantially increased via maintaining destination authenticity and excellence in hospitality (along with the services offered), making tourism experience a crucial element of sustained destination competitiveness (Heath 2003). Heath, like many researchers before – illustrated through his model, which had people as the center of attention – explained the importance of competitiveness as a critical factor in determining the success and sustainability of a tourism destination (Goffi and Cucculelli 2014). The goal of any destination should be to offer unique, extraordinary and memorable tourism experiences to the target tourists in order to maintain a sustainable tourism competitive advantage (Vilić and Dujaković 2016).

The destination's attributes and their effect

A destination's distinct attributes (if and when perceived by the tourist) prompt experience creation of a certain nature and scope (Oh et al. 2007). These experiences in a destination may be fashioned and built to enable the comprehension of its past, or to depict a visited place via the attributed meaning of its 'actual' name (history, aesthetics, geography and culture). The experiences could also be shaped by the attitude of the host community towards the tourism activities in a particular region (Trauer and Ryan 2005; Gao 2017).

Governments are also well aware of the significance held by tourism in their respective regional as well as national economies. Their keen interest in keeping track of the competitiveness may be considered as evidence pointing towards their understanding of this key concept. This becomes more evident when researchers point out the relationship between destination competitiveness and the ability of a destination to attract potential tourists by offering a more satisfying tourism experience than its competitors and as a result – pointed out by the researchers – the subsequent improvement in the local as well as the national economies (Dwyer and Chulwon 2003; Ritchie and Crouch 2003; Gândara and Chim-Miki 2016).

The uniqueness of each destination and finding the USP (unique selling proposition)

One tourism destination is inherently different from another one in various ways. Therefore, the evaluation of destinations has been proven to be a difficult task due to this difference in nature. As in most cases, certain features are unique to a particular destination and, ergo, cannot be generalized (Prebensen et al. 2014; Vilić and Dujaković 2016). Hence, in order to analyze tourism destinations, it is important to understand the level of satisfaction and the quality of experience generated among visitors as a parameter of destination competitiveness development in the global tourism market (Vilić and Dujaković 2016). For example, the emphasis on tourism experience has swung away from the family, in which the main focus of travel was relaxation with a basic goal of producing a special moment to be shared among all the family members (Ritchie and Crouch 2003). Another aspect, no less important, is the decision regarding which aspect of a tourism destination needs to be prioritized in order to provide a certain competitive edge over its contemporary rivals (in this scenario, other similar destinations). It becomes very important to differentiate between what constitutes and needs to be promoted as a core element and what needs to be a secondary tangential component at a destination. For instance, majestic scenery will be the core element whereas the hotels, restaurants, signage, etc., may be considered as elements that are peripheral in nature (Ritchie and Crouch 2003). Both of these components together add up to give the sum tourism experience at a destination.

Case study 49.1: Kerala; a holistic experience

Kerala, the southernmost state of the Indian subcontinent, has been an important trading center for many centuries. Trade and commerce have had a vital role to play in how this southern state has been culturally evolving throughout history. Hence, it should not come as a surprise to learn that it is one of the most important tourism destinations for a modern tourist visiting the country. The state has a vast amount of natural resources, along with the other cultural resources, to offer to a wide audience. The foreign influence in architecture, cuisine, art, literature, philosophy, etc., is evident in many parts of the state. For example, the Jewish synagogue of Mattanchery (Fort Kochi) stands as a reminder of the region's trade and relations with the Middle East. And the Chinese fishing nets on the backwaters of Cochin stand as testimony to the exchange of technology that happened well back in the pages of history.

The present scenario is much different from that of earlier days. In recent years, the regional tourism governing body, the Department of Tourism, Government of Kerala, along with the Kerala Tourism Development Corporation (KTDC), has played its part in making sure that the state always has a competitive edge over its contemporary competitors. The state tourism board has gone to great lengths to promote, market and manage the destination in a sustainable manner. They have been carrying out initiatives to maintain a competitive advantage over the other Indian states, such as Rajasthan, Maharashtra and Madhya Pradesh, etc. The aforementioned states have individual tourism destinations whereas Kerala stands apart by offering the entire state as a destination for a holistic tourism experience. One of the recent initiatives taken by the state is to promote responsible tourism (Responsible Tourism Mission) in the state. The mission adopts tourism as a tool for the development of village and local communities, eliminating poverty and giving importance to the empowerment of women (Department of Tourism 2017). From tourists' perspectives, it provides an opportunity for them to explore the less-explored life of the villages of Kumarakom, Kovalam, Vaikom, Thekkady, Vythiri, etc. The tourists get an opportunity to witness artisans at work and even find themselves tempted to have a shot at it. The unique small-scale traditional industries and the skilled artisans become the core element of the tourism product that is being offered.

Grihasthali was a scheme initiated to facilitate the conversion of heritage buildings such as homes which are more than 50 years old into tourist accommodations. The scheme was not only to promote the unique heritage tourism product but also to offer a taste of the authentic Kerala lifestyle. For example, Chengara House, one of the famous mansions in Northern Kerala, was restored (infrastructural) to its former glory and converted into a heritage resort (Smitha Vasudevan 2008).

The Kumbalangi Tourism Village (Kochi) is another case from Kerala, where the unique tourism experience acts as the catalyst for a competitive tourism product. The initiatives taken to convert the tiny island village of Kumbalangi into a model fishing village and tourism spot have proven to be successful in providing a unique experience of the rural lifestyle in Kerala and, more importantly, tourism has given a new livelihood to many inhabitants of the village.

The destination, which is primarily a fishing village, owing to the rich aquatic life, is surrounded by backwaters and the Chinese fishing nets on the banks. The homestays inside the village are in the residences of the local community, which have been converted to provide an extra room or two for their guests. The tourists are encouraged to have their daily meals with the family (with whom they have sought accommodation) and experience the various aspects of life in a village. The meals comprise dishes from the local cuisine (mainly seafood), seasoned and spiced to the preference of the guests. During the daytime, tourists are able to roam around in the village, trying their hand at fishing and making the local handicrafts out of coconut products.

Conclusion

Like any other field of study, tourism destination competitiveness is also found to be a highly debated topic. Research studies until the present time suggest that there is not a definitive standard method to assess the competitiveness of any particular destination, but rather a number of methods (chosen according to the scenario in question) are used to complete the task. The characteristics of each tourism product are found to be varying in nature and each product has its own unique selling point. Researchers found it difficult to incorporate this variety into a single definitive model, which led to the existence of multiple models pertaining to each scenario at different tourism destinations. In the case of the various models proposed, one of the most prominent commonly accepted themes is the significance of the overall tourism experience in each scenario. This further illustrates the significance of subjectivity that is inherent in the concept of destination competitiveness. Quality experiences at a destination should be given equal (if not greater) attention, while at the same time the more objective variables such as tourism expenditure, arrival counts, etc., are also of great significance. These experiences should be sound enough to justify traveling to a particular destination. The case study above will suffice to show how tourism products, which are oriented on the overall tourism experience, point us towards the direction of sustainable and responsible tourism practices.

The good experiences at a destination need to be fostered and the stakeholders need to be educated about how to nourish and improve such experiences. There needs to be a structure for the comprehension of various aspects associated with the different levels of the concept concerning the different groups of stakeholders. The importance of analyzing tourism experience in relation to the competitiveness of a destination, and vice versa, will prove to be of even greater significance in time as the scenarios show, like the aforementioned case, and will decide whether a tourism destination and practices are sustainable or not.

References

Binkhorst, E. and Dekker, T. D. (2009). Agenda for co-creation tourism experience research, *Journal of Hospitality Marketing and Management*, 18(2-3):311–327. DOI: 10.1080/19368620802594193

Buhalis, D. (2000). Marketing the competitive destination of the future, *Tourism Management*, 21 (1):97–116, DOI: 10.1016/s0261-5177(99)00095-3

Cracolici, M. and Nijkamp, P. (2009). The attractiveness and competitiveness of tourist destinations: A study of southern Italian regions, *Tourism Management*, 30(3):336–344. DOI: 10.1016/j.tourman.2008.07.006

Crouch, G. and Ritchie, J. (1995), Destination competitiveness and the role of the tourism enterprise. Istanbul, Turkey, Proceedings of the Fourth Annual Business Congress.

Crouch, G. I. and Ritchie, J. (1999). Tourism, competitiveness, and societal prosperity, *Journal of Business Research*, 44(3):137–152. DOI: 10.1016/s0148-2963(97)00196-3

Csikszentmihalyi, M. (1990) Flow: The psychology of optimal experience, *Journal of Leisure Research*, 24 (1):93–94. DOI: 10.1080/00222216.1992.11969876

Cutler, S. and Carmichael, B. (2010). The dimensions of the tourist experience. In M. Morgan, P. Lugosi, and J. B. Ritchie (eds.), *The Tourism and Leisure Experience: Consumer and Managerial Perspectives* (pp. 3–26). Bristol: Channel View.

d'Harteserre, A. (2000) Lessons in managerial destination competitiveness in the case of Foxwoods Casino resort, *Tourism Management*, 21(1):23–32. DOI: 10.1016/s0261-5177(99)00097-7

Darmer, P. and Sundbo, J. (2008). Introduction to experience creation. In P. Darmer and J. Sundbo (eds.), *Creating Experiences in the Experience Economy* (pp. 1–12). Cheltenham: Edward Elgar.

Department of Tourism. (2017). 'Responsible tourism mission'. Kerala. [Online] Available at: www.keralatourism.org/responsible-tourism/ [Accessed: April 20, 2019].

Dwyer, L. and Chulwon, K. (2003), Destination competitiveness: Determinants and indicators, *Current Issues in Tourism*, 6(5):369–414. DOI: 10.1080/13683500308667962

Echtner, C. M. and Ritchie, J. B. (1993), The measurement of destination image: An empirical assessment, *Journal of Travel Research*, 31(4):3–13. DOI: 10.1177/004728759303100402

Enright, M. J. and Newton, J. (2004), Tourism destination competitiveness: A quantitative approach, *Tourism Management*, 25(6):777–788. DOI: 10.1016/j.tourman.2004.06.008

Gândara, J. M. and Chim-Miki, A. F. (2016). Destination evaluation through the prioratization of the competitiveness pillars: The case of Brazil. In A. Artal-Tur and M. Kozak (eds.), *Destination Competitiveness, the Environment and Sustainability: Challenges and Cases* (pp. 24–39). Oxfordshire: CABI.

Gao, L. (2017). The meaning of a destination experience in Zhouzhuang, China. In N. Scott, J. Gao, and J. Ma (eds.), *Visitor Experience Design* (pp. 201–219). Oxfordshire: CABI.

Goffi, G. and Cucculelli, M. (2014), Components of destination competitiveness. The case of small tourism destinations in Italy, *International Journal of Tourism Policy*, 5(4):296–326. DOI: 10.1504/ijtp.2014.068035

Hassan, S. (2000), Determinants of market competitiveness in an environmentally sustainable tourism industry, *Journal of Travel Research*, 38(3):239–245. DOI: 10.1177/004728750003800305

Heath, E. (2003), Towards a model to enhance destination competitiveness: A Southern African perspective, *Journal of Hospitality and Tourism Management*, January, 10(2):124–141.

Kim, J.-H., Ritchie, J., and McCormick, B. (2010), Development of a scale to measure memorable, *Journal of Travel Research*, 51(1):12–25. DOI: 10.1177/0047287510385467

Kotler, P., Haider, D. H., and Rein, I. (1993). *Marketing Places: Attracting Investment, Industry, and Tourism to Cities, States, and Nations*. New York: Free Press.

Kozak, M. and Rimmington, M. (1998), Benchmarking: Destination attractiveness and small hospitality business performance, *International Journal of Contemporary Hospitality Management*, 10(5):184–188. DOI: 10.1108/09596119810227767

Larsen, S. (2007), Aspects of a psychology of the tourist experience, *Scandinavian Journal of Hospitality*, 7 (1):7–18. DOI: http://doi.org/10.1080/15022250701226014

Murphy, P., Pritchard, M., and Smith, B. (2000), The destination product and its impact on traveller perceptions, *Tourism Management*, 21(1):43–52. DOI: 10.1016/s0261-5177(99)00080-1

Novais, M. A., Ruhanen, L., and Arcodia, C. (2017), Destination competitiveness: A phenomenographic study, *Tourism Management*, 64(C):324–334. DOI: 10.1016/j.tourman.2017.08.014

Oh, H., Fiore, A., and Jeoung, M. (2007), Measuring experience economy concepts: Tourism applications, *Journal of Travel Research*, 46(2):119–132. DOI: 10.1177/0047287507304039

Pine, B. and Gilmore, J. (1999). *The Experience Economy: Work Is Theatre and Every Business a Stage*. Boston, Massachusetts: Harvard Business School Press.

Prebensen, N. K., Chen, J. S., and Uysal, M. (2014). Co-creation of tourist experience: Scope, definition and structure. In N. K. Prebensen, J. S. Chen, and M. Uysal (eds.), *Creating Experience Value in Tourism* (pp. 1–10). Boston: CABI.

Ritchie, J. and Crouch, G. (1993). *Competitiveness in International Tourism: A Framework for Understanding and Analysis*. San Carlos de Bariloche, Argentina: Association Internationale d'Experts Scientifique de Tourisme.

Ritchie, J. and Crouch, G. (2012). *Competitiveness and Tourism*. Massachusetts: Edward Elgar.

Ritchie, J. B. and Crouch, G. (2000), The competitive destination: A sustainability perspective, *Tourism Management*, 21(1):1–7.

Ritchie, J. B. and Crouch, G. I. (2003). *The Competitive Destination: A Sustainable Tourism Perspective*. Cambridge: CABI.

Scott, N. and Le, D. (2017). Tourism experience: A review. In N. Scott, J. Gao, and J. Ma (eds.), *Visitor Experience Design* (pp. 30–49). Oxford: CABI.

Trauer, B. and Ryan, C. (2005), Destination image, romance and place experience: An application of intimacy theory in tourism, *Tourism Management*, 26(4):481–491. DOI: 10.1016/j.tourman.2004.02.014

Tung, V. and Ritchie, J. (2011), Exploring the essence of memorable tourism experiences, *Annals of Tourism Research*, 38(4):1367–1386. DOI: 10.1016/j.annals.2011.03.009

Tyson, K. W. M. (1997). *Competition in the 21st Century*. Boca Raton: St Lucie Press.

Vasudevan, S. (2008), The role of internal stakeholders in destination branding: Observations from Kerala tourism, *Place Branding and Public Diplomacy*, 4(4):331–335. DOI: 10.1057/pb.2008.24

Vilić, S. and Dujaković, T. (2016), Tourist experience as the basis for destination competitiveness: A case study of Ohrid, Belgrade, SITCON 2016 - Singidunum International Tourism Conference.

World Economic Forum. (2007). 'The travel and tourism competitiveness report 2007'. Geneva. [online] Available at: generis.com/_res/file/4247/49/0/WEF_TravelTourismCompetitiveness_Report_2007.pdf [Accessed: March 20, 2019].

World Economic Forum. (2017). 'The travel and tourism competitiveness report 2017'. Geneva. [online] Available at: www3.weforum.org/docs/WEF_TTCR_2017_web_0401.pdf [Accessed: March 20, 2019].

50

Senior tourism and customer experience

Links and opportunities

Adela Balderas-Cejudo and George W. Leeson

Introduction

The age structure of the population is projected to change significantly in the coming years and undergo profound variation. As fertility declines and life expectancy rises, the proportion of the population above a certain age rises as well. "This phenomenon, known as population ageing, is occurring throughout the world" (UN 2017a: 11). By 2070, the EU's share of the total world population is forecast to shrink to 4.5%, and its dependency ratios will be the second-highest globally among large countries (European Commission 2017).

According to the 2018 Ageing Report by the European Commission (2017) most continents are forecasted to experience significant aging of their populations between 2015–70. Future demographic change is particularly pronounced in China and Japan, where the old-age dependency ratios are projected to reach 53.3% and 69.6% respectively in 2070. The UN projections show that Europe is currently the oldest continent in the world when looking at the "very-old-age dependency ratio" – the ratio of people over 80 years old to the working-age population – and will remain so by 2070. The number of older persons in the world is projected to be 1.4 billion in 2030 and 2.1 billion in 2050 and could rise to 3.1 billion in 2100. Over the next few decades, a further increase in the population of older persons is almost inevitable, given the size of the cohorts born in recent decades (UN 2015). As projections show, the age distribution of the population shifting towards the elderly is a current trend (see Figure 50.1). Adapting to a future aging population requires not only demographic changes but also political, social and economic ones. There is an increasing awareness among academic and business communities about the growing power of the senior (or silver) market, the population segment whose growing size is aided by demographic trends (Nimrod 2008).

These global population changes have formed one of the principal demographic trends that are now affecting the tourism industry. Academic researchers and destination managers are now beginning to recognize the importance of older people as a growing market segment of the tourism and hospitality industry (Patterson et al. 2017). In terms of their market potential, seniors will soon become one of the largest prospective market segments for the hospitality and travel industries, contributing significantly to the leisure sector (Huang

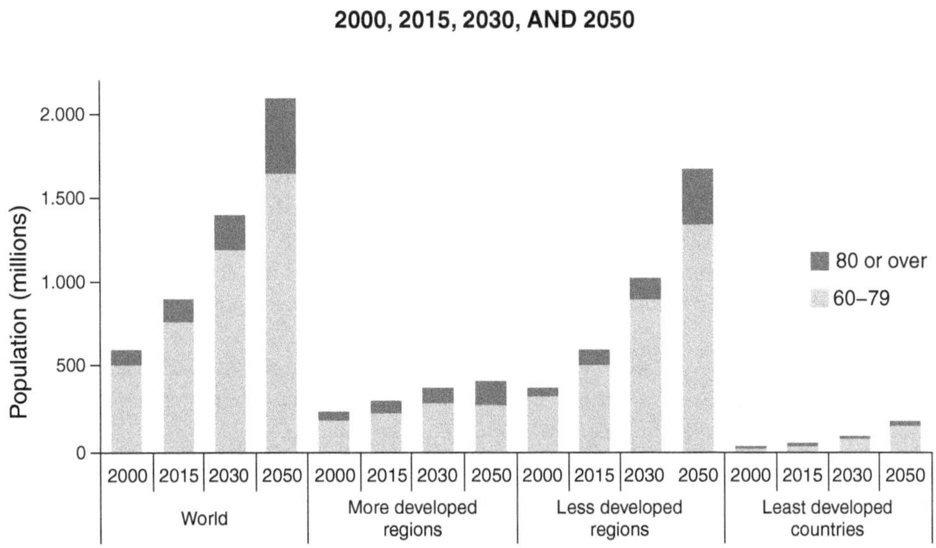

Figure 50.1 Population aged 60–79 and aged 80 years and over by development group

Source: United Nations (2015). World Population Prospects. The 2015 Revision

and Tsai 2003; Balderas-Cejudo et al. 2016). Nevertheless, in an era of fast-changing consumer profiles and behaviors, companies must strive for a thorough understanding of what consumers want and are willing to pay for. The travel industry has been criticized for failing to recognize the diversity of travel preferences within the senior travel market (Javalgi et al. 1992).

Part of the problem underlying this perception of the travel industry stems from an inaccurate and misunderstood stereotypical view of elderly people (Horneman et al. 2002). However, seniors do not travel the same way as the previous generations. Their needs, wants and expectations are completely different. A more sophisticated traveler demands tourism products and hospitality experiences that are not typically associated with the older adult tourists of previous generations (Patterson et al. 2017).

Hospitality in a new aging reality: the power of customer value

Competing in a global market has become increasingly difficult and only the creation of value and long-lasting competitive advantages seems to offer an opportunity for survival and also for differentiation. The study of customer value has become one of increasing interest to researchers and managers (Slater and Narver 2000) and is regarded as the next source of competitive advantage. Furthermore, creating a strong customer experience is now a leading management objective (Lemon and Verhoef 2016). Early on, Abbott (1955) focused on the broader notion that "what people really desire are not products but satisfying experiences" (Abbott 1955: 40). Following Gentile et al. (2007), the concept of customer experience was first conceived in the mid-1980s when, along with the mainstream literature on consumer behavior that deemed customers as rational decision-makers, a new experiential approach offered an original view of consumer behavior (Holbrook and Hirschman 1982). The concept of customer experience came more relevantly to the fore in the 1990s with Pine and Gilmore's book the *Experience Economy* (1999); the authors presented the "experiences" as a new economic offering which emerges as the next step after commodities, goods and services, in what they call the progression of economic value (Gentile et al. 2007).

In the contemporary and challenging hospitality industry, superior customer experiences are essential to gaining customer loyalty and achieving a competitive advantage. The tourism and hospitality industry is often regarded as the more experience-based service industry. Selling services, however, is not enough to differentiate service delivery offerings (Gilmore and Pine 2002). Customer satisfaction is the leading criterion for determining the quality that is actually delivered to customers through the product/service and by the accompanying servicing (Vavra 1997).

According to Patterson and Pegg (2009), it has become generally accepted that most people travel for the leisure experiences that they actively pursue. Cooper et al. (1998) stated that "most tourism throughout the world is a leisure activity" (p. 11).

Following Choo et al. (2018), the key to providing superior service is learning to understand through social interactions and responding to the focused customer's changing expectations (Parasuraman et al. 1991). Events and trends affecting the tourism industry are becoming more complex and fast-paced. It is essential to acknowledge that customer experience does not improve until it becomes a top priority and a company's work processes, systems, and structure change to reflect that (Meyer and Schwager 2007).

Experience-based service encompasses the total customer experience quality, including an integrated dimension of product experience, outcome focus, a moment of truth, and peace of mind in the hospitality industry (Khan et al. 2015). A focus on experiences leads to an emphasis on the emotional aspects of consumer decision-making rather than only rational cognition in consumption choices (Jennings et al. 2009). The customer experience originates from a set of interactions between a customer and a product, a company, a service or part of its organization, which provoke a reaction (LaSalle and Britton 2003; Shaw and Ivens 2005). This experience is strictly personal and implies the customer's involvement at different levels, not only rational but emotional, sensorial, physical and spiritual (LaSalle and Britton 2003). Hence the customer experience construct is holistic in nature and involves the customer's cognitive, affective, emotional, social and physical responses to the retailer (Verhoef et al. 2009) and it reflects the customer's journey through all interactions with the firm; pre-consumption, consumption and post-consumption (Kandampully and Solnet 2015). Moreover, following Verhoef et al. (2009), the customer experience encompasses the total experience, including the search, purchase, consumption, and after-sale phases of the experience, and may involve multiple retail channels. According to a survey by Gartner (2014) 89% of firms expect to compete primarily on the basis of customer experience, and it is now one of the key strategies adopted by hospitality firms. While technology leads to innovation among globally connected customers, service employees continue to hold the key that transforms simple hospitality interactions into memorable customer experiences (Kandampully et al. 2018).

It is key to highlight that such a shift in mindset will have an immediate and notable impact on the type of holidays undertaken and the destinations that are chosen, not only by younger travelers but also by older people (Patterson and Pegg 2009). A new generation of retirees has emerged who are "hungry to go off the beaten path" resulting in the adventure travel business now growing to a $500 million segment and at a rate of 30% per year. This is driven partly by, "... more and more retirees with time, money, and a yen for the exotic" (Symonds 1998: 102).

The older generations are increasingly becoming attracted to a variety of tourism experiences and are wanting to experience and discover the world, and this is especially true of baby boomers (born between 1946 and 1964) who have been found to have different attitudes and lifestyles compared to previous generations of retirees (Faranda and Schmidt 2000; Patterson and Balderas-Cejudo 2018; Balderas-Cejudo 2019).

A deeper awareness and comprehension of the behavior, preferences and trends of senior travelers – being a potential and evolving market segment – is absolutely essential so as to offer innovative, efficient and appropriate services adapted to their requirements, thereby fulfilling the expectations and needs of an emerging market segment.

Understanding this reality, anticipating the forthcoming demographic changes and developing appropriate and proactive policies to address these shifts and challenges must be recognized as major issues (Harper 2014).

The silver market: aging travelers in an aging world

All over the world, societies are rapidly changing, shaped by declining fertility rates, the lengthening of life expectancy, urbanization and migration. The rise of the "silver economy" will generate a non-negligible impact on world societies/economies (EC 2014).

On a global scale, the world is becoming increasingly aware of the growing impact that older adults will have on the tourism industry in the decades ahead (Goeldner 1992). The scholarly literature has extensively examined older tourists and their travel behavior, especially their travel-related activities, socio-demographic characteristics, preferences and motivations. Older people are not only growing rapidly in absolute numbers but have also become substantially healthier. In a phenomenon referred to by demographers and health specialists as the "compression of morbidity", the length of healthy old-age appears to be increasing. Part of this trend can be attributed to increases in life expectancy and part to shorter and later periods of illness. The net effect is an increase in the number of years lived at old age without major health problems (Bloom et al. 2010). This increase in life expectancy will bring a series of outcomes and undoubtedly will allow seniors to travel more, and later in life. In 1999, over 593 million international travelers were aged 60 years and over. This level of tourism activity accounted for approximately a third of the total amount spent on holidays in that year. By 2050, this figure is projected to grow to exceed two billion trips per annum (World Tourism Organization 2001). This increased propensity to travel will expand much more in the future due to a more active generation of seniors who will benefit from higher life expectancy; higher disposable income; improved health standards; familiarity with traveling; and a change in attitude towards lifetime savings and disposable time and what it should be used for (Möller et al. 2007). Living longer, healthier and wealthier than previous generations makes this group one of the largest prospective market segments for the hospitality and travel industries, industries which are contributing significantly to the leisure sector (Huang and Tsai 2003; Jang and Wu 2006).

Following Patterson and Pegg (2009), it has become generally accepted that most people travel for the leisure experiences that they actively pursue. Cooper et al. (1998) stated that "most tourism throughout the world is a leisure activity" (p. 11).

According to Meiners and Seeberger (2010), marketing to seniors now means that the objectives and measures of a company must be adjusted to the specific current and potential requirements and needs of the senior market (Grosskopf 1998; Nieschlag et al. 2002). Following Patterson et al. (2017), tourism and hospitality operators need to avoid treating seniors and baby boomers born from the end of World War II onwards (between 1946 and 1964) as "older people" and appeal to their "forever young" attitudes. The current perception of "old" is not as it was perceived previously, which was related to a person's chronological age and has now been found to be an unreliable predictor of consumer behavior or a person's physical health (Ruys and Wei 2001). Seniors and baby boomers prefer memorable experiences where they can learn and broaden their minds through their travels. "They are the biggest participants of volunteer-related travel who look for spiritual or experiential aspects of travel as much as the young market do" (Fraser 2012: 6).

In the coming decades, major shifts in the leisure and tourism environment will be seen, reflecting changing consumer values, political forces, environmental changes and the explosive growth of information and communication technology (Dwyer et al. 2009).

Case study 50.1: Seniors and travel experiences: gastronomic tourism in the Basque Country

The Organization for Economic Co-operation and Development (OECD) has a particular interest in food and the tourism experience because it seeks to promote policies that will improve the economic and social well-being of people around the world (Richards 2012).

The European Commission in their ESCAPE project (ENAT 2015) focused on the development and promotion of transnational exchanges for seniors (aged 55 years and older) in five European countries during the low season, namely in Bulgaria, Cyprus, France, Italy and Portugal. A questionnaire was administered to 894 respondents of whom 41.5% were men and 58.5% were women. The results showed that gastronomy was the third most favored theme (774 replies), behind nature and culture (858 replies) and health and well-being (780 replies). It is key to highlight that the importance of access to gastronomic activities increased until the age of 80 and then it slowly began to decline.

The marketing opportunities for gastronomy tourism are being rapidly developed by countries that place great importance on this issue (Sormaz et al. 2016). Trends associated with fashion, famous chefs, TV shows, mass communication and social media have played a key role in providing food with a new and different status. Our daily lives are now immersed in culinary influences, with cooking shows being designed for all targets. Food experiences have become part of the distinctive intangible culture of places and are part of the "stockpile of knowledge, traditions, memories and images" (Scott 2010) that are used to create distinction in an increasingly crowded marketplace (Richards 2012).

Gastronomy has become a central part of tourism and is becoming a special interest area for older travelers. Tourists are becoming food savvy in search of new and special and authentic cuisines, and senior travelers are not an exception. An example of this is the Basque Country in Spain. The Basque Country is located in the Western Pyrenees, straddling the border between France and Spain on the coast of the Bay of Biscay. Tourism in the Basque Autonomous Community has increased considerably in recent years with a 56% increase in the influx of tourists from 2008 to 2018 (Eustat).

The Basque Country's cuisine is one of its highest-rated and best-known strengths. Eating well is part of Basque culture. One good example of this can be seen in Donostia-San Sebastián, the city with one of the largest concentrations of Michelin stars per square meter in the world.

In an internal research project developed by Basquetour (the public company of the Ministry for Tourism, Trade and Consumer of the Basque Government) on senior tourism as a strategic segment (Balderas-Cejudo and Rivera-Hernaez 2016), no statistical source was allowed to assess the behavior of this segment at a global level, but taking its data on tourist behavior in the Basque Country as a sample, it could be seen that the average stay of tourists aged between 50–59 years is almost 6% longer than that of the average tourist, while their total expenditure is 18% higher.

Also, the average stay of tourists aged 60 or over is 25% longer and their total expenditure is also 18% higher than the average. The Basque Country has shown, then, a greater tendency to specialize in senior tourism than the rest of Spain, gastronomy being one of the main factors of attraction. One of the main reasons for traveling to the Basque Country is because gastronomy is part of the experience. Many older travelers are experienced travelers who are more educated, with more free time, and they show a greater willingness to travel during the off-peak season and stay for longer periods. Senior travelers have greater purchasing power and value quality in services. Many seniors plan their vacations specifically to incorporate culinary trips and are particularly interested in sampling local food products and tasting local recipes. They also request more authentic experiences and authenticity in food is highly valued by these experienced travelers. Seniors are eager to enjoy the tourist experience, wanting to optimize their time, their money and their freedom.

In regard to the future of senior travel, it is likely that many seniors are experienced travelers, in contrast to their parents and grandparents. Thus, because older travelers are more experienced, they will have clearer ideas of what they want, making it more difficult to surprise them, and will demand "personalized" and often exotic location offerings to meet their travel needs (Patterson and Balderas-Cejudo 2018).

Many of today's seniors are considered to be quite distinct from previous generations in that they actively seek out ad hoc leisure opportunities while on work-related travel, as well as on trips for pleasure more generally. This only serves to add to the appeal of this niche segment for the tourism and hospitality market (Voigt et al. 2010).

There is increasing competition in the tourism and hospitality industries – between destinations worldwide (both established markets and new markets), between destinations domestically and between firms within a particular destination. The greater their knowledge of these trends relating to tourism development in general, and the senior market segment in particular, the greater the capacity of all stakeholders to contribute to new and innovative services and to achieve competitive advantage for their organizations and their regions.

Conclusions

The hospitality industry is dynamic in that the guest dictates the pace and type of service and increasing competitiveness in the industry has resulted in satisfactory service being the minimum expectation of guests (Crick and Spencer 2010). The demand for leisure and tourism activities has grown steadily in our society, and these are now regarded as important aspects of life for enhancing psychological and physical well-being (Janke et al. 2006), and to assist in the achievement of a successful retirement (Silverstein and Parker 2002). As for the decision-making process in regard to tourist behavior, changes that are occurring through the evolution of communication and technology have been dramatic, creating many new opportunities as well as posing other threats (Buhalis 2000).

Over time, the way in which the senior market is perceived has undergone considerable changes; from considering older people as a homogeneous market during the 80s and 90s to the gradual acceptance of its diversity and complexity over the last 30 years, which has been emphasized in the consumer marketing literature (Moschis 1991). The increasing attention that is paid to this market is a worldwide phenomenon and requires appropriate exploration, not only because of seniors' consumption characteristics but also because of their insights and the way they feel, act and react. Many scholars state that this generation of seniors will become the most significant consumer group of the coming decades, as they are considered the richest generation in the world (Meiners and Seeberger 2010). The relevance of the senior segment for this industry, as part of a continuously aging population with a high economic status, has made this segment a very attractive group for the sector (Losada et al. 2019) and as a direct consequence of global aging patterns, older travelers have become a significant proportion of annual total holiday spending (Sie et al. 2015). Nevertheless, it is essential to highlight that because older people are often stereotyped as part of the past, they can be overlooked in the surge towards the future (WHO 2015). Today, many seniors are active, enjoy flexible schedules, take frequent and longer holidays, and outspend the youth market on leisure activities (Baloglu and Shoemaker 2001; Huang and Tsai 2003; Nimrod and Rotem 2012). Many have the time to travel, are willing to spend a significant amount of their savings doing so (Fleischer and Pizam 2002) and will demand products and services that deliver value, cater to their needs, preferences and insights and provide memorable experiences. To cater for the diversity of needs of this growing market, what is needed is a more comprehensive understanding of seniors and their travel behavior and knowledge about their motivations for leisure travel, so that tourism products and services can be tailored to meet their current and future preferences and needs when traveling. Accordingly, delivering value for customers should become a central theme in business in general and in the hospitality sector in particular. Pine et al. (1999) argued that creating a distinctive customer experience can provide enormous economic value for firms. The sources of superior customer value can be identified from organizational resources and capabilities, such as human resource management, innovation, knowledge management, organizational culture and structure (Weinstein and Pohlman 1998; Mittal and Sheth 2001).

Berry et al. stated that (2002: 1) in recent years managers and companies have become increasingly aware of the need to create value for their customers in the form of experiences. Nevertheless, they have often proceeded as if managing experiences meant providing entertainment or being engagingly creative. A customer service experience can alter a customer's entire perception of an organization (Chen et al. 2013). A deeper understanding of this evolving segment will be essential through innovative thinking and also innovation in service, technology, continuous improvement, the delivery of that customer experience and a profound ability to both anticipate and invent new ways of delighting this emerging and sensitive to experience market segment. Creating innovative ways to address some of the gaps identified, as well as recognizing the need to understand the wants of an aging population without stereotypes and the need to expand knowledge of the trends underpinning tourism development, will increase the capacity of stakeholders to formulate strategies to achieve competitive advantage for their organizations.

Stakeholders will need to provide for seniors' travel motivations, which include the need for social interaction, special events, memorable experiences, cultural amenities, educational offerings and a desire for self-fulfillment. These will be important motives to be taken into account in order to offer products and services that seniors seek and value. A more experiential senior traveler will seek authenticity, self-improvement and new experiences (Patterson and Balderas-Cejudo 2018).

At this critical moment, in an age of customers' empowerment, the key to meeting the new challenges and opportunities confronting the hospitality industry and providing that total and memorable experience is for stakeholders to look beyond their customers' expectations and become deeply aware of their needs, thoughts, emotions, insights, fears and states of mind, so as to provide an unforgettable and memorable experience.

References

Abbott, L. (1955). *Quality and Competition*. New York: Columbia University Press.

Balderas-Cejudo, A. (2019). *Senior Tourism: Determinants, Motivations and Behaviour in a Globalized and Evolving Market Segment*. Madrid: ESIC. ISBN 978-84-17513-51-1. https://www.unebook.es/es/libro/senior-tourism-determinats-motivations-and-behaviour-in-a-globalized-and-evolving-market-segment_249641

Balderas-Cejudo, A., Rivera-Hernaez, O., and Patterson, I. (2016). The strategic impact of country of origin on senior tourism demand: The need to balance global and local strategies. *Journal of Population Ageing*, 9(4): 345–373. DOI: 10.1007/s12062-016-9146-2

Baloglu, S., and Shoemaker, S. (2001). Prediction of senior travelers' motorcoach use from demographic, psychological, and psychographic characteristics. *Journal of Travel Research*, 40(1): 12–18. DOI: 10.1177/004728750104000103

Berry, L. L., Carbone, L. P., and Haeckel, S. H. (2002). Managing the total customer experience. *MIT Sloan Management Review*, 43(3): 85–89. Boston: Harvard Business School Press.

Bloom, D. E., Canning, D., and Fink, G. (2010). Implications of population ageing for economic growth. *Oxford Review of Economic Policy*, 26(4): 583–612. www.jstor.org/stable/43664646

Buhalis, D. (2000). Marketing the competitive destination of the future. *Tourism Management*, 21(1): 97–116. DOI: 10.1016/S0261-5177(99)00095-3

Chen, K. H., Liu, H. H., and Chang, F. H. (2013). Essential customer service factors and the segmentation of older visitors within wellness tourism based on hot springs hotels. *International Journal of Hospitality Management*, 35: 122–132. DOI: 10.1016/j.ijhm.2013.05.013

Choo, P. W., Tan, C. L., and Yeo, S. F. (2018). A review of customer experience quality measurement in Malaysian heritage hotels. *Global Business and Management Research*, 10(1): 379–395.

Cooper, C., Fletcher, J., Gilbert, D., and Wanhill, S. (1998). *Tourism: Principles and Practice*. Harlow: Prentice Hall.

Crick, A. P., and Spencer, A. (2010). Hospitality quality: New directions and new challenges. *International Journal of Contemporary Hospitality Management*, 23(4): 463–478.

Dwyer, L., Edwards, D., Mistilis, N., Roman, C., and Scott, N. (2009). Destination and enterprise management for a tourism future. *Tourism Management*, 30(1): 63–74.

ENAT (European Network for Accessible Tourism). (2015). European senior citizens' actions to promote exchange in tourism – ESCAPE. https://www.accessibletourism.org/?i=enat.en.enat_projects_and_good_practices.1883

European Commission. (2014). Europe, the best destination for Seniors. http://epp.eurostat.ec.europa.eu/statistics_explained/index.php [Accessed: March 28, 2019].

European Commission. (2017). 2018 ageing report by the European Commission. https://ec.europa.eu/info/sites/info/files/economy-finance/ip065_en.pdf [Accessed: May 2, 2019].

Eustat. http://en.eustat.eus/estadisticas/tema_159/opt_0/tipo_1/ti_Population/temas.html [Accessed: September 28, 2019].

Faranda, W. T., and Schmidt, S. L. (2000). Segmentation and the senior traveller: Implications for today's and tomorrow's aging consume. *Journal of Travel and Tourism Marketing*, 8(2): 3–27. DOI: 10.1300/J073v08n02_02

Fleischer, A., and Pizam, A. (2002). Tourism constraints among Israeli seniors. *Annals of Tourism Research*, 29(1): 106–123.

Fraser, J. E. (2012). Advanced sense of adventure. *The Sydney Morning Herald*. www.smh.com.au/travel/advanced-sense-of-adventure-20120607-1zxvt.html [Accessed: March 30, 2019].

Gartner. (2014). Importance of customer experience is on the rise: Marketing is on the hook. www.gartner.com/doc/2853519/importance-customer-experience-rise-marketing [Accessed: March 16, 2019].

Gentile, C., Spiller, N., and Noci, G. (2007). How to sustain the customer experience: An overview of experience components that co-create value with the customer. *European Management Journal*, 25(5): 395–410. DOI: 10.1016/j.emj.2007.08.005

Gilmore, J. H., and Pine, B. J. (2002). Differentiating hospitality operations via experiences: Why selling services is not enough. *The Cornell Hotel and Restaurant Administration Quarterly*, 43(3): 87–96. DOI: 10.1016/S0010-8804(02)80022-2

Goeldner, C. R. (1992). Trends in North American tourism. *American Behaviouralist Scientist*, 36: 144–154.

Grosskopf, A. K. (1998) Der Markt der älteren Konsumenten. *Werbeforschung & Praxis*, 43(1): 20–25.

Harper, S. (2014). Introduction: Conceptualizing social policy for the twenty-first-century demography. In Harper, S., and Hamblin, K. (Eds.). *International Handbook on Ageing and Public Policy* (pp. 1–9). Cheltenham: Edward Elgar Publishing.

Holbrook, M. B., and Hirschman, E. C. (1982). The experiential aspects of consumption: Consumer fantasies, feelings, and fun. *Journal of Consumer Research*, 9(2): 132–140.

Horneman, L., Carter, R. W., Wei, S., and Ruys, H. (2002). Profiling the senior traveler: An Australian perspective. *Journal of Travel Research*, 41(1): 23–37. DOI: 10.1177/004728750204100104

Huang, L., and Tsai, H. T. (2003). The study of senior traveler behavior in Taiwan. *Tourism Management*, 24(5): 561–574. DOI: 10.1016/S0261-5177(03)00008-6

Jang, S. (Shawn), and Wu, C.-M. E. (2006). Senior's travel, motivation and the influential factors: An examination of examination of Taiwanese seniors. *Tourism Management*, 27(2): 306–316. DOI: 10.1016/j.tourman.2004.11.006

Janke, M., Davey, A., and Kleiber, D. (2006). Modeling change in older adults' leisure activities. *Leisure Sciences*, 28(3): 285–303. DOI: 10.1080/01490400600598145

Javalgi, R. G., Thomas, E. G., and Rao, S. R. (1992). Consumer behavior in the US pleasure travel marketplace: An analysis of senior and nonsenior travellers. *Journal of Travel Research*, 31(2): 14–19. DOI: 10.1177/004728759203100203

Jennings, G., Lee, Y. S., Ayling, A., Lunny, B., Cater, C., and Ollenburg, C. (2009). Quality tourism experiences:(re) constructions and (re) interpretations by adventure travellers and providers. *Journal of Hospitality and Leisure Marketing*, 18: 294–310. DOI: 10.1080/19368620802594169

Kandampully, J., and Solnet, D. (2015). *Service Management: Principles for Hospitality and Tourism*. Dubuque, IA: Kendall Hunt Publishing.

Kandampully, J., Zhang, T., and Jaakkola, E. (2018). Customer experience management in hospitality: A literature synthesis, new understanding and research agenda. *International Journal of Contemporary Hospitality Management*, 30(1): 21–56. DOI: 10.1108/IJCHM-10-2015-0549

Khan, I., Garg, R. J., and Rahman, Z. (2015). Customer service experience in hotel operations: An empirical analysis. *Procedia-Social and Behavioral Sciences*, 189: 266–274. New York: Leadership. McGraw-Hill.

LaSalle, D., and Britton, T. A. (2003). *Priceless: Turning Ordinary Products into Extraordinary Experiences*. Boston: Harvard Business School Press.

Lemon, K. N., and Verhoef, P. C. (2016). Understanding customer experience throughout the customer journey. *Journal of Marketing*, 80(6): 69–96. DOI: 10.1509/jm.15.0420

Losada, N., Alen, E., Cotos-Yanez, T. R., and Dominguez, T. (2019). Spatial heterogeneity in Spain for senior travel behavior. *Tourism Management*, 70: 444–452. DOI: 10.1016/j.tourman.2018.09.011

Meiners, N. H., and Seeberger, B. (2010). Marketing to senior citizens: Challenges and opportunities. *The Journal of Social, Political, and Economic Studies*, 35(3): 293–328.

Meyer, C., and Schwager, A. (2007). Understanding customer experience. *Harvard Business Review*, 85(2): 116–126.

Mittal, B., and Sheth, J. N. (2001). *Value Space: Winning the Battle for Market Leadership*. New York: McGraw-Hill.

Möller, C., Weiermair, K., and Wintersberger, E. (2007). The changing travel behaviour of Austria's ageing population and its impact on tourism. *Tourism Review*, 62(3/4): 15–20. DOI: 10.1108/16605370780000316

Moschis, G. P. (1991). Marketing to older adults: An overview and assessment of present knowledge and practice. *Journal of Services Marketing*, 5(2): 33–41. DOI: 10.1108/08876049110035521

Nieschlag, R., Dichtl, E., and Hörschgen, H. (2002). Marketing, 19. *Aufl., Berlin*, 7.

Nimrod, G. (2008). Retirement and tourism themes in retirees' narratives. *Annals of Tourism Research*, 35 (4): 859–878. DOI: 10.1016/j.annals.2008.06.001

Nimrod, G., and Rotem, A. (2012). An exploration of the innovation theory of successful ageing among older tourists. *Ageing and Society*, 32(3): 379–404. DOI: 10.1017/S0144686X1100033X

Parasuraman, A., Berry, L. L., and Zeithaml, V. A. (1991). Understanding customer expectations of service. *Sloan Management Review*, 32(3): 39–49.

Patterson, I., and Balderas-Cejudo, A. (2018). Continuing and emerging trends of senior tourism: A review of the literature. *Journal of Population Ageing*. DOI: 10.1007/s12062-018-9228-4

Patterson, I., and Pegg, S. (2009). Marketing the leisure experience to baby boomers and older tourists. *Journal of Hospitality Marketing & Management*, 18(2–3): 254–272.

Patterson, I., Sie, L., Balderas-Cejudo, A., and Rivera-Hernaez, O. (2017). Changing trends in the baby boomer travel market: Importance of memorable experiences. *Journal of Hospitality Marketing and Management*, 26(4): 347–360. DOI: 10.1080/19368623.2017.1255162

Pine, B. J., Pine, J., and Gilmore, J. H. (1999). *The Experience Economy: Work Is Theatre & Every Business a Stage*. Boston: Harvard Business Press.

Richards, G. (2012). Food and the tourism experience: Major findings and policy orientations. In Dodd, D. (Ed.). *Food and the Tourism Experience* (pp. 13–46). Paris: OECD.

Ruys, H., and Wei, S. (2001). Senior tourism. In Douglas, N., Douglas, N. and Derrett, R. (Eds.), *Special Interest Tourism*. Australia: John Wiley and Sons.

Scott, A. J. (2010). Cultural economy and the creative field of the city, Geografiska Annaler: Series B. *Human Geography*, 92: 115–130. DOI: 10.1111/j.1468-0467.2010.00337.x

Shaw, C., and Ivens, J. (2005). *Building Great Customer Experiences*. New York: Macmillan.

Sie, L., Patterson, I., and Pegg, S. (2015). Towards an understanding of older adult educational tourism through the development of a three-phase integrated framework. *Current Issues in Tourism*. DOI: 10.1080/13683500.2015.1021303

Silverstein, M., and Parker, M. G. (2002). Leisure activities and quality of life among the oldest old in Sweden. *Research on Aging*, 24(5): 528–547. DOI: 10.1177/0164027502245003

Slater, S. F., and Narver, J. C. (2000). Intelligence generation and superior customer value. *Journal of the Academy of Marketing Science*, 28(1): 120–127. DOI: 10.1177/0092070300281011

Sormaz, U., Akmese, H., Gunes, E., and Aras, S. (2016). Gastronomy in tourism. *Procedia Economics and Finance*, 39: 725–730.

Symonds, W. C. (1998), July 20). Far from the tour bus crowd: Track gorillas in Uganda, explore a rainforest, or pedal across France. *Business Week*, 3587: 102.

United Nations, Department of Economic and Social Affairs, Population Division. (2015). *World Population Ageing 2015* (ST/ESA/SER.A/390).

United Nations, Department of Economic and Social Affairs, Population Division. (2017a). World Population Prospects: The 2017 Revision, Key Findings and Advance Tables. Working Paper No. ESA/P/WP/248.

United Nations, Department of Economic and Social Affairs, Population Division. (2017b). World Population Ageing 2017 – Highlights (ST/ESA/SER.A/397).

Vavra, T. G. (1997). *Improving Your Measurement of Customer Satisfaction: A Guide to Creating, Conducting, Analyzing, and Reporting Customer Satisfaction Measurement Programs*. Milwaukee, WI: ASQ Quality Press.

Verhoef, P. C., Lemon, K. N., Parasuraman, A., Roggeveen, A., Tsiros, M., and Schlesinger, L. A. (2009). Customer experience creation: Determinants, dynamics and management strategies. *Journal of Retailing*, 85(1): 31–41. DOI: 10.1016/j.jretai.2008.11.001

Voigt, C., Howat, G., and Brown, G. (2010). Hedonic and eudaimonic experiences among wellness tourists: An exploratory enquiry. *Annals of Leisure Research*, 13(3): 541–562. DOI: 10.1080/11745398.2010.9686862

Weinstein, A., and Pohlman, R. A. (1998). Customer value: A new paradigm for marketing management. *Advance in Business Studies*, 6(10): 89–97.

World Health Organization. (2015). *World Report on Ageing and Health*. Geneva: WHO.

World Tourism Organization. (2001). *Tourism 2020 vision: Global forecasts and profiles of market segments*. Madrid: World Tourism Organization (UNWTO). http://www.unwto.org/facts/eng/vision.htm

51

Crisis management and tourism experiences

Adiyukh Berbekova and Melissa A. Baker

Introduction

Crises and disasters are an unfortunate part of human history with no place and no organization immune to crises. They can happen everywhere and are often sudden and unexpected in nature. Crisis incidents can strongly influence the tourism industry of the destination and individual hospitality and tourism firms and can bring about several negative impacts. Therefore, within the globalized world the significance of effective crisis management has increased. The wide usage of social media also affects the life cycle of the crisis. Crisis management may come in the form of terrorist attacks, such as those that have taken place in Sri Lanka at a number of hotels (CBS 2019), the effects of Hurricane Katrina on the city of New Orleans and the tourism economy or negative online reviews that are seen by a wide audience that can have a detrimental effect on the hospitality or tourism firm (Baker 2017). In today's global environment, it is critical to have a plan to manage crises, to effectively manage during the crisis and to critically evaluate the event post crisis. Thus, crisis management is a crucial function for hospitality and tourism organizations of different levels. The firm's responses at every stage of crisis development will affect customers' experiences and influence the organization's further existence.

Therefore, this chapter discusses this vital topic for future hospitality and tourism leaders. First, the chapter discusses key definitions of crisis management. Second, it discusses the importance of crisis management plans. Next, leadership and the importance of communication are discussed. Finally, the stages of crisis management are defined including the pre-crisis planning phase, the response phase and the evaluation phase. This chapter seeks to inform readers of the importance of crisis management and how to most effectively manage crises.

Crisis management: definitions and types

There are numerous studies concerning the definitions of crisis management and its features (Faulkner 2003; Glaesser 2006) and crisis management issues seem to go viral almost every day.

Crisis definitions

There is no commonly accepted and used definition in the literature for crisis management. According to Faulkner (2003), a crisis is any action or failure to act that interferes with a firm's ongoing functions, the acceptable attainment of its objectives or its viability or survival, or that has a detrimental personal effect as perceived by the majority of its employees, clients or constituents. Alternatively, Glaesser (2006) offers another definition that points out the ambiguous nature of crises where a crisis is an undesired, extraordinary, often unexpected and timely limited process with ambivalent development possibilities. It demands immediate decisions and countermeasures in order to influence positive further development for the organization (destination) and to limit the negative consequences as much as possible. Thus, a crisis is a negative event of an exceptional nature, which leads to deviation from an organization's normal activities. Nonetheless, the ambiguity of the crisis shows that such a negative event can be a source of new opportunities for an organization if handled properly. In most cases, the possible beginning and end of the crisis cannot be predicted. This makes it nearly impossible for tourism organizations to be fully prepared for all types of potential crises. However, it is important to learn from each case regarding the causes, consequences, and possible measures that need to be taken to address similar situations in the future (Paraskevas and Quek 2019). The crises can also be characterized by the high level of uncertainty concerning the type of crises that a company or destination might deal with. Time compression (Stafford, Yu and Alex 2002) or time pressure refers to the amount of "available time" (Glaesser 2006:14) during which the major steps to overcome the crisis must be executed. This is a critical issue for managers to consider, in that there is a narrow window of time to address the issue.

Crisis management definition

Crisis management, on the other hand, refers to an organization's pre-established activities and guidelines for preparing for and responding to significant catastrophic events or incidents (i.e., fires, earthquakes, severe storms, workplace violence, kidnappings, bomb threats, acts of terrorism, etc.) in a safe and effective manner (Lockwood 2005). In addition, many types of service failures are also considered types of crisis management (Su, Stepchenkova and Kirilenko 2019) as they are potential factors that disrupt an organization's normal work. In other words, crisis management entails how an organization or destination deals with negative events. Crisis management has to be considered as part of the overall management of a hospitality or tourism organization or destination. Therefore, it becomes part of a company's strategy and corresponds with its goals and objectives. According to Mitroff, Shrivastava and Udwadia (1987) and Glaesser (2006) there are two crisis management strategies: proactive or offensive and reactive or defensive. A proactive handling strategy is directed towards diminishing the risk of the crisis occurring. It is understandable that the elimination of the cause of the crisis, in the case of natural disasters for example, is impossible. In this case, an offensive crisis strategy is used to accept and capitalize on the event (Glaesser 2006) properly. A reactive strategy for handling crises involves response actions from the organization's side only when the negative event has happened. It is an ad hoc strategy. This type of crisis strategy aims to avoid the negative situation as long as possible and let it develop without any intervention. Obviously, such indifferent attitudes towards a crisis can lead to negative consequences, such as a negative image of the destination or loss of credibility (Paraskevas and Quek 2019). One twenty-year study of

Fortune 500 companies' crisis management preparedness found that companies that employed a proactive strategy faced a significantly lower number of crises and were financially more successful. In addition, on average these organizations stayed in business longer in contrast to the companies that either had an ad hoc strategy to handle crises or no strategy (Mitroff and Alpaslan 2003).

Types of crises

For effective crisis management, a tourism organization must be aware of the possible types of negative events. One way to categorize crises is by those that are natural versus human-induced (Glaesser 2006). A broader classification is provided by Henderson (2007), who categorizes crises as: "economic, political, socio-cultural, environmental, technological, commercial and terrorism". The economic crises refer to economic deterioration, inflation, unstable exchange rates, etc., which can all influence both the tourism industry of the destination and separate hospitality and tourism organizations. The environmental crises are the natural catastrophes and disasters: tsunamis, earthquakes, volcano eruptions, tornadoes, floods, typhoons and so on. Technological crises are the ones that developed because of a technological failure. For example, the Chernobyl crisis in Ukraine in 1986. The conflicts initiated by the clashes between representatives of different social groups or nations (nationalities) refer to socio-cultural conflicts. The political causes of the crises can be: wars, sanctions, abortive international relations and revolutions (Henderson 2007). Commercial crises are the ones concerned with government intervention in an organization's activity with novel financial regulations. The management decisions which lead to commercial failures also refer to internal threats.

The most common types of crises in today's tourism environment are planned attacks, natural disasters and service failure. First, crises can involve planned attacks on a business. Such examples would include the planned terrorist attack at the Ariana Grande concert in Manchester, England in 2017. A second type would be natural disasters such as the ongoing volcanic eruptions, earthquakes, and flooding that occurred in Hawaii, which had significant impacts on Hawaii's tourism industry (CNN 2018). A third type is service failures. For example, in 2017 United Airlines was at the center of a severe crisis when the company's poor response to an overbooking incident turned into a disaster and caused the company's stock to drop one billion dollars in value (also see Table 51.1). Thus, service failures and recoveries should be an essential part of crisis management for service industries, as these can also have detrimental effects (Laws 2001).

Crisis management plans

The effectiveness of crisis management depends on the diligence given to and the effectiveness of the crisis management plan, crisis communication and crisis leadership. Crisis management plans demonstrate management's commitment to provide reasonable care to protect guests and can successfully be used as a marketing tool (Rittichainuwat 2013). The major goal of a crisis management plan is to reduce the possibility of any negative event and thereby gain an advantage in the fight against crises. It also serves as a strategic initiative. A crisis management plan is developed as a guideline for dealing with crisis situations and should be adaptable to any negative events of any kind. The plan is a working document for a company and therefore is updated and changed constantly as the firm, industry or business environment is ever-changing. Thus, in order to develop an efficient plan, the organization should be aware of its main objectives: both long-term and short-term.

Table 51.1 Crisis Management Cases

Year	*Location*	*Crisis*
2001	United States	9/11 Terrorist attacks
2002–2003	Hong Kong China	SARS (Severe acute respiratory syndrome) outbreak
2004	Asia	Asian Tsunami. Indonesia, Sri Lanka, India, Thailand and the Maldives were affected
2005	New Orleans, United States	Hurricane Katrina
2011	Egypt	Revolution in Egypt. The political crisis continued until 2014, affecting the number of tourist arrivals in the country
2015	Tunisia	Terrorist attacks: Bardo National Museum and Sousse Beach resort were targeted
2016	Istanbul	Terrorist attack: In a tourist district near the famous Blue Mosque
2017	Manchester, England	Terrorist attack: At the Ariana Grande concert
2017	United States	Service failure: Overbooking incident on United Express Flight 3411
2017	Virgin Islands	Service failure: After Hurricane Irma hit St. John, Marriott sent a ship to rescue only the guests of the property, leaving other travelers behind
2018	Hawaii	Volcano eruptions: Affected tourists and residents
2019	Sri Lanka	Terrorist attacks: Three luxury hotels in the capital city of Colombo; the Shangri La, the Cinnamon Grand and the Kingsbury were targeted.

Sources: BBC 2004; BBC 2016; BBC 2019; CBS 2017; CNN 2018; NBC 2019; NCBI 2004

The content of a crisis management plan includes clarified crisis management objectives, the completed analysis of possible threats for an organization, the list of the stakeholders involved and actions for managing the negative event and bringing a company to a normal state. It has to be noted that a peculiar feature of a crisis management plan is that it has to contain several strategies for dealing with crises of different origins (Sönmez and Backman 1992) such as those that are terrorist attacks, natural disasters or service failures. In addition, the duration of the crisis can also have differing impacts on the firm. Prolonged crises, which are typically human-induced, can continue for years and have a variable effect on the organization. Similarly, firms need to consider the scope of the crisis.

Once the organization has familiarized itself with potential crisis types the planning begins. Crisis management planning follows three stages: generic planning, contingency planning and preventive planning (Glaesser 2006). The generic plan can be understood as the "draft" of the contingency plan. Generic planning determines the goals of crisis management, its fundamental requirements and the available resources, in order to deal with the crisis. For that purpose, during this stage of planning all of the divisions and subdivisions of the organization are analyzed through the prism of their potential in a crisis situation (Glaesser 2006). Another aspect of generic planning consists in identifying the role of different individuals. The next step of crisis management planning is the formation of the contingency plan (Glaesser 2006). The contingency plan includes the review and consideration of the possible crises that can befall the organization. In accordance with the completed review, several action plans are developed. Moreover, for each of these scenarios a particular plan is developed, considering all required aspects: the necessary data about the negative event, the activities to be undertaken, contact information concerning the responsible staff and the responsibilities of the management and employees. These may include events that are more likely, such as a snowstorm turning off the power at a hotel, to those that are less likely, such as the bombing at the Boston Marathon event. The preventive planning phase starts when there is an imminent negative event and it is clear that it will affect the organization or destination. A preventive plan is based on one of the optional contingency plans. In regard to the nature of the oncoming crisis, one of the developed scenarios is chosen. When preventive planning is organized, therefore, not only the obtained general information about the type of crisis is considered but also the specific characteristics that the crisis reveals while it is unfolding (Glaesser 2006).

In summary, a crisis management plan has to be practical, which means that the amount of money spent on the foundation of the crisis management team and the external and internal analysis of the company's strengths and weaknesses, etc., has to be less than the losses that may be incurred by the company without a crisis management plan.

Crisis leadership

Crisis leadership can strongly affect how well organizations recover from the negative event (Lockwood 2005). The leader is responsible for running the whole process of overcoming the crisis including implementation of crisis management and crisis communication plans, managing the crisis management team, and sustaining a company's image and reputation (Lockwood 2005; Watkins and Bazerman 2003). According to Boin et al. (2005) there are five crucial functions which managers ought to fulfill during a crisis: sense-making, decision-making, meaning-making, terminating and learning. Sense-making refers to a leader's ability to determine all the parameters of a negative event. The leader is in charge of decoding the warning signals of the coming crisis situation and uses them to interpret the crisis. Decision-making as a leadership function involves not only the processes of making decisions and delegating them if needed, but also the coordination of all the different groups involved. Meaning-making is dedicated to providing the correct information on the occurred crisis and giving meaning to the unfolding crisis in such a way that efforts to manage it are enhanced (Boin et al. 2005). The terminating task of leadership deals with returning from crisis to routine. Finally, the learning function includes the attempts of the leader to take a lesson from the event and apply it to future crisis management planning (Boin et al. 2005).

Crisis communication

Crisis communication represents, perhaps, the most important part of overall crisis management in the era of rapid technological development. Communication plays a key role in the processes of restoration and recovery (Regester and Larkin 2005; Ritchie et al. 2004) as companies are judged as harshly on the communication about the crisis as they will be about the crisis itself (Anthonissen 2008). The crisis communication plan is normally implemented during the crisis phase. Crisis communication involves factual information on the current state of a crisis, its influences and taken and planned actions. In order to be effective, the connection with media, stakeholders and customers has to be organized within a certain time frame. For most tourism-related crises, communication within the first 24 hours is the most critical (Glaesser 2006).

In addition, crisis communication happens both externally and internally and the significance of the internal communication cannot be underestimated (Coombs 2014). The internal communication during a crisis involves the connections between the managers, the crisis management team and the employees. Successful internal communication helps to keep employees informed and supports them during the stressful crisis situation (Coombs 2014). Successful crisis communication must comply with certain principles, including being current and up-to-date. Second, a crisis communication plan requires the company's management to think about potential disaster scenarios, regardless of how painful they can be, and ensure that these are communicated across the different constituents (Anthonissen 2008), stressing that the value of every communication action increases when it is done on time and, most importantly, to the point. Lastly, the communication must show that the interests of the victims and the involved actors are of great importance.

Stages of crisis management

Crisis management can be examined as a process of several phases which coincide with the stages of crisis development. While there are different ways to categorize the stages, it is generally agreed that the key stages involve the Pre-crisis phase, the Response phase and the Evaluation phase.

Pre-crisis phase

The pre-crisis phase is dedicated to preparing for a possible crisis. It is concerned mainly with the prevention of the negative events, and is therefore responsible for the strategic scanning of the internal and external environment in order to investigate the risks and hazards for an organization. For example, a foodservice establishment will always have to consider food-borne illnesses as potential crises. However, they may also need to consider other risks involved in their business, such as natural or man-made disasters. Even though the majority of crises are unpredictable and sudden in nature, it is possible to detect major threats (Beirman 2003; Faulkner 2003; Glaesser 2006). A tourist destination should pay closer attention to past experiences and identify the areas which are most prone to negative events. It is within this stage that the crisis management plan is developed, the crisis management team formed and responsibilities are allocated to the employees (Pforr and Hosie 2008). To increase understanding and the fulfillment of designated duties during crisis time, it is recommended to carry out workshops and regular training for the employees (Glaesser 2006). All in all, the outcome of this phase of crisis management has to be the full preparedness of a hospitality or tourism organization for possible crises. It can be concluded that this phase is of vital importance, as the effectiveness of a response to a crisis largely depends on how the company was prepared for it. As Argenti (2002), who interviewed several CEOs of different companies after the 9/11 terrorist attacks, points out, even if organizations' crisis management plan cannot cover all possible crisis scenarios, having one in place will greatly help.

Response phase

The response phase occurs as the actual crisis is happening and involves the implementation of the developed crisis management plan. The first steps in the response phase are to identify what kind of crisis it is, what the cause was and what scope it has (Faulkner 2003; Rousaki and Alcott 2006; Siaw and Loosemore 2006). By monitoring the crisis, management can more effectively and efficiently discover the characteristics of the crisis and determine the most suitable contingency crisis management plan. The activities identified in the plan should be strictly and thoroughly followed; however, the randomness of the crisis's evolution should be considered as well (Pforr and Hosie 2009). Therefore, crisis leadership can serve as a critical element in the successful implementation of the plan. In other words, the leader of the crisis management team is in charge of the decision-making process and of how smoothly things go, as well as the implementation of the chosen plan and the execution of all necessary activities (Beirman and Walbeek 2011). Similarly, crisis communication becomes essential in this stage of the crisis's development. Crisis communication has to be organized in three main directions: media, customers (tourists or potential tourists) and stakeholders (Beirman 2003). Information concerning the current state of the company (destination) and the actions which were taken, or are about to be taken, to cope with the crisis event needs to be shared. The flow of information from the tourism organization or destination authorities has to be regulated by appointing one spokesperson (Beirman 2003). This will help to avoid confusion and misunderstanding while cooperating with the mass media. As a part of the crisis management plan in the response phase the prescribed marketing strategy is applied. The ideal result of this stage of crisis management is to overcome the crisis with no or minimum negative consequences for a hospitality or tourism organization.

Evaluation phase

The major aim of the recovery stage is to restore normal operations in the organization. The crisis management team ought to carry out activities for the fastest recovery of an organization's operations. A detailed and in-depth analysis is carried out concerning the response actions and the performance of the personnel, as well as evaluating how strictly the crisis management plan was followed (Laws, Prideaux and Chon 2007; Pforr and Hosie 2009). The marketing strategy which began with the response phase continues. The strategy can be possibly modified depending on the extent of the consequences (Beirman 2003). Furthermore, an evaluation of the effectiveness of the implementation of the crisis management plan and the management team's response and performance is assessed. This will involve feedback to identify the mistakes made during the emergency phase, to point out the weaknesses of the crisis management plan, and how to improve future preparation processes with special attention paid to the consequences of the crisis. Consequences should be clearly observed and analyzed to determine whether they had short- or long-lasting effects on the tourism industry of the destination and how heavily the organization will be affected. At the same time, tourism firms should take into account the ambiguity of the crisis and identify which opportunities it has brought with it. It is important to identify whether there is a possibility to use them in new developments for an organization (Beirman 2003; Beirman and Walbeek 2011; Pforr and Hosie 2008; Faulkner 2003). The recovery and evaluation phase plays an important role in crisis management. It helps to reflect all of the work that was done on overcoming the crisis. It is also time for learning from the failures and the successes that were achieved while responding to the crisis situation.

Case study 51.1

Imagine that you are a general manager of a hotel located in a region with a high level of political instability.

1. What is the type of crisis your hotel may face?
2. Identify the crisis management strategy you would employ and explain why this would be the best strategy. Be sure to indicate the main elements you would include in the crisis management plan.

Case study 51.2

After overbooking a flight, United Airlines had airport security request that a passenger leave the airplane. After he refused, airport security physically removed him against his will. The video of the passenger's rough treatment was recorded and posted to YouTube. The video went viral, causing a PR disaster for United Airlines.

Research the case and provide an assessment of United Airlines' crisis management.

1. What was the crisis that came about during this event?
2. What were the negative impacts?
3. How did United Airlines respond?
4. How would you assess their response? What advice would you give them to more effectively manage overbooking issues in the future?

Conclusion

This chapter shows that hospitality and tourism organizations and destinations are vulnerable to significant crises and failures and why it is so important for firms to have crisis management plans to more effectively manage the tourist experience. The chapter discusses the types of crisis and how the crisis management plan is a critical strategic initiative. The crisis management plan should correspond with the long- and short-term goals of an organization. The plan covers all of the necessary aspects of crisis management such as: the analysis of possible hazards, the partners and stakeholders involved and a crisis communication plan. The crisis management plan works as a guideline for handling a negative event, rather than being a step-by-step procedure. The uncertainty of a crisis has to be taken into consideration and the organization should be ready to adjust the prescribed plan according to arising circumstances. Nonetheless, it is highly recommended to have a prepared crisis management plan in order to improve the chances of successful restoration.

References

Anthonissen, P. (2008). *Crisis Communication. Practical PR Strategies for Reputation Management and Company Survival*. London: Kogan Page.

Argenti, P. (2002). 'Crisis communication lessons from 9/11'. *Harvard Business Review*, 80(12): 103–119.

Baker, M. A. (2017). 'Electronic customer relationship management and customer satisfaction'. In: Dixit, S. K. (ed.). *The Routledge Handbook of Consumer Behaviour in Hospitality and Tourism* (pp. 396–403). Routledge. DOI: 10.4324/9781315659657

BBC. (2004), Retrieved from: http://news.bbc.co.uk/onthisday/hi/dates/stories/december/26/newsid_4631000/4631713.stm. [Accessed: April 29th, 2019].

BBC. (2016). Retrieved from: www.bbc.com/news/world-europe-35290760. [Accessed: April 29th, 2019].

BBC. (2019). Retrieved from: www.bbc.com/news/world-africa-13313370. [Accessed: April 29th, 2019].

Beirman, D. (2003). *Restoring Tourist Destinations in Crisis. A Strategic Marketing Approach.* UK. Cambridge: CABI Publishing.

Beirman, D., and Walbeek, B.,(2011). Tourism risk, crisis and recovery management guide. [Online]. Available from: www.uts.edu.au/sites/default/files/20160329-Beirman-Risk-Crisis-Recovery-Tour-Wholesalers.pdf. [Accessed: April 29th, 2019].

Boin, A. et al., (2005). *The Politics of Crisis Management: Public Leadership under Pressure.* New-York: Cambridge University Press.

CBS. (2017). Retrieved from: www.cbsnews.com/news/marriott-hurricane-irma-rescue-st-thomas-stranded-tourists/. [Accessed: April 29th, 2019].

CBS. (2019). Retrieved from: www.cbsnews.com/live-news/sri-lanka-bombings-death-toll-easter-attack-churches-hotels-live-updates-22-04-2019/. [Accessed: April 29th, 2019].

CNN. (2018). Retrieved from www.cnn.com/2018/09/27/us/hawaii-lane-eruption-august-spending-wtc/index.html.[Accessed: April 29th, 2019].

Coombs, T. W. (2014). 'Crisis management and communications'. Available from: www.instituteforpr.org/crisis-management-communications/. [Accessed: April 29th, 2019].

Faulkner, B. (2003). 'Towards a framework for tourism disaster management'. In: Faulkner, B. et al., (ed.). *Progressing Tourism Research* (pp. 244–269). Great Britain: Cromwell Press.

Glaesser, D. (2006). *Crisis Management in the Tourism Industry.* Second edition. Burlington: Elsevier.

Henderson, J. C. (2007). *Tourism Crises: Causes, Consequences, Management.* Oxford: Elsevier.

Laws, E. (2001). 'Service problems, service crises and service disasters'. *Journal of Quality Assurance in Hospitality and Tourism*, 2(1-2): 119–129. DOI: 10.1300/J162v02n01_08

Laws, E., Prideaux, B., and Chon, K. (2007). *Crisis Management in Tourism.* Wallingford: CABI International.

Lockwood, N., (2005). Crisis management in today's business environment: HR's strategic role. SHRM (The Society for Human Resource Management). www.shrm.org/hr-today/news/hr-magazine/Documents/1205RQuartpdf.pdf. [Accessed: April 5, 2019].

Mitroff, I. I., and Alpaslan, M. C. (2003). 'Preparing for evil'. *Harvard Business Review*, 81(12): 5–11.

Mitroff, I. I., Shrivastava, P., and Udwadia, F. (1987). 'Effective crisis management'. *The Academy of Management Executive*, 1(4): 283–292. DOI: 10.5465/ame.1987.4275639

NBC. (2019). Retrieved from www.nbcnews.com/news/world/deaths-injuries-confirmed-after-explosions-heard-u-k-concert-featuring-n763286. [Accessed: April 5th, 2019].

NCBI. (2004). Retrieved from www.ncbi.nlm.nih.gov/books/NBK92479/.[Accessed: April 5th, 2019].

Paraskevas, A., and Quek, M. (2019). 'When castro seized the Hilton: Risk and crisis management lessons from the past'. *Tourism Management*, 70: 419–429. DOI: 10.1016/j.tourman.2018.09.007

Pforr, C., and Hosie, P. (2009). *Crisis Management in the Tourism Industry: Beating the Odds?* Ashgate: Routledge.

Pforr, C., and Hosie, P. J. (2008). 'Crisis management in tourism'. *Journal of Travel & Tourism Marketing*, 23(2-4): 249–264. DOI: 10.1300/J073v23n02_19

Regester, M., and Larkin, J. (2005). *Risk Issues and Crisis Management. A Case-book of Best Practice.* London: Kogan Page.

Ritchie, B. W. et al., (2004). 'Crisis communication and recovery for the tourism industry'. *Journal of Travel & Tourism Marketing*, 15(2-3): 199–216. DOI: 10.1300/J073v15n02_11

Rittichainuwat, B. N. (2013). 'Tourists' and tourism suppliers' perceptions toward crisis management on tsunami'. *Tourism Management*, 34: 112–121. DOI: 10.1016/j.tourman.2012.03.018

Rousaki, B., and Alcott, P. (2006). 'Exploring the crisis readiness perceptions of hotel managers in the UK'. *Tourism and Hospitality Research*, 7(1): 27–38. DOI: 10.1057/palgrave.thr.6050030

Siaw, K. T., and Loosemore, M. (2006). 'Terrorism prevention, preparedness, and response in built facilities'. *Facilities*, 24(5): 157–176. DOI: 10.1108/02632770610665766

Sönmez, S., and Backman, S. J. (1992). 'Crisis management in tourist destinations'. *Visions in Leisure and Business*, 11(3): 25–33.

Stafford, G., Yu, L., and Alex, K. A. (2002). 'Crisis management and recovery: How Washington D. C. hotels responded to terrorism'. *Cornell Hotel and Restaurant Administration Quarterly*, 43(5): 27–40. DOI: 10.1016/S0010-8804(02)80054-4

Su, L., Stepchenkova, S., and Kirilenko, A. P. (2019). 'Online public response to a service failure incident: Implications for crisis communications'. *Tourism Management*, 73: 1–12. DOI: 10.1016/j.tourman.2019.01.011

Watkins, M. D., and Bazerman, M. H. (2003). 'Predictable surprises: The disasters you should have seen coming'. *Harvard Business Review*, 81(3): 72–80.

52

Conclusion

Some reflections on experiential tourism management and marketing

Saurabh Kumar Dixit

The aim of this *Handbook* is to develop theoretical and applied evidence for embracing and implementing management and marketing approaches and to address the challenges and seize the opportunities in the arena of tourism experiences. This chapter will be highlighting the key notions and multifaceted nature of experiential tourism management and marketing underscored in different contributions included in this volume. The authors in this handbook have reflected on a number of the most contemporary themes relating to experiential tourism management and marketing. Not all of the chapters focus on 'tourism experience' per se, but instead on the complex inter-relationships in tourism experience, marketing, management and co-creation. In most of the chapters the authors have carried out diverse case studies in order to highlight the theme of discussion in the chapter. The purpose of this final chapter is to reflect the central concerns and significant foci prompted in the 52 chapters of this *Handbook*.

Tourism experience: prospective and reflective views

Experiences are considered as the core of all tourism activities. Visitors travel in search of memorable/rewarding destinations, besides learning about the peculiar features of host cultures and discovering unique travel experiences. Therefore it is of the utmost importance to explore the ways in which tourists experience the places and people they visit during the study of the tourism consumption experience. This book validates the premise that various theoretical lenses can construct an improved understanding of tourism experiences. Chapters 1, 3, 4 and 5 offer the theoretical foundation of the tourism experience from diverse perspectives, the prime motivators for a memorable tourism experience, the typology and classification of tourism experiences and the notion of tourism experience and its relationship with customer experience. Chapter 2 uses systematic review techniques to provide a comprehensive outlook of origins and concerns by examining the theoretical frameworks under which tourism experience has been investigated by different researchers during the years 2009–2018. The volume also highlights important paradigms of tourism experience design in its different chapters. Chapters 11, 12, 17 and 21 revisit the dimensions involved in the designing of an effective tourism experience. They offer vital reference points for experience designers to ensure appropriate solutions for different problems, experiential choice and the theory of experiential dissonance. Chapter 10 offers a critique of the concept of the experience economy in tourism. The value offered to consumers in the experience economy is not limited to the core benefits associated with the products but rather it includes a whole collection of value-added activities needed to stage, market and deliver the complete tourism experience. Chapter 13 elaborates the *experiencescape* as an amalgam of both the physical and social elements that affect the overall experience of an individual.

Prosumption of tourism experience

The creation and co-creation of experiences, as theoretical constructs, situates the tourist as a dynamic component in consuming and producing value (Dabholkar et al. 2000), which further leads to consumer engagements in designing the experience. This phenomenon is named prosumption by Toffler (1980). Prosumption therefore includes both production and consumption rather than focusing on either production or consumption. The modern tourists facilitated by the internet are digitally connected and well-informed, which enables them to pursue authentic, customized, environmentally friendly tourism experiences. Present-day travelers tend to immerse themselves in designing their own travel experiences. Prosumption results when tourists help themselves (Meuter et al. 2000) and also encourage others to become actively involved in the production and consumption of experiences (Toffler 1980). Chapters 14, 34, 44 and 45 deliberate on the different perspectives of how travelers' involvement contributes to the creation of their overall experience. The chapters emphasize the contribution of traveler perception, travel motivation, image, co-creation through service-dominant logic and social media platforms in the formation of travelers' whole travel experience. Chapter 41 advocates that sustainable rural tourism experiences (SRTE) can be achieved through co-creation experiences by focusing on active participation and a high level of interaction between the tourist and host communities/physical environment. Chapter 47 explores what motivates Asian/Chinese tourists to visit the Arctic region through a case study on Chinese face, to understand the cultural co–creation of tourist experiences.

Co-production of tourism experience

Several chapters of the handbook deal with experience co-production and recognize the active role of tourists and the industry in the formation of tourism experiences. Employees are one of the main means of making contented customers and generating higher profits. Therefore, understanding the effect of the roles and contributions of service employees on customer experience is necessary, which is explained in Chapter 8. Chapter 16 offers *The City Hospitality Experience Model*, which helps to define the critical touchpoints in the tourists' journey. Memorable experiences are created through tourism offerings, atmosphere and hospitable behavior. Chapter 19 illustrates how storylines offer a framework for developing opportunities for visitors to co-create their own tourism experience. At a visitor attraction, if storytelling is complemented by physical resources such as museums, national parks, heritage monuments, etc., it will surely add value to the tourism experience of the visitors. Chapter 46 presents Albergo Diffuso, an Italian model of combined hospitality businesses that offers a framework that creates and delivers value to its customers. Chapter 48 explores the notion of emotional labor that plays out between the employee and the customer. There is a lack of more nuanced takes on emotional labor in the tourism sharing economy, so this chapter offers such a take on emotional labor, highlighting the emotional labor of being an Airbnb host as a case.

Sensory tourism experience

Tourism experience is a very complex phenomenon whereby the overall experience is associated with multiple interpretations of social, environmental and activity components. Traditionally, experience has been defined as a personal occurrence, with highly emotional significance being obtained from the consumption of products and services (Holbrook and Hirschman 1982), and emotions that are translated into senses. The *Handbook* also posits in its different chapters that hedonic experiences play a vital role in shaping travelers' experiences at different stages of their visit. Chapter 6 appraises the body of knowledge on the role and measurement of emotions in tourism experience formation. Chapter 9 enlightens readers with the linkages of emotional experiences to perceived overall image, satisfaction and loyalty. Chapter 42 investigates the effects of green experiential quality and its dimensions, green emotional attachment, green image, green switching experience and green experiential satisfaction on green loyalty in a green restaurant setting.

Memorable tourism experience

Today, there is an increasing shift among tourism service providers from providing services to offering memorable experiences (Ariffin and Maghzi 2012). A memorable tourism experience (MTE) is a tourism experience that is positively remembered and recalled after the event has occurred (Kim et al. 2012). Providing consumers with memorable tourism experiences is key to being competitive in the tourism industry (Sthapit and Jiménez-Barreto 2018). Chapter 7 of this handbook offers a framework to help in understanding the central characteristics of memorable tourism experiences through a holistic approach. Chapter 15 highlights how extraordinary tourism experiences can be intellectually engaging and transformative. Two cases, an oyster experience on the West Coast of Sweden and a guided tour in Kisumu, Kenya, are used to demonstrate how learning takes place for tourists engaged in extraordinary experiences. Chapter 27 presents the notion of memorable tourism experiences, their fundamentals and their perceived importance. It also outlines the role of managers as facilitators and providers of means that help create memorable experiences through the process of co-creation and customization. Chapter 28 explores the categories and components of memorable gastronomic experiences.

Gastronomic tourism experiences

Destinations nowadays are using food and related events to not only fulfill the basic dietary requirements of tourists, but also market a unique gastronomic experience to highlight local cultures, places and people (Dixit 2019). Gastronomic tourism therefore spawns prospects to explore, taste and experience the foods and beverages of a destination, in order to contribute deeper experience to the local tourism product (Dixit 2019). Chapter 28 emphasizes the endeavors to use gastronomy to market the tourism experiences of a destination. The chapter also emphasizes the quest for authenticity in gastronomic tourism experiences. Chapters 18 and 35 explore the wine tourism experiences of tourists. Chapter 18, through a qualitative case study, explores trends in experience design implemented by wineries in Victoria, Australia, with the aim of attracting millennials as tourists. Chapter 35 describes how local producers have employed enhanced traditional services (e.g. visits and tastings) by integrating new technologies into the winery experience, so as to overcome physical barriers (e.g. allowing tourists to experience the vineyard during all seasons), to provide content and to increase visitors' experience. Chapter 26 integrates Kano's theory as it fits within the more comprehensive customer journey applied to the Oktoberfest beer tent experience.

Experience intensification via technology

Modern travelers are transforming their travel experiences and are endeavoring to make them more existent by clicking pictures, buying mementos and gifts and posting online travel reviews (Dong and Siu 2013). Therefore, social media sites and other technological platforms have become popular tools that contribute to the intensification of travel experiences through photographs, website design/postings, messaging and other user-generated contents. Chapters 32 to 38 of the handbook deal with the different facets of technological experiential intensification. Chapter 32 explores how the emergence of the sharing economy revolutionized the tourism industry and profoundly changed the structure of accommodation and transportation supply. Chapter 33 displays a gradual convergence of travel behavior activities from the pre-trip and the post-trip stages to the during-trip. Chapters 34 and 37 deliberate on the emergence and growth of digital media and the role of digital narratives as media to co-construct travel experiences. Chapter 35 analyzes how technology is able to activate the narrative facet of wine and contribute to tourists' memorable experiences. Chapter 36 deals about presentation of information about the destinations on website pages. The chapter also emphasizes that the use of appropriate pictures and colors leads to a greater amount of attention from potential tourists, making them important in website design. Chapter 38 explains how destination managers may operate during the traveler journey using new technologies (VR, AR and artificial intelligence).

Marketing tourism experiences

Tourism experience is a fairly topical arena of interest for marketing scholars and yet has produced a multifaceted and fascinating array of conceptualizations. The experiential marketing approach can be an immensely worthwhile means of augmenting understanding of traveler behavior. The broad objective of this handbook is to provide a comprehensive and critical overview of the theoretical, methodological and practical issues in tourism experiential marketing. Chapter 22 highlights case study methodology in order to understand whether and how the Accor Hotel chain operates through revenue management techniques that use customer relationship management (CRM) for the creation of personalized offers. Chapter 23 introduces a framework of tourist behavior to explicate the effect of experiential marketing on tourist behavior. Chapter 24 propounds a model of tourist engagement and its effect on experience, brand identification and loyalty. Chapter 25 deliberates the concept of relationship marketing as an efficient system for tourism marketing through the relational exchange happening between tourists and the organizations. As branding and experience have long been in the focus of hotel businesses, Chapter 29 sheds light on the emergence of the concept brand experience and its design in targeting different customer segments. Chapter 30 educates readers on the critical management of service failures and recoveries in managing tourism experiences. Chapter 31 explores experiencing negative incidents and places that leads to traumatizing, disturbing or unsettling travel experiences. This chapter further spells out the phenomenon of dark tourism experience marketing and addresses criticisms about it.

Future research directions

It is significant to note that experience in tourism is quite complex and different from other consumptions. Academic researchers in tourism have spawned numerous interesting and broad frameworks relating to tourism experience. In this section, we will identify several themes in the contemporary body of knowledge of tourism experience which will pave the way for future studies in the area of experiential research in tourism and hospitality. Chapter 20 of the handbook enlightens readers on the performativity and embodiment of an Arctic animal-based tourism experience, the husky safari. Animal-based tourism experiences have rarely been analyzed through embodiment. The chapter concludes that there is further need to examine tourism experiences co-created through emotional, affective and embodied engagement between humans and non-human agents such as robots, autonomous vehicles and other forms of artificial intelligence. Chapter 39 elaborates how visitor experience could be designed, staged and managed in ways that result in less or no negative impacts without deteriorating the ultimate experience for tourists. Chapter 40 observes the connection between identity and experience, concerning the experiential needs of tourists, on one hand, and the emancipation of the territories and their local communities, on the other hand. Chapter 43 also investigates guest and host perspectives in sustainable tourism experience formation. Chapter 49 affirms the interlinkage between destination competitiveness and tourism experiences. Chapter 50 presents a review that sheds light on the needs and wants of seniors to shape their lasting travel experiences. Chapter 51 serves as an introduction to the pertinent theme of crisis management and tourism experiences.

Each chapter draws together authors from different backgrounds and interdisciplinary interests and research methodologies. As a consequence, the volume provides a model of the way researchers can work together to illuminate an area of study and provide multiple representations and interpretations of the subject. Moreover, the book demonstrates intradisciplinary, transdisciplinary and interdisciplinary approaches and collaborations. This exhaustive handbook has attempted to convey an enhanced understanding of tourism experience in an unambiguous manner. As we have grasped through different chapters, tourism experience has a rich and complex assemblage of approaches which can be professed from several perspectives. The chapters also emphasize the need of further investigation in this progressing field. Besides the important imminent issues revealed across the *Handbook*, other paths identified for experiential tourism scholars are to place issues of sustainability prominently on the research agenda. Innovative technology prevalent in the tourism industry converts intangible tourism products into tangible offerings and service providers have now become more innovative in creating quality experiences for travelers. This trend definitely requires an in-depth understanding of how tourists deal with both the service provider and the business environment. The final research issue is that of the reflexivity of researchers on tourism experiences.

I hope that this *Handbook* will garner a greater interest and dialogue on the assorted dimensions of experiential tourism research and lay down a base for a much larger research contribution from its readers.

References

Ariffin, A. A. M., and Maghzi, A. (2012). A preliminary study on customer expectations of hotel hospitality: Influences of personal and hotel factors. *International Journal of Hospitality Management*, 31(1): 191–198. DOI: 10.1016/j.ijhm.2011.04.012

Dabholkar, P. A., Shepherd, C. D., and Thorpe, D. I. (2000). A comprehensive framework for service quality: An investigation of critical conceptual and measurement issues through a longitudinal study. *Journal of Retailing*, 76(2): 139–173. DOI: 10.1016/S0022-4359(00)00029-4

Dixit, S. K. (2019). Gastronomic tourism: A theoretical construct. In: Dixit, S. K. (ed.) *The Routledge Handbook of Gastronomic Tourism* (pp. 13–23). London: Routledge. DOI: 10.4324/9781315147628

Dong, P., and Siu, N. Y.-M. (2013). Servicescape elements, customer predispositions and service experience: The case of theme park visitors. *Tourism Management*, 36: 541–551. DOI: 10.1016/j.tourman.2012.09.004

Holbrook, M. B., and Hirschman, E. C. (1982). The experiential aspects of consumption: Consumer fantasies, feelings, and fun. *Journal of Consumer Research*, 9(2): 132–140. DOI: 10.1086/208906

Kim, J.-H., Ritchie, J., and McCormick, B. (2012). Development of a scale to measure memorable tourism experiences. *Journal of Travel Research*, 51(1): 12–25. DOI: 10.1177/0047287510385467

Meuter, M., Ostrom, A., Roundtree, R., and Bitner, M. (2000). Self-service technologies: Understanding customer satisfaction with technology-based service encounters. *Journal of Marketing*, 64 (3): 50–64. DOI: 10.1509/jmkg.64.3.50.18024

Sthapit, E., and Jiménez-Barreto, J. (2018). Exploring tourists' memorable hospitality experiences: An Airbnb perspective. *Tourism Management Perspectives*, 28: 83–92. DOI: 10.1016/j.tmp.2018.08.006

Toffler, A. (1980). *The Third Wave*. New York, NY: Bantam Books.

Index

For Product Safety Concerns and Information please contact our EU representative GPSR@taylorandfrancis.com
Taylor & Francis Verlag GmbH, Kaufingerstraße 24, 80331 München, Germany

www.ingramcontent.com/pod-product-compliance
Lightning Source LLC
LaVergne TN
LVHW080759011026
849160LV00015B/1489

* 9 7 8 0 3 6 7 4 9 2 7 5 5 *